Practice Problems
for the Environmental Engineering PE Exam

A Companion to the *Environmental Engineering Reference Manual*

Third Edition

Michael R. Lindeburg, PE

The Power to Pass®
www.ppi2pass.com

Professional Publications, Inc. • Belmont, California

PRACTICE PROBLEMS FOR THE ENVIRONMENTAL ENGINEERING PE EXAM

Third Edition

Current printing of this edition: 2

Printing History

date	edition number	printing number	update
Feb 2013	2	4	Minor revisions.
Dec 2014	3	1	New edition. Copyright update.
Sep 2016	3	2	Minor corrections. Minor cover update.

PPI
1250 Fifth Avenue
Belmont, CA 94002
(650) 593-9119
ppi2pass.com

ISBN: 978-1-59126-476-7

Library of Congress Control Number: 2014951241

F E D C B A

Topics

Background

Flow of Fluids

Water Treatment

Ventilation

Combustion

Solid Waste

Health Safety

Systems Mgmt

Where do I find help
solving these Practice Problems?

Practice Problems for the Environmental Engineering PE Exam presents complete, step-by-step solutions for more than 625 problems to help you prepare for the Environmental PE exam. You can find all the background information, including charts and tables of data, that you need to solve these problems in the *Environmental Engineering Reference Manual*.

The *Environmental Engineering Reference Manual* may be obtained from PPI at **ppi2pass.com** or from your favorite retailer.

Table of Contents

Preface to the Third Edition

I know from first-hand experience that the easiest type of engineering book to write is a collection of solved problems. Relatively easiest, I should say, because no engineering book is easy to write, and it is essentially impossible to write a flawless engineering book. But, given that proviso, a non-engineering cataloger could assemble groups of solved problems from various sources and almost instantly become an "author."

When budding authors send me book proposals and even entire manuscripts that consist of photocopies of page after page of problems and solutions borrowed from other sources, I decline to publish them. Aside from ignoring the obvious intellectual property ownership issues, such wannabes have lost sight of the purpose of writing books: to entertain, compile information, and teach. Books written merely to start a royalty income stream or to add lines to one's resume are not difficult to recognize when one is a publisher.

As an author as well as a publisher, I can tell you that, while this book is a collection of solved problems, it is not one of those easy books to write. This book has a purpose: to support the *Environmental Engineering Reference Manual* and to provide you with the practice needed to get you ready for the Professional Engineering licensing exam in that discipline.

The third edition of the *Environmental Engineering Reference Manual* contains no end-of-chapter practice problems, no problem assignments, no homework. The task of providing the problems to support the chapters in that book falls to this book. This decoupling eliminates any mismatch between problem statements and solutions in different editions or printings of the two books, but it does not reduce the banes of engineering publishing: relevance, difficulty level, consistency, cross-referencing, completeness, and linearity.

For this third edition, 255 new problems have been incorporated. Among these are new problems written for chapters that in the last edition were devoid of practice problems: Environmental Pollutants, Disposition of Hazardous Materials, Environmental Remediation, Organic Chemistry, Instrumentation and Digital Numbering Systems, Engineering Law, and Engineering Ethics. But don't think for one minute that I focused only on these new problems. All existing problems were reviewed, and some of the problem statements have been changed to include additional information in order to more closely simulate the exam and to eliminate the need to make assumptions that might get you off track. Suggestions for improvements from readers of previous editions have been incorporated, as have corrections of typos and, of course, errata. All in all, the completely new problems and the revised existing problems will be of significant benefit to you as you study for your exam.

I've tried to anticipate your questions by adding strategic commentary throughout many solutions. My daily dose of fan mail (as I like to view the correspondence) brings questions that show me how to improve logic and flow. In these solutions, you won't find any professor chalkboard talk such as "the interested reader may prove the expression in his/her own time," or "it is patently obvious what the next seven steps should be, so we'll skip them." Although I might leave out some interim steps (e.g., omitting the steps to solve for the roots of a quadratic or higher-order equation, which is something your calculator can do), for the most part, the solutions are complete down to the algebraic manipulations.

The chapters in this edition parallel those in the *Environmental Engineering Reference Manual*. This book contains numerous references to the equations, tables, and figures in the *Reference Manual*, and this book uses the same nomenclature, data, and methods. Though independent, these two books intentionally complement each other. Because this book is essentially an independent collection of solved practice problems that you can take with you anywhere, even if you don't have the *Reference Manual* with you while you review this book's solutions, you should be able to follow the solution steps in most of the chapters. Except where the intended equation is obvious (e.g., $A = \pi r^2$ and $v = Q/A$), the actual equation is given in variable format before the numbers are plugged in.

As usual, I've already begun thinking about the next edition of this book. Even before it arrives, however, you can keep up with changes to this book by logging onto PPI's website at **ppi2pass.com/errata**.

I have done my best to make *Practice Problems for the Environmental Engineering PE Exam* and the *Environmental Engineering Reference Manual* the most useful books in your library. Now, it's up to you.

Michael R. Lindeburg, PE

Acknowledgments

Every new edition is a lot of work, for someone. Sometimes, as when new content is needed, the work of bringing out a new edition falls to the author; sometimes the work falls to the publisher's editorial and production staff, as when existing content is reformatted and reorganized; and sometimes the work falls to one or more subject matter experts, as when specialized knowledge is required to integrate changes in standards, codes, and federal legislation. Sometimes, one player shoulders a disproportionate share of the load; sometimes one player gets off easier than another.

I know enough about environmental engineering to know how little I know. So, for the most critical chapters, I asked several environmental experts from industry and academia to fill in holes in my outline. Their contributions ranged from small sections to entire chapters. When their contributions were submitted to me, I edited every sentence and checked every calculation. However, the genius of the chapters remains theirs. This is how 9 of the 58 chapters of the first edition of this book were created. The following experts were invaluable in helping me prepare the first edition of the manuscript in an accurate and timely manner.

> Jeffrey S. Mueller, MS, PE, CHP: Biological Effects of Radiation, Shielding (see Chap. 48, Chap. 49)

> R. Wane Schneiter, PhD, PE, DEE: Air Quality/Pollution (see Chap. 38) plus technical review of Protection of Wetlands (see Chap. 45)

> James R. Sheetz, PE, DEE: Risk Analysis, Emergency Response, Protection of the Wetlands, Toxicology, Industrial Hygiene, Health and Safety (see Chap. 11 [part], Chap. 28, Chap. 45, Chap. 46, Chap. 47, Chap. 51)

Now, I'd like to introduce you to the "I couldn't have done it without you" crew, the talented team at PPI.

Editing and Production: Tom Bergstrom, David Chu, Nicole Evans, Hilary Flood, Kate Hayes, Tyler Hayes, Julia Lopez, Scott Marley, Ellen Nordman, Heather Turbeville, and Ian A. Walker

Management: Sarah Hubbard, director of product development and implementation; Cathy Schrott, production services manager; and Jenny Lindeburg King, Chelsea Logan, Magnolia Molcan, and Julia White, editorial project managers

This edition incorporates the comments, questions, suggestions, and errata submitted by many people who have used the previous edition for their own preparations. As an author, I am humbled to know that these individuals have read the previous edition in such detail as to notice typos, illogic, and other errata, and that they subsequently took the time to share their observations with me. Their suggestions have been incorporated into this edition, and their attention to detail will benefit you and all future readers. The following is a partial list (in alphabetical order) of some of those who have improved this book through their comments.

> Robert DeRosier, Michael Fleming, Samuel Haffey, Christopher Hardee, Joy Jenkins, Alia Johnson, Richard Mestan, and Walter Trinkala

This edition shares a common developmental heritage with its previous editions, and I have not forgotten those that submitted errata for them. For this edition, though, there isn't a single contributor that I intentionally excluded. Still, I could have slipped up and forgotten to mention you. I hope you'll let me know if you should have been credited, but were inadvertently left out. I'd appreciate the opportunity to list your name in the next printing of this edition.

Near the end of the acknowledgments, after mentioning a lot of people who contributed to the book and, therefore, could be blamed for a variety of types of errors, it is common for an author to say something like, "I take responsibility for all of the errors you find in this book." Or, "All of the mistakes are mine." This is certainly true, given the process of publishing, since the author sees and approves the final version before his/her book goes to the printer. You would think that after more than 35 years of writing, I would have figured out how to write something without making a mistake and how to proofread without missing those blunders that are so obvious to readers. However, such perfection continues to elude me. So, yes, the finger points straight at me.

All I can say instead is that I'll do my best to respond to any suggestions and errata that you report through PPI's website, **ppi2pass.com/errata**. I'd love to see your name in the acknowledgments for the next edition.

Thank you, everyone!

Michael R. Lindeburg, PE

Codes Used to Prepare This Book

PPI lists on its website the dates of the codes, standards, and regulations on which NCEES has announced the exams are based. However, these postings do not contain any announcements specifically about the environmental exam. The conclusion you and I must reach from such an omission is that the exam is not sensitive to changes in standards, regulations, or announcements in the Federal Register.

How to Use This Book

This book is primarily a companion to the *Environmental Engineering Reference Manual*. As a collection of solved problems, there are a few, but not very many, ways to use it.

I envisioned this book (with its included problem statements) being taken to work, on the bus or train, on business trips, even on weekend pleasure getaways to the beach. I figured that it would "carry" a lot easier than the big *Environmental Engineering Reference Manual*, which I knew, before I started to write, would be a big book. Though I don't think you'll be taking this book on any backpacking trips, you might still find yourself working problems in the cafeteria during your lunch break.

The really big issue is whether you actually work the practice problems or not. Some people think they can read a problem statement, think about it for about ten seconds, read the solution, and then say "Yes, that's what I was thinking of, and that's what I would have done." Sadly, these people find out too late that the human brain doesn't learn very efficiently that way. Under pressure, they find they know and remember little. For real learning, you have to spend some time with a stubby pencil.

There are so many places where you can get messed up solving a problem. Maybe it's in the use of your calculator, like pushing log instead of ln, or forgetting to set the angle to radians instead of degrees, and so on. Maybe it's rusty math. What is e^x, anyway?

How do you factor a polynomial? Maybe it's in finding the data needed or the proper unit conversion. Maybe it's just trying to find out if that funky building code equation expects L to be in feet or inches. These things take time. And you have to make the mistakes once so that you don't make them again.

If you do decide to get your hands dirty and actually work these problems, you'll have to decide how much reliance you place on the solutions. It's tempting to turn to a solution when you get slowed down by details or stumped by the subject material. You'll probably be tempted to maximize the number of problems you solve by spending as little time as possible on each one. I want you to struggle a little bit more than that.

Studying a new subject is analogous to using a machete to cut a path through a dense jungle. By doing the work, you develop pathways that weren't there before. It's a lot different from just looking at the route on a map. You actually get nowhere by looking at a map. But cut that path once, and you're in business.

So do the problems. All of them. Don't look at the answers until you've sweated a little. And let's not have any whining. Please.

$$0.1713 \cdot 10^{-8} \frac{Btu}{hr \, ft^2 \cdot R^4} \cdot \frac{hr \; 0.2931 \, W}{1 \, Btu} \cdot \frac{10.7639 \, ft^2}{1 \, m^2} \cdot \frac{1 \, R}{5/9^4 \, K}$$

$$0.9$$

1 Systems of Units

PRACTICE PROBLEMS

1. Most nearly, what is 250°F converted to degrees Celsius?

(A) 115°C

(B) 121°C

(C) 124°C

(D) 420°C

$$\frac{5}{9}\left(250-32\right)$$

2. Most nearly, what is the Stefan-Boltzmann constant $(0.1713 \times 10^{-8}$ Btu/hr-ft^2-°R$^4)$ converted from English to SI units?

(A) 5.14×10^{-10} W/m^2·K^4

(B) 0.95×10^{-8} W/m^2·K^4

(C) 5.67×10^{-8} W/m^2·K^4

(D) 7.33×10^{-6} W/m^2·K^4

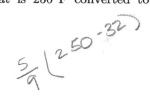

1 Btu/hr = 0.2931 W

10.7639 ft^2 = 1 m^2

1 R = 5/9 K

3. Approximately how many U.S. tons (2000 lbm per ton) of coal with a heating value of 13,000 Btu/lbm must be burned to provide as much energy as a complete nuclear conversion of 1 g of coal? (Hint: Use Einstein's equation: $E = mc^2$.)

(A) 1.7 tons

(B) 14 tons

(C) 780 tons

(D) 3300 tons

$c = 3 \times 10^8$ m/s

$m = 1g = 0.001$ Kg

$E = 0.001$ Kg $\cdot (3 \cdot 10^8$ m/s)^2

$E = 9 \cdot 10^{13}$ J

$9 \cdot 10^{13}$ J $\cdot 0.000948\ \dfrac{Btu}{J} = 8.532 \cdot 10^{10}$ Btu

$$\dfrac{8.532 \cdot 10^{10}\ Btu}{13000\ \frac{Btu}{lbm} \cdot 2000\ \frac{lbm}{ton}} = 3281.5\ tons$$

SOLUTIONS

1. The conversion to degrees Celsius is

$$T_{°C} = \tfrac{5}{9}(T_{°F} - 32°F)$$
$$= \left(\tfrac{5}{9}\right)(250°F - 32°F)$$
$$= \boxed{121.1°C \quad (121°C)}$$

The answer is (B).

2. The Stefan-Boltzmann constant represents a certain amount of energy (in Btu). Since it has hr-ft^2-°R^4 in the denominator, the energy is reported on a per hour basis and represents power. The energy is an areal value because it is reported per square foot. Since a square meter is larger than a square foot, the energy must be multiplied by a number larger than 1.0 in order to put it into a per square meter basis. (For example, on any given day, the energy that can be derived from a square meter of sunlight is greater than can be derived from a square foot of sunlight.) 1/0.3048 is the conversion between feet and meters, and it is larger than 1.0.

The Stefan-Boltzmann constant is reported on a per degree basis, and the same logic applies. One Kelvin (K) is a larger (longer) interval on the temperature scale than one degree Rankine. The amount of energy per Kelvin is larger than the amount of energy per degree Rankine. Therefore, the energy must be multiplied by a number greater than 1.0 to report it on a "per Kelvin" basis. This requires multiplying by $^9/_5$. (An incorrect conversion will result if the unit in the denominator is thought of as an actual temperature. Any given numerical value of temperature in degrees Rankine is larger than the same temperature in Kelvins. However, this problem does not require a temperature conversion. It requires a unit conversion.)

Use the following conversion factors.

$$1\text{ Btu/hr} = 0.2931\text{ W}$$
$$1\text{ ft} = 0.3048\text{ m}$$
$$1°\text{R} = \tfrac{5}{9}\text{ K}$$

Performing the conversion gives

$$\sigma = \left(0.1713 \times 10^{-8} \ \frac{\text{Btu}}{\text{hr-ft}^2\text{-}^\circ\text{R}^4}\right)\left(0.2931 \ \frac{\text{W-hr}}{\text{Btu}}\right)$$

$$\times \left(\frac{1 \ \text{ft}}{0.3048 \ \text{m}}\right)^2\left(\frac{1^\circ\text{R}}{\frac{5}{9}\text{K}}\right)^4$$

$$= \boxed{5.67 \times 10^{-8} \ \text{W/m}^2\text{-K}^4}$$

The answer is (C).

3. The energy produced from the nuclear conversion of any quantity of mass is

$$E = mc^2$$

The speed of light, c, is 3×10^8 m/s.

For a mass of 1 g (0.001 kg),

$$E = mc^2$$
$$= (0.001 \ \text{kg})\left(3 \times 10^8 \ \frac{\text{m}}{\text{s}}\right)^2$$
$$= 9 \times 10^{13} \ \text{J}$$

Convert to U.S. customary units with the conversion 1 Btu = 1055.1 J.

$$E = \frac{9 \times 10^{13} \ \text{J}}{1055.1 \ \frac{\text{J}}{\text{Btu}}}$$
$$= 8.53 \times 10^{10} \ \text{Btu}$$

The number of tons of 13,000 Btu/lbm coal is

$$\frac{8.53 \times 10^{10} \ \text{Btu}}{\left(13,000 \ \frac{\text{Btu}}{\text{lbm}}\right)\left(2000 \ \frac{\text{lbm}}{\text{ton}}\right)} = \boxed{3281 \ \text{tons} \quad (3300 \ \text{tons})}$$

The answer is (D).

2 Properties of Areas

PRACTICE PROBLEMS

1. Where is the x-coordinate of the centroid of the area most nearly located?

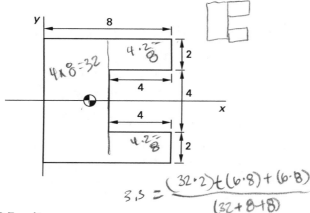

(A) 2.7 units

(B) 2.9 units

(C) 3.1 units

(D) 3.3 units

2. Replace the distributed load with three concentrated loads, and indicate the points of application.

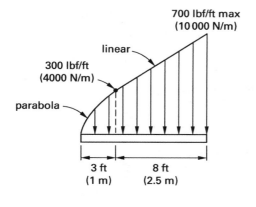

3. Where is the centroidal moment of inertia about an axis parallel to the x-axis most nearly located?

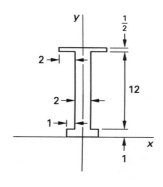

(A) 160 units^4

(B) 290 units^4

(C) 570 units^4

(D) 740 units^4

4. A rectangular 4 in $\times$ 2 in area has a 2 in diameter hole in its geometric center. What is most nearly the moment of inertia about the y-axis?

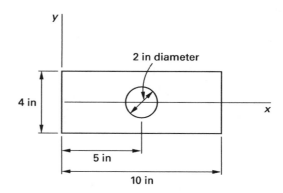

(A) 1030 in^4

(B) 1150 in^4

(C) 1250 in^4

(D) 1370 in^4

5. What is most nearly the moment of inertia about the x-axis of the area OAB?

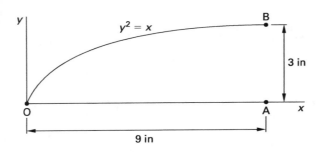

(A) 32 in^4

(B) 67 in^4

(C) 70 in^4

(D) 76 in^4

6. An annular flat ring has an outer diameter of 4 in and an inner diameter of 2 in. What is most nearly the radius of gyration of the ring about the y-axis?

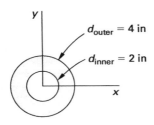

(A) 0.89 in

(B) 1.1 in

(C) 1.3 in

(D) 2.2 in

SOLUTIONS

1. The area is divided into three basic shapes.

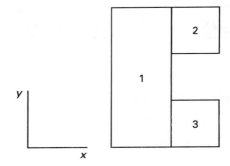

First, calculate the areas of the basic shapes.

$$A_1 = (4)(8) = 32 \text{ units}^2$$
$$A_2 = (4)(2) = 8 \text{ units}^2$$
$$A_3 = (4)(2) = 8 \text{ units}^2$$

Next, find the x-components of the centroids of the basic shapes.

$$x_{c,1} = 2 \text{ units}$$
$$x_{c,2} = 6 \text{ units}$$
$$x_{c,3} = 6 \text{ units}$$

Finally, use Eq. 2.5.

$$x_c = \frac{\sum A_i x_{c,i}}{\sum A_i}$$

$$= \frac{\begin{array}{c}(32 \text{ units}^2)(2 \text{ units}) + (8 \text{ units}^2)(6 \text{ units}) \\ + (8 \text{ units}^2)(6 \text{ units})\end{array}}{32 \text{ units}^2 + 8 \text{ units}^2 + 8 \text{ units}^2}$$

$$= \boxed{3.33 \text{ units} \quad (3.3 \text{ units})}$$

The answer is (D).

2. *Customary U.S. Solution*

The parabolic shape is

$$f(x) = \left(300 \ \frac{\text{lbf}}{\text{ft}}\right)\sqrt{\frac{x}{3}}$$

First, use Eq. 2.3 to find the concentrated load given by the area.

$$A = \int f(x)\,dx = \int_{0\text{ ft}}^{3\text{ ft}} 300\sqrt{\frac{x}{3}}\,dx$$

$$= \left(\frac{300\ \frac{\text{lbf}}{\text{ft}}}{\sqrt{3}}\right)\left(\frac{x^{3/2}}{\frac{3}{2}}\bigg|_{0\text{ ft}}^{3\text{ ft}}\right)$$

$$= \left(\frac{300\ \frac{\text{lbf}}{\text{ft}}}{\frac{3}{2}\sqrt{3}}\right)(3^{3/2} - 0^{3/2})$$

$$= \boxed{600\text{ lbf}} \quad \text{[first concentrated load]}$$

Next, from Eq. 2.4,

$$dA = f(x)\,dx = 300\sqrt{\frac{x}{3}}$$

Finally, use Eq. 2.1 to find the location, x_c, of the concentrated load from the left end.

$$x_c = \frac{\int x\,dA}{A} = \frac{1}{600\text{ lbf}}\int_{0\text{ ft}}^{3\text{ ft}} 300x\sqrt{\frac{x}{3}}\,dx$$

$$= \frac{300}{600\sqrt{3}}\int_{0\text{ ft}}^{3\text{ ft}} x^{3/2}\,dx$$

$$= \frac{300x^{5/2}}{600\sqrt{3}\left(\frac{5}{2}\right)}\bigg|_{0\text{ ft}}^{3\text{ ft}}$$

$$= \left(\frac{300\ \frac{\text{lbf}}{\text{ft}}}{(600\text{ lbf})\sqrt{3}\left(\frac{5}{2}\right)}\right)(3^{5/2} - 0^{5/2})$$

$$= \boxed{1.8\text{ ft}} \quad \text{[location]}$$

Alternative solution for the parabola:

Use App. 2.A.

$$A = \frac{2bh}{3} = \frac{(2)\left(300\ \frac{\text{lbf}}{\text{ft}}\right)(3\text{ ft})}{3} = \boxed{600\text{ lbf}}$$

The centroid is located at a distance from the left end of

$$\frac{3h}{5} = \frac{(3)(3\text{ ft})}{5} = \boxed{1.8\text{ ft}}$$

The concentrated load for the triangular shape is the area from App. 2.A.

$$A = \frac{bh}{2} = \frac{\left(700\ \frac{\text{lbf}}{\text{ft}} - 300\ \frac{\text{lbf}}{\text{ft}}\right)(8\text{ ft})}{2}$$

$$= \boxed{1600\text{ lbf}} \quad \text{[second concentrated load]}$$

From App. 2.A, the location of the concentrated load from the right end is

$$\frac{h}{3} = \frac{8\text{ ft}}{3} = \boxed{2.67\text{ ft}}$$

The concentrated load for the rectangular shape is the area from App. 2.A.

$$A = bh = \left(300\ \frac{\text{lbf}}{\text{ft}}\right)(8\text{ ft})$$

$$= \boxed{2400\text{ lbf}} \quad \text{[third concentrated load]}$$

From App. 2.A, the location of the concentrated load from the right end is

$$\frac{h}{2} = \frac{8\text{ ft}}{2} = \boxed{4\text{ ft}}$$

SI Solution

The parabolic shape is

$$f(x) = \left(4000\ \frac{\text{N}}{\text{m}}\right)\sqrt{x}$$

First, use Eq. 2.3 to find the concentrated load given by the area.

$$A = \int f(x)\,dx = \int_{0\text{ m}}^{1\text{ m}} 4000\sqrt{x}\,dx = \frac{4000x^{3/2}}{\frac{3}{2}}\bigg|_{0\text{ m}}^{1\text{ m}}$$

$$= \left(\frac{4000\ \frac{\text{N}}{\text{m}}}{\frac{3}{2}}\right)(1^{3/2} - 0^{3/2})$$

$$= \boxed{2666.7\text{ N}} \quad \text{[first concentrated load]}$$

Next, from Eq. 2.4,

$$dA = f(x)\,dx = 4000\sqrt{x}\,dx$$

Finally, use Eq. 2.1 to find the location, x_c, of the concentrated load from the left end.

$$
\begin{aligned}
x_c &= \frac{\int x\, dA}{A} = \frac{1}{2666.7 \text{ N}} \int_{0 \text{ m}}^{1 \text{ m}} 4000 x \sqrt{x}\, dx \\
&= \frac{4000}{2666.7} \int_{0 \text{ m}}^{1 \text{ m}} x^{3/2}\, dx \\
&= \frac{4000 x^{5/2}}{(2666.7)\left(\frac{5}{2}\right)} \bigg|_{0 \text{ m}}^{1 \text{ m}} \\
&= \left(\frac{4000\, \frac{\text{N}}{\text{m}}}{(2666.7 \text{ N})\left(\frac{5}{2}\right)} \right)(1^{5/2} - 0^{5/2}) \\
&= \boxed{0.60 \text{ m}} \qquad \text{[location]}
\end{aligned}
$$

Alternative solution for the parabola:

Use App. 2.A.

$$
\begin{aligned}
A &= \frac{2bh}{3} = \frac{(2)\left(4000\, \frac{\text{N}}{\text{m}}\right)(1 \text{ m})}{3} \\
&= \boxed{2666.7 \text{ N}}
\end{aligned}
$$

The centroid is located at a distance from the left end of

$$
\frac{3h}{5} = \frac{(3)(1 \text{ m})}{5} = \boxed{0.6 \text{ m}}
$$

The concentrated load for the triangular shape is the area from App. 2.A.

$$
\begin{aligned}
A &= \frac{bh}{2} = \frac{\left(10\,000\, \frac{\text{N}}{\text{m}} - 4000\, \frac{\text{N}}{\text{m}}\right)(2.5 \text{ m})}{2} \\
&= \boxed{7500 \text{ N}} \quad \text{[second concentrated load]}
\end{aligned}
$$

From App. 2.A, the location of the concentrated load from the right end is

$$
\frac{h}{3} = \frac{2.5 \text{ m}}{3} = \boxed{0.83 \text{ m}}
$$

The concentrated load for the rectangular shape is the area from App. 2.A.

$$
\begin{aligned}
A = bh &= \left(4000\, \frac{\text{N}}{\text{m}}\right)(2.5 \text{ m}) \\
&= \boxed{10\,000 \text{ N}} \quad \text{[third concentrated load]}
\end{aligned}
$$

From App. 2.A, the location of the concentrated load from the right end is

$$
\frac{h}{2} = \frac{2.5 \text{ m}}{2} = \boxed{1.25 \text{ m}}
$$

3. The area is divided into three basic shapes.

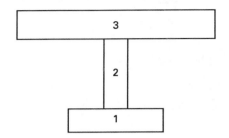

First, calculate the areas of the basic shapes.

$$
\begin{aligned}
A_1 &= (4)(1) = 4 \text{ units}^2 \\
A_2 &= (2)(12) = 24 \text{ units}^2 \\
A_3 &= (6)(0.5) = 3 \text{ units}^2
\end{aligned}
$$

Next, find the y-components of the centroids of the basic shapes.

$$
\begin{aligned}
y_{c,1} &= 0.5 \text{ units} \\
y_{c,2} &= 7 \text{ units} \\
y_{c,3} &= 13.25 \text{ units}
\end{aligned}
$$

From Eq. 2.6, the centroid of the area is

$$
\begin{aligned}
y_c &= \frac{\sum A_i y_{c,i}}{\sum A_i} = \frac{(4)(0.5) + (24)(7) + (3)(13.25)}{4 + 24 + 3} \\
&= 6.77 \text{ units}
\end{aligned}
$$

From App. 2.A, the moment of inertia of basic shape 1 about its own centroid is

$$
I_{cx,1} = \frac{bh^3}{12} = \frac{(4)(1)^3}{12} = 0.33 \text{ units}^4
$$

The moment of inertia of basic shape 2 about its own centroid is

$$
I_{cx,2} = \frac{bh^3}{12} = \frac{(2)(12)^3}{12} = 288 \text{ units}^4
$$

The moment of inertia of basic shape 3 about its own centroid is

$$I_{cx,3} = \frac{bh^3}{12} = \frac{(6)(0.5)^3}{12} = 0.063 \text{ units}^4$$

From the parallel axis theorem, Eq. 2.20, the moment of inertia of basic shape 1 about the centroidal axis of the section is

$$I_{x,1} = I_{cx,1} + A_1 d_1^2 = 0.33 + (4)(6.77 - 0.5)^2$$
$$= 157.6 \text{ units}^4$$

The moment of inertia of basic shape 2 about the centroidal axis of the section is

$$I_{x,2} = I_{cx,2} + A_2 d_2^2 = 288 + (24)(7.0 - 6.77)^2$$
$$= 289.3 \text{ units}^4$$

The moment of inertia of basic shape 3 about the centroidal axis of the section is

$$I_{x,3} = I_{cx,3} + A_3 d_3^2 = 0.063 + (3)(13.25 - 6.77)^2$$
$$= 126.0 \text{ units}^4$$

The total moment of inertia about the centroidal axis of the section is

$$I_x = I_{x,1} + I_{x,2} + I_{x,3}$$
$$= 157.6 \text{ units}^4 + 289.3 \text{ units}^4 + 126.0 \text{ units}^4$$
$$= \boxed{572.9 \text{ units}^4 \quad (570 \text{ units}^4)}$$

The answer is (C).

4. This is a composite area. Divide the composite area into two basic shapes, a rectangle and a circle.

From App. 2.A, the moment of inertia of the rectangle with respect to an edge (in this case, the y-axis) is

$$I_{y,\text{rectangle}} = \frac{bh^3}{3} = \frac{(4 \text{ in})(10 \text{ in})^3}{3} = 1333.33 \text{ in}^4$$

From App. 2.A, the centroidal moment of inertia of the circle is

$$I_{c,\text{circle}} = \frac{\pi r^4}{4} = \frac{\pi (1 \text{ in})^4}{4} = 0.785 \text{ in}^4$$

From the parallel axis theorem, the moment of inertia of the circle with respect to the y-axis is

$$I_{y,\text{circle}} = I_{c,\text{circle}} + d_2 A = 0.785 \text{ in}^4 + (5 \text{ in})^2 \pi (1 \text{ in})^2$$
$$= 79.285 \text{ in}^4$$

The moment of inertia about the y-axis is

$$I_{y,\text{composite area}} = I_{y,\text{rectangle}} - I_{y,\text{circle}}$$
$$= 1333.33 \text{ in}^4 - 79.285 \text{ in}^4$$
$$= \boxed{1254.05 \text{ in}^4 \quad (1250 \text{ in}^4)}$$

The answer is (C).

5. Determine dA, which is the shaded area within the curve parallel to the x-axis.

$$dA = (9 - x)dy$$

Using Eq. 2.17, calculate the moment of inertia with respect to the y-axis from $y = 0$ in to $y = 3$ in.

$$I_x = \int y^2 \, dA = \int_{0 \text{ in}}^{3 \text{ in}} y^2 (9 - x) \, dy$$
$$= \int_{0 \text{ in}}^{3 \text{ in}} y^2 (9 - y^2) \, dy$$
$$= 9 \int_{0 \text{ in}}^{3 \text{ in}} y^2 \, dy - \int_{0 \text{ in}}^{3 \text{ in}} y^4 \, dy$$
$$= (9) \left(\frac{y^3}{3} \Big|_{0 \text{ in}}^{3 \text{ in}} \right) - \frac{y^5}{5} \Big|_{0 \text{ in}}^{3 \text{ in}}$$
$$= \boxed{32.4 \text{ in}^4 \quad (32 \text{ in}^4)}$$

The answer is (A).

Background
and Support

6. Due to symmetry, the moment of inertia and radius of gyration about the y-axis will be the same as for the x-axis. From App. 2.A, the moment of inertia of the annular ring is

$$I = I_{outer} - I_{inner} = \frac{\pi r_{outer}^4}{4} - \frac{\pi r_{inner}^4}{4}$$

$$= \frac{\pi \left(\dfrac{4 \text{ in}}{2}\right)^4}{4} - \frac{\pi \left(\dfrac{2 \text{ in}}{2}\right)^4}{4}$$

$$= 11.78 \text{ in}^4$$

The area of the composite area a is

$$A = \pi r_{outer}^2 - \pi r_{inner}^2 = \pi \left(\frac{4 \text{ in}}{2}\right)^2 - \pi \left(\frac{2 \text{ in}}{2}\right)^2$$

$$= 9.42 \text{ in}^2$$

From Eq. 2.25, the radius of gyration of the annular ring about the y-axis is

$$k = \sqrt{\frac{I}{A}} = \sqrt{\frac{11.78 \text{ in}^4}{9.42 \text{ in}^2}}$$

$$= \boxed{1.12 \text{ in} \quad (1.1 \text{ in})}$$

The answer is (B).

3 Algebra

PRACTICE PROBLEMS

Roots of Quadratic Equations

1. What are the roots of the quadratic equation $x^2 - 7x - 44 = 0$?

(A) −11, 4

(B) −7.5, 3.5

(C) 7.5, −3.5

(D) 11, −4

Logarithm Identities

2. What is most nearly the value of x that satisfies the expression $17.3 = e^{1.1x}$?

(A) 0.17

(B) 2.6

(C) 5.8

(D) 15

Series

3. What is most nearly the following sum?

$$\sum_{j=1}^{5} \left((j+1)^2 - 1 \right)$$

(A) 15

(B) 24

(C) 35

(D) 85

Logarithms

4. If a quantity increases by 0.1% of its current value every 0.1 sec, the doubling time is most nearly

(A) 14 sec

(B) 69 sec

(C) 690 sec

(D) 69,000 sec

SOLUTIONS

1. *method 1:* Use inspection.

Assuming that the roots are integers, there are only a few ways to arrive at 44 by multiplication: 1×44, 2×22, and 4×11. Of these, the difference of 4 and 11 is −7. Therefore, the quadratic equation can be factored into $(x - 11)(x + 4) = 0$. The roots are +11 and −4.

method 2: Use the quadratic formula, where $a = 1$, $b = -7$, and $c = -44$.

$$x_1, x_2 = \frac{-b \pm \sqrt{b^2 - 4ac}}{2a}$$

$$= \frac{-(-7) \pm \sqrt{(-7)^2 - (4)(1)(-44)}}{(2)(1)}$$

$$= \frac{7 \pm 15}{2}$$

$$= -4, 11$$

method 3: Use completing the square.

$$x^2 - 7x - 44 = 0$$

$$x^2 - 7x = 44$$

$$\left(x - \frac{7}{2} \right)^2 = 44 + \left(\frac{7}{2} \right)^2$$

$$\left(x - \frac{7}{2} \right)^2 = 56.25$$

$$x - 3.5 = \sqrt{56.25}$$

$$x = \pm 7.5 + 3.5$$

$$= \boxed{11, -4}$$

The answer is (D).

2. Take the natural logarithm of both sides, using the identity $\log_b b^n = n$.

$$17.3 = e^{1.1x}$$

$$\ln 17.3 = \ln e^{1.1x}$$

$$= 1.1x$$

$$x = \frac{\ln 17.3}{1.1}$$

$$= \boxed{\ln e^{1.1x} \quad (2.6)}$$

The answer is (B).

3. Let $S_n = (j+1)^2 - 1$.

For $j = 1$,

$$S_1 = (1+1)^2 - 1 = 3$$

For $j = 2$,

$$S_2 = (2+1)^2 - 1 = 8$$

For $j = 3$,

$$S_3 = (3+1)^2 - 1 = 15$$

For $j = 4$,

$$S_4 = (4+1)^2 - 1 = 24$$

For $j = 5$,

$$S_5 = (5+1)^2 - 1 = 35$$

Substituting the above expressions gives

$$\sum_{j=1}^{5}\left((j+1)^2 - 1\right) = \sum_{j=1}^{5}S_j$$
$$= S_1 + S_2 + S_3 + S_4 + S_5$$
$$= 3 + 8 + 15 + 24 + 35$$
$$= \boxed{85}$$

The answer is (D).

4. Let n represent the number of elapsed periods of 0.1 sec, and let y_n represent the amount present after n periods.

y_0 represents the initial quantity.

$$y_1 = 1.001 y_0$$
$$y_2 = 1.001 y_1 = (1.001)(1.001 y_0) = (1.001)^2 y_0$$

Therefore, by induction,

$$y_n = (1.001)^n y_0$$

The expression for a doubling of the original quantity is

$$2 y_0 = y_n$$

Substitute for y_n.

$$2 y_0 = (1.001)^n y_0$$
$$2 = (1.001)^n$$

Take the logarithm of both sides.

$$\log 2 = \log(1.001)^n = n \log 1.001$$

Solve for n.

$$n = \frac{\log 2}{\log 1.001} = 693.5$$

Since each period is 0.1 sec, the time is

$$t = n(0.1 \text{ sec}) = (693.5)(0.1 \text{ sec})$$
$$= \boxed{69.35 \text{ sec} \quad (69 \text{ sec})}$$

The answer is (B).

4 Linear Algebra

PRACTICE PROBLEMS

Determinants

1. What is most nearly the determinant of matrix $\mathbf{A}$?

$$\mathbf{A} = \begin{bmatrix} 8 & 2 & 0 & 0 \\ 2 & 8 & 2 & 0 \\ 0 & 2 & 8 & 2 \\ 0 & 0 & 2 & 4 \end{bmatrix}$$

(A) 459

(B) 832

(C) 1552

(D) 1776

Simultaneous Linear Equations

2. Use Cramer's rule to solve for the values of x, y, and z that simultaneously satisfy the following equations.

$$x + y = -4$$
$$x + z - 1 = 0$$
$$2z - y + 3x = 4$$

(A) $(x, y, z) = (3, 2, 1)$

(B) $(x, y, z) = (-3, -1, 2)$

(C) $(x, y, z) = (3, -1, -3)$

(D) $(x, y, z) = (-1, -3, 2)$

SOLUTIONS

1. Expand by cofactors of the first row since there are two zeros in that row.

$$|\mathbf{A}| = 8 \begin{vmatrix} 8 & 2 & 0 \\ 2 & 8 & 2 \\ 0 & 2 & 4 \end{vmatrix} - 2 \begin{vmatrix} 2 & 0 & 0 \\ 2 & 8 & 2 \\ 0 & 2 & 4 \end{vmatrix} + 0 - 0$$

By first row:

$$\begin{vmatrix} 8 & 2 & 0 \\ 2 & 8 & 2 \\ 0 & 2 & 4 \end{vmatrix} = (8)\big((8)(4) - (2)(2)\big) - (2)\big((2)(4) - (2)(0)\big)$$

$$= (8)(28) - (2)(8)$$

$$= 208$$

By first row:

$$\begin{vmatrix} 2 & 0 & 0 \\ 2 & 8 & 2 \\ 0 & 2 & 4 \end{vmatrix} = (2)\big((8)(4) - (2)(2)\big)$$

$$= 56$$

$$|\mathbf{A}| = (8)(208) - (2)(56) = \boxed{1552}$$

The answer is (C).

2. Rearrange the equations.

$$x + y = -4$$
$$x + z = 1$$
$$3x - y + 2z = 4$$

Write the set of equations in matrix form: $\mathbf{AX} = \mathbf{B}$.

$$\begin{bmatrix} 1 & 1 & 0 \\ 1 & 0 & 1 \\ 3 & -1 & 2 \end{bmatrix} \begin{bmatrix} x \\ y \\ z \end{bmatrix} = \begin{bmatrix} -4 \\ 1 \\ 4 \end{bmatrix}$$

Find the determinant of the matrix $\mathbf{A}$.

$$|\mathbf{A}| = \begin{vmatrix} 1 & 1 & 0 \\ 1 & 0 & 1 \\ 3 & -1 & 2 \end{vmatrix}$$

$$= 1 \begin{vmatrix} 0 & 1 \\ -1 & 2 \end{vmatrix} - 1 \begin{vmatrix} 1 & 0 \\ -1 & 2 \end{vmatrix} + 3 \begin{vmatrix} 1 & 0 \\ 0 & 1 \end{vmatrix}$$

$$= (1)\big((0)(2) - (1)(-1)\big)$$
$$\quad - (1)\big((1)(2) - (-1)(0)\big)$$
$$\quad + (3)\big((1)(1) - (0)(0)\big)$$
$$= (1)(1) - (1)(2) + (3)(1)$$
$$= 1 - 2 + 3$$
$$= 2$$

Find the determinant of the substitutional matrix $\mathbf{A_1}$.

$$|\mathbf{A_1}| = \begin{vmatrix} -4 & 1 & 0 \\ 1 & 0 & 1 \\ 4 & -1 & 2 \end{vmatrix}$$

$$= -4 \begin{vmatrix} 0 & 1 \\ -1 & 2 \end{vmatrix} - 1 \begin{vmatrix} 1 & 0 \\ -1 & 2 \end{vmatrix} + 4 \begin{vmatrix} 1 & 0 \\ 0 & 1 \end{vmatrix}$$

$$= (-4)\big((0)(2) - (1)(-1)\big)$$
$$\quad - (1)\big((1)(2) - (-1)(0)\big)$$
$$\quad + (4)\big((1)(1) - (0)(0)\big)$$
$$= (-4)(1) - (1)(2) + (4)(1)$$
$$= -4 - 2 + 4$$
$$= -2$$

Find the determinant of the substitutional matrix $\mathbf{A_2}$.

$$|\mathbf{A_2}| = \begin{vmatrix} 1 & -4 & 0 \\ 1 & 1 & 1 \\ 3 & 4 & 2 \end{vmatrix}$$

$$= 1 \begin{vmatrix} 1 & 1 \\ 4 & 2 \end{vmatrix} - 1 \begin{vmatrix} -4 & 0 \\ 4 & 2 \end{vmatrix} + 3 \begin{vmatrix} -4 & 0 \\ 1 & 1 \end{vmatrix}$$

$$= (1)\big((1)(2) - (4)(1)\big)$$
$$\quad - (1)\big((-4)(2) - (4)(0)\big)$$
$$\quad + (3)\big((-4)(1) - (1)(0)\big)$$
$$= (1)(-2) - (1)(-8) + (3)(-4)$$
$$= -2 + 8 - 12$$
$$= -6$$

Find the determinant of the substitutional matrix $\mathbf{A_3}$.

$$|\mathbf{A_3}| = \begin{vmatrix} 1 & 1 & -4 \\ 1 & 0 & 1 \\ 3 & -1 & 4 \end{vmatrix}$$

$$= 1 \begin{vmatrix} 0 & 1 \\ -1 & 4 \end{vmatrix} - 1 \begin{vmatrix} 1 & -4 \\ -1 & 4 \end{vmatrix} + 3 \begin{vmatrix} 1 & -4 \\ 0 & 1 \end{vmatrix}$$

$$= (1)\big((0)(4) - (-1)(1)\big)$$
$$\quad - (1)\big((1)(4) - (-1)(-4)\big)$$
$$\quad + (3)\big((1)(1) - (0)(-4)\big)$$
$$= (1)(1) - (1)(0) + (3)(1)$$
$$= 1 - 0 + 3$$
$$= 4$$

Use Cramer's rule.

$$x = \frac{|\mathbf{A_1}|}{|\mathbf{A}|} = \frac{-2}{2} = \boxed{-1}$$

$$y = \frac{|\mathbf{A_2}|}{|\mathbf{A}|} = \frac{-6}{2} = \boxed{-3}$$

$$z = \frac{|\mathbf{A_3}|}{|\mathbf{A}|} = \frac{4}{2} = \boxed{2}$$

The answer is (D).

5 Vectors

PRACTICE PROBLEMS

1. What are the dot products for the following vector pairs?

(a) $\mathbf{V}_1 = 2\mathbf{i} + 3\mathbf{j}$; $\mathbf{V}_2 = 5\mathbf{i} - 2\mathbf{j}$

   (A) -4

   (B) 4

   (C) 8

   (D) 11

(b) $\mathbf{V}_1 = 1\mathbf{i} + 4\mathbf{j}$; $\mathbf{V}_2 = 9\mathbf{i} - 3\mathbf{j}$

   (A) -23

   (B) -11

   (C) -3

   (D) 33

(c) $\mathbf{V}_1 = 7\mathbf{i} - 3\mathbf{j}$; $\mathbf{V}_2 = 3\mathbf{i} + 4\mathbf{j}$

   (A) -21

   (B) 9

   (C) 11

   (D) 21

(d) $\mathbf{V}_1 = 2\mathbf{i} - 3\mathbf{j} + 6\mathbf{k}$; $\mathbf{V}_2 = 8\mathbf{i} + 2\mathbf{j} - 3\mathbf{k}$

   (A) -12

   (B) -8

   (C) 37

   (D) 40

(e) $\mathbf{V}_1 = 6\mathbf{i} + 2\mathbf{j} + 3\mathbf{k}$; $\mathbf{V}_2 = \mathbf{i} + \mathbf{k}$

   (A) -11

   (B) 9

   (C) 11

   (D) 13

2. (a) The angle between the vectors in Prob. 1(a) is most nearly

   (A) $60°$

   (B) $70°$

   (C) $80°$

   (D) $100°$

(b) The angle between the vectors in Prob. 1(b) is most nearly

   (A) $70°$

   (B) $85°$

   (C) $90°$

   (D) $95°$

(c) The angle between the vectors in Prob. 1(c) is most nearly

   (A) $73°$

   (B) $76°$

   (C) $100°$

   (D) $120°$

3. (a) The cross product for the vector pair in Prob. 1(a) is

   (A) $-19\mathbf{k}$

   (B) $4\mathbf{k}$

   (C) $11\mathbf{k}$

   (D) $19\mathbf{k}$

(b) The cross product for the vector pair in Prob. 1(b) is

   (A) $-39\mathbf{k}$

   (B) $-23\mathbf{k}$

   (C) $-3\mathbf{k}$

   (D) $33\mathbf{k}$

(c) The cross product for the vector pair in Prob. 1(c) is most nearly

  (A) $-9\mathbf{k}$

  (B) $11\mathbf{k}$

  (C) $21\mathbf{k}$

  (D) $37\mathbf{k}$

(d) The cross product for the vector pair in Prob. 1(d) is most nearly

  (A) $-24\mathbf{i} - 2\mathbf{j} + 10\mathbf{k}$

  (B) $-3\mathbf{i} + 54\mathbf{j} + 28\mathbf{k}$

  (C) $3\mathbf{i} - 42\mathbf{j} + 28\mathbf{k}$

  (D) $21\mathbf{i} - 42\mathbf{j} - 20\mathbf{k}$

(e) The cross product for the vector pair in Prob. 1(e) is most nearly

  (A) $-2\mathbf{i} - 3\mathbf{j} + \mathbf{k}$

  (B) $-9\mathbf{j} + 2\mathbf{k}$

  (C) $2\mathbf{i} - 3\mathbf{j} - 2\mathbf{k}$

  (D) $-5\mathbf{i} - 9\mathbf{j} + 4\mathbf{k}$

SOLUTIONS

1. (a) The dot product is

$$\mathbf{V}_1 \cdot \mathbf{V}_2 = \mathbf{V}_{1x}\mathbf{V}_{2x} + \mathbf{V}_{1y}\mathbf{V}_{2y}$$
$$= (2)(5) + (3)(-2)$$
$$= \boxed{4}$$

The answer is (B).

(b) The dot product is

$$\mathbf{V}_1 \cdot \mathbf{V}_2 = \mathbf{V}_{1x}\mathbf{V}_{2x} + \mathbf{V}_{1y}\mathbf{V}_{2y}$$
$$= (1)(9) + (4)(-3)$$
$$= \boxed{-3}$$

The answer is (C).

(c) The dot product is

$$\mathbf{V}_1 \cdot \mathbf{V}_2 = \mathbf{V}_{1x}\mathbf{V}_{2x} + \mathbf{V}_{1y}\mathbf{V}_{2y}$$
$$= (7)(3) + (-3)(4)$$
$$= \boxed{9}$$

The answer is (B).

(d) The dot product is

$$\mathbf{V}_1 \cdot \mathbf{V}_2 = \mathbf{V}_{1x}\mathbf{V}_{2x} + \mathbf{V}_{1y}\mathbf{V}_{2y} + \mathbf{V}_{1z}\mathbf{V}_{2z}$$
$$= (2)(8) + (-3)(2) + (6)(-3)$$
$$= \boxed{-8}$$

The answer is (B).

(e) The dot product is

$$\mathbf{V}_1 \cdot \mathbf{V}_2 = \mathbf{V}_{1x}\mathbf{V}_{2x} + \mathbf{V}_{1y}\mathbf{V}_{2y} + \mathbf{V}_{1z}\mathbf{V}_{2z}$$
$$= (6)(1) + (2)(0) + (3)(1)$$
$$= \boxed{9}$$

The answer is (B).

2. (a) The angle between the vectors is

$$\cos\phi = \frac{\mathbf{V}_1 \cdot \mathbf{V}_2}{|\mathbf{V}_1||\mathbf{V}_2|} = \frac{4}{\sqrt{(2)^2 + (3)^2}\sqrt{(5)^2 + (-2)^2}}$$

$$= 0.206$$

$$\phi = \arccos 0.206 = \boxed{78.1° \quad (80°)}$$

The answer is (C).

(b) The angle between the vectors is

$$\cos\phi = \frac{\mathbf{V}_1 \cdot \mathbf{V}_2}{|\mathbf{V}_1||\mathbf{V}_2|} = \frac{-3}{\sqrt{(1)^2 + (4)^2}\sqrt{(9)^2 + (-3)^2}}$$

$$= -0.0767$$

$$\phi = \arccos(-0.0767) = \boxed{94.4° \quad (95°)}$$

The answer is (D).

(c) The angle between the vectors is

$$\cos\phi = \frac{\mathbf{V}_1 \cdot \mathbf{V}_2}{|\mathbf{V}_1||\mathbf{V}_2|} = \frac{9}{\sqrt{(7)^2 + (-3)^2}\sqrt{(3)^2 + (4)^2}}$$

$$= 0.236$$

$$\phi = \arccos 0.236 = \boxed{76.3° \quad (76°)}$$

The answer is (B).

3. (a) The cross product is

$$\mathbf{V}_1 \times \mathbf{V}_2 = \begin{vmatrix} \mathbf{i} & \mathbf{V}_{1x} & \mathbf{V}_{2x} \\ \mathbf{j} & \mathbf{V}_{1y} & \mathbf{V}_{2y} \\ \mathbf{k} & \mathbf{V}_{1z} & \mathbf{V}_{2z} \end{vmatrix}$$

$$= \begin{vmatrix} \mathbf{i} & 2 & 5 \\ \mathbf{j} & 3 & -2 \\ \mathbf{k} & 0 & 0 \end{vmatrix}$$

Expand by the third row.

$$\mathbf{V}_1 \times \mathbf{V}_2 = \mathbf{k}\begin{vmatrix} 2 & 5 \\ 3 & -2 \end{vmatrix} = ((2)(-2) - (3)(5))\mathbf{k}$$

$$= \boxed{-19\mathbf{k}}$$

The answer is (A).

(b) The cross product is

$$\mathbf{V}_1 \times \mathbf{V}_2 = \begin{vmatrix} \mathbf{i} & 1 & 9 \\ \mathbf{j} & 4 & -3 \\ \mathbf{k} & 0 & 0 \end{vmatrix}$$

Expand by the third row.

$$\mathbf{V}_1 \times \mathbf{V}_2 = \mathbf{k}\begin{vmatrix} 1 & 9 \\ 4 & -3 \end{vmatrix} = ((1)(-3) - (4)(9))\mathbf{k}$$

$$= \boxed{-39\mathbf{k}}$$

The answer is (A).

(c) The cross product is

$$\mathbf{V}_1 \times \mathbf{V}_2 = \begin{vmatrix} \mathbf{i} & \mathbf{V}_{1x} & \mathbf{V}_{2x} \\ \mathbf{j} & \mathbf{V}_{1y} & \mathbf{V}_{2y} \\ \mathbf{k} & \mathbf{V}_{1z} & \mathbf{V}_{2z} \end{vmatrix} = \begin{vmatrix} \mathbf{i} & 7 & 3 \\ \mathbf{j} & -3 & 4 \\ \mathbf{k} & 0 & 0 \end{vmatrix}$$

Expand by the third row.

$$\mathbf{V}_1 \times \mathbf{V}_2 = \mathbf{k}\begin{vmatrix} 7 & 3 \\ -3 & 4 \end{vmatrix} = ((7)(4) - (-3)(3))\mathbf{k}$$

$$= \boxed{37\mathbf{k}}$$

The answer is (D).

(d) The cross product is

$$\mathbf{V}_1 \times \mathbf{V}_2 = \begin{vmatrix} \mathbf{i} & \mathbf{V}_{1x} & \mathbf{V}_{2x} \\ \mathbf{j} & \mathbf{V}_{1y} & \mathbf{V}_{2y} \\ \mathbf{k} & \mathbf{V}_{1z} & \mathbf{V}_{2z} \end{vmatrix} = \begin{vmatrix} \mathbf{i} & 2 & 8 \\ \mathbf{j} & -3 & 2 \\ \mathbf{k} & 6 & -3 \end{vmatrix}$$

Expand by the first column.

$$\mathbf{V}_1 \times \mathbf{V}_2 = \mathbf{i}\begin{vmatrix} -3 & 2 \\ 6 & -3 \end{vmatrix} - \mathbf{j}\begin{vmatrix} 2 & 8 \\ 6 & -3 \end{vmatrix} + \mathbf{k}\begin{vmatrix} 2 & 8 \\ -3 & 2 \end{vmatrix}$$

$$= \boxed{-3\mathbf{i} + 54\mathbf{j} + 28\mathbf{k}}$$

The answer is (B).

Background and Support

Background
and Support

(e) The cross product is

$$\mathbf{V}_1 \times \mathbf{V}_2 = \begin{vmatrix} \mathbf{i} & V_{1x} & V_{2x} \\ \mathbf{j} & V_{1y} & V_{2y} \\ \mathbf{k} & V_{1z} & V_{2z} \end{vmatrix} = \begin{vmatrix} \mathbf{i} & 6 & 1 \\ \mathbf{j} & 2 & 0 \\ \mathbf{k} & 3 & 1 \end{vmatrix}$$

Expand by the second row.

$$\mathbf{V}_1 \times \mathbf{V}_2 = -\mathbf{j}\begin{vmatrix} 6 & 1 \\ 3 & 1 \end{vmatrix} + (2)\begin{vmatrix} \mathbf{i} & 1 \\ \mathbf{k} & 1 \end{vmatrix}$$
$$= \boxed{2\mathbf{i} - 3\mathbf{j} - 2\mathbf{k}}$$

The answer is (C).

6 Trigonometry

PRACTICE PROBLEMS

1. A 5 lbm (5 kg) block sits on a 20° incline without slipping. (a) Draw the free-body diagram with respect to the axes parallel and perpendicular to the surface of the incline. (b) What is most nearly the magnitude of the frictional force (holding the block stationary) on the block?

(A) 1.7 lbf (17 N)

(B) 3.4 lbf (33 N)

(C) 4.7 lbf (46 N)

(D) 5.0 lbf (49 N)

2. Complete the following calculation related to a catenary cable.

$$S = c\left(\cosh\frac{a}{h} - 1\right) = (245 \text{ ft})\left(\cosh\frac{50 \text{ ft}}{245 \text{ ft}} - 1\right)$$

(A) 3.9 ft

(B) 4.5 ft

(C) 5.1 ft

(D) 7.4 ft

3. Part of a turn-around area in a parking lot is shaped as a circular segment. The segment has a central angle of 120° and a radius of 75 ft. If the circular segment is to receive a special surface treatment, what area will be treated?

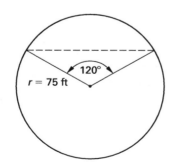

(A) 49 ft²

(B) 510 ft²

(C) 3500 ft²

(D) 5900 ft²

SOLUTIONS

1. (a) The free-body diagram is

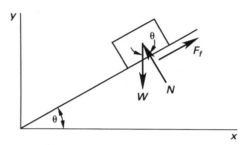

Customary U.S. Solution

(b) The mass of the block is $m = 5$ lbm. The angle of inclination is $\theta = 20°$. The weight is

$$W = \frac{mg}{g_c} = \frac{(5 \text{ lbm})\left(32.2 \frac{\text{ft}}{\text{sec}^2}\right)}{32.2 \frac{\text{lbm-ft}}{\text{lbf-sec}^2}} = 5 \text{ lbf}$$

The frictional force is

$$F_f = W\sin\theta = (5 \text{ lbf})\sin 20°$$
$$= \boxed{1.71 \text{ lbf} \quad (1.7 \text{ lbf})}$$

The answer is (A).

SI Solution

(b) The mass of the block is $m = 5$ kg. The angle of inclination is $\theta = 20°$. The gravitational force is

$$W = mg = (5 \text{ kg})\left(9.81 \frac{\text{m}}{\text{s}^2}\right) = 49.1 \text{ N}$$

The frictional force is

$$F_f = W\sin\theta = (49.1 \text{ N})\sin 20°$$
$$= \boxed{16.8 \text{ N} \quad (17 \text{ N})}$$

The answer is (A).

Background
and Support

2. This calculation contains a hyperbolic cosine function.

$$S = c\left(\cosh\frac{a}{h} - 1\right)$$

$$= (245 \text{ ft})\left(\cosh\frac{50 \text{ ft}}{245 \text{ ft}} - 1\right)$$

$$= (245 \text{ ft})(1.0209 - 1)$$

$$= \boxed{5.12 \text{ ft} \quad (5.1 \text{ ft})}$$

The answer is (C).

3. From App. 7.A, the area of a circular segment is

$$A = \tfrac{1}{2}r^2(\phi - \sin\phi)$$

Since ϕ appears in the expression by itself, it must be expressed in radians.

$$\phi = \frac{(120°)(2\pi)}{360°} = 2.094 \text{ rad}$$

$$A = \tfrac{1}{2}r^2(\phi - \sin\phi) = \left(\tfrac{1}{2}\right)(75 \text{ ft})^2(2.094 \text{ rad} - \sin 120°)$$

$$= \boxed{3453.7 \text{ ft}^2 \quad (3500 \text{ ft}^2)}$$

The answer is (C).

7 Analytic Geometry

PRACTICE PROBLEMS

1. The diameter of a sphere and the base of a cone are equal. What approximate percentage of that diameter must the cone's height be so that both volumes are equal?

(A) 133%

(B) 150%

(C) 166%

(D) 200%

2. The distance between the entrance and exit points on a horizontal circular roadway curve is 747 ft. The radius of the curve is 400 ft. The central angle between the entrance and exit points is most nearly

(A) 0.27°

(B) 54°

(C) 110°

(D) 340°

3. A vertical parabolic roadway crest curve starts deviating from a constant grade at station 103 (i.e., 10,300 ft from an initial benchmark). At sta 103+62, the curve is 2.11 ft lower than the tangent (i.e., from the straight line extension of the constant grade). Approximately how far will the curve be from the tangent at sta 103+87?

(A) 1.1 ft

(B) 1.5 ft

(C) 3.0 ft

(D) 4.2 ft

4. A pile driving hammer emits 143 W of sound power with each driving stroke. Assume isotropic emission and disregard reflected power. What is most nearly the maximum areal sound power density at the ground when 10.7 m of pile remains to be driven?

(A) 0.030 W/m^2

(B) 0.10 W/m^2

(C) 0.53 W/m^2

(D) 1.1 W/m^2

SOLUTIONS

1. Let d be the diameter of the sphere and the base of the cone. Use App. 7.B.

The volume of the sphere is

$$V_{\text{sphere}} = \tfrac{4}{3}\pi r^3 = \tfrac{4}{3}\pi\left(\frac{d}{2}\right)^3$$
$$= \tfrac{1}{6}\pi d^3$$

The volume of the circular cone is

$$V_{\text{cone}} = \tfrac{1}{3}\pi r^2 h = \tfrac{1}{3}\pi\left(\frac{d}{2}\right)^2 h$$
$$= \tfrac{1}{12}\pi d^2 h$$

Since the volume of the sphere and cone are equal,

$$V_{\text{cone}} = V_{\text{sphere}}$$
$$\tfrac{1}{12}\pi d^2 h = \tfrac{1}{6}\pi d^3$$
$$h = 2d$$

The height of the cone must be $\boxed{200\%}$ of the diameter of the sphere.

The answer is (D).

2. Horizontal roadway curves are circular arcs. The circumference (perimeter) of an entire circle with a radius of 400 ft is

$$p = 2\pi r = (2\pi)(400 \text{ ft}) = 2513.3 \text{ ft}$$

From a ratio of curve length to angles,

$$\phi = \left(\frac{747 \text{ ft}}{2513.3 \text{ ft}}\right)(360°) = \boxed{107° \quad (110°)}$$

The answer is (C).

Background and Support

3. Vertical roadway curves are parabolic arcs. Parabolas are second-degree polynomials. Deviations, y, from a baseline are proportional to the square of the separation distance. That is, $y \propto x^2$.

$$\frac{y_1}{y_2} = \left(\frac{x_1}{x_2}\right)^2$$

$$y_2 = y_1 \left(\frac{x_2}{x_1}\right)^2 = (2.11 \text{ ft})\left(\frac{87 \text{ ft}}{62 \text{ ft}}\right)^2$$

$$= \boxed{4.15 \text{ ft} \quad (4.2 \text{ ft})}$$

The answer is (D).

4. The power is emitted isotropically, spherically, in all directions. The maximum sound power will occur at the surface, adjacent to the pile. The surface area of a sphere with a radius of 10.7 m is

$$A = 4\pi r^2 = (4\pi)(10.7 \text{ m})^2 = 1438.7 \text{ m}^2$$

The areal sound power density is

$$\rho_S = \frac{P}{A} = \frac{143 \text{ W}}{1438.7 \text{ m}^2}$$

$$= \boxed{0.0994 \text{ W/m}^2 \quad (0.10 \text{ W/m}^2)}$$

The answer is (B).

Differential Calculus

PRACTICE PROBLEMS

1. Find all minima, maxima, and inflection points for

$$y = x^3 - 9x^2 - 3$$

(A) inflection at $x = -3$
maximum at $x = 0$
minimum at $x = -6$

(B) inflection at $x = 3$
maximum at $x = 0$
minimum at $x = 6$

(C) inflection at $x = 3$
maximum at $x = 6$
minimum at $x = 0$

(D) inflection at $x = 0$
maximum at $x = -3$
minimum at $x = 3$

2. The equation for the elevation above mean sea level of a sag vertical roadway curve is

$$y(x) = 0.56x^2 - 3.2x + 708.28$$

y is measured in feet, and x is measured in 100 ft stations past the beginning of the curve. What is most nearly the elevation of the turning point (i.e., the lowest point on the curve)?

(A) 702 ft

(B) 704 ft

(C) 705 ft

(D) 706 ft

3. A car drives on a highway with a legal speed limit of 100 km/h. The fuel usage, Q (in liters per 100 kilometers driven), of a car driven at speed v (in km/h) is

$$Q(v) = \frac{1750v}{v^2 + 6700}$$

At what approximate legal speed should the car travel in order to maximize the fuel efficiency?

(A) 82 km/h

(B) 87 km/h

(C) 93 km/h

(D) 100 km/h

4. A chemical feed storage tank is needed with a volume of 3000 ft^3 (gross of fittings). The tank will be formed as a circular cylinder with barrel length, L, capped by two hemispherical ends of radius, r. The manufacturing cost per unit area of hemispherical ends is double that of the cylinder. The dimensions that will minimize the manufacturing cost are most nearly

(A) radius $= 4^1/_2$ ft; cylinder barrel length $= 42$ ft

(B) radius $= 5$ ft; cylinder barrel length $= 31^1/_2$ ft

(C) radius $= 5^1/_2$ ft; cylinder barrel length $= 22^1/_2$ ft

(D) radius $= 6$ ft; cylinder barrel length $= 18^1/_2$ ft

SOLUTIONS

1. Determine the critical points by taking the first derivative of the function and setting it equal to zero.

$$\frac{dy}{dx} = 3x^2 - 18x = 3x(x - 6)$$

$$3x(x - 6) = 0$$

$$x(x - 6) = 0$$

The critical points are located at $x = 0$ and $x = 6$.

Determine the inflection points by setting the second derivative equal to zero. Take the second derivative.

$$\frac{d^2y}{dx^2} = \left(\frac{d}{dx}\right)\left(\frac{dy}{dx}\right)$$

$$= \frac{d}{dx}(3x^2 - 18x)$$

$$= 6x - 18$$

Set the second derivative equal to zero.

$$\frac{d^2y}{dx^2} = 0 = 6x - 18 = (6)(x - 3)$$

$$(6)(x - 3) = 0$$

$$x - 3 = 0$$

$$x = 3$$

The inflection point is at $\boxed{x = 3.}$

Determine the local maximum and minimum by substituting the critical points into the expression for the second derivative.

At the critical point $x = 0$,

$$\left.\frac{d^2y}{dx^2}\right|_{x=0} = (6)(x - 3) = (6)(0 - 3) = -18$$

Since $-18 < 0$, $\boxed{x = 0}$ is a local maximum.

At the critical point $x = 6$,

$$\left.\frac{d^2y}{dx^2}\right|_{x=6} = (6)(x - 3) = (6)(6 - 3) = 18$$

Since $18 > 0$, $\boxed{x = 6}$ is a local minimum.

The answer is (B).

2. Set the derivative of the curve's equation to zero.

$$\frac{dy(x)}{dx} = \frac{d}{dx}(0.56x^2 - 3.2x + 708.28) = 1.12x - 3.2$$

$$x_c = \frac{3.2}{1.12} = 2.857 \text{ sta}$$

Insert x_c into the elevation equation.

$$y_{min} = y(x_c) = 0.56x^2 - 3.2x + 708.28$$

$$= (0.56)(2.857)^2 - (3.2)(2.857) + 708.28$$

$$= \boxed{703.71 \text{ ft} \quad (704 \text{ ft})}$$

The answer is (B).

3. Use the quotient rule to calculate the derivative.

$$\mathbf{D}\left(\frac{f(x)}{g(x)}\right) = \frac{g(x)\mathbf{D}f(x) - f(x)\mathbf{D}g(x)}{\left(g(x)\right)^2}$$

$$\frac{dQ(v)}{dv} = \frac{d}{dv}\left(\frac{1750v}{v^2 + 6700}\right)$$

$$= \frac{(v^2 + 6700)(1750) - (1750v)(2v)}{(v^2 + 6700)^2}$$

Combining terms and simplifying,

$$\frac{dQ(v)}{dv} = \frac{-1750v^2 + 11,725,000}{v^4 + 13,400v^2 + 44,890,000}$$

Set the derivative of the fuel consumption equation to zero. In order for the derivative to be zero, the numerator must be zero.

$$-1750v^2 + 11,725,000 = 0$$

$$v = \sqrt{\frac{11,725,000}{1750}} = 81.85 \quad (82 \text{ km/h})$$

Maximizing the fuel efficiency is the same as minimizing the fuel usage. It is not known if setting $dQ(v)/dt = 0$ results in a minimum or maximum. While using $d^2 Q(v)/dt^2$ is possible, it is easier just to plot the points.

v	Q(v)
82 km/h	10.689 L
87 km/h	10.669 L
93 km/h	10.603 L
100 km/h	10.479 L

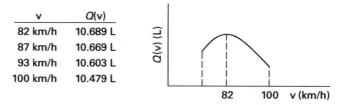

The answer is (B).

Clearly, $Q(82 \text{ km/h})$ is a maximum, and the minimum fuel usage occurs at the endpoint of the range, at $\boxed{100 \text{ km/h.}}$

The answer is (D).

4. The volume of the tank will be the combined volume of a cylinder and a sphere. The cylinder and sphere have the same radius, r.

$$V = \pi r^2 L + \tfrac{4}{3}\pi r^3 = 3000 \text{ ft}^3$$

For any given cost per unit area (arbitrarily selected as $\$1/\text{ft}^2$), the cost function is

$$C(r, L) = \left(1 \; \frac{\$}{\text{ft}^2}\right)(A_{\text{cylinder}} + 2A_{\text{sphere}}) = 2\pi r L + 2(4\pi r^2)$$

$$= 2\pi r L + 8\pi r^2$$

Solve the volume equation for barrel length, L.

$$L = \frac{3000 - \tfrac{4}{3}\pi r^3}{\pi r^2}$$

Substitute L into the cost equation to get a cost function of a single variable.

$$C(r) = 2\pi r L + 8\pi r^2 = 2\pi r\left(\frac{3000 - \tfrac{4}{3}\pi r^3}{\pi r^2}\right) + 8\pi r^2$$

$$= \frac{6000}{r} - \frac{8\pi r^2}{3} + 8\pi r^2$$

$$= \frac{6000}{r} + \frac{16\pi r^2}{3}$$

Find the optimal value of r by setting the first derivative of the cost function equal to zero.

$$\frac{dC(r)}{dr} = \frac{d}{dr}\left(\frac{6000}{r} + \frac{16\pi r^2}{3}\right) = \frac{-6000}{r^2} + \frac{32\pi r}{3} = 0$$

$$\frac{6000}{r^2} = \frac{32\pi r}{3}$$

Cross multiply, and solve for the optimal value of radius, r.

$$32\pi r^3 = 18{,}000$$

$$r = \sqrt[3]{\frac{18{,}000}{32\pi}}$$

$$= \boxed{5.636 \text{ ft} \quad (5\tfrac{1}{2} \text{ ft})}$$

Calculate the optimal value of the barrel length, L.

$$L = \frac{3000 - \tfrac{4}{3}\pi r^3}{\pi r^2}$$

$$= \frac{3000 - \tfrac{4}{3}\pi(5.636 \text{ ft})^3}{\pi(5.636 \text{ ft})^2}$$

$$= \boxed{22.55 \text{ ft} \quad (22\tfrac{1}{2} \text{ ft})}$$

The answer is (C).

 **Integral Calculus**

PRACTICE PROBLEMS

1. Find the indefinite integrals.

(a) $\int \sqrt{1-x}\, dx$

    (A) $-\frac{2}{3}(1-x)^{3/2} + C$

    (B) $-\frac{1}{2}(1-x)^{-1/2} + C$

    (C) $\frac{3}{2}(1-x)^{3/2} + C$

    (D) $2(1-x)^{3/2} + C$

(b) $\int \frac{x}{x^2+1}\, dx$

    (A) $\frac{1}{x}\ln|(x^2+1)| + C$

    (B) $\frac{1}{4}\ln|(x^2+1)| + C$

    (C) $\frac{1}{2}\ln|(x^2+1)| + C$

    (D) $\ln|(x^2+1)| + C$

(c) $\int \frac{x^2}{x^2+x-6}\, dx$

    (A) $\ln|(x+3)| + \frac{4}{5}\ln|(x+2)| + C$

    (B) $\ln|(x-3)| + \frac{4}{5}\ln|(x-2)| + C$

    (C) $x - \frac{5}{9}\ln|(x-3)| + \frac{4}{5}\ln|(x-2)| + C$

    (D) $x - \frac{9}{5}\ln|(x+3)| + \frac{4}{5}\ln|(x-2)| + C$

2. Calculate the definite integrals.

(a) $\int_1^3 (x^2 + 4x)\, dx$

    (A) $16\frac{1}{2}$

    (B) $24\frac{2}{3}$

    (C) 27

    (D) $42\frac{1}{3}$

(b) $\int_{-2}^2 (x^3 + 1)\, dx$

    (A) -4

    (B) -2

    (C) 0

    (D) 4

(c) $\int_1^2 (4x^3 - 3x^2)\, dx$

    (A) 8

    (B) 16

    (C) 24

    (D) 32

3. Find the area bounded by $x = 1$, $x = 3$, $y + x + 1 = 0$, and $y = 6x - x^2$.

    (A) $7\frac{1}{2}$

    (B) $13\frac{1}{3}$

    (C) $21\frac{1}{3}$

    (D) $25\frac{1}{2}$

4. The velocity profile of a fluid experiencing laminar flow in a pipe of radius R is

$$v(r) = v_{\max}\left(1 - \left(\frac{r}{R}\right)^2\right)$$

r is the distance from the centerline. $v_{\max}$ is the centerline velocity.

(a) What is the volumetric flow rate?

    (A) $\dfrac{\pi v_{\max} R^2}{2}$

    (B) $\pi v_{\max} R^2$

    (C) $\dfrac{3\pi v_{\max} R^2}{2}$

    (D) $2\pi v_{\max} R^2$

(b) What is the average velocity?

    (A) $\dfrac{v_{\max}}{6}$

    (B) $\dfrac{v_{\max}}{4}$

    (C) $\dfrac{v_{\max}}{3}$

    (D) $\dfrac{v_{\max}}{2}$

5. Find a_0 (the first term of a Fourier series approximation, corresponding to the waveform's average value) for the two waveforms shown.

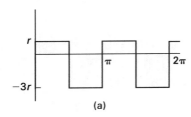

(a)

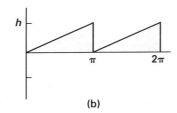

(b)

(a) For waveform (a),

   (A) $-r$

   (B) $-\frac{1}{2}r$

   (C) r

   (D) $2r$

(b) For waveform (b),

   (A) $-\dfrac{h}{2}$

   (B) $\dfrac{h}{2}$

   (C) h

   (D) $2h$

6. For each of the two waveforms shown, determine if its Fourier series is of type A, B, or C.

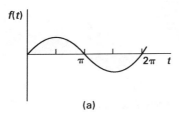

(a)

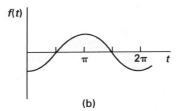

(b)

type A: $f(t) = a_0 + a_2 \cos 2t + b_2 \sin 2t$
$$+ \, a_4 \cos 4t + b_4 \sin 4t + \cdots$$

type B: $f(t) = a_0 + b_1 \sin t + b_2 \sin 2t + b_3 \sin 3t + \cdots$

type C: $f(t) = a_0 + a_1 \cos t + a_2 \cos 2t + a_3 \cos 3t + \cdots$

(a) For waveform (a),

   (A) type A

   (B) type B

   (C) type C

   (D) both type A and type C

(b) For waveform (b),

   (A) type A

   (B) type B

   (C) type C

   (D) both type A and type C

SOLUTIONS

1. (a) The indefinite integral is

$$\int \sqrt{1-x}\,dx = \int (1-x)^{1/2}\,dx$$
$$= \boxed{-\tfrac{2}{3}(1-x)^{3/2} + C}$$

The answer is (A).

(b) The indefinite integral is

$$\int \frac{x}{x^2+1}\,dx = \tfrac{1}{2}\int \frac{2x}{x^2+1}\,dx$$
$$= \boxed{\tfrac{1}{2}\ln|(x^2+1)| + C}$$

The answer is (C).

(c) The indefinite integral is

$$\frac{x^2}{x^2+x-6} = \frac{x^2-x-x-6+6}{x^2+x-6}$$
$$= 1 - \frac{x-6}{x^2+x-6}$$
$$= 1 - \frac{x-6}{(x+3)(x-2)}$$
$$= 1 - \frac{\tfrac{9}{5}}{x+3} + \frac{\tfrac{4}{5}}{x-2}$$

$$\int \frac{x^2}{x^2+x-6}\,dx = \int \left(1 - \frac{\tfrac{9}{5}}{x+3} + \frac{\tfrac{4}{5}}{x-2}\right) dx$$
$$= \int dx - \int \frac{\tfrac{9}{5}}{x+3}\,dx + \int \frac{\tfrac{4}{5}}{x-2}\,dx$$
$$= \boxed{x - \tfrac{9}{5}\ln|(x+3)| + \tfrac{4}{5}\ln|(x-2)| + C}$$

The answer is (D).

2. (a) The definite integral is

$$\int_1^3 (x^2+4x)\,dx = \left(\frac{x^3}{3} + 2x^2\right)\Bigg|_1^3$$
$$= \frac{(3)^3}{3} + (2)(3)^2 - \left(\frac{(1)^3}{3} + (2)(1)^2\right)$$
$$= \boxed{24\tfrac{2}{3}}$$

The answer is (B).

(b) The definite integral is

$$\int_{-2}^2 (x^3+1)\,dx = \left(\frac{x^4}{4} + x\right)\Bigg|_{-2}^2$$
$$= \frac{(2)^4}{4} + 2 - \left(\frac{(-2)^4}{4} + (-2)\right)$$
$$= \boxed{4}$$

The answer is (D).

(c) The definite integral is

$$\int_1^2 (4x^3 - 3x^2)\,dx = \left(x^4 - x^3\right)\Bigg|_1^2$$
$$= (2)^4 - (2)^3 - \left((1)^4 - (1)^3\right)$$
$$= \boxed{8}$$

The answer is (A).

3. The bounded area is shown.

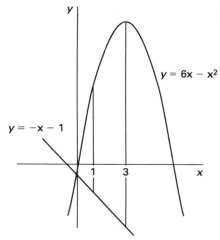

$$\text{area} = \int_1^3 \left((6x - x^2) - (-x - 1)\right) dx$$
$$= \int_1^3 (-x^2 + 7x + 1)\,dx$$
$$= \left(-\frac{x^3}{3} + \tfrac{7}{2}x^2 + x\right)\Bigg|_1^3$$
$$= -\frac{(3)^3}{3} + \left(\tfrac{7}{2}\right)(3)^2 + 3 - \left(-\frac{(1)^3}{3} + \left(\tfrac{7}{2}\right)(1)^2 + 1\right)$$
$$= \boxed{21\tfrac{1}{3}}$$

The answer is (C).

4. (a) Divide the circular internal area of the pipe into small annular rings. The radius at the ring is r; the differential thickness of the ring is dr; the differential area of the ring is $dA = 2\pi r\, dr$; and the velocity is $\mathrm{v}(r)$. The volumetric flow rate through the annular ring is

$$dQ = \mathrm{v}(r)\, dA = 2\pi r \mathrm{v}_{\max}\left(1 - \left(\frac{r}{R}\right)^2\right) dr$$

$$Q = \int_{r=0}^{r=R} 2\pi r \mathrm{v}_{\max}\left(1 - \left(\frac{r}{R}\right)^2\right) dr$$

$$= 2\pi \mathrm{v}_{\max} \int_{r=0}^{r=R} \left(r - \frac{r^3}{R^2}\right) dr$$

$$= 2\pi \mathrm{v}_{\max} \left.\left(\frac{r^2}{2} - \frac{r^4}{4R^2}\right)\right|_0^R$$

$$= 2\pi \mathrm{v}_{\max} \left(\frac{R^2}{2} - \frac{R^4}{4R^2} - 0 - 0\right)$$

$$= \boxed{\frac{\pi \mathrm{v}_{\max} R^2}{2}}$$

The answer is (A).

(b) The average velocity is

$$\overline{\mathrm{v}} = \frac{Q}{A} = \frac{\dfrac{\pi \mathrm{v}_{\max} R^2}{2}}{\pi R^2} = \boxed{\frac{\mathrm{v}_{\max}}{2}}$$

The answer is (D).

5. (a) For waveform (a),

$$a_0 = \frac{1}{2\pi} \int_0^{2\pi} f(t)\, dt = \frac{1}{\pi} \int_0^{\pi} f(t)\, dt$$

$$= \frac{1}{\pi}\left(r\left(\frac{\pi}{2}\right) + (-3r)\left(\frac{\pi}{2}\right)\right)$$

$$= \boxed{-r}$$

The answer is (A).

(b) For waveform (b),

$$a_0 = \frac{1}{2\pi} \int_0^{2\pi} f(t)\, dt = \frac{1}{\pi} \int_0^{\pi} f(t)\, dt$$

$$= \left(\frac{1}{\pi}\right)\left(\tfrac{1}{2}\pi h\right)$$

$$= \boxed{\frac{h}{2}}$$

The answer is (B).

6. (a) For waveform (a),

Since $f(t) = -f(-t)$, it is $\boxed{\text{type B.}}$

The answer is (B).

(b) For waveform (b),

Since $f(t) = f(-t)$, it is $\boxed{\text{type C.}}$

The answer is (C).

10 Differential Equations

PRACTICE PROBLEMS

1. Solve the following differential equation for y.

$$y'' - 4y' - 12y = 0$$

(A) $A_1 e^{6x} + A_2 e^{-2x}$

(B) $A_1 e^{-6x} + A_2 e^{2x}$

(C) $A_1 e^{6x} + A_2 e^{2x}$

(D) $A_1 e^{-6x} + A_2 e^{-2x}$

2. Solve the following differential equation for y.

$$y' - y = 2xe^{2x} \quad y(0) = 1$$

(A) $y = 2e^{-2x}(x-1) + 3e^{-x}$

(B) $y = 2e^{2x}(x-1) + 3e^{x}$

(C) $y = -2e^{-2x}(x-1) + 3e^{-x}$

(D) $y = 2e^{2x}(x-1) + 3e^{-x}$

3. The oscillation exhibited by the top story of a certain building in free motion is given by the following differential equation.

$$x'' + 2x' + 2x = 0 \quad x(0) = 0 \quad x'(0) = 1$$

(a) What is x as a function of time?

(A) $e^{-2t}\sin t$

(B) $e^{t}\sin t$

(C) $e^{-t}\sin t$

(D) $e^{-t}\sin t + e^{-t}\cos t$

(b) The building's fundamental natural frequency of vibration is most nearly

(A) $^1/_2$

(B) 1

(C) $\sqrt{2}$

(D) 2

(c) The amplitude of oscillation is most nearly

(A) 0.32

(B) 0.54

(C) 1.7

(D) 6.6

(d) What is x as a function of time if a lateral wind load is applied with a form of $\sin t$?

(A) $\frac{6}{5}e^{-t}\sin t + \frac{2}{5}e^{-t}\cos t$

(B) $\frac{6}{5}e^{t}\sin t + \frac{2}{5}e^{t}\cos t$

(C) $\frac{2}{5}e^{t}\sin t + \frac{6}{5}e^{-t}\cos t + \frac{2}{5}\sin t - \frac{1}{5}\cos t$

(D) $\frac{2}{5}e^{-t}\cos t + \frac{6}{5}e^{-t}\sin t - \frac{2}{5}\cos t + \frac{1}{5}\sin t$

4. A 90 lbm (40 kg) bag of a chemical is accidentally dropped in an aerating lagoon. The chemical is water soluble and nonreacting. The lagoon is 120 ft (35 m) in diameter and filled to a depth of 10 ft (3 m). The aerators circulate and distribute the chemical evenly throughout the lagoon.

Water enters the lagoon at a rate of 30 gal/min (115 L/min). Fully mixed water is pumped into a reservoir at a rate of 30 gal/min (115 L/min). The established safe concentration of this chemical is 1 ppb (part per billion).

(a) The volume of the lagoon at time t is most nearly

(A) 28,000 ft^3 (720 m^3)

(B) 110,000 ft^3 (2900 m^3)

(C) 140,000 ft^3 (3700 m^3)

(D) 230,000 ft^3 (5600 m^3)

(b) The volumetric flow rate out of the lagoon is most nearly

(A) 2.3 ft^3/min (0.066 m^3/min)

(B) 3.6 ft^3/min (0.10 m^3/min)

(C) 4.0 ft^3/min (0.12 m^3/min)

(D) 4.6 ft^3/min (0.13 m^3/min)

(c) The initial mass of the water in the lagoon is most nearly

(A) 1.8×10^6 lbm (7.2×10^5 kg)

(B) 7.1×10^6 lbm (2.9×10^6 kg)

(C) 9.0×10^6 lbm (3.7×10^6 kg)

(D) 14×10^7 lbm (5.6×10^6 kg)

(d) The final mass of chemicals at a concentration of 1 ppb is most nearly

(A) 0.002 lbm (0.0007 kg)

(B) 0.007 lbm (0.003 kg)

(C) 0.009 lbm (0.004 kg)

(D) 0.014 lbm (0.006 kg)

(e) The number of days it will take for the concentration of the discharge water to reach this level is most nearly

(A) 25 days

(B) 50 days

(C) 100 days

(D) 200 days

5. A tank contains 100 gal (100 L) of brine made by dissolving 60 lbm (60 kg) of salt in pure water. Salt water with a concentration of 1 lbm/gal (1 kg/L) enters the tank at a rate of 2 gal/min (2 L/min). A well-stirred mixture is drawn from the tank at a rate of 3 gal/min (3 L/min). The mass of salt in the tank after one hour is most nearly

(A) 13 lbm (13 kg)

(B) 37 lbm (37 kg)

(C) 43 lbm (43 kg)

(D) 51 lbm (51 kg)

SOLUTIONS

1. Obtain the characteristic equation by replacing each derivative with a polynomial term of equal degree.

$$r^2 - 4r - 12 = 0$$

Factor the characteristic equation.

$$(r - 6)(r + 2) = 0$$

The roots are $r_1 = 6$ and $r_2 = -2$.

Since the roots are real and distinct, the solution is

$$y = A_1 e^{r_1 x} + A_2 e^{r_2 x} = \boxed{A_1 e^{6x} + A_2 e^{-2x}}$$

The answer is (A).

2. The equation is a first-order linear differential equation of the form

$$y' + p(x)y = g(x)$$
$$p(x) = -1$$
$$g(x) = 2xe^{2x}$$

The integration factor $u(x)$ is

$$u(x) = \exp\left(\int p(x)\,dx\right) = \exp\left(\int (-1)\,dx\right) = e^{-x}$$

The closed form of the solution is

$$y = \frac{1}{u(x)}\left(\int u(x)g(x)\,dx + C\right)$$
$$= \frac{1}{e^{-x}}\left(\int (e^{-x})(2xe^{2x})\,dx + C\right)$$
$$= e^x\left(2(xe^x - e^x) + C\right)$$
$$= e^x\left(2e^x(x - 1) + C\right)$$
$$= 2e^{2x}(x - 1) + Ce^x$$

Apply the initial condition $y(0) = 1$ to obtain the integration constant, C.

$$y(0) = 2e^{(2)(0)}(0 - 1) + Ce^0 = 1$$
$$(2)(1)(-1) + C(1) = 1$$
$$-2 + C = 1$$
$$C = 3$$

Substituting in the value for the integration constant, C, the solution is

$$y = 2e^{2x}(x-1) + 3e^x$$

The answer is (B).

3. (a) The differential equation is a homogeneous second-order linear differential equation with constant coefficients. Write the characteristic equation.

$$r^2 + 2r + 2 = 0$$

This is a quadratic equation of the form $ar^2 + br + c = 0$, where $a = 1$, $b = 2$, and $c = 2$.

Solve for r.

$$r = \frac{-b \pm \sqrt{b^2 - 4ac}}{2a} = \frac{-2 \pm \sqrt{(2)^2 - (4)(1)(2)}}{(2)(1)}$$

$$= \frac{-2 \pm \sqrt{4 - 8}}{2}$$

$$= -1 \pm \sqrt{-1}$$

$$= -1 \pm i$$

$$r_1 = -1 + i, \text{ and } r_2 = -1 - i$$

Since the roots are imaginary and of the form $\alpha + i\omega$ and $\alpha - i\omega$, where $\alpha = -1$ and $\omega = 1$, the general form of the solution is

$$x(t) = A_1 e^{\alpha t} \cos \omega t + A_2 e^{\alpha t} \sin \omega t$$

$$= A_1 e^{-1t} \cos t + A_2 e^{-1t} \sin t$$

$$= A_1 e^{-t} \cos t + A_2 e^{-t} \sin t$$

Apply the initial conditions, $x(0) = 0$ and $x'(0) = 1$, to solve for A_1 and A_2.

First, apply the initial condition, $x(0) = 0$.

$$x(t) = A_1 e^0 \cos 0 + A_2 e^0 \sin 0 = 0$$

$$A_1 (1)(1) + A_2 (1)(0) = 0$$

$$A_1 = 0$$

Substituting, the solution of the differential equation becomes

$$x(t) = A_2 e^{-t} \sin t$$

To apply the second initial condition, take the first derivative.

$$x'(t) = \frac{d}{dt}(A_2 e^{-t} \sin t) = A_2 \frac{d}{dt}(e^{-t} \sin t)$$

$$= A_2 \left(\sin t \frac{d}{dt}(e^{-t}) + e^{-t} \frac{d}{dt} \sin t \right)$$

$$= A_2 \left(\sin t(-e^{-t}) + e^{-t}(\cos t) \right)$$

$$= A_2 (e^{-t})(-\sin t + \cos t)$$

Apply the initial condition, $x'(0) = 1$.

$$x(0) = A_2 e^0 (-\sin 0 + \cos 0) = 1$$

$$A_2 (1)(0 + 1) = 1$$

$$A_2 = 1$$

The solution is

$$x(t) = A_2 e^{-t} \sin t = (1) e^{-t} \sin t$$

$$= e^{-t} \sin t$$

The answer is (C).

(b) To determine the natural frequency, set the damping term to zero. The equation has the form

$$x'' + 2x = 0$$

This equation has a general solution of the form

$$x(t) = x_0 \cos \omega t + \left(\frac{v_0}{\omega} \right) \sin \omega t$$

ω is the natural frequency. Given the equation $x'' + 2x = 0$, the characteristic equation is

$$r^2 + 2 = 0$$

$$r = \sqrt{-2}$$

$$= \pm \sqrt{2} i$$

Since the roots are imaginary and of the form $\alpha + i\omega$ and $\alpha - i\omega$, where $\alpha = 0$ and $\omega = \sqrt{2}$, the general form of the solution is

$$x(t) = A_1 e^{\alpha t} \cos \omega t + A_2 e^{\alpha t} \sin \omega t$$

$$= A_1 e^{0t} \cos \sqrt{2} t + A_2 e^{0t} \sin \sqrt{2} t$$

$$= A_1 (1) \cos \sqrt{2} t + A_2 (1) \sin \sqrt{2} t$$

$$= A_1 \cos \sqrt{2} t + A_2 \sin \sqrt{2} t$$

Apply the initial conditions, $x(0) = 0$ and $x'(0) = 1$, to solve for A_1 and A_2. Applying the initial condition $x(0) = 0$ gives

$$x(0) = A_1 \cos\big((\sqrt{2})(0)\big) + A_2 \sin\big((\sqrt{2})(0)\big) = 0$$
$$A_1 \cos 0 + A_2 \sin 0 = 0$$
$$A_1(1) + A_2(0) = 0$$
$$A_1 = 0$$

Substituting, the solution of the differential equation becomes

$$x(t) = A_2 \sin \sqrt{2}t$$

To apply the second initial condition, take the first derivative.

$$x'(t) = \frac{d}{dt}(A_2 \sin \sqrt{2}t)$$
$$= A_2\sqrt{2}\cos\sqrt{2}t$$

Apply the second initial condition, $x'(0) = 1$.

$$x'(0) = A_2\sqrt{2}\cos\big((\sqrt{2})(0)\big) = 1$$
$$A_2\sqrt{2}\cos(0) = 1$$
$$A_2(\sqrt{2})(1) = 1$$
$$A_2\sqrt{2} = 1$$
$$A_2 = \frac{1}{\sqrt{2}} = \frac{\sqrt{2}}{2}$$

Substituting, the undamped solution becomes

$$x(t) = \frac{\sqrt{2}}{2}\sin\sqrt{2}t$$

This is of the form

$$x(t) = A\sin\omega t$$

Therefore, the undamped natural frequency is $\omega = \boxed{\sqrt{2}.}$

The answer is (C).

(c) The amplitude of the oscillation is the maximum displacement.

Take the derivative of the solution, $x(t) = e^{-t}\sin t$.

$$x'(t) = \frac{d}{dt}(e^{-t}\sin t) = \sin t \frac{d}{dt}(e^{-t}) + e^{-t}\frac{d}{dt}\sin t$$
$$= \sin t(-e^{-t}) + e^{-t}\cos t$$
$$= e^{-t}(\cos t - \sin t)$$

The maximum displacement occurs at $x'(t) = 0$.

Since $e^{-t} \neq 0$ except as t approaches infinity,

$$\cos t - \sin t = 0$$
$$\tan t = 1$$
$$t = \tan^{-1}(1)$$
$$= 0.785 \text{ rad}$$

At $t = 0.785$ rad, the displacement is maximum. Substitute into the orginal solution to obtain a value for the maximum displacement.

$$x(0.785) = e^{-0.785}\sin 0.785 = 0.322$$

The amplitude is $\boxed{0.322 \ (0.32).}$

The answer is (A).

(d) (An alternative solution using Laplace transforms follows this solution.) The application of a lateral wind load with the form $\sin t$ revises the differential equation to the form

$$x'' + 2x' + 2x = \sin t$$

Express the solution as the sum of the complementary x_c and particular x_p solutions.

$$x(t) = x_c(t) + x_p(t)$$

From part (a),

$$x_c(t) = A_1 e^{-t}\cos t + A_2 e^{-t}\sin t$$

The general form of the particular solution is given by

$$x_p(t) = x^s(A_3 \cos t + A_4 \sin t)$$

Determine the value of s; check to see if the terms of the particular solution solve the homogeneous equation.

Examine the term $A_3 \cos t$.

Take the first derivative.

$$\frac{d}{dx}(A_3 \cos t) = -A_3 \sin t$$

Take the second derivative.

$$\frac{d}{dx}\left(\frac{d}{dx}(A_3 \cos t)\right) = \frac{d}{dx}(-A_3 \sin t)$$
$$= -A_3 \cos t$$

Substitute the terms into the homogeneous equation.

$$x'' + 2x' + 2x = -A_3 \cos t + (2)(-A_3 \sin t)$$
$$+ (2)(-A_3 \cos t)$$
$$= A_3 \cos t - 2A_3 \sin t$$
$$\neq 0$$

Except for the trival solution $A_3 = 0$, the term $A_3 \cos t$ does not solve the homogeneous equation.

Examine the second term, $A_4 \sin t$.

Take the first derivative.

$$\frac{d}{dx}(A_4 \sin t) = A_4 \cos t$$

Take the second derivative.

$$\frac{d}{dx}\left(\frac{d}{dx}(A_4 \sin t)\right) = \frac{d}{dx}(A_4 \cos t) = -A_4 \sin t$$

Substitute the terms into the homogeneous equation.

$$x'' + 2x' + 2x = -A_4 \sin t + (2)(A_4 \cos t)$$
$$+ (2)(A_4 \sin t)$$
$$= A_4 \sin t + 2A_4 \cos t$$
$$\neq 0$$

Except for the trivial solution $A_4 = 0$, the term $A_4 \sin t$ does not solve the homogeneous equation.

Neither of the terms satisfies the homogeneous equation, so $s = 0$. Therefore, the particular solution is of the form

$$x_p(t) = A_3 \cos t + A_4 \sin t$$

Use the method of undetermined coefficients to solve for A_3 and A_4. Take the first derivative.

$$x_p'(t) = \frac{d}{dx}(A_3 \cos t + A_4 \sin t)$$
$$= -A_3 \sin t + A_4 \cos t$$

Take the second derivative.

$$x_p''(t) = \frac{d}{dx}\left(\frac{d}{dx}(A_3 \cos t + A_4 \sin t)\right)$$
$$= \frac{d}{dx}(-A_3 \sin t + A_4 \cos t)$$
$$= -A_3 \cos t - A_4 \sin t$$

Substitute the expressions for the derivatives into the differential equation.

$$x'' + 2x' + 2x = (-A_3 \cos t - A_4 \sin t)$$
$$+ (2)(-A_3 \sin t + A_4 \cos t)$$
$$+ (2)(A_3 \cos t + A_4 \sin t)$$
$$= \sin t$$

Rearranging terms gives

$$(-A_3 + 2A_4 + 2A_3)\cos t$$
$$+ (-A_4 - 2A_3 + 2A_4)\sin t = \sin t$$
$$(A_3 + 2A_4)\cos t + (-2A_3 + A_4)\sin t = \sin t$$

Equating coefficients gives

$$A_3 + 2A_4 = 0$$
$$-2A_3 + A_4 = 1$$

Multiplying the first equation by 2 and adding equations gives

$$2A_3 + 4A_4 = 0$$
$$\underline{+ (-2A_3 + A_4) = 1}$$
$$5A_4 = 1 \text{ or } A_4 = \tfrac{1}{5}$$

From the first equation for $A_4 = \frac{1}{5}$, $A_3 + (2)(\frac{1}{5}) = 0$, and $A_3 = -\frac{2}{5}$.

Substituting for the coefficients, the particular solution becomes

$$x_p(t) = -\tfrac{2}{5}\cos t + \tfrac{1}{5}\sin t$$

Combining the complementary and particular solutions gives

$$x(t) = x_c(t) + x_p(t)$$
$$= A_1 e^{-t}\cos t + A_2 e^{-t}\sin t - \tfrac{2}{5}\cos t + \tfrac{1}{5}\sin t$$

Apply the initial conditions to solve for the coefficients A_1 and A_2, then apply the first initial condition, $x(0) = 0$.

$$x(t) = A_1 e^0 \cos 0 + A_2 e^0 \sin 0$$
$$\tfrac{2}{5}\cos 0 + \tfrac{1}{5}\sin 0 = 0$$
$$A_1(1)(1) + A_2(1)(0) + \left(-\tfrac{2}{5}\right)(1) + \left(\tfrac{1}{5}\right)(0) = 0$$
$$A_1 - \tfrac{2}{5} = 0$$
$$A_1 = \tfrac{2}{5}$$

Substituting for A_1, the solution becomes

$$x(t) = \tfrac{2}{5}e^{-t}\cos t + A_2 e^{-t}\sin t - \tfrac{2}{5}\cos t + \tfrac{1}{5}\sin t$$

Take the first derivative.

$$x'(t) = \frac{d}{dx}\left(\begin{array}{c} \left(\tfrac{2}{5}e^{-t}\cos t + A_2 e^{-t}\sin t\right) \\ + \left(-\tfrac{2}{5}\cos t + \tfrac{1}{5}\sin t\right) \end{array} \right)$$

$$= \left(\tfrac{2}{5}\right)(-e^{-t}\cos t - e^{-t}\sin t)$$
$$+ A_2(-e^{-t}\sin t + e^{-t}\cos t)$$
$$+ \left(-\tfrac{2}{5}\right)(-\sin t) + \tfrac{1}{5}\cos t$$

Apply the second initial condition, $x'(0) = 1$.

$$x'(0) = \left(\tfrac{2}{5}\right)(-e^0\cos 0 - e^0\sin 0)$$
$$+ A_2(-e^0\sin 0 + e^0\cos 0)$$
$$+ \left(-\tfrac{2}{5}\right)(-\sin 0) + \tfrac{1}{5}\cos 0$$
$$= 1$$

$$\left(\tfrac{2}{5}\right)\big(-(1)(1) - (1)(0)\big) + A_2\big(-(1)(0) + (1)(1)\big)$$
$$+\left(-\tfrac{2}{5}\right)(0) + \left(\tfrac{1}{5}\right)(1) = 1$$
$$\left(\tfrac{2}{5}\right)(-1) + A_2(1) + \left(\tfrac{1}{5}\right) = 1$$
$$A_2 = \tfrac{6}{5}$$

Substituting for A_2, the solution becomes

$$\boxed{x(t) = \tfrac{2}{5}e^{-t}\cos t + \tfrac{6}{5}e^{-t}\sin t - \tfrac{2}{5}\cos t + \tfrac{1}{5}\sin t}$$

The answer is (D).

(d) *Alternate solution:*

Use the Laplace transform method.

$$x'' + 2x' + 2x = \sin t$$
$$\mathcal{L}(x'') + 2\mathcal{L}(x') + 2\mathcal{L}(x) = \mathcal{L}(\sin t)$$
$$s^2\mathcal{L}(x) - 1 + 2s\mathcal{L}(x) + 2\mathcal{L}(x) = \frac{1}{s^2 + 1}$$
$$\mathcal{L}(x)(s^2 + 2s + 2) - 1 = \frac{1}{s^2 + 1}$$
$$\mathcal{L}(x) = \frac{1}{s^2 + 2s + 2} + \frac{1}{(s^2 + 1)(s^2 + 2s + 2)}$$
$$= \frac{1}{(s+1)^2 + 1} + \frac{1}{(s^2 + 1)(s^2 + 2s + 2)}$$

Use partial fractions to expand the second term.

$$\frac{1}{(s^2 + 1)(s^2 + 2s + 2)} = \frac{A_1 + B_1 s}{s^2 + 1} + \frac{A_2 + B_2 s}{s^2 + 2s + 2}$$

Cross multiply.

$$A_1 s^2 + 2A_1 s + 2A_1 + B_1 s^3 + 2B_1 s^2$$
$$= \frac{+ A_2 s^2 + A_2 + B_2 s^3 + B_2 s}{(s^2 + 1)(s^2 + 2s + 2)}$$
$$= \frac{\begin{array}{c} s^3(B_1 + B_2) + s^2(A_1 + A_2 + 2B_1) \\ + s(2A_1 + 2B_1 + B_2) + 2A_1 + A_2 \end{array}}{(s^2 + 1)(s^2 + 2s + 2)}$$

Compare numerators to obtain the following four simultaneous equations.

$$\begin{array}{rcl} B_1 + B_2 &=& 0 \\ A_1 + A_2 + 2B_1 &=& 0 \\ 2A_1 + 2B_1 + B_2 &=& 0 \\ 2A_1 + A_2 &=& 1 \end{array}$$

Use Cramer's rule to find A_1.

$$A_1 = \frac{\begin{vmatrix} 0 & 0 & 1 & 1 \\ 0 & 1 & 2 & 0 \\ 0 & 0 & 2 & 1 \\ 1 & 1 & 0 & 0 \end{vmatrix}}{\begin{vmatrix} 0 & 0 & 1 & 1 \\ 1 & 1 & 2 & 0 \\ 2 & 0 & 2 & 1 \\ 2 & 1 & 0 & 0 \end{vmatrix}} = \frac{-1}{-5} = \frac{1}{5}$$

The rest of the coefficients are found similarly.

$$A_1 = \tfrac{1}{5}$$
$$A_2 = \tfrac{3}{5}$$
$$B_1 = -\tfrac{2}{5}$$
$$B_2 = \tfrac{2}{5}$$

Then,

$$\mathcal{L}(x) = \frac{1}{(s+1)^2 + 1} + \frac{\frac{1}{5}}{s^2 + 1} + \frac{-\frac{2}{5}s}{s^2 + 1}$$

$$+ \frac{\frac{3}{5}}{s^2 + 2s + 2} + \frac{\frac{2}{5}s}{s^2 + 2s + 2}$$

Take the inverse transform.

$$x(t) = \mathcal{L}^{-1}\{\mathcal{L}(x)\}$$

$$= e^{-t}\sin t + \frac{1}{5}\sin t - \frac{2}{5}\cos t + \frac{3}{5}e^{-t}\sin t$$

$$+ \frac{2}{5}(e^{-t}\cos t - e^{-t}\sin t)$$

$$= \boxed{\frac{6}{5}e^{-t}\sin t + \frac{2}{5}e^{-t}\cos t + \frac{1}{5}\sin t - \frac{2}{5}\cos t}$$

The answer is (D).

4. *Customary U.S. Solution*

(a) The differential equation is

$$m'(t) = a(t) - \frac{m(t)o(t)}{V(t)}$$

$a(t) = $ rate of addition of chemical

$m(t) = $ mass of chemical at time t

$o(t) = $ volumetric flow out of the lagoon, 30 gpm

$V(t) = $ volume in the lagoon at time t

Water flows into the lagoon at a rate of 30 gpm, and a water-chemical mix flows out of the lagoon at a rate of 30 gpm. Therefore, the volume of the lagoon at time t is equal to the initial volume.

$$V(t) = \left(\frac{\pi}{4}\right)(\text{diameter of lagoon})^2(\text{depth of lagoon})$$

$$= \left(\frac{\pi}{4}\right)(120 \text{ ft})^2(10 \text{ ft})$$

$$= \boxed{113{,}097 \text{ ft}^3 \quad (110{,}000 \text{ ft}^3)}$$

The answer is (B).

(b) Using a conversion factor of 7.48 gal/ft^3 gives

$$o(t) = \frac{30 \frac{\text{gal}}{\text{min}}}{7.48 \frac{\text{gal}}{\text{ft}^3}} = \boxed{4.01 \text{ ft}^3/\text{min} \quad (4.0 \text{ ft}^3/\text{min})}$$

The answer is (C).

(c) Substituting into the general form of the differential equation gives

$$m'(t) = a(t) - \frac{m(t)o(t)}{V(t)}$$

$$= 0 - m(t)\left(\frac{4.01 \frac{\text{ft}^3}{\text{min}}}{113{,}097 \text{ ft}^3}\right)$$

$$= -\left(\frac{3.55 \times 10^{-5}}{\text{min}}\right)m(t)$$

$$m'(t) + \left(\frac{3.55 \times 10^{-5}}{\text{min}}\right)m(t) = 0$$

The differential equation of the problem has the following characteristic equation.

$$r + \frac{3.55 \times 10^{-5}}{\text{min}} = 0$$

$$r = -3.55 \times 10^{-5}/\text{min}$$

The general form of the solution is

$$m(t) = Ae^{rt}$$

Substituting the root, r, gives

$$m(t) = Ae^{(-3.55 \times 10^{-5}/\text{min})t}$$

Apply the initial condition $m(0) = 90$ lbm at time $t = 0$.

$$m(0) = Ae^{(-3.55 \times 10^{-5}/\text{min})(0)} = 90 \text{ lbm}$$

$$Ae^0 = 90 \text{ lbm}$$

$$A = 90 \text{ lbm}$$

Therefore,

$$m(t) = (90 \text{ lbm})e^{(-3.55 \times 10^{-5}/\text{min})t}$$

Solve for t.

$$\frac{m(t)}{90 \text{ lbm}} = e^{(-3.55 \times 10^{-5}/\text{min})t}$$

$$\ln\left(\frac{m(t)}{90 \text{ lbm}}\right) = \ln\left(e^{(-3.55 \times 10^{-5}/\text{min})t}\right)$$

$$= \left(\frac{-3.55 \times 10^{-5}}{\text{min}}\right)t$$

$$t = \frac{\ln\left(\frac{m(t)}{90 \text{ lbm}}\right)}{\frac{-3.55 \times 10^{-5}}{\text{min}}}$$

The initial mass of the water in the lagoon is

$$m_i = V\rho = (113{,}097 \text{ ft}^3)\left(62.4 \; \frac{\text{lbm}}{\text{ft}^3}\right)$$
$$= \boxed{7.06 \times 10^6 \text{ lbm} \quad (7.1 \times 10^6 \text{ lbm})}$$

The answer is (B).

(d) The final mass of chemicals at a concentration of 1 ppb is

$$m_f = \frac{7.06 \times 10^6 \text{ lbm}}{1 \times 10^9}$$
$$= \boxed{7.06 \times 10^{-3} \text{ lbm} \quad (0.007 \text{ lbm})}$$

The answer is (B).

(e) Find the time required to achieve a mass of 7.06×10^{-3} lbm.

$$t = \left(\frac{\ln\left(\dfrac{m(t)}{90 \text{ lbm}}\right)}{\dfrac{-3.55 \times 10^{-5}}{\text{min}}}\right)\left(\frac{1 \text{ hr}}{60 \text{ min}}\right)\left(\frac{1 \text{ day}}{24 \text{ hr}}\right)$$

$$= \left(\frac{\ln\left(\dfrac{7.06 \times 10^{-3} \text{ lbm}}{90 \text{ lbm}}\right)}{\dfrac{-3.55 \times 10^{-5}}{\text{min}}}\right)\left(\frac{1 \text{ hr}}{60 \text{ min}}\right)\left(\frac{1 \text{ day}}{24 \text{ hr}}\right)$$

$$= \boxed{185 \text{ days} \quad (200 \text{ days})}$$

The answer is (D).

SI Solution

(a) The differential equation is

$$m'(t) = a(t) - \frac{m(t)o(t)}{V(t)}$$

$a(t) = $ rate of addition of chemical
$m(t) = $ mass of chemical at time t
$o(t) = $ volumetric flow out of the lagoon, 115 L/min
$V(t) = $ volume in the lagoon at time t

Water flows into the lagoon at a rate of 115 L/min, and a water-chemical mix flows out of the lagoon at a rate of 115 L/min. Therefore, the volume of the lagoon at time t is equal to the initial volume.

$$V(t) = \left(\frac{\pi}{4}\right)(\text{diameter of lagoon})^2(\text{depth of lagoon})$$
$$= \left(\frac{\pi}{4}\right)(35 \text{ m})^2(3 \text{ m})$$
$$= \boxed{2886 \text{ m}^3 \quad (2900 \text{ m}^3)}$$

The answer is (B).

(b) Using a conversion factor of 1000 L/m³ gives

$$o(t) = \frac{115 \; \dfrac{\text{L}}{\text{min}}}{1000 \; \dfrac{\text{L}}{\text{m}^3}} = \boxed{0.115 \text{ m}^3/\text{min} \quad (0.12 \text{ m}^3/\text{min})}$$

The answer is (C).

(c) Substituting into the general form of the differential equation gives

$$m'(t) = a(t) - \frac{m(t)o(t)}{V(t)}$$
$$= 0 - m(t)\left(\frac{0.115 \; \dfrac{\text{m}^3}{\text{min}}}{2886 \text{ m}^3}\right)$$
$$= -\left(\frac{3.985 \times 10^{-5}}{\text{min}}\right)m(t)$$

$$m'(t) + \left(\frac{3.985 \times 10^{-5}}{\text{min}}\right)m(t) = 0$$

The differential equation of the problem has the following characteristic equation.

$$r + \frac{3.985 \times 10^{-5}}{\text{min}} = 0$$
$$r = -3.985 \times 10^{-5}/\text{min}$$

The general form of the solution is

$$m(t) = Ae^{rt}$$

Substituting in for the root, r, gives

$$m(t) = Ae^{(-3.985 \times 10^{-5}/\text{min})t}$$

Apply the initial condition $m(0) = 40$ kg at time $t = 0$.

$$m(0) = Ae^{(-3.985 \times 10^{-5}/\text{min})(0)} = 40 \text{ kg}$$
$$Ae^0 = 40 \text{ kg}$$
$$A = 40 \text{ kg}$$

Therefore,

$$m(t) = (40 \text{ kg})e^{(-3.985 \times 10^{-5}/\text{min})t}$$

Solve for t.

$$\frac{m(t)}{40 \text{ kg}} = e^{(-3.985 \times 10^{-5}/\text{min})t}$$

$$\ln\left(\frac{m(t)}{40 \text{ kg}}\right) = \ln\left(e^{(-3.985 \times 10^{-5}/\text{min})t}\right)$$

$$= \left(\frac{-3.985 \times 10^{-5}}{\text{min}}\right)t$$

$$t = \frac{\ln\left(\frac{m(t)}{40 \text{ kg}}\right)}{\frac{-3.985 \times 10^{-5}}{\text{min}}}$$

The initial mass of water in the lagoon is

$$m_i = V\rho = (2886 \text{ m}^3)\left(1000 \frac{\text{kg}}{\text{m}^3}\right)$$

$$= \boxed{2.886 \times 10^6 \text{ kg} \quad (2.9 \times 10^6 \text{ kg})}$$

The answer is (B).

(d) The final mass of chemicals at a concentration of 1 ppb is

$$m_f = \frac{2.886 \times 10^6 \text{ kg}}{1 \times 10^9} = \boxed{2.886 \times 10^{-3} \text{ kg} \quad (0.003 \text{ kg})}$$

The answer is (B).

(e) Find the time required to achieve a mass of 2.886×10^{-3} kg.

$$t = \left(\frac{\ln\left(\frac{m(t)}{40 \text{ kg}}\right)}{\frac{-3.985 \times 10^{-5}}{\text{min}}}\right)\left(\frac{1 \text{ h}}{60 \text{ min}}\right)\left(\frac{1 \text{ d}}{24 \text{ h}}\right)$$

$$= \left(\frac{\ln\left(\frac{2.886 \times 10^{-3} \text{ kg}}{40 \text{ kg}}\right)}{\frac{-3.985 \times 10^{-5}}{\text{min}}}\right)\left(\frac{1 \text{ h}}{60 \text{ min}}\right)\left(\frac{1 \text{ d}}{24 \text{ h}}\right)$$

$$= \boxed{166 \text{ days} \quad (200 \text{ days})}$$

The answer is (D).

5. Let

$$m(t) = \text{mass of salt in tank at time } t$$
$$m_0 = 60 \text{ mass units (lbm or kg)}$$
$$m'(t) = \text{rate at which salt content is changing}$$

Two mass units of salt enter each minute, and three volumes leave each minute. The amount of salt leaving each minute is

$$\left(3 \frac{\text{vol}}{\text{min}}\right)\left(\text{concentration in } \frac{\text{mass}}{\text{vol}}\right)$$

$$= \left(3 \frac{\text{vol}}{\text{min}}\right)\left(\frac{\text{salt mass}}{\text{volume}}\right)$$

$$= \left(3 \frac{\text{vol}}{\text{min}}\right)\left(\frac{m(t)}{100 - t}\right)$$

$$m'(t) = 2 - (3)\left(\frac{m(t)}{100 - t}\right) \text{ or } m'(t) + \frac{3m(t)}{100 - t}$$

$$= 2 \text{ mass units/min}$$

This is a first-order linear differential equation. The integrating factor is

$$m = \exp\left(3 \int \frac{dt}{100 - t}\right)$$

$$= \exp\left((3)\left(-\ln(100 - t)\right)\right)$$

$$= (100 - t)^{-3}$$

$$m(t) = (100 - t)^3 \left(2 \int \frac{dt}{(100 - t)^3} + k\right)$$

$$= 100 - t + k(100 - t)^3$$

$m = 60$ mass units at $t = 0$, so $k = -0.00004$.

$$m(t) = 100 - t - (0.00004)(100 - t)^3$$

At $t = 60$ min,

$$m = 100 - 60 \text{ min} - (0.00004)(100 - 60 \text{ min})^3$$

$$= \boxed{37.44 \text{ (37) mass units}}$$

The answer is (B).

11 Probability and Statistical Analysis of Data

PRACTICE PROBLEMS

Probability

1. Four military recruits whose respective shoe sizes are 7, 8, 9, and 10 report to the supply clerk to be issued boots. The supply clerk selects one pair of boots in each of the four required sizes and hands them at random to the recruits.

(a) Use exhaustive enumeration to determine the probability that all recruits will receive boots of an incorrect size.

(A) 0.25

(B) 0.38

(C) 0.45

(D) 0.61

(b) The probability that exactly three recruits will receive boots of the correct size is most nearly

(A) 0

(B) 0.063

(C) 0.17

(D) 0.25

Probability Distributions

2. The time taken by a toll taker to collect the toll from vehicles crossing a bridge is an exponential distribution with a mean of 23 sec. The probability that a random vehicle will be processed in 25 sec or more (i.e., will take longer than 25 sec) is most nearly

(A) 0.17

(B) 0.25

(C) 0.34

(D) 0.52

3. The number of cars entering a toll plaza on a bridge during the hour after midnight follows a Poisson distribution with a mean of 20.

(a) The probability that exactly 17 cars will pass through the toll plaza during that hour on any given night is most nearly

(A) 0.08

(B) 0.12

(C) 0.16

(D) 0.23

(b) The percent probability that three or fewer cars will pass through the toll plaza at that hour on any given night is most nearly

(A) 0.00032%

(B) 0.0019%

(C) 0.079%

(D) 0.11%

4. A survey field crew measures one leg of a traverse four times. The following results are obtained.

repetition	measurement	direction
1	1249.529	forward
2	1249.494	backward
3	1249.384	forward
4	1249.348	backward

The crew chief is under orders to obtain readings with confidence limits of 90%.

(a) How many readings are acceptable?

(A) No readings are acceptable.

(B) Two readings are acceptable.

(C) Three readings are acceptable.

(D) All four readings are acceptable.

(b) How many readings are unacceptable?

(A) No readings are unacceptable.

(B) One reading is unacceptable.

(C) Two readings are unacceptable.

(D) All four readings are unacceptable.

(c) Explain how to determine which readings are unacceptable.

(A) Readings inside the 90% confidence limits are unacceptable.

(B) Readings outside the 90% confidence limits are unacceptable.

(C) Readings outside the upper 90% confidence limit are unacceptable.

(D) Readings outside the lower 90% confidence limit are unacceptable.

(d) The most probable value of the distance is most nearly

(A) 1249.399

(B) 1249.410

(C) 1249.439

(D) 1249.452

(e) The error in the most probable value (at 90% confidence) is most nearly

(A) 0.08

(B) 0.11

(C) 0.14

(D) 0.19

(f) If the distance is one side of a square traverse whose sides are all equal, the most probable closure error is most nearly

(A) 0.14

(B) 0.20

(C) 0.28

(D) 0.35

(g) The most probable error of part (f) expressed as a fraction is most nearly

(A) 1:17,600

(B) 1:14,200

(C) 1:12,500

(D) 1:10,900

(h) Define accuracy and distinguish it from precision.

(A) If an experiment can be repeated with identical results, the results are considered accurate.

(B) If an experiment has a small bias, the results are considered precise.

(C) If an experiment is precise, it cannot also be accurate.

(D) If an experiment is unaffected by experimental error, the results are accurate.

(i) Which of the following is an example of a systematic error?

(A) measuring river depth as a motorized ski boat passes by

(B) using a steel tape that is too short to measure consecutive distances

(C) locating magnetic north near a large iron ore deposit along an overland route

(D) determining local wastewater BOD after a toxic spill

Statistical Analysis

5. Two resistances, the meter resistor and a shunt resistor, are connected in parallel in an ammeter. Most of the current passing through the meter goes through the shunt resistor. In order to determine the accuracy of the resistance of shunt resistors being manufactured for a line of ammeters, a manufacturer tests a sample of 100 shunt resistors. The numbers of shunt resistors with the resistance indicated (to the nearest hundredth of an ohm) are as follows.

$0.200 \ \Omega$, 1; $0.210 \ \Omega$, 3; $0.220 \ \Omega$, 5; $0.230 \ \Omega$, 10; $0.240 \ \Omega$, 17; $0.250 \ \Omega$, 40; $0.260 \ \Omega$, 13; $0.270 \ \Omega$, 6; $0.280 \ \Omega$, 3; $0.290 \ \Omega$, 2.

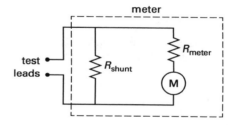

(a) The mean resistance is most nearly

(A) $0.235 \ \Omega$

(B) $0.247 \ \Omega$

(C) $0.251 \ \Omega$

(D) $0.259 \ \Omega$

(b) The sample standard deviation is most nearly

(A) $0.0003 \ \Omega$

(B) $0.010 \ \Omega$

(C) $0.016 \ \Omega$

(D) $0.24 \ \Omega$

(c) The median resistance is most nearly

(A) $0.22 \ \Omega$

(B) $0.24 \ \Omega$

(C) $0.25 \ \Omega$

(D) $0.26 \ \Omega$

(d) The sample variance is most nearly

(A) $0.00027 \ \Omega^2$

(B) $0.0083 \ \Omega^2$

(C) $0.0114 \ \Omega^2$

(D) $0.0163 \ \Omega^2$

6. California law requires a statistical analysis of the average speed driven by motorists on a road prior to the use of radar speed control. The following speeds (all in mph) were observed in a random sample of 40 cars.

44, 48, 26, 25, 20, 43, 40, 42, 29, 39, 23, 26, 24, 47, 45, 28, 29, 41, 38, 36, 27, 44, 42, 43, 29, 37, 34, 31, 33, 30, 42, 43, 28, 41, 29, 36, 35, 30, 32, 31

(a) Tabulate the frequency distribution and the cumulative frequency distribution of the data.

(b) Draw the frequency histogram.

(c) Draw the frequency polygon.

(d) Draw the cumulative frequency graph.

(e) The upper quartile speed is most nearly

(A) 30 mph

(B) 35 mph

(C) 40 mph

(D) 45 mph

(f) The mean speed is most nearly

(A) 31 mph

(B) 33 mph

(C) 35 mph

(D) 37 mph

(g) The standard deviation of the sample data is most nearly

(A) 2.1 mph

(B) 6.1 mph

(C) 6.8 mph

(D) 7.4 mph

(h) The sample standard deviation is most nearly

(A) 7.5 mph

(B) 18 mph

(C) 35 mph

(D) 56 mph

(i) The sample variance is most nearly

(A) $60 \ \text{mi}^2/\text{hr}^2$

(B) $320 \ \text{mi}^2/\text{hr}^2$

(C) $1200 \ \text{mi}^2/\text{hr}^2$

(D) $3100 \ \text{mi}^2/\text{hr}^2$

7. A spot speed study is conducted for a stretch of roadway. During a normal day, the speeds were found to be normally distributed with a mean of 46 and a standard deviation of 3.

(a) The 50th percentile speed is most nearly

(A) 39

(B) 43

(C) 46

(D) 49

(b) The 85th percentile speed is most nearly

(A) 47

(B) 48

(C) 49

(D) 52

(c) The upper two-standard deviation speed is most nearly

(A) 47

(B) 49

(C) 51

(D) 52

(d) The daily average speeds for the same stretch of roadway on consecutive normal days were determined by sampling 25 vehicles each day. The upper two-standard deviation average speed is most nearly

(A) 46

(B) 47

(C) 52

(D) 54

8. The diameters of bolt holes drilled in structural steel members are normally distributed with a mean of 0.502 in and a standard deviation of 0.005 in. Holes are out of specification if their diameters are less than 0.497 in or more than 0.507 in.

(a) The probability that a hole chosen at random will be out of specification is most nearly

(A) 0.16

(B) 0.22

(C) 0.32

(D) 0.68

(b) The probability that two holes out of a sample of 15 will be out of specification is most nearly

(A) 0.07

(B) 0.12

(C) 0.15

(D) 0.32

9. The length of a project's critical path is 43.83 days with a variance of 10.53 days^2. The length of the project is distributed normally.

(a) What is most nearly the probability of the project finishing in less than 42 days?

(A) 0.16

(B) 0.29

(C) 0.37

(D) 0.44

(b) Without using the standard normal table, what is most nearly the probability of the project finishing in less than 42 days?

(A) 0.16

(B) 0.29

(C) 0.37

(D) 0.44

10. The conductive transient temperature profile within an infinite solid block (uniform initial temperature of T_0, diffusivity of α) whose outer surface is maintained at a temperature T_s is

$$\frac{T(x, t) - T_s}{T_0 - T_s} = \text{erf}\left(\frac{x}{2\sqrt{\alpha t}}\right)$$

A large, smooth block of aluminum (diffusivity of 4.6×10^{-5} m^2/s) initially at 18°C is placed on a 400°C flat, smooth surface. Approximately what will be the temperature in the block 8 cm from the plane of contact after 10 minutes?

(A) 210°C

(B) 270°C

(C) 300°C

(D) 340°C

11. A researcher wants to know if a sample of 20 normally distributed executive salaries has a standard deviation greater than \$150,000. The average of the sample is \$400,000. The sample standard deviation is \$195,000. At approximately what confidence level does a standard deviation greater than \$150,000 become likely?

(A) 50%

(B) 95%

(C) 97%

(D) 99%

Reliability

12. A mechanical component exhibits a negative exponential failure distribution with a mean time to failure of 1000 hr. The maximum operating time such that the reliability remains above 99% is most nearly

(A) 3.3 hr

(B) 5.6 hr

(C) 8.1 hr

(D) 10 hr

13. An electro-mechanical system is a fully redundant, 1-out-of-3 system. The mean service (repair) rate is 0.1 repairs per hour, and the mean failure rate is one failure per 300 hours. The operational availability of the system is most nearly

(A) 77%

(B) 89%

(C) 98%

(D) 99%

14. A system with three components in parallel has a mean time to failure of 10,000 hours. The reliabilities of the first and second components are 0.50 and 0.75, respectively. The system is required to operate for 1000 hours. The minimum reliability of the third component to satisfy the previously mentioned conditions is most nearly

(A) 0.2

(B) 0.3

(C) 0.4

(D) 0.6

15. A serial system consists of five identical components. The failure history of a number of these components has been recorded. What is most nearly the mean time to failure for the system?

elapsed time, t (years)	number of failures, f
0	0
1	0
2	0
3	0
4	1
5	2
6	2
7	2
8	3
9	3
10	3

(A) 1.7

(B) 2.5

(C) 3.2

(D) 18

16. The mean time between failures of a metal cutting lathe is 2770 hours. Assuming an exponential distribution, the reliability of the lathe after operating for 700 hours is most nearly

(A) 10%

(B) 23%

(C) 56%

(D) 78%

Hypothesis Testing

17. 100 bearings were tested to failure. The average life was 1520 hours, and the sample standard deviation was 120 hours. The manufacturer wants to claim an average 1600 hour life. Evaluate using confidence limits of 95% and 99%.

(A) The claim is accurate at both 95% and 99% confidence.

(B) The claim is inaccurate only at 95% confidence.

(C) The claim is inaccurate only at 99% confidence.

(D) The claim is inaccurate at both 95% and 99% confidence.

Curve Fitting

18. (a) What is most nearly the best equation for a straight line passing through the points given?

x	y
400	370
800	780
1250	1210
1600	1560
2000	1980
2500	2450
4000	3950

(A) $y = 0.276x + 259.6$

(B) $y = 0.768x + 62.8$

(C) $y = 0.994x - 25.0$

(D) $y = 1.210x - 114.0$

(b) The correlation coefficient is most nearly

(A) 0.284

(B) 0.501

(C) 0.537

(D) 1.000

19. What is most nearly the best equation for a line passing through the points given?

s	t
20	43
18	141
16	385
14	1099

(A) $\ln t = -22.80 + 1.324s$

(B) $\ln t = 5.57 - 0.00924s$

(C) $\ln t = 14.53 - 0.536s$

(D) $\ln t = 7.56 - 0.854s$

20. The number of vehicles lining up behind a flashing railroad crossing has been observed for five trains of different lengths, as given. What is most nearly the mathematical formula that relates the two variables?

no. of cars in train, x	no. of vehicles, y
2	14.8
5	18.0
8	20.4
12	23.0
27	29.9

(A) $y = -26.18 + 52.71 \log x$

(B) $y = 2.48 + 21.32 \log x$

(C) $y = 7.46 + 15.6 \log x$

(D) $y = 9.57 + 13.20 \log x$

21. The following yield data are obtained from five identical treatment plants.

treatment plant	average temperature, T	average yield, Y
1	207.1	92.30
2	210.3	92.58
3	200.4	91.56
4	201.1	91.63
5	203.4	91.83

(a) Develop a linear equation to correlate the yield and average temperature.

(A) $Y = -0.04\,T + 100$

(B) $Y = 0.019\,T + 88$

(C) $Y = 0.11\,T + 70$

(D) $Y = 0.44\,T + 1.2$

(b) The correlation coefficient is most nearly

(A) 0.80

(B) 0.87

(C) 0.90

(D) 1.00

22. The following data are obtained from a soil compaction test. What is most nearly the nonlinear formula that relates the two variables?

x	y
−1	0
0	1
1	1.4
2	1.7
3	2
4	2.2
5	2.4
6	2.6
7	2.8
8	3

(A) $y = \sqrt{x - 1}$

(B) $y = \sqrt{x}$

(C) $y = \sqrt{x + 0.5}$

(D) $y = \sqrt{x + 1}$

Quantitative Risk Analysis

23. What is the difference between risk analysis and risk management?

(A) Risk analysis is typically performed by engineers, and risk management is typically performed by risk managers.

(B) Risk analysis is primarily a technical evaluation, while risk management is primarily a political (or management) decision.

(C) Risk analysis is a fact-based process and has no uncertainties associated with it, while risk management involves a great many uncertainties and judgmental factors.

(D) Risk analysis can be applied to many different situations and conditions, but risk management is limited to industrial processes.

24. What is the sequence of steps in the risk analysis process?

I.   hazard identification

II. toxicity assessment

III. risk characterization

IV. exposure assessment

V.   fate and transport modeling

VI. process hazard analysis

(A) I, III, IV, V

(B) II, IV, V, VI

(C) I, II, IV, III

(D) I, II, V, VI

25. The objective of an initial screening of hazardous chemicals at a remediation site is to

(A) identify the carcinogenic and noncarcinogenic chemicals

(B) identify the potential release mechanisms at the site

(C) identify the potential pathways of exposure

(D) identify the chemicals or their surrogates that will account for 99% of the risk

26. A child has been exposed for 5 years to a drinking water well that is contaminated with heptachlor at a concentration of 1.3 μg / L. The slope factor (oral) for heptachlor is 4.5 $(mg / kg \cdot d)^{-1}$. For a standard factor of 1.0 L/d water consumption and 10 kg body mass, and assuming heptachlor is carcinogenic, the excess risk of cancer over the child's lifetime is most nearly

(A) 3.4×10^{-4}

(B) 4.2×10^{-5}

(C) 3.6×10^{-6}

(D) 9.7×10^{-6}

27. An adult has been continuously exposed to two noncarcinogens for 2 years by the inhalation route, as shown below.

chemical	exposure (mg/m^3)	RfD, inhalation (mg/kg·d)
toluene	0.02	1.4
propylene glycol	0.33	0.572

Standard factors are 20 m^3/d of air breathed and 70 kg body mass. The hazard index for exposure to these two chemicals is most nearly

(A) 0.17

(B) 0.34

(C) 0.68

(D) 1.24

SOLUTIONS

1. (a) There are 4! = 24 different possible outcomes. By enumeration, there are 9 completely wrong combinations.

$$p\{\text{all wrong}\} = \frac{9}{24} = \boxed{0.375 \quad (0.38)}$$

	sizes				
correct →	7	8	9	10	all wrong
	7	8	9	10	
	7	8	10	9	
	7	9	8	10	
	7	9	10	8	
	7	10	8	9	
	7	10	9	8	
	8	9	10	7	X
	8	9	7	10	
	8	10	9	7	
	8	10	7	9	X
	8	7	9	10	
	8	7	10	9	X
	9	10	7	8	X
	9	10	8	7	X
	9	7	10	8	X
	9	7	8	10	
	9	8	7	10	
	9	8	10	7	
	10	7	8	9	X
	10	7	9	8	
	10	8	7	9	
	10	8	9	7	
	10	9	8	7	X
	10	9	7	8	X

(left margin label: sizes issued)

The answer is (B).

(b) If three recruits get the correct size, the fourth recruit will also since there will be only one pair remaining.

$$p\{\text{exactly } 3\} = \boxed{0}$$

The answer is (A).

2. For an exponential distribution function, the mean is

$$\mu = \frac{1}{\lambda}$$

Using Eq. 11.41, for a mean of 23,

$$\mu = 23 = \frac{1}{\lambda}$$

$$\lambda = \frac{1}{23} = 0.0435$$

Using Eq. 11.40, for an exponential distribution function,

$$p = F(x) = 1 - e^{-\lambda x}$$
$$p\{X < x\} = 1 - p$$
$$p\{X > x\} = 1 - F(x)$$
$$= 1 - (1 - e^{-\lambda x})$$
$$= e^{-\lambda x}$$

The probability of a random vehicle being processed in 25 sec or more is

$$p\{x > 25\} = e^{-(0.0435)(25)}$$
$$= \boxed{0.337 \quad (0.34)}$$

The answer is (C).

3. (a) The distribution is a Poisson distribution with an average of $\lambda = 20$.

The probability for a Poisson distribution is given by Eq. 11.63.

$$p\{x\} = f(x)$$
$$= \frac{e^{-\lambda} \lambda^x}{x!}$$

Therefore, the probability of 17 cars is

$$p\{x = 17\} = f(17)$$
$$= \frac{e^{-20} 20^{17}}{17!}$$
$$= \boxed{0.076 \quad (0.08)}$$

The answer is (A).

(b) The probability of three or fewer cars is

$$p\{x \le 3\} = p\{x = 0\} + p\{x = 1\} + p\{x = 2\}$$
$$+ p\{x = 3\}$$
$$= f(0) + f(1) + f(2) + f(3)$$
$$= \frac{e^{-20} 20^0}{0!} + \frac{e^{-20} 20^1}{1!}$$
$$+ \frac{e^{-20} 20^2}{2!} + \frac{e^{-20} 20^3}{3!}$$
$$= 2 \times 10^{-9} + 4.1 \times 10^{-8}$$
$$+ 4.12 \times 10^{-7} + 2.75 \times 10^{-6}$$
$$= 3.2 \times 10^{-6}$$
$$= \boxed{0.0000032 \quad (0.00032\%)}$$

The answer is (A).

4. (a) Find the average using Eq. 11.68.

$$\overline{x} = \frac{\sum x_i}{n}$$
$$= \frac{1249.529 + 1249.494 + 1249.384 + 1249.348}{4}$$
$$= 1249.439$$

Since the sample population is small, use Eq. 11.75 to find the sample standard deviation.

$$s = \sqrt{\frac{\sum(x_i - \overline{x})^2}{n - 1}}$$

$$= \sqrt{\frac{\begin{array}{c}(1249.529 - 1249.439)^2 \\ + (1249.494 - 1249.439)^2 \\ + (1249.384 - 1249.439)^2 \\ + (1249.348 - 1249.439)^2\end{array}}{4 - 1}}$$
$$= 0.08647$$

From Table 11.2 or App. 11.A, two-tail 90% confidence limits are located $1.645s$ from $\overline{x}$.

$$1249.439 \pm (1.645)(0.08647) = 1249.439 \pm 0.142$$

Therefore, (1249.297, 1249.581) are the 90% confidence limits.

By observation, $\boxed{\text{all four readings}}$ fall within the 90% confidence limits.

The answer is (D).

(b) $\boxed{\text{No readings}}$ are unacceptable.

The answer is (A).

(c) $\boxed{\text{Readings outside the 90\% confidence limits are unacceptable.}}$

The answer is (B).

(d) The unbiased estimate of the most probable distance is $\boxed{1249.439.}$

The answer is (C).

(e) The error in the most probable value for the 90% confidence range is $\boxed{0.142\ (0.14).}$

The answer is (C).

(f) If the surveying crew places a marker, measures a distance x, places a second marker, and then measures the same distance x back to the original marker, the ending point should coincide with the original marker. If, due to measurement errors, the ending and starting points do not coincide, the difference is the closure error.

In this example, the survey crew moves around the four sides of a square, so there are two measurements in the x-direction and two measurements in the y-direction. If the errors E_1 and E_2 are known for two measurements, x_1 and x_2, the error associated with the sum or difference $x_1 \pm x_2$ is

$$E\{x_1 \pm x_2\} = \sqrt{E_1^2 + E_2^2}$$

In this case, the error in the x-direction is

$$E_x = \sqrt{(0.1422)^2 + (0.1422)^2} = 0.2011$$

The error in the y-direction is calculated the same way and is also 0.2011. E_x and E_y are combined by the Pythagorean theorem to yield

$$E_{\text{closure}} = \sqrt{(0.2011)^2 + (0.2011)^2}$$
$$= \boxed{0.2844\quad (0.28)}$$

The answer is (C).

(g) In surveying, error may be expressed as a fraction of one or more legs of the traverse. Assume that the total of all four legs is to be used as the basis.

$$\frac{0.2844}{(4)(1249)} = \boxed{\frac{1}{17,567}}\quad (1{:}17{,}600)$$

The answer is (A).

(h) An experiment is $\boxed{\text{accurate}}$ if it is unchanged by experimental error. Precision is concerned with the repeatability of the experimental results. If an experiment is repeated with identical results, the experiment is said to be precise. However, it is possible to have a highly precise experiment with a large bias.

The answer is (D).

(i) A systematic error is one that is always present and is unchanged from sample to sample. For example, a steel tape that is 0.02 ft $\boxed{\text{too short}}$ to measure consecutive distances introduces a systematic error.

The answer is (B).

5. (a) For convenience, tabulate the frequency-weighted values of R and R^2.

R	f	fR	fR^2
0.200	1	0.200	0.0400
0.210	3	0.360	0.1323
0.220	5	1.100	0.2420
0.230	10	2.300	0.5290
0.240	17	4.080	0.9792
0.250	40	10.000	2.5000
0.260	13	3.380	0.8788
0.270	6	1.620	0.4374
0.280	3	0.840	0.2352
0.290	2	0.580	0.1682
	100	24.730	6.1421

The mean resistance is

$$\overline{R} = \frac{\sum fR}{\sum f} = \frac{24.730\ \Omega}{100} = \boxed{0.2473\ \Omega\quad (0.247\ \Omega)}$$

The answer is (B).

(b) The sample standard deviation is given by Eq. 11.75.

$$s = \sqrt{\frac{\sum fR^2 - \dfrac{\left(\sum fR\right)^2}{n}}{n-1}} = \sqrt{\frac{6.1421\ \Omega^2 - \dfrac{(24.73\ \Omega)^2}{100}}{99}}$$
$$= \boxed{0.0163\ \Omega\quad (0.016\ \Omega)}$$

The answer is (C).

(c) The 50th and 51st values are both 0.25 Ω. The median is $\boxed{0.25\ \Omega.}$

The answer is (C).

(d) The sample variance is

$$s^2 = (0.0163\ \Omega)^2 = \boxed{0.0002656\ \Omega^2\quad (0.00027\ \Omega^2)}$$

The answer is (A).

6. (a) Tabulate the frequency distribution and the cumulative frequency distribution of the data.

The lowest speed is 20 mph and the highest speed is 48 mph; therefore, the range is 28 mph. Choose 10 cells with a width of 3 mph.

midpoint	interval (mph)	frequency	cumulative frequency	cumulative percent
21	20–22	1	1	3
24	23–25	3	4	10
27	26–28	5	9	23
30	29–31	8	17	43
33	32–34	3	20	50
36	35–37	4	24	60
39	38–40	3	27	68
42	41–43	8	35	88
45	44–46	3	38	95
48	47–49	2	40	100

(b) Draw the frequency histogram.

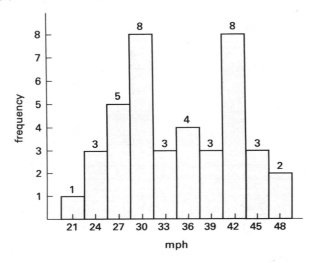

(c) Draw the frequency polygon.

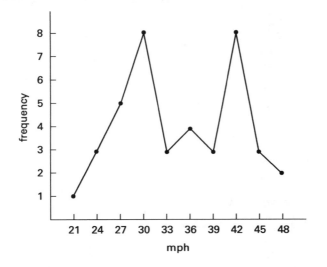

(d) Use the table in part (a).

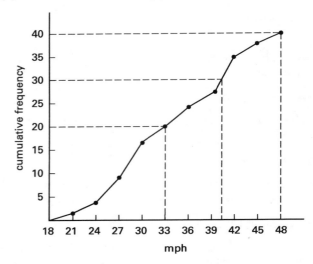

(e) From the cumulative frequency graph in part (d), the upper quartile speed occurs at 30 cars or 75%, which corresponds to approximately $\boxed{40 \text{ mph.}}$

The answer is (C).

(f) Calculate the following quantities.

$$\sum x_i = 1390 \text{ mi/hr}$$

$$n = 40$$

The mean is computed using Eq. 11.68.

$$\bar{x} = \frac{\sum x_i}{n} = \frac{1390 \, \frac{\text{mi}}{\text{hr}}}{40}$$

$$= \boxed{34.75 \text{ mi/hr} \quad (35 \text{ mph})}$$

The answer is (C).

(g) The standard deviation of the sample data is given by Eq. 11.74.

$$\sigma = \sqrt{\frac{\sum x^2}{n} - \mu^2}$$

$$\sum x^2 = 50{,}496 \text{ mi}^2/\text{hr}^2$$

Use the sample mean as an unbiased estimator of the population mean, μ.

$$\sigma = \sqrt{\frac{\sum x^2}{n} - \mu^2} = \sqrt{\frac{50{,}496 \, \frac{\text{mi}^2}{\text{hr}^2}}{40} - \left(34.75 \, \frac{\text{mi}}{\text{hr}}\right)^2}$$

$$= \boxed{7.405 \text{ mi/hr} \quad (7.4 \text{ mph})}$$

The answer is (D).

(h) The sample standard deviation is

$$s = \sqrt{\frac{\sum x^2 - \dfrac{\left(\sum x\right)^2}{n}}{n-1}}$$

$$= \sqrt{\frac{50{,}496 \ \dfrac{\text{mi}^2}{\text{hr}^2} - \dfrac{\left(1390 \ \dfrac{\text{mi}}{\text{hr}}\right)^2}{40}}{40 - 1}}$$

$$= \boxed{7.5 \ \text{mi/hr} \quad (7.5 \ \text{mph})}$$

The answer is (A).

(i) The sample variance is given by the square of the sample standard deviation.

$$s^2 = \left(7.5 \ \frac{\text{mi}}{\text{hr}}\right)^2$$

$$= \boxed{56.25 \ \text{mi}^2/\text{hr}^2 \quad (60 \ \text{mi}^2/\text{hr}^2)}$$

The answer is (A).

7. (a) The 50th percentile speed is the median speed, $\boxed{46,}$ which for a symmetrical normal distribution is the mean speed.

The answer is (C).

(b) The 85th percentile speed is the speed that is exceeded by only 15% of the measurements. Since this is a normal distribution, App. 11.A can be used. 15% in the upper tail corresponds to 35% between the mean and the 85th percentile. From App. 11.A, this occurs at approximately 1.04σ. The 85th percentile speed is

$$x_{85\%} = \mu + 1.04\sigma = 46 + (1.04)(3)$$

$$= \boxed{49.12 \quad (49)}$$

The answer is (C).

(c) The upper 2σ speed is

$$x_{2\sigma} = \mu + 2\sigma = 46 + (2)(3)$$

$$= \boxed{52.0 \quad (52)}$$

The answer is (D).

(d) According to the central limit theorem, the mean of the average speeds is the same as the distribution mean, and the standard deviation of sample means (from Eq. 11.83) is

$$s_{\overline{x}} = \frac{\sigma_x}{\sqrt{n}} = \frac{3}{\sqrt{25}}$$

$$= 0.6$$

The upper two-standard deviation average speed is

$$\overline{x}_{2\sigma} = \mu + 2\sigma_{\overline{x}} = 46 + (2)(0.6)$$

$$= \boxed{47.2 \quad (47)}$$

The answer is (B).

8. (a) From Eq. 11.45,

$$z = \frac{x_0 - \mu}{\sigma}$$

$$z_{\text{upper}} = \frac{0.507 \ \text{in} - 0.502 \ \text{in}}{0.005 \ \text{in}} = +1$$

From App. 11.A, the area outside $z = +1$ is

$$0.5 - 0.3413 = 0.1587$$

Since these are symmetrical limits, $z_{\text{lower}} = -1$.

$$\text{total fraction defective} = (2)(0.1587)$$

$$= \boxed{0.3174 \quad (0.32)}$$

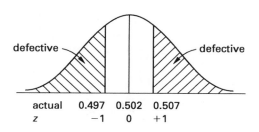

The answer is (C).

(b) This is a binomial problem.

$$p = p\{\text{defective}\} = 0.3174$$

$$q = 1 - p = 0.6826$$

From Eq. 11.28,

$$p\{x\} = f(x) = \binom{n}{x}\hat{p}^x\hat{q}^{n-x}$$

$$f(2) = \binom{15}{2}(0.3174)^2(0.6826)^{13}$$

$$= \left(\frac{15!}{13!2!}\right)(0.3174)^2(0.6826)^{13}$$

$$= \boxed{0.0739 \quad (0.07)}$$

The answer is (A).

9. (a) The standard deviation is

$$\sigma = \sqrt{\sigma^2} = \sqrt{10.53 \ \text{days}^2} = 3.245 \ \text{days}$$

Calculate the standard normal variable.

$$z = \left|\frac{\bar{x} - \mu}{\sigma}\right| = \left|\frac{42 \text{ days} - 43.83 \text{ days}}{3.245 \text{ days}}\right| = 0.564$$

From App. 11.A, the area under the normal curve between $z = 0.5$ and $z = 0.564$ is 0.2137 (interpolating between $z = 0.56$ and $z = 0.57$). The area under the curve between $z = 0.564$ and $z = +\infty$ (equivalent to the area under the curve between $t = -\infty$ and $t = 42$ days) is

$$p(t < 42 \text{ days}) - p(z < 0.564) = 0.5 - 0.2137$$
$$= \boxed{0.2863 \quad (0.29)}$$

The answer is (B).

(b) The standard normal variable was found to be 0.564 in part (a). From Eq. 11.57,

$$p\{0 < z < z_0\} = \tfrac{1}{2}\text{erf}\left(\frac{z_0}{\sqrt{2}}\right) = \tfrac{1}{2}\text{erf}\left(\frac{0.564}{\sqrt{2}}\right)$$
$$= \tfrac{1}{2}\text{erf}(0.3988)$$
$$\approx \tfrac{1}{2}(0.4284)$$
$$= 0.2142$$

The area in the tail of the standard normal distribution is

$$p(t < 42 \text{ days}) = p(z < 0.564) = 0.5 - 0.2142$$
$$= \boxed{0.2858 \quad (0.29)}$$

The answer is (B).

10.   The depth of the unknown temperature is

$$x = \frac{8 \text{ cm}}{100 \ \frac{\text{cm}}{\text{m}}} = 0.08 \text{ m}$$

$$\frac{T(x, t) - T_s}{T_0 - T_s} = \text{erf}\left(\frac{x}{2\sqrt{\alpha t}}\right)$$

$$\frac{T - 400°\text{C}}{25°\text{C} - 400°\text{C}} = \text{erf}\left(\frac{0.08 \text{ m}}{2\sqrt{\left(4.6 \times 10^{-5} \ \frac{\text{m}^2}{\text{s}}\right)(10 \text{ min})} \times \left(60 \ \frac{\text{s}}{\text{min}}\right)}\right)$$

$$\frac{T - 400°\text{C}}{-375°\text{C}} = \text{erf}(0.2408)$$

From App. 11.D, $\text{erf}(0.24) = 0.2657$.

$$\frac{T - 400°\text{C}}{-375°\text{C}} = 0.2657$$
$$T = \boxed{300.4°\text{C} \quad (300°\text{C})}$$

The answer is (C).

11.   The chi-squared statistic for this hypothesis test is

$$\chi^2 = \frac{(n-1)s^2}{\sigma^2} = \frac{(20-1)(\$195,000)^2}{(\$150,000)^2} = 32.11$$

From App. 11.B, with $20 - 1 = 19$ degrees of freedom, $\chi^2 = 32.11$ at approximately $\alpha = 0.03$ (3%). A standard deviation greater than \$150,000 can be supported with a $100\% - 3\% = \boxed{97\%}$ confidence limit, but not a higher confidence limit.

The answer is (C).

12.   Using Eq. 11.60,

$$\lambda = \frac{1}{\text{MTTF}} = \frac{1}{1000 \text{ hr}} = 0.001 \text{ hr}^{-1}$$

Using Eq. 11.61, the reliability function is

$$R\{t\} = e^{-\lambda t} = e^{-0.001t}$$

Since the reliability is greater than 99%,

$$e^{-0.001t} > 0.99$$
$$\ln e^{-0.001t} > \ln 0.99$$
$$-0.001t > \ln 0.99$$
$$t < -1000 \ln 0.99$$
$$t < 10.05 \quad (10)$$

The maximum operating time such that the reliability remains above 99% is $\boxed{10 \text{ hr.}}$

The answer is (D).

13.   The mean time to repair (MTTR) is the reciprocal of the mean service (repair) rate.

$$\text{MTTR} = \frac{1}{\mu} = \frac{1}{0.1 \ \frac{\text{repairs}}{\text{hr}}} = 10 \text{ hr/repair}$$

From Table 11.1, the mean time to failure (MTTF) of a 1-out-of-3 fully redundant system of identical components is 11/6 times the inverse of the mean failure rate.

$$\text{MTTF} = \left(\frac{11}{6}\right)\left(\frac{1}{\lambda}\right) = \frac{\frac{11}{6}}{\frac{1}{300 \text{ hr}}}$$
$$= 550 \text{ hr}$$

The mean time between failures (MTBF) is the sum of the mean time to failure and the mean time to repair.

$$\text{MTBF} = \text{MTTR} + \text{MTTF} = 10 \text{ hr} + 550 \text{ hr} = 560 \text{ hr}$$

The availability of the system is the ratio of the mean time between failures and the mean time to failure.

$$A_o = \frac{\text{MTTF}}{\text{MTBF}} = \frac{550 \text{ hr}}{560 \text{ hr}} \times 100\%$$
$$= \boxed{98\%}$$

The answer is (C).

14. The total required reliability of the system is a function of the mean time to failure and the 1000 hour operational requirement.

$$R_t = e^{-(t/\text{MTTF})}$$
$$= e^{-(1000 \text{ hr}/10{,}000 \text{ hr})}$$
$$= 0.90$$

The total reliability of the system can also be calculated as a function of the reliabilities of all three components.

$$R_t = 1 - (1 - R_1)(1 - R_2)(1 - R_3)$$
$$0.90 = 1 - (1 - 0.50)(1 - 0.75)(1 - R_3)$$
$$R_3 = \boxed{0.2}$$

The answer is (A).

15. The mean time to failure (MTTF) of the system is a function of the mean failure rate for the component and the number of components, n.

$$\text{MTTF} = \left(\frac{1}{\lambda}\right)\left(1 + \frac{1}{2} + \frac{1}{3} + \cdots + \frac{1}{n}\right)$$

The mean failure rate of one of the identical components is computed based on the table provided in the problem.

$$\lambda = \frac{\text{total no. of failures}}{\text{total yrs operated}} = \frac{\sum\limits_{i=1}^{\text{yr } 10} f_i}{\sum\limits_{i=1}^{\text{yr } 10} t_i f_i}$$

$$= \frac{1 + 2 + 2 + 2 + 3 + 3 + 3}{\begin{aligned}&(0 \text{ yr})(0) + (1 \text{ yr})(0) + (2 \text{ yr})(0) + (3 \text{ yr})(0) \\ &+ (4 \text{ yr})(1) + (5 \text{ yr})(2) + (6 \text{ yr})(2) + (7 \text{ yr})(2) \\ &+ (8 \text{ yr})(3) + (9 \text{ yr})(3) + (10 \text{ yr})(3)\end{aligned}}$$

$$= 0.13 \text{ failures/yr}$$

The mean time to failure is calculated for the five-component system.

$$\text{MTTF} = \left(\frac{1}{\lambda}\right)\left(1 + \frac{1}{2} + \frac{1}{3} + \frac{1}{4} + \frac{1}{5}\right)$$
$$= \left(\frac{1}{0.13}\right)\left(1 + \frac{1}{2} + \frac{1}{3} + \frac{1}{4} + \frac{1}{5}\right)$$
$$= \boxed{17.6 \quad (18)}$$

The answer is (D).

16. The failure (hazard) rate is

$$\lambda = \frac{1}{\text{mean time between failures}} = \frac{1}{2770 \text{ hr}}$$
$$= 3.61 \times 10^{-4} \text{ hr}^{-1}$$

The exponential reliability is

$$R(t) = e^{-\lambda t}$$
$$R(700 \text{ hr}) = e^{-(3.61 \times 10^{-4} \text{ hr}^{-1})(700 \text{ hr})}$$
$$= \boxed{0.7766 \quad (78\%)}$$

The answer is (D).

17. This is a typical hypothesis test of two sample population means. The two populations are the original population the manufacturer used to determine the 1600 hr average life value and the new population the sample was taken from. The mean ($\bar{x} = 1520$ hr) of the sample and its sample standard deviation ($s = 120$ hr) are known. The mean of the population is 1600 hr.

The standard deviation of the average lifetime population is

$$\sigma_{\bar{x}} = \frac{s}{\sqrt{n}} = \frac{120 \text{ hr}}{\sqrt{100}} = 12 \text{ hr}$$

The manufacturer can be reasonably sure that the claim of a 1600 hr average life is justified if 1600 hr is near the average test life of 1520 hr. "Reasonably sure" must be evaluated based on acceptable probability of being incorrect. If the manufacturer is willing to be wrong with a 5% probability, then a 95% confidence level is required.

Since the direction of bias is known, a one-tailed test is required. To determine if the mean has shifted downward, test the hypothesis that 1600 hr is within the 95% limit of a distribution with a mean of 1520 hr and a standard deviation of 12 hr. From a standard normal table, 5% of a standard normal distribution is outside of $z = 1.645$. Therefore, the 95% confidence limit is

$$1520 \text{ hr} + (1.645)(12 \text{ hr}) = 1540 \text{ hr}$$

The manufacturer can be 95% certain that the average lifetime of the bearings is less than 1600 hr.

If the manufacturer is willing to be wrong with a probability of only 1%, then a 99% confidence limit is required. From the normal table, $z = 2.33$, and the 99% confidence limit is

$$1520 \text{ hr} + (2.33)(12 \text{ hr}) = 1548 \text{ hr}$$

The manufacturer can be 99% certain that the average bearing life is less than 1600 hr.

Therefore, the claim is inaccurate at both 95% and 99% confidence.

The answer is (D).

18. (a) Plot the data points to determine if the relationship is linear.

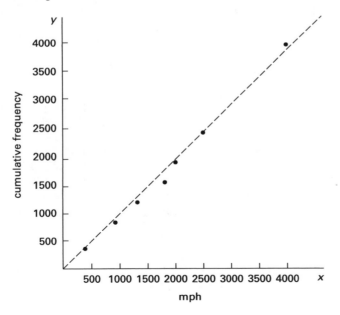

The data appear to be essentially linear. The slope, m, and the y-intercept, b, can be determined using linear regression.

The individual terms are

$$n = 7$$

$$\sum x_i = 400 + 800 + 1250 + 1600 + 2000$$
$$+ 2500 + 4000$$
$$= 12{,}550$$

$$\left(\sum x_i\right)^2 = (12{,}550)^2 = 1.575 \times 10^8$$

From Eq. 11.68,

$$\bar{x} = \frac{\sum x_i}{n} = \frac{12{,}550}{7} = 1792.9$$

$$\sum x_i^2 = (400)^2 + (800)^2 + (1250)^2 + (1600)^2$$
$$+ (2000)^2 + (2500)^2 + (4000)^2$$
$$= 3.117 \times 10^7$$

Similarly,

$$\sum y_i = 370 + 780 + 1210 + 1560 + 1980$$
$$+ 2450 + 3950$$
$$= 12{,}300$$

$$\left(\sum y_i\right)^2 = (12{,}300)^2 = 1.513 \times 10^8$$

$$\bar{y} = \frac{\sum y_i}{n} = \frac{12{,}300}{7} = 1757.1$$

$$\sum y_i^2 = (370)^2 + (780)^2 + (1210)^2 + (1560)^2$$
$$+ (1980)^2 + (2450)^2 + (3950)^2$$
$$= 3.017 \times 10^7$$

Also,

$$\sum x_i y_i = (400)(370) + (800)(780) + (1250)(1210)$$
$$+ (1600)(1560) + (2000)(1980)$$
$$+ (2500)(2450) + (4000)(3950)$$
$$= 3.067 \times 10^7$$

Using Eq. 11.89, the slope is

$$m = \frac{n\sum x_i y_i - \sum x_i \sum y_i}{n\sum x_i^2 - \left(\sum x_i\right)^2}$$

$$= \frac{(7)(3.067 \times 10^7) - (12{,}550)(12{,}300)}{(7)(3.117 \times 10^7) - (12{,}550)^2}$$

$$= 0.994$$

Using Eq. 11.90, the y-intercept is

$$b = \bar{y} - m\bar{x} = 1757.1 - (0.994)(1792.9)$$

$$= -25.0$$

The least squares equation of the line is

$$y = mx + b$$

$$= \boxed{0.994x - 25.0}$$

The answer is (C).

(b) Using Eq. 11.91, the correlation coefficient is

$$r = \frac{n\sum x_i y_i - \sum x_i \sum y_i}{\sqrt{\left(n\sum x_i^2 - \left(\sum x_i\right)^2\right)\left(n\sum y_i^2 - \left(\sum y_i\right)^2\right)}}$$

$$= \frac{(7)(3.067 \times 10^7) - (12{,}500)(12{,}300)}{\sqrt{\begin{array}{c}\left((7)(3.117 \times 10^7) - (12{,}500)^2\right) \\ \times \left((7)(3.017 \times 10^7) - (12{,}300)^2\right)\end{array}}}$$

$$\approx \boxed{1.000}$$

The answer is (D).

19. Plotting the data shows that the relationship is nonlinear.

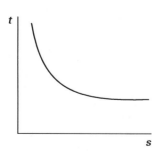

This appears to be an exponential with the form

$$t = ae^{bs}$$

Take the natural log of both sides.

$$\ln t = \ln ae^{bs} = \ln a + \ln e^{bs}$$

$$= \ln a + bs$$

But, $\ln a$ is just a constant, c.

$$\ln t = c + bs$$

Make the transformation $R = \ln t$.

$$R = c + bs$$

s	R
20	3.76
18	4.95
16	5.95
14	7.00

This is linear.

$$n = 4$$

$$\sum s_i = 20 + 18 + 16 + 14 = 68$$

$$\bar{s} = \frac{\sum s}{n} = \frac{68}{4} = 17$$

$$\sum s_i^2 = (20)^2 + (18)^2 + (16)^2 + (14)^2 = 1176$$

$$\left(\sum s_i\right)^2 = (68)^2 = 4624$$

$$\sum R_i = 3.76 + 4.95 + 5.95 + 7.00 = 21.66$$

$$\bar{R} = \frac{\sum R_i}{n} = \frac{21.66}{4} = 5.415$$

$$\sum R_i^2 = (3.76)^2 + (4.95)^2 + (5.95)^2 + (7.00)^2$$

$$= 123.04$$

$$\left(\sum R_i\right)^2 = (21.66)^2 = 469.16$$

$$\sum s_i R_i = (20)(3.76) + (18)(4.95) + (16)(5.95)$$

$$+ (14)(7.00)$$

$$= 357.5$$

The slope, b, of the transformed line is

$$b = \frac{n\sum s_i R_i - \sum s_i \sum R_i}{n\sum s_i^2 - \left(\sum s_i\right)^2} = \frac{(4)(357.5) - (68)(21.66)}{(4)(1176) - (68)^2}$$

$$= -0.536$$

The intercept is

$$c = \overline{R} - b\overline{s} = 5.415 - (-0.536)(17) = 14.527$$

The transformed equation is

$$R = c + bs = 14.527 - 0.536s$$

$$\boxed{\ln t = 14.527 - 0.536s \quad (14.53 - 0.536s)}$$

The answer is (C).

20. The first step is to graph the data.

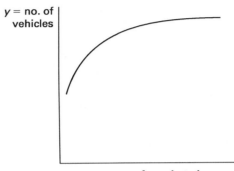

y = no. of vehicles

x = no. of cars in train

It is assumed that the relationship between the variables has the form $y = a + b\log x$. Therefore, the variable change $z = \log x$ is made, resulting in the following set of data.

z	y
0.301	14.8
0.699	18.0
0.903	20.4
1.079	23.0
1.431	29.9

$$\sum z_i = 4.413$$

$$\sum y_i = 106.1$$

$$\sum z_i^2 = 4.6082$$

$$\sum y_i^2 = 2382.2$$

$$\left(\sum z_i\right)^2 = 19.475$$

$$\left(\sum y_i\right)^2 = 11{,}257.2$$

$$\overline{z} = 0.8826$$

$$\overline{y} = 21.22$$

$$\sum z_i y_i = 103.06$$

$$n = 5$$

Using Eq. 11.89, the slope is

$$m = \frac{n\sum z_i y_i - \sum z_i \sum y_i}{n\sum z_i^2 - \left(\sum z_i\right)^2} = \frac{(5)(103.06) - (4.413)(106.1)}{(5)(4.6082) - 19.475}$$

$$= 13.20$$

The y-intercept is

$$b = \overline{y} - m\overline{z} = 21.22 - (13.20)(0.8826) = 9.57$$

The resulting equation is

$$y = 9.57 + 13.20z$$

The relationship between x and y is approximately

$$\boxed{y = 9.57 + 13.20\log x}$$

(This is not an optimal correlation, as better correlation coefficients can be obtained if other assumptions about the form of the equation are made. For example, $y = 9.1 + 4\sqrt{x}$ has a better correlation coefficient.)

The answer is (D).

21. (a) Plot the data to verify that they are linear.

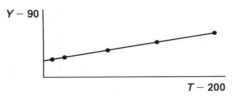

$Y - 90$

$T - 200$

x $T - 200$	y $Y - 90$
7.1	2.30
10.3	2.58
0.4	1.56
1.1	1.63
3.4	1.83

step 1: Calculate the following quantities.

$$\sum x_i = 22.3 \qquad\qquad \sum y_i = 9.9$$

$$\sum x_i^2 = 169.43 \qquad\qquad \sum y_i^2 = 20.39$$

$$\left(\sum x_i\right)^2 = 497.29 \qquad \left(\sum y_i\right)^2 = 98.01$$

$$\overline{x} = \frac{22.3}{5} = 4.46 \qquad\qquad \overline{y} = 1.98$$

$$\sum x_i y_i = 51.54$$

step 2: From Eq. 11.89, the slope is

$$m = \frac{n\sum x_i y_i - \sum x_i \sum y_i}{n\sum x_i^2 - \left(\sum x_i\right)^2} = \frac{(5)(51.54) - (22.3)(9.9)}{(5)(169.43) - 497.29}$$

$$= 0.1055$$

step 3: From Eq. 11.90, the y-intercept is

$$b = \overline{y} - m\overline{x} = 1.98 - (0.1055)(4.46) = 1.509$$

The equation of the line is

$$y = mx + b = 0.1055x + 1.509$$

$$Y - 90 = (0.1055)(T - 200) + 1.509$$

$$\boxed{Y = 0.1055\,T + 70.409 \quad (0.11\,T + 70)}$$

The answer is (C).

(b) *step 4:* Use Eq. 11.91 to get the correlation coefficient.

$$r = \frac{n\sum x_i y_i - \sum x_i \sum y_i}{\sqrt{\left(n\sum x_i^2 - \left(\sum x_i\right)^2\right)\left(n\sum y_i^2 - \left(\sum y_i\right)^2\right)}}$$

$$= \frac{(5)(51.54) - (22.3)(9.9)}{\sqrt{\left((5)(169.43) - 497.29\right)\left((5)(20.39) - 98.01\right)}}$$

$$= \boxed{0.995 \quad (1.00)}$$

The answer is (D).

22. Plot the data to see if they are linear.

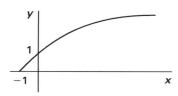

This looks like it could be of the form

$$y = a + b\sqrt{x}$$

However, when x is negative (as is the first point), the function is imaginary. Try shifting the curve to the right, replacing x with $x + 1$.

$$y = a + bz$$
$$z = \sqrt{x + 1}$$

z	y
0	0
1	1
1.414	1.4
1.732	1.7
2	2
2.236	2.2
2.45	2.4
2.65	2.6
2.83	2.8
3	3

Since $y \approx z$, the relationship is

$$\boxed{y = \sqrt{x + 1}}$$

In this problem, the answer was found accidentally. Usually, regression would be necessary.

The answer is (D).

23. Risk analysis is primarily a technical evaluation, while risk management is primarily a political (or management) decision.

The answer is (B).

24. The sequence of steps in the risk analysis process is (1) hazard identification, (2) toxicity assessment, (3) exposure assessment, and (4) risk characterization.

The answer is (C).

25. The objective of an initial screening of hazardous chemicals at a remediation site is to identify the chemicals or their surrogates that will account for 99% of the risk.

The answer is (D).

26. Use Eq. 11.94.

$$CDI_{a/w} = \frac{C(CR)(EF)(ED)}{(BW)(AT)}$$

$$CDI_w = \frac{\left(1.3\,\frac{\mu g}{L}\right)\left(1.0\,\frac{L}{d}\right)\left(365\,\frac{d}{yr}\right)(5\,yr)}{(10\,kg)(70\,yr)\left(365\,\frac{d}{yr}\right)\left(1000\,\frac{\mu g}{mg}\right)}$$

$$= 9.2857 \times 10^{-6}\ mg/kg \cdot d$$

Use Eq. 11.97.

$$R_c = (\text{CDI})(\text{SF})$$
$$= \left(9.2857 \times 10^{-6} \ \frac{\text{mg}}{\text{kg·d}}\right)\left(4.5 \ \frac{\text{kg·d}}{\text{mg}}\right)$$
$$\boxed{= 4.18 \times 10^{-5} \quad (4.2 \times 10^{-5})}$$

The answer is (B).

27. Use Eq. 11.94.

$$\text{CDI}_{a/w} = \frac{C(\text{CR})(\text{EF})(\text{ED})}{(\text{BW})(\text{AT})}$$

For toluene,

$$\text{CDI}_a = \frac{\left(0.02 \ \frac{\text{mg}}{\text{m}^3}\right)\left(20 \ \frac{\text{m}^3}{\text{d}}\right)\left(365 \ \frac{\text{d}}{\text{yr}}\right)(2 \ \text{yr})}{(70 \ \text{kg})(2 \ \text{yr})\left(365 \ \frac{\text{d}}{\text{yr}}\right)}$$
$$= 5.714 \times 10^{-3} \ \text{mg/kg·d}$$

Use Eq. 11.96.

$$\text{HI} = \frac{\text{CDI}}{\text{RfD}}$$
$$= \frac{5.714 \times 10^{-3} \ \frac{\text{mg}}{\text{kg·d}}}{1.4 \ \frac{\text{mg}}{\text{kg·d}}}$$
$$= 4.08 \times 10^{-3}$$

For propylene glycol,

$$\text{CDI}_a = \frac{\left(0.33 \ \frac{\text{mg}}{\text{m}^3}\right)\left(20 \ \frac{\text{m}^3}{\text{d}}\right)\left(365 \ \frac{\text{d}}{\text{yr}}\right)(2 \ \text{yr})}{(70 \ \text{kg})(2 \ \text{yr})\left(365 \ \frac{\text{d}}{\text{yr}}\right)}$$
$$= 9.429 \times 10^{-2} \ \text{mg/kg·d}$$

$$\text{HI} = \frac{\text{CDI}}{\text{RfD}} = \frac{9.429 \times 10^{-2} \ \frac{\text{mg}}{\text{kg·d}}}{0.572 \ \frac{\text{mg}}{\text{kg·d}}} = 0.165$$

The total HI is

$$4.08 \times 10^{-3} + 0.165 = \boxed{0.169 \quad (0.17)}$$

The answer is (A).

12 Numerical Analysis

PRACTICE PROBLEMS

1. A function is given as $y = 3x^{0.93} + 4.2$. What is most nearly the percent relative error if the value of y at $x = 2.7$ is found by using straight-line interpolation between $x = 2$ and $x = 3$?

(A) 0.06%

(B) 0.18%

(C) 2.5%

(D) 5.4%

2. Given the following data points, estimate y by straight-line interpolation for $x = 2.75$.

x	y
1	4
2	6
3	2
4	−14

(A) 2.1

(B) 2.4

(C) 2.7

(D) 3.0

3. Using the bisection method, find all of the roots of $f(x) = 0$ to the nearest 0.000005.

$$f(x) = x^3 + 2x^2 + 8x - 2$$

4. The increase in concentration of mixed-liquor suspended solids (MLSS) in an activated sludge aeration tank as a function of time is given in the table. Use a second-order Lagrangian interpolation to estimate the MLSS after 16 min of aeration.

t (min)	MLSS (mg/L)
0	0
10	227
15	362
20	517
22.5	602
30	901

(A) 350 mg/L

(B) 390 mg/L

(C) 540 mg/L

(D) 640 mg/L

SOLUTIONS

1. The actual value at $x = 2.7$ is

$$y(x) = 3x^{0.93} + 4.2$$
$$y(2.7) = (3)(2.7)^{0.93} + 4.2$$
$$= 11.756$$

At $x = 3$,

$$y(3) = (3)(3)^{0.93} + 4.2 = 12.534$$

At $x = 2$,

$$y(2) = (3)(2)^{0.93} + 4.2 = 9.916$$

Use straight-line interpolation.

$$\frac{x_2 - x}{x_2 - x_1} = \frac{y_2 - y}{y_2 - y_1}$$
$$\frac{3 - 2.7}{3 - 2} = \frac{12.534 - y}{12.534 - 9.916}$$
$$y = 11.749$$

The relative error is

$$\frac{\text{actual value} - \text{predicted value}}{\text{actual value}} = \frac{11.756 - 11.749}{11.756}$$
$$= \boxed{0.0006 \quad (0.06\%)}$$

The answer is (A).

2. Let $x_1 = 2$; therefore, from the table of data points, $y_1 = 6$. Let $x_2 = 3$; therefore, from the table of data points, $y_2 = 2$.

Let $x = 2.75$. By straight-line interpolation,

$$\frac{x_2 - x}{x_2 - x_1} = \frac{y_2 - y}{y_2 - y_1}$$
$$\frac{3 - 2.75}{3 - 2} = \frac{2 - y}{2 - 6}$$

$$\boxed{y = 3.0}$$

The answer is (D).

3. Use the equation $f(x) = x^3 + 2x^2 + 8x - 2$ to try to find an interval in which there is a root.

x	$f(x)$
0	-2
1	9

A root exists in the interval $[0, 1]$.

Try $x = \frac{1}{2}(0 + 1) = 0.5$.

$$f(0.5) = (0.5)^3 + (2)(0.5)^2 + (8)(0.5) - 2 = 2.625$$

A root exists in $[0, 0.5]$.

Try $x = 0.25$.

$$f(0.25) = (0.25)^3 + (2)(0.25)^2 + (8)(0.25) - 2 = 0.1406$$

A root exists in $[0, 0.25]$.

Try $x = 0.125$.

$$f(0.125) = (0.125)^3 + (2)(0.125)^2 + (8)(0.125) - 2$$
$$= -0.967$$

A root exists in $[0.125, 0.25]$.

Try $x = \frac{1}{2}(0.125 + 0.25) = 0.1875$.

Continuing,

$$f(0.1875) = -0.42 \quad [0.1875, 0.25]$$
$$f(0.21875) = -0.144 \quad [0.21875, 0.25]$$
$$f(0.234375) = -0.002 \quad \text{[This is close enough.]}$$

One root is $x_1 \approx \boxed{0.234375 \ (0.234).}$

Try to find the other two roots. Use long division to factor the polynomial.

$$
\begin{array}{r}
x^2 + 2.234375x + 8.52368 \\
x - 0.234375{\overline{\smash{\big)}\,x^3 + 2x^2 + 8x - 2}} \\
\underline{-(x^3 - 0.234375x^2)} \\
2.234375x^2 + 8x - 2 \\
\underline{-(2.234375x^2 - 0.52368x)} \\
8.52368x - 2 \\
\underline{-(8.52368x - 1.9977)} \\
\approx 0
\end{array}
$$

Use the quadratic equation to find the roots of $x^2 + 2.234375x + 8.52368$.

$$x_2, x_3 = \frac{-2.234375 \pm \sqrt{(2.234375)^2 - (4)(1)(8.52368)}}{(2)(1)}$$
$$= \boxed{-1.117189 \pm i2.697327} \quad \text{[both imaginary]}$$

4. Choose the three data points that bracket $t = 16$ as closely as possible. These three points are $t_0 = 10$, $t_1 = 15$, and $t_2 = 20$.

	$i = 0$	$i = 1$	$i = 2$
$k = 0$: $S_0(16) = -0.08$	$\left(\dfrac{16 - 10}{10 - 10}\right)$	$\left(\dfrac{16 - 15}{10 - 15}\right)$	$\left(\dfrac{16 - 20}{10 - 20}\right)$
$k = 1$: $S_1(16) = 0.96$	$\left(\dfrac{16 - 10}{15 - 10}\right)$	$\left(\dfrac{16 - 15}{15 - 15}\right)$	$\left(\dfrac{16 - 20}{15 - 20}\right)$
$k = 2$: $S_2(16) = 0.12$	$\left(\dfrac{16 - 10}{20 - 10}\right)$	$\left(\dfrac{16 - 15}{20 - 15}\right)$	$\left(\dfrac{16 - 20}{20 - 20}\right)$

Use Eq. 12.3.

$$S = S(10)S_0(16) + S(15)S_1(16) + S(20)S_2(16)$$
$$= \left(227 \ \frac{\text{mg}}{\text{L}}\right)(-0.08) + \left(362 \ \frac{\text{mg}}{\text{L}}\right)(0.96)$$
$$+ \left(517 \ \frac{\text{mg}}{\text{L}}\right)(0.12)$$
$$= \boxed{391 \ \text{mg/L} \quad (390 \ \text{mg/L})}$$

The answer is (B).

13 Energy, Work, and Power

PRACTICE PROBLEMS

Energy

1. A solid, cast-iron sphere (density of 0.256 lbm/in^3 (7090 kg/m^3)) of 10 in (25 cm) diameter travels without friction at 30 ft/sec (9 m/s) horizontally. Its kinetic energy is most nearly

- (A) 900 ft-lbf (1.2 kJ)
- (B) 1200 ft-lbf (1.6 kJ)
- (C) 1600 ft-lbf (2.0 kJ)
- (D) 1900 ft-lbf (2.3 kJ)

Work

2. The work done when a balloon carries a 12 lbm (5.2 kg) load to 40,000 ft (12 000 m) height is most nearly

- (A) 2.4×10^5 ft-lbf (300 kJ)
- (B) 4.8×10^5 ft-lbf (610 kJ)
- (C) 7.7×10^5 ft-lbf (980 kJ)
- (D) 9.9×10^5 ft-lbf (1.3 MJ)

3. What is most nearly the compression (deflection) if a 100 lbm (50 kg) weight is dropped from 8 ft (2 m) onto a spring with a stiffness of 33.33 lbf/in (5.837×10^3 N/m)?

- (A) 27 in (0.67 m)
- (B) 34 in (0.85 m)
- (C) 39 in (0.90 m)
- (D) 45 in (1.1 m)

4. A punch press flywheel with a moment of inertia of 483 lbm-ft^2 (20 kg·m^2) operates at 300 rpm. What is most nearly the speed in rpm to which the wheel will be reduced after a sudden punching requiring 4500 ft-lbf (6100 J) of work?

- (A) 160 rpm
- (B) 190 rpm
- (C) 220 rpm
- (D) 280 rpm

5. A force of 550 lbf (2500 N) making a 40° angle (upward) from the horizontal pushes a box 20 ft (6 m) across the floor. The work done is most nearly

- (A) 2200 ft-lbf (3.0 kJ)
- (B) 3700 ft-lbf (5.2 kJ)
- (C) 4200 ft-lbf (6.0 kJ)
- (D) 8400 ft-lbf (12 kJ)

6. A 1000 ft long (300 m long) cable has a mass of 2 lbm/ft (3 kg/m) and is fully suspended from a winding drum down into a vertical shaft. The work done to rewind the cable is most nearly

- (A) 0.50×10^6 ft-lbf (0.6 MJ)
- (B) 0.75×10^6 ft-lbf (0.9 MJ)
- (C) 1×10^6 ft-lbf (1.3 MJ)
- (D) 2×10^6 ft-lbf (2.6 MJ)

Power

7. Approximately what volume in ft^3 (m^3) of water can be pumped to a 130 ft (40 m) height in 1 hr by a 7 hp (5 kW) pump? Assume 85% efficiency.

- (A) 1500 ft^3 (40 m^3)
- (B) 1800 ft^3 (49 m^3)
- (C) 2000 ft^3 (54 m^3)
- (D) 2400 ft^3 (65 m^3)

8. The power in horsepower (kW) that is required to lift a 3300 lbm (1500 kg) mass 250 ft (80 m) vertically in 14 sec is most nearly

- (A) 40 hp (30 kW)
- (B) 70 hp (53 kW)
- (C) 90 hp (68 kW)
- (D) 110 hp (84 kW)

SOLUTIONS

1. *Customary U.S. Solution*

Since there is no friction, there is no rotation. The sphere slides.

$$E_{\text{kinetic}} = \tfrac{1}{2}\left(\frac{m}{g_c}\right)v^2 = \tfrac{1}{2}\left(\frac{V\rho}{g_c}\right)v^2$$

$$= \left(\tfrac{1}{2}\right)\left(\tfrac{4}{3}\pi r^3\right)\left(\frac{\rho}{g_c}\right)v^2$$

$$= \tfrac{2}{3}\pi r^3\left(\frac{\rho}{g_c}\right)v^2$$

$$= \tfrac{2}{3}\pi\left(\frac{10\text{ in}}{2}\right)^3\left(\frac{0.256\ \frac{\text{lbm}}{\text{in}^3}}{32.2\ \frac{\text{lbm-ft}}{\text{lbf-sec}^2}}\right)\left(30\ \frac{\text{ft}}{\text{sec}}\right)^2$$

$$= \boxed{1873\text{ ft-lbf}\quad(1900\text{ ft-lbf})}$$

The answer is (D).

SI Solution

Since there is no friction, there is no rotation. The sphere slides.

$$E_{\text{kinetic}} = \tfrac{1}{2}mv^2 = \tfrac{1}{2}(\rho V)v^2$$

$$= \tfrac{1}{2}\rho\left(\tfrac{4}{3}\pi r^3\right)v^2$$

$$= \tfrac{2}{3}\pi r^3\rho v^2$$

$$= \tfrac{2}{3}\pi\left(\frac{0.25\text{ m}}{2}\right)^3\left(7090\ \frac{\text{kg}}{\text{m}^3}\right)\left(9\ \frac{\text{m}}{\text{s}}\right)^2$$

$$= \boxed{2349\text{ J}\quad(2.3\text{ kJ})}$$

The answer is (D).

2. *Customary U.S. Solution*

From Eq. 13.11(b) and Eq. 13.12, the work done by the balloon is

$$W = \Delta E_{\text{potential}} = \frac{mg\Delta h}{g_c}$$

$$= \frac{(12\text{ lbm})\left(32.2\ \frac{\text{ft}}{\text{sec}^2}\right)(40{,}000\text{ ft})}{32.2\ \frac{\text{lbm-ft}}{\text{lbf-sec}^2}}$$

$$= \boxed{4.8\times10^5\text{ ft-lbf}}$$

The answer is (B).

SI Solution

From Eq. 13.11(a) and Eq. 13.12, the work done by the balloon is

$$W = \Delta E_{\text{potential}} = mg\Delta h$$

$$= \frac{(5.2\text{ kg})\left(9.81\ \frac{\text{m}}{\text{s}^2}\right)(12\,000\text{ m})}{1000\ \frac{\text{J}}{\text{kJ}}}$$

$$= \boxed{612.1\text{ kJ}\quad(610\text{ kJ})}$$

The answer is (B).

3. *Customary U.S. Solution*

Equate the potential energy to the energy of the spring.

$$\Delta E_{\text{potential}} = \Delta E_{\text{spring}}$$

$$W(\Delta h + \Delta x) = \tfrac{1}{2}k(\Delta x)^2$$

Rearranging and using $W = m(g/g_c)$,

$$\tfrac{1}{2}k(\Delta x)^2 - W\Delta x - W\Delta h = 0$$

$$\left(\tfrac{1}{2}\right)\left(33.33\ \frac{\text{lbf}}{\text{in}}\right)(\Delta x)^2$$

$$- (100\text{ lbm})\left(\frac{32.2\ \frac{\text{ft}}{\text{sec}^2}}{32.2\ \frac{\text{lbm-ft}}{\text{lbf-sec}^2}}\right)\Delta x$$

$$- (100\text{ lbm})\left(\frac{32.2\ \frac{\text{ft}}{\text{sec}^2}}{32.2\ \frac{\text{lbm-ft}}{\text{lbf-sec}^2}}\right)(8\text{ ft})\left(12\ \frac{\text{in}}{\text{ft}}\right) = 0$$

$$16.665(\Delta x)^2 - 100\Delta x = 9600$$

Complete the square.

$$(\Delta x)^2 - 6\Delta x = 576$$

$$(\Delta x - 3)^2 = 576 + 9$$

$$\Delta x - 3 = \pm\sqrt{585} = \pm24.2$$

$$\Delta x = \boxed{27.2\text{ in}\quad(27\text{ in})}$$

The answer is (A).

SI Solution

Equate the potential energy to the energy of the spring.

$$\Delta E_{\text{potential}} = \Delta E_{\text{spring}}$$

$$mg(\Delta h + \Delta x) = \tfrac{1}{2}k(\Delta x)^2$$

Rearrange.

$$\tfrac{1}{2}k(\Delta x)^2 - mg\Delta x - mg\Delta h = 0$$

$$\left(\tfrac{1}{2}\right)\left(5.837\times 10^3\ \tfrac{\text{N}}{\text{m}}\right)(\Delta x)^2$$

$$- (50\text{ kg})\left(9.81\ \tfrac{\text{m}}{\text{s}^2}\right)\Delta x$$

$$- (50\text{ kg})\left(9.81\ \tfrac{\text{m}}{\text{s}^2}\right)(2\text{ m}) = 0$$

$$2918.5(\Delta x)^2 - 490.5\Delta x - 981.0 = 0$$

$$(\Delta x)^2 - 0.1681\Delta x = 0.3361$$

Complete the square.

$$(\Delta x - 0.08403)^2 = 0.3361 + (0.08403)^2$$

$$= 0.3432$$

$$\Delta x - 0.08403 = \pm\sqrt{0.3432} = \pm 0.5858$$

$$\Delta x = \boxed{0.6699\text{ m} \quad (0.67\text{ m})}$$

The answer is (A).

4. *Customary U.S. Solution*

The work done by the wheel is equal to the change in the rotational energy.

$$W_{\text{done by wheel}} = \Delta E_{\text{rotational}}$$

$$= \tfrac{1}{2}\left(\frac{I}{g_c}\right)\omega_{\text{initial}}^2 - \tfrac{1}{2}\left(\frac{I}{g_c}\right)\omega_{\text{final}}^2$$

The final angular velocity is found from

$$\omega_{\text{final}} = \sqrt{\omega_{\text{initial}}^2 - \frac{2g_c W}{I}} = 2\pi f$$

The final speed is

$$n_{\text{final,rpm}} = f_{\text{final,Hz}}\left(60\ \tfrac{\text{sec}}{\text{min}}\right) = \frac{\omega_{\text{final}}\left(60\ \tfrac{\text{sec}}{\text{min}}\right)}{2\pi}$$

$$= \left(\frac{1}{2\pi}\right)\left(60\ \tfrac{\text{sec}}{\text{min}}\right)$$

$$\times \sqrt{\begin{array}{c}\left(\left(2\pi\ \tfrac{\text{rad}}{\text{rev}}\right)\left(\dfrac{300\ \tfrac{\text{rev}}{\text{min}}}{60\ \tfrac{\text{sec}}{\text{min}}}\right)\right)^2 \\[4pt] \dfrac{(2)\left(32.2\ \tfrac{\text{lbm-ft}}{\text{lbf-sec}^2}\right)}{} \\[4pt] -\dfrac{\times (4500\text{ ft-lbf})}{483\text{ lbm-ft}^2}\end{array}}$$

$$= \boxed{187.8\text{ rpm} \quad (190\text{ rpm})}$$

The answer is (B).

SI Solution

The work done by the wheel is equal to the change in the rotational energy.

$$W_{\text{done by wheel}} = \Delta E_{\text{rotational}}$$

$$= \tfrac{1}{2}I\omega_{\text{initial}}^2 - \tfrac{1}{2}I\omega_{\text{final}}^2$$

The final angular velocity is found from

$$\omega_{\text{final}} = \sqrt{\omega_{\text{initial}}^2 - \frac{2W}{I}} = 2\pi f$$

The final speed is

$$n_{\text{final,rpm}} = f_{\text{final,Hz}}\left(60\ \tfrac{\text{s}}{\text{min}}\right) = \frac{\omega_{\text{final}}\left(60\ \tfrac{\text{s}}{\text{min}}\right)}{2\pi}$$

$$= \left(\frac{1}{2\pi}\right)\left(60\ \tfrac{\text{s}}{\text{min}}\right)\sqrt{\begin{array}{c}\left(\left(2\pi\ \tfrac{\text{rad}}{\text{rev}}\right)\left(\dfrac{300\ \tfrac{\text{rev}}{\text{min}}}{60\ \tfrac{\text{s}}{\text{min}}}\right)\right)^2 \\[4pt] -\dfrac{(2)(6100\text{ J})}{20\text{ kg}\cdot\text{m}^2}\end{array}}$$

$$= \boxed{185.4\text{ rpm} \quad (190\text{ rpm})}$$

The answer is (B).

5. *Customary U.S. Solution*

The work done is

$$W_{\text{done on box}} = F_x\Delta x = F(\cos\theta)\Delta x$$

$$= (550\text{ lbf})(\cos 40°)(20\text{ ft})$$

$$= \boxed{8426\text{ ft-lbf} \quad (8400\text{ ft-lbf})}$$

The answer is (D).

SI Solution

The work done is

$$W_{\text{done on box}} = F_x\Delta x = F(\cos\theta)\Delta x$$

$$= \frac{(2500\text{ N})(\cos 40°)(6\text{ m})}{1000\ \tfrac{\text{J}}{\text{kJ}}}$$

$$= \boxed{11.49\text{ kJ} \quad (12\text{ kJ})}$$

The answer is (D).

6. *Customary U.S. Solution*

The work done to rewind the cable is

$$W_{\text{to rewind cable}} = \int_0^l F\,dh = \int_0^l ((l-h)w)\,dh$$
$$= \tfrac{1}{2}wl^2$$
$$= \left(\tfrac{1}{2}\right)\left(2\,\frac{\text{lbf}}{\text{ft}}\right)(1000\text{ ft})^2$$
$$= \boxed{1 \times 10^6 \text{ ft-lbf}}$$

The answer is (C).

SI Solution

The work done to rewind the cable is

$$W_{\text{to rewind cable}} = \int_0^l F\,dh = \int_0^l ((l-h)m_lg)\,dh$$
$$= \tfrac{1}{2}m_lgl^2$$
$$= \left(\tfrac{1}{2}\right)\left(3\,\frac{\text{kg}}{\text{m}}\right)\left(9.81\,\frac{\text{m}}{\text{s}^2}\right)(300\text{ m})^2$$
$$= \boxed{1.32 \times 10^6 \text{ J} \quad (1.3 \text{ MJ})}$$

The answer is (C).

7. *Customary U.S. Solution*

The volume of water is found from the work performed.

$$P_{\text{actual}}\,\Delta t = W_{\text{done by pump}}$$
$$\eta P_{\text{ideal}}\,\Delta t = \Delta E_{\text{potential}}$$
$$= \frac{mg\Delta h}{g_c}$$
$$= \frac{(\rho V)g\Delta h}{g_c}$$
$$V = \frac{\eta P_{\text{ideal}}\,\Delta t}{\dfrac{\rho g\Delta h}{g_c}}$$
$$= \frac{(0.85)(7\text{ hp})\left(550\,\dfrac{\text{ft-lbf}}{\text{hp-sec}}\right)(1\text{ hr})\left(3600\,\dfrac{\text{sec}}{\text{hr}}\right)}{\dfrac{\left(62.4\,\dfrac{\text{lbm}}{\text{ft}^3}\right)\left(32.2\,\dfrac{\text{ft}}{\text{sec}^2}\right)(130\text{ ft})}{32.2\,\dfrac{\text{lbm-ft}}{\text{lbf-sec}^2}}}$$
$$= \boxed{1452 \text{ ft}^3 \quad (1500 \text{ ft}^3)}$$

The answer is (A).

SI Solution

The volume of water is found from the work performed.

$$P_{\text{actual}}\,\Delta t = W_{\text{done by pump}}$$
$$\eta P_{\text{ideal}}\,\Delta t = \Delta E_{\text{potential}}$$
$$= mg\Delta h$$
$$= (\rho V)g\Delta h$$
$$V = \frac{\eta P_{\text{ideal}}\,\Delta t}{\rho g\Delta h}$$
$$= \frac{(0.85)(5\text{ kW})\left(1000\,\dfrac{\text{W}}{\text{kW}}\right)(1\text{ h})\left(3600\,\dfrac{\text{s}}{\text{h}}\right)}{\left(1000\,\dfrac{\text{kg}}{\text{m}^3}\right)\left(9.81\,\dfrac{\text{m}}{\text{s}^2}\right)(40\text{ m})}$$
$$= \boxed{39.0 \text{ m}^3 \quad (40 \text{ m}^3)}$$

The answer is (A).

8. *Customary U.S. Solution*

The work required to lift the mass is

$$W = P\Delta t = \frac{mg\Delta h}{g_c}$$
$$P = \frac{mg\Delta h}{g_c\Delta t}$$
$$= \frac{(3300\text{ lbm})\left(32.2\,\dfrac{\text{ft}}{\text{sec}^2}\right)(250\text{ ft})}{\left(32.2\,\dfrac{\text{lbm-ft}}{\text{sec}^2\text{-lbf}}\right)(14\text{ sec})\left(550\,\dfrac{\text{ft-lbf}}{\text{hp-sec}}\right)}$$
$$= \boxed{107 \text{ hp} \quad (110 \text{ hp})}$$

The answer is (D).

SI Solution

The work required to lift the mass is

$$W = P\Delta t = mg\Delta h$$
$$P = \frac{mg\Delta h}{\Delta t} = \frac{(1500\text{ kg})\left(9.81\,\dfrac{\text{m}}{\text{s}^2}\right)(80\text{ m})}{(14\text{ s})\left(1000\,\dfrac{\text{W}}{\text{kW}}\right)}$$
$$= \boxed{84.1 \text{ kW} \quad (84 \text{ kW})}$$

The answer is (D).

14 Fluid Properties

PRACTICE PROBLEMS

(Use $g = 32.2$ ft/sec^2 or 9.81 m/s^2 unless told to do otherwise in the problem.)

Pressure

1. Atmospheric pressure is 14.7 lbf/in^2 (101.3 kPa). What is most nearly the absolute pressure in a tank if a gauge reads 8.7 lbf/in^2 (60 kPa) vacuum?

(A) 4 psia (27 kPa)

(B) 6 psia (41 kPa)

(C) 8 psia (55 kPa)

(D) 10 psia (68 kPa)

Viscosity

2. Air is considered to be an ideal gas with a specific gas constant of 53.3 ft-lbf/lbm-°R (287 J/kg·K). What is most nearly the kinematic viscosity of air at 80°F (27°C) and 70 psia (480 kPa)?

(A) 3.5×10^{-5} ft^2/sec (3.0×10^{-6} m^2/s)

(B) 4.0×10^{-5} ft^2/sec (4.0×10^{-6} m^2/s)

(C) 5.0×10^{-5} ft^2/sec (5.0×10^{-6} m^2/s)

(D) 6.0×10^{-5} ft^2/sec (6.0×10^{-6} m^2/s)

Solutions

3. An 8% (by volume) solution, a 10% solution, and a 20% solution of nitric acid are combined to produce 100 mL of a 12% solution. The 8% solution contributes half of the total volume of nitric acid contributed by the 10% and 20% solutions. The volume of 10% acid solution is most nearly

(A) 20 mL

(B) 30 mL

(C) 50 mL

(D) 80 mL

Properties of Mixtures

4. A blend contains equal (weights) masses of two non-reacting oils. The two oils have kinematic viscosities of 1 cSt and 1000 cSt, respectively. What is most nearly the kinematic viscosity of the mixture?

(A) 7 cSt

(B) 20 cSt

(C) 30 cSt

(D) 200 cSt

SOLUTIONS

1. *Customary U.S. Solution*

$$p_{gage} = -8.7 \text{ lbf/in}^2$$
$$p_{atmospheric} = 14.7 \text{ lbf/in}^2$$

The relationship between absolute, gage, and atmospheric pressure is

$$p_{absolute} = p_{gage} + p_{atmospheric} = -8.7 \frac{\text{lbf}}{\text{in}^2} + 14.7 \frac{\text{lbf}}{\text{in}^2}$$
$$= \boxed{6 \text{ lbf/in}^2 \quad (6 \text{ psia})}$$

The answer is (B).

SI Solution

$$p_{gage} = -60 \text{ kPa}$$
$$p_{atmospheric} = 101.3 \text{ kPa}$$

The relationship between absolute, gage, and atmospheric pressure is

$$p_{absolute} = p_{gage} + p_{atmospheric} = -60 \text{ kPa} + 101.3 \text{ kPa}$$
$$= \boxed{41.3 \text{ kPa} \quad (41 \text{ kPa})}$$

The answer is (B).

2. *Customary U.S. Solution*

From App. 14.D, for air at 14.7 psia and 80°F, the absolute viscosity (independent of pressure) is $\mu = 3.869 \times 10^{-7}$ lbf-sec/ft^2.

Determine the density of air at 70 psia and 80°F.

$$\rho = \frac{p}{RT} = \frac{\left(70 \frac{\text{lbf}}{\text{in}^2}\right)\left(12 \frac{\text{in}}{\text{ft}}\right)^2}{\left(53.3 \frac{\text{ft-lbf}}{\text{lbm-°R}}\right)(80°F + 460°)}$$
$$= 0.350 \text{ lbm/ft}^3$$

The kinematic viscosity, ν, is

$$\nu = \frac{\mu g_c}{\rho} = \frac{\left(3.869 \times 10^{-7} \frac{\text{lbf-sec}}{\text{ft}^2}\right)\left(32.2 \frac{\text{lbm-ft}}{\text{lbf-sec}^2}\right)}{0.350 \frac{\text{lbm}}{\text{ft}^3}}$$
$$= \boxed{3.56 \times 10^{-5} \text{ ft}^2/\text{sec} \quad (3.6 \times 10^{-5} \text{ ft}^2/\text{sec})}$$

The answer is (A).

SI Solution

From App. 14.E, for air at 480 kPa and 27°C, the absolute viscosity (independent of pressure) is $\mu = 1.854 \times 10^{-5}$ Pa·s.

Determine the density of air at 480 kPa and 27°C.

$$\rho = \frac{p}{RT} = \frac{(480 \text{ kPa})\left(1000 \frac{\text{Pa}}{\text{kPa}}\right)}{\left(287 \frac{\text{J}}{\text{kg·K}}\right)(27°C + 273°)}$$
$$= 5.575 \text{ kg/m}^3$$

The kinematic viscosity, ν, is

$$\nu = \frac{\mu}{\rho} = \frac{1.854 \times 10^{-5} \text{ Pa·s}}{5.575 \frac{\text{kg}}{\text{m}^3}}$$
$$= \boxed{3.33 \times 10^{-6} \text{ m}^2/\text{s} \quad (3.0 \times 10^{-6} \text{ m}^2/\text{s})}$$

The answer is (A).

3. Let

$$x = \text{volume of 8\% solution}$$
$$0.08x = \text{volume of nitric acid contributed by 8\%}$$
$$\text{solution}$$
$$y = \text{volume of 10\% solution}$$
$$0.10y = \text{volume of nitric acid contributed by 10\%}$$
$$\text{solution}$$
$$z = \text{volume of 20\% solution}$$
$$0.20z = \text{volume of nitric acid contributed by 20\%}$$
$$\text{solution}$$

The three conditions that must be satisfied are

$$x + y + z = 100 \text{ mL}$$
$$0.08x + 0.10y + 0.20z = (0.12)(100 \text{ mL}) = 12 \text{ mL}$$
$$0.08x = \left(\tfrac{1}{2}\right)(0.10y + 0.20z)$$

Simplifying these equations,

$$x + y + z = 100$$
$$4x + 5y + 10z = 600$$
$$8x - 5y - 10z = 0$$

Adding the second and third equations gives

$$12x = 600$$
$$x = 50 \text{ mL}$$

Work with the first two equations to get

$$y + z = 100 - 50 = 50$$
$$5y + 10z = 600 - (4)(50) = 400$$

Multiplying the top equation by -5 and adding to the bottom equation,

$$5z = 150$$
$$z = 30 \text{ mL}$$

From the first equation,

$$y = \boxed{20 \text{ mL}}$$

The answer is (A).

4. The gravimetric fraction of each oil is $G = 0.50$.
From Eq. 14.43, the two viscosity blending indexes are

$$
\begin{aligned}
\text{VBI}_{1\,\text{cSt}} &= 10.975 + 14.534 \times \ln\big(\ln(\nu_{i,\text{cSt}} + 0.8)\big) \\
&= 10.975 + 14.534 \times \ln\big(\ln(1 \text{ cSt} + 0.8)\big) \\
&= 3.2518
\end{aligned}
$$

$$
\begin{aligned}
\text{VBI}_{1000\,\text{cSt}} &= 10.975 + 14.534 \times \ln\big(\ln(\nu_{i,\text{cSt}} + 0.8)\big) \\
&= 10.975 + 14.534 \times \ln\big(\ln(1000 \text{ cSt} + 0.8)\big) \\
&= 39.0657
\end{aligned}
$$

Using Eq. 14.44, the mixture VBI is

$$
\begin{aligned}
\text{VBI}_{\text{mixture}} &= \sum_i G_i \times \text{VBI}_i \\
&= (0.50)(3.2518) + (0.50)(39.0657) \\
&= 21.1588
\end{aligned}
$$

From Eq. 14.45, the viscosity of the mixture is

$$
\begin{aligned}
\nu_{\text{mixture,cSt}} &= \exp\Big(\exp\Big(\frac{\text{VBI}_{\text{mixture}} - 10.975}{14.534}\Big)\Big) - 0.8 \\
&= \exp\Big(\exp\Big(\frac{21.1588 - 10.975}{14.534}\Big)\Big) - 0.8 \\
&= \boxed{6.70 \text{ cSt} \quad (7 \text{ cSt})}
\end{aligned}
$$

The answer is (A).

Flow of Fluids

15 Fluid Statics

PRACTICE PROBLEMS

(Use $g = 32.2$ ft/sec^2 or 9.81 m/s^2 unless told to do otherwise in the problem.)

1. A 4000 lbm (1800 kg) blimp contains 10,000 lbm (4500 kg) of hydrogen (specific gas constant = 766.5 ft-lbf/lbm-°R (4124 J/kg·K)) at 56°F (13°C) and 30.2 in Hg (770 mm Hg). If the hydrogen and air are in thermal and pressure equilibrium, what is most nearly the blimp's lift (lifting force)?

(A) 7.6×10^3 lbf (3.4×10^4 N)

(B) 1.2×10^4 lbf (5.3×10^4 N)

(C) 1.3×10^5 lbf (5.7×10^5 N)

(D) 1.7×10^5 lbf (7.7×10^5 N)

2. A hollow 6 ft (1.8 m) diameter sphere floats half-submerged in seawater. The mass of concrete that is required as an external anchor to just submerge the sphere completely is most nearly

(A) 2700 lbm (1200 kg)

(B) 4200 lbm (1900 kg)

(C) 5500 lbm (2500 kg)

(D) 6300 lbm (2700 kg)

3. Water removed from Lake Superior (elevation, 601 ft above mean sea level; water density, 62.4 lbm/ft^3) is transported by tanker ship to the Atlantic Ocean (elevation, 0 ft) through 16 Seaway locks. A tanker's displacement is 32,000 tonnes when loaded, and 5100 tonnes when empty. Each lock is 766 ft long and 80 ft wide. Water pumped from each lock flows to the Atlantic Ocean. Compared to a passage from Lake Superior to the Atlantic Ocean when empty, what is most nearly the change in water loss from Lake Superior when a ship passes through the locks fully loaded?

(A) 27,000 tonnes less loss

(B) no change in loss

(C) 27,000 tonnes additional loss

(D) 54,000 tonnes additional loss

SOLUTIONS

1. *Customary U.S. Solution*

The lift (lifting force) of the hydrogen-filled blimp, F_{lift}, is equal to the difference between the buoyant force, F_b, and the weight of the hydrogen contained in the blimp, W_H.

$$F_{\text{lift}} = F_b - W_H - W_{\text{blimp}}$$

The weight of the hydrogen is calculated from the mass of hydrogen.

$$W_H = \frac{m_H g}{g_c} = \frac{(10{,}000 \text{ lbm})\left(32.2 \, \dfrac{\text{ft}}{\text{sec}^2}\right)}{32.2 \, \dfrac{\text{lbm-ft}}{\text{lbf-sec}^2}}$$

$$= 10{,}000 \text{ lbf}$$

The buoyant force is equal to the weight of the displaced air. The volume of the air displaced is equal to the volume of hydrogen enclosed in the blimp.

The absolute temperature of the hydrogen is

$$T = 56°\text{F} + 460° = 516°\text{R}$$

The pressure of the hydrogen is

$$p = \frac{(30.2 \text{ in Hg})\left(12 \, \dfrac{\text{in}}{\text{ft}}\right)^2}{2.036 \, \dfrac{\text{in Hg}}{\dfrac{\text{lbf}}{\text{in}^2}}} = 2136 \text{ lbf/ft}^2$$

Compute the volume of hydrogen from the ideal gas law.

$$V_H = \frac{m_H R T}{p} = \frac{(10{,}000 \text{ lbm})\left(766.5 \, \dfrac{\text{ft-lbf}}{\text{lbm-°R}}\right)(516°\text{R})}{2136 \, \dfrac{\text{lbf}}{\text{ft}^2}}$$

$$= 1.85 \times 10^6 \text{ ft}^3$$

Since the volume of the hydrogen contained in the blimp is equal to the air displaced, the air displaced can be computed from the ideal gas equation. Since the air and hydrogen are in thermal and pressure equilibrium, the temperature and pressure are equal to the values given for the hydrogen.

For air, $R = 53.35$ ft-lbf/lbm-°R.

$$m_{air} = \frac{pV_H}{RT} = \frac{\left(2136 \; \frac{lbf}{ft^2}\right)\left(1.85 \times 10^6 \; ft^3\right)}{\left(53.35 \; \frac{ft\text{-}lbf}{lbm\text{-}°R}\right)(516°R)}$$

$$= 1.435 \times 10^5 \; lbm$$

The buoyant force is equal to the weight of the air.

$$F_b = W_{air} = \frac{m_{air}g}{g_c} = \frac{\left(1.435 \times 10^5 \; lbm\right)\left(32.2 \; \frac{ft}{sec^2}\right)}{32.2 \; \frac{lbm\text{-}ft}{lbf\text{-}sec^2}}$$

$$= 1.435 \times 10^5 \; lbf$$

The lift (lifting force) is

$$F_{lift} = F_b - W_H - W_{blimp}$$

$$= 1.435 \times 10^5 \; lbf - 10{,}000 \; lbf - 4000 \; lbf$$

$$= \boxed{1.295 \times 10^5 \; lbf \quad (1.3 \times 10^5 \; lbf)}$$

The answer is (C).

SI Solution

The lift (lifting force) of the hydrogen-filled blimp, F_{lift}, is equal to the difference between the buoyant force, F_b, and the weight of the hydrogen contained in the blimp, W_H.

$$F_{lift} = F_b - W_H - W_{blimp}$$

The weight of the hydrogen is calculated from the mass of hydrogen.

$$W_H = m_H g$$

$$= (4500 \; kg)\left(9.81 \; \frac{m}{s^2}\right)$$

$$= 44\,145 \; N$$

The buoyant force is equal to the weight of the displaced air. The volume of the air displaced is equal to the volume of hydrogen enclosed in the blimp.

The absolute temperature of the hydrogen is

$$T = 13°C + 273° = 286K$$

The absolute pressure of the hydrogen is

$$p = \frac{(770 \; mm \; Hg)\left(133.4 \; \frac{kPa}{m}\right)}{1000 \; \frac{mm}{m}}$$

$$= 102.7 \; kPa$$

Compute the volume of hydrogen from the ideal gas law.

$$V_H = \frac{m_H RT}{p} = \frac{(4500 \; kg)\left(4124 \; \frac{J}{kg\cdot K}\right)(286K)}{(102.7 \; kPa)\left(1000 \; \frac{Pa}{kPa}\right)}$$

$$= 5.168 \times 10^4 \; m^3$$

Since the volume of the hydrogen contained in the blimp is equal to the air displaced, the air displaced can be computed from the ideal gas equation. Since the air and hydrogen are assumed to be in thermal and pressure equilibrium, the temperature and pressure are equal to the values given for the hydrogen.

For air, $R = 287.03$ J/kg·K.

$$m_{air} = \frac{pV_H}{RT} = \frac{(102.7 \; kPa)\left(1000 \; \frac{Pa}{kPa}\right)\left(5.168 \times 10^4 \; m^3\right)}{\left(287.03 \; \frac{J}{kg\cdot K}\right)(286K)}$$

$$= 6.465 \times 10^4 \; kg$$

The buoyant force is equal to the weight of the air.

$$F_b = W_{air} = m_{air}g = (6.465 \times 10^4 \; kg)\left(9.81 \; \frac{m}{s^2}\right)$$

$$= 6.34 \times 10^5 \; N$$

The lift (lifting force) is

$$F_{lift} = F_b - W_H - W_{blimp}$$

$$= 6.34 \times 10^5 \; N - 44\,145 \; N - (1800 \; kg)\left(9.81 \; \frac{m}{s^2}\right)$$

$$= \boxed{5.7 \times 10^5 \; N}$$

The answer is (C).

2. *Customary U.S. Solution*

The weight of the sphere is equal to the weight of the displaced volume of water when floating.

The buoyant force is given by

$$F_b = \frac{\rho g V_{displaced}}{g_c}$$

Since the sphere is half submerged,

$$W_{sphere} = \frac{1}{2}\left(\frac{\rho g V_{sphere}}{g_c}\right)$$

For seawater, $\rho = 64.0$ lbm/ft^3.

The volume of the sphere is

$$V_{sphere} = \frac{\pi}{6}d^3 = \left(\frac{\pi}{6}\right)(6 \text{ ft})^3$$
$$= 113.1 \text{ ft}^3$$

The weight of the sphere is

$$W_{sphere} = \frac{1}{2}\left(\frac{\rho g V_{sphere}}{g_c}\right)$$
$$= \left(\frac{1}{2}\right)\left(\frac{\left(64.0 \frac{\text{lbm}}{\text{ft}^3}\right)\left(32.2 \frac{\text{ft}}{\text{sec}^2}\right)(113.1 \text{ ft}^3)}{32.2 \frac{\text{lbm-ft}}{\text{lbf-sec}^2}}\right)$$
$$= 3619 \text{ lbf}$$

The equilibrium equation for a fully submerged sphere and anchor can be solved for the concrete volume.

$$W_{sphere} + W_{concrete} = (V_{sphere} + V_{concrete})\rho_{water}$$
$$W_{sphere} + \rho_{concrete}V_{concrete}\left(\frac{g}{g_c}\right)$$
$$= (V_{sphere} + V_{concrete})\rho_{water}\left(\frac{g}{g_c}\right)$$
$$3619 \text{ lbf} + \left(150 \frac{\text{lbm}}{\text{ft}^3}\right)V_{concrete}\left(\frac{32.2 \frac{\text{ft}}{\text{sec}^2}}{32.2 \frac{\text{lbm-ft}}{\text{lbf-sec}^2}}\right)$$
$$= (113.1 \text{ ft}^3 + V_{concrete})\left(64.0 \frac{\text{lbm}}{\text{ft}^3}\right)\left(\frac{32.2 \frac{\text{ft}}{\text{sec}^2}}{32.2 \frac{\text{lbm-ft}}{\text{lbf-sec}^2}}\right)$$
$$V_{concrete} = 42.09 \text{ ft}^3$$
$$m_{concrete} = \rho_{concrete}V_{concrete}$$
$$= \left(150 \frac{\text{lbm}}{\text{ft}^3}\right)(42.09 \text{ ft}^3)$$
$$= \boxed{6314 \text{ lbm} \quad (6300 \text{ lbm})}$$

The answer is (D).

SI Solution

The weight of the sphere is equal to the weight of the displaced volume of water when floating.

The buoyant force is given by

$$F_b = \rho g V_{displaced}$$

Since the sphere is half submerged,

$$W_{sphere} = \frac{1}{2}\rho g V_{sphere}$$

For seawater, $\rho = 1025$ kg/m^3.

The volume of the sphere is

$$V_{sphere} = \frac{\pi}{6}d^3 = \left(\frac{\pi}{6}\right)(1.8 \text{ m})^3 = 3.054 \text{ m}^3$$

The weight of the sphere required is

$$W_{sphere} = \frac{1}{2}\rho g V_{sphere}$$
$$= \left(\frac{1}{2}\right)\left(1025 \frac{\text{kg}}{\text{m}^3}\right)\left(9.81 \frac{\text{m}}{\text{s}^2}\right)(3.054 \text{ m}^3)$$
$$= 15\,354 \text{ N}$$

The equilibrium equation for a fully submerged sphere and anchor can be solved for the concrete volume.

$$W_{sphere} + W_{concrete} = (V_{sphere} + V_{concrete})\rho_{water}$$
$$W_{sphere} + \rho_{concrete}g V_{concrete} = g(V_{sphere} + V_{concrete})\rho_{water}$$
$$15\,354 \text{ N} + \left(2400 \frac{\text{kg}}{\text{m}^3}\right)\left(9.81 \frac{\text{m}}{\text{s}^2}\right)V_{concrete}$$
$$= (3.054 \text{ m}^3 + V_{concrete})\left(1025 \frac{\text{kg}}{\text{m}^3}\right)\left(9.81 \frac{\text{m}}{\text{s}^2}\right)$$
$$V_{concrete} = 1.138 \text{ m}^3$$
$$m_{concrete} = \rho_{concrete}V_{concrete}$$
$$= \left(2400 \frac{\text{kg}}{\text{m}^3}\right)(1.138 \text{ m}^3)$$
$$= \boxed{2731 \text{ kg} \quad (2700 \text{ kg})}$$

The answer is (D).

3. From Archimedes' principle, each tonne of water carried in the tanker displaces a tonne of water in the lock. So, each tonne of water transported out of the lake results in a tonne less of lock loss. Compared to an empty tanker, the net result is zero.

The answer is (B).

16 Fluid Flow Parameters

PRACTICE PROBLEMS

Use the following values unless told to do otherwise in the problem:

$$g = 32.2 \text{ ft/sec}^2 \, (9.81 \text{ m/s}^2)$$

$$\rho_{\text{water}} = 62.4 \text{ lbm/ft}^3 \, (1000 \text{ kg/m}^3)$$

$$p_{\text{atmospheric}} = 14.7 \text{ psia} \, (101.3 \text{ kPa})$$

Hydraulic Radius

1. A 10 in (25 cm) composition pipe is compressed by a tree root into an elliptical cross section until its inside height is only 7.2 in (18 cm). What is its approximate hydraulic radius when flowing half full?

(A) 2.2 in (5.5 cm)

(B) 2.7 in (6.9 cm)

(C) 3.2 in (8.1 cm)

(D) 4.5 in (11.4 cm)

2. A pipe with an inside diameter of 18.812 in contains water to a depth of 15.7 in. What is most nearly the hydraulic radius? (Work in customary U.S. units only.)

(A) 4.4 in

(B) 5.1 in

(C) 5.7 in

(D) 6.5 in

Pipe Ratings

3. A class IV, 60 in diameter C76 concrete pipe has a $D_{0.01}$ rating of 2000 lbf/ft² and a D_{ultimate} rating of 3000 lbf/ft². Approximately what vertical line force must be applied in three-point loading to a pipe section 8 ft long in order to induce a concrete crack at least 1 ft long?

(A) 80,000 lbf

(B) 120,000 lbf

(C) 240,000 lbf

(D) 680,000 lbf

SOLUTIONS

1. *Customary U.S. Solution*

The perimeter of the pipe is

$$p = \pi d = \pi(10 \text{ in}) = 31.42 \text{ in}$$

If the pipe is flowing half full, the wetted perimeter becomes

$$\text{wetted perimeter} = \tfrac{1}{2}p = \left(\tfrac{1}{2}\right)(31.42 \text{ in}) = 15.71 \text{ in}$$

The ellipse will have a minor axis, b, equal to one-half the height of the compressed pipe or

$$b = \frac{7.2 \text{ in}}{2} = 3.6 \text{ in}$$

When the pipe is compressed, the perimeter of the pipe will remain constant. The perimeter of an ellipse is given by

$$p \approx 2\pi\sqrt{\tfrac{1}{2}(a^2 + b^2)}$$

Solve for the major axis.

$$a = \sqrt{2\left(\frac{p}{2\pi}\right)^2 - b^2} = \sqrt{(2)\left(\frac{31.42 \text{ in}}{2\pi}\right)^2 - (3.6 \text{ in})^2}$$
$$= 6.09 \text{ in}$$

The flow area or area of the ellipse is given by

$$\text{flow area} = \tfrac{1}{2}\pi ab = \tfrac{1}{2}\pi(6.09 \text{ in})(3.6 \text{ in})$$
$$= 34.4 \text{ in}^2$$

The hydraulic radius is

$$r_h = \frac{\text{area in flow}}{\text{wetted perimeter}} = \frac{34.4 \text{ in}^2}{15.71 \text{ in}} = \boxed{2.19 \text{ in} \quad (2.2 \text{ in})}$$

The answer is (A).

SI Solution

The perimeter of the pipe is

$$p = \pi d = \pi(25 \text{ cm}) = 78.54 \text{ cm}$$

If the pipe is flowing half full, the wetted perimeter becomes

$$\text{wetted perimeter} = \tfrac{1}{2}p = \left(\tfrac{1}{2}\right)(78.54 \text{ cm}) = 39.27 \text{ cm}$$

Assume the compressed pipe is an elliptical cross section. The ellipse will have a minor axis, b, equal to one-half the height of the compressed pipe or

$$b = \frac{18 \text{ cm}}{2} = 9 \text{ cm}$$

When the pipe is compressed, the perimeter of the pipe will remain constant. The perimeter of an ellipse is given by

$$p \approx 2\pi\sqrt{\tfrac{1}{2}(a^2 + b^2)}$$

Solve for the major axis.

$$a = \sqrt{2\left(\frac{p}{2\pi}\right)^2 - b^2} = \sqrt{(2)\left(\frac{78.54 \text{ cm}}{2\pi}\right)^2 - (9 \text{ cm})^2}$$

$$= 15.2 \text{ cm}$$

The flow area or area of the ellipse is given by

$$\text{flow area} = \tfrac{1}{2}\pi ab = \tfrac{1}{2}\pi(15.2 \text{ cm})(9 \text{ cm})$$

$$= 214.9 \text{ cm}^2$$

The hydraulic radius is

$$r_h = \frac{\text{area in flow}}{\text{wetted perimeter}} = \frac{214.9 \text{ cm}^2}{39.27 \text{ cm}}$$

$$= \boxed{5.47 \text{ cm} \quad (5.5 \text{ cm})}$$

The answer is (A).

2. *method 1:* Use App. 7.A for a circular segment.

$$r = \frac{D}{2} = \frac{18.812 \text{ in}}{2} = 9.406 \text{ in}$$

$$\phi = 2 \arccos \frac{r - D}{r}$$

$$= 2 \arccos \frac{9.406 \text{ in} - 3.112 \text{ in}}{9.406 \text{ in}}$$

$$= 1.675 \text{ rad}$$

$$\sin \phi = 0.9946$$

$$A_1 = \tfrac{1}{2}r^2(\phi - \sin \phi)$$

$$= \left(\tfrac{1}{2}\right)(9.406 \text{ in})^2(1.675 - 0.9946)$$

$$= 30.1 \text{ in}^2$$

$$A_{\text{total}} = A_1 + A_2 = \frac{\pi}{4}D^2 = \left(\frac{\pi}{4}\right)(18.812 \text{ in})^2$$

$$= 277.95 \text{ in}^2$$

$$A_2 = A_{\text{total}} - A_1 = 277.95 \text{ in}^2 - 30.1 \text{ in}^2$$

$$= 247.85 \text{ in}^2$$

$$s_1 = r\phi = (9.406 \text{ in})(1.675) = 15.76 \text{ in}$$

$$s_{\text{total}} = s_1 + s_2 = \pi D = \pi(18.812 \text{ in})$$

$$= 59.1 \text{ in}$$

$$s_2 = s_{\text{total}} - s_1 = 59.1 \text{ in} - 15.76 \text{ in} = 43.34 \text{ in}$$

$$r_h = \frac{A_2}{s_2} = \frac{247.85 \text{ in}^2}{43.34 \text{ in}} = \boxed{5.719 \text{ in} \quad (5.7 \text{ in})}$$

method 2: Use App. 16.A.

$$\frac{d}{D} = \frac{15.7 \text{ in}}{18.812 \text{ in}} = 0.83$$

From App. 16.A, $r_h/D = 0.3041$.

$$r_h = (0.3041)(18.812 \text{ in})$$

$$= \boxed{5.72 \text{ in} \quad (5.7 \text{ in})}$$

The answer is (C).

3. The definition of $D_{0.01}$ is the vertical force applied in three-point loading required to cause a crack at least 1 ft long. The dimensions used to calculate the D-load rating of 2000 lbf/ft^2 are pipe length and pipe diameter. Rearranging Eq. 16.35, the required line force is

$$F = D_{0.01} D_{\text{ft}} L_{\text{ft}} = \frac{\left(2000 \dfrac{\text{lbf}}{\text{ft}^2}\right)(60 \text{ in})(8 \text{ ft})}{12 \dfrac{\text{in}}{\text{ft}}}$$

$$= \boxed{80,000 \text{ lbf}}$$

The answer is (A).

17 Fluid Dynamics

PRACTICE PROBLEMS

(Use $g = 32.2$ ft/sec^2 (9.81 m/s^2) and 60°F (16°C) water unless told to do otherwise in the problem.)

Basic Concepts

1. The capacity of a municipal, old, pressurized water supply pipe system must be doubled without increasing the average velocity. A replacement new pipe will have 40% less friction than the existing pipe. Approximately what increase in pipe diameter is required?

(A) 20%

(B) 30%

(C) 40%

(D) 100%

2. A pressurized water pipe system will be modified such that the pipe diameter will be doubled and the average flow velocity halved. Fluid properties are unchanged. When these changes are made, the Reynolds number will

(A) halve

(B) double

(C) quadruple

(D) remain the same

Conservation of Energy

3. 5 ft^3/sec (130 L/s) of water flows through a schedule-40 steel pipe that changes gradually in diameter from 6 in at point A to 18 in at point B. Point B is 15 ft (4.6 m) higher than point A. The respective pressures at points A and B are 10 psia (70 kPa) and 7 psia (48.3 kPa). All minor losses are insignificant. The velocity and direction of flow at point A are most nearly

(A) 3.2 ft/sec (1 m/s); from A to B

(B) 25 ft/sec (7 m/s); from A to B

(C) 3.2 ft/sec (1 m/s); from B to A

(D) 25 ft/sec (7 m/s); from B to A

4. Points A and B are separated by 3000 ft of new 6 in schedule-40 steel pipe. 750 gal/min of 60°F water flows from point A to point B. Point B is 60 ft above point A. Approximately what must be the pressure at point A if the pressure at B must be 50 psig?

(A) 90 psig

(B) 100 psig

(C) 120 psig

(D) 170 psig

5. A pipe network connects junctions A, B, C, and D as shown. All pipe sections have a Hazen-Williams C-value of 150. Water can be added and removed at any of the junctions to achieve the flows listed. Water flows from point A to point D. No flows are backward. All minor losses are insignificant. For simplicity, use the nominal pipe diameters.

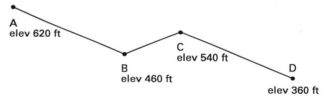

pipe section	length	diameter (in)	flow
A to B	20,000 ft	6	120 gal/min
B to C	10,000 ft	6	160 gal/min
C to D	30,000 ft	4	120 gal/min

(a) The friction loss from A to B is most nearly

(A) 0.1 ft

(B) 20 ft

(C) 40 ft

(D) 60 ft

(b) The velocity head in pipe AB is most nearly

(A) 0.007 ft

(B) 0.018 ft

(C) 0.020 ft

(D) 0.030 ft

Flow of Fluids

(c) The friction loss from B to C is most nearly

(A) 0.3 ft

(B) 20 ft

(C) 30 ft

(D) 50 ft

(d) The friction loss from C to D is most nearly

(A) 240 ft

(B) 300 ft

(C) 310 ft

(D) 340 ft

(e) Assume the static pressure at point A is 20 psig. The pressure at point B is most nearly

(A) 30 psig

(B) 60 psig

(C) 80 psig

(D) 90 psig

(f) Assume the static pressure at point A is 20 psig. The pressure at point C is most nearly

(A) 20 psig

(B) 30 psig

(C) 40 psig

(D) 50 psig

(g) Assume the static pressure at point A is 20 psig. The pressure at point D is most nearly

(A) 10 psig

(B) 12 psig

(C) 18 psig

(D) 20 psig

(h) If the minimum static pressure anywhere in the system is 20 psig, the pressure at point A is most nearly

(A) 14 psig

(B) 17 psig

(C) 24 psig

(D) 30 psig

(i) If the minimum static pressure anywhere in the system is 20 psig, the elevation of the hydraulic grade line at point A referenced to point D is most nearly

(A) 280 ft

(B) 300 ft

(C) 480 ft

(D) 600 ft

Friction Loss

6. Based on the formulas commonly used by engineers, is friction head loss ever proportional to velocity, instead of velocity squared?

(A) yes, in laminar flow

(B) yes, for non-Newtonian fluids

(C) yes, in smooth pipes

(D) no, never

7. Water flows quietly through a pipe network consisting of 50 ft of new 3 in diameter, schedule-40 steel pipe, four 45° standard elbows, and a fully open gate valve. All fittings are flanged. The elevation of the network discharge is 20 ft lower than the elevation of the inlet. The element that contributes the most to specific energy loss is the

(A) four elbows

(B) gate valve

(C) pipe friction (excluding the fittings)

(D) elevation change

8. 300 ft of 18 in high-density polyethylene (HDPE) pipe ($C = 120$) and 400 ft of 14 in HDPE pipe ($C = 120$) are currently joined in series as part of a pipe network. The length of 16 in diameter HDPE pipe ($C = 120$) that can replace the two pipes without increasing the pumping power required is most nearly

(A) 640 ft

(B) 790 ft

(C) 840 ft

(D) 940 ft

9. A 10 in diameter reinforced concrete pressure pipe ($C = 100$) carries water flowing at 4 ft/sec. What is most nearly the theoretical friction loss per 100 ft of pipe length?

(A) 1.0 ft

(B) 4.2 ft

(C) 10 ft

(D) 14 ft

10. A pressurized supply line ($C = 140$) brings cold water to 600 residential connections. The line is 17,000 ft long and has a diameter of 10 in. The elevation is 1000 ft at the start of the line and 850 ft at the delivery end of the line. The average flow rate is 1.1 gal/min per residence, and the peaking factor is 2.5. The minimum required pressure at the delivery end is 60 psig. Approximately what must be the pressure at the start of the line during peak flow?

(A) 9.4 psig

(B) 40 psig

(C) 95 psig

(D) 110 psig

11. 1.5 ft³/sec (40 L/s) of 70°F (20°C) water flows through 1200 ft (355 m) of 6 in (nominal) diameter new schedule-40 steel pipe. The friction loss is most nearly

(A) 4 ft (1.2 m)

(B) 18 ft (5.2 m)

(C) 36 ft (9.5 m)

(D) 70 ft (21 m)

12. 500 gal/min (30 L/s) of 100°F (40°C) water flows through 300 ft (90 m) of 6 in schedule-40 pipe. The pipe contains two 6 in flanged steel elbows, two full-open gate valves, a full-open 90° angle valve, and a swing check valve. The discharge is located 20 ft (6 m) higher than the entrance. The pressure difference between the two ends of the pipe is most nearly

(A) 12 psi (78 kPa)

(B) 21 psi (140 kPa)

(C) 45 psi (310 kPa)

(D) 87 psi (600 kPa)

13. 70°F (20°C) air is flowing at 60 ft/sec (18 m/s) through 300 ft (90 m) of 6 in schedule-40 pipe. The pipe contains two 6 in (0.15 m) flanged steel elbows, two full-open gate valves, a full-open 90° angle valve, and a swing check valve. The discharge is located 20 ft (6 m) higher than the entrance, and the average air density is 0.075 lbm/ft³. The pressure difference between the two ends of the pipe is most nearly

(A) 0.26 psi (1.8 kPa)

(B) 0.49 psi (3.2 kPa)

(C) 1.5 psi (10 kPa)

(D) 13 psi (90 kPa)

Reservoirs

14. Three reservoirs (A, B, and C) are interconnected with a common junction (point D) at elevation 25 ft above an arbitrary reference point. The water levels for reservoirs A, B, and C are at elevations of 50 ft, 40 ft, and 22 ft, respectively. The pipe from reservoir A to the junction is 800 ft of 3 in (nominal) steel pipe. The pipe from reservoir B to the junction is 500 ft of 10 in (nominal) steel pipe. The pipe from reservoir C to the junction is 1000 ft of 4 in (nominal) steel pipe. All pipes are schedule-40 with a friction factor of 0.02. All minor losses and velocity heads can be neglected. The direction of flow and the pressure at point D are most nearly

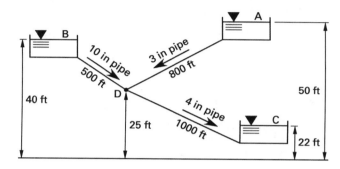

(A) out of reservoir B; 500 psf

(B) out of reservoir B; 930 psf

(C) into reservoir B; 1100 psf

(D) into reservoir B; 1260 psf

Water Hammer

15. A cast-iron pipe with expansion joints throughout has an inside diameter of 24 in (600 mm) and a wall thickness of 0.75 in (20 mm). The pipe's modulus of elasticity is 20×10^6 psi (140 GPa). The pipeline is 500 ft (150 m) long. 70°F (20°C) water is flowing at 6 ft/sec (2 m/s).

(a) The modulus of elasticity of the cast-iron pipe is most nearly

(A) 2.1×10^5 lbf/in² (1.5×10^9 Pa)

(B) 2.6×10^5 lbf/in² (1.8×10^9 Pa)

(C) 2.8×10^5 lbf/in² (2.0×10^9 Pa)

(D) 3.1×10^5 lbf/in² (2.2×10^9 Pa)

(b) The speed of sound in the pipe is most nearly

(A) 330 ft/sec (100 m/s)

(B) 1500 ft/sec (400 m/s)

(C) 4000 ft/sec (1200 m/s)

(D) 4500 ft/sec (1300 m/s)

(c) If a valve is closed instantaneously, the pressure increase experienced in the pipe will be most nearly

(A) 48 psi (330 kPa)

(B) 140 psi (970 kPa)

(C) 320 psi (2.5 MPa)

(D) 470 psi (3.2 MPa)

(d) If the pipe is 500 ft (150 m) long, over what approximate length of time must the valve be closed to create a pressure equivalent to instantaneous closure?

(A) 0.25 sec

(B) 0.68 sec

(C) 1.6 sec

(D) 2.1 sec

Parallel Pipe Systems

16. 8 MGD (millions of gallons per day) (350 L/s) of 70°F (20°C) water flows into the new schedule-40 steel pipe network shown. Minor losses are insignificant.

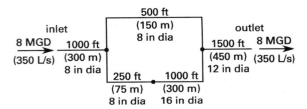

(a) The quantity of water flowing in the upper branch is most nearly

(A) 1.2 ft^3/sec (0.034 m^3/s)

(B) 2.9 ft^3/sec (0.081 m^3/s)

(C) 4.1 ft^3/sec (0.11 m^3/s)

(D) 5.3 ft^3/sec (0.15 m^3/s)

(b) The energy loss per unit mass between the inlet and the outlet is most nearly

(A) 120 ft-lbf/lbm (0.37 kJ/kg)

(B) 300 ft-lbf/lbm (0.90 kJ/kg)

(C) 480 ft-lbf/lbm (1.4 kJ/kg)

(D) 570 ft-lbf/lbm (1.7 kJ/kg)

Pipe Networks

17. A single-loop pipe network is shown. The distance between each junction is 1000 ft. All junctions are on the same elevation. All pipes have a Hazen-Williams C-value of 100. The volumetric flow rates are to be determined to within 2 gal/min. Start by assuming the following flows.

A to D: 300 gal/min

D to C: 100 gal/min

B to C: 200 gal/min

A to B: 400 gal/min

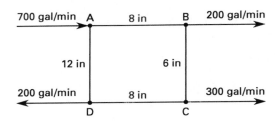

The flow rate between junctions B and C is most nearly

(A) 36 gal/min from B to C

(B) 58 gal/min from B to C

(C) 84 gal/min from C to B

(D) 110 gal/min from C to B

18. A double-loop pipe network is shown. The distance between each junction is 1000 ft. The water temperature is 60°F. Elevations and some pressure are known for the junctions. All pipes have a Hazen-Williams C-value of 100. Start by assuming the following flows.

A to F: 500 gal/min

A to B: 300 gal/min

B to E: 700 gal/min

C to B: 400 gal/min

C to D: 600 gal/min

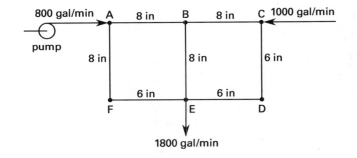

point	pressure	elevation
A		200 ft
B		150 ft
C	40 psig	300 ft
D		150 ft
E		200 ft
F		150 ft

(a) After one iteration, the corrections for loops 1 and 2, respectively, will be most nearly

    (A) +100 gal/min, −116 gal/min

    (B) +108 gal/min, −205 gal/min

    (C) +112 gal/min, −196 gal/min

    (D) +120 gal/min, −210 gal/min

(b) After two iterations, the corrections for loops 1 and 2, respectively, will most nearly be

    (A) −70 gal/min, −9 gal/min

    (B) −50 gal/min, −4 gal/min

    (C) −10 gal/min, −6 gal/min

    (D) 10 gal/min, 9 gal/min

(c) After three iterations, the corrections for loops 1 and 2, respectively, will most nearly be

    (A) 0 gal/min, −8 gal/min

    (B) −0.23 gal/min, −7 gal/min

    (C) −0.16 gal/min, −8 gal/min

    (D) −0.09 gal/min, −10 gal/min

(d) The flow rate between junctions B and E is most nearly

    (A) 540 gal/min

    (B) 620 gal/min

    (C) 810 gal/min

    (D) 980 gal/min

(e) When pressure at C is 40 psig, the friction head is most nearly

    (A) 10 ft

    (B) 40 ft

    (C) 90 ft

    (D) 110 ft

(f) The pressure at point D is most nearly

    (A) 51 psig

    (B) 73 psig

    (C) 96 psig

    (D) 120 psig

(g) If the pump receives water at 20 psig (140 kPa), the hydraulic power required is most nearly

    (A) 14 hp

    (B) 28 hp

    (C) 35 hp

    (D) 67 hp

19. The water distribution network shown consists of class A cast-iron pipe (specific roughness of 0.0008 ft) installed 1.5 years ago. 1.5 MGD of 50°F water enters at junction A and leaves at junction B. The minimum acceptable pressure at point D is 40 psig (280 kPa).

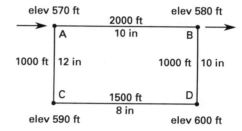

For parts (a) through (e), solve as a pipe network problem.

(a) The friction coefficient between junctions A and B is most nearly

    (A) 6×10^{-6}

    (B) 9×10^{-6}

    (C) 12×10^{-6}

    (D) 26×10^{-6}

(b) The friction coefficient between junctions B and D is most nearly

    (A) 2×10^{-6}

    (B) 6×10^{-6}

    (C) 12×10^{-6}

    (D) 26×10^{-6}

(c) The friction coefficient between junctions D and C is most nearly

    (A) 6×10^{-6}

    (B) 8×10^{-6}

    (C) 12×10^{-6}

    (D) 26×10^{-6}

(d) The friction coefficient between junctions C and A is most nearly

(A) 1×10^{-6}

(B) 2×10^{-6}

(C) 5×10^{-6}

(D) 6×10^{-6}

(e) The flow rate between junctions A and B is most nearly

(A) 320 gal/min

(B) 480 gal/min

(C) 590 gal/min

(D) 660 gal/min

For parts (f) through (h), solve as a parallel pipe problem. Assume $v_{max} = 5$ ft/sec.

(f) The flow rate between junctions A and B is most nearly

(A) 320 gal/min

(B) 480 gal/min

(C) 590 gal/min

(D) 660 gal/min

(g) Use a Darcy friction factor of 0.021. The pressure at point B is most nearly

(A) 32 psig

(B) 48 psig

(C) 57 psig

(D) 88 psig

(h) Use a Darcy friction factor of 0.021. The pressure at point A is most nearly

(A) 39 psig

(B) 55 psig

(C) 69 psig

(D) 110 psig

Discharge Through an Orifice

20. The velocity of discharge from a fire hose is 50 ft/sec (15 m/s). The hose is oriented 45° from the horizontal. Disregarding air friction, the maximum range of the discharge is most nearly

(A) 45 ft (14 m)

(B) 78 ft (23 m)

(C) 91 ft (27 m)

(D) 110 ft (33 m)

21. A full cylindrical tank that is 40 ft (12 m) high has a constant diameter of 20 ft. The tank has a 4 in (100 mm) diameter hole in its bottom. The coefficient of discharge for the hole is 0.98. Approximately how long will it take for the water level to drop from 40 ft to 20 ft (12 m to 6 m)?

(A) 950 sec

(B) 1200 sec

(C) 1450 sec

(D) 1700 sec

Siphons

22. A 24 in diameter siphon is used to transfer irrigation water from a water distribution canal to an irrigation ditch for a field of row crops below. The elevation of the water in the canal is 320 ft MSL (relative to mean sea level), and the elevation of the stilling basin for the row crops is 305 ft MSL. Counting fittings and minor losses, the siphon has a total equivalent length of 42 ft. The siphon is constructed of corrugated metal pipe with a standard galvanized surface. Counting entrance and exit losses, what is most nearly the total head loss experienced by the water?

(A) 7.0 ft

(B) 11 ft

(C) 15 ft

(D) 23 ft

Venturi Meters

23. A venturi meter with an 8 in diameter throat is installed in a 12 in diameter water line. The venturi is perfectly smooth, so that the discharge coefficient is 1.00. An attached mercury manometer registers a 4 in differential. The volumetric flow rate is most nearly

(A) 1.7 ft³/sec

(B) 5.2 ft³/sec

(C) 6.4 ft³/sec

(D) 18 ft³/sec

24. 60°F (15°C) benzene (specific gravity at 60°F (15°C) of 0.885) flows through an 8 in/3.5 in (200 mm/90 mm) venturi meter whose coefficient of discharge is 0.99. A mercury manometer indicates a 4 in difference in the heights of the mercury columns. The volumetric flow rate of the benzene is most nearly

(A) 1.2 ft³/sec (34 L/s)

(B) 9.1 ft³/sec (250 L/s)

(C) 13 ft³/sec (360 L/s)

(D) 27 ft³/sec (760 L/s)

Orifice Meters

25. A sharp-edged orifice meter with a 0.2 ft diameter opening is installed in a 1 ft diameter pipe. 70°F water approaches the orifice at 2 ft/sec. The indicated pressure drop across the orifice meter is most nearly

(A) 5.9 psi

(B) 13 psi

(C) 22 psi

(D) 47 psi

26. A mercury manometer is used to measure a pressure difference across an orifice meter in a water line. The difference in mercury levels is 7 in (17.8 cm). The pressure differential is most nearly

(A) 1.7 psi (12 kPa)

(B) 3.2 psi (22 kPa)

(C) 7.9 psi (55 kPa)

(D) 23 psi (160 kPa)

27. A sharp-edged ISA orifice is used in a schedule-40 steel 12 in (300 mm inside diameter) water line. (Figure 17.28 is applicable.) The water temperature is 70°F (20°C), and the flow rate is 10 ft^3/sec (250 L/s). The differential pressure change across the orifice (to the vena contracta) should be approximately 25 ft (7.5 m). The smallest orifice that can be used is most nearly

(A) 5.5 in (14 cm)

(B) 7.3 in (19 cm)

(C) 8.1 in (20 cm)

(D) 8.9 in (23 cm)

Impulse-Momentum

28. A pipe necks down from 24 in at point A to 12 in at point B. 8 ft^3/sec of 60°F water flows from point A to point B. The pressure head at point A is 20 ft. Friction is insignificant over the distance between points A and B. The magnitude and direction of the resultant force on the water are most nearly

(A) 2900 lbf, toward A

(B) 3500 lbf, toward A

(C) 2900 lbf, toward B

(D) 3500 lbf, toward B

29. A 2 in (50 mm) diameter horizontal water jet has an absolute velocity (with respect to a stationary point) of 40 ft/sec (12 m/s) as it strikes a curved blade. The blade is moving horizontally away with an absolute velocity of 15 ft/sec (4.5 m/s). Water is deflected 60° from the horizontal. The force on the blade is most nearly

(A) 18 lbf (80 N)

(B) 26 lbf (110 N)

(C) 35 lbf (160 N)

(D) 47 lbf (210 N)

Pumps

30. 2000 gal/min (125 L/s) of brine with a specific gravity of 1.2 passes through an 85% efficient pump. The centerlines of the pump's 12 in inlet and 8 in outlet are at the same elevation. The inlet suction gauge indicates 6 in (150 mm) of mercury below atmospheric. The discharge pressure gauge is located 4 ft (1.2 m) above the centerline of the pump's outlet and indicates 20 psig (138 kPa). All pipes are schedule-40. The input power to the pump is most nearly

(A) 12 hp (8.9 kW)

(B) 36 hp (26 kW)

(C) 52 hp (39 kW)

(D) 87 hp (65 kW)

Turbines

31. 100 ft^3/sec (2.6 m^3/s) of water passes through a horizontal turbine. The water's pressure is reduced from 30 psig (210 kPa) to 5 psig (35 kPa) vacuum. Disregarding friction, velocity, and other factors, the power generated is most nearly

(A) 110 hp (82 kW)

(B) 380 hp (280 kW)

(C) 730 hp (540 kW)

(D) 920 hp (640 kW)

Drag

32. A refrigeration truck is driven at 65 mph into a 15 mph headwind. The frontal area of the truck is 100 ft^2. The coefficient of drag is 0.5. When calculating the drag force on the truck, the velocity that should be used is most nearly

(A) 50 mph

(B) 65 mph

(C) 73 mph

(D) 80 mph

33. A dish-shaped antenna faces directly into a 60 mph wind. The projected area of the antenna is 0.8 ft^2; the coefficient of drag, C_D, is 1.2; and the density of air is 0.076 lbm/ft^3. The total amount of drag force experienced by the antenna is most nearly

(A) 9.0 lbf

(B) 10 lbf

(C) 14 lbf

(D) 16 lbf

34. A 30 ft long smooth, straight, round log has been floated downstream in a 6 ft deep flume by upstream loggers. The outside diameter of the log is 12 in, and the saturated specific gravity of the wood is 0.72 relative to 50°F water. The 60°F flume water flows with an average velocity of 10 ft/sec. At the downstream collection point, one end of the log is chained to an excavator crane arm directly over the flume and lifted until the log is approximately vertical. The lower 2 ft of the log remain submerged in the flume. The upper end of the log is free to rotate, but the upper end cannot translate. Neglect buoyancy effects. What is most nearly the angle from the vertical that the log will deflect in the moving water?

(A) 5°

(B) 7°

(C) 11°

(D) 14°

35. A car traveling through 70°F (20°C) air has the following characteristics.

frontal area	28 ft^2 (2.6 m^2)
mass	3300 lbm (1500 kg)
drag coefficient	0.42
rolling resistance	1% of weight
engine thermal efficiency	28%
fuel heating value	115,000 Btu/gal (32 MJ/L)

For parts (a) through (e), assume the car is traveling at 55 mi/hr (90 km/h).

(a) The velocity of the car is most nearly

(A) 0.9 ft/sec (0.3 m/s)

(B) 14 ft/sec (4 m/s)

(C) 45 ft/sec (14 m/s)

(D) 80 ft/sec (25 m/s)

(b) The drag on the car is most nearly

(A) 1 lbf (5 N)

(B) 3 lbf (12 N)

(C) 30 lbf (130 N)

(D) 90 lbf (410 N)

(c) The total resisting force on the car is most nearly

(A) 60 lbf (280 N)

(B) 120 lbf (560 N)

(C) 3300 lbf (15 000 N)

(D) 3400 lbf (15 500 N)

(d) The power consumed by the car is most nearly

(A) 6 Btu/sec (7000 W)

(B) 13 Btu/sec (14 000 W)

(C) 22 Btu/sec (24 000 W)

(D) 350 Btu/sec (38 000 W)

(e) Considering only the drag and rolling resistance, the fuel consumption of the car is most nearly

(A) 0.026 gal/mi (0.062 L/km)

(B) 0.038 gal/mi (0.087 L/km)

(C) 0.051 gal/mi (0.12 L/km)

(D) 0.13 gal/mi (0.30 L/km)

For parts (f) through (j), assume the car is traveling at 65 mi/hr (105 km/h).

(f) The drag on the car is most nearly

(A) 30 lbf (130 N)

(B) 90 lbf (410 N)

(C) 120 lbf (560 N)

(D) 124 lbf (570 N)

(g) The total resisting force on the car is most nearly

(A) 120 lbf (560 N)

(B) 150 lbf (670 N)

(C) 155 lbf (710 N)

(D) 160 lbf (720 N)

(h) The power consumed by the car is most nearly

(A) 15 Btu/sec (16 000 W)

(B) 16 Btu/sec (18 000 W)

(C) 19 Btu/sec (21 000 W)

(D) 24 Btu/sec (26 000 W)

(i) The fuel consumption of the car is most nearly

(A) 0.03 gal/mi (0.08 L/km)

(B) 0.07 gal/mi (0.16 L/km)

(C) 0.10 gal/mi (0.23 L/km)

(D) 0.11 gal/mi (0.25 L/km)

(j) What is the approximate percentage increase in fuel consumption at 65 mi/hr (105 km/h) compared to 55 mi/hr (90 km/h)?

(A) 10%

(B) 20%

(C) 30%

(D) 40%

Similarity

36. A 1/20th airplane model is tested in a wind tunnel at full velocity and temperature. What is the approximate ratio of the wind tunnel pressure to normal ambient pressure?

(A) 5

(B) 10

(C) 20

(D) 40

37. 68°F (20°C) castor oil (kinematic viscosity at 68°F (20°C) of 1110×10^{-5} ft^2/sec (103×10^{-5} m^2/s)) flows through a pump whose impeller turns at 1000 rpm. A similar pump twice the first pump's size is tested with 68°F (20°C) air. Theoretically, what should be the approximate speed of the second pump's impeller to ensure similarity?

(A) 3.6 rpm

(B) 88 rpm

(C) 250 rpm

(D) 1600 rpm

SOLUTIONS

1. The decrease in friction will affect pumping power required, but it will not affect capacity. Capacity is a function of velocity and area only. The velocity is unchanged.

$$\frac{Q_2}{Q_1} = \frac{vA_2}{vA_1} = \frac{\frac{\pi v d_2^2}{4}}{\frac{\pi v d_1^2}{4}} = \frac{d_2^2}{d_1^2}$$

$$= 2$$

$$d_2 = \sqrt{2}d_1 \approx \boxed{1.41 d_1 \quad (40\% \text{ increase})}$$

The answer is (C).

2. The initial Reynolds number is

$$\mathrm{Re}_1 = \frac{Dv}{\nu}$$

The Reynolds number after modifications are made will be

$$\mathrm{Re}_2 = \frac{(2D)\left(\frac{v}{2}\right)}{\nu} = \frac{Dv}{\nu} = \mathrm{Re}_1$$

The Reynolds number will $\boxed{\text{remain the same.}}$

The answer is (D).

3. *Customary U.S. Solution*

For schedule-40 pipe,

$$D_A = 0.5054 \text{ ft}$$

$$D_B = 1.4063 \text{ ft}$$

Let point A be at zero elevation.

The total energy at point A from Bernoulli's equation is

$$E_{t,A} = E_p + E_v + E_z = \frac{p_A}{\rho} + \frac{v_A^2}{2g_c} + \frac{z_A g}{g_c}$$

At point A, the diameter is 6 in. The velocity at point A is

$$\dot{V} = v_A A_A = v_A \left(\frac{\pi}{4}\right) D_A^2$$

$$v_A = \left(\frac{4}{\pi}\right)\left(\frac{\dot{V}}{D_A^2}\right) = \left(\frac{4}{\pi}\right)\left(\frac{5 \frac{\text{ft}^3}{\text{sec}}}{(0.5054 \text{ ft})^2}\right)$$

$$= \boxed{24.9 \text{ ft/sec} \quad (25 \text{ ft/sec})}$$

$$p_A = \left(10 \frac{\text{lbf}}{\text{in}^2}\right)\left(12 \frac{\text{in}}{\text{ft}}\right)^2 = 1440 \text{ lbf/ft}^2$$

$$z_A = 0$$

For water, $\rho \approx 62.4 \text{ lbm/ft}^3$.

$$E_{t,A} = \frac{p_A}{\rho} + \frac{v_A^2}{2g_c} + \frac{z_A g}{g_c}$$

$$= \frac{1440 \frac{\text{lbf}}{\text{ft}^2}}{62.4 \frac{\text{lbm}}{\text{ft}^3}} + \frac{\left(24.9 \frac{\text{ft}}{\text{sec}}\right)^2}{(2)\left(32.2 \frac{\text{lbm-ft}}{\text{lbf-sec}^2}\right)} + 0$$

$$= 32.7 \text{ ft-lbf/lbm}$$

Similarly, the total energy at point B is

$$v_B = \left(\frac{4}{\pi}\right)\left(\frac{\dot{V}}{D_B^2}\right) = \left(\frac{4}{\pi}\right)\left(\frac{5 \frac{\text{ft}^3}{\text{sec}}}{(1.4063 \text{ ft})^2}\right)$$

$$= 3.22 \text{ ft/sec}$$

$$p_B = \left(7 \frac{\text{lbf}}{\text{in}^2}\right)\left(12 \frac{\text{in}}{\text{ft}}\right)^2 = 1008 \text{ lbf/ft}^2$$

$$z_B = 15 \text{ ft}$$

$$E_{t,B} = \frac{p_B}{\rho} + \frac{v_B^2}{2g_c} + \frac{z_B g}{g_c}$$

$$= \frac{1008 \frac{\text{lbf}}{\text{ft}^2}}{62.4 \frac{\text{lbm}}{\text{ft}^3}} + \frac{\left(3.22 \frac{\text{ft}}{\text{sec}}\right)^2}{(2)\left(32.2 \frac{\text{lbm-ft}}{\text{lbf-sec}^2}\right)}$$

$$+ \frac{(15 \text{ ft})\left(32.2 \frac{\text{ft}}{\text{sec}^2}\right)}{32.2 \frac{\text{lbm-ft}}{\text{lbf-sec}^2}}$$

$$= 31.3 \text{ ft-lbf/lbm}$$

Since $E_{t,A} > E_{t,B}$, the flow is from point A to point B.

The answer is (B).

SI Solution

Let point A be at zero elevation.

The total energy at point A from Bernoulli's equation is

$$E_{t,A} = E_p + E_v + E_z = \frac{p_A}{\rho} + \frac{v_A^2}{2} + z_A g$$

At point A, from App. 16.C, the diameter is 154 mm (0.154 m). The velocity at point A is

$$\dot{V} = v_A A_A = v_A \left(\frac{\pi}{4}\right) D_A^2$$

$$v_A = \left(\frac{4}{\pi}\right)\left(\frac{\dot{V}}{D_A^2}\right) = \left(\frac{4}{\pi}\right)\left(\frac{130 \frac{\text{L}}{\text{s}}}{(0.154 \text{ m})^2 \left(1000 \frac{\text{L}}{\text{m}^3}\right)}\right)$$

$$= \boxed{6.98 \text{ m/s} \quad (7 \text{ m/s})}$$

$$p_A = 70 \text{ kPa} \quad (70\,000 \text{ Pa})$$

$$z_A = 0$$

For water, $\rho = 1000 \text{ kg/m}^3$.

$$E_{t,A} = \frac{p_A}{\rho} + \frac{v_A^2}{2} + z_A g$$

$$= \frac{70\,000 \text{ Pa}}{1000 \frac{\text{kg}}{\text{m}^3}} + \frac{\left(6.98 \frac{\text{m}}{\text{s}}\right)^2}{2} + 0$$

$$= 94.36 \text{ J/kg}$$

Similarly, at B, the diameter is 429 mm. The total energy at point B is

$$v_B = \left(\frac{4}{\pi}\right)\left(\frac{\dot{V}}{D_B^2}\right)$$

$$= \left(\frac{4}{\pi}\right)\left(\frac{130 \frac{\text{L}}{\text{s}}}{(0.429 \text{ m})^2 \left(1000 \frac{\text{L}}{\text{m}^3}\right)}\right)$$

$$= 0.90 \text{ m/s}$$

$$p_B = 48.3 \text{ kPa} \quad (48\,300 \text{ Pa})$$

$$z_B = 4.6 \text{ m}$$

$$E_{t,B} = \frac{p_B}{\rho} + \frac{v_B^2}{2} + z_B g$$

$$= \frac{48\,300 \text{ Pa}}{1000 \frac{\text{kg}}{\text{m}^3}} + \frac{\left(0.90 \frac{\text{m}}{\text{s}}\right)^2}{2}$$

$$+ (4.6 \text{ m})\left(9.81 \frac{\text{m}}{\text{s}^2}\right)$$

$$= 93.8 \text{ J/kg}$$

Since $E_{t,A} > E_{t,B}$, the flow is from point A to point B.

The answer is (B).

4.

$$\dot{V} = \frac{750 \frac{\text{gal}}{\text{min}}}{\left(7.4805 \frac{\text{gal}}{\text{ft}^3}\right)\left(60 \frac{\text{sec}}{\text{min}}\right)}$$

$$= 1.671 \text{ ft}^3/\text{sec}$$

From App. 16.B, $D = 0.5054 \text{ ft}$, and $A = 0.2006 \text{ ft}^2$.

$$v = \frac{\dot{V}}{A} = \frac{1.671 \frac{\text{ft}^3}{\text{sec}}}{0.2006 \text{ ft}^2} = 8.33 \text{ ft/sec}$$

For 60°F water, from App. 14.A,

$$\rho = 62.37 \ \text{lbm/ft}^3$$

$$\nu = 1.217 \times 10^{-5} \ \text{ft}^2/\text{sec}$$

$$\text{Re} = \frac{\text{v}D}{\nu} = \frac{\left(8.33 \ \frac{\text{ft}}{\text{sec}}\right)(0.5054 \ \text{ft})}{1.217 \times 10^{-5} \ \frac{\text{ft}^2}{\text{sec}}} = 3.46 \times 10^5$$

The specific weight is

$$\gamma = \frac{\rho g_c}{g} = \frac{\left(62.37 \ \frac{\text{lbm}}{\text{ft}^3}\right)\left(32.2 \ \frac{\text{lbm-ft}}{\text{lbf-sec}^2}\right)}{32.2 \ \frac{\text{ft}}{\text{sec}^2}} = 62.37 \ \text{lbf/ft}^3$$

For steel,

$$\epsilon = 0.0002$$

$$\frac{\epsilon}{D} = \frac{0.0002 \ \text{ft}}{0.5054 \ \text{ft}} \approx 0.0004$$

$$f = 0.0175$$

From Eq. 17.22,

$$h_f = \frac{fL\text{v}^2}{2Dg} = \frac{(0.0175)(3000 \ \text{ft})\left(8.33 \ \frac{\text{ft}}{\text{sec}}\right)^2}{(2)(0.5054 \ \text{ft})\left(32.2 \ \frac{\text{ft}}{\text{sec}^2}\right)}$$

$$= 111.9 \ \text{ft}$$

Use the Bernoulli equation. Since velocity is the same at points A and B, it may be omitted.

$$\frac{p_1}{\gamma_1} = \frac{p_2}{\gamma_2} + (z_2 - z_1) + h_f$$

$$\frac{\left(12 \ \frac{\text{in}}{\text{ft}}\right)^2 p_1}{62.37 \ \frac{\text{lbf}}{\text{ft}^3}} = \frac{\left(50 \ \frac{\text{lbf}}{\text{in}^2}\right)\left(12 \ \frac{\text{in}}{\text{ft}}\right)^2}{62.37 \ \frac{\text{lbf}}{\text{ft}^3}} + 60 \ \text{ft} + 111.9 \ \text{ft}$$

$$p_1 = \boxed{124.5 \ \text{lbf/in}^2 \quad (120 \ \text{psig})}$$

The answer is (C).

5. (a) From Eq. 17.29, the friction loss from A to B is

$$h_{f,\text{ft},\text{A-B}} = \frac{10.44 L_{\text{ft}} \, Q_{\text{gpm}}^{1.85}}{C^{1.85} d_{\text{in}}^{4.87}}$$

$$= \frac{(10.44)(20,000 \ \text{ft})\left(120 \ \frac{\text{gal}}{\text{min}}\right)^{1.85}}{(150)^{1.85}(6 \ \text{in})^{4.87}}$$

$$= \boxed{22.4 \ \text{ft} \quad (20 \ \text{ft})}$$

The answer is (B).

(b) Calculate the velocity head.

$$\text{v} = \frac{\dot{V}}{A} = \frac{120 \ \frac{\text{gal}}{\text{min}}}{\left(\frac{\pi}{4}\right)\left(\frac{6 \ \text{in}}{12 \ \frac{\text{in}}{\text{ft}}}\right)^2 \left(7.4805 \ \frac{\text{gal}}{\text{ft}^3}\right)\left(60 \ \frac{\text{sec}}{\text{min}}\right)}$$

$$= 1.36 \ \text{ft/sec}$$

$$h_\text{v} = \frac{\text{v}^2}{2g} = \frac{\left(1.36 \ \frac{\text{ft}}{\text{sec}}\right)^2}{(2)\left(32.2 \ \frac{\text{ft}}{\text{sec}^2}\right)}$$

$$= \boxed{0.029 \ \text{ft} \quad (0.030 \ \text{ft})}$$

The answer is (D).

(c) Velocity heads are low and can be disregarded. The friction loss from B to C is

$$h_{f,\text{B-C}} = \frac{(10.44)(10,000 \ \text{ft})\left(160 \ \frac{\text{gal}}{\text{min}}\right)^{1.85}}{(150)^{1.85}(6 \ \text{in})^{4.87}}$$

$$= \boxed{19.10 \ \text{ft} \quad (20 \ \text{ft})}$$

The answer is (B).

(d) For C to D,

$$h_{f,\text{C-D}} = \frac{(10.44)(30,000 \ \text{ft})\left(120 \ \frac{\text{gal}}{\text{min}}\right)^{1.85}}{(150)^{1.85}(4 \ \text{in})^{4.87}}$$

$$= \boxed{242.4 \ \text{ft} \quad (240 \ \text{ft})}$$

The answer is (A).

(e) Assume a pressure of 20 psig at point A.

$$h_{p,\text{A}} = \frac{\left(20 \ \frac{\text{lbf}}{\text{in}^2}\right)\left(12 \ \frac{\text{in}}{\text{ft}}\right)^2}{62.4 \ \frac{\text{lbf}}{\text{ft}^3}}$$

$$= 46.2 \ \text{ft}$$

Flow of Fluids

From the Bernoulli equation, ignoring velocity head,

$$h_{p,A} + z_A = h_{p,B} + z_B + h_{f,A\text{-}B}$$
$$46.2 \text{ ft} + 620 \text{ ft} = h_{p,B} + 460 \text{ ft} + 22.4 \text{ ft}$$
$$h_{p,B} = 183.8 \text{ ft}$$

$$p_B = \gamma h_{p,B} = \frac{\left(62.4 \dfrac{\text{lbf}}{\text{ft}^3}\right)(183.8 \text{ ft})}{\left(12 \dfrac{\text{in}}{\text{ft}}\right)^2}$$
$$= \boxed{79.6 \text{ lbf/in}^2 \quad (80 \text{ psig})}$$

The answer is (C).

(f) For B to C,

$$h_{p,B} + z_B = h_{p,C} + z_C + h_{f,B\text{-}C}$$
$$183.8 \text{ ft} + 460 \text{ ft} = h_{p,C} + 540 \text{ ft} + 19.10 \text{ ft}$$
$$h_{p,C} = 84.7 \text{ ft}$$

$$p_C = \gamma h_{p,C} = \frac{\left(62.4 \dfrac{\text{lbf}}{\text{ft}^3}\right)(84.7 \text{ ft})}{\left(12 \dfrac{\text{in}}{\text{ft}}\right)^2}$$
$$= \boxed{36.7 \text{ lbf/in}^2 \quad (40 \text{ psig})}$$

The answer is (C).

(g) For C to D,

$$h_{p,C} + z_C = h_{p,D} + z_D + h_{f,C\text{-}D}$$
$$84.7 \text{ ft} + 540 \text{ ft} = h_{p,D} + 360 \text{ ft} + 242.4 \text{ ft}$$
$$h_{p,D} = 22.3 \text{ ft}$$

$$p_D = \gamma h_{p,D} = \frac{\left(62.4 \dfrac{\text{lbf}}{\text{ft}^3}\right)(22.3 \text{ ft})}{\left(12 \dfrac{\text{in}}{\text{ft}}\right)^2}$$
$$= \boxed{9.7 \text{ lbf/in}^2 \quad (10 \text{ psig})}$$

The answer is (A).

(h) $p_D = 9.7 \text{ lbf/in}^2$ is too low; therefore, add $20 \text{ lbf/in}^2 - 9.7 \text{ lbf/in}^2 = 10.3 \text{ lbf/in}^2$ (psig) to each point.

$$p_A = 20.0 \frac{\text{lbf}}{\text{in}^2} + 10.3 \frac{\text{lbf}}{\text{in}^2} = \boxed{30.3 \text{ lbf/in}^2 \quad (30 \text{ psig})}$$

$$p_B = 79.6 \frac{\text{lbf}}{\text{in}^2} + 10.3 \frac{\text{lbf}}{\text{in}^2} = 89.9 \text{ lbf/in}^2 \quad (\text{psig})$$

$$p_C = 36.7 \frac{\text{lbf}}{\text{in}^2} + 10.3 \frac{\text{lbf}}{\text{in}^2} = 47.0 \text{ lbf/in}^2 \quad (\text{psig})$$

$$p_D = 9.7 \frac{\text{lbf}}{\text{in}^2} + 10.3 \frac{\text{lbf}}{\text{in}^2} = 20.0 \text{ lbf/in}^2 \quad (\text{psig})$$

The answer is (D).

(i) The elevation of the hydraulic grade line above point D is the sum of the potential and static heads.

$$\Delta h_{A\text{-}D} = z_A - z_D + h_{p,A} - h_{p,D}$$
$$= z_A - z_D + \frac{p_A - p_D}{\gamma}$$
$$= 620 \text{ ft} - 360 \text{ ft}$$
$$+ \frac{\left(30.3 \dfrac{\text{lbf}}{\text{in}^2} - 20 \dfrac{\text{lbf}}{\text{in}^2}\right)\left(12 \dfrac{\text{in}}{\text{ft}}\right)^2}{62.4 \dfrac{\text{lbf}}{\text{ft}^3}}$$
$$= \boxed{283.8 \text{ ft} \quad (280 \text{ ft})}$$

The answer is (A).

6. The Darcy equation is applicable to fluids in the laminar and turbulent regions.

$$h_f = \frac{fLv^2}{2Dg}$$

In laminar flow in circular pipes, the friction factor is

$$f = \frac{64}{\text{Re}} = \frac{64\nu}{Dv}$$

Combining these two equations,

$$h_f = \frac{fLv^2}{2Dg} = \frac{\left(\dfrac{64\nu}{Dv}\right)Lv^2}{2Dg} = \frac{32\nu Lv}{D^2 g}$$

This is the Hagen-Poiseuille equation.

$$\boxed{\text{Friction head loss in laminar flow is} \\ \text{proportional to velocity.}}$$

The answer is (A).

7. From App. 17.D, the equivalent length of four flanged 3 in elbows is

$$L_{e,\text{elbows}} = (4)(2.6 \text{ ft}) = 10.4 \text{ ft}$$

From App. 17.D, the equivalent length of the open gate valve is 2.8 ft.

The equivalent length of the pipe without fittings is 50 ft.

The elevation change is 20 ft, but this distance is not an equivalent length of pipe. Taking terms from the Bernoulli equation,

$$\Delta z = \frac{fL_e v^2}{2Dg}$$

$$L_e = \frac{2Dg\Delta z}{fv^2}$$

The flow rate is not given, so the velocity must be estimated. Since the water is said to flow quietly, the velocity is most likely less than 10 ft/sec. Since the velocity is not known, an exact determination of the friction factor, f, is not possible. Use a value of 0.02, which is appropriate for steel pipe and turbulent flow. Using nominal values for a quick estimate, the equivalent length of pipe equal to the elevation drop is

$$L_e = \frac{2Dg\Delta z}{fv^2} = \frac{(2)(3\text{ in})\left(32.2 \frac{\text{ft}}{\text{sec}^2}\right)(20\text{ ft})}{(0.02)\left(10 \frac{\text{ft}}{\text{sec}}\right)^2\left(12 \frac{\text{in}}{\text{ft}}\right)}$$

$$= 161 \text{ ft}$$

If the velocity was 15 ft/sec, the equivalent length would be 71 ft. If the velocity was less than 10 ft/sec, the equivalent length would be even larger than 161 ft. Using the actual diameter of the pipe would also increase the equivalent length.

The answer is (D).

8. The pumping power will be the same if the head loss due to friction is the same. From Eq. 17.29, the friction loss in a section of pipe is

$$h_f = \frac{10.44LQ^{1.85}}{C^{1.85}d^{4.87}}$$

The flow rate is the same in all pipe sections, as is the Hazen-Williams roughness coefficient. These terms and the constant term cancel out.

$$\frac{300\text{ ft}}{(18\text{ in})^{4.87}} + \frac{400\text{ ft}}{(14\text{ in})^{4.87}} = \frac{L}{(16\text{ in})^{4.87}}$$

$$L = \boxed{935\text{ ft}\quad(940\text{ ft})}$$

The answer is (D).

9. Using App. 17.E, the friction loss is approximately 10 ft per 1000 ft of pipe, or approximately $\boxed{1.0\text{ ft per }100\text{ ft}}$ of pipe.

The answer is (A).

10. The total flow rate is

$$\dot{V} = (600\text{ res})(2.5)\left(1.1 \frac{\frac{\text{gal}}{\text{min}}}{\text{res}}\right) = 1650\text{ gpm}$$

Write the energy equation, Eq. 17.64, in terms of specific weight, disregarding the velocity head (which is assumed to be small and constant in the line).

$$\frac{p_1}{\gamma} + z_1 = \frac{p_2}{\gamma} + z_2 + \frac{10.44LQ^{1.85}}{C^{1.85}d^{4.87}}$$

$$\frac{p_{1,\text{psig}}\left(12 \frac{\text{in}}{\text{ft}}\right)^2}{62.4 \frac{\text{lbf}}{\text{ft}^3}} + 1000\text{ ft}$$

$$= \frac{\left(60 \frac{\text{lbf}}{\text{in}^2}\right)\left(12 \frac{\text{in}}{\text{ft}}\right)^2}{62.4 \frac{\text{lbf}}{\text{ft}^3}} + 850\text{ ft}$$

$$+ \frac{(10.44)(17{,}000\text{ ft})\left(1650 \frac{\text{gal}}{\text{min}}\right)^{1.85}}{(140)^{1.85}(10\text{ in})^{4.87}}$$

$$p_1 = \boxed{94.5\text{ psig}\quad(95\text{ psig})}$$

The answer is (C).

11. *Customary U.S. Solution*

For 6 in schedule-40 pipe, the internal diameter, D, is 0.5054 ft. The internal area is 0.2006 ft^2.

The velocity, v, is calculated from the volumetric flow, $\dot{V}$, and the flow area, A, by

$$v = \frac{\dot{V}}{A} = \frac{1.5 \frac{\text{ft}^3}{\text{sec}}}{0.2006\text{ ft}^2}$$

$$= 7.48\text{ ft/sec}$$

Use App. 14.A. For water at 70°F, the kinematic viscosity, ν, is 1.059×10^{-5} ft^2/sec.

Calculate the Reynolds number.

$$\text{Re} = \frac{Dv}{\nu} = \frac{(0.5054\text{ ft})\left(7.48 \frac{\text{ft}}{\text{sec}}\right)}{1.059 \times 10^{-5} \frac{\text{ft}^2}{\text{sec}}}$$

$$= 3.57 \times 10^5$$

Since Re > 2100, the flow is turbulent. The friction loss coefficient can be determined from the Moody diagram.

Flow of Fluids

For new steel pipe, the specific roughness, ϵ, is 0.0002 ft.

The relative roughness is

$$\frac{\epsilon}{D} = \frac{0.0002 \text{ ft}}{0.5054 \text{ ft}} = 0.0004$$

From the Moody diagram with Re = 3.57×10^5 and $\epsilon/D = 0.0004$, the friction factor, f, is 0.0174.

Use Darcy's equation to compute the frictional loss.

$$h_f = \frac{fLv^2}{2Dg} = \frac{(0.0174)(1200 \text{ ft})\left(7.48 \frac{\text{ft}}{\text{sec}}\right)^2}{(2)(0.5054 \text{ ft})\left(32.2 \frac{\text{ft}}{\text{sec}^2}\right)}$$

$$= \boxed{35.9 \text{ ft} \quad (36 \text{ ft})}$$

The answer is (C).

SI Solution

For 6 in pipe, the internal diameter is 154.1 mm, and the internal area is 186.5×10^{-4} m^2.

The velocity, v, is calculated from the volumetric flow, $\dot{V}$, and the flow area, A, by

$$v = \frac{\dot{V}}{A} = \frac{40 \frac{\text{L}}{\text{s}}}{(186.5 \times 10^{-4} \text{ m}^2)\left(1000 \frac{\text{L}}{\text{m}^3}\right)}$$

$$= 2.145 \text{ m/s}$$

From App. 14.B, for water at 20°C, the kinematic viscosity is

$$\nu = \frac{\mu}{\rho} = \frac{1.0050 \times 10^{-3} \text{ Pa·s}}{998.23 \frac{\text{kg}}{\text{m}^3}}$$

$$= 1.007 \times 10^{-6} \text{ m}^2/\text{s}$$

Calculate the Reynolds number.

$$\text{Re} = \frac{D\text{v}}{\nu} = \frac{(154.1 \text{ mm})\left(2.145 \frac{\text{m}}{\text{s}}\right)}{\left(1.007 \times 10^{-6} \frac{\text{m}^2}{\text{s}}\right)\left(1000 \frac{\text{mm}}{\text{m}}\right)}$$

$$= 3.282 \times 10^5$$

Since Re > 2100, the flow is turbulent. The friction loss coefficient can be determined from the Moody diagram.

For new steel pipe, the specific roughness, ϵ, is 6.0×10^{-5} m.

The relative roughness is

$$\frac{\epsilon}{D} = \frac{6.0 \times 10^{-5} \text{ m}}{0.1541 \text{ m}} = 0.0004$$

From the Moody diagram with Re = 3.28×10^5 and $\epsilon/D = 0.0004$, the friction factor, f, is 0.0175.

Use Darcy's equation to compute the frictional loss.

$$h_f = \frac{fLv^2}{2Dg} = \frac{(0.0175)(355 \text{ m})\left(2.145 \frac{\text{m}}{\text{s}}\right)^2}{(2)(0.1541 \text{ m})\left(9.81 \frac{\text{m}}{\text{s}^2}\right)}$$

$$= \boxed{9.45 \text{ m} \quad (9.5 \text{ m})}$$

The answer is (C).

12. *Customary U.S. Solution*

For 6 in schedule-40 pipe, the internal diameter, D, is 0.5054 ft. The internal area is 0.2006 ft^2.

Convert the volumetric flow rate from gal/min to ft^3/sec.

$$\dot{V} = \frac{500 \frac{\text{gal}}{\text{min}}}{\left(60 \frac{\text{sec}}{\text{min}}\right)\left(7.4805 \frac{\text{gal}}{\text{ft}^3}\right)} = 1.114 \text{ ft}^3/\text{sec}$$

The velocity is

$$v = \frac{\dot{V}}{A} = \frac{1.114 \frac{\text{ft}^3}{\text{sec}}}{0.2006 \text{ ft}^2} = 5.55 \text{ ft/sec}$$

Use App. 14.A. For water at 100°F, the kinematic viscosity, ν, is 0.739×10^{-5} ft^2/sec, and the density is 62.00 lbm/ft^2.

Calculate the Reynolds number.

$$\text{Re} = \frac{D\text{v}}{\nu} = \frac{(0.5054 \text{ ft})\left(5.55 \frac{\text{ft}}{\text{sec}}\right)}{0.739 \times 10^{-5} \frac{\text{ft}^2}{\text{sec}}}$$

$$= 3.80 \times 10^5$$

Since Re > 2100, the flow is turbulent. The friction loss coefficient can be determined from the Moody diagram.

For new steel pipe, the specific roughness, ϵ, is 0.0002 ft.

The relative roughness is

$$\frac{\epsilon}{D} = \frac{0.0002 \text{ ft}}{0.5054 \text{ ft}} = 0.0004$$

From the Moody diagram with Re = 3.80×10^5 and $\epsilon/D = 0.0004$, the friction factor, f, is 0.0173.

Use App. 17.D. The equivalent lengths of the valves and fittings are

standard radius elbow	2×8.9 ft =	17.8 ft
gate valve (fully open)	2×3.2 ft =	6.4 ft
90° angle valve (fully open)	1×63.0 ft =	63.0 ft
swing check valve	1×63.0 ft =	63.0 ft
		150.2 ft

The equivalent pipe length is the sum of the straight run of pipe and the equivalent length of pipe for the valves and fittings.

$$L_e = L + L_{\text{fittings}} = 300 \text{ ft} + 150.2 \text{ ft}$$
$$= 450.2 \text{ ft}$$

Use Darcy's equation to compute the frictional loss.

$$h_f = \frac{fL_e v^2}{2Dg}$$

$$= \frac{(0.0173)(450.2 \text{ ft})\left(5.55 \frac{\text{ft}}{\text{sec}}\right)^2}{(2)(0.5054 \text{ ft})\left(32.2 \frac{\text{ft}}{\text{sec}^2}\right)}$$

$$= 7.37 \text{ ft}$$

The head loss is the sum of the head losses through the pipe, valves, and fittings and the change in elevation.

$$\Delta h = h_f + \Delta z = 7.37 \text{ ft} + 20 \text{ ft}$$
$$= 27.37 \text{ ft}$$

The pressure difference between the entrance and discharge is

$$\Delta p = \gamma \Delta h = \rho \Delta h \times \frac{g}{g_c}$$

$$= \frac{\left(62.0 \frac{\text{lbm}}{\text{ft}^3}\right)(27.37 \text{ ft})}{\left(12 \frac{\text{in}}{\text{ft}}\right)^2} \times \frac{32.2 \frac{\text{ft}}{\text{sec}^2}}{32.2 \frac{\text{lbm-ft}}{\text{lbf-sec}^2}}$$

$$= \boxed{11.8 \text{ lbf/in}^2 \quad (12 \text{ psi})}$$

The answer is (A).

SI Solution

For 6 in pipe, the internal diameter is 154.1 mm (0.1541 m). The internal area is 186.5×10^{-4} m².

The velocity, v, is

$$v = \frac{\dot{V}}{A} = \frac{30 \frac{\text{L}}{\text{s}}}{(186.5 \times 10^{-4} \text{ m}^2)\left(1000 \frac{\text{L}}{\text{m}^3}\right)}$$

$$= 1.61 \text{ m/s}$$

Use App. 14.B. For water at 40°C, the kinematic viscosity, ν, is 6.611×10^{-7} m²/s, and the density is 992.25 kg/m³.

Calculate the Reynolds number.

$$\text{Re} = \frac{Dv}{\nu} = \frac{(0.1541 \text{ m})\left(1.61 \frac{\text{m}}{\text{s}}\right)}{6.611 \times 10^{-7} \frac{\text{m}^2}{\text{s}}}$$

$$= 3.75 \times 10^5$$

Since Re > 2100, the flow is turbulent. The friction loss coefficient can be determined from the Moody diagram.

For new steel pipe, the specific roughness, ϵ, is 6.0×10^{-5} m.

The relative roughness is

$$\frac{\epsilon}{D} = \frac{6.0 \times 10^{-5} \text{ m}}{0.1541 \text{ m}} = 0.0004$$

From the Moody diagram with Re = 3.75×10^5 and $\epsilon/D = 0.0004$, the friction factor, f, is 0.0173.

Use App. 17.D. The equivalent lengths of the valves and fittings are

standard radius elbow	$2 \times (8.9 \text{ ft})\left(0.3048 \frac{\text{m}}{\text{ft}}\right) =$	5.4 m
gate valve (fully open)	$2 \times (3.2 \text{ ft})\left(0.3048 \frac{\text{m}}{\text{ft}}\right) =$	2.0 m
90° angle valve (fully open)	$1 \times (63.0 \text{ ft})\left(0.3048 \frac{\text{m}}{\text{ft}}\right) =$	19.2 m
swing check valve	$1 \times (63.0 \text{ ft})\left(0.3048 \frac{\text{m}}{\text{ft}}\right) =$	19.2 m
		45.8 m

The equivalent pipe length is the sum of the straight run of pipe and the equivalent length of pipe for the valves and fittings.

$$L_e = L + L_{\text{fittings}} = 90 \text{ m} + 45.8 \text{ m}$$
$$= 135.8 \text{ m}$$

Flow of Fluids

Use Darcy's equation to compute the frictional loss.

$$h_f = \frac{fL_e v^2}{2Dg} = \frac{(0.0173)(135.8 \text{ m})\left(1.61 \ \frac{\text{m}}{\text{s}}\right)^2}{(2)(0.1541 \text{ m})\left(9.81 \ \frac{\text{m}}{\text{s}^2}\right)}$$

$$= 2.01 \text{ m}$$

The total head loss is the sum of the head losses through the pipe, valves, and fittings and the change in elevation.

$$\Delta h = h_f + \Delta z = 2.01 \text{ m} + 6 \text{ m}$$

$$= 8.01 \text{ m}$$

The pressure difference between the entrance and discharge is

$$\Delta p = \rho \Delta h g = \left(992.25 \ \frac{\text{kg}}{\text{m}^3}\right)(8.01 \text{ m})\left(9.81 \ \frac{\text{m}}{\text{s}^2}\right)$$

$$= \boxed{77\,969 \text{ Pa} \quad (78 \text{ kPa})}$$

The answer is (A).

13. *Customary U.S. Solution*

For 6 in schedule-40 pipe, the internal diameter, D, is 0.5054 ft. The internal area is 0.2006 ft^2.

Use App. 14.D. For air at 70°F and atmospheric pressure, the kinematic viscosity is 15.83×10^{-5} ft^2/sec.

Calculate the Reynolds number.

$$\text{Re} = \frac{D\text{v}}{\nu} = \frac{(0.5054 \text{ ft})\left(60 \ \frac{\text{ft}}{\text{sec}}\right)}{15.83 \times 10^{-5} \ \frac{\text{ft}^2}{\text{sec}}}$$

$$= 1.92 \times 10^5$$

Since Re > 2100, the flow is turbulent. The friction loss coefficient can be determined from the Moody diagram.

For new steel pipe, the specific roughness, ϵ, is 0.0002 ft.

The relative roughness is

$$\frac{\epsilon}{D} = \frac{0.0002 \text{ ft}}{0.5054 \text{ ft}} = 0.0004$$

From the Moody diagram with Re = 1.92×10^5 and $\epsilon/D = 0.0004$, the friction factor, f, is 0.0184.

Use App. 17.D. The equivalent lengths of the valves and fittings are

standard radius elbow	2×8.9 ft = 17.8 ft
gate valve (fully open)	2×3.2 ft = 6.4 ft
90° angle valve (fully open)	1×63.0 ft = 63.0 ft
swing check valve	1×63.0 ft = 63.0 ft
	150.2 ft

The equivalent pipe length is the sum of the straight run of pipe and the equivalent lengths of pipe for the valves and fittings.

$$L_e = L + L_{\text{fittings}} = 300 \text{ ft} + 150.2 \text{ ft}$$

$$= 450.2 \text{ ft}$$

Use Darcy's equation to compute the frictional loss.

$$h_f = \frac{fL_e v^2}{2Dg} = \frac{(0.0184)(450.2 \text{ ft})\left(60 \ \frac{\text{ft}}{\text{sec}}\right)^2}{(2)(0.5054 \text{ ft})\left(32.2 \ \frac{\text{ft}}{\text{sec}^2}\right)}$$

$$= 916.2 \text{ ft}$$

The head loss is the sum of the head losses through the pipe, valves, and fittings and the change in elevation.

$$\Delta h = h_f + \Delta z = 916.2 \text{ ft} + 20 \text{ ft}$$

$$= 936.2 \text{ ft}$$

The pressure difference between the entrance and discharge is

$$\Delta p = \gamma \Delta h = \rho \Delta h \times \frac{g}{g_c}$$

$$= \frac{\left(0.075 \ \frac{\text{lbm}}{\text{ft}^3}\right)(936.2 \text{ ft})}{\left(12 \ \frac{\text{in}}{\text{ft}}\right)^2} \times \frac{32.2 \ \frac{\text{ft}}{\text{sec}^2}}{32.2 \ \frac{\text{lbm-ft}}{\text{lbf-sec}^2}}$$

$$= \boxed{0.49 \text{ lbf/in}^2 \quad (0.49 \text{ psi})}$$

The answer is (B).

SI Solution

Use App. 16.C. For 6 in pipe, the internal diameter, D, is 154.1 mm (0.1541 m), and the internal area is 186.5×10^{-4} m^2.

Use App. 14.E. For air at 20°C, the kinematic viscosity, ν, is 1.512×10^{-5} m^2/s.

Calculate the Reynolds number.

$$\mathrm{Re} = \frac{D\mathrm{v}}{\nu} = \frac{(0.1541 \text{ m})\left(18 \, \frac{\text{m}}{\text{s}}\right)}{1.512 \times 10^{-5} \, \frac{\text{m}^2}{\text{s}}}$$

$$= 1.83 \times 10^5$$

Since Re > 2100, the flow is turbulent. The friction loss coefficient can be determined from the Moody diagram.

For new steel pipe, the specific roughness, ϵ, is 6.0×10^{-5} m.

The relative roughness is

$$\frac{\epsilon}{D} = \frac{6.0 \times 10^{-5}}{0.1541 \text{ m}} = 0.0004$$

From the Moody diagram with Re = 1.83×10^5 and $\epsilon/D = 0.0004$, the friction factor, f, is 0.0185.

Compute the equivalent lengths of the valves and fittings. (Convert from App. 17.D.)

standard radius elbow	$2 \times (8.9 \text{ ft})\left(0.3048 \, \frac{\text{m}}{\text{ft}}\right) =$	5.4 m
gate valve (fully open)	$2 \times (3.2 \text{ ft})\left(0.3048 \, \frac{\text{m}}{\text{ft}}\right) =$	2.0 m
90° angle valve (fully open)	$1 \times (63.0 \text{ ft})\left(0.3048 \, \frac{\text{m}}{\text{ft}}\right) =$	19.2 m
swing check valve	$1 \times (63.0 \text{ ft})\left(0.3048 \, \frac{\text{m}}{\text{ft}}\right) =$	19.2 m
		45.8 m

The equivalent pipe length is the sum of the straight run of pipe and the equivalent lengths of pipe for the valves and fittings.

$$L_e = L + L_{\text{fittings}} = 90 \text{ m} + 45.8 \text{ m}$$

$$= 135.8 \text{ m}$$

Use Darcy's equation to compute the frictional loss.

$$h_f = \frac{fL_e\mathrm{v}^2}{2Dg} = \frac{(0.0185)(135.8 \text{ m})\left(18 \, \frac{\text{m}}{\text{s}}\right)^2}{(2)(0.1541 \text{ m})\left(9.81 \, \frac{\text{m}}{\text{s}^2}\right)} = 269.2 \text{ m}$$

The head loss is the sum of the head losses through the pipe, valves, and fittings and the change in elevation.

$$\Delta h = h_f + \Delta z = 269.2 \text{ m} + 6 \text{ m}$$

$$= 275.2 \text{ m}$$

The density of the air, ρ, is approximately 1.20 kg/m³.

The pressure difference between the entrance and discharge is

$$\Delta p = \rho\Delta hg = \left(1.20 \, \frac{\text{kg}}{\text{m}^3}\right)(275.2 \text{ m})\left(9.81 \, \frac{\text{m}}{\text{s}^2}\right)$$

$$= \boxed{3240 \text{ Pa} \quad (3.2 \text{ kPa})}$$

The answer is (B).

14. Assume that flows from reservoirs A and B are toward D and then toward C. From continuity,

$$\dot{V}_{\text{A-D}} + \dot{V}_{\text{B-D}} = \dot{V}_{\text{D-C}}$$

$$A_A\mathrm{v}_{\text{A-D}} + A_B\mathrm{v}_{\text{B-D}} - A_C\mathrm{v}_{\text{D-C}} = 0$$

From App. 16.B, for schedule-40 pipe,

$$A_A = 0.05134 \text{ ft}^2 \quad D_A = 0.2557 \text{ ft}$$

$$A_B = 0.5476 \text{ ft}^2 \quad D_B = 0.8350 \text{ ft}$$

$$A_C = 0.08841 \text{ ft}^2 \quad D_C = 0.3355 \text{ ft}$$

$$0.05134\mathrm{v}_{\text{A-D}} + 0.5476\mathrm{v}_{\text{B-D}} - 0.08841\mathrm{v}_{\text{D-C}} = 0 \quad [\text{Eq. I}]$$

Ignoring the velocity heads, the conservation of energy equation between A and D is

$$z_A = \frac{p_D}{\gamma} + z_D + h_{f,\text{A-D}}$$

$$50 \text{ ft} = \frac{p_D}{62.4 \, \frac{\text{lbf}}{\text{ft}^3}} + 25 \text{ ft} + \frac{(0.02)(800 \text{ ft})\mathrm{v}_{\text{A-D}}^2}{(2)(0.2557 \text{ ft})\left(32.2 \, \frac{\text{ft}}{\text{sec}^2}\right)}$$

$$\mathrm{v}_{\text{A-D}} = \mathrm{v}_{\text{A-D}} = \sqrt{25.73 - 0.0165p_D} \quad [\text{Eq. II}]$$

Similarly, for B–D,

$$40 \text{ ft} = \frac{p_D}{62.4 \, \frac{\text{lbf}}{\text{ft}^3}} + 25 \text{ ft} + \frac{(0.02)(500 \text{ ft})\mathrm{v}_{\text{B-D}}^2}{(2)(0.8350 \text{ ft})\left(32.2 \, \frac{\text{ft}}{\text{sec}^2}\right)}$$

$$\mathrm{v}_{\text{B-D}} = \sqrt{80.66 - 0.0862p_D} \quad [\text{Eq. III}]$$

For D–C,

$$22 \text{ ft} = \frac{p_D}{62.4 \, \frac{\text{lbf}}{\text{ft}^3}} + 25 \text{ ft} - \frac{(0.02)(1000 \text{ ft})\mathrm{v}_{\text{D-C}}^2}{(2)(0.3355 \text{ ft})\left(32.2 \, \frac{\text{ft}}{\text{sec}^2}\right)}$$

$$\mathrm{v}_{\text{D-C}} = \sqrt{3.24 + 0.0173p_D} \quad [\text{Eq. IV}]$$

Equations I, II, III, and IV must be solved simultaneously. To do this, assume a value for p_D. This value then determines all three velocities in Eqs. II, III, and IV. These velocities are substituted into Eq. I. A trial and error solution yields

$$v_{A-D} = 3.21 \text{ ft/sec}$$

$$v_{B-D} = 0.408 \text{ ft/sec}$$

$$v_{D-C} = 4.40 \text{ ft/sec}$$

$$\boxed{p_D = 933.8 \text{ lbf/ft}^2 \quad (930 \text{ psf})}$$

$$\boxed{\text{Flow is from B to D.}}$$

The answer is (B).

15. *Customary U.S. Solution*

(a) For water at 70°F, $\rho = 62.3$ lbm/ft^3, and $E_{water} = 320 \times 10^3$ lbf/in^2.

For cast-iron pipe, $E_{pipe} = 20 \times 10^6$ lbf/in^2. From Eq. 17.210, the composite modulus of elasticity of the pipe and water is

$$E = \frac{E_{water} t_{pipe} E_{pipe}}{t_{pipe} E_{pipe} + c_P D_{pipe} E_{water}}$$

$$= \frac{\left(320 \times 10^3 \, \frac{\text{lbf}}{\text{in}^2}\right)(0.75 \text{ in})\left(20 \times 10^6 \, \frac{\text{lbf}}{\text{in}^2}\right)}{(0.75 \text{ in})\left(20 \times 10^6 \, \frac{\text{lbf}}{\text{in}^2}\right)}$$

$$+ (1)(24 \text{ in})\left(320 \times 10^3 \frac{\text{lbf}}{\text{in}^2}\right)$$

$$= \boxed{2.12 \times 10^5 \text{ lbf/in}^2 \quad (2.1 \times 10^5 \text{ lbf/in}^2)}$$

The answer is (A).

(b) Using the value found in part (a) and from Eq. 17.210, the speed of sound in the pipe is

$$a = \sqrt{\frac{E g_c}{\rho}}$$

$$= \sqrt{\frac{\left(2.12 \times 10^5 \, \frac{\text{lbf}}{\text{in}^2}\right)\left(12 \, \frac{\text{in}}{\text{ft}}\right)^2\left(32.2 \, \frac{\text{lbm-ft}}{\text{lbf-sec}^2}\right)}{62.3 \, \frac{\text{lbm}}{\text{ft}^3}}}$$

$$= \boxed{3972 \text{ ft/sec} \quad (4000 \text{ ft/sec})}$$

The answer is (C).

(c) The maximum pressure is given by Eq. 17.208(b).

$$\Delta p = \frac{\rho a \Delta v}{g_c}$$

$$= \frac{\left(62.3 \, \frac{\text{lbm}}{\text{ft}^3}\right)\left(3972 \, \frac{\text{ft}}{\text{sec}}\right)\left(6 \, \frac{\text{ft}}{\text{sec}}\right)}{\left(32.2 \, \frac{\text{lbm-ft}}{\text{lbf-sec}^2}\right)\left(12 \, \frac{\text{in}}{\text{ft}}\right)^2}$$

$$= \boxed{320.2 \text{ lbf/in}^2 \quad (320 \text{ psi})}$$

The answer is (C).

(d) The length of time the pressure is constant at the valve is

$$t = \frac{2L}{a} = \frac{(2)(500 \text{ ft})}{3972 \, \frac{\text{ft}}{\text{sec}}} = \boxed{0.25 \text{ sec}}$$

The answer is (A).

SI Solution

(a) For water at 20°C, $\rho = 998.2$ kg/m^3, and $E_{water} = 2.2 \times 10^9$ Pa.

For cast-iron pipe, $E_{pipe} = 1.4 \times 10^{11}$ Pa. From Eq. 17.210, the composite modulus of elasticity of the pipe and water is

$$E = \frac{E_{water} t_{pipe} E_{pipe}}{t_{pipe} E_{pipe} + c_P D_{pipe} E_{water}}$$

$$= \frac{(2.2 \times 10^9 \text{ Pa})(0.02 \text{ m})(1.4 \times 10^{11} \text{ Pa})}{(0.02 \text{ m})(1.4 \times 10^{11} \text{ Pa}) + (1)(0.6 \text{ m})(2.2 \times 10^9 \text{ Pa})}$$

$$= \boxed{1.5 \times 10^9 \text{ Pa}}$$

The answer is (A).

(b) Using the value found in part (a) and from Eq. 17.210, the speed of sound in the pipe is

$$a = \sqrt{\frac{E}{\rho}} = \sqrt{\frac{1.50 \times 10^9 \text{ Pa}}{998.2 \, \frac{\text{kg}}{\text{m}^3}}}$$

$$= \boxed{1226 \text{ m/s} \quad (1200 \text{ m/s})}$$

The answer is (C).

(c) The maximum pressure is given by Eq. 17.208(a).

$$\Delta p = \rho a \Delta v$$

$$= \left(998.2 \, \frac{\text{kg}}{\text{m}^3}\right)\left(1226 \, \frac{\text{m}}{\text{s}}\right)\left(2 \, \frac{\text{m}}{\text{s}}\right)$$

$$= \boxed{2.45 \times 10^6 \text{ Pa} \quad (2.5 \text{ MPa})}$$

The answer is (C).

(d) The length of time the pressure is constant at the valve is

$$t = \frac{2L}{a} = \frac{(2)(150 \text{ m})}{1225 \ \frac{\text{m}}{\text{s}}} = \boxed{0.25 \text{ s}}$$

The answer is (A).

16. *Customary U.S. Solution*

(a) First, it is necessary to collect data on schedule-40 pipe and water. The fluid viscosity, pipe dimensions, and other parameters can be found in various appendices in Chap. 14 and Chap. 16. At 70°F water, $\nu = 1.059 \times 10^{-5} \text{ ft}^2/\text{sec}$.

From Table 17.2, $\epsilon = 0.0002$ ft. From App. 16.B,

$$
\begin{array}{lll}
\text{8 in pipe} & D = 0.6651 \text{ ft} \\
 & A = 0.3474 \text{ ft}^2 \\
\text{12 in pipe} & D = 0.9948 \text{ ft} \\
 & A = 0.7773 \text{ ft}^2 \\
\text{16 in pipe} & D = 1.25 \text{ ft} \\
 & A = 1.2272 \text{ ft}^2
\end{array}
$$

The flow quantity is converted from gallons per minute to cubic feet per second.

$$\dot{V} = \frac{(8 \text{ MGD})\left(10^6 \ \frac{\frac{\text{gal}}{\text{day}}}{\text{MGD}}\right)}{\left(24 \ \frac{\text{hr}}{\text{day}}\right)\left(60 \ \frac{\text{min}}{\text{hr}}\right)\left(7.4805 \ \frac{\text{gal}}{\text{ft}^3}\right)\left(60 \ \frac{\text{sec}}{\text{min}}\right)}$$
$$= 12.378 \text{ ft}^3/\text{sec}$$

For the inlet pipe, the velocity is

$$\text{v} = \frac{\dot{V}}{A} = \frac{12.378 \ \frac{\text{ft}^3}{\text{sec}}}{0.3474 \text{ ft}^2} = 35.63 \text{ ft/sec}$$

The Reynolds number is

$$\text{Re} = \frac{D\text{v}}{\nu} = \frac{(0.6651 \text{ ft})\left(35.63 \ \frac{\text{ft}}{\text{sec}}\right)}{1.059 \times 10^{-5} \ \frac{\text{ft}^2}{\text{sec}}}$$
$$= 2.24 \times 10^6$$

The relative roughness is

$$\frac{\epsilon}{D} = \frac{0.0002 \text{ ft}}{0.6651 \text{ ft}} = 0.0003$$

From the Moody diagram, $f = 0.015$.

Equation 17.23(b) is used to calculate the frictional energy loss.

$$E_{f,1} = h_f \times \frac{g}{g_c} = \frac{fL\text{v}^2}{2Dg_c}$$
$$= \frac{(0.015)(1000 \text{ ft})\left(35.63 \ \frac{\text{ft}}{\text{sec}}\right)^2}{(2)(0.6651 \text{ ft})\left(32.2 \ \frac{\text{lbm-ft}}{\text{lbf-sec}^2}\right)}$$
$$= 444.6 \text{ ft-lbf/lbm}$$

For the outlet pipe, the velocity is

$$\text{v} = \frac{\dot{V}}{A} = \frac{12.378 \ \frac{\text{ft}^3}{\text{sec}}}{0.7773 \text{ ft}^2} = 15.92 \text{ ft/sec}$$

The Reynolds number is

$$\text{Re} = \frac{D\text{v}}{\nu} = \frac{(0.9948 \text{ ft})\left(15.92 \ \frac{\text{ft}}{\text{sec}}\right)}{1.059 \times 10^{-5} \ \frac{\text{ft}^2}{\text{sec}}}$$
$$= 1.5 \times 10^6$$

The relative roughness is

$$\frac{\epsilon}{D} = \frac{0.0002 \text{ ft}}{0.9948 \text{ ft}} = 0.0002$$

From the Moody diagram, $f = 0.014$.

Equation 17.23(b) is used to calculate the frictional energy loss.

$$E_{f,2} = h_f \times \frac{g}{g_c} = \frac{fL\text{v}^2}{2Dg_c}$$
$$= \frac{(0.014)(1500 \text{ ft})\left(15.92 \ \frac{\text{ft}}{\text{sec}}\right)^2}{(2)(0.9948 \text{ ft})\left(32.2 \ \frac{\text{lbm-ft}}{\text{lbf-sec}^2}\right)}$$
$$= 83.1 \text{ ft-lbf/lbm}$$

Assume a 50% split through the two branches. In the upper branch, the velocity is

$$\text{v} = \frac{\dot{V}}{A} = \frac{\left(\frac{1}{2}\right)\left(12.378 \ \frac{\text{ft}^3}{\text{sec}}\right)}{0.3474 \text{ ft}^2} = 17.82 \text{ ft/sec}$$

The Reynolds number is

$$Re = \frac{Dv}{\nu} = \frac{(0.6651 \text{ ft})\left(17.82 \ \frac{\text{ft}}{\text{sec}}\right)}{1.059 \times 10^{-5} \ \frac{\text{ft}^2}{\text{sec}}}$$

$$= 1.1 \times 10^6$$

The relative roughness is

$$\frac{\epsilon}{D} = \frac{0.0002 \text{ ft}}{0.6651 \text{ ft}} = 0.0003$$

From the Moody diagram, $f = 0.015$.

For the 16 in pipe in the lower branch, the velocity is

$$v = \frac{\dot{V}}{A} = \frac{\left(\frac{1}{2}\right)\left(12.378 \ \frac{\text{ft}^3}{\text{sec}}\right)}{1.2272 \text{ ft}^2} = 5.04 \text{ ft/sec}$$

The Reynolds number is

$$Re = \frac{Dv}{\nu} = \frac{(1.25 \text{ ft})\left(5.04 \ \frac{\text{ft}}{\text{sec}}\right)}{1.059 \times 10^{-5} \ \frac{\text{ft}^2}{\text{sec}}}$$

$$= 5.95 \times 10^5$$

The relative roughness is

$$\frac{\epsilon}{D} = \frac{0.0002 \text{ ft}}{1.25 \text{ ft}} = 0.00016$$

From the Moody diagram, $f = 0.015$.

These values of f for the two branches are fairly insensitive to changes in $\dot{V}$, so they will be used for the rest of the problem in both branches.

Equation 17.23(b) is used to calculate the frictional energy loss in the upper branch.

$$E_{f,\text{upper}} = h_f \times \frac{g}{g_c} = \frac{fLv^2}{2Dg_c}$$

$$= \frac{(0.015)(500 \text{ ft})\left(17.81 \ \frac{\text{ft}}{\text{sec}}\right)^2}{(2)(0.6651 \text{ ft})\left(32.2 \ \frac{\text{lbm-ft}}{\text{lbf-sec}^2}\right)}$$

$$= 55.5 \text{ ft-lbf/lbm}$$

To calculate a loss for any other flow in the upper branch,

$$E_{f,\text{upper 2}} = E_{f,\text{upper}}\left(\frac{\dot{V}}{\left(\frac{1}{2}\right)\left(12.378 \ \frac{\text{ft}^3}{\text{sec}}\right)}\right)^2$$

$$= \left(55.5 \ \frac{\text{ft-lbf}}{\text{lbm}}\right)\left(\frac{\dot{V}}{6.189 \ \frac{\text{ft}^3}{\text{sec}}}\right)^2$$

$$= 1.45 \dot{V}^2$$

Similarly, for the lower branch, in the 8 in section,

$$E_{f,\text{lower,8 in}} = h_f \times \frac{g}{g_c} = \frac{fLv^2}{2Dg_c}$$

$$= \frac{(0.015)(250 \text{ ft})\left(17.81 \ \frac{\text{ft}}{\text{sec}}\right)^2}{(2)(0.6651 \text{ ft})\left(32.2 \ \frac{\text{lbm-ft}}{\text{lbf-sec}^2}\right)}$$

$$= 27.8 \text{ ft-lbf/lbm}$$

For the lower branch, in the 16 in section,

$$E_{f,\text{lower,16 in}} = h_f \times \frac{g}{g_c} = \frac{fLv^2}{2Dg_c}$$

$$= \frac{(0.015)(1000 \text{ ft})\left(5.04 \ \frac{\text{ft}}{\text{sec}}\right)^2}{(2)(1.25 \text{ ft})\left(32.2 \ \frac{\text{lbm-ft}}{\text{lbf-sec}^2}\right)}$$

$$= 4.7 \text{ ft-lbf/lbm}$$

The total loss in the lower branch is

$$E_{f,\text{lower}} = E_{f,\text{lower,8 in}} + E_{f,\text{lower,16 in}}$$

$$= 27.8 \ \frac{\text{ft-lbf}}{\text{lbm}} + 4.7 \ \frac{\text{ft-lbf}}{\text{lbm}}$$

$$= 32.5 \text{ ft-lbf/lbm}$$

To calculate a loss for any other flow in the lower branch,

$$E_{f,\text{lower 2}} = E_{f,\text{lower}}\left(\frac{\dot{V}}{\left(\frac{1}{2}\right)\left(12.378 \ \frac{\text{ft}^3}{\text{sec}}\right)}\right)^2$$

$$= \left(32.5 \ \frac{\text{ft-lbf}}{\text{lbm}}\right)\left(\frac{\dot{V}}{6.189 \ \frac{\text{ft}^3}{\text{sec}}}\right)^2$$

$$= 0.85 \dot{V}^2$$

Let x be the fraction flowing in the upper branch. Then, because the friction losses are equal,

$$E_{f,\text{upper 2}} = E_{f,\text{lower 2}}$$

$$1.45x^2 = (0.85)(1-x)^2$$

$$x = 0.434$$

$$\dot{V}_{\text{upper}} = (0.434)\left(12.378 \ \frac{\text{ft}^3}{\text{sec}}\right)$$

$$= \boxed{5.372 \ \text{ft}^3/\text{sec} \quad (5.4 \ \text{ft}^3/\text{sec})}$$

The answer is (D).

(b) $\qquad \dot{V}_{\text{lower}} = (1 - 0.432)\left(12.378 \ \frac{\text{ft}^3}{\text{sec}}\right)$

$$= 7.03 \ \text{ft}^3/\text{sec}$$

$$E_{f,\text{total}} = E_{f,1} + E_{f,\text{lower 2}} + E_{f,2}$$

$$E_{f,\text{lower 2}} = 0.85 \ \dot{V}_{\text{lower}}^2$$

$$= (0.85)\left(7.03 \ \frac{\text{ft}^3}{\text{sec}}\right)^2$$

$$= 42.0 \ \text{ft}$$

$$E_{f,\text{total}} = 444.6 \ \frac{\text{ft-lbf}}{\text{lbm}} + 42.0 \ \frac{\text{ft-lbf}}{\text{lbm}} + 83.1 \ \frac{\text{ft-lbf}}{\text{lbm}}$$

$$= \boxed{569.7 \ \text{ft-lbf/lbm} \quad (570 \ \text{ft-lbf/lbm})}$$

The answer is (D).

SI Solution

(a) First, it is necessary to collect data on schedule-40 pipe and water. The fluid viscosity, pipe dimensions, and other parameters can be found in various appendices in Chap. 14 and Chap. 16. At 20°C water, $\nu = 1.007 \times 10^{-6} \ \text{m}^2/\text{s}$.

From Table 17.2, $\epsilon = 6 \times 10^{-5}$ m. From App. 16.C,

8 in pipe	$D = 202.7$ mm	
	$A = 322.75 \ \text{cm}^2$	
12 in pipe	$D = 303.2$ mm	
	$A = 721.9 \times 10^{-4} \ \text{m}^2$	
16 in pipe	$D = 381$ mm	
	$A = 1140 \times 10^{-4} \ \text{m}^2$	

For the inlet pipe, the velocity is

$$v = \frac{\dot{V}}{A} = \frac{\left(350 \ \frac{\text{L}}{\text{s}}\right)\left(100 \ \frac{\text{cm}}{\text{m}}\right)^2}{(322.75 \ \text{cm}^2)\left(1000 \ \frac{\text{L}}{\text{m}^3}\right)} = 10.85 \ \text{m/s}$$

The Reynolds number is

$$\text{Re} = \frac{D v}{\nu} = \frac{(0.2027 \ \text{m})\left(10.85 \ \frac{\text{m}}{\text{s}}\right)}{1.007 \times 10^{-6} \ \frac{\text{m}^2}{\text{s}}}$$

$$= 2.18 \times 10^6$$

The relative roughness is

$$\frac{\epsilon}{D} = \frac{6 \times 10^{-5} \ \text{m}}{0.2027 \ \text{m}} = 0.0003$$

From the Moody diagram, $f = 0.015$.

Equation 17.23(a) is used to calculate the frictional energy loss.

$$E_{f,1} = h_f g = \frac{f L v^2}{2D}$$

$$= \frac{(0.015)(300 \ \text{m})\left(10.85 \ \frac{\text{m}}{\text{s}}\right)^2}{(2)(0.2027 \ \text{m})}$$

$$= 1307 \ \text{J/kg}$$

For the outlet pipe, the velocity is

$$v = \frac{\dot{V}}{A} = \frac{350 \ \frac{\text{L}}{\text{s}}}{(721.9 \times 10^{-4} \ \text{m}^2)\left(1000 \ \frac{\text{L}}{\text{m}^3}\right)} = 4.848 \ \text{m/s}$$

The Reynolds number is

$$\text{Re} = \frac{D v}{\nu} = \frac{(0.3032 \ \text{m})\left(4.848 \ \frac{\text{m}}{\text{s}}\right)}{1.007 \times 10^{-6} \ \frac{\text{m}^2}{\text{s}}}$$

$$= 1.46 \times 10^6$$

The relative roughness is

$$\frac{\epsilon}{D} = \frac{6 \times 10^{-5} \ \text{m}}{0.3032 \ \text{m}} = 0.0002$$

From the Moody diagram, $f = 0.014$.

Equation 17.23(a) is used to calculate the frictional energy loss.

$$E_{f,2} = h_f g = \frac{f L v^2}{2D}$$

$$= \frac{(0.014)(450 \ \text{m})\left(4.848 \ \frac{\text{m}}{\text{s}}\right)^2}{(2)(0.3032 \ \text{m})}$$

$$= 244.2 \ \text{J/kg}$$

Assume a 50% split through the two branches. In the upper branch, the velocity is

$$v = \frac{\dot{V}}{A} = \frac{\left(\frac{1}{2}\right)\left(350 \; \frac{L}{s}\right)}{(322.7 \times 10^{-4} \; m^2)\left(1000 \; \frac{L}{m^3}\right)} = 5.423 \; m/s$$

The Reynolds number is

$$Re = \frac{Dv}{\nu} = \frac{(0.2027 \; m)\left(5.423 \; \frac{m}{s}\right)}{1.007 \times 10^{-6} \; \frac{m^2}{s}}$$

$$= 1.1 \times 10^6$$

The relative roughness is

$$\frac{\epsilon}{D} = \frac{6 \times 10^{-5} \; m}{0.2027 \; m} = 0.0003$$

From the Moody diagram, $f = 0.015$.

For the 16 in pipe in the lower branch, the velocity is

$$v = \frac{\dot{V}}{A} = \frac{\left(\frac{1}{2}\right)\left(350 \; \frac{L}{s}\right)}{(1140 \times 10^{-4} \; m^2)\left(1000 \; \frac{L}{m^3}\right)} = 1.535 \; m/s$$

The Reynolds number is

$$Re = \frac{Dv}{\nu} = \frac{(0.381 \; m)\left(1.535 \; \frac{m}{s}\right)}{1.007 \times 10^{-6} \; \frac{m^2}{s}}$$

$$= 5.81 \times 10^5$$

The relative roughness is

$$\frac{\epsilon}{D} = \frac{6 \times 10^{-5} \; m}{0.381 \; m} = 0.00016$$

From the Moody diagram, $f = 0.015$.

These values of f for the two branches are fairly insensitive to changes in $\dot{V}$, so they will be used for the rest of the problem in both branches.

Equation 17.23(a) is used to calculate the frictional energy loss in the upper branch.

$$E_{f,\text{upper}} = h_f g = \frac{fLv^2}{2D}$$

$$= \frac{(0.015)(150 \; m)\left(5.423 \; \frac{m}{s}\right)^2}{(2)(0.2027 \; m)}$$

$$= 163.2 \; J/kg$$

To calculate a loss for any other flow in the upper branch,

$$E_{f,\text{upper 2}} = E_{f,\text{upper}}\left(\frac{\dot{V}}{\left(\frac{1}{2}\right)\left(0.350 \; \frac{m^3}{s}\right)}\right)^2$$

$$= \left(163.2 \; \frac{J}{kg}\right)\left(\frac{\dot{V}}{0.175 \; \frac{m^3}{s}}\right)^2$$

$$= 5329\dot{V}^2$$

Similarly, for the lower branch, in the 8 in section,

$$E_{f,\text{lower,8 in}} = h_f g = \frac{fLv^2}{2D} = \frac{(0.015)(75 \; m)\left(5.423 \; \frac{m}{s}\right)^2}{(2)(0.2027 \; m)}$$

$$= 81.61 \; J/kg$$

For the lower branch, in the 16 in section,

$$E_{f,\text{lower,16 in}} = h_f g = \frac{fLv^2}{2D} = \frac{(0.015)(300 \; m)\left(1.585 \; \frac{m}{s}\right)^2}{(2)(0.381 \; m)}$$

$$= 14.84 \; J/kg$$

The total loss in the lower branch is

$$E_{f,\text{lower}} = E_{f,\text{lower,8 in}} + E_{f,\text{lower,16 in}}$$

$$= 81.61 \; \frac{J}{kg} + 14.84 \; \frac{J}{kg}$$

$$= 96.45 \; J/kg$$

To calculate a loss for any other flow in the lower branch,

$$E_{f,\text{lower 2}} = E_{f,\text{lower}}\left(\frac{\dot{V}}{\left(\frac{1}{2}\right)\left(0.350 \; \frac{m^3}{s}\right)}\right)^2$$

$$= \left(96.45 \; \frac{J}{kg}\right)\left(\frac{\dot{V}}{0.175 \; \frac{m^3}{s}}\right)^2$$

$$= 3149\dot{V}^2$$

Let x be the fraction flowing in the upper branch. Then, because the friction losses are equal,

$$E_{f,\text{upper 2}} = E_{f,\text{lower 2}}$$

$$5329x^2 = (3149)(1-x)^2$$

$$x = 0.435$$

$$\dot{V}_{upper} = (0.435)\left(0.350 \ \frac{m^3}{s}\right)$$

$$= \boxed{0.15 \ m^3/s}$$

The answer is (D).

(b)
$$\dot{V}_{lower} = (1 - 0.435)\left(0.350 \ \frac{m^3}{s}\right)$$

$$= 0.198 \ m^3/s$$

$$E_{f,total} = E_{f,1} + E_{f,lower\,2} + E_{f,2}$$

$$E_{f,lower\,2} = 3149\dot{V}_{lower}^2$$

$$= (3149)\left(0.198 \ \frac{m^3}{s}\right)^2$$

$$= 123.5 \ J/kg$$

$$E_{f,total} = 1307 \ \frac{J}{kg} + 123.5 \ \frac{J}{kg} + 244.2 \ \frac{J}{kg}$$

$$= \boxed{1675 \ J/kg \quad (1.7 \ kJ/kg)}$$

The answer is (D).

17. *steps 1, 2, and 3:*

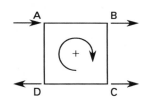

step 4: There is only one loop: ABCD.

step 5: Use Eq. 17.132.

$$K' = \frac{10.44L}{d^{4.87}C^{1.85}}$$

pipe AB: $\quad K' = \dfrac{(10.44)(1000 \ ft)}{(8 \ in)^{4.87}(100)^{1.85}}$

$$= 8.33 \times 10^{-5}$$

pipe BC: $\quad K' = \dfrac{(10.44)(1000 \ ft)}{(6 \ in)^{4.87}(100)^{1.85}}$

$$= 3.38 \times 10^{-4}$$

pipe CD: $\quad K' = 8.33 \times 10^{-5} \quad$ [same as AB]

pipe DA: $\quad K' = \dfrac{(10.44)(1000 \ ft)}{(12 \ in)^{4.87}(100)^{1.85}}$

$$= 1.16 \times 10^{-5}$$

step 6: Assume the flows are as shown in the following illustration.

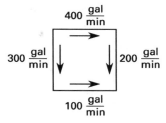

step 7: Use Eq. 17.140.

$$\delta = \frac{-\sum K'\dot{V}_a^n}{n\sum|K'\dot{V}_a^{n-1}|}$$

$$= \frac{-\left(\begin{array}{l}(8.33 \times 10^{-5})\left(400 \ \frac{gal}{min}\right)^{1.85} \\[2mm] + (3.38 \times 10^{-4})\left(200 \ \frac{gal}{min}\right)^{1.85} \\[2mm] - (8.33 \times 10^{-5})\left(100 \ \frac{gal}{min}\right)^{1.85} \\[2mm] - (1.16 \times 10^{-5})\left(300 \ \frac{gal}{min}\right)^{1.85}\end{array}\right)}{(1.85)\left(\begin{array}{l}(8.33 \times 10^{-5})\left(400 \ \frac{gal}{min}\right)^{0.85} \\[2mm] + (3.38 \times 10^{-4})\left(200 \ \frac{gal}{min}\right)^{0.85} \\[2mm] + (8.33 \times 10^{-5})\left(100 \ \frac{gal}{min}\right)^{0.85} \\[2mm] + (1.16 \times 10^{-5})\left(300 \ \frac{gal}{min}\right)^{0.85}\end{array}\right)}$$

$$= \frac{-10.67 \ \frac{gal}{min}}{(1.85)\left(4.98 \times 10^{-2} \ \frac{gal}{min}\right)}$$

$$= -116 \ gal/min$$

step 8: The adjusted flows are shown.

step 7: $\delta = -24$ gal/min

step 8: The adjusted flows are shown.

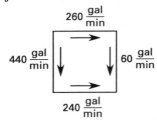

step 7: $\delta = -2$ gal/min [small enough]

step 8: The final adjusted flows are shown.

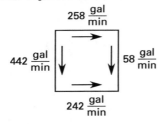

The answer is (B).

18. (a) This is a Hardy Cross problem. The pressure at point C does not change the solution procedure.

step 1: The Hazen-Williams roughness coefficient is given.

step 2: Choose clockwise as positive.

step 3: Nodes are already numbered.

step 4: Choose the loops as shown.

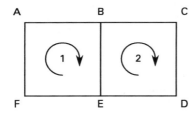

step 5: $\dot{V}$ is in gal/min, so use Eq. 17.132. d is in inches. L is in feet.

Each pipe has the same length.

$$K' = \frac{10.44L}{d^{4.87}C^{1.85}}$$

$$K'_{8\,\text{in}} = \frac{(10.44)(1000\ \text{ft})}{(8\ \text{in})^{4.87}(100)^{1.85}} = 8.33 \times 10^{-5}$$

$$K'_{6\,\text{in}} = \frac{(10.44)(1000\ \text{ft})}{(6\ \text{in})^{4.87}(100)^{1.85}} = 3.38 \times 10^{-4}$$

$$K'_{\text{CE}} = 2K'_{6\,\text{in}} = 6.76 \times 10^{-4}$$

$$K'_{\text{EA}} = K'_{6\,\text{in}} + K'_{8\,\text{in}} = 4.21 \times 10^{-4}$$

step 6: Assume the flows shown.

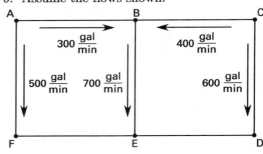

$$AB = 300\ \text{gal/min}$$

$$BE = 700\ \text{gal/min}$$

$$AF = AE = 500\ \text{gal/min}$$

$$CF = CE = 600\ \text{gal/min}$$

$$CB = 400\ \text{gal/min}$$

step 7: If the elevations are included as part of the head loss,

$$\sum h = \sum K' \dot{V}_a^n + \delta \sum nK' \dot{V}_a^{n-1} + z_2 - z_1 = 0$$

However, since the loop closes on itself, $z_2 = z_1$, and the elevations can be omitted.

First iteration:

Loop 1:

$$\delta_1 = \frac{-\left(\begin{array}{c}(8.33 \times 10^{-5})(300)^{1.85} \\ + (8.33 \times 10^{-5})(700)^{1.85} \\ - (4.21 \times 10^{-4})(500)^{1.85}\end{array}\right)}{(1.85)\left(\begin{array}{c}(8.33 \times 10^{-5})(300)^{0.85} \\ + (8.33 \times 10^{-5})(700)^{0.85} \\ + (4.21 \times 10^{-4})(500)^{0.85}\end{array}\right)}$$

$$= \frac{-(-22.97)}{0.213}$$

$$= \boxed{+108\ \text{gal/min}}$$

Loop 2:

$$\delta_2 = \frac{-\left(\begin{array}{c}(6.76 \times 10^{-4})(600)^{1.85} \\ - (8.33 \times 10^{-5})(700)^{1.85} \\ - (8.33 \times 10^{-5})(400)^{1.85}\end{array}\right)}{(1.85)\left(\begin{array}{c}(6.76 \times 10^{-4})(600)^{0.85} \\ + (8.33 \times 10^{-5})(700)^{0.85} \\ + (8.33 \times 10^{-5})(400)^{0.85}\end{array}\right)}$$

$$= \frac{-72.52}{0.353}$$

$$= \boxed{-205\ \text{gal/min}}$$

The answer is (B).

Flow of Fluids

(b) *Second iteration:*

AB: $300 +$ 108 $= 408$ gal/min

BE: $700 +$ 108 $-(-205) = 1013$ gal/min

AE: $500 -$ 108 $= 392$ gal/min

CE: $600 +$ (-205) $= 395$ gal/min

CB: $400 -$ (-205) $= 605$ gal/min

Loop 1:

$$\delta_1 = \frac{-9.48}{0.205} = \boxed{-46 \text{ gal/min} \quad (-50 \text{ gal/min})}$$

Loop 2:

$$\delta_2 = \frac{-1.08}{0.292} = \boxed{-3.7 \text{ gal/min} \quad (-4 \text{ gal/min})}$$

The answer is (B).

(c) *Third iteration:*

AB: $408 + (-46)$ $= 362$ gal/min

BE: $1013 + (-46)$ $-(-4) = 971$ gal/min

AE: $392 - (-46)$ $= 438$ gal/min

CE: $395 + (-4)$ $= 391$ gal/min

CB: $605 - (-4)$ $= 609$ gal/min

Loop 1:

$$\delta_1 = \frac{-0.066}{0.213} = \boxed{-0.31 \text{ gal/min} \quad (0 \text{ gal/min})}$$

Loop 2:

$$\delta_2 = \frac{-2.4}{0.29} = \boxed{-8.3 \text{ gal/min} \quad (-8 \text{ gal/min})}$$

The answer is (A).

(d) Use the following flows.

AB: $362 +$ 0 $= 362$ gal/min

BE: $971 +$ 0 $-(-8) = \boxed{\begin{array}{l} 979 \text{ gal/min} \\ (980 \text{ gal/min}) \end{array}}$

AE: $438 -$ 0 $= 438$ gal/min

CE: $391 +$ (-8) $= 383$ gal/min

CB: $609 -$ (-8) $= 617$ gal/min

The answer is (D).

(e) The friction loss in each section is

$$h_{f,\text{AB}} = (8.33 \times 10^{-5})(362)^{1.85} = 4.5 \text{ ft}$$

$$h_{f,\text{BE}} = (8.33 \times 10^{-5})(979)^{1.85} = 28.4 \text{ ft}$$

$$h_{f,\text{AF}} = (8.33 \times 10^{-5})(438)^{1.85} = 6.4 \text{ ft}$$

$$h_{f,\text{FE}} = (3.38 \times 10^{-4})(438)^{1.85} = 26.0 \text{ ft}$$

$$h_{f,\text{CD}} = h_{f,\text{DE}} = (3.38 \times 10^{-4})(383)^{1.85} = 20.3 \text{ ft}$$

$$h_{f,\text{CB}} = (8.33 \times 10^{-5})(617)^{1.85} = 12.1 \text{ ft}$$

The pressure at C is 40 psig. Water density at 60°F is 62.37 lbf/ft³.

$$h_C = \frac{p}{\gamma} = \frac{\left(40 \frac{\text{lbf}}{\text{in}^2}\right)\left(12 \frac{\text{in}}{\text{ft}}\right)^2}{62.37 \frac{\text{lbf}}{\text{ft}^3}} = \boxed{92.4 \text{ ft} \quad (90 \text{ ft})}$$

The answer is (C).

(f) Use the energy continuity equation between adjacent points. The sign of the friction head depends on the loop and flow directions.

$$h = h_C + z_C - z - h_f$$

$$h_D = 92.4 \text{ ft} + 300 \text{ ft} - 150 \text{ ft} - 20.3 \text{ ft} = 222.1 \text{ ft}$$

$$h_E = 222.1 \text{ ft} + 150 \text{ ft} - 200 \text{ ft} - 20.3 \text{ ft} = 151.8 \text{ ft}$$

$$h_F = 151.8 \text{ ft} + 200 \text{ ft} - 150 \text{ ft} + 26 \text{ ft} = 227.8 \text{ ft}$$

$$h_B = 92.4 \text{ ft} + 300 \text{ ft} - 150 \text{ ft} - 12.1 \text{ ft} = 230.3 \text{ ft}$$

$$h_A = 230.3 \text{ ft} + 150 \text{ ft} - 200 \text{ ft} + 4.5 \text{ ft} = 184.8 \text{ ft}$$

Using $p = \gamma h$,

$$p_A = \frac{\left(62.37 \frac{\text{lbf}}{\text{ft}^3}\right)(184.8 \text{ ft})}{\left(12 \frac{\text{in}}{\text{ft}}\right)^2} = 80.0 \text{ lbf/in}^2 \quad (80.0 \text{ psig})$$

$$p_B = \frac{\left(62.37 \frac{\text{lbf}}{\text{ft}^3}\right)(230.3 \text{ ft})}{\left(12 \frac{\text{in}}{\text{ft}}\right)^2} = 99.7 \text{ psig}$$

$$p_C = 40 \text{ psig} \quad [\text{given}]$$

$$p_D = \frac{\left(62.37 \frac{\text{lbf}}{\text{ft}^3}\right)(222.1 \text{ ft})}{\left(12 \frac{\text{in}}{\text{ft}}\right)^2} = \boxed{96.2 \text{ psig} \quad (96 \text{ psig})}$$

$$p_E = \frac{\left(62.37 \frac{\text{lbf}}{\text{ft}^3}\right)(151.8 \text{ ft})}{\left(12 \frac{\text{in}}{\text{ft}}\right)^2} = 65.7 \text{ psig}$$

$$p_F = \frac{\left(62.37 \frac{\text{lbf}}{\text{ft}^3}\right)(227.8 \text{ ft})}{\left(12 \frac{\text{in}}{\text{ft}}\right)^2} = 98.7 \text{ psig}$$

The answer is (C).

Flow of Fluids

(g) The pressure increase across the pump is

$$\Delta p = \left(80 \ \frac{\text{lbf}}{\text{in}^2} - 20 \ \frac{\text{lbf}}{\text{in}^2}\right)\left(12 \ \frac{\text{in}}{\text{ft}}\right)^2 = 8640 \ \text{lbf/ft}^2$$

Use Table 18.5 to find the hydraulic horsepower.

$$P = \frac{\Delta p Q}{2.468 \times 10^5}$$

$$= \frac{\left(8640 \ \frac{\text{lbf}}{\text{ft}^2}\right)\left(800 \ \frac{\text{gal}}{\text{min}}\right)}{2.468 \times 10^5 \ \frac{\text{lbf-gal}}{\text{min-ft}^2\text{-hp}}}$$

$$= \boxed{28 \ \text{hp}}$$

The answer is (B).

19. (a) *Pipe network solution*

step 1: Use the Darcy equation since Hazen-Williams coefficients are not given (or, assume *C*-values based on the age of the pipe). (See Eq. 17.133.)

step 2: Clockwise is positive.

step 3: All nodes are lettered.

step 4: There is only one loop.

step 5: $\epsilon = 0.0008$ ft. Assume full turbulence. For class-A cast-iron pipe, $D = 10.1$ in.

$$\frac{\epsilon}{D} = \frac{0.0008 \ \text{ft}}{\dfrac{10.1 \ \text{in}}{12 \ \frac{\text{in}}{\text{ft}}}} = 0.00095 \quad [\text{use } 0.001]$$

$f \approx 0.020$ for full turbulence. (See Fig. 17.4.)

$$K' = \frac{1.251 \times 10^{-7} fL}{D^5}$$

The friction coefficient between junctions A and B is

$$K'_{AB} = (1.251 \times 10^{-7})\left(\frac{(0.02)(2000 \ \text{ft})}{\left(\dfrac{10.1 \ \text{in}}{12 \ \frac{\text{in}}{\text{ft}}}\right)^5}\right)$$

$$= \boxed{11.8 \times 10^{-6} \quad (12 \times 10^{-6})}$$

The answer is (C).

(b) The friction coefficient between junctions B and D is

$$K'_{BD} = (1.251 \times 10^{-7})\left(\frac{(0.02)(1000 \ \text{ft})}{\left(\dfrac{10.1 \ \text{in}}{12 \ \frac{\text{in}}{\text{ft}}}\right)^5}\right)$$

$$= \boxed{5.92 \times 10^{-6} \quad (6 \times 10^{-6})}$$

The answer is (B).

(c) The friction coefficient between junctions D and C is

$$K'_{DC} = (1.251 \times 10^{-7})\left(\frac{(0.02)(1500 \ \text{ft})}{\left(\dfrac{8.13 \ \text{in}}{12 \ \frac{\text{in}}{\text{ft}}}\right)^5}\right)$$

$$= \boxed{26.3 \times 10^{-6} \quad (26 \times 10^{-6})}$$

The answer is (D).

(d) The friction coefficient between junctions C and A is

$$K'_{CA} = (1.251 \times 10^{-7})\left(\frac{(0.02)(1000 \ \text{ft})}{\left(\dfrac{12.12 \ \text{in}}{12 \ \frac{\text{in}}{\text{ft}}}\right)^5}\right)$$

$$= \boxed{2.38 \times 10^{-6} \quad (2 \times 10^{-6})}$$

The answer is (B).

(e) *step 6:* Assume $\dot{V}_{AB} = 1$ MGD.

$$\dot{V}_{ACDB} = 0.5 \ \text{MGD}$$

$$\frac{\dot{V}_{AB}}{\dot{V}_{ACDB}} = \frac{1 \ \text{MGD}}{0.5 \ \text{MGD}} = 2$$

Convert $\dot{V}$ to gal/min.

$$\dot{V}_{AB} = \frac{(1 \ \text{MGD})\left(1 \times 10^6 \ \frac{\text{gal}}{\text{MG}}\right)}{\left(24 \ \frac{\text{hr}}{\text{day}}\right)\left(60 \ \frac{\text{min}}{\text{hr}}\right)} = 694 \ \text{gal/min}$$

$$\dot{V}_{ACDB} = \frac{(0.5 \ \text{MGD})\left(1 \times 10^6 \ \frac{\text{gal}}{\text{MG}}\right)}{\left(24 \ \frac{\text{hr}}{\text{day}}\right)\left(60 \ \frac{\text{min}}{\text{hr}}\right)} = 347 \ \text{gal/min}$$

step 7: There is only one loop.

$$694 \ \frac{\text{gal}}{\text{min}} = (2)\left(347 \ \frac{\text{gal}}{\text{min}}\right)$$

$$\left(694 \ \frac{\text{gal}}{\text{min}}\right)^2 = (4)\left(347 \ \frac{\text{gal}}{\text{min}}\right)^2$$

$$\delta = \frac{\begin{aligned}&(-1)(1 \times 10^{-6})(347)^2 \\ &\times \big((11.8)(4) - 5.92 - 26.3 - 2.38\big)\end{aligned}}{\begin{aligned}&(2)(1 \times 10^{-6})(347) \\ &\times \big((11.8)(2) + 5.92 + 26.3 + 2.38\big)\end{aligned}}$$

$$= -37.6 \ \text{gal/min} \quad [\text{use} -38 \ \text{gal/min}]$$

step 8:

$$\dot{V}_{AB} = 694 \ \frac{\text{gal}}{\text{min}} + \left(-38 \ \frac{\text{gal}}{\text{min}}\right) = 656 \ \text{gal/min}$$

$$\dot{V}_{ACDB} = 347 \ \frac{\text{gal}}{\text{min}} - \left(-38 \ \frac{\text{gal}}{\text{min}}\right) = 385 \ \text{gal/min}$$

Repeat step 7.

$$\frac{\dot{V}_{AB}}{\dot{V}_{ACDB}} = \frac{656 \ \frac{\text{gal}}{\text{min}}}{385 \ \frac{\text{gal}}{\text{min}}} = 1.7$$

$$(1.7)^2 = 2.90$$

$$656 \ \frac{\text{gal}}{\text{min}} = (1.70)\left(385 \ \frac{\text{gal}}{\text{min}}\right)$$

$$\left(656 \ \frac{\text{gal}}{\text{min}}\right)^2 = (2.90)\left(385 \ \frac{\text{gal}}{\text{min}}\right)^2$$

$$\delta = \frac{\begin{aligned}&(-1)(1 \times 10^{-6})(385)^2 \\ &\times \big((11.8)(2.90) - 5.92 - 26.3 - 2.38\big)\end{aligned}}{\begin{aligned}&(2)(1 \times 10^{-6})(385) \\ &\times \big((11.8)(1.70) + 5.92 + 26.3 + 2.38\big)\end{aligned}}$$

$$= 1.34 \quad (1.3)$$

$$\boxed{\dot{V}_{AB} = 656 \ \text{gal/min} \quad (660 \ \text{gal/min})}$$

$$\dot{V}_{ACDB} = 385 \ \text{gal/min}$$

Check the Reynolds number in leg AB to verify that $f = 0.02$ (from part (a)) was a good choice.

$$A_{10 \, \text{in pipe}} = \left(\frac{\pi}{4}\right)\left(\frac{10.1 \ \text{in}}{12 \ \frac{\text{in}}{\text{ft}}}\right)^2 = 0.5564 \quad [\text{cast-iron pipe}]$$

The flow rate is

$$\frac{656 \ \frac{\text{gal}}{\text{min}}}{\left(7.4805 \ \frac{\text{gal}}{\text{ft}^3}\right)\left(60 \ \frac{\text{sec}}{\text{min}}\right)} = 1.46 \ \text{ft}^3/\text{sec}$$

$$\text{v} = \frac{\dot{V}}{A} = \frac{1.46 \ \frac{\text{ft}^3}{\text{sec}}}{0.5564 \ \text{ft}^2} = 2.62 \ \text{ft/sec} \quad [\text{reasonable}]$$

For 50°F water,

$$\nu = 1.410 \times 10^{-5} \ \text{ft}^2/\text{sec}$$

$$\text{Re} = \frac{\text{v}D}{\nu} = \frac{\left(2.62 \ \frac{\text{ft}}{\text{sec}}\right)\left(\frac{10.1 \ \text{in}}{12 \ \frac{\text{in}}{\text{ft}}}\right)}{1.410 \times 10^{-5} \ \frac{\text{ft}^2}{\text{sec}}}$$

$$= 1.56 \times 10^5 \quad [\text{turbulent}]$$

The assumption of full turbulence made in step 5 is justified.

The answer is (D).

(f) *Alternative closed-form (parallel pipe) solution*

Use the Darcy equation. $\epsilon = 0.0008$ ft. The relative roughness is

$$\frac{\epsilon}{D} = \frac{0.0008 \ \text{ft}}{\frac{10.1 \ \text{in}}{12 \ \frac{\text{in}}{\text{ft}}}} = 0.00095 \quad [\text{use } 0.001]$$

$\text{v}_{\text{max}} = 5$ ft/sec, and the temperature is 50°F. The Reynolds number is

$$\text{Re} = \frac{\text{v}D}{\nu} = \frac{\left(5 \ \frac{\text{ft}}{\text{sec}}\right)\left(\frac{10.1 \ \text{in}}{12 \ \frac{\text{in}}{\text{ft}}}\right)}{1.410 \times 10^{-5} \ \frac{\text{ft}^2}{\text{sec}}} = 2.98 \times 10^5$$

From the Moody diagram, $f \approx 0.0205$.

$$h_{f,AB} = \frac{fL\text{v}^2}{2Dg} = \frac{fL\dot{V}^2}{2DA^2 g}$$

$$= \frac{(0.0205)(2000 \ \text{ft}) \, \dot{V}_{AB}^2}{(2)\left(\frac{10.1 \ \text{in}}{12 \ \frac{\text{in}}{\text{ft}}}\right)(0.556 \ \text{ft}^2)^2\left(32.2 \ \frac{\text{ft}}{\text{sec}^2}\right)}$$

$$= 2.447 \, \dot{V}_{AB}^2$$

$$h_{f,ACDB} = \frac{fLv^2}{2Dg} = \frac{fL\dot{V}^2}{2DA^2g}$$

$$= \frac{(0.0205)(1000 \text{ ft})\dot{V}^2_{ACDB}}{(2)\left(\dfrac{12.12 \text{ in}}{12 \frac{\text{in}}{\text{ft}}}\right)(0.801 \text{ ft}^2)^2\left(32.2 \frac{\text{ft}}{\text{sec}^2}\right)}$$

$$+ \frac{(0.0205)(1500 \text{ ft})\dot{V}^2_{ACDB}}{(2)\left(\dfrac{8.13 \text{ in}}{12 \frac{\text{in}}{\text{ft}}}\right)(0.360 \text{ ft}^2)^2\left(32.2 \frac{\text{ft}}{\text{sec}^2}\right)}$$

$$+ \frac{(0.0205)(1000 \text{ ft})\dot{V}^2_{ACDB}}{(2)\left(\dfrac{10.1 \text{ in}}{12 \frac{\text{in}}{\text{ft}}}\right)(0.556 \text{ ft}^2)^2\left(32.2 \frac{\text{ft}}{\text{sec}^2}\right)}$$

$$= 0.4912\,\dot{V}^2_{ACDB} + 5.438\,\dot{V}^2_{ACDB} + 1.223\,\dot{V}^2_{ACDB}$$

$$= 7.152\,\dot{V}^2_{ACDB}$$

$$h_{f,AB} = h_{f,ACDB}$$

$$2.447\,\dot{V}^2_{AB} = 7.152\,\dot{V}^2_{ACDB}$$

$$\dot{V}_{AB} = \sqrt{\frac{7.152}{2.447}}\,\dot{V}_{ACDB} = 1.71\,\dot{V}_{ACDB} \quad \text{[Eq. I]}$$

The total flow rate is

$$\frac{1.5 \text{ MGD}}{\left(24 \frac{\text{hr}}{\text{day}}\right)\left(60 \frac{\text{min}}{\text{hr}}\right)} = 1041.7 \text{ gal/min}$$

$$\dot{V}_{AB} + \dot{V}_{ACDB} = 1041.7 \text{ gal/min} \quad \text{[Eq. II]}$$

Solving Eqs. I and II simultaneously,

$$\dot{V}_{AB} = \boxed{657.3 \text{ gal/min} \quad (660 \text{ gal/min})}$$

$$\dot{V}_{ACDB} = 384.4 \text{ gal/min}$$

This answer is insensitive to the $v_{max} = 5$ ft/sec assumption. A second iteration using actual velocities from these flow rates does not change the answer.

The same technique can be used with the Hazen-Williams equation and an assumed value of C. If $C = 100$ is used, then

$$\dot{V}_{AB} = 725.4 \text{ gal/min}$$

$$\dot{V}_{ACDB} = 316.3 \text{ gal/min}$$

The answer is (D).

(g)

$$h_{f,AB} = \frac{fLv^2}{2Dg} = \frac{(0.021)(2000 \text{ ft})\left(2.62 \frac{\text{ft}}{\text{sec}}\right)^2}{(2)\left(\dfrac{10.1 \text{ in}}{12 \frac{\text{in}}{\text{ft}}}\right)\left(32.2 \frac{\text{ft}}{\text{sec}^2}\right)}$$

$$= 5.32 \text{ ft}$$

For leg BD,

$$v = \frac{\dot{V}}{A} = \frac{\left(385 \frac{\text{gal}}{\text{min}}\right)\left(0.002228 \frac{\text{ft}^3\text{-min}}{\text{sec-gal}}\right)}{0.5564 \text{ ft}^2}$$

$$= 1.54 \text{ ft/sec}$$

$$h_{f,DB} = \frac{fLv^2}{2Dg}$$

$$= \frac{(0.021)(1000 \text{ ft})\left(1.54 \frac{\text{ft}}{\text{sec}}\right)^2}{(2)\left(\dfrac{10.1 \text{ in}}{12 \frac{\text{in}}{\text{ft}}}\right)\left(32.2 \frac{\text{ft}}{\text{sec}^2}\right)}$$

$$= 0.9 \text{ ft}$$

At 50°F, $\gamma = 62.4$ lbf/ft^3. From the Bernoulli equation (omitting the velocity term),

$$\frac{p_B}{\gamma} + z_B + h_{f,DB} = \frac{p_D}{\gamma} + z_D$$

$$p_B = \left(\frac{62.4 \frac{\text{lbf}}{\text{ft}^3}}{\left(12 \frac{\text{in}}{\text{ft}}\right)^2}\right)$$

$$\times \left(\frac{\left(40 \frac{\text{lbf}}{\text{in}^2}\right)\left(12 \frac{\text{in}}{\text{ft}}\right)^2}{62.4 \frac{\text{lbf}}{\text{ft}^3}}\right.$$

$$\left. + 600 \text{ ft} - 0.9 \text{ ft} - 580 \text{ ft}\right)$$

$$= \boxed{48.3 \text{ lbf/in}^2 \quad (48 \text{ psig})}$$

The answer is (B).

(h) The pressure at point A is

$$\frac{p_A}{\gamma} + z_A = \frac{p_B}{\gamma} + z_B + h_{f,AB}$$

$$p_A = \left(\frac{62.4 \ \frac{\text{lbf}}{\text{ft}^3}}{\left(12 \ \frac{\text{in}}{\text{ft}}\right)^2}\right)$$

$$\times \left(\frac{\left(48.3 \ \frac{\text{lbf}}{\text{in}^2}\right)\left(12 \ \frac{\text{in}}{\text{ft}}\right)^2}{62.4 \ \frac{\text{lbf}}{\text{ft}^3}} + 580 \ \text{ft} + 5.32 \ \text{ft} - 570 \ \text{ft}\right)$$

$$= \boxed{54.9 \ \text{lbf/in}^2 \quad (55 \ \text{psig})}$$

The answer is (B).

20. *Customary U.S. Solution*

Use projectile equations.

The maximum range of the discharge is given by

$$R = v_o^2\left(\frac{\sin 2\phi}{g}\right) = \left(50 \ \frac{\text{ft}}{\text{sec}}\right)^2\left(\frac{\sin(2)(45°)}{32.2 \ \frac{\text{ft}}{\text{sec}^2}}\right)$$

$$= \boxed{77.64 \ \text{ft} \quad (78 \ \text{ft})}$$

The answer is (B).

SI Solution

Use projectile equations.

The maximum range of the discharge is given by

$$R = v_o^2\left(\frac{\sin 2\phi}{g}\right) = \left(15 \ \frac{\text{m}}{\text{s}}\right)^2\left(\frac{\sin(2)(45°)}{9.81 \ \frac{\text{m}}{\text{s}^2}}\right)$$

$$= \boxed{22.94 \ \text{m} \quad (23 \ \text{m})}$$

The answer is (B).

21.

$$A_o = \left(\frac{\pi}{4}\right)\left(\frac{4 \ \text{in}}{12 \ \frac{\text{in}}{\text{ft}}}\right)^2 = 0.08727 \ \text{ft}^2$$

$$A_t = \left(\frac{\pi}{4}\right)(20 \ \text{ft})^2 = 314.16 \ \text{ft}^2$$

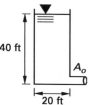

The time it takes to drop from 40 ft to 20 ft is given by Eq. 17.83.

$$t = \frac{2A_t(\sqrt{z_1} - \sqrt{z_2})}{C_d A_o \sqrt{2g}}$$

$$= \frac{(2)(314.16 \ \text{ft}^2)(\sqrt{40 \ \text{ft}} - \sqrt{20 \ \text{ft}})}{(0.98)(0.08727 \ \text{ft}^2)\sqrt{(2)\left(32.2 \ \frac{\text{ft}}{\text{sec}^2}\right)}}$$

$$= \boxed{1696 \ \text{sec} \quad (1700 \ \text{sec})}$$

The answer is (D).

22. For low velocities, the total energy head of the water in the canal is 320 ft. Similarly, the total energy head of the water in the irrigation ditch is 305 ft. The head loss is 320 ft − 305 ft = $\boxed{15 \ \text{ft.}}$

The answer is (C).

23.
$$C_d = 1.00 \quad \text{[given]}$$

$$F_{va} = \frac{1}{\sqrt{1 - \left(\frac{D_2}{D_1}\right)^4}} = \frac{1}{\sqrt{1 - \left(\frac{8 \ \text{in}}{12 \ \text{in}}\right)^4}}$$

$$= 1.116$$

$$A_2 = \left(\frac{\pi}{4}\right)\left(\frac{8 \ \text{in}}{12 \ \frac{\text{in}}{\text{ft}}}\right)^2 = 0.3491 \ \text{ft}^2$$

The specific weight of mercury is 0.491 lbf/in³; the specific weight of water is 0.0361 lbf/in³.

$$p_1 - p_2 = \Delta(\gamma h)$$

$$= \left(\left(0.491 \ \frac{\text{lbf}}{\text{in}^3}\right)(4 \ \text{in}) - \left(0.0361 \ \frac{\text{lbf}}{\text{in}^3}\right)(4 \ \text{in})\right)\left(12 \ \frac{\text{in}}{\text{ft}}\right)^2$$

$$= 262.0 \ \text{lbf/ft}^2$$

From Eq. 17.154,

$$\dot{V} = F_{va}C_dA_2\sqrt{\frac{2g(p_1 - p_2)}{\gamma}}$$

$$= (1.116)(1)(0.3491 \text{ ft}^2)$$

$$\times \sqrt{\frac{(2)\left(32.2 \dfrac{\text{ft}}{\text{sec}^2}\right)\left(262 \dfrac{\text{lbf}}{\text{ft}^2}\right)}{62.4 \dfrac{\text{lbf}}{\text{ft}^3}}}$$

$$= \boxed{6.406 \text{ ft}^3/\text{sec} \quad (6.4 \text{ ft}^3/\text{sec})}$$

The answer is (C).

24. *Customary U.S. Solution*

The volumetric flow rate of benzene through the venturi meter is given by

$$\dot{V} = C_fA_2\sqrt{\frac{2g(\rho_m - \rho)h}{\rho}}$$

The density of mercury, ρ_m, at 60°F is approximately 848 lbm/ft³.

The density of the benzene at 60°F is

$$\rho = (SG)\rho_{\text{water}} = (0.885)\left(62.4 \frac{\text{lbm}}{\text{ft}^3}\right) = 55.22 \text{ lbm/ft}^3$$

The throat area is

$$A_2 = \frac{\pi D_2^2}{4} = \frac{\pi\left(\dfrac{3.5 \text{ in}}{12 \dfrac{\text{in}}{\text{ft}}}\right)^2}{4} = 0.0668 \text{ ft}^2$$

$$\beta = \frac{3.5 \text{ in}}{8 \text{ in}} = 0.4375$$

$$C_f = \frac{C_d}{\sqrt{1 - \beta^4}} = \frac{0.99}{\sqrt{1 - (0.4375)^4}} = 1.00865$$

Find the volumetric flow of benzene.

$$\dot{V} = C_fA_2\sqrt{\frac{2g(\rho_m - \rho)h}{\rho}}$$

$$= (1.00865)(0.0668 \text{ ft}^2)$$

$$\times \sqrt{\frac{(2)\left(32.2 \dfrac{\text{ft}}{\text{sec}^2}\right)\left(848 \dfrac{\text{lbm}}{\text{ft}^3} - 55.22 \dfrac{\text{lbm}}{\text{ft}^3}\right)}{\left(55.22 \dfrac{\text{lbm}}{\text{ft}^3}\right)\left(12 \dfrac{\text{in}}{\text{ft}}\right)} \times (4 \text{ in})}$$

$$= \boxed{1.181 \text{ ft}^3/\text{sec} \quad (1.2 \text{ ft}^3/\text{sec})}$$

The answer is (A).

SI Solution

The volumetric flow rate of benzene through the venturi meter is given by

$$\dot{V} = C_fA_2\sqrt{\frac{2g(\rho_m - \rho)h}{\rho}}$$

ρ_m is the density of mercury at 15°C; ρ_m is approximately 13 600 kg/m³.

The density of the benzene at 15°C is

$$\rho = (SG)\rho_{\text{water}} = (0.885)\left(1000 \frac{\text{kg}}{\text{m}^3}\right)$$

$$= 885 \text{ kg/m}^3$$

The throat area is

$$A_2 = \frac{\pi D_2^2}{4} = \frac{\pi(0.09 \text{ m})^2}{4} = 0.0064 \text{ m}^2$$

$$\beta = \frac{9 \text{ cm}}{20 \text{ cm}} = 0.45$$

$$C_f = \frac{C_d}{\sqrt{1 - \beta^4}} = \frac{0.99}{\sqrt{1 - (0.45)^4}} = 1.01094$$

Find the volumetric flow of benzene.

$$\dot{V} = C_fA_2\sqrt{\frac{2g(\rho_m - \rho)h}{\rho}}$$

$$= (1.01094)(0.0064 \text{ m}^2)$$

$$\times \sqrt{\frac{(2)\left(9.81 \dfrac{\text{m}}{\text{s}^2}\right) \times \left(13\,600 \dfrac{\text{kg}}{\text{m}^3} - 885 \dfrac{\text{kg}}{\text{m}^3}\right)(0.1 \text{ m})}{885 \dfrac{\text{kg}}{\text{m}^3}}}$$

$$= \boxed{0.0344 \text{ m}^3/\text{s} \quad (34 \text{ L/s})}$$

The answer is (A).

25. From App. 14.A, for 70°F water,

$$\nu = 1.059 \times 10^{-5} \text{ ft}^2/\text{sec}$$

$$\gamma = 62.3 \text{ lbf/ft}^3$$

$$D_o = 0.2 \text{ ft}$$

$$v_o = v\left(\frac{D}{D_o}\right)^2 = \left(2 \frac{\text{ft}}{\text{sec}}\right)\left(\frac{1 \text{ ft}}{0.2 \text{ ft}}\right)^2$$

$$= 50 \text{ ft/sec}$$

$$\mathrm{Re} = \frac{D_o \mathrm{v}_o}{\nu} = \frac{(0.2\ \mathrm{ft})\left(50\ \dfrac{\mathrm{ft}}{\mathrm{sec}}\right)}{1.059 \times 10^{-5}\ \dfrac{\mathrm{ft}^2}{\mathrm{sec}}} = 9.44 \times 10^5$$

$$A_o = \left(\frac{\pi}{4}\right)(0.2\ \mathrm{ft})^2 = 0.0314\ \mathrm{ft}^2$$

$$A = \left(\frac{\pi}{4}\right)(1\ \mathrm{ft})^2 = 0.7854\ \mathrm{ft}^2$$

$$\frac{A_o}{A} = \frac{0.0314\ \mathrm{ft}^2}{0.7854\ \mathrm{ft}^2} = 0.040$$

From Fig. 17.28,

$$C_f \approx 0.60$$

$$\dot{V} = A\mathrm{v} = (0.7854\ \mathrm{ft}^2)\left(2\ \frac{\mathrm{ft}}{\mathrm{sec}}\right) = 1.571\ \mathrm{ft}^3/\mathrm{sec}$$

From Eq. 17.162(b), substituting $\gamma = \rho g / g_c$,

$$p_p - p_o = \left(\frac{\gamma}{2g}\right)\left(\frac{\dot{V}}{C_f A_o}\right)^2$$

$$= \frac{\left(\dfrac{62.3\ \dfrac{\mathrm{lbf}}{\mathrm{ft}^3}}{(2)\left(32.2\ \dfrac{\mathrm{ft}}{\mathrm{sec}^2}\right)}\right)\left(\dfrac{1.571\ \dfrac{\mathrm{ft}^3}{\mathrm{sec}}}{(0.60)(0.0314\ \mathrm{ft}^2)}\right)^2}{\left(12\ \dfrac{\mathrm{in}}{\mathrm{ft}}\right)^2}$$

$$= \boxed{46.7\ \mathrm{lbf/in}^2 \quad (47\ \mathrm{psi})}$$

The answer is (D).

26. *Customary U.S. Solution*

The densities of mercury and water are

$$\rho_{\mathrm{mercury}} = 848\ \mathrm{lbm/ft}^3$$

$$\rho_{\mathrm{water}} = 62.4\ \mathrm{lbm/ft}^3$$

The manometer tube is filled with water above the mercury column. The pressure differential across the orifice meter is

$$\Delta p = p_1 - p_2 = (\rho_{\mathrm{mercury}} - \rho_{\mathrm{water}})h \times \frac{g}{g_c}$$

$$= \frac{\left(848\ \dfrac{\mathrm{lbm}}{\mathrm{ft}^3} - 62.4\ \dfrac{\mathrm{lbm}}{\mathrm{ft}^3}\right)(7\ \mathrm{in})}{12\ \dfrac{\mathrm{in}}{\mathrm{ft}}} \times \frac{32.2\ \dfrac{\mathrm{ft}}{\mathrm{sec}^2}}{32.2\ \dfrac{\mathrm{lbm\text{-}ft}}{\mathrm{lbf\text{-}sec}^2}}$$

$$= \boxed{458.3\ \mathrm{lbf/ft}^2 \quad (3.2\ \mathrm{psi})}$$

The answer is (B).

SI Solution

The densities of mercury and water are

$$\rho_{\mathrm{mercury}} = 13\,600\ \mathrm{kg/m}^3$$

$$\rho_{\mathrm{water}} = 1000\ \mathrm{kg/m}^3$$

The manometer tube is filled with water above the mercury column. The pressure differential across the orifice meter is given by

$$\Delta p = p_1 - p_2 = (\rho_{\mathrm{mercury}} - \rho_{\mathrm{water}})hg$$

$$= \left(13\,600\ \frac{\mathrm{kg}}{\mathrm{m}^3} - 1000\ \frac{\mathrm{kg}}{\mathrm{m}^3}\right)(0.178\ \mathrm{m})\left(9.81\ \frac{\mathrm{m}}{\mathrm{s}^2}\right)$$

$$= \boxed{22\,002\ \mathrm{Pa} \quad (22\ \mathrm{kPa})}$$

The answer is (B).

27. *Customary U.S. Solution*

Use App. 16.B. For 12 in pipe,

$$D = 0.99483\ \mathrm{ft}$$

$$A = 0.7773\ \mathrm{ft}^2$$

The velocity is

$$\mathrm{v} = \frac{\dot{V}}{A} = \frac{10\ \dfrac{\mathrm{ft}^3}{\mathrm{sec}}}{0.7773\ \mathrm{ft}^2} = 12.87\ \mathrm{ft/sec}$$

For water at 70°F, $\nu = 1.059 \times 10^{-5}\ \mathrm{ft}^2/\mathrm{sec}$.

The Reynolds number in the pipe is

$$\mathrm{Re} = \frac{\mathrm{v}D}{\nu} = \frac{\left(12.87\ \dfrac{\mathrm{ft}}{\mathrm{sec}}\right)(0.99483\ \mathrm{ft})}{1.059 \times 10^{-5}\ \dfrac{\mathrm{ft}^2}{\mathrm{sec}}}$$

$$= 1.21 \times 10^6 \quad [\text{fully turbulent}]$$

Flow through the orifice will have a higher Reynolds number and will also be turbulent.

The volumetric flow rate through a sharp-edged orifice is

$$\dot{V} = C_f A_o \sqrt{\frac{2g(\rho_m - \rho)h}{\rho}} = C_f A_o \sqrt{\frac{2g_c(p_1 - p_2)}{\rho}}$$

Rearranging,

$$C_f A_o = \frac{\dot{V}}{\sqrt{\dfrac{2g_c(p_1 - p_2)}{\rho}}}$$

The maximum head loss must not exceed 25 ft.

$$\frac{\dfrac{g_c}{g} \times (p_1 - p_2)}{\rho} = 25 \text{ ft}$$

$$\frac{g_c(p_1 - p_2)}{\rho} = (25 \text{ ft})g$$

Substituting,

$$C_f A_o = \frac{10 \ \dfrac{\text{ft}^3}{\text{sec}}}{\sqrt{(2)\left(32.2 \ \dfrac{\text{ft}}{\text{sec}^2}\right)(25 \text{ ft})}} = 0.249 \text{ ft}^2$$

Both C_f and A_o depend on the orifice diameter. For a 7 in diameter orifice,

$$A_o = \frac{\pi D_o^2}{4} = \frac{\pi \left(\dfrac{7 \text{ in}}{12 \ \frac{\text{in}}{\text{ft}}}\right)^2}{4} = 0.267 \text{ ft}^2$$

$$\frac{A_o}{A_1} = \frac{0.267 \text{ ft}^2}{0.7773 \text{ ft}^2} = 0.343$$

From Fig. 17.28, for $A_o/A_1 = 0.343$ and fully turbulent flow,

$$C_f = 0.645$$

$$C_f A_o = (0.645)(0.267 \text{ ft}^2) = 0.172 \text{ ft}^2 < 0.249 \text{ ft}^2$$

Therefore, a 7 in diameter orifice is too small.

Try a 9 in diameter orifice.

$$A_o = \frac{\pi D_o^2}{4} = \frac{\pi \left(\dfrac{9 \text{ in}}{12 \ \frac{\text{in}}{\text{ft}}}\right)^2}{4} = 0.442 \text{ ft}^2$$

$$\frac{A_o}{A_1} = \frac{0.442 \text{ ft}^2}{0.7773 \text{ ft}^2} = 0.569$$

From Fig. 17.28, for $A_o/A_1 = 0.569$ and fully turbulent flow,

$$C_f = 0.73$$

$$C_f A_o = (0.73)(0.442 \text{ ft}^2) = 0.323 \text{ ft}^2 > 0.249 \text{ ft}^2$$

Therefore, a 9 in orifice is too large.

Interpolating gives

$$D_o = 7 \text{ in} + \frac{(9 \text{ in} - 7 \text{ in})(0.249 \text{ ft}^2 - 0.172 \text{ ft}^2)}{0.323 \text{ ft}^2 - 0.172 \text{ ft}^2}$$

$$= 8.0 \text{ in}$$

Further iterations yield

$$D_o \approx \boxed{8.1 \text{ in}}$$

$$C_f A_o = 0.243 \text{ ft}^2$$

The answer is (C).

SI Solution

For 300 mm inside diameter pipe, $D = 0.3$ m.

The velocity is

$$v = \frac{\dot{V}}{A} = \frac{\dot{V}}{\dfrac{\pi D^2}{4}} = \frac{250 \ \dfrac{\text{L}}{\text{s}}}{\left(\dfrac{\pi (0.3 \text{ m})^2}{4}\right)\left(1000 \ \dfrac{\text{L}}{\text{m}^3}\right)}$$

$$= 3.54 \text{ m/s}$$

From App. 14.B, for water at 20°C, $\nu = 1.007 \times 10^{-6} \text{ m}^2/\text{s}$.

The Reynolds number in the pipe is

$$\text{Re} = \frac{vD}{\nu} = \frac{\left(3.54 \ \dfrac{\text{m}}{\text{s}}\right)(0.3 \text{ m})}{1.007 \times 10^{-6} \ \dfrac{\text{m}^2}{\text{s}}}$$

$$= 1.05 \times 10^6 \quad [\text{fully turbulent}]$$

Flow through the orifice will have a higher Reynolds number and also be turbulent.

The volumetric flow rate through a sharp-edged orifice is

$$\dot{V} = C_f A_o \sqrt{\frac{2g(\rho_m - \rho)h}{\rho}} = C_f A_o \sqrt{\frac{2(p_1 - p_2)}{\rho}}$$

Rearranging,

$$C_f A_o = \frac{\dot{V}}{\sqrt{\dfrac{2(p_1 - p_2)}{\rho}}}$$

The maximum head loss must not exceed 7.5 m.

$$\frac{p_1 - p_2}{g\rho} = 7.5 \text{ m}$$

$$\frac{p_1 - p_2}{\rho} = (7.5 \text{ m})g$$

Substituting,

$$C_f A_o = \frac{0.25 \ \dfrac{\text{m}^3}{\text{s}}}{\sqrt{(2)\left(9.81 \ \dfrac{\text{m}}{\text{s}^2}\right)(7.5 \text{ m})}} = 0.021 \text{ m}^2$$

Both C_f and A_o depend on the orifice diameter. For an 18 cm diameter orifice,

$$A_o = \frac{\pi D_o^2}{4} = \frac{\pi(0.18 \text{ m})^2}{4} = 0.0254 \text{ m}^2$$

$$\frac{A_o}{A_1} = \frac{0.0254 \text{ m}^2}{0.0707 \text{ m}^2} = 0.359$$

From Fig. 17.28, for $A_o/A_1 = 0.359$ and fully turbulent flow,

$$C_f = 0.65$$

$$C_f A_o = (0.65)(0.0254 \text{ m}^2) = 0.0165 \text{ m}^2 < 0.021 \text{ m}^2$$

Therefore, an 18 cm diameter orifice is too small.

Try a 23 cm diameter orifice.

$$A_o = \frac{\pi D_o^2}{4} = \frac{\pi(0.23 \text{ m})^2}{4} = 0.0415 \text{ m}^2$$

$$\frac{A_o}{A_1} = \frac{0.0415 \text{ m}^2}{0.0707 \text{ m}^2} = 0.587$$

From Fig. 17.28, for $A_o/A_1 = 0.587$ and fully turbulent flow,

$$C_f = 0.73$$

$$C_f A_o = (0.73)(0.0415 \text{ m}^2) = 0.0303 \text{ m}^2 > 0.021 \text{ m}^2$$

Therefore, a 23 cm orifice is too large.

Interpolating gives

$$D_o = 18 \text{ cm}$$
$$+ (23 \text{ cm} - 18 \text{ cm})\left(\frac{0.021 \text{ m}^2 - 0.0165 \text{ m}^2}{0.0303 \text{ m}^2 - 0.0165 \text{ m}^2}\right)$$
$$= 19.6 \text{ cm}$$

Further iteration yields

$$D_o = \boxed{20 \text{ cm}}$$
$$C_f = 0.675$$
$$C_f A_o = 0.021 \text{ m}^2$$

The answer is (C).

28.

$$A_A = \left(\frac{\pi}{4}\right)\left(\frac{24 \text{ in}}{12 \frac{\text{in}}{\text{ft}}}\right)^2 = 3.142 \text{ ft}^2$$

$$A_B = \left(\frac{\pi}{4}\right)\left(\frac{12 \text{ in}}{12 \frac{\text{in}}{\text{ft}}}\right)^2 = 0.7854 \text{ ft}^2$$

$$v_A = \frac{\dot{V}}{A} = \frac{8 \frac{\text{ft}^3}{\text{sec}}}{3.142 \text{ ft}^2} = 2.546 \text{ ft/sec}$$

$$p_A = \gamma h = \left(62.4 \frac{\text{lbf}}{\text{ft}^3}\right)(20 \text{ ft}) = 1248 \text{ lbf/ft}^2$$

$$v_B = \frac{\dot{V}}{A} = \frac{8 \frac{\text{ft}^3}{\text{sec}}}{0.7854 \text{ ft}^2} = 10.19 \text{ ft/sec}$$

Using the Bernoulli equation to solve for p_B,

$$p_B = p_A + \left(\frac{v_A^2}{2g} - \frac{v_B^2}{2g}\right)\gamma$$

$$= 1248 \frac{\text{lbf}}{\text{ft}^2} + \left(\frac{\left(2.546 \frac{\text{ft}}{\text{sec}}\right)^2 - \left(10.19 \frac{\text{ft}}{\text{sec}}\right)^2}{(2)\left(32.2 \frac{\text{ft}}{\text{sec}^2}\right)}\right)$$

$$\times \left(62.4 \frac{\text{lbf}}{\text{ft}^3}\right)$$

$$= 1153.7 \text{ lbf/ft}^2$$

With $\theta = 0°$, from Eq. 17.202(b),

$$F_x = p_B A_B - p_A A_A + \frac{\dot{m}(v_B - v_A)}{g_c}$$

$$= \left(1153.7 \frac{\text{lbf}}{\text{ft}^2}\right)(0.7854 \text{ ft}^2) - \left(1248 \frac{\text{lbf}}{\text{ft}^2}\right)(3.142 \text{ ft}^2)$$

$$+ \frac{\left(\left(8 \frac{\text{ft}^3}{\text{sec}}\right)\left(62.4 \frac{\text{lbm}}{\text{ft}^3}\right)\right)\left(10.19 \frac{\text{ft}}{\text{sec}} - 2.546 \frac{\text{ft}}{\text{sec}}\right)}{32.2 \frac{\text{lbm-ft}}{\text{lbf-sec}^2}}$$

$$= \boxed{-2897 \text{ lbf (2900 lbf) on the fluid (toward A)}}$$

$$F_y = 0$$

The answer is (A).

29. *Customary U.S. Solution*

The mass flow rate of the water is

$$\dot{m} = \rho \dot{V} = \rho v A = \frac{\rho v \pi D^2}{4}$$

$$= \frac{\left(62.4 \ \frac{\text{lbm}}{\text{ft}^3}\right)\left(40 \ \frac{\text{ft}}{\text{sec}}\right)\pi \left(\frac{2 \ \text{in}}{12 \ \frac{\text{in}}{\text{ft}}}\right)^2}{4}$$

$$= 54.45 \ \text{lbm/sec}$$

The effective mass flow rate of the water is

$$\dot{m}_{\text{eff}} = \left(\frac{v - v_b}{v}\right)\dot{m}$$

$$= \left(\frac{40 \ \frac{\text{ft}}{\text{sec}} - 15 \ \frac{\text{ft}}{\text{sec}}}{40 \ \frac{\text{ft}}{\text{sec}}}\right)\left(54.45 \ \frac{\text{lbm}}{\text{sec}}\right)$$

$$= 34.0 \ \text{lbm/sec}$$

The force in the (horizontal) x-direction is

$$F_x = \frac{\dot{m}_{\text{eff}}(v - v_b)(\cos\theta - 1)}{g_c}$$

$$= \frac{\left(34.0 \ \frac{\text{lbm}}{\text{sec}}\right)\left(40 \ \frac{\text{ft}}{\text{sec}} - 15 \ \frac{\text{ft}}{\text{sec}}\right)(\cos 60° - 1)}{32.2 \ \frac{\text{lbm-ft}}{\text{lbf-sec}^2}}$$

$$= -13.2 \ \text{lbf} \quad [\text{acting to the left}]$$

The force in the (vertical) y-direction is

$$F_y = \frac{\dot{m}_{\text{eff}}(v - v_b)\sin\theta}{g_c}$$

$$= \frac{\left(34.0 \ \frac{\text{lbm}}{\text{sec}}\right)\left(40 \ \frac{\text{ft}}{\text{sec}} - 15 \ \frac{\text{ft}}{\text{sec}}\right)(\sin 60°)}{32.2 \ \frac{\text{lbm-ft}}{\text{lbf-sec}^2}}$$

$$= 22.9 \ \text{lbf} \quad [\text{acting upward}]$$

The net resultant force is

$$F = \sqrt{F_x^2 + F_y^2} = \sqrt{(-13.2 \ \text{lbf})^2 + (22.9 \ \text{lbf})^2}$$

$$= \boxed{26.4 \ \text{lbf} \quad (26 \ \text{lbf})}$$

The answer is (B).

SI Solution

The mass flow rate of the water is

$$\dot{m} = \rho \dot{V} = \rho v A = \frac{\rho v \pi D^2}{4}$$

$$= \frac{\left(1000 \ \frac{\text{kg}}{\text{m}^3}\right)\left(12 \ \frac{\text{m}}{\text{s}}\right)\pi(0.05 \ \text{m})^2}{4}$$

$$= 23.56 \ \text{kg/s}$$

The effective mass flow rate of the water is

$$\dot{m}_{\text{eff}} = \left(\frac{v - v_b}{v}\right)\dot{m}$$

$$= \left(\frac{12 \ \frac{\text{m}}{\text{s}} - 4.5 \ \frac{\text{m}}{\text{s}}}{12 \ \frac{\text{m}}{\text{s}}}\right)\left(23.56 \ \frac{\text{kg}}{\text{s}}\right)$$

$$= 14.73 \ \text{kg/s}$$

The force in the (horizontal) x-direction is

$$F_x = \dot{m}_{\text{eff}}(v - v_b)(\cos\theta - 1)$$

$$= \left(14.73 \ \frac{\text{kg}}{\text{s}}\right)\left(12 \ \frac{\text{m}}{\text{s}} - 4.5 \ \frac{\text{m}}{\text{s}}\right)(\cos 60° - 1)$$

$$= -55.2 \ \text{N} \quad [\text{acting to the left}]$$

The force in the (vertical) y-direction is

$$F_y = \dot{m}_{\text{eff}}(v - v_b)\sin\theta$$

$$= \left(14.73 \ \frac{\text{kg}}{\text{s}}\right)\left(12 \ \frac{\text{m}}{\text{s}} - 4.5 \ \frac{\text{m}}{\text{s}}\right)(\sin 60°)$$

$$= 95.7 \ \text{N} \quad [\text{acting upward}]$$

The net resultant force is

$$F = \sqrt{F_x^2 + F_y^2} = \sqrt{(-55.2 \ \text{N})^2 + (95.7 \ \text{N})^2}$$

$$= \boxed{110.5 \ \text{N} \quad (110 \ \text{N})}$$

The answer is (B).

30. *Customary U.S. Solution*

For schedule-40 pipe,

$$D_i = 0.9948 \ \text{ft}$$

$$A_i = 0.7773 \ \text{ft}^2$$

$$v = \frac{\dot{V}}{A_i} = \frac{2000 \ \frac{\text{gal}}{\text{min}}}{(0.7773 \ \text{ft}^2)\left(7.4805 \ \frac{\text{gal}}{\text{ft}^3}\right)\left(60 \ \frac{\text{sec}}{\text{min}}\right)}$$

$$= 5.73 \ \text{ft/sec}$$

The pressures are in terms of gage pressure, and the density of mercury is 0.491 lbm/in³.

$$p_i = \left(14.7 \ \frac{\text{lbf}}{\text{in}^2} - \frac{(6 \ \text{in})\left(0.491 \ \frac{\text{lbm}}{\text{in}^3}\right)\left(32.2 \ \frac{\text{ft}}{\text{sec}^2}\right)}{32.2 \ \frac{\text{lbm-ft}}{\text{lbf-sec}^2}}\right)$$

$$\times \left(12 \ \frac{\text{in}}{\text{ft}}\right)^2$$

$$= 1692.6 \ \text{lbf/ft}^2$$

$$E_{ti} = \frac{p_i}{\rho} + \frac{v_i^2}{2g_c} + \frac{z_i g}{g_c}$$

Since the pump inlet and outlet are at the same elevation, use $\Delta z = 0$. $\rho = (\text{SG})\rho_{\text{water}}$.

$$E_{ti} = \frac{p_i}{(\text{SG})\rho_{\text{water}}} + \frac{v_i^2}{2g_c} + 0$$

$$= \frac{1692.6 \ \frac{\text{lbf}}{\text{ft}^2}}{(1.2)\left(62.4 \ \frac{\text{lbm}}{\text{ft}^3}\right)} + \frac{\left(5.73 \ \frac{\text{ft}}{\text{sec}}\right)^2}{(2)\left(32.2 \ \frac{\text{lbm-ft}}{\text{lbf-sec}^2}\right)}$$

$$= 23.11 \ \text{ft-lbf/lbm}$$

Calculate the total head at the inlet.

$$h_{ti} = E_{ti} \times \frac{g_c}{g} = 23.11 \ \frac{\text{ft-lbf}}{\text{lbm}} \times \frac{32.2 \ \frac{\text{lbm-ft}}{\text{lbf-sec}^2}}{32.2 \ \frac{\text{ft}}{\text{sec}^2}}$$

$$= 23.11 \ \text{ft}$$

At the outlet side of the pump,

$$D_o = 0.6651 \ \text{ft}$$

$$A_o = 0.3474 \ \text{ft}^2$$

$$v_o = \frac{Q}{A_o} = \frac{2000 \ \frac{\text{gal}}{\text{min}}}{(0.3474 \ \text{ft}^2)\left(7.4805 \ \frac{\text{gal}}{\text{ft}^3}\right)\left(60 \ \frac{\text{sec}}{\text{min}}\right)}$$

$$= 12.83 \ \text{ft/sec}$$

The pressures are in terms of gage pressure. The gauge is located 4 ft above the pump outlet, which adds 4 ft of pressure head at the pump outlet.

$$p_o = \left(14.7 \ \frac{\text{lbf}}{\text{in}^2} + 20 \ \frac{\text{lbf}}{\text{in}^2}\right)\left(12 \ \frac{\text{in}}{\text{ft}}\right)^2$$

$$+ 4 \ \text{ft}\left(\frac{(1.2)\left(62.4 \ \frac{\text{lbm}}{\text{ft}^3}\right)\left(32.2 \ \frac{\text{ft}}{\text{sec}^2}\right)}{32.2 \ \frac{\text{lbm-ft}}{\text{lbf-sec}^2}}\right)$$

$$= 5296 \ \text{lbf/ft}^2$$

$$E_{to} = \frac{p_o}{\rho} + \frac{v_o^2}{2g_c} + \frac{z_o g}{g_c}$$

Since the pump inlet and outlet are at the same elevation, $\Delta z = 0$. $\rho = (\text{SG})\rho_{\text{water}}$.

$$E_{to} = \frac{p_o}{(\text{SG})\rho_{\text{water}}} + \frac{v_o^2}{2g_c} + 0$$

$$= \frac{5296 \ \frac{\text{lbf}}{\text{ft}^2}}{(1.2)\left(62.4 \ \frac{\text{lbm}}{\text{ft}^3}\right)} + \frac{\left(12.83 \ \frac{\text{ft}}{\text{sec}}\right)^2}{(2)\left(32.2 \ \frac{\text{lbm-ft}}{\text{lbf-sec}^2}\right)}$$

$$= 73.28 \ \text{ft-lbf/lbm}$$

Calculate the total head at the outlet.

$$h_{to} = E_{to} \times \frac{g_c}{g} = 73.28 \ \frac{\text{ft-lbf}}{\text{lbm}} \times \frac{32.2 \ \frac{\text{lbm-ft}}{\text{lbf-sec}^2}}{32.2 \ \frac{\text{ft}}{\text{sec}^2}}$$

$$= 73.28 \ \text{ft}$$

Compute the total head across the pump.

$$\Delta h = h_{to} - h_{ti} = 73.28 \ \text{ft} - 23.11 \ \text{ft} = 50.17 \ \text{ft}$$

The mass flow rate is

$$\dot{m} = \rho \dot{V} = (\text{SG})\rho_{\text{water}} \ \dot{V}$$

$$= \frac{(1.2)\left(62.4 \ \frac{\text{lbm}}{\text{ft}^3}\right)\left(2000 \ \frac{\text{gal}}{\text{min}}\right)}{\left(7.4805 \ \frac{\text{gal}}{\text{ft}^3}\right)\left(60 \ \frac{\text{sec}}{\text{min}}\right)}$$

$$= 333.7 \ \text{lbm/sec}$$

The power input to the pump is

$$P = \frac{\Delta h \dot{m} \times \frac{g}{g_c}}{\eta}$$

$$= \frac{(50.17 \ \text{ft})\left(333.7 \ \frac{\text{lbm}}{\text{sec}}\right) \times \frac{32.2 \ \frac{\text{ft}}{\text{sec}^2}}{32.2 \ \frac{\text{lbm-ft}}{\text{lbf-sec}^2}}}{(0.85)\left(550 \ \frac{\text{ft-lbf}}{\text{hp-sec}}\right)}$$

$$= \boxed{35.8 \ \text{hp} \quad (36 \ \text{hp})}$$

(It is not necessary to use absolute pressures as has been done in this solution.)

The answer is (B).

Flow of Fluids

SI Solution

For schedule-40 pipe,

$$D_i = 303.2 \text{ mm}$$

$$A_i = 0.0722 \text{ m}^2$$

$$v = \frac{\dot{V}}{A_i} = \frac{0.125 \ \frac{\text{m}^3}{\text{s}}}{0.0722 \text{ m}^2} = 1.73 \text{ m/s}$$

The pressures are in terms of gage pressure, and the density of mercury is $13\,600 \text{ kg/m}^3$.

$$p_i = 1.013 \times 10^5 \text{ Pa} - (0.15 \text{ m})\left(13\,600 \ \frac{\text{kg}}{\text{m}^3}\right)\left(9.81 \ \frac{\text{m}}{\text{s}^2}\right)$$

$$= 8.13 \times 10^4 \text{ Pa}$$

$$E_{ti} = \frac{p}{\rho} + \frac{v_i^2}{2} + z_i g$$

Since the pump inlet and outlet are at the same elevation, $\Delta z = 0$. $\rho = (\text{SG})\rho_{\text{water}}$.

$$E_{ti} = \frac{p}{(\text{SG})\rho_{\text{water}}} + \frac{v_i^2}{2} + 0$$

$$= \frac{8.13 \times 10^4 \text{ Pa}}{(1.2)\left(1000 \ \frac{\text{kg}}{\text{m}^3}\right)} + \frac{\left(1.73 \ \frac{\text{m}}{\text{s}}\right)^2}{2}$$

$$= 69.2 \text{ J/kg}$$

The total head at the inlet is

$$h_{ti} = \frac{E_{ti}}{g} = \frac{69.2 \ \frac{\text{J}}{\text{kg}}}{9.81 \ \frac{\text{m}}{\text{s}^2}} = 7.05 \text{ m}$$

Assume the pipe nominal diameter is equal to the internal diameter. On the outlet side of the pump,

$$D_i = 202.7 \text{ mm}$$

$$A_o = 0.0323 \text{ m}^2$$

$$v_o = \frac{\dot{V}}{A_o} = \frac{0.125 \ \frac{\text{m}^3}{\text{s}}}{0.0323 \text{ m}^2} = 3.87 \text{ m/s}$$

The pressures are in terms of gage pressure. The gauge is located 1.2 m above the pump outlet, which adds 1.2 m of pressure head at the pump outlet.

$$p_o = 1.013 \times 10^5 \text{ Pa} + 138 \times 10^3 \text{ Pa}$$

$$+ (1.2 \text{ m})\left((1.2)\left(1000 \ \frac{\text{kg}}{\text{m}^3}\right)\left(9.81 \ \frac{\text{m}}{\text{s}^2}\right)\right)$$

$$= 2.53 \times 10^5 \text{ Pa}$$

$$E_{to} = \frac{p_o}{\rho} + \frac{v_o^2}{2} + z_o g$$

Since the pump inlet and outlet are at the same elevation, $\Delta z = 0$. $\rho = (\text{SG})\rho_{\text{water}}$.

$$E_{to} = \frac{p_o}{(\text{SG})\rho_{\text{water}}} + \frac{v_o^2}{2} + 0$$

$$= \frac{2.53 \times 10^5 \text{ Pa}}{(1.2)\left(1000 \ \frac{\text{kg}}{\text{m}^3}\right)} + \frac{\left(3.87 \ \frac{\text{m}}{\text{s}}\right)^2}{2}$$

$$= 218.3 \text{ J/kg}$$

The total head at the outlet is

$$h_{to} = \frac{E_{to}}{g} = \frac{218.3 \ \frac{\text{J}}{\text{kg}}}{9.81 \ \frac{\text{m}}{\text{s}^2}} = 22.25 \text{ m}$$

The total head across the pump is

$$\Delta h = h_{to} - h_{ti} = 22.25 \text{ m} - 7.05 \text{ m} = 15.2 \text{ m}$$

The mass flow rate is

$$\dot{m} = \rho \dot{V} = (\text{SG})\rho_{\text{water}} Q$$

$$= (1.2)\left(1000 \ \frac{\text{kg}}{\text{m}^3}\right)\left(0.125 \ \frac{\text{m}^3}{\text{s}}\right)$$

$$= 150 \text{ kg/s}$$

The power input to the pump is

$$P = \frac{\Delta h \dot{m} g}{\eta} = \frac{(15.2 \text{ m})\left(150 \ \frac{\text{kg}}{\text{s}}\right)\left(9.81 \ \frac{\text{m}}{\text{s}^2}\right)}{0.85}$$

$$= \boxed{26\,314 \text{ W} \quad (26 \text{ kW})}$$

(It is not necessary to use absolute pressures as has been done in this solution.)

The answer is (B).

31. *Customary U.S. Solution*

The mass flow rate is

$$\dot{m} = \dot{V}\rho = \left(100 \ \frac{\text{ft}^3}{\text{sec}}\right)\left(62.4 \ \frac{\text{lbm}}{\text{ft}^3}\right)$$
$$= 6240 \ \text{lbm/sec}$$

The head loss across the horizontal turbine is

$$h_{\text{loss}} = \frac{\Delta p}{\rho} \times \frac{g_c}{g}$$

$$= \frac{\left(30 \ \frac{\text{lbf}}{\text{in}^2} - \left(-5 \ \frac{\text{lbf}}{\text{in}^2}\right)\right)\left(12 \ \frac{\text{in}}{\text{ft}}\right)^2}{62.4 \ \frac{\text{lbm}}{\text{ft}^3}}$$

$$\times \frac{32.2 \ \frac{\text{lbm-ft}}{\text{lbf-sec}^2}}{32.2 \ \frac{\text{ft}}{\text{sec}^2}}$$

$$= 80.77 \ \text{ft}$$

From Table 18.5, the power developed by the turbine is

$$P = \dot{m}h_{\text{loss}} \times \frac{g}{g_c}$$

$$= \frac{\left(6240 \ \frac{\text{lbm}}{\text{sec}}\right)(80.77 \ \text{ft})}{550 \ \frac{\text{ft-lbf}}{\text{hp-sec}}} \times \frac{32.2 \ \frac{\text{ft}}{\text{sec}^2}}{32.2 \ \frac{\text{lbm-ft}}{\text{lbf-sec}^2}}$$

$$= \boxed{916 \ \text{hp} \quad (920 \ \text{hp})}$$

The answer is (D).

SI Solution

The mass flow rate is

$$\dot{m} = \dot{V}\rho = \left(2.6 \ \frac{\text{m}^3}{\text{s}}\right)\left(1000 \ \frac{\text{kg}}{\text{m}^3}\right) = 2600 \ \text{kg/s}$$

The head loss across the horizontal turbine is

$$h_{\text{loss}} = \frac{\Delta p}{\rho g} = \frac{(210 \ \text{kPa} - (-35 \ \text{kPa}))\left(1000 \ \frac{\text{Pa}}{\text{kPa}}\right)}{\left(1000 \ \frac{\text{kg}}{\text{m}^3}\right)\left(9.81 \ \frac{\text{m}}{\text{s}^2}\right)}$$

$$= 25.0 \ \text{m}$$

From Table 18.5, the power developed by the turbine is

$$P = \dot{m}h_{\text{loss}}g = \left(2600 \ \frac{\text{kg}}{\text{s}}\right)(25.0 \ \text{m})\left(9.81 \ \frac{\text{m}}{\text{s}^2}\right)$$

$$= \boxed{637\,650 \ \text{W} \quad (640 \ \text{kW})}$$

The answer is (D).

32. The speed of the truck relative to the ground is 65 mph. However, power is also required to overcome oncoming wind. The speed of the truck relative to the wind is 65 mph + 15 mph = $\boxed{80 \ \text{mph.}}$ If the truck was stationary, the wind would exert a force on the truck that would push the truck backward, relative to the ground. The frictional forces between the tires and ground, as well as within the parking brake systems, perform work while preventing this motion.

The answer is (D).

33. From Eq. 17.218(b), the drag force on the antenna is

$$F_D = \frac{C_D A \rho \text{v}^2}{2g_c}$$

$$= \frac{(1.2)(0.8 \ \text{ft}^2)\left(0.076 \ \frac{\text{lbm}}{\text{ft}^3}\right)}{(2)\left(32.2 \ \frac{\text{lbm-ft}}{\text{lbf-sec}^2}\right)}$$

$$\times \left(\frac{\left(60 \ \frac{\text{mi}}{\text{hr}}\right)\left(5280 \ \frac{\text{ft}}{\text{mi}}\right)}{3600 \ \frac{\text{sec}}{\text{hr}}}\right)^2$$

$$= \boxed{8.77 \ \text{lbf} \quad (9.0 \ \text{lbf})}$$

The answer is (A).

34. From App. 14.A, the density and kinematic viscosity of 60°F water are

$$\rho = 62.37 \ \text{lbm/ft}^3$$
$$\nu = 1.217 \times 10^{-5} \ \text{ft}^2/\text{sec}$$

The Reynolds number is

$$\text{Re} = \frac{D\text{v}}{\nu} = \frac{(12 \ \text{in})\left(10 \ \frac{\text{ft}}{\text{sec}}\right)}{\left(1.217 \times 10^{-5} \ \frac{\text{ft}^2}{\text{sec}}\right)\left(12 \ \frac{\text{in}}{\text{ft}}\right)}$$

$$= 8.22 \times 10^5$$

Flow of Fluids

The projected area of the submerged portion of the log is

$$A = L_{\text{submerged}} D = \frac{(2\ \text{ft})(12\ \text{in})}{12\ \frac{\text{in}}{\text{ft}}} = 2\ \text{ft}^2$$

From Fig. 17.53, the drag coefficient, C_D, is approximately 0.35.

The total drag force on the log is

$$F_D = \frac{C_D A \rho v^2}{2g_c}$$

$$= \frac{(0.35)(2\ \text{ft}^2)\left(62.37\ \frac{\text{lbm}}{\text{ft}^3}\right)\left(10\ \frac{\text{ft}}{\text{sec}}\right)^2}{(2)\left(32.2\ \frac{\text{lbm-ft}}{\text{lbf-sec}^2}\right)}$$

$$= 67.8\ \text{lbf}$$

The volume of the log is

$$V = LA = L\pi r^2$$

$$= (30\ \text{ft})\left(12\ \frac{\text{in}}{\text{ft}}\right)\pi\left(\frac{12\ \text{in}}{2}\right)^2$$

$$= 40{,}715\ \text{in}^3$$

The density of 50°F water is 62.41 lbm/ft³. The weight of the entire log is

$$W = \frac{\rho V g}{g_c} = \frac{(\text{SG})\rho_w V g}{g_c}$$

$$= \frac{(0.72)\left(62.41\ \frac{\text{lbm}}{\text{ft}^3}\right)(40{,}715\ \text{in}^3)\left(32.2\ \frac{\text{ft}}{\text{sec}^2}\right)}{\left(32.2\ \frac{\text{lbm-ft}}{\text{lbf-sec}^2}\right)\left(12\ \frac{\text{in}}{\text{ft}}\right)^3}$$

$$= 1058.8\ \text{lbf}$$

Choose the x- and y-directions as being perpendicular and parallel to the log, respectively. For small deflection angles, the drag force acts perpendicular to the log at half the submerged depth, $2\ \text{ft}/2 = 1\ \text{ft}$ from the lower, free end, and $30\ \text{ft} - 1\ \text{ft} = 29\ \text{ft}$ from the suspended, upper end. For a small deflection angle of θ, the x-component of the drag force is

$$F_{D,x} = F_D \cos\theta$$

The log weight acts vertically downward at the midpoint of the log, $30\ \text{ft}/2 = 15\ \text{ft}$ from the suspended end. For a small deflection angle of θ, the x-component of the log weight is

$$W_x = W \sin\theta$$

Neglect buoyancy. Take moments about the suspended end.

$$\sum M = 0:\ F_{D,x}(29\ \text{ft}) - W_x(15\ \text{ft}) = 0$$
$$F_D \cos\theta(29\ \text{ft}) - W\sin\theta(15\ \text{ft}) = 0$$
$$(67.8\ \text{lbf})\cos\theta(29\ \text{ft}) - (1058.8\ \text{lbf})\sin\theta(15\ \text{ft}) = 0$$

$$\frac{\sin\theta}{\cos\theta} = \tan\theta = \frac{(67.8\ \text{lbf})(29\ \text{ft})}{(1058.8\ \text{lbf})(15\ \text{ft})} = 0.1238$$

$$\theta = \arctan 0.1238 = \boxed{7.06°\quad (7°)}$$

The answer is (B).

35. *Customary U.S. Solution*

(a) For air at 70°F,

$$\rho = \frac{p}{RT} = \frac{\left(14.7\ \frac{\text{lbf}}{\text{in}^2}\right)\left(12\ \frac{\text{in}}{\text{ft}}\right)^2}{\left(53.35\ \frac{\text{ft-lbf}}{\text{lbm-°R}}\right)(70°\text{F} + 460°)}$$

$$= 0.0749\ \text{lbm/ft}^3$$

$$v = \frac{\left(55\ \frac{\text{mi}}{\text{hr}}\right)\left(5280\ \frac{\text{ft}}{\text{mi}}\right)}{3600\ \frac{\text{sec}}{\text{hr}}}$$

$$= \boxed{80.67\ \text{ft/sec}\quad (80\ \text{ft/sec})}$$

The answer is (D).

(b) The drag on the car is

$$F_D = \frac{C_D A \rho v^2}{2g_c}$$

$$= \frac{(0.42)(28\ \text{ft}^2)\left(0.0749\ \frac{\text{lbm}}{\text{ft}^3}\right)\left(80.67\ \frac{\text{ft}}{\text{sec}}\right)^2}{(2)\left(32.2\ \frac{\text{lbm-ft}}{\text{lbf-sec}^2}\right)}$$

$$= \boxed{89.0\ \text{lbf}\quad (90\ \text{lbf})}$$

The answer is (D).

(c) The total resisting force is

$$F = F_D + \text{rolling resistance}$$

$$= 89.0\ \text{lbf} + (0.01)(3300\ \text{lbm}) \times \frac{g}{g_c}$$

$$= 89.0\ \text{lbf} + (0.01)(3300\ \text{lbm}) \times \frac{32.2\ \frac{\text{ft}}{\text{sec}^2}}{32.2\ \frac{\text{lbm-ft}}{\text{lbf-sec}^2}}$$

$$= \boxed{122.0\ \text{lbf}\quad (120\ \text{lbf})}$$

The answer is (B).

(d) The power consumed is

$$P = Fv = \frac{(122.0 \text{ lbf})\left(80.67 \frac{\text{ft}}{\text{sec}}\right)}{778 \frac{\text{ft-lbf}}{\text{Btu}}}$$

$$= \boxed{12.65 \text{ Btu/sec} \quad (13 \text{ Btu/sec})}$$

The answer is (B).

(e) The energy available from the fuel is

$$E_A = (\text{engine thermal efficiency})(\text{fuel heating value})$$

$$= (0.28)\left(115{,}000 \frac{\text{Btu}}{\text{gal}}\right)$$

$$= 32{,}200 \text{ Btu/gal}$$

The fuel consumption at 55 mi/hr is

$$\frac{P}{E_A v} = \frac{\left(12.65 \frac{\text{Btu}}{\text{sec}}\right)\left(3600 \frac{\text{sec}}{\text{hr}}\right)}{\left(32{,}200 \frac{\text{Btu}}{\text{gal}}\right)\left(55 \frac{\text{mi}}{\text{hr}}\right)}$$

$$= \boxed{0.0257 \text{ gal/mi} \quad (0.026 \text{ gal/mi})}$$

The answer is (A).

(f) At 65 mi/hr,

$$v = \frac{\left(65 \frac{\text{mi}}{\text{hr}}\right)\left(5280 \frac{\text{ft}}{\text{mi}}\right)}{3600 \frac{\text{sec}}{\text{hr}}} = 95.33 \text{ ft/sec}$$

The drag on the car is

$$F_D = \frac{C_D A \rho v^2}{2g_c}$$

$$= \frac{(0.42)(28 \text{ ft}^2)\left(0.0749 \frac{\text{lbm}}{\text{ft}^3}\right)\left(95.33 \frac{\text{ft}}{\text{sec}}\right)^2}{(2)\left(32.2 \frac{\text{lbm-ft}}{\text{lbf-sec}^2}\right)}$$

$$= \boxed{124.3 \text{ lbf} \quad (124 \text{ lbf})}$$

The answer is (D).

(g) The total resisting force is

$$F = F_D + \text{rolling resistance}$$

$$= 124.3 \text{ lbf} + (0.01)(3300 \text{ lbm}) \times \frac{g}{g_c}$$

$$= 124.3 \text{ lbf} + (0.01)(3300 \text{ lbm}) \times \frac{32.2 \frac{\text{ft}}{\text{sec}^2}}{32.2 \frac{\text{lbm-ft}}{\text{lbf-sec}^2}}$$

$$= \boxed{157.3 \text{ lbf} \quad (160 \text{ lbf})}$$

The answer is (D).

(h) The power consumed by the car is

$$P = Fv = \frac{(157.3 \text{ lbf})\left(95.33 \frac{\text{ft}}{\text{sec}}\right)}{778 \frac{\text{ft-lbf}}{\text{Btu}}}$$

$$= \boxed{19.27 \text{ Btu/sec} \quad (19 \text{ Btu/sec})}$$

The answer is (C).

(i) The fuel consumption at 65 mi/hr is

$$\frac{P}{E_A v} = \frac{\left(19.27 \frac{\text{Btu}}{\text{sec}}\right)\left(3600 \frac{\text{sec}}{\text{hr}}\right)}{\left(32{,}200 \frac{\text{Btu}}{\text{gal}}\right)\left(65 \frac{\text{mi}}{\text{hr}}\right)}$$

$$= \boxed{0.0331 \text{ gal/mi} \quad (0.03 \text{ gal/mi})}$$

The answer is (A).

(j) The relative difference between the fuel consumptions at 55 mi/hr and 65 mi/hr is

$$\frac{0.0331 \frac{\text{gal}}{\text{mi}} - 0.0257 \frac{\text{gal}}{\text{mi}}}{0.0257 \frac{\text{gal}}{\text{mi}}} = \boxed{0.288 \quad (30\%)}$$

The answer is (C).

SI Solution

(a) For air at 20°C,

$$\rho = \frac{p}{RT} = \frac{1.013 \times 10^5 \text{ Pa}}{\left(287.03 \frac{\text{J}}{\text{kg·K}}\right)(20°\text{C} + 273°)}$$

$$= 1.205 \text{ kg/m}^3$$

$$v = \frac{\left(90 \frac{\text{km}}{\text{h}}\right)\left(1000 \frac{\text{m}}{\text{km}}\right)}{3600 \frac{\text{s}}{\text{h}}}$$

$$= \boxed{25.0 \text{ m/s} \quad (25 \text{ m/s})}$$

The answer is (D).

(b) The drag on the car is

$$F_D = \frac{C_D A \rho v^2}{2} = \frac{(0.42)(2.6 \text{ m}^2)\left(1.205 \frac{\text{kg}}{\text{m}^3}\right)\left(25.0 \frac{\text{m}}{\text{s}}\right)^2}{2}$$

$$= \boxed{411.2 \text{ N} \quad (410 \text{ N})}$$

The answer is (D).

(c) The total resisting force is

$$F = F_D + \text{rolling resistance} \times g$$
$$= 411.2 \text{ N} + (0.01)(1500 \text{ kg})g$$
$$= 411.2 \text{ N} + (0.01)(1500 \text{ kg})\left(9.81 \ \frac{\text{m}}{\text{s}^2}\right)$$
$$= \boxed{558.4 \text{ N} \quad (560 \text{ N})}$$

The answer is (B).

(d) The power required is

$$P = F\text{v} = (558.4 \text{ N})\left(25 \ \frac{\text{m}}{\text{s}}\right)$$
$$= \boxed{13\,960 \text{ W} \quad (14\,000 \text{ W})}$$

The answer is (B).

(e) The energy available from the fuel is

$$E_A = (\text{engine thermal efficiency})(\text{fuel heating value})$$
$$= (0.28)\left(32 \times 10^6 \ \frac{\text{J}}{\text{L}}\right)$$
$$= 8.96 \times 10^6 \text{ J/L}$$

The fuel consumption at 90 km/h is

$$\frac{P}{E_A \text{v}} = \frac{(13\,960 \text{ W})\left(3600 \ \frac{\text{s}}{\text{h}}\right)}{\left(8.96 \times 10^6 \ \frac{\text{J}}{\text{L}}\right)\left(90 \ \frac{\text{km}}{\text{h}}\right)}$$
$$= \boxed{0.0623 \text{ L/km} \quad (0.062 \text{ L/km})}$$

The answer is (A).

(f) At 105 km/h,

$$\text{v} = \frac{\left(105 \ \frac{\text{km}}{\text{h}}\right)\left(1000 \ \frac{\text{m}}{\text{km}}\right)}{3600 \ \frac{\text{s}}{\text{h}}}$$
$$= 29.2 \text{ m/s}$$

The drag on the car is

$$F_D = \frac{C_D A \rho \text{v}^2}{2} = \frac{(0.42)(2.6 \text{ m}^2)\left(1.205 \ \frac{\text{kg}}{\text{m}^3}\right)\left(29.2 \ \frac{\text{m}}{\text{s}}\right)^2}{2}$$
$$= \boxed{561.0 \text{ N} \quad (570 \text{ N})}$$

The answer is (D).

(g) The total resisting force is

$$F = F_D + \text{rolling resistance} \times g$$
$$= 561.0 \text{ N} + (0.01)(1500 \text{ kg})g$$
$$= 561.0 \text{ N} + (0.01)(1500 \text{ kg})\left(9.81 \ \frac{\text{m}}{\text{s}^2}\right)$$
$$= \boxed{708.2 \text{ N} \quad (720 \text{ N})}$$

The answer is (D).

(h) The power consumed by the car is

$$P = F\text{v} = (708.2 \text{ N})\left(29.2 \ \frac{\text{m}}{\text{s}}\right)$$
$$= \boxed{20\,679 \text{ W} \quad (21\,000 \text{ W})}$$

The answer is (C).

(i) The fuel consumption at 105 km/h is

$$\frac{P}{E_A \text{v}} = \frac{(20\,679 \text{ W})\left(3600 \ \frac{\text{s}}{\text{h}}\right)}{\left(8.96 \times 10^6 \ \frac{\text{J}}{\text{L}}\right)\left(105 \ \frac{\text{km}}{\text{h}}\right)}$$
$$= \boxed{0.0791 \text{ L/km} \quad (0.08 \text{ L/km})}$$

The answer is (A).

(j) The relative difference between the fuel consumptions at 90 km/h and 105 km/h is

$$\frac{0.0791 \ \frac{\text{L}}{\text{km}} - 0.0623 \ \frac{\text{L}}{\text{km}}}{0.0623 \ \frac{\text{L}}{\text{km}}} = \boxed{0.270 \quad (30\%)}$$

The answer is (C).

36. To ensure similarity between the model and the true conditions of the full-scale airplane, the Reynolds numbers must be equal.

$$\left(\frac{\text{v}L}{\nu}\right)_{\text{model}} = \left(\frac{\text{v}L}{\nu}\right)_{\text{true}}$$

Use the absolute viscosity.

$$\mu = \frac{\rho \nu}{g_c}$$
$$\nu = \frac{\mu g_c}{\rho}$$

$$\left(\frac{\text{v}L\rho}{\mu g_c}\right)_{\text{model}} = \left(\frac{\text{v}L\rho}{\mu g_c}\right)_{\text{true}}$$

Recall that the absolute viscosity is independent of pressure, so $\mu_{\text{model}} = \mu_{\text{true}}$.

Since g_c is a constant,

$$\left(\frac{vL\rho}{\mu}\right)_{\text{model}} = \left(\frac{vL\rho}{\mu}\right)_{\text{true}}$$

Assume the air behaves as an ideal gas.

$$\rho = \frac{p}{RT}$$

$$\left(\frac{vLp}{\mu RT}\right)_{\text{model}} = \left(\frac{vLp}{\mu RT}\right)_{\text{true}}$$

Since the tunnel operates with air at true velocity and temperature,

$$R_{\text{model}} = R_{\text{true}}$$

$$v_{\text{model}} = v_{\text{true}}$$

$$T_{\text{model}} = T_{\text{true}}$$

$$\left(\frac{Lp}{\mu}\right)_{\text{model}} = \left(\frac{Lp}{\mu}\right)_{\text{true}}$$

Therefore,

$$(Lp)_{\text{model}} = (Lp)_{\text{true}}$$

Since the scale of the model is 1/20,

$$L_{\text{model}} = \frac{L_{\text{true}}}{20}$$

Substituting gives

$$(Lp)_{\text{model}} = (Lp)_{\text{true}}$$

$$\left(\frac{L_{\text{true}}}{20}\right)p_{\text{model}} = L_{\text{true}}p_{\text{true}}$$

$$\boxed{p_{\text{model}} = 20p_{\text{true}}}$$

The answer is (C).

37. To ensure similarity between the two impellers, the Reynolds numbers, Re, must be equal.

$$\text{Re}_{\text{oil}} = \text{Re}_{\text{air}}$$

$$\frac{v_{\text{oil}}D_{\text{oil}}}{\nu_{\text{oil}}} = \frac{v_{\text{air}}D_{\text{air}}}{\nu_{\text{air}}}$$

$$\frac{v_{\text{oil}}}{v_{\text{air}}} = \left(\frac{\nu_{\text{oil}}}{\nu_{\text{air}}}\right)\left(\frac{D_{\text{air}}}{D_{\text{oil}}}\right)$$

v is the tangential velocity, and D is the impeller diameter.

$$v \propto nD$$

$$\frac{v_{\text{oil}}}{v_{\text{air}}} = \left(\frac{n_{\text{oil}}}{n_{\text{air}}}\right)\left(\frac{D_{\text{oil}}}{D_{\text{air}}}\right)$$

Therefore,

$$\left(\frac{\nu_{\text{oil}}}{\nu_{\text{air}}}\right)\left(\frac{D_{\text{air}}}{D_{\text{oil}}}\right) = \left(\frac{n_{\text{oil}}}{n_{\text{air}}}\right)\left(\frac{D_{\text{oil}}}{D_{\text{air}}}\right)$$

$$n_{\text{air}} = n_{\text{oil}}\left(\frac{\nu_{\text{air}}}{\nu_{\text{oil}}}\right)\left(\frac{D_{\text{oil}}}{D_{\text{air}}}\right)^2$$

Since the air impeller is twice the size of the oil impeller, $D_{\text{air}} = 2D_{\text{oil}}$.

$$n_{\text{air}} = n_{\text{oil}}\left(\frac{\nu_{\text{air}}}{\nu_{\text{oil}}}\right)\left(\frac{D_{\text{oil}}}{D_{\text{air}}}\right)^2$$

$$= n_{\text{oil}}\left(\frac{\nu_{\text{air}}}{\nu_{\text{oil}}}\right)\left(\frac{D_{\text{oil}}}{2D_{\text{oil}}}\right)^2$$

$$= \tfrac{1}{4}n_{\text{oil}}\left(\frac{\nu_{\text{air}}}{\nu_{\text{oil}}}\right)$$

Customary U.S. Solution

From App. 14.D, for air at 68°F, $\nu = 15.72 \times 10^{-5}$ ft²/sec.

For castor oil at 68°F, $\nu = 1110 \times 10^{-5}$ ft²/sec (given).

$$n_{\text{air}} = \tfrac{1}{4}n_{\text{oil}}\left(\frac{\nu_{\text{air}}}{\nu_{\text{oil}}}\right)$$

$$= \left(\tfrac{1}{4}\right)\left(1000 \; \frac{\text{rev}}{\text{min}}\right)\left(\frac{15.72 \times 10^{-5} \; \frac{\text{ft}^2}{\text{sec}}}{1110 \times 10^{-5} \; \frac{\text{ft}^2}{\text{sec}}}\right)$$

$$= \boxed{3.54 \text{ rpm}}$$

The answer is (A).

SI Solution

From App. 14.E, for air at 20°C, $\nu = 1.512 \times 10^{-5}$ m²/s.

For castor oil at 20°C, $\nu = 103 \times 10^{-5}$ m²/s (given).

$$n_{\text{air}} = \tfrac{1}{4}n_{\text{oil}}\left(\frac{\nu_{\text{air}}}{\nu_{\text{oil}}}\right)$$

$$= \left(\tfrac{1}{4}\right)\left(1000 \; \frac{\text{rev}}{\text{min}}\right)\left(\frac{1.512 \times 10^{-5} \; \frac{\text{m}^2}{\text{s}}}{103 \times 10^{-5} \; \frac{\text{m}^2}{\text{s}}}\right)$$

$$= \boxed{3.67 \text{ rpm}}$$

The answer is (A).

18 Hydraulic Machines

PRACTICE PROBLEMS

1. Two centrifugal pumps used in a water pumping application have the characteristic curves shown. The pumps operate in parallel and discharge into a common header against a head of 40 ft. What is most nearly the discharge rate of the pumps operating in parallel?

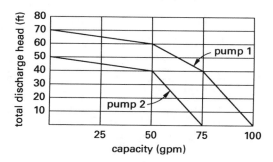

(A) 30 gpm

(B) 50 gpm

(C) 75 gpm

(D) 130 gpm

2. An electric motor drives a pump in a gasoline (specific gravity of 0.7) transfer network. The system and pump curves are defined by the points in the given table. What is most nearly the minimum horsepower for the motor?

head (ft)	volume	
	system curve (gpm)	pump curve (gpm)
10	0	1500
20	500	1200
30	1000	1000
40	1200	500
50	1500	0

(A) 4.0 hp

(B) 5.3 hp

(C) 5.5 hp

(D) 7.6 hp

3. A pump intended for occasional use in normal ambient conditions has an overall hydraulic efficiency of 0.85 and is required to develop a hydraulic horsepower of 4.5 hp. The pump is driven by an electric motor with a service factor of 1.80 and an electrical efficiency of 90%. The smallest NEMA standard motor size suitable for this application is

(A) 3 hp

(B) 5 hp

(C) 8 hp

(D) 10 hp

4. In a valve test bed, the fluid flow rate was measured as 800 gpm. The specific gravity of the fluid was 1.2. The pressure of the fluid one pipe diameter upstream of the valve was measured as 10 psig. The pressure ten pipe diameters downstream of the valve was measured as 0.1 psig. The pressure at the vena contracta was measured as −3 psig. The valve's pressure recovery factor for liquids is most nearly

(A) 0.3

(B) 0.6

(C) 0.8

(D) 0.9

5. Two centrifugal pumps used in a water pumping application have the characteristic curves shown. The pumps operate in series and have a combined discharge of 50 gpm. What is most nearly the total discharge head?

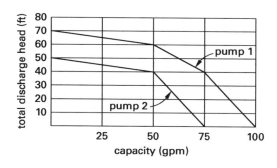

Flow of Fluids

(A) 24 ft

(B) 50 ft

(C) 80 ft

(D) 100 ft

6. 2000 gal/min of 60°F thickened sludge with a specific gravity of 1.2 flows through a pump with an inlet diameter of 12 in and an outlet diameter of 8 in. The centerlines of the inlet and outlet are at the same elevation. The inlet pressure is 8 in of mercury (vacuum). A discharge pressure gauge located 4 ft above the pump discharge centerline reads 20 psig. The pump efficiency is 85%. All pipes are schedule-40. The input power of the pump is most nearly

(A) 26 hp

(B) 31 hp

(C) 37 hp

(D) 53 hp

7. 1.25 ft³/sec (35 L/s) of 70°F (21°C) water is pumped from the bottom of a tank through 700 ft (230 m) of 4 in (102.3 mm) schedule-40 steel pipe. The line includes a 50 ft (15 m) rise in elevation, two right-angle elbows, a wide-open gate valve, and a swing check valve. All fittings and valves are regular screwed. The inlet pressure is 50 psig (345 kPa), and a working pressure of 20 psig (140 kPa) is needed at the end of the pipe. The hydraulic power for this pumping application is most nearly

(A) 16 hp (13 kW)

(B) 23 hp (17 kW)

(C) 49 hp (37 kW)

(D) 66 hp (50 kW)

8. 80 gal/min (5 L/s) of 80°F (27°C) water is lifted 12 ft (4 m) vertically by a pump through a total length of 50 ft (15 m) of a 2 in (5.1 cm) diameter smooth rubber hose. The discharge end of the hose is submerged in 8 ft (2.5 m) of water as shown. The head added by the pump is most nearly

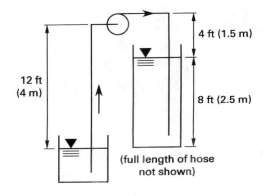

(full length of hose not shown)

(A) 10 ft (3.0 m)

(B) 13 ft (4.0 m)

(C) 22 ft (6.6 m)

(D) 31 ft (9.3 m)

9. A 20 hp motor drives a centrifugal pump. The pump discharges 60°F (16°C) water at 12 ft/sec (4 m/s) into a 6 in (15.2 cm) steel schedule-40 line. The inlet is 8 in (20.3 cm) schedule-40 steel pipe. The pump suction is 5 psi (35 kPa) below standard atmospheric pressure. The friction and fitting head loss in the system is 10 ft (3.3 m). The pump efficiency is 70%. The suction and discharge lines are at the same elevation. The maximum height above the pump inlet that water is available with that velocity at standard atmospheric pressure is most nearly

(A) 28 ft (6.9 m)

(B) 37 ft (11 m)

(C) 49 ft (15 m)

(D) 81 ft (25 m)

10. An electrically driven pump is used to fill a tank on a hill from a lake below. The flow rate is 10,000 gal/hr (10.5 L/s) of 60°F (16°C) water. The atmospheric pressure is 14.7 psia (101 kPa). The pump is 12 ft (4 m) above the lake, and the tank surface level is 350 ft (115 m) above the pump. The suction and discharge lines are 4 in (10.2 cm) diameter schedule-40 steel pipe. The equivalent length of the inlet line between the lake and the pump is 300 ft (100 m). The total equivalent length between the lake and the tank is 7000 ft (2300 m), including all fittings, bends, screens, and valves. The cost of electricity is $0.04 per kW-hr. The overall efficiency of the pump and motor set is 70%.

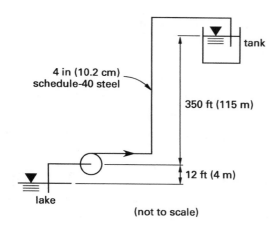

(not to scale)

(a) The velocity in the pipe is most nearly

(A) 0.5 ft/sec (0.2 m/s)

(B) 4 ft/sec (1.0 m/s)

(C) 6 ft/sec (2.0 m/s)

(D) 10 ft/sec (5.0 m/s)

(b) The Reynolds number is most nearly

(A) 1.2×10^5

(B) 1.8×10^5

(C) 4.0×10^5

(D) 6.0×10^5

(c) The friction head is most nearly

(A) 28 ft (10 m)

(B) 120 ft (40 m)

(C) 290 ft (95 m)

(D) 950 ft (310 m)

(d) The hydraulic horsepower is most nearly

(A) 2.7 hp (2.2 kW)

(B) 5.4 hp (4.4 kW)

(C) 19 hp (15 kW)

(D) 20 hp (16 kW)

(e) The cost to operate the pump for one hour is most nearly

(A) $0.1

(B) $1

(C) $3

(D) $6

(f) The motor power required is approximately

(A) 10 hp (7.5 kW)

(B) 30 hp (25 kW)

(C) 50 hp (40 kW)

(D) 75 hp (60 kW)

(g) The NPSHA for this application is most nearly

(A) 4 ft (1.2 m)

(B) 8 ft (2.4 m)

(C) 12 ft (3.6 m)

(D) 16 ft (4.5 m)

11. A town with a stable, constant population of 10,000 produces sewage at the average rate of 100 gallons per capita day (gpcd), with peak flows of 250 gpcd. The pipe to the pumping station is 5000 ft in length and has a C-value of 130. The elevation drop along the length is 48 ft. Minor losses in infiltration are insignificant. The pump's maximum suction lift is 10 ft.

(a) If all whole-inch pipe diameters are available and the pipe flows 100% full under gravity flow, the minimum pipe diameter required is

(A) 8 in

(B) 12 in

(C) 14 in

(D) 18 in

(b) If constant-speed pumps are used, the minimum number of pumps (disregarding spares and backups) that should be used is

(A) 2

(B) 3

(C) 4

(D) 5

(c) If variable-speed pumps are used, the minimum number of pumps (disregarding spares and backups) that should be used is

(A) 1

(B) 2

(C) 3

(D) 4

(d) If three constant-speed pumps are used, with a fourth as backup, and the pump-motor set efficiency is 60%, the motor power required is most nearly

(A) 2 hp

(B) 3 hp

(C) 5 hp

(D) 8 hp

(e) If two variable-speed pumps are used, with a third as backup, and the pump-motor set efficiency is 80%, the motor power required is most nearly

(A) 3 hp

(B) 8 hp

(C) 12 hp

(D) 18 hp

(f) Which of the following are valid ways of controlling sump pump on-off cycles?

I.   detecting sump levels

II.   detecting pressure in the sump

III.   detecting incoming flow rates

IV.   using fixed run times

V.   detecting outgoing flow rates

VI.   operating manually

   (A) I, II, and III

   (B) I, II, IV, and V

   (C) I, III, and V

   (D) I, II, IV, V, and VI

12. A pump transfers 3.5 MGD of filtered water from the clear well of a 10 ft × 20 ft (plan) rapid sand filter to a higher elevation. The pump efficiency is 85%, and the motor driving the pump has an efficiency of 90%. Minor losses are insignificant. Refer to the illustration shown for additional information.

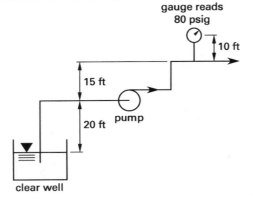

(a) The static suction lift is most nearly

(A) 15 ft

(B) 20 ft

(C) 35 ft

(D) 40 ft

(b) The static discharge head is most nearly

(A) 15 ft

(B) 20 ft

(C) 35 ft

(D) 40 ft

(c) Based on the information given, the approximate total dynamic head is most nearly

(A) 45 ft

(B) 185 ft

(C) 210 ft

(D) 230 ft

(d) The required motor power is most nearly

(A) 50 hp

(B) 100 hp

(C) 150 hp

(D) 200 hp

13. Gasoline with a specific gravity of 0.7 and kinematic viscosity of 6×10^{-6} ft^2/sec (5.6×10^{-10} m^2/s) is transferred from a tanker to a storage tank. The interior of the storage tank is maintained at atmospheric pressure by a vapor-recovery system. The free surface in the storage tank is 60 ft (20 m) above the tanker's free surface. The pipe consists of 500 ft (170 m) of 3 in (7.62 cm) schedule-40 steel pipe with six flanged elbows and two wide-open gate valves. The pump and motor both have individual efficiencies of 88%. Electricity costs $0.045 per kW-hr. The pump's performance data (based on cold, clear water) are known.

flow rate (gpm (L/s))	head (ft (m))
0 (0)	127 (42)
100 (6.3)	124 (41)
200 (12)	117 (39)
300 (18)	108 (36)
400 (24)	96 (32)
500 (30)	80 (27)
600 (36)	55 (18)

(a) The total equivalent length of pipe and fittings is most nearly

(A) 500 ft (169 m)

(B) 510 ft (170 m)

(C) 525 ft (178 m)

(D) 530 ft (180 m)

(b) If the flow rate is 100 gal/min (6.3 L/s), the velocity in the flow pipe is most nearly

(A) 2.8 ft/sec (0.86 m/s)

(B) 3.4 ft/sec (1.0 m/s)

(C) 4.3 ft/sec (1.3 m/s)

(D) 5.1 ft/sec (1.6 m/s)

(c) The friction head loss is most nearly

(A) 2.6 ft (1.0 m)

(B) 6.5 ft (2.2 m)

(C) 9.1 ft (3.1 m)

(D) 11 ft (3.8 m)

(d) The transfer rate is most nearly

(A) 150 gal/min (9.2 L/s)

(B) 180 gal/min (11 L/s)

(C) 200 gal/min (12 L/s)

(D) 230 gal/min (14 L/s)

(e) The total cost of operating the pump for one hour is most nearly

(A) $0.20

(B) $0.80

(C) $1.30

(D) $2.70

14. The pressure of 37 gal/min (65 L/s) of 80°F (27°C) SAE 40 oil is increased from 1 atm to 40 psig (275 kPa). The hydraulic power required is most nearly

(A) 0.45 hp (9 kW)

(B) 0.9 hp (18 kW)

(C) 1.8 hp (36 kW)

(D) 3.6 hp (72 kW)

15. A double-suction water pump moving 300 gal/sec (1.1 kL/s) turns at 900 rpm. The pump adds 20 ft (7 m) of head to the water. The specific speed is most nearly

(A) 3000 rpm (52 rpm)

(B) 6000 rpm (100 rpm)

(C) 9000 rpm (160 rpm)

(D) 12,000 rpm (210 rpm)

16. A two-stage centrifugal pump draws water from an inlet 10 ft below its eye. Each stage of the pump adds 150 ft of head. What is the approximate maximum suggested speed for this application?

(A) 900 rpm

(B) 1200 rpm

(C) 1700 rpm

(D) 2000 rpm

17. 100 gal/min (6.3 L/s) of pressurized hot water at 281°F and 80 psia (138°C and 550 kPa) is drawn through 30 ft (10 m) of 1.5 in (3.81 cm) schedule-40 steel pipe into a tank pressurized to a constant 2 psig (14 kPa). The inlet and outlet are both 20 ft (6 m) below the surface of the water when the tank is full. The inlet line contains a square mouth inlet, two wide-open gate valves, and two long-radius elbows. All components are regular screwed. The pump's NPSHR is 10 ft (3 m) for this application. The kinematic viscosity of 281°F (138°C) water is 0.239×10^{-5} ft²/sec (0.222×10^{-6} m²/s), and the vapor pressure is 50.02 psia (3.431 bar). Will the pump cavitate?

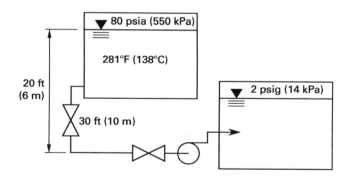

(A) yes; NPSHA = 4 ft (1.2 m)

(B) yes; NPSHA = 9 ft (2.7 m)

(C) no; NPSHA = 24 ft (7.2 m)

(D) no; NPSHA = 68 ft (21 m)

18. The velocity of the tip of a marine propeller is 4.2 times the boat velocity. The propeller is located 8 ft (3 m) below the surface. The temperature of the seawater is 68°F (20°C). The density of seawater is approximately 64.0 lbm/ft³ (1024 kg/m³), and the salt content is 2.5% by weight. The practical maximum boat velocity, as limited strictly by cavitation, is most nearly

(A) 9.1 ft/sec (2.7 m/s)

(B) 12 ft/sec (3.8 m/s)

(C) 15 ft/sec (4.5 m/s)

(D) 22 ft/sec (6.6 m/s)

19. The inlet of a centrifugal water pump is 7 ft (2.3 m) above the free surface from which it draws. The suction point is a submerged pipe. The suction line consists of 12 ft (4 m) of 2 in (5.08 cm) schedule-40 steel pipe and contains one long-radius elbow and one check valve. The discharge line is 2 in (5.08 cm) schedule-40 steel pipe and includes two long-radius elbows and an 80 ft (27 m) run. The discharge is 20 ft (6.3 m) above the free surface and is a jet to the open atmosphere. All components are regular screwed. The water temperature is 70°F (21°C). Use the following pump curve data.

flow rate (gpm (L/s))	head (ft (m))
0 (0)	110 (37)
10 (0.6)	108 (36)
20 (1.2)	105 (35)
30 (1.8)	102 (34)
40 (2.4)	98 (33)
50 (3.2)	93 (31)
60 (3.6)	87 (29)
70 (4.4)	79 (26)
80 (4.8)	66 (22)
90 (5.7)	50 (17)

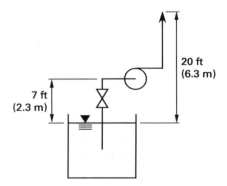

(a) The flow rate is most nearly

(A) 44 gal/min (2.9 L/s)

(B) 69 gal/min (4.5 L/s)

(C) 82 gal/min (5.5 L/s)

(D) 95 gal/min (6.2 L/s)

(b) What can be said about the use of this pump in this installation?

(A) A different pump should be used.

(B) The pump is operating near its most efficient point.

(C) Pressure fluctuations could result from surging.

(D) Overloading will not be a problem.

20. A pump was intended to run at 1750 rpm when driven by a 0.5 hp (0.37 kW) motor. The required power rating of a motor that will turn the pump at 2000 rpm is most nearly

(A) 0.25 hp (0.19 kW)

(B) 0.45 hp (0.34 kW)

(C) 0.65 hp (0.49 kW)

(D) 0.75 hp (0.55 kW)

21. A centrifugal pump running at 1400 rpm has the curve shown. The pump will be installed in an existing pipeline with known head requirements given by the formula $H = 30 + 2Q^2$. H is the system head in feet of water. Q is the flow rate in cubic feet per second.

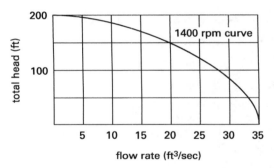

(a) If the pump is turned at 1400 rpm, the flow rate will most nearly be

(A) 2000 gal/min

(B) 3500 gal/min

(C) 4000 gal/min

(D) 4500 gal/min

(b) The power required to drive the pump is most nearly

(A) 190 hp

(B) 210 hp

(C) 230 hp

(D) 260 hp

(c) If the pump is turned at 1200 rpm, the flow rate will most nearly be

(A) 2000 gal/min

(B) 3500 gal/min

(C) 4000 gal/min

(D) 4500 gal/min

22. A horizontal turbine reduces 100 ft³/sec of water from 30 psia to 5 psia. Friction is negligible. The power developed is most nearly

(A) 350 hp

(B) 500 hp

(C) 650 hp

(D) 800 hp

23. 1000 ft³/sec of 60°F water flows from a high reservoir through a hydroelectric turbine installation, exiting 625 ft lower. The head loss due to friction is 58 ft. The turbine efficiency is 89%. The power developed in the turbines is most nearly

(A) 40 kW

(B) 18 MW

(C) 43 MW

(D) 71 MW

24. Water at 500 psig and 60°F (3.5 MPa and 16°C) drives a 250 hp (185 kW) turbine at 1750 rpm against a back pressure of 30 psig (210 kPa). The water discharges through a 4 in (100 mm) diameter nozzle at 35 ft/sec (10.5 m/s). The water is deflected 80° by a single blade moving directly away at 10 ft/sec (3 m/s).

(a) The specific speed is most nearly

(A) 4 (17)

(B) 25 (85)

(C) 75 (260)

(D) 230 (770)

(b) The total force acting on a single blade is most nearly

(A) 100 lbf (450 N)

(B) 140 lbf (570 N)

(C) 160 lbf (720 N)

(D) 280 lbf (1300 N)

25. A Francis-design hydraulic reaction turbine with 22 in (560 mm) diameter blades runs at 610 rpm. The turbine develops 250 hp (185 kW) when 25 ft³/sec (700 L/s) of water flow through it. The pressure head at the turbine entrance is 92.5 ft (28.2 m). The elevation of the turbine above the tailwater level is 5.26 ft (1.75 m). The inlet and outlet velocities are both 12 ft/sec (3.6 m/s).

(a) The effective head is most nearly

(A) 90 ft (27 m)

(B) 95 ft (29 m)

(C) 100 ft (31 m)

(D) 105 ft (35 m)

(b) The overall turbine efficiency is most nearly

(A) 81%

(B) 88%

(C) 93%

(D) 96%

(c) If the effective head is 225 ft (70 m), the turbine speed will most nearly be

(A) 600 rpm

(B) 920 rpm

(C) 1100 rpm

(D) 1400 rpm

(d) If the effective head is 225 ft (70 m), the horsepower developed will most nearly be

(A) 560 hp (420 kW)

(B) 630 hp (470 kW)

(C) 750 hp (560 kW)

(D) 840 hp (640 kW)

(e) If the effective head is 225 ft (70 m), the flow rate will most nearly be

(A) 25 ft³/sec (700 L/s)

(B) 38 ft³/sec (1100 L/s)

(C) 56 ft³/sec (1600 L/s)

(D) 64 ft³/sec (1800 L/s)

SOLUTIONS

1. Theoretically, when operating in parallel, each pump performs as if the other pump isn't present. The capacities of each pump at 40 ft discharge head are cumulative: 50 gpm for pump 2 and 75 gpm for pump 1.

$$Q_{\text{parallel}} = Q_2 + Q_1 = 50 \ \frac{\text{gal}}{\text{min}} + 75 \ \frac{\text{gal}}{\text{min}}$$
$$= \boxed{125 \text{ gpm} \quad (130 \text{ gpm})}$$

The answer is (D).

2. The system and pump curves intersect at 1000 gpm and 30 ft. From Table 18.5, the hydraulic horsepower is

$$\text{WHP} = \frac{h_A Q(\text{SG})}{3956} = \frac{(30 \text{ ft})\left(1000 \ \frac{\text{gal}}{\text{min}}\right)(0.7)}{3956 \ \frac{\text{ft-gal}}{\text{hp-min}}}$$
$$= \boxed{5.31 \text{ hp} \quad (5.3 \text{ hp})}$$

This is the minimum power that the electric motor can produce.

The answer is (B).

3. The motor efficiency is not used because NEMA motor power ratings are motor power output ratings. The motor is intended to occasional use, so the service factor can be confidently included.

From Eq. 18.11, the smallest suitable motor size is

$$\text{BHP} = \frac{\text{WHP}}{\eta_p(\text{SF})} = \frac{4.5 \text{ hp}}{(0.85)(1.80)} = 2.94 \text{ hp}$$

From Table 18.7, the smallest NEMA standard motor size with a rating greater than 2.94 hp is $\boxed{3 \text{ hp.}}$

The answer is (A).

4. It is usually assumed that the lowest pressure will be found at the vena contracta. The liquid pressure recovery factor, F_L, of a valve is basically (the square root of) the ratio of the actual pressure loss to the maximum pressure loss.

$$F_L = \sqrt{\frac{p_1 - p_2}{(p_1 - p_{\text{vena contracta}})\text{SG}}}$$
$$= \sqrt{\frac{10 \ \frac{\text{lbf}}{\text{in}^2} - 0.1 \ \frac{\text{lbf}}{\text{in}^2}}{\left(10 \ \frac{\text{lbf}}{\text{in}^2} - \left(-3 \ \frac{\text{lbf}}{\text{in}^2}\right)\right)(1.2)}}$$
$$= \boxed{0.796 \quad (0.8)}$$

The answer is (C).

5. When operated in series, the second pump receives water at the rate of the first pump's discharge, so both pumps experience the same flow rate. The second pump adds pressure head to the first pump's pressurization, so the discharge heads are cumulative. At 50 gpm, the discharge heads are 60 ft for pump 1 and 40 ft for pump 2, respectively.

$$h_{A,\text{series}} = h_{A,1} + h_{A,2} = 60 \text{ ft} + 40 \text{ ft} = \boxed{100 \text{ ft}}$$

The answer is (D).

6. The flow rate is

$$\dot{V} = \frac{2000 \ \frac{\text{gal}}{\text{min}}}{\left(7.4805 \ \frac{\text{gal}}{\text{ft}^3}\right)\left(60 \ \frac{\text{sec}}{\text{min}}\right)} = 4.456 \text{ ft}^3/\text{sec}$$

From App. 16.B,

$$12 \text{ in}: D_1 = 0.9948 \text{ ft} \qquad A_1 = 0.7773 \text{ ft}^2$$
$$8 \text{ in}: D_2 = 0.6651 \text{ ft} \qquad A_2 = 0.3473 \text{ ft}^2$$

$$p_1 = \left(14.7 \ \frac{\text{lbf}}{\text{in}^2} - (8 \text{ in})\left(0.491 \ \frac{\text{lbf}}{\text{in}^3}\right)\right)\left(12 \ \frac{\text{in}}{\text{ft}}\right)^2$$
$$= 1551.2 \text{ lbf/ft}^2$$

$$p_2 = \left(14.7 \ \frac{\text{lbf}}{\text{in}^2} + 20 \ \frac{\text{lbf}}{\text{in}^2}\right)\left(12 \ \frac{\text{in}}{\text{ft}}\right)^2$$
$$+ (4 \text{ ft})(1.2)\left(62.4 \ \frac{\text{lbf}}{\text{ft}^3}\right)$$
$$= 5296.3 \text{ lbf/ft}^2$$

$$\text{v}_1 = \frac{\dot{V}}{A_1} = \frac{4.456 \ \frac{\text{ft}^3}{\text{sec}}}{0.7773 \text{ ft}^2} = 5.73 \text{ ft/sec}$$

$$\text{v}_2 = \frac{\dot{V}}{A_2} = \frac{4.456 \ \frac{\text{ft}^3}{\text{sec}}}{0.3473 \text{ ft}^2} = 12.83 \text{ ft/sec}$$

From Eq. 18.8, the total heads (in feet of sludge) at points 1 and 2 are

$$h_{t,1} = h_{t,s} = \frac{p_1}{\gamma} + \frac{\text{v}_1^2}{2g} = \frac{1551.2 \ \frac{\text{lbf}}{\text{ft}^2}}{\left(62.4 \ \frac{\text{lbf}}{\text{ft}^3}\right)(1.2)} + \frac{\left(5.73 \ \frac{\text{ft}}{\text{sec}}\right)^2}{(2)\left(32.2 \ \frac{\text{ft}}{\text{sec}^2}\right)}$$
$$= 21.23 \text{ ft}$$

$$h_{t,2} = h_{t,d} = \frac{p_2}{\gamma} + \frac{\text{v}_2^2}{2g} = \frac{5296.3 \ \frac{\text{lbf}}{\text{ft}^2}}{\left(62.4 \ \frac{\text{lbf}}{\text{ft}^3}\right)(1.2)} + \frac{\left(12.83 \ \frac{\text{ft}}{\text{sec}}\right)^2}{(2)\left(32.2 \ \frac{\text{ft}}{\text{sec}^2}\right)}$$
$$= 73.29 \text{ ft}$$

The pump must add 73.29 ft − 21.23 ft = 52.06 ft of head (sludge head).

The power required is given in Table 18.5.

$$P_{ideal} = \frac{\Delta p \dot{V}}{550} = \frac{\Delta h \gamma \dot{V}}{550} = \frac{\Delta h (SG) \gamma_w \dot{V}}{550}$$

$$= \frac{(52.06 \text{ ft})(1.2)\left(62.4 \frac{\text{lbf}}{\text{ft}^3}\right)\left(4.456 \frac{\text{ft}^3}{\text{sec}}\right)}{550 \frac{\text{ft-lbf}}{\text{hp-sec}}}$$

$$= 31.58 \text{ hp}$$

The input horsepower is

$$P_{in} = \frac{P_{ideal}}{\eta} = \frac{31.58 \text{ hp}}{0.85} = \boxed{37.15 \text{ hp} \quad (37 \text{ hp})}$$

The answer is (C).

7. *Customary U.S. Solution*

From App. 16.B, data for 4 in schedule-40 steel pipe are

$$D = 0.3355 \text{ ft}$$
$$A = 0.08841 \text{ ft}^2$$

The velocity in the pipe is

$$v = \frac{\dot{V}}{A} = \frac{1.25 \frac{\text{ft}^3}{\text{sec}}}{0.08841 \text{ ft}^2} = 14.139 \text{ ft/sec}$$

From App. 17.D, typical equivalent lengths for schedule-40, screwed steel fittings for 4 in pipes are

$$90° \text{ elbow: } 13 \text{ ft}$$
$$\text{gate valve: } 2.5 \text{ ft}$$
$$\text{check valve: } 38 \text{ ft}$$

The total equivalent length is

$$(2)(13 \text{ ft}) + (1)(2.5 \text{ ft}) + (1)(38 \text{ ft}) = 66.5 \text{ ft}$$

At 70°F, from App. 14.A, the density of water is 62.3 lbm/ft³, and the kinematic viscosity of water, ν, is 1.059×10^{-5} ft²/sec. The Reynolds number is

$$Re = \frac{Dv}{\nu} = \frac{(0.3355 \text{ ft})\left(14.139 \frac{\text{ft}}{\text{sec}}\right)}{1.059 \times 10^{-5} \frac{\text{ft}^2}{\text{sec}}}$$

$$= 4.479 \times 10^5$$

From App. 17.A, for steel, $\epsilon = 0.0002$ ft. So,

$$\frac{\epsilon}{D} = \frac{0.0002 \text{ ft}}{0.3355 \text{ ft}} \approx 0.0006$$

From App. 17.B, the friction factor is $f = 0.01835$.

The friction head is given by Eq. 18.5.

$$h_f = \frac{fLv^2}{2Dg}$$

$$= \frac{(0.01835)(700 \text{ ft} + 66.5 \text{ ft})\left(14.139 \frac{\text{ft}}{\text{sec}}\right)^2}{(2)(0.3355 \text{ ft})\left(32.2 \frac{\text{ft}}{\text{sec}^2}\right)}$$

$$= 130.1 \text{ ft}$$

The total dynamic head is given by Eq. 18.8. Point 1 is taken as the bottom of the supply tank. Point 2 is taken as the end of the discharge pipe.

$$h = \frac{(p_2 - p_1)g_c}{\rho g} + \frac{v_2^2 - v_1^2}{2g} + z_2 - z_1$$

$$v_1 \approx 0$$

$$z_2 - z_1 = 50 \text{ ft} \quad \text{[given as rise in elevation]}$$

The outlet and inlet pressures are

$$p_2 = 20 \text{ psig}$$
$$p_1 = 50 \text{ psig}$$

The pressure head added by the pump is

$$h = \frac{(p_2 - p_1)g_c}{\rho g} + \frac{v_2^2 - v_1^2}{2g} + z_2 - z_1$$

$$= \frac{\left(20 \frac{\text{lbf}}{\text{in}^2} - 50 \frac{\text{lbf}}{\text{in}^2}\right)\left(12 \frac{\text{in}}{\text{ft}}\right)^2\left(32.2 \frac{\text{lbm-ft}}{\text{lbf-sec}^2}\right)}{\left(62.3 \frac{\text{lbm}}{\text{ft}^3}\right)\left(32.2 \frac{\text{ft}}{\text{sec}^2}\right)}$$

$$+ \frac{\left(14.139 \frac{\text{ft}}{\text{sec}}\right)^2}{(2)\left(32.2 \frac{\text{ft}}{\text{sec}^2}\right)} + 50 \text{ ft}$$

$$= -16.2 \text{ ft}$$

The head added is

$$h_A = h + h_f$$
$$= -16.2 \text{ ft} + 130.1 \text{ ft}$$
$$= 113.9 \text{ ft}$$

The mass flow rate is

$$\dot{m} = \rho \dot{V}$$

$$= \left(62.3 \frac{\text{lbm}}{\text{ft}^3}\right)\left(1.25 \frac{\text{ft}^3}{\text{sec}}\right)$$

$$= 77.875 \text{ lbm/sec}$$

From Table 18.5, the hydraulic horsepower is

$$\text{WHP} = \frac{h_A \dot{m}}{550} \times \frac{g}{g_c}$$

$$= \frac{(113.9 \text{ ft})\left(77.875 \dfrac{\text{lbm}}{\text{sec}}\right)}{550 \dfrac{\text{ft-lbf}}{\text{hp-sec}}} \times \frac{32.2 \dfrac{\text{ft}}{\text{sec}^2}}{32.2 \dfrac{\text{lbm-ft}}{\text{lbf-sec}^2}}$$

$$= \boxed{16.13 \text{ hp} \quad (16 \text{ hp})}$$

The answer is (A).

SI Solution

From App. 16.C, data for 4 in schedule-40 steel pipe are

$$D = 102.3 \text{ mm}$$
$$A = 82.19 \times 10^{-4} \text{ m}^2$$

The velocity in the pipe is

$$v = \frac{\dot{V}}{A} = \frac{35 \dfrac{\text{L}}{\text{s}}}{(82.19 \times 10^{-4} \text{ m}^2)\left(1000 \dfrac{\text{L}}{\text{m}^3}\right)} = 4.26 \text{ m/s}$$

From App. 17.D, typical equivalent lengths for schedule-40, screwed steel fittings for 4 in pipes are

90° elbow: 13 ft

gate valve: 2.5 ft

check valve: 38 ft

The total equivalent length is

$$(2)(13 \text{ ft}) + (1)(2.5 \text{ ft}) + (1)(38 \text{ ft}) = 66.5 \text{ ft}$$
$$(66.5 \text{ ft})\left(0.3048 \dfrac{\text{m}}{\text{ft}}\right) = 20.27 \text{ m}$$

At 21°C, from App. 14.B, the water properties are

$$\rho = 998 \text{ kg/m}^3$$
$$\mu = 0.9827 \times 10^{-3} \text{ Pa·s}$$
$$\nu = \frac{\mu}{\rho} = \frac{0.9827 \times 10^{-3} \text{ Pa·s}}{998 \dfrac{\text{kg}}{\text{m}^3}}$$
$$= 9.85 \times 10^{-7} \text{ m}^2/\text{s}$$

The Reynolds number is

$$\text{Re} = \frac{Dv}{\nu} = \frac{(102.3 \text{ mm})\left(4.26 \dfrac{\text{m}}{\text{s}}\right)}{\left(9.85 \times 10^{-7} \dfrac{\text{m}^2}{\text{s}}\right)\left(1000 \dfrac{\text{mm}}{\text{m}}\right)}$$

$$= 4.424 \times 10^5$$

From Table 17.2, for steel, $\epsilon = 6 \times 10^{-5}$ m.

$$\frac{\epsilon}{D} = \frac{(6 \times 10^{-5} \text{ m})\left(1000 \dfrac{\text{mm}}{\text{m}}\right)}{102.3 \text{ mm}}$$
$$= 0.0006$$

From App. 17.B, the friction factor is $f = 0.01836$.

From Eq. 18.5, the friction head is

$$h_f = \frac{fLv^2}{2Dg}$$

$$= \frac{(0.01836)(230 \text{ m} + 20.27 \text{ m})\left(4.26 \dfrac{\text{m}}{\text{s}}\right)^2\left(1000 \dfrac{\text{mm}}{\text{m}}\right)}{(2)(102.3 \text{ mm})\left(9.81 \dfrac{\text{m}}{\text{s}^2}\right)}$$

$$= 41.5 \text{ m}$$

The total dynamic head is given by Eq. 18.8. Point 1 is taken as the bottom of the supply tank. Point 2 is taken as the end of the discharge pipe.

$$h = \frac{p_2 - p_1}{\rho g} + \frac{v_2^2 - v_1^2}{2g} + z_2 - z_1$$

$$v_1 \approx 0$$

$$z_2 - z_1 = 15 \text{ m} \quad \text{[given as rise in elevation]}$$

The difference between outlet and inlet pressure is

$$p_2 - p_1 = 140 \text{ kPa} - 345 \text{ kPa} = -205 \text{ kPa}$$

$$h = \frac{(-205 \text{ kPa})\left(1000 \dfrac{\text{Pa}}{\text{kPa}}\right)}{\left(998 \dfrac{\text{kg}}{\text{m}^3}\right)\left(9.81 \dfrac{\text{m}}{\text{s}^2}\right)}$$

$$+ \frac{\left(4.26 \dfrac{\text{m}}{\text{s}}\right)^2}{(2)\left(9.81 \dfrac{\text{m}}{\text{s}^2}\right)} + 15 \text{ m}$$

$$= -5.0 \text{ m}$$

The head added by the pump is

$$h_A = h + h_f = -5.0 \text{ m} + 41.5 \text{ m} = 36.5 \text{ m}$$

The mass flow rate is

$$\dot{m} = \rho \dot{V} = \frac{\left(998 \dfrac{\text{kg}}{\text{m}^3}\right)\left(35 \dfrac{\text{L}}{\text{s}}\right)}{1000 \dfrac{\text{L}}{\text{m}^3}}$$

$$= 34.93 \text{ kg/s}$$

From Table 18.6, the hydraulic power is

$$\text{WkW} = \frac{9.81 h_A \dot{m}}{1000} = \frac{\left(9.81 \frac{\text{m}}{\text{s}^2}\right)(36.5 \text{ m})\left(34.93 \frac{\text{kg}}{\text{s}}\right)}{1000 \frac{\text{W}}{\text{kW}}}$$

$$= \boxed{12.51 \text{ kW} \quad (13 \text{ kW})}$$

The answer is (A).

8. *Customary U.S. Solution*

The area of the rubber hose is

$$A = \frac{\pi D^2}{4} = \frac{\pi \left(\dfrac{2 \text{ in}}{12 \frac{\text{in}}{\text{ft}}}\right)^2}{4} = 0.0218 \text{ ft}^2$$

The velocity of water in the hose is

$$\text{v} = \frac{\dot{V}}{A} = \frac{80 \frac{\text{gal}}{\text{min}}}{(0.0218 \text{ ft}^2)\left(7.4805 \frac{\text{gal}}{\text{ft}^3}\right)\left(60 \frac{\text{sec}}{\text{min}}\right)}$$

$$= 8.176 \text{ ft/sec}$$

At 80°F from App. 14.A, the kinematic viscosity of water is $\nu = 0.93 \times 10^{-5} \text{ ft}^2/\text{sec}$.

The Reynolds number is

$$\text{Re} = \frac{\text{v}D}{\nu} = \frac{\left(8.176 \frac{\text{ft}}{\text{sec}}\right)(2 \text{ in})}{\left(0.93 \times 10^{-5} \frac{\text{ft}^2}{\text{sec}}\right)\left(12 \frac{\text{in}}{\text{ft}}\right)}$$

$$= 1.47 \times 10^5$$

Since the rubber hose is smooth, from App. 17.B, the friction factor is $f = 0.0166$.

From Eq. 18.5, the friction head is

$$h_f = \frac{f L \text{v}^2}{2Dg} = \frac{(0.0166)(50 \text{ ft})\left(8.176 \frac{\text{ft}}{\text{sec}}\right)^2\left(12 \frac{\text{in}}{\text{ft}}\right)}{(2)(2 \text{ in})\left(32.2 \frac{\text{ft}}{\text{sec}^2}\right)}$$

$$= 5.17 \text{ ft}$$

Neglecting entrance and exit losses, the head added by the pump is

$$h_A = h_f + h_z = 5.17 \text{ ft} + 12 \text{ ft} - 4 \text{ ft}$$

$$= \boxed{13.17 \text{ ft} \quad (13 \text{ ft})}$$

The answer is (B).

SI Solution

The area of the rubber hose is

$$A = \frac{\pi D^2}{4} = \frac{\pi (5.1 \text{ cm})^2}{(4)\left(100 \frac{\text{cm}}{\text{m}}\right)^2} = 0.00204 \text{ m}^2$$

The velocity of water in the hose is

$$\text{v} = \frac{\dot{V}}{A} = \frac{5 \frac{\text{L}}{\text{s}}}{(0.00204 \text{ m}^2)\left(1000 \frac{\text{L}}{\text{m}^3}\right)} = 2.45 \text{ m/s}$$

At 27°C from App. 14.B, the kinematic viscosity of water is $\nu = 8.60 \times 10^{-7} \text{ m}^2/\text{s}$.

The Reynolds number is

$$\text{Re} = \frac{\text{v}D}{\nu} = \frac{\left(2.45 \frac{\text{m}}{\text{s}}\right)(5.1 \text{ cm})}{\left(8.60 \times 10^{-7} \frac{\text{m}^2}{\text{s}}\right)\left(100 \frac{\text{cm}}{\text{m}}\right)} = 1.45 \times 10^5$$

Since the rubber hose is smooth, from App. 17.B, the friction factor is $f \approx 0.0166$.

From Eq. 18.5, the friction head is

$$h_f = \frac{f L \text{v}^2}{2Dg} = \frac{(0.0166)(15 \text{ m})\left(2.45 \frac{\text{m}}{\text{s}}\right)^2\left(100 \frac{\text{cm}}{\text{m}}\right)}{(2)(5.1 \text{ cm})\left(9.81 \frac{\text{m}}{\text{s}^2}\right)}$$

$$= 1.49 \text{ m}$$

Neglecting entrance and exit losses, the head added by the pump is

$$h_A = h_f + h_z$$

$$= 1.49 \text{ m} + 4 \text{ m} - 1.5 \text{ m}$$

$$= \boxed{3.99 \text{ m} \quad (4.0 \text{ m})}$$

The answer is (B).

9. *Customary U.S. Solution*

From App. 16.B, the diameters (inside) for 8 in and 6 in schedule-40 steel pipe are

$$D_1 = 7.981 \text{ in}$$

$$D_2 = 6.065 \text{ in}$$

At 60°F from App. 14.A, the density of water is 62.37 lbm/ft³.

Flow of Fluids

The mass flow rate through 6 in pipe is

$$\dot{m} = A_2 v_2 \rho = \frac{\pi \left(\frac{(6.065 \text{ in})^2}{4} \right) \left(12 \frac{\text{ft}}{\text{sec}} \right) \left(62.37 \frac{\text{lbm}}{\text{ft}^3} \right)}{\left(12 \frac{\text{in}}{\text{ft}} \right)^2}$$

$$= 150.2 \text{ lbm/sec}$$

The inlet (suction) pressure is

$$(14.7 \text{ psia} - 5 \text{ psig}) \left(12 \frac{\text{in}}{\text{ft}} \right)^2 = 1397 \text{ lbf/ft}^2$$

From Table 18.5, the head added by the pump is

$$h_A = \frac{550(\text{BHP})\eta}{\dot{m}} \times \frac{g_c}{g}$$

$$= \frac{\left(550 \frac{\text{ft-lbf}}{\text{hp-sec}} \right)(20 \text{ hp})(0.70)}{150.2 \frac{\text{lbm}}{\text{sec}}} \times \frac{32.2 \frac{\text{lbm-ft}}{\text{lbf-sec}^2}}{32.2 \frac{\text{ft}}{\text{sec}^2}}$$

$$= 51.26 \text{ ft}$$

At 1 (pump inlet):

$$p_1 = 1397 \text{ lbf/ft}^2 \quad [\text{absolute}]$$

$$z_1 = 0$$

$$v_1 = \frac{v_2 A_2}{A_1} = v_2 \left(\frac{D_2}{D_1} \right)^2 = \left(12 \frac{\text{ft}}{\text{sec}} \right) \left(\frac{6.065 \text{ in}}{7.981 \text{ in}} \right)^2$$

$$= 6.93 \text{ ft/sec}$$

At 2 (pump outlet):

$$p_2 \quad [\text{unknown}]$$

$$v_2 = 12 \text{ ft/sec} \quad [\text{given}]$$

$$z_2 = z_1 = 0$$

$$h_{f,1-2} = 0$$

Let z_3 be the additional head above atmospheric. From Eq. 18.8(b), the head added by the pump is

$$h_A = \frac{(p_2 - p_1)g_c}{\rho g} + \frac{v_2^2 - v_1^2}{2g} + z_2 - z_1 + h_{f,1-2}$$

$$51.26 \text{ ft} = \frac{\left(p_2 - 1397 \frac{\text{lbf}}{\text{ft}^2} \right) \left(32.2 \frac{\text{lbm-ft}}{\text{lbf-sec}^2} \right)}{\left(62.37 \frac{\text{lbm}}{\text{ft}^3} \right) \left(32.2 \frac{\text{ft}}{\text{sec}^2} \right)}$$

$$+ \frac{\left(12 \frac{\text{ft}}{\text{sec}} \right)^2 - \left(6.93 \frac{\text{ft}}{\text{sec}} \right)^2}{(2) \left(32.2 \frac{\text{ft}}{\text{sec}^2} \right)}$$

$$+ 0 - 0 + 0$$

$$p_2 = 4501.2 \text{ lbf/ft}^2$$

At 3 (discharge):

$$p_3 = \left(14.7 \frac{\text{lbf}}{\text{in}^2} \right) \left(12 \frac{\text{in}}{\text{ft}} \right)^2 = 2117 \text{ lbf/ft}^2$$

$$v_3 = 12 \text{ ft/sec} \quad [\text{given}]$$

$$z_3 \quad [\text{unknown}]$$

$$h_{f,2-3} = 10 \text{ ft}$$

$$h_A = 0 \quad [\text{no pump between points 2 and 3}]$$

$$h_A = \frac{(p_3 - p_2)g_c}{\rho g} + \frac{v_3^2 - v_2^2}{2g} + z_3 - z_2 + h_{f,2-3}$$

$$0 = \frac{\left(2117 \frac{\text{lbf}}{\text{ft}^2} - 4501.2 \frac{\text{lbf}}{\text{ft}^2} \right) \left(32.2 \frac{\text{lbm-ft}}{\text{lbf-sec}^2} \right)}{\left(62.37 \frac{\text{lbm}}{\text{ft}^3} \right) \left(32.2 \frac{\text{ft}}{\text{sec}^2} \right)}$$

$$+ \frac{\left(12 \frac{\text{ft}}{\text{sec}} \right)^2 - \left(12 \frac{\text{ft}}{\text{sec}} \right)^2}{(2) \left(32.2 \frac{\text{ft}}{\text{sec}^2} \right)}$$

$$+ z_3 - 0 + 10 \text{ ft}$$

$$z_3 = \boxed{28.2 \text{ ft} \quad (28 \text{ ft})}$$

z_3 could have been found directly without determining the intermediate pressure, p_2. This method is illustrated in the SI solution.

The answer is (A).

SI Solution

From App. 16.C, the inside diameters for 8 in and 6 in steel schedule-40 pipe are

$$D_1 = 202.7 \text{ mm}$$

$$D_2 = 154.1 \text{ mm}$$

At 16°C from App. 14.B, the density of water is 998.83 kg/m³.

The mass flow rate through the 6 in pipe is

$$\dot{m} = A_2 v_2 \rho = \frac{\pi \left(\frac{(154.1 \text{ mm})^2}{4} \right) \left(4 \frac{\text{m}}{\text{s}} \right) \left(998.83 \frac{\text{kg}}{\text{m}^3} \right)}{\left(1000 \frac{\text{mm}}{\text{m}} \right)^2}$$

$$= 74.5 \text{ kg/s}$$

The inlet (suction) pressure is

$$101.3 \text{ kPa} - 35 \text{ kPa} = 66.3 \text{ kPa}$$

From Table 18.6, the head added by the pump is

$$h_A = \frac{1000(\text{BkW})\eta}{9.81\dot{m}}$$

$$= \frac{\left(1000 \; \frac{\text{W}}{\text{kW}}\right)(20 \; \text{hp})\left(0.7457 \; \frac{\text{kW}}{\text{hp}}\right)(0.70)}{\left(9.81 \; \frac{\text{m}}{\text{s}^2}\right)\left(74.5 \; \frac{\text{kg}}{\text{s}}\right)}$$

$$= 14.28 \; \text{m}$$

At 1 (pump inlet):

$$p_1 = 66.3 \; \text{kPa}$$

$$z_1 = 0$$

$$v_1 = v_2\left(\frac{A_2}{A_1}\right) = v_2\left(\frac{D_2}{D_1}\right)^2 = \left(4 \; \frac{\text{m}}{\text{s}}\right)\left(\frac{154.1 \; \text{mm}}{202.7 \; \text{mm}}\right)^2$$

$$= 2.31 \; \text{m/s}$$

At 2 (pump outlet):

$$p_2 = 101.3 \; \text{kPa}$$

$$v_2 = 4 \; \text{m/s} \quad \text{[given]}$$

From Eq. 18.8(a), the head added by the pump is

$$h_A = \frac{p_2 - p_1}{\rho g} + \frac{v_2^2 - v_1^2}{2g} + z_2 - z_1 + h_f + z_3$$

$$14.28 \; \text{m} = \frac{(101.3 \; \text{kPa} - 66.3 \; \text{kPa})\left(1000 \; \frac{\text{Pa}}{\text{kPa}}\right)}{\left(998.83 \; \frac{\text{kg}}{\text{m}^3}\right)\left(9.81 \; \frac{\text{m}}{\text{s}^2}\right)}$$

$$+ \frac{\left(4 \; \frac{\text{m}}{\text{s}}\right)^2 - \left(2.31 \; \frac{\text{m}}{\text{s}}\right)^2}{(2)\left(9.81 \; \frac{\text{m}}{\text{s}^2}\right)}$$

$$+ 0 - 0 + 3.3 \; \text{m} + z_3$$

$$z_3 = \boxed{6.86 \; \text{m} \quad (6.9 \; \text{m})}$$

The answer is (A).

10. *Customary U.S. Solution*

(a) The flow rate is

$$\dot{V} = \left(10,000 \; \frac{\text{gal}}{\text{hr}}\right)\left(0.1337 \; \frac{\text{ft}^3}{\text{gal}}\right) = 1337 \; \text{ft}^3/\text{hr}$$

From App. 16.B, for 4 in schedule-40 steel pipe,

$$D = 0.3355 \; \text{ft}$$

$$A = 0.08841 \; \text{ft}^2$$

The velocity in the pipe is

$$v = \frac{\dot{V}}{A} = \frac{1337 \; \frac{\text{ft}^3}{\text{hr}}}{(0.08841 \; \text{ft}^2)\left(3600 \; \frac{\text{sec}}{\text{hr}}\right)}$$

$$= \boxed{4.20 \; \text{ft/sec} \quad (4 \; \text{ft/sec})}$$

The answer is (B).

(b) From App. 14.A, the kinematic viscosity of water at 60°F is

$$\nu = 1.217 \times 10^{-5} \; \text{ft}^2/\text{sec}$$

$$\rho = 62.37 \; \text{lbm/ft}^3$$

The Reynolds number is

$$\text{Re} = \frac{Dv}{\nu} = \frac{(0.3355 \; \text{ft})\left(4.20 \; \frac{\text{ft}}{\text{sec}}\right)}{1.217 \times 10^{-5} \; \frac{\text{ft}^2}{\text{sec}}}$$

$$= \boxed{1.16 \times 10^5 \quad (1.2 \times 10^5)}$$

The answer is (A).

(c) From App. 17.A, for welded and seamless steel, $\epsilon = 0.0002$ ft.

$$\frac{\epsilon}{D} = \frac{0.0002 \; \text{ft}}{0.3355 \; \text{ft}} \approx 0.0006$$

From App. 17.B, the friction factor, f, is 0.0205. The 7000 ft of equivalent length includes the pipe between the lake and the pump. The friction head is

$$h_f = \frac{fLv^2}{2Dg} = \frac{(0.0205)(7000 \; \text{ft})\left(4.20 \; \frac{\text{ft}}{\text{sec}}\right)^2}{(2)(0.3355 \; \text{ft})\left(32.2 \; \frac{\text{ft}}{\text{sec}^2}\right)}$$

$$= \boxed{117.2 \; \text{ft} \quad (120 \; \text{ft})}$$

The answer is (B).

(d) The head added by the pump is

$$h_A = h_f + h_z = 117.2 \; \text{ft} + 12 \; \text{ft} + 350 \; \text{ft}$$

$$= 479.2 \; \text{ft}$$

From Table 18.5, the hydraulic horsepower is

$$\text{WHP} = \frac{h_A Q(\text{SG})}{3956} = \frac{(479.2 \; \text{ft})\left(10,000 \; \frac{\text{gal}}{\text{hr}}\right)(1)}{\left(3956 \; \frac{\text{ft-gal}}{\text{hp-min}}\right)\left(60 \; \frac{\text{min}}{\text{hr}}\right)}$$

$$= \boxed{20.2 \; \text{hp} \quad (20 \; \text{hp})}$$

The answer is (D).

(e) From Eq. 18.15, the electrical horsepower is

$$\text{EHP} = \frac{\text{WHP}}{\eta} = \frac{20.2 \text{ hp}}{0.7}$$
$$= 28.9 \text{ hp}$$

At \$0.04/kW-hr, power costs for 1 hr are

$$(28.9 \text{ hp})\left(0.7457 \frac{\text{kW}}{\text{hp}}\right)(1 \text{ hr})\left(0.04 \frac{\$}{\text{kW-hr}}\right)$$
$$= \boxed{\$0.86 \text{ per hour} \quad (\$1 \text{ per hour})}$$

The answer is (B).

(f) The motor horsepower, EHP, is 28.9 hp. Select the next higher standard motor size. Use a $\boxed{30 \text{ hp motor.}}$

The answer is (B).

(g) From Eq. 18.4(b),

$$h_\text{atm} = \frac{p_\text{atm}}{\rho} \times \frac{g_c}{g} = \frac{\left(14.7 \frac{\text{lbf}}{\text{in}^2}\right)\left(12 \frac{\text{in}}{\text{ft}}\right)^2}{62.37 \frac{\text{lbm}}{\text{ft}^3}} \times \frac{32.2 \frac{\text{lbm-ft}}{\text{lbf-sec}^2}}{32.2 \frac{\text{ft}}{\text{sec}^2}}$$
$$= 33.94 \text{ ft}$$

The friction loss through 300 ft is provided.

$$h_{f(s)} = \left(\frac{300 \text{ ft}}{7000 \text{ ft}}\right)h_f = \left(\frac{300 \text{ ft}}{7000 \text{ ft}}\right)(117.2 \text{ ft})$$
$$= 5.0 \text{ ft}$$

From App. 14.A, the vapor pressure head at 60°F is 0.59 ft.

From Eq. 18.30(a), the NPSHA is

$$\text{NPSHA} = h_\text{atm} + h_{z(s)} - h_{f(s)} - h_\text{vp}$$
$$= 33.94 \text{ ft} - 12 \text{ ft} - 5.0 \text{ ft} - 0.59 \text{ ft}$$
$$= \boxed{16.35 \text{ ft} \quad (16 \text{ ft})}$$

The answer is (D).

SI Solution

(a) From App. 16.C, for 4 in schedule-40 steel pipe,

$$D = 102.3 \text{ mm}$$
$$A = 82.19 \times 10^{-4} \text{ m}^2$$

The velocity in the pipe is

$$\text{v} = \frac{\dot{V}}{A} = \frac{10.5 \frac{\text{L}}{\text{s}}}{(82.19 \times 10^{-4} \text{ m}^2)\left(1000 \frac{\text{L}}{\text{m}^3}\right)}$$
$$= \boxed{1.28 \text{ m/s} \quad (1.0 \text{ m/s})}$$

The answer is (B).

(b) From App. 14.B, at 16°C the water data are

$$\rho = 998.83 \text{ kg/m}^3$$
$$\mu = 1.1261 \times 10^{-3} \text{ Pa·s}$$

The Reynolds number is

$$\text{Re} = \frac{\rho \text{v} D}{\mu} = \frac{\left(998.83 \frac{\text{kg}}{\text{m}^3}\right)\left(1.28 \frac{\text{m}}{\text{s}}\right)(102.3 \text{ mm})}{(1.1261 \times 10^{-3} \text{ Pa·s})\left(1000 \frac{\text{mm}}{\text{m}}\right)}$$
$$= \boxed{1.16 \times 10^5 \quad (1.2 \times 10^5)}$$

The answer is (A).

(c) From Table 17.2, for welded and seamless steel, $\epsilon = 6.0 \times 10^{-5}$ m.

$$\frac{\epsilon}{D} = \frac{(6.0 \times 10^{-5} \text{ m})\left(1000 \frac{\text{mm}}{\text{m}}\right)}{102.3 \text{ mm}} \approx 0.0006$$

From App. 17.B, the friction factor is $f = 0.0205$.

From Eq. 18.5, the friction head is

$$h_f = \frac{fL\text{v}^2}{2Dg} = \frac{(0.0205)(2300 \text{ m})\left(1.28 \frac{\text{m}}{\text{s}}\right)^2\left(1000 \frac{\text{mm}}{\text{m}}\right)}{(2)(102.3 \text{ mm})\left(9.81 \frac{\text{m}}{\text{s}^2}\right)}$$
$$= \boxed{38.5 \text{ m} \quad (40 \text{ m})}$$

The answer is (B).

(d) The head added by the pump is

$$h_A = h_f + h_z = 38.5 \text{ m} + 4 \text{ m} + 115 \text{ m} = 157.5 \text{ m}$$

From Table 18.6, the hydraulic power is

$$\text{WkW} = \frac{9.81 h_A Q(\text{SG})}{1000}$$
$$= \frac{\left(9.81 \frac{\text{m}}{\text{s}^2}\right)(157.5 \text{ m})\left(10.5 \frac{\text{L}}{\text{s}}\right)(1)}{1000 \frac{\text{W·L}}{\text{kW·kg}}}$$
$$= \boxed{16.22 \text{ kW} \quad (16 \text{ kW})}$$

The answer is (D).

(e) From Eq. 18.15, the electrical power is

$$\text{EHP} = \frac{\text{WHP}}{\eta} = \frac{16.22 \text{ kW}}{0.7} = 23.2 \text{ kW}$$

At \$0.04/kW·h, power costs for 1 h are

$$(23.2 \text{ kW})(1 \text{ h})\left(0.04 \frac{\$}{\text{kW·h}}\right) = \boxed{\$0.93 \quad (\$1) \text{ per hour}}$$

The answer is (B).

(f) The required motor power is 23.2 kW. Select the next higher standard motor size. Use a 25 kW motor.

The answer is (B).

(g) From Eq. 18.4(a),

$$h_{atm} = \frac{p}{\rho g} = \frac{(101 \text{ kPa})\left(1000 \frac{\text{Pa}}{\text{kPa}}\right)}{\left(998.83 \frac{\text{kg}}{\text{m}^3}\right)\left(9.81 \frac{\text{m}}{\text{s}^2}\right)} = 10.31 \text{ m}$$

The prorated friction loss through 100 m is

$$h_{f(s)} = \left(\frac{100 \text{ m}}{2300 \text{ m}}\right)h_f = \left(\frac{100 \text{ m}}{2300 \text{ m}}\right)(38.5 \text{ m}) = 1.67 \text{ m}$$

Interpolating from App. 14.B, the vapor pressure at 16°C is 1.894 kPa.

From Eq. 18.4(a),

$$h_{vp} = \frac{p_{vp}}{g\rho} = \frac{(1.894 \text{ kPa})\left(1000 \frac{\text{Pa}}{\text{kPa}}\right)}{\left(9.81 \frac{\text{m}}{\text{s}^2}\right)\left(998.83 \frac{\text{kg}}{\text{m}^3}\right)}$$

$$= 0.19 \text{ m}$$

From Eq. 18.30(a), the NPSHA is

$$\begin{aligned}\text{NPSHA} &= h_{atm} + h_{z(s)} - h_{f(s)} - h_{vp} \\ &= 10.31 \text{ m} - 4 \text{ m} - 1.67 \text{ m} - 0.19 \text{ m} \\ &= \boxed{4.45 \text{ m} \quad (4.5 \text{ m})}\end{aligned}$$

The answer is (D).

11. (a) Sewers are usually gravity-flow (open channel) systems. $h_f = \Delta z = 48$ ft since $\Delta p = 0$ and $\Delta v = 0$.

$$Q = \frac{\left(250 \frac{\text{gal}}{\text{person-day}}\right)(10{,}000 \text{ people})}{\left(24 \frac{\text{hr}}{\text{day}}\right)\left(60 \frac{\text{min}}{\text{hr}}\right)}$$

$$= 1736 \text{ gal/min}$$

Solving for d from Eq. 17.31,

$$d_{in}^{4.87} = \frac{10.44 L_{ft} Q_{gpm}^{1.85}}{C^{1.85} h_f}$$

$$= \frac{(10.44)(5000 \text{ ft})\left(1736 \frac{\text{gal}}{\text{min}}\right)^{1.85}}{(130)^{1.85}(48 \text{ ft})}$$

$$= 131{,}462$$

$$d = \boxed{11.24 \text{ in} \quad \text{[round to 12 in minimum]}}$$

The answer is (B).

(b) Without having a specific pump curve, the number of pumps can only be specified based on general rules. Use the *Ten States' Standards* (TSS), which states:

- No station will have less than two identical pumps.
- Capacity must be met with one pump out of service.
- Provision must be made in order to alternate pumps automatically.

Two pumps are required, plus spares.

The answer is (A).

(c) With a variable speed pump, it will be possible to adjust to the wide variations in flow (100–250 gpcd). It may be possible to operate with one pump. However, TSS still requires two.

The answer is (B).

(d) With three constant speed pumps,

$$Q = \frac{1736 \frac{\text{gal}}{\text{min}}}{3} = 579 \text{ gal/min at maximum capacity}$$

To get to the pump, the sewage descended 48 ft under the influence of gravity. The pump only has to lift the sewage 10 ft. From Table 18.5, assuming specific gravity ≈ 1.00,

$$\text{rated motor power} = \frac{h_A Q(\text{SG})}{3956\eta} = \frac{(10 \text{ ft})\left(579 \frac{\text{gal}}{\text{min}}\right)(1)}{\left(3956 \frac{\text{gal-ft}}{\text{min-hp}}\right)(0.60)}$$

$$= \boxed{2.44 \text{ hp} \quad (3 \text{ hp})}$$

The answer is (B).

(e) With two variable-speed pumps,

$$Q = \frac{1736 \frac{\text{gal}}{\text{min}}}{2} = 868 \text{ gal/min}$$

$$\text{rated motor power} = \frac{h_A Q(\text{SG})}{3956\eta} = \frac{(10 \text{ ft})\left(868 \frac{\text{gal}}{\text{min}}\right)(1)}{\left(3956 \frac{\text{gal-ft}}{\text{min-hp}}\right)(0.80)}$$

$$= \boxed{2.74 \text{ hp} \quad (3 \text{ hp})}$$

The answer is (A).

(f) Incoming flow rate (choke III) cannot be controlled. It is independent of sump level. All other options are valid.

The answer is (D).

12. (a) The static suction lift, $h_{p(s)}$, is 20 ft.

The answer is (B).

(b) The static discharge head, $h_{p(d)}$, is $\boxed{15 \text{ ft.}}$

The answer is (A).

(c) There is no pipe size specified, so h_v cannot be calculated. Even so, v is typically in the 5–10 ft/sec range, and $h_v \approx 0$.

Since pipe lengths are not given, assume $h_f \approx 0$.

$$20 \text{ ft} + 15 \text{ ft} + \frac{\left(80 \frac{\text{lbf}}{\text{in}^2}\right)\left(12 \frac{\text{in}}{\text{ft}}\right)^2}{62.4 \frac{\text{lbf}}{\text{ft}^3}} + 10 \text{ ft}$$

$$= \boxed{229.6 \text{ ft } (230 \text{ ft}) \text{ of water}}$$

The answer is (D).

(d) The mass flow rate is

$$\frac{(3.5 \text{ MGD})\left(62.4 \frac{\text{lbm}}{\text{ft}^3}\right)\left(10^6 \frac{\text{gal}}{\text{MG}}\right)}{\left(7.4805 \frac{\text{gal}}{\text{ft}^3}\right)\left(24 \frac{\text{hr}}{\text{day}}\right)} = 337.9 \text{ lbm/sec}$$

$$\times \left(60 \frac{\text{min}}{\text{hr}}\right)\left(60 \frac{\text{sec}}{\text{min}}\right)$$

The rated motor output power does not depend on the motor efficiency. The motor produces what it is rated to produce. From Table 18.5,

$$P = \frac{h_A \dot{m}}{550 \eta_{\text{pump}}} \times \frac{g}{g_c}$$

$$= \frac{(229.6 \text{ ft})\left(337.9 \frac{\text{lbm}}{\text{sec}}\right)}{\left(550 \frac{\text{ft-lbf}}{\text{hp-sec}}\right)(0.85)} \times \frac{32.2 \frac{\text{ft}}{\text{sec}^2}}{32.2 \frac{\text{lbm-ft}}{\text{lbf-sec}^2}}$$

$$= 166.0 \text{ hp}$$

$$\boxed{\text{Use a 200 hp motor.}}$$

The answer is (D).

13. *Customary U.S. Solution*

(a) From App. 16.B, for 3 in schedule-40 steel pipe,

$$D = 0.2557 \text{ ft}$$

$$A = 0.05134 \text{ ft}^2$$

From App. 17.D, the equivalent lengths for various fittings are

flanged elbow, $L_e = 4.4$ ft

wide-open gate valve, $L_e = 2.8$ ft

The total equivalent length of pipe and fittings is

$$L_e = 500 \text{ ft} + (6)(4.4 \text{ ft}) + (2)(2.8 \text{ ft})$$

$$= \boxed{532 \text{ ft } (530 \text{ ft})}$$

The answer is (D).

(b) The flow rate is 100 gal/min.

The velocity in the pipe is

$$\text{v} = \frac{\dot{V}}{A} = \frac{100 \frac{\text{gal}}{\text{min}}}{(0.05134 \text{ ft}^2)\left(7.4805 \frac{\text{gal}}{\text{ft}^3}\right)\left(60 \frac{\text{sec}}{\text{min}}\right)}$$

$$= \boxed{4.34 \text{ ft/sec } (4.3 \text{ ft/sec})}$$

The answer is (C).

(c) The Reynolds number is

$$\text{Re} = \frac{\text{v}D}{\nu} = \frac{\left(4.34 \frac{\text{ft}}{\text{sec}}\right)(0.2557 \text{ ft})}{6 \times 10^{-6} \frac{\text{ft}^2}{\text{sec}}} = 1.85 \times 10^5$$

From App. 17.A, $\epsilon = 0.0002$ ft.

$$\frac{\epsilon}{D} = \frac{0.0002 \text{ ft}}{0.2557 \text{ ft}} \approx 0.0008$$

From the friction factor table, $f \approx 0.0204$.

For higher flow rates, f approaches 0.0186. Since the chosen flow rate was almost the lowest, $f = 0.0186$ should be used.

From Eq. 18.5, the friction head loss is

$$h_f = \frac{fL\text{v}^2}{2Dg} = \frac{(0.0186)(532 \text{ ft})\left(4.34 \frac{\text{ft}}{\text{sec}}\right)^2}{(2)(0.2557 \text{ ft})\left(32.2 \frac{\text{ft}}{\text{sec}^2}\right)}$$

$$= \boxed{11.3 \text{ ft } (11 \text{ ft}) \text{ of gasoline}}$$

The answer is (D).

(d) The friction head loss neglects the small velocity head. The other system points can be found using Eq. 18.43.

$$\frac{h_{f_1}}{h_{f_2}} = \left(\frac{Q_1}{Q_2}\right)^2$$

$$h_{f_2} = h_{f_1}\left(\frac{Q_2}{100 \frac{\text{gal}}{\text{min}}}\right)^2 = (11.3 \text{ ft})\left(\frac{Q_2}{100 \frac{\text{gal}}{\text{min}}}\right)^2$$

$$= 0.00113 Q_2^2$$

Q (gal/min)	h_f (ft)	$h_f + 60$ (ft)
100	11.3	71.3
200	45.2	105.2
300	101.7	161.7
400	180.8	240.8
500	282.5	342.5
600	406.8	466.8

The pump's characteristic curve is independent of the liquid's specific gravity. Plot the system and pump curves.

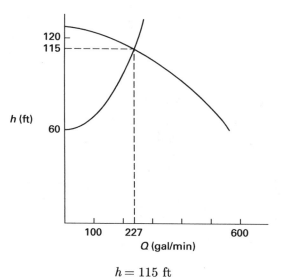

$$h = 115 \text{ ft}$$

The transfer rate is

$$\boxed{Q = 227 \text{ gal/min} \quad (230 \text{ gal/min})}$$

(This value could be used to determine a new friction factor.)

The answer is (D).

(e) From Table 18.5, the electrical power supplied to the motor is

$$\text{EHP} = \frac{h_A Q (\text{SG})}{3956 \eta_{\text{pump}} \eta_{\text{motor}}} = \frac{(115 \text{ ft})\left(227 \dfrac{\text{gal}}{\text{min}}\right)(0.7)}{\left(3956 \dfrac{\text{ft-gal}}{\text{hp-min}}\right)(0.88)(0.88)}$$

$$= 5.96 \text{ hp}$$

The cost per hour is

$$(5.96 \text{ hp})\left(0.7457 \dfrac{\text{kW}}{\text{hp}}\right)(1 \text{ hr})\left(0.045 \dfrac{\$}{\text{kW-hr}}\right) = \boxed{\$0.20}$$

The answer is (A).

SI Solution

(a) From App. 16.C, for 3 in schedule-40 pipe,

$$D = 77.92 \text{ mm}$$

$$A = 47.69 \times 10^{-4} \text{ m}^2$$

From App. 17.D, the equivalent lengths for various fittings are

$$\text{flanged elbow,} \quad L_e = 4.4 \text{ ft}$$

$$\text{wide-open gate valve,} \quad L_e = 2.8 \text{ ft}$$

The total equivalent length of pipe and fittings is

$$L_e = 170 \text{ m} + \big((6)(4.4 \text{ ft}) + (2)(2.8 \text{ ft})\big)\left(0.3048 \dfrac{\text{m}}{\text{ft}}\right)$$

$$= \boxed{179.8 \text{ m} \quad (180 \text{ m})}$$

The answer is (D).

(b) The flow rate is 6.3 L/s. The velocity in the pipe is

$$\text{v} = \frac{\dot{V}}{A} = \frac{6.3 \dfrac{\text{L}}{\text{s}}}{(47.69 \times 10^{-4} \text{ m}^2)\left(1000 \dfrac{\text{L}}{\text{m}^3}\right)}$$

$$= \boxed{1.32 \text{ m/s} \quad (1.3 \text{ m/s})}$$

The answer is (C).

(c) The Reynolds number is

$$\text{Re} = \frac{\text{v}D}{\nu} = \frac{\left(1.32 \dfrac{\text{m}}{\text{s}}\right)(77.92 \text{ mm})}{\left(5.6 \times 10^{-7} \dfrac{\text{m}^2}{\text{s}}\right)\left(1000 \dfrac{\text{mm}}{\text{m}}\right)} = 1.84 \times 10^5$$

From Table 17.2, $\epsilon = 6.0 \times 10^{-5}$ m.

$$\frac{\epsilon}{D} = \frac{(6.0 \times 10^{-5} \text{ m})\left(1000 \dfrac{\text{mm}}{\text{m}}\right)}{77.92 \text{ mm}} \approx 0.0008$$

From the friction factor table (see App. 17.B), $f = 0.0204$.

For higher flow rates, f approaches 0.0186. Since the chosen flow rate was almost the lowest, $f = 0.0186$ should be used.

From Eq. 18.5, the friction head loss is

$$h_f = \frac{f L \text{v}^2}{2Dg} = \frac{(0.0186)(179.8 \text{ m})\left(1.32 \dfrac{\text{m}}{\text{s}}\right)^2\left(1000 \dfrac{\text{mm}}{\text{m}}\right)}{(2)(77.92 \text{ mm})\left(9.81 \dfrac{\text{m}}{\text{s}^2}\right)}$$

$$= \boxed{3.8 \text{ m of gasoline}}$$

The answer is (D).

(d) The friction head loss neglects the small velocity head. The other system points can be found using Eq. 18.43.

$$\frac{h_{f1}}{h_{f2}} = \left(\frac{Q_1}{Q_2}\right)^2$$

$$h_{f2} = h_{f1}\left(\frac{Q_2}{Q_1}\right)^2 = (3.80 \text{ m})\left(\frac{Q_2}{6.3 \dfrac{\text{L}}{\text{s}}}\right)^2$$

$$= 0.0957 Q_2^2$$

Q (L/s)	h_f (m)	$h_f + 20$ (m)
6.3	3.80	23.80
12	13.78	33.78
18	31.0	51.0
24	55.1	75.1
30	86.1	106.1
36	124.0	144.0

The pump's characteristic curve is independent of the liquid's specific gravity. Plot the system and pump curves.

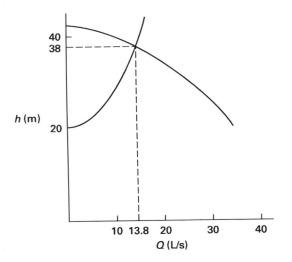

$$h = 38.0 \text{ m}$$

The transfer rate is

$$\boxed{Q = 13.6 \text{ L/s} \quad (14 \text{ L/s})}$$

(This value could be used to determine a new friction factor.)

The answer is (D).

(e) From Table 18.6, the electric power delivered to the motor is

$$\text{EkW} = \frac{9.81 h_A Q(\text{SG})}{1000 \eta_{\text{pump}} \eta_{\text{motor}}}$$

$$= \frac{\left(9.81 \frac{\text{m}}{\text{s}^2}\right)(38.0 \text{ m})\left(13.8 \frac{\text{L}}{\text{s}}\right)(0.7)}{\left(1000 \frac{\text{W}}{\text{kW}}\right)(0.88)(0.88)}$$

$$= 4.65 \text{ kW}$$

The cost per hour is

$$(4.65 \text{ kW})(1 \text{ h})\left(0.045 \frac{\$}{\text{kW·h}}\right) = \boxed{\$0.21 \quad (\$0.20)}$$

The answer is (A).

14. *Customary U.S. Solution*

From Table 18.5, the hydraulic horsepower is

$$\text{WHP} = \frac{\Delta p Q}{1714}$$

$$\Delta p = p_d - p_s$$

The absolute pressures are

$$p_d = 40 \text{ psig} + 14.7 \text{ psia} = 54.7 \text{ psia}$$

$$p_s = 1 \text{ atm} = 14.7 \text{ psia}$$

$$\Delta p = p_d - p_s = 54.7 \text{ psia} - 14.7 \text{ psia} = 40 \text{ psia}$$

$$\text{WHP} = \frac{\left(40 \frac{\text{lbf}}{\text{in}^2}\right)\left(37 \frac{\text{gal}}{\text{min}}\right)}{1714 \frac{\text{lbf-gal}}{\text{in}^2\text{-min-hp}}} = \boxed{0.863 \text{ hp} \quad (0.9 \text{ hp})}$$

The answer is (B).

SI Solution

From Table 18.6, the hydraulic kilowatts are

$$\text{WkW} = \frac{\Delta p Q}{1000}$$

$$\Delta p = p_d - p_s$$

The absolute pressures are

$$p_d = 275 \text{ kPa} + 101.3 \text{ kPa} = 376.3 \text{ kPa}$$

$$p_s = 1 \text{ atm} = 101.3 \text{ kPa}$$

$$\Delta p = p_d - p_s = 376.3 \text{ kPa} - 101.3 \text{ kPa} = 275 \text{ kPa}$$

$$\text{WkW} = \frac{(275 \text{ kPa})\left(65 \frac{\text{L}}{\text{s}}\right)}{1000 \frac{\text{W}}{\text{kW}}} = \boxed{17.88 \text{ kW} \quad (18 \text{ kW})}$$

The answer is (B).

15. *Customary U.S. Solution*

From Eq. 18.28(b), the specific speed is

$$n_s = \frac{n\sqrt{Q}}{h_A^{0.75}}$$

For a double-suction pump, Q in the preceding equation is half of the full flow rate.

$$n_s = \frac{\left(900 \frac{\text{rev}}{\text{min}}\right)\sqrt{\left(\tfrac{1}{2}\right)\left(300 \frac{\text{gal}}{\text{sec}}\right)\left(60 \frac{\text{sec}}{\text{min}}\right)}}{(20 \text{ ft})^{0.75}}$$

$$= \boxed{9028 \text{ rpm} \quad (9000 \text{ rpm})}$$

The answer is (C).

SI Solution

From Eq. 18.28(a), the specific speed is

$$n_s = \frac{n\sqrt{\dot{V}}}{h_A^{0.75}}$$

For a double-suction pump, $\dot{V}$ in the preceding equation is half of the full flow rate.

$$n_s = \frac{\left(900\ \frac{\text{rev}}{\text{min}}\right)\sqrt{\left(\frac{1}{2}\right)\left(1.1\ \frac{\text{kL}}{\text{s}}\right)\left(1\ \frac{\text{m}^3}{\text{kL}}\right)}}{(7\ \text{m})^{0.75}}$$

$$= \boxed{155.1\ \text{rpm} \quad (160\ \text{rpm})}$$

The answer is (C).

16. *Customary U.S. Solution*

This problem is solved graphically using charts of maximum suction lift from *Standards of the Hydraulic Institute.*

Each stage adds 150 ft of head and the suction lift is 10 ft. Therefore, for a single-suction pump, $n \approx$ $\boxed{2050\ \text{rpm}\ (2000\ \text{rpm}).}$

The answer is (D).

17. *Customary U.S. Solution*

From App. 16.B, for 1.5 in schedule-40 steel pipe,

$$D = 0.1342\ \text{ft}$$

$$A = 0.01414\ \text{ft}^2$$

The velocity in the pipe is

$$\text{v} = \frac{\dot{V}}{A} = \frac{100\ \frac{\text{gal}}{\text{min}}}{(0.01414\ \text{ft}^2)\left(7.4805\ \frac{\text{gal}}{\text{ft}^3}\right)\left(60\ \frac{\text{sec}}{\text{min}}\right)}$$

$$= 15.76\ \text{ft/sec}$$

From App. 17.D, the equivalent lengths for screwed steel fittings are

inlet (square mouth): $L_e = 3.1$ ft

long radius 90° elbow: $L_e = 3.4$ ft

wide-open gate valves: $L_e = 1.2$ ft

The total equivalent length is

$$30\ \text{ft} + 3.1\ \text{ft} + (2)(3.4\ \text{ft}) + (2)(1.2\ \text{ft}) = 42.3\ \text{ft}$$

From App. 17.A, for steel, $\epsilon = 0.0002$ ft.

$$\frac{\epsilon}{D} = \frac{0.0002\ \text{ft}}{0.1342\ \text{ft}} = 0.0015$$

At 281°F, $\nu = 0.239 \times 10^{-5}$ ft²/sec. The Reynolds number is

$$\text{Re} = \frac{D\text{v}}{\nu} = \frac{(0.1342\ \text{ft})\left(15.76\ \frac{\text{ft}}{\text{sec}}\right)}{0.239 \times 10^{-5}\ \frac{\text{ft}^2}{\text{sec}}}$$

$$= 8.85 \times 10^5$$

From App. 17.B, the friction factor is $f = 0.022$.

From Eq. 18.5, the friction head is

$$h_f = \frac{fL\text{v}^2}{2Dg} = \frac{(0.022)(42.3\ \text{ft})\left(15.76\ \frac{\text{ft}}{\text{sec}}\right)^2}{(2)(0.1342\ \text{ft})\left(32.2\ \frac{\text{ft}}{\text{sec}^2}\right)}$$

$$= 26.74\ \text{ft}$$

The density of the liquid is the reciprocal of the specific volume, taken from App. 29.B at 281°F.

$$\rho = \frac{1}{v_f} = \frac{1}{0.01727\ \frac{\text{ft}^3}{\text{lbm}}} = 57.9\ \text{lbm/ft}^3$$

From Eq. 18.4(b),

$$h_{\text{vp}} = \frac{p_{\text{vapor}}}{\rho} \times \frac{g_c}{g} = \frac{\left(50.02\ \frac{\text{lbf}}{\text{in}^2}\right)\left(12\ \frac{\text{in}}{\text{ft}}\right)^2}{57.9\ \frac{\text{lbm}}{\text{ft}^3}} \times \frac{32.2\ \frac{\text{lbm-ft}}{\text{lbf-sec}^2}}{32.2\ \frac{\text{ft}}{\text{sec}^2}}$$

$$= 124.4\ \text{ft}$$

From Eq. 18.4(b), the pressure head is

$$h_p = \frac{p}{\rho} \times \frac{g_c}{g} = \frac{\left(80\ \frac{\text{lbf}}{\text{in}^2}\right)\left(12\ \frac{\text{in}}{\text{ft}}\right)^2}{57.9\ \frac{\text{lbm}}{\text{ft}^3}} \times \frac{32.2\ \frac{\text{lbm-ft}}{\text{lbf-sec}^2}}{32.2\ \frac{\text{ft}}{\text{sec}^2}}$$

$$= 199.0\ \text{ft}$$

From Eq. 18.30, the NPSHA is

$$\text{NPSHA} = h_p + h_{z(s)} - h_{f(s)} - h_{\text{vp}}$$

$$= 199.0\ \text{ft} + 20\ \text{ft} - 26.74\ \text{ft} - 124.4\ \text{ft}$$

$$= \boxed{67.9\ \text{ft} \quad (68\ \text{ft})}$$

Since NPSHR = 10 ft, $\boxed{\text{the pump will not cavitate.}}$

(A pump may not be needed in this configuration.)

The answer is (D).

SI Solution

From App. 16.C, for 1.5 in schedule-40 steel pipe,

$$D = 40.89 \text{ mm}$$
$$A = 13.13 \times 10^{-4} \text{ m}^2$$

The velocity in the pipe is

$$v = \frac{\dot{V}}{A} = \frac{6.3 \frac{\text{L}}{\text{s}}}{(13.13 \times 10^{-4} \text{ m}^2)\left(1000 \frac{\text{L}}{\text{m}^3}\right)}$$
$$= 4.80 \text{ m/s}$$

From App. 17.D, the equivalent lengths for screwed steel fittings are

inlet (square mouth): $L_e = 3.1$ ft

long radius 90° elbow: $L_e = 3.4$ ft

wide-open gate valves: $L_e = 1.2$ ft

The total equivalent length is

$$10 \text{ m} + \left(\begin{array}{c} 3.1 \text{ ft} + (2)(3.4 \text{ ft}) \\ + (2)(1.2 \text{ ft}) \end{array}\right)\left(0.3048 \frac{\text{m}}{\text{ft}}\right) = 13.75 \text{ m}$$

From Table 17.2, for steel, $\epsilon = 6.0 \times 10^{-5}$ m.

$$\frac{\epsilon}{D} = \frac{(6.0 \times 10^{-5} \text{ m})\left(1000 \frac{\text{mm}}{\text{m}}\right)}{40.89 \text{ mm}}$$
$$\approx 0.0015$$

At 138°C, $\nu = 0.222 \times 10^{-6}$ m^2/s. The Reynolds number is

$$\text{Re} = \frac{Dv}{\nu} = \frac{(40.89 \text{ mm})\left(4.80 \frac{\text{m}}{\text{s}}\right)}{\left(0.222 \times 10^{-6} \frac{\text{m}^2}{\text{s}}\right)\left(1000 \frac{\text{mm}}{\text{m}}\right)}$$
$$= 8.84 \times 10^5$$

From App. 17.B, the friction factor is $f = 0.022$.

From Eq. 18.5, the friction head is

$$h_f = \frac{fLv^2}{2Dg}$$

$$= \frac{(0.022)(13.75 \text{ m})\left(4.80 \frac{\text{m}}{\text{s}}\right)^2\left(1000 \frac{\text{mm}}{\text{m}}\right)}{(2)(40.89 \text{ mm})\left(9.81 \frac{\text{m}}{\text{s}^2}\right)}$$

$$= 8.69 \text{ m}$$

The density of the liquid is the reciprocal of the specific volume. From App. 29.N at 281°F (138°C),

$$\rho = \frac{1}{v_f} = \frac{(1)\left(3.281 \frac{\text{ft}}{\text{m}}\right)^3}{\left(0.01727 \frac{\text{ft}^3}{\text{lbm}}\right)\left(2.205 \frac{\text{lbm}}{\text{kg}}\right)} = 927.5 \text{ kg/m}^3$$

From Eq. 18.4(a),

$$h_{\text{vp}} = \frac{p_{\text{vapor}}}{\rho g} = \frac{(3.431 \text{ bar})\left(1 \times 10^5 \frac{\text{Pa}}{\text{bar}}\right)}{\left(927.5 \frac{\text{kg}}{\text{m}^3}\right)\left(9.81 \frac{\text{m}}{\text{s}^2}\right)}$$

$$= 37.71 \text{ m}$$

From Eq. 18.4(a), the pressure head is

$$h_p = \frac{p}{\rho g} = \frac{(550 \text{ kPa})\left(1000 \frac{\text{Pa}}{\text{kPa}}\right)}{\left(927.5 \frac{\text{kg}}{\text{m}^3}\right)\left(9.81 \frac{\text{m}}{\text{s}^2}\right)}$$

$$= 60.45 \text{ m}$$

From Eq. 18.30, the NPSHA is

$$\text{NPSHA} = h_p + h_{z(s)} - h_{f(s)} - h_{\text{vp}}$$
$$= 60.45 \text{ m} + 6 \text{ m} - 8.69 \text{ m} - 37.71 \text{ m}$$
$$= \boxed{20.05 \text{ m} \quad (21 \text{ m})}$$

Since NPSHR is 3 m, $\boxed{\text{the pump will not cavitate.}}$

(A pump may not be needed in this configuration.)

The answer is (D).

18. The solvent is the water (fresh), and the solution is the seawater. Since seawater contains approximately $2\frac{1}{2}\%$ salt (NaCl) by weight, 100 lbm of seawater will yield 2.5 lbm salt and 97.5 lbm water. The molecular weight of salt is $23.0 + 35.5 = 58.5$ lbm/lbmol. The number of moles of salt in 100 lbm of seawater is

$$n_{\text{salt}} = \frac{m}{\text{MW}} = \frac{2.5 \text{ lbm}}{58.5 \frac{\text{lbm}}{\text{lbmol}}} = 0.043 \text{ lbmol}$$

Similarly, water's molecular weight is 18.016 lbm/lbmol. The number of moles of water is

$$n_{\text{water}} = \frac{97.5 \text{ lbm}}{18.016 \frac{\text{lbm}}{\text{lbmol}}} = 5.412 \text{ lbmol}$$

The mole fraction of water is

$$\frac{5.412 \text{ lbmol}}{5.412 \text{ lbmol} + 0.043 \text{ lbmol}} = 0.992$$

Customary U.S. Solution

Cavitation will occur when

$$h_{\text{atm}} - h_{\text{v}} < h_{\text{vp}}$$

The density of seawater is 64.0 lbm/ft^3.

From Eq. 18.4(b), the atmospheric head is

$$h_{\text{atm}} = \frac{p}{\rho} \times \frac{g_c}{g} = \frac{\left(14.7 \, \frac{\text{lbf}}{\text{in}^2}\right)\left(12 \, \frac{\text{in}}{\text{ft}}\right)^2}{64.0 \, \frac{\text{lbm}}{\text{ft}^3}} \times \frac{32.2 \, \frac{\text{lbm-ft}}{\text{lbf-sec}^2}}{32.2 \, \frac{\text{ft}}{\text{sec}^2}}$$

$$= 33.075 \text{ ft} \quad [\text{ft of seawater}]$$

$h_{\text{depth}} = 8 \text{ ft} \quad [\text{given}]$

From Eq. 18.6, the velocity head is

$$h_{\text{v}} = \frac{v_{\text{propeller}}^2}{2g} = \frac{(4.2 v_{\text{boat}})^2}{(2)\left(32.2 \, \frac{\text{ft}}{\text{sec}^2}\right)} = 0.2739 v_{\text{boat}}^2$$

From App. 29.A, the vapor pressure of freshwater at 68°F is $p_{\text{vp}} = 0.3393$ psia.

From App. 14.A, the density of water at 68°F is 62.32 lbm/ft^3. Raoult's law predicts the actual vapor pressure of the solution.

$$p_{\text{vapor,solution}} = \left(\begin{array}{c}\text{mole fraction} \\ \text{of the solvent}\end{array}\right) p_{\text{vapor,solvent}}$$

$$p_{\text{vapor,seawater}} = (0.992)\left(0.3393 \, \frac{\text{lbf}}{\text{in}^2}\right) = 0.3366 \text{ lbf/in}^2$$

From Eq. 18.4(b), the vapor pressure head is

$$h_{\text{vapor,seawater}} = \frac{p}{\rho} \times \frac{g_c}{g}$$

$$= \frac{\left(0.3366 \, \frac{\text{lbf}}{\text{in}^2}\right)\left(12 \, \frac{\text{in}}{\text{ft}}\right)^2}{64.0 \, \frac{\text{lbm}}{\text{ft}^3}} \times \frac{32.2 \, \frac{\text{lbm-ft}}{\text{lbf-sec}^2}}{32.2 \, \frac{\text{ft}}{\text{sec}^2}}$$

$$= 0.7574 \text{ ft}$$

Solve for the boat velocity.

$$8 \text{ ft} + 33.075 \text{ ft}$$

$$- 0.2739 v_{\text{boat}}^2 = 0.7574 \text{ ft}$$

$$v_{\text{boat}} = \boxed{12.13 \text{ ft/sec} \quad (12 \text{ ft/sec})}$$

The answer is (B).

SI Solution

Cavitation will occur when

$$h_{\text{atm}} - h_{\text{v}} < h_{\text{vp}}$$

The density of seawater is 1024 kg/m^3.

From Eq. 18.4(a), the atmospheric head is

$$h_{\text{atm}} = \frac{p}{\rho g} = \frac{(101.3 \text{ kPa})\left(1000 \, \frac{\text{Pa}}{\text{kPa}}\right)}{\left(1024 \, \frac{\text{kg}}{\text{m}^3}\right)\left(9.81 \, \frac{\text{m}}{\text{s}^2}\right)} = 10.08 \text{ m}$$

$h_{\text{depth}} = 3 \text{ m} \quad [\text{given}]$

From Eq. 18.6, the velocity head is

$$h_{\text{v}} = \frac{v_{\text{propeller}}^2}{2g} = \frac{(4.2 v_{\text{boat}})^2}{(2)\left(9.81 \, \frac{\text{m}}{\text{s}^2}\right)} = 0.899 v_{\text{boat}}^2$$

The vapor pressure of 20°C freshwater is

$$p_{\text{vp}} = (0.02339 \text{ bar})\left(100 \, \frac{\text{kPa}}{\text{bar}}\right) = 2.339 \text{ kPa}$$

From App. 14.B, the density of water at 20°C is 998.23 kg/m^3. Raoult's law predicts the actual vapor pressure of the solution.

$$p_{\text{vapor,solution}} = p_{\text{vapor,solvent}}\left(\begin{array}{c}\text{mole fraction} \\ \text{of the solvent}\end{array}\right)$$

The solvent is the freshwater and the solution is the seawater.

The mole fraction of water is 0.992.

$$p_{\text{vapor,seawater}} = (2.339 \text{ kPa})(0.992) = 2.320 \text{ kPa}$$

From Eq. 18.4(a), the vapor pressure head is

$$h_{\text{vapor,seawater}} = \frac{p}{g\rho} = \frac{(2.320 \text{ kPa})\left(1000 \, \frac{\text{Pa}}{\text{kPa}}\right)}{\left(9.81 \, \frac{\text{m}}{\text{s}^2}\right)\left(1024 \, \frac{\text{kg}}{\text{m}^3}\right)} = 0.231 \text{ m}$$

Solve for the boat velocity.

$$3 \text{ m} + 10.08 \text{ m} - 0.899 v_{\text{boat}}^2 = 0.231 \text{ m}$$

$$v_{\text{boat}} = \boxed{3.78 \text{ m/s} \quad (3.8 \text{ m/s})}$$

The answer is (B).

19. *Customary U.S. Solution*

(a) From App. 17.D, the equivalent lengths of various screwed steel fittings are

> inlet: $L_e = 8.5$ ft   [essentially a reentrant inlet]
>
> check valve: $L_e = 19$ ft
>
> long radius elbow: $L_e = 3.6$ ft

Flow of Fluids

The total equivalent length of the 2 in line is

$$L_e = 12 \text{ ft} + 8.5 \text{ ft} + 19 \text{ ft} + (3)(3.6 \text{ ft}) + 80 \text{ ft}$$

$$= 130.3 \text{ ft}$$

From App. 16.B, for 2 in schedule-40 pipe,

$$D = 0.1723 \text{ ft}$$

$$A = 0.0233 \text{ ft}^2$$

Since the flow rate is unknown, it must be assumed in order to find velocity. Assume 90 gal/min.

$$\dot{V} = \frac{90 \dfrac{\text{gal}}{\text{min}}}{\left(7.4805 \dfrac{\text{gal}}{\text{ft}^3}\right)\left(60 \dfrac{\text{sec}}{\text{min}}\right)} = 0.2005 \text{ ft}^3/\text{sec}$$

The velocity is

$$v = \frac{\dot{V}}{A} = \frac{0.2005 \dfrac{\text{ft}^3}{\text{sec}}}{0.0233 \text{ ft}^2} = 8.605 \text{ ft/sec}$$

From App. 14.A, the kinematic viscosity of water at 70°F is $\nu = 1.059 \times 10^{-5}$ ft^2/sec.

The Reynolds number is

$$Re = \frac{Dv}{\nu} = \frac{(0.1723 \text{ ft})\left(8.605 \dfrac{\text{ft}}{\text{sec}}\right)}{1.059 \times 10^{-5} \dfrac{\text{ft}^2}{\text{sec}}} = 1.4 \times 10^5$$

From Table 17.2 or App. 17.A, the specific roughness of steel pipe is $\epsilon = 0.0002$ ft.

$$\frac{\epsilon}{D} = \frac{0.0002 \text{ ft}}{0.1723 \text{ ft}} \approx 0.0012$$

From App. 17.B, $f = 0.022$. At 90 gal/min, the friction loss in the line from Eq. 18.5 is

$$h_f = \frac{fLv^2}{2Dg} = \frac{(0.022)(130.3 \text{ ft})\left(8.605 \dfrac{\text{ft}}{\text{sec}}\right)^2}{(2)(0.1723 \text{ ft})\left(32.2 \dfrac{\text{ft}}{\text{sec}^2}\right)} = 19.1 \text{ ft}$$

From Eq. 18.6, the velocity head at 90 gal/min is

$$h_v = \frac{v^2}{2g} = \frac{\left(8.605 \dfrac{\text{ft}}{\text{sec}}\right)^2}{(2)\left(32.2 \dfrac{\text{ft}}{\text{sec}^2}\right)} = 1.1 \text{ ft}$$

In general, the friction head and velocity head are proportional to v^2 and Q^2.

$$h_f = (19.1 \text{ ft})\left(\frac{Q_2}{90 \dfrac{\text{gal}}{\text{min}}}\right)^2$$

$$h_v = (1.1 \text{ ft})\left(\frac{Q_2}{90 \dfrac{\text{gal}}{\text{min}}}\right)^2$$

(The 7 ft suction lift is included in the 20 ft static discharge head.)

The equation for the total system head is

$$h = h_z + h_v + h_f = 20 \text{ ft} + (1.1 \text{ ft} + 19.1 \text{ ft})\left(\frac{Q_2}{90 \dfrac{\text{gal}}{\text{min}}}\right)^2$$

Q_2 (gal/min)	system head, h (ft)
0	20.0
10	20.2
20	21.0
30	22.2
40	24.0
50	26.2
60	29.0
70	32.2
80	36.0
90	40.2
100	44.9
110	50.2

The intersection point of the system curve and the pump curve defines the operating flow rate.

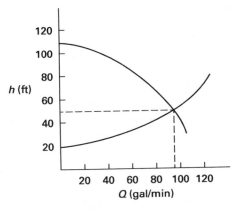

The flow rate is $\boxed{95 \text{ gal/min.}}$

The answer is (D).

(b) The intersection point is not in an efficient range for the pump because it is so far down on the system curve that the pumping efficiency will be low.

$\boxed{\text{A different pump should be used.}}$

The answer is (A).

SI Solution

(a) Use the equivalent lengths of various screwed steel fittings from the customary U.S. solution. The total equivalent length of 5.08 cm schedule-40 pipe is

$$L_e = 4 \text{ m} + \left(8.5 \text{ ft} + 19 \text{ ft} + (3)(3.6 \text{ ft})\right)\left(0.3048 \ \frac{\text{m}}{\text{ft}}\right)$$
$$+ 27 \text{ m}$$
$$= 42.67 \text{ m}$$

From App. 16.C, for 2 in schedule-40 pipe,

$$D = 52.50 \text{ mm}$$
$$A = 21.65 \times 10^{-4} \text{ m}^2$$

Since the flow rate is unknown, it must be assumed in order to find velocity. Assume 6 L/s.

$$\dot{V} = \frac{6 \ \frac{\text{L}}{\text{s}}}{1000 \ \frac{\text{L}}{\text{m}^3}} = 6 \times 10^{-3} \text{ m}^3/\text{s}$$

The velocity is

$$\text{v} = \frac{\dot{V}}{A} = \frac{6 \times 10^{-3} \ \frac{\text{m}^3}{\text{s}}}{21.65 \times 10^{-4} \text{ m}^2} = 2.77 \text{ m/s}$$

From App. 14.B, the kinematic viscosity of water at 21°C is approximately $\nu = 9.849 \times 10^{-7} \text{ m}^2/\text{s}$.

The Reynolds number is

$$\text{Re} = \frac{\text{v}D}{\nu} = \frac{\left(2.77 \ \frac{\text{m}}{\text{s}}\right)(52.50 \text{ mm})}{\left(9.849 \times 10^{-7} \ \frac{\text{m}^2}{\text{s}}\right)\left(1000 \ \frac{\text{mm}}{\text{m}}\right)}$$
$$= 1.48 \times 10^5$$

From Table 17.2, the specific roughness of steel pipe is $\epsilon = 6.0 \times 10^{-5}$ m.

$$\frac{\epsilon}{D} = \frac{\left(6.0 \times 10^{-5} \text{ m}\right)\left(1000 \ \frac{\text{mm}}{\text{m}}\right)}{52.50 \text{ mm}} \approx 0.0011$$

From App. 17.B, $f = 0.022$. At 6 L/s, the friction loss in the line from Eq. 18.5 is

$$h_f = \frac{fL\text{v}^2}{2Dg} = \frac{(0.022)(42.67 \text{ m})\left(2.77 \ \frac{\text{m}}{\text{s}}\right)^2\left(1000 \ \frac{\text{mm}}{\text{m}}\right)}{(2)(52.50 \text{ mm})\left(9.81 \ \frac{\text{m}}{\text{s}^2}\right)}$$
$$= 6.99 \text{ m}$$

At 6 L/s, the velocity head from Eq. 18.6 is

$$h_\text{v} = \frac{\text{v}^2}{2g} = \frac{\left(2.77 \ \frac{\text{m}}{\text{s}}\right)^2}{(2)\left(9.81 \ \frac{\text{m}}{\text{s}^2}\right)} = 0.39 \text{ m}$$

In general, the friction head and velocity head are proportional to v^2 and Q^2.

$$h_f = (6.99 \text{ m})\left(\frac{Q_2}{6 \ \frac{\text{L}}{\text{s}}}\right)^2$$

$$h_\text{v} = (0.39 \text{ m})\left(\frac{Q_2}{6 \ \frac{\text{L}}{\text{s}}}\right)^2$$

(The 2.3 m suction lift is included in the 6.3 m static discharge head.)

The equation for the total system head is

$$h = h_z + h_\text{v} + h_f = 6.3 \text{ m} + (0.39 \text{ m} + 6.99 \text{ m})\left(\frac{Q_2}{6 \ \frac{\text{L}}{\text{s}}}\right)^2$$

Q_2 (L/s)	h (m)
0	6.3
0.6	6.37
1.2	6.60
1.8	6.96
2.4	7.48
3.2	8.40
3.6	8.96
4.4	10.27
4.8	11.02
5.7	12.96
6.0	13.68
6.5	14.96
7.0	16.35
7.5	17.83

The intersection point of the system curve and the pump curve defines the operating flow rate.

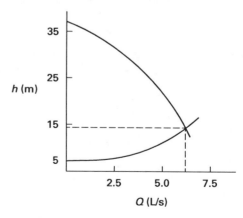

The flow rate is $\boxed{6.2 \text{ L/s.}}$

The answer is (D).

(b) The intersection point is not in an efficient range of the pump because it is so far down on the system curve that the pumping efficiency will be low.

> A different pump should be used.

The answer is (A).

20. From Eq. 18.53,

$$P_2 = P_1 \left(\frac{\rho_2 n_2^3 D_2^5}{\rho_1 n_1^3 D_1^5} \right) = P_1 \left(\frac{n_2}{n_1} \right)^3 \quad [\rho_2 = \rho_1 \text{ and } D_2 = D_1]$$

Customary U.S. Solution

$$P_2 = (0.5 \text{ hp}) \left(\frac{2000 \frac{\text{rev}}{\text{min}}}{1750 \frac{\text{rev}}{\text{min}}} \right)^3 = \boxed{0.75 \text{ hp}}$$

The answer is (D).

SI Solution

$$P_2 = (0.37 \text{ kW}) \left(\frac{2000 \frac{\text{rev}}{\text{min}}}{1750 \frac{\text{rev}}{\text{min}}} \right)^3 = \boxed{0.55 \text{ kW}}$$

The answer is (D).

21. (a) Random values of Q are chosen, and the corresponding values of H are determined by the formula $H = 30 + 2Q^2$.

Q (ft³/sec)	H (ft)
0	30
2.5	42.5
5	80
7.5	142.5
10	230
15	480
20	830
25	1280
30	1830

The intersection of the system curve and the 1400 rpm pump curve defines the operating point at that rpm.

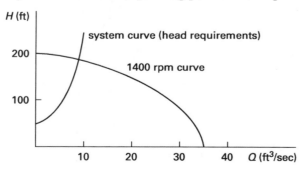

From the intersection of the graphs, at 1400 rpm the flow rate is approximately 9 ft³/sec, and the corresponding head is $30 + (2)(9)^2 \approx 192$ ft.

$$Q = \left(9 \frac{\text{ft}^3}{\text{sec}} \right) \left(7.4805 \frac{\text{gal}}{\text{ft}^3} \right) \left(60 \frac{\text{sec}}{\text{min}} \right)$$
$$= \boxed{4039 \text{ gal/min} \quad (4000 \text{ gal/min})}$$

The answer is (C).

(b) From Table 18.5, the hydraulic horsepower is

$$\text{WHP} = \frac{h_A \dot{V}(\text{SG})}{8.814} = \frac{(192 \text{ ft}) \left(9 \frac{\text{ft}^3}{\text{sec}} \right) (1)}{8.814 \frac{\text{ft}^4}{\text{hp-sec}}} = 196 \text{ hp}$$

From Eq. 18.28(b), the specific speed is

$$n_s = \frac{n\sqrt{Q}}{h_A^{0.75}} = \frac{\left(1400 \frac{\text{rev}}{\text{min}} \right) \sqrt{4039 \frac{\text{gal}}{\text{min}}}}{(192 \text{ ft})^{0.75}} = 1725$$

From Fig. 18.8 with curve E, $\eta \approx 86\%$.

The minimum motor power should be

$$\frac{196 \text{ hp}}{0.86} = \boxed{228 \text{ hp} \quad (230 \text{ hp})}$$

The answer is (C).

(c) From Eq. 18.42,

$$Q_2 = Q_1 \left(\frac{n_2}{n_1} \right) = \left(4039 \frac{\text{gal}}{\text{min}} \right) \left(\frac{1200 \frac{\text{rev}}{\text{min}}}{1400 \frac{\text{rev}}{\text{min}}} \right)$$
$$= \boxed{3462 \text{ gal/min} \quad (3500 \text{ gal/min})}$$

The answer is (B).

22. Since turbines are essentially pumps running backward, use Table 18.5.

$$\Delta p = \left(30 \frac{\text{lbf}}{\text{in}^2} - 5 \frac{\text{lbf}}{\text{in}^2} \right) \left(12 \frac{\text{in}}{\text{ft}} \right)^2$$
$$= 3600 \text{ lbf/ft}^2$$

$$P = \frac{\Delta p \dot{V}}{550} = \frac{\left(3600 \frac{\text{lbf}}{\text{ft}^2} \right) \left(100 \frac{\text{ft}^3}{\text{sec}} \right)}{550 \frac{\text{ft-lbf}}{\text{hp-sec}}}$$
$$= \boxed{654.5 \text{ hp} \quad (650 \text{ hp})}$$

The answer is (C).

23. The flow rate is

$$\dot{m} = \rho \dot{V} = \left(62.4 \ \frac{\text{lbm}}{\text{ft}^3}\right)\left(1000 \ \frac{\text{ft}^3}{\text{sec}}\right)$$

$$= 6.24 \times 10^4 \ \text{lbm/sec}$$

The head available for work is

$$\Delta h = 625 \ \text{ft} - 58 \ \text{ft} = 567 \ \text{ft}$$

Use Table 18.5. The turbine efficiency is 89%. So, the power is

$$P = \frac{h_A \dot{m}}{550} \times \frac{g}{g_c}$$

$$= \frac{(0.89)\left(6.24 \times 10^4 \ \frac{\text{lbm}}{\text{sec}}\right)(567 \ \text{ft})}{550 \ \frac{\text{ft-lbf}}{\text{hp-sec}}} \times \frac{32.2 \ \frac{\text{ft}}{\text{sec}^2}}{32.2 \ \frac{\text{lbm-ft}}{\text{lbf-sec}^2}}$$

$$= 57{,}253 \ \text{hp}$$

Convert from hp to kW.

$$P = (57{,}253 \ \text{hp})\left(0.7457 \ \frac{\text{kW}}{\text{hp}}\right)$$

$$= \boxed{4.27 \times 10^4 \ \text{kW} \quad (43 \ \text{MW})}$$

The answer is (C).

24. *Customary U.S. Solution*

(a) From App. 14.A, the density of water at 60°F is 62.37 lbm/ft³. From Eq. 18.4(b), the head dropped is

$$h = \frac{\Delta p}{\rho} \times \frac{g_c}{g}$$

$$= \frac{\left(500 \ \frac{\text{lbf}}{\text{in}^2} - 30 \ \frac{\text{lbf}}{\text{in}^2}\right)\left(12 \ \frac{\text{in}}{\text{ft}}\right)^2}{62.37 \ \frac{\text{lbm}}{\text{ft}^3}} \times \frac{32.2 \ \frac{\text{lbm-ft}}{\text{lbf-sec}^2}}{32.2 \ \frac{\text{ft}}{\text{sec}^2}}$$

$$= 1085 \ \text{ft}$$

From Eq. 18.56(b), the specific speed of a turbine is

$$n_s = \frac{n\sqrt{P}}{h_t^{1.25}} = \frac{\left(1750 \ \frac{\text{rev}}{\text{min}}\right)\sqrt{250 \ \text{hp}}}{(1085 \ \text{ft})^{1.25}} = \boxed{4.443 \quad (4)}$$

The answer is (A).

(b) The flow rate, $\dot{V}$, is

$$\dot{V} = A\text{v} = \left(\frac{\pi}{4}\right)\left(\frac{4 \ \text{in}}{12 \ \frac{\text{in}}{\text{ft}}}\right)^2\left(35 \ \frac{\text{ft}}{\text{sec}}\right)$$

$$= 3.054 \ \text{ft}^3/\text{sec}$$

Since the analysis is for a single blade, not the entire turbine, only a portion of the water will catch up with the blade. From Eq. 17.195, the flow rate, considering the blade's movement away at 10 ft/sec, is

$$\dot{V}' = \frac{\left(35 \ \frac{\text{ft}}{\text{sec}} - 10 \ \frac{\text{ft}}{\text{sec}}\right)\left(3.054 \ \frac{\text{ft}^3}{\text{sec}}\right)}{35 \ \frac{\text{ft}}{\text{sec}}}$$

$$= 2.181 \ \text{ft}^3/\text{sec}$$

From Eq. 17.196(b) and Eq. 17.197(b), the forces in the x-direction and the y-direction are

$$F_x = \left(\frac{\dot{V}'\rho}{g_c}\right)(\text{v}_j - \text{v}_b)(\cos\theta - 1)$$

$$= \left(\frac{\left(2.181 \ \frac{\text{ft}^3}{\text{sec}}\right)\left(62.37 \ \frac{\text{lbm}}{\text{ft}^3}\right)}{32.2 \ \frac{\text{lbm-ft}}{\text{lbf-sec}^2}}\right)$$

$$\times \left(35 \ \frac{\text{ft}}{\text{sec}} - 10 \ \frac{\text{ft}}{\text{sec}}\right)(\cos 80° - 1)$$

$$= -87.27 \ \text{lbf}$$

$$F_y = \left(\frac{\dot{V}'\rho}{g_c}\right)(\text{v}_j - \text{v}_b)\sin\theta$$

$$= \left(\frac{\left(2.181 \ \frac{\text{ft}^3}{\text{sec}}\right)\left(62.37 \ \frac{\text{lbm}}{\text{ft}^3}\right)}{32.2 \ \frac{\text{lbm-ft}}{\text{lbf-sec}^2}}\right)$$

$$\times \left(35 \ \frac{\text{ft}}{\text{sec}} - 10 \ \frac{\text{ft}}{\text{sec}}\right)\sin 80°$$

$$= 104.0 \ \text{lbf}$$

The total force acting on the blade is

$$R = \sqrt{F_x^2 + F_y^2} = \sqrt{(-87.27 \ \text{lbf})^2 + (104.0 \ \text{lbf})^2}$$

$$= \boxed{135.8 \ \text{lbf} \quad (140 \ \text{lbf})}$$

The answer is (B).

SI Solution

(a) From App. 14.B, the density of water at 16°C is 998.83 kg/m³. From Eq. 18.4(a), the head dropped is

$$h = \frac{\Delta p}{\rho g}$$

$$= \frac{(3.5 \ \text{MPa})\left(1 \times 10^6 \ \frac{\text{Pa}}{\text{MPa}}\right) - (210 \ \text{kPa})\left(1000 \ \frac{\text{Pa}}{\text{kPa}}\right)}{\left(998.83 \ \frac{\text{kg}}{\text{m}^3}\right)\left(9.81 \ \frac{\text{m}}{\text{s}^2}\right)}$$

$$= 335.8 \ \text{m}$$

From Eq. 18.56(a), the specific speed of a turbine is

$$n_s = \frac{n\sqrt{P}}{h_t^{1.25}} = \frac{\left(1750 \ \frac{\text{rev}}{\text{min}}\right)\sqrt{185 \text{ kW}}}{(335.8 \text{ m})^{1.25}}$$

$$= \boxed{16.56 \quad (17)}$$

The answer is (A).

(b) The flow rate, $\dot{V}$, is

$$\dot{V} = A\text{v} = \left(\frac{\pi\left(\dfrac{100 \text{ mm}}{1000 \ \frac{\text{mm}}{\text{m}}}\right)^2}{4}\right)\left(10.5 \ \frac{\text{m}}{\text{s}}\right)$$

$$= 0.08247 \text{ m}^3/\text{s}$$

Since the analysis is for a single blade, not the entire turbine, only a portion of the water will catch up with the blade. From Eq. 17.195, the flow rate, considering the blade's movement away at 3 m/s, is

$$\dot{V}' = \frac{\left(10.5 \ \frac{\text{m}}{\text{s}} - 3 \ \frac{\text{m}}{\text{s}}\right)\left(0.08247 \ \frac{\text{m}^3}{\text{s}}\right)}{10.5 \ \frac{\text{m}}{\text{s}}}$$

$$= 0.05891 \text{ m}^3/\text{s}$$

From Eq. 17.196(a) and Eq. 17.197(a), the forces in the x-direction and the y-direction are

$$F_x = \dot{V}'\rho(\text{v}_j - \text{v}_b)(\cos\theta - 1)$$

$$= \left(0.05891 \ \frac{\text{m}^3}{\text{s}}\right)\left(998.83 \ \frac{\text{kg}}{\text{m}^3}\right)$$

$$\times \left(10.5 \ \frac{\text{m}}{\text{s}} - 3 \ \frac{\text{m}}{\text{s}}\right)(\cos 80° - 1)$$

$$= -364.7 \text{ N}$$

$$F_y = \dot{V}'\rho(\text{v}_j - \text{v}_b)\sin\theta$$

$$= \left(0.05891 \ \frac{\text{m}^3}{\text{s}}\right)\left(998.83 \ \frac{\text{kg}}{\text{m}^3}\right)$$

$$\times \left(10.5 \ \frac{\text{m}}{\text{s}} - 3 \ \frac{\text{m}}{\text{s}}\right)\sin 80°$$

$$= 434.6 \text{ N}$$

The total force acting on the blade is

$$R = \sqrt{F_x^2 + F_y^2} = \sqrt{(-364.7 \text{ N})^2 + (434.6 \text{ N})^2}$$

$$= \boxed{567.3 \text{ N} \quad (570 \text{ N})}$$

The answer is (B).

25. *Customary U.S. Solution*

(a) The total effective head is due to the pressure head, velocity head, and tailwater head.

$$h_{\text{eff}} = h_p + h_v - h_{z,\text{tailwater}} = h_p + \frac{\text{v}^2}{2g} - h_{z,\text{tailwater}}$$

$$= 92.5 \text{ ft} + \frac{\left(12 \ \frac{\text{ft}}{\text{sec}}\right)^2}{(2)\left(32.2 \ \frac{\text{ft}}{\text{sec}^2}\right)} - (-5.26 \text{ ft})$$

$$= \boxed{100 \text{ ft}}$$

The answer is (C).

(b) From Table 18.5, the theoretical hydraulic horsepower is

$$P_{\text{th}} = \frac{h_A\dot{V}(\text{SG})}{8.814} = \frac{(100 \text{ ft})\left(25 \ \frac{\text{ft}^3}{\text{sec}}\right)(1)}{8.814 \ \frac{\text{ft}^4}{\text{hp-sec}}}$$

$$= 283.6 \text{ hp}$$

The overall turbine efficiency is

$$\eta = \frac{P_{\text{brake}}}{P_{\text{th}}} = \frac{250 \text{ hp}}{283.6 \text{ hp}} = \boxed{0.882 \quad (88\%)}$$

The answer is (B).

(c) From Eq. 18.43,

$$n_2 = n_1\sqrt{\frac{h_2}{h_1}} = \left(610 \ \frac{\text{rev}}{\text{min}}\right)\sqrt{\frac{225 \text{ ft}}{100 \text{ ft}}}$$

$$= \boxed{915 \text{ rpm} \quad (920 \text{ rpm})}$$

The answer is (B).

(d) Combine Eq. 18.43 and Eq. 18.44.

$$P_2 = P_1\left(\frac{n_2}{n_1}\right)^3 = P_1\left(\left(\frac{h_2}{h_1}\right)^{1/2}\right)^3$$

$$= P_1\left(\frac{h_2}{h_1}\right)^{3/2}$$

$$= (250 \text{ hp})\left(\frac{225 \text{ ft}}{100 \text{ ft}}\right)^{3/2}$$

$$= \boxed{843.8 \text{ hp} \quad (840 \text{ hp})}$$

The answer is (D).

(e) From Eq. 18.43,

$$Q_2 = Q_1\sqrt{\frac{h_2}{h_1}} = \left(25 \ \frac{\text{ft}^3}{\text{sec}}\right)\sqrt{\frac{225 \text{ ft}}{100 \text{ ft}}}$$

$$= \boxed{37.5 \text{ ft}^3/\text{sec} \quad (38 \text{ ft}^3/\text{sec})}$$

The answer is (B).

SI Solution

(a) The total effective head is due to the pressure head, velocity head, and tailwater head.

$$h_{\text{eff}} = h_p + h_v - h_{z,\text{tailwater}} = h_p + \frac{v^2}{2g} - h_{z,\text{tailwater}}$$

$$= 28.2 \text{ m} + \frac{\left(3.6 \dfrac{\text{m}}{\text{s}}\right)^2}{(2)\left(9.81 \dfrac{\text{m}}{\text{s}^2}\right)} - (-1.75 \text{ m})$$

$$= \boxed{30.61 \text{ m} \quad (31 \text{ m})}$$

The answer is (C).

(b) From Table 18.6, the theoretical hydraulic kilowatts are

$$P_{\text{th}} = \frac{9.81 h_A Q(\text{SG})}{1000}$$

$$= \frac{\left(9.81 \dfrac{\text{m}}{\text{s}^2}\right)(30.61 \text{ m})\left(700 \dfrac{\text{L}}{\text{s}}\right)(1)}{1000 \dfrac{\text{W}}{\text{kW}}}$$

$$= 210.2 \text{ kW}$$

The overall turbine efficiency is

$$\eta = \frac{P_{\text{brake}}}{P_{\text{th}}} = \frac{185 \text{ kW}}{210.2 \text{ kW}} = \boxed{0.88 \quad (88\%)}$$

The answer is (B).

(c) From Eq. 18.43,

$$n_2 = n_1 \sqrt{\frac{h_2}{h_1}} = \left(610 \frac{\text{rev}}{\text{min}}\right) \sqrt{\frac{70 \text{ m}}{30.61 \text{ m}}}$$

$$= \boxed{922 \text{ rpm} \quad (920 \text{ rpm})}$$

The answer is (B).

(d) Combine Eq. 18.43 and Eq. 18.44.

$$P_2 = P_1 \left(\frac{n_2}{n_1}\right)^3 = P_1 \left(\left(\frac{h_2}{h_1}\right)^{1/2}\right)^3 = P_1 \left(\frac{h_2}{h_1}\right)^{3/2}$$

$$= (185 \text{ kW}) \left(\frac{70 \text{ m}}{30.61 \text{ m}}\right)^{3/2}$$

$$= \boxed{639.8 \text{ kW} \quad (640 \text{ kW})}$$

The answer is (D).

(e) From Eq. 18.43,

$$Q_2 = Q_1 \sqrt{\frac{h_2}{h_1}} = \left(700 \frac{\text{L}}{\text{s}}\right) \sqrt{\frac{70 \text{ m}}{30.61 \text{ m}}}$$

$$= \boxed{1059 \text{ L/s} \quad (1100 \text{ L/s})}$$

The answer is (B).

Flow of Fluids

19 Open Channel Flow

PRACTICE PROBLEMS

Rectangular Channels

1. A wooden flume ($n = 0.012$) with a rectangular cross section is 2 ft wide. The flume carries 3 ft³/sec of water down a 1% slope. What is the depth of flow?

(A) 0.3 ft

(B) 0.4 ft

(C) 0.5 ft

(D) 0.6 ft

2. A rectangular open channel is to be constructed with smooth concrete on a slope of 0.08. The design flow rate is 17 m³/s. The channel design must be optimum. What are the channel dimensions?

(A) 0.6 m deep; 1.2 m wide

(B) 0.8 m deep; 1.6 m wide

(C) 1.0 m deep; 2.0 m wide

(D) 1.2 m deep; 2.4 m wide

3. (*Time limit: one hour*) A dam's outfall structure has three rectangular inlets sized 1 ft high × 2 ft wide as shown. The orifice coefficient for each inlet is 0.7. The vertical riser joins with a 100 ft long square box culvert ($n = 0.013$) placed on a 0.05 slope. The box culvert drains into a tailwater basin whose water level remains constant at the top of the culvert opening. During steady-state operation, the water level in the vertical barrel of the structure is 4 ft above the top of the box culvert. The orifice coefficient of the box culvert entrance is ⅔. What are the dimensions of the box culvert?

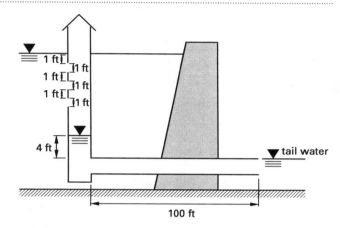

(A) 1.5 ft × 1.5 ft

(B) 2.0 ft × 2.0 ft

(C) 2.5 ft × 2.5 ft

(D) 4.0 ft × 4.0 ft

Circular Channels

4. A 24 in diameter pipe ($n = 0.013$) was installed 30 years ago on a 0.001 slope. Recent tests indicate that the full-flow capacity of the pipe is 6.0 ft³/sec.

(a) What was the original full-flow capacity?

(A) 4.1 ft³/sec

(B) 6.3 ft³/sec

(C) 6.7 ft³/sec

(D) 7.2 ft³/sec

(b) What was the original full-flow velocity?

(A) 1.8 ft/sec

(B) 2.3 ft/sec

(C) 2.7 ft/sec

(D) 3.7 ft/sec

(c) What is the current full-flow velocity?

(A) 1.9 ft/sec

(B) 2.4 ft/sec

(C) 3.1 ft/sec

(D) 5.8 ft/sec

(d) What is the current Manning coefficient?

(A) 0.011

(B) 0.013

(C) 0.016

(D) 0.024

5. A circular sewer is to be installed on a 1% grade. Its Manning coefficient is $n = 0.013$. The maximum full-flow capacity is to be 3.5 ft^3/sec.

(a) What size pipe should be recommended?

(A) 8 in

(B) 12 in

(C) 16 in

(D) 18 in

(b) Assuming that the diameter chosen is 12 in, what is the full-flow capacity?

(A) 2.7 ft^3/sec

(B) 3.3 ft^3/sec

(C) 3.6 ft^3/sec

(D) 4.5 ft^3/sec

(c) What is the full-flow velocity?

(A) 2.9 ft/sec

(B) 3.3 ft/sec

(C) 4.1 ft/sec

(D) 4.6 ft/sec

(d) What is the depth of flow when the flow is 0.7 ft^3/sec?

(A) 4 in

(B) 7 in

(C) 9 in

(D) 12 in

(e) What minimum velocity will prevent solids from settling out in the sewer?

(A) 1.0–1.5 ft/sec

(B) 1.5–2.0 ft/sec

(C) 2.0–2.5 ft/sec

(D) 2.5–3.0 ft/sec

6. A 4 ft diameter concrete storm drain on a 0.02 slope carries water at a depth of 1.5 ft. Manning's roughness coefficient for a full storm drain is $n_{full} = 0.013$, but n varies with depth.

(a) What is the velocity of the water in the storm drain?

(A) 3 ft/sec

(B) 6 ft/sec

(C) 8 ft/sec

(D) 11 ft/sec

(b) What is the maximum velocity that can be achieved in the storm drain?

(A) 8 ft/sec

(B) 13 ft/sec

(C) 17 ft/sec

(D) 21 ft/sec

(c) What is the maximum capacity of the storm drain?

(A) 130 ft^3/sec

(B) 160 ft^3/sec

(C) 190 ft^3/sec

(D) 210 ft^3/sec

7. A circular storm sewer ($n = 0.012$) is being designed to carry a peak flow of 5 ft^3/sec. To allow for excess capacity, the depth at peak flow is to be 75% of the sewer diameter. The sewer is to be installed in a bed with a slope of 2%. What is the required sewer diameter?

(A) 12 in

(B) 16 in

(C) 20 in

(D) 24 in

8. (*Time limit: one hour*) Flow in a 6 ft diameter, newly formed concrete culvert is 150 ft^3/sec at a point where the depth of flow is 3 ft. Disregard the variation in roughness with depth.

(a) What is the hydraulic radius?

(A) 1.1 ft

(B) 1.5 ft

(C) 2.3 ft

(D) 3.0 ft

(b) What is the slope of the pipe?

(A) 0.004

(B) 0.008

(C) 0.010

(D) 0.030

(c) What type of flow occurs?

(A) subcritical

(B) critical

(C) supercritical

(D) choked

(d) What is the hydraulic radius if the pipe flows full?

(A) 1.5 ft

(B) 3.0 ft

(C) 4.0 ft

(D) 6.0 ft

(e) What is the flow quantity if the culvert flows full?

(A) 100 ft^3/sec

(B) 120 ft^3/sec

(C) 200 ft^3/sec

(D) 300 ft^3/sec

(f) If the flow is 150 ft^3/sec, what is the critical depth?

(A) 1.1 ft

(B) 1.5 ft

(C) 3.3 ft

(D) 4.3 ft

(g) What is the critical slope?

(A) 0.002

(B) 0.003

(C) 0.004

(D) 0.05

(h) If the exit depth is less than critical, which culvert flow types are most probable?

(A) 1, 5

(B) 1, 2, 3

(C) 1, 3, 5

(D) 5 only

(i) What is the Froude number for a depth of flow of 3 ft?

(A) 0.86

(B) 0.99

(C) 1.22

(D) 2.84

(j) After the water exits the circular culvert, it flows at a velocity of 15 ft/sec and a depth of 2 ft in a rectangular channel. Assuming a hydraulic jump is possible, what will be the depth after the jump?

(A) 2.5 ft

(B) 3.0 ft

(C) 3.3 ft

(D) 4.4 ft

Trapezoidal Channels

9. The trapezoidal channel shown has a Manning coefficient of $n = 0.013$ and is laid at a slope of 0.002. The depth of flow is 2 ft. What is the flow rate?

(A) 120 ft^3/sec

(B) 150 ft^3/sec

(C) 180 ft^3/sec

(D) 210 ft^3/sec

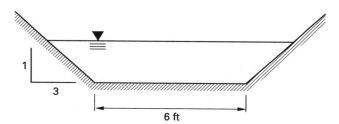

10. A trapezoidal open channel is to be constructed with smooth concrete on a slope of 0.08. The design flow rate is 17 m^3/s. The channel design must be optimum. What are most nearly the channel dimensions?

(A) depth 0.8 m; base 1.0 m

(B) depth 1.0 m; base 1.2 m

(C) depth 1.2 m; base 1.4 m

(D) depth 1.4 m; base 1.8 m

Weirs

11. A sharp-crested rectangular weir with two end contractions is 5 ft wide. The weir height is 6 ft. The head over the weir is 0.43 ft. What is the flow rate?

(A) 3.9 ft^3/sec

(B) 4.3 ft^3/sec

(C) 4.6 ft^3/sec

(D) 6.4 ft^3/sec

12. A weir is constructed with a trapezoidal opening. The base of the opening is 18 in wide, and the sides have a slope of 4:1 (vertical:horizontal). The depth of flow over the weir is 9 in. What is the rate of discharge?

(A) 1.8 ft^3/sec

(B) 3.3 ft^3/sec

(C) 5.1 ft^3/sec

(D) 8.9 ft^3/sec

13. (*Time limit: one hour*) A 17 ft × 20 ft rectangular tank is fed from the bottom and drains through a double-constricted rectangular weir into a trough. The weir opening is to be designed using the following parameters.

discharge rate:	2 MGD minimum
	4 MGD average
	8 MGD maximum
minimum tank freeboard:	4 in
minimum head over weir:	10 in
maximum weir elevation:	590 ft
tank bottom elevation:	580 ft

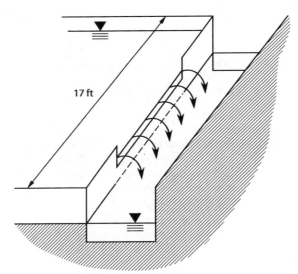

(a) What is the required weir opening length?

(A) 1.2 ft

(B) 1.4 ft

(C) 2.1 ft

(D) 2.7 ft

(b) What is the surface elevation in the tank at maximum flow?

(A) 581.9 ft

(B) 589.8 ft

(C) 592.6 ft

(D) 593.1 ft

(c) What is the surface elevation in the tank at minimum flow?

(A) 590.8 ft

(B) 591.2 ft

(C) 591.7 ft

(D) 592.0 ft

Hydraulic Jumps and Drops

14. Water flowing in a rectangular channel 5 ft wide experiences a hydraulic jump. The depth of flow upstream from the jump is 1 ft. The depth of flow downstream of the jump is 2.4 ft. What quantity is flowing?

(A) 39 ft^3/sec

(B) 45 ft^3/sec

(C) 52 ft^3/sec

(D) 57 ft^3/sec

15. A hydraulic jump forms at the toe of a spillway. The water surface levels are 0.2 ft and 6 ft above the apron before and after the jump, respectively. The velocity before the jump is 54.7 ft/sec. What is the energy loss in the jump?

(A) 41 ft-lbf/lbm

(B) 58 ft-lbf/lbm

(C) 65 ft-lbf/lbm

(D) 92 ft-lbf/lbm

16. Water flowing in a rectangular channel 6 ft wide experiences a hydraulic jump. The depth of flow upstream from the jump is 1.2 ft. The depth of flow downstream of the jump is 3.8 ft.

(a) What quantity is flowing?

(A) 75 ft³/sec

(B) 115 ft³/sec

(C) 180 ft³/sec

(D) 210 ft³/sec

(b) What is the energy loss in the jump?

(A) 1 ft-lbf/lbm

(B) 3 ft-lbf/lbm

(C) 5 ft-lbf/lbm

(D) 12 ft-lbf/lbm

Spillways

17. A spillway operates with 2 ft of head. The toe of the spillway is 40 ft below the crest of the spillway. The spillway discharge coefficient is 3.5.

(a) What is the discharge per foot of crest?

(A) 9.9 ft³/sec

(B) 11 ft³/sec

(C) 15 ft³/sec

(D) 27 ft³/sec

(b) What is the depth of flow at the toe?

(A) 0.19 ft

(B) 0.35 ft

(C) 0.78 ft

(D) 1.4 ft

18. 10,000 ft³/sec of water flows over a 100 ft wide spillway and continues down a constant-width discharge chute on a 5% grade. The discharge chute has a Manning coefficient of $n = 0.012$.

(a) What is the normal depth of the water flowing down the chute?

(A) 1.9 ft

(B) 2.2 ft

(C) 2.7 ft

(D) 3.5 ft

(b) What is the critical depth of the water flowing down the chute?

(A) 3.8 ft

(B) 5.1 ft

(C) 6.8 ft

(D) 7.7 ft

(c) What type of flow occurs on the chute?

(A) tranquil

(B) subcritical

(C) rapid

(D) nonsteady

(d) If a hydraulic jump forms at the bottom of the 5% slope where it joins with a horizontal apron, what is the depth after the jump?

(A) 9.8 ft

(B) 12 ft

(C) 16 ft

(D) 21 ft

19. A spillway has a crest length of 60 ft. The stilling basin below the crest has a width of 60 ft and a level bottom. The spillway discharge coefficient is 3.7. The head loss in the chute is 20% of the difference in level between the reservoir surface and the stilling basin bottom.

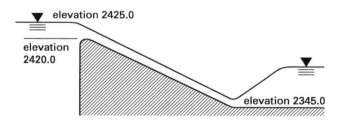

(a) What is the flow rate over the spillway?

(A) 1800 ft³/sec

(B) 2100 ft³/sec

(C) 2500 ft³/sec

(D) 2900 ft³/sec

(b) What is the depth of flow at the toe of the spillway?

(A) 0.65 ft

(B) 1.2 ft

(C) 1.6 ft

(D) 2.0 ft

(c) What tailwater depth is required to cause a hydraulic jump to form at the toe?

(A) 4 ft

(B) 7 ft

(C) 10 ft

(D) 13 ft

(d) What energy is lost in the hydraulic jump?

(A) 30 ft-lbf/lbm

(B) 50 ft-lbf/lbm

(C) 75 ft-lbf/lbm

(D) 100 ft-lbf/lbm

Parshall Flumes

20. A Parshall flume has a throat width of 6 ft. The upstream head measured from the throat floor is 18 in.

(a) What is the flow rate?

(A) 41 ft^3/sec

(B) 46 ft^3/sec

(C) 54 ft^3/sec

(D) 63 ft^3/sec

(b) Which of the following is NOT normally considered to be a feature of Parshall flumes?

(A) They are self-cleansing.

(B) They can be operated by unskilled personnel.

(C) They have low head loss.

(D) They can be placed in temporary installations.

Backwater Curves

21. A rectangular channel 8 ft wide carries a flow of 150 ft^3/sec. The channel slope is 0.0015, and the Manning coefficient is $n = 0.015$. A weir installed across the channel raises the depth at the weir to 6 ft.

(a) What is the normal depth of flow?

(A) 2.9 ft

(B) 3.3 ft

(C) 3.6 ft

(D) 4.5 ft

(b) What is the distance between the weir and the point where the depth is 5 ft?

(A) 610 ft

(B) 840 ft

(C) 1100 ft

(D) 1700 ft

(c) What is the distance between the points where the depth is 4 ft and 5 ft?

(A) 700 ft

(B) 800 ft

(C) 950 ft

(D) 1100 ft

Culverts

22. A 42 in diameter concrete culvert is 250 ft long and laid at a slope of 0.006. The culvert entrance is flush and square-edged. The tailwater level at the outlet is just above the crown of the barrel, and the headwater is 5.0 ft above the crown of the culvert's inlet. What is the capacity?

(A) 60 ft^3/sec

(B) 80 ft^3/sec

(C) 100 ft^3/sec

(D) 120 ft^3/sec

23. (*Time limit: one hour*) A dam spills 2000 m³/s of water over its crest as shown. The crest height is 30 m. The width of the dam is 150 m. The depth at point B is 0.6 m. The stilling basin, constructed of rough concrete, is rectangular in shape, 150 m in width, and has a slope of 0.002.

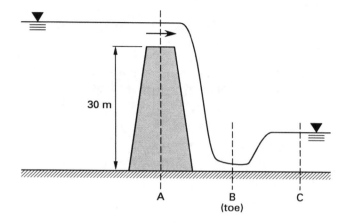

(a) What is the head over the crest?

(A) 0.8 m

(B) 1.2 m

(C) 2.2 m

(D) 3.3 m

(b) What is the total upstream specific energy head?

(A) 32.2 m

(B) 34.1 m

(C) 36.1 m

(D) 38.0 m

(c) What is the total specific energy head at point B?

(A) 25.8 m

(B) 32.2 m

(C) 34.1 m

(D) 43.3 m

(d) What is the friction head loss between point A and point B?

(A) 2.2 m

(B) 4.2 m

(C) 8.3 m

(D) 9.8 m

(e) What is the normal downstream depth?

(A) 2.0 m

(B) 2.6 m

(C) 5.6 m

(D) 10.0 m

(f) What is the critical depth?

(A) 2.0 m

(B) 2.6 m

(C) 7.0 m

(D) 8.3 m

(g) At what point will a hydraulic jump occur?

(A) on the spillway slope

(B) at point B

(C) downstream of point B

(D) nowhere (a hydraulic jump will not occur)

(h) The stilling basin feeds into a natural channel composed of graded material from silt to cobble size. What is the maximum velocity in this channel to prevent erosion?

(A) 1.2 m/s

(B) 6.7 m/s

(C) 15 m/s

(D) 26 m/s

(i) If the natural channel has the same geometry as the stilling basin (i.e., rectangular with a width of 150 m), what slope is required to maintain the maximum velocity and prevent erosion?

(A) 0.00001

(B) 0.00004

(C) 0.0002

(D) 0.002

(j) Assuming that the slope in the natural channel remains at 0.002, what will be the friction loss over a 1000 m reach?

(A) 0.1 m

(B) 0.6 m

(C) 1.5 m

(D) 2.0 m

SOLUTIONS

1.

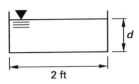

$$R = \frac{2d}{2+2d} = \frac{d}{1+d}$$

From Eq. 19.13,

$$Q = \left(\frac{1.49}{n}\right) A R^{2/3} \sqrt{S}$$

$$3 \; \frac{\text{ft}^3}{\text{sec}} = \left(\frac{1.49}{0.012}\right)(2d)\left(\frac{d}{d+1}\right)^{2/3} \sqrt{0.01}$$

$$0.120805 = \left(\frac{d^{5/2}}{d+1}\right)^{2/3}$$

$$0.042 = \frac{d^{5/2}}{d+1}$$

By trial and error, $d = \boxed{0.314 \text{ ft } (0.3 \text{ ft}).}$

The answer is (A).

2. From App. 19.A, $n = 0.011$. $d = w/2$ for an optimum channel.

$$A = wd = \frac{w^2}{2}$$

$$P = w + 2\left(\frac{w}{2}\right) = 2w$$

$$R = \frac{A}{P} = \frac{\dfrac{w^2}{2}}{2w} = \frac{w}{4}$$

(The units conversion factor of 1.49 is not needed in problems with SI units.)

$$17 \; \frac{\text{m}^3}{\text{s}} = \left(\frac{w^2}{2}\right)\left(\frac{1}{0.011}\right)\left(\frac{w}{4}\right)^{2/3} \sqrt{0.08}$$

$$w = \boxed{1.57 \text{ m} \quad (1.6 \text{ m})}$$

$$d = \boxed{0.79 \text{ m} \quad (0.8 \text{ m})}$$

The answer is (B).

3. Find the flow of the culvert.

$$Q_o = C_d A_o \sqrt{2gh}$$

$$Q_1 = (0.7)(2 \text{ ft}^2)\sqrt{(2)\left(32.2 \; \frac{\text{ft}}{\text{sec}^2}\right)(1.5 \text{ ft})}$$

$$= (11.235)\sqrt{1.5} \quad [\text{top inlet}]$$

$$Q_1 + Q_2 + Q_3 = (11.235)\left(\sqrt{1.5} + \sqrt{3.5} + \sqrt{5.5}\right)$$

$$= 61.13 \text{ ft}^3/\text{sec}$$

Culvert size (and hence R) is unknown. To get a trial value, initially disregard entrance loss and barrel friction.

$$H = 4 \text{ ft} + (100 \text{ ft})(0.05) = 9 \text{ ft}$$

$$\text{v} = \sqrt{2gh} = \sqrt{(2)\left(32.2 \; \frac{\text{ft}}{\text{sec}^2}\right)(9 \text{ ft})}$$

$$= 24.07 \text{ ft/sec}$$

$$A = \frac{Q}{C_d \text{v}} = \frac{61.13 \; \dfrac{\text{ft}^3}{\text{sec}}}{\left(\dfrac{2}{3}\right)\left(24.07 \; \dfrac{\text{ft}}{\text{sec}}\right)} = 3.81 \text{ ft}^2$$

$$\text{box size} = \sqrt{3.81 \text{ ft}^2} = 1.95 \text{ ft}$$

(Round to $2^1/_4 \text{ ft}^2$ to account for friction.)

Due to the hydrostatic pressure, assume the culvert box flows full.

Use Eq. 19.101. $k_e = 0.50$ for a square-edged entrance.

$$R = \frac{(2.25 \text{ ft})^2}{(4)(2.25 \text{ ft})} = 0.5625 \text{ ft}$$

$$\text{v} = \sqrt{\frac{H}{\dfrac{1+k_e}{2g} + \dfrac{n^2 L}{2.21 R^{4/3}}}}$$

$$= \sqrt{\frac{9 \text{ ft}}{\dfrac{1+0.5}{(2)\left(32.2 \; \dfrac{\text{ft}}{\text{sec}^2}\right)} + \dfrac{(0.013)^2(100 \text{ ft})}{(2.21)(0.5625 \text{ ft})^{4/3}}}}$$

$$= 15.05 \text{ ft/sec}$$

$$\text{box size} = \sqrt{\frac{61.13 \; \dfrac{\text{ft}^3}{\text{sec}}}{\left(\dfrac{2}{3}\right)\left(15.05 \; \dfrac{\text{ft}}{\text{sec}}\right)}}$$

$$= \boxed{2.47 \text{ ft} \quad (2.5 \text{ ft or more})}$$

The answer is (C).

4. (a) From Eq. 19.13,

$$Q_o = \left(\frac{1.49}{n}\right)AR^{2/3}\sqrt{S}$$

$$= \left(\frac{1.49}{0.013}\right)\left(\frac{\pi}{4}\right)\left(\frac{24\text{ in}}{12\frac{\text{in}}{\text{ft}}}\right)^2\left(\frac{24\text{ in}}{\left(12\frac{\text{in}}{\text{ft}}\right)(4)}\right)^{2/3}$$

$$\times\sqrt{0.001}$$

$$= \boxed{7.173\text{ ft}^3/\text{sec}\quad(7.2\text{ ft}^3/\text{sec})}$$

The answer is (D).

(b) The original full-flow capacity is

$$v_o = \frac{Q_o}{A} = \frac{7.173\frac{\text{ft}^3}{\text{sec}}}{\left(\frac{\pi}{4}\right)(2\text{ ft})^2}$$

$$= \boxed{2.28\text{ ft/sec}\quad(2.3\text{ ft/sec})}$$

The answer is (B).

(c) The current full-flow velocity is

$$v_{\text{current}} = \frac{Q_{\text{current}}}{A}$$

$$= \frac{6.0\frac{\text{ft}^3}{\text{sec}}}{\left(\frac{\pi}{4}\right)(2\text{ ft})^2}$$

$$= \boxed{1.91\text{ ft/sec}\quad(1.9\text{ ft/sec})}$$

The answer is (A).

(d) Since Q is inversely proportional to n,

$$n_{\text{current}} = \left(\frac{Q_o}{Q_{\text{current}}}\right)n_o = \left(\frac{7.173\frac{\text{ft}^3}{\text{sec}}}{6.0\frac{\text{ft}^3}{\text{sec}}}\right)(0.013)$$

$$= \boxed{0.01554\quad(0.016)}$$

The answer is (C).

5. (a) From Eq. 19.16,

$$D = 1.335\left(\frac{nQ}{\sqrt{S}}\right)^{3/8}$$

$$= (1.335)\left(\frac{\left(3.5\frac{\text{ft}^3}{\text{sec}}\right)(0.013)}{\sqrt{0.01}}\right)^{3/8}$$

$$= 0.99\text{ ft}$$

$$\boxed{\text{Use 12 in pipe.}}$$

The answer is (B).

(b) From Eq. 19.13,

$$D = \frac{12\text{ in}}{12\frac{\text{in}}{\text{ft}}} = 1\text{ ft}$$

$$R = \frac{D}{4}$$

$$Q = \left(\frac{1.49}{n}\right)AR^{2/3}\sqrt{S}$$

$$= \left(\frac{1.49}{0.013}\right)\left(\frac{\pi}{4}\right)(1\text{ ft})^2\left(\frac{1\text{ ft}}{4}\right)^{2/3}\sqrt{0.01}$$

$$= \boxed{3.57\text{ ft}^3/\text{sec}\quad(3.6\text{ ft}^3/\text{sec})}$$

The answer is (C).

(c) The full-flow velocity is

$$v = \frac{Q}{A} = \frac{3.57\frac{\text{ft}^3}{\text{sec}}}{\left(\frac{\pi}{4}\right)(1\text{ ft})^2}$$

$$= \boxed{4.55\text{ ft/sec}\quad(4.6\text{ ft/sec})}$$

The answer is (D).

(d) The depth of flow is calculated as follows.

$$\frac{Q}{Q_{\text{full}}} = \frac{0.7\frac{\text{ft}^3}{\text{sec}}}{3.57\frac{\text{ft}^3}{\text{sec}}} \approx 0.2$$

From App. 19.C, letting n vary with depth,

$$\frac{d}{D} \approx 0.35$$

$$d = (0.35)(12\text{ in})$$

$$= \boxed{4.2\text{ in}\quad(4\text{ in})}$$

The answer is (A).

(e) To be self-cleansing, $v = \boxed{2.0\text{--}2.5\text{ ft/sec.}}$

The answer is (C).

6. *Full flowing:*

$$R = \frac{D}{4} = \frac{4\text{ ft}}{4} = 1\text{ ft}$$

From Eq. 19.12 and Eq. 19.13,

$$v = \left(\frac{1.49}{n}\right)R^{2/3}\sqrt{S}$$

$$= \left(\frac{1.49}{0.013}\right)(1\text{ ft})^{2/3}\sqrt{0.02}$$

$$= 16.21\text{ ft/sec}$$

$$Q = Av = \left(\frac{\pi}{4}\right)(4\text{ ft})^2\left(16.21\frac{\text{ft}}{\text{sec}}\right)$$

$$= 203.7\text{ ft}^3/\text{sec}$$

Actual:

$$\frac{d}{D} = \frac{1.5 \text{ ft}}{4 \text{ ft}} = 0.375$$

From App. 19.C, assuming that n varies with depth,

$$\frac{\text{v}}{\text{v}_{\text{full}}} = 0.68$$

(a) $\text{v} = (0.68)\left(16.21 \ \frac{\text{ft}}{\text{sec}}\right) = \boxed{11 \text{ ft/sec}}$

The answer is (D).

(b) From App. 19.C, the maximum value of $\text{v}/\text{v}_{\text{full}}$ is 1.04 (at $d/D \approx 0.9$).

$$\text{v}_{\text{max}} = (1.04)\left(16.21 \ \frac{\text{ft}}{\text{sec}}\right)$$
$$= \boxed{16.86 \text{ ft/sec} \quad (17 \text{ ft/sec})}$$

The answer is (C).

(c) Similarly, $Q_{\text{max}}/Q_{\text{full}} = 1.02$ (at $d/D \approx 0.96$).

$$Q_{\text{max}} = (1.02)\left(203.7 \ \frac{\text{ft}^3}{\text{sec}}\right)$$
$$= \boxed{207.8 \text{ ft}^3/\text{sec} \quad (210 \text{ ft}^3/\text{sec})}$$

The answer is (D).

7. From App. 16.A, for $Q/Q_{\text{full}} = 0.75$,

$$A = 0.6318 D^2$$
$$R = 0.3017 D$$

From Eq. 19.13,

$$Q = \left(\frac{1.49}{n}\right) A R^{2/3} \sqrt{S}$$
$$5 \ \frac{\text{ft}^3}{\text{sec}} = \left(\frac{1.49}{0.012}\right)(0.6318 D^2)(0.3017 D)^{2/3} \sqrt{0.02}$$
$$1.0 = D^{8/3}$$
$$D = \boxed{1 \text{ ft} \quad (12 \text{ in})}$$

The answer is (A).

(Appendix 19.C can also be used to solve this problem.)

8. (a) For a circular channel flowing half-full, the hydraulic radius is

$$R = \frac{D}{4} = \frac{6 \text{ ft}}{4} = \boxed{1.5 \text{ ft}}$$

The answer is (B).

(b) From App. 19.A, $n = 0.012$. From Eq. 19.13,

$$Q = \left(\frac{1.49}{n}\right) A R^{2/3} \sqrt{S}$$

Rearranging and recognizing that the culvert flows half full,

$$S = \left(\frac{Qn}{1.49 A R^{2/3}}\right)^2$$
$$= \left(\frac{\left(150 \ \frac{\text{ft}^3}{\text{sec}}\right)(0.012)}{(1.49)(0.5)\left(\dfrac{\pi(6 \text{ ft})^2}{4}\right)(1.5 \text{ ft})^{2/3}}\right)^2$$
$$= \boxed{0.0043 \quad (0.004)}$$

The answer is (A).

(c) Equation 19.78 could be used, or the Froude number could be calculated from Eq. 19.79, but App. 19.D is more convenient for use with circular channels. Entering with $Q = 150 \text{ ft}^3/\text{sec}$ and $D = 6 \text{ ft}$, $d_c = 3.3 \text{ ft}$. The actual flow depth is less than the critical depth; therefore, the flow is $\boxed{\text{supercritical.}}$

If the Froude number is used,

$$\text{v} = \frac{Q}{A} = \frac{150 \ \frac{\text{ft}^3}{\text{sec}}}{\left(\dfrac{1}{2}\right)\left(\dfrac{\pi}{4}\right)(6 \text{ ft})^2} = 10.61 \text{ ft/sec}$$

$$L = \frac{\pi D}{8} = \frac{\pi(6 \text{ ft})}{8} = 2.36 \text{ ft}$$

$$\text{Fr} = \frac{\text{v}}{\sqrt{gL}} = \frac{10.61 \ \frac{\text{ft}}{\text{sec}}}{\sqrt{\left(32.2 \ \dfrac{\text{ft}}{\text{sec}^2}\right)(2.36 \text{ ft})}} = 1.22$$

$\text{Fr} > 1.0$, so the flow is $\boxed{\text{supercritical.}}$

The answer is (C).

(d) For a circular channel flowing full, the hydraulic radius is the same as for half-full flow.

$$R = \frac{D}{4} = \frac{6 \text{ ft}}{4} = \boxed{1.5 \text{ ft}}$$

The answer is (A).

(e) From Eq. 19.13,

$$Q = \left(\frac{1.49}{n}\right) A R^{2/3} \sqrt{S}$$
$$= \left(\frac{1.49}{0.012}\right)\left(\frac{\pi(6 \text{ ft})^2}{4}\right)(1.5 \text{ ft})^{2/3} \sqrt{0.0043}$$
$$= \boxed{301.7 \text{ ft}^3/\text{sec} \quad (300 \text{ ft}^3/\text{sec})}$$

The answer is (D).

(f) This part was solved in part (c): $d_c = \boxed{3.3 \text{ ft.}}$

The answer is (C).

(g) From Eq. 19.13, use $d = d_c$.

$$Q = \left(\frac{1.49}{n}\right) A R^{2/3} \sqrt{S}$$

R is calculated using the formula in Table 19.2.

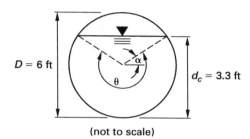

(not to scale)

$$\sin \alpha = \frac{0.3}{3}$$

$$\alpha = 5.74°$$

$$\theta = 180° + 2\alpha = 180° + (2)(5.74°) = 191.48°$$

Convert to radians.

$$\theta = (191.48°)\left(\frac{2\pi \text{ rad}}{360°}\right) = 3.34 \text{ rad}$$

$$R = \tfrac{1}{4}\left(1 - \frac{\sin \theta}{\theta}\right) D = \left(\tfrac{1}{4}\right)\left(1 - \frac{\sin(3.34 \text{ rad})}{3.34 \text{ rad}}\right)(6 \text{ ft})$$

$$= 1.59 \text{ ft}$$

$$A = \tfrac{1}{8}(\theta - \sin \theta) D^2 = \left(\tfrac{1}{8}\right)(3.34 \text{ rad} - \sin(3.34 \text{ rad}))(6 \text{ ft})^2$$

$$= 15.9 \text{ ft}^2$$

$$S = \left(\frac{Qn}{(1.49 A R^{2/3})}\right)^2 = \left(\frac{\left(150 \frac{\text{ft}^3}{\text{sec}}\right)(0.012)}{(1.49)(15.9 \text{ ft}^2)(1.59 \text{ ft})^{2/3}}\right)^2$$

$$= \boxed{0.00311 \quad (0.003)}$$

The answer is (B).

(h) The pipe flows partially full, and the exit is not submerged. This eliminates types 4 and 6 and leaves types 1, 2, 3, and 5. Type 2 is eliminated because the flow is not at critical depth at the outlet. Type 3 is tranquil flow throughout, so it is also eliminated. Distinguishing between types $\boxed{1 \text{ and } 5}$ cannot be done without further information about the entrance conditions.

The answer is (A).

(i)
$$v = \left(\frac{1.49}{n}\right) R^{2/3} \sqrt{S}$$

$$= \left(\frac{1.49}{0.012}\right)(1.5 \text{ ft})^{2/3} \sqrt{0.0043}$$

$$= 10.67 \text{ ft/sec}$$

The characteristic length is the hydraulic depth. For a circular channel flowing half full,

$$L = D_h = \frac{\pi D}{8} = \frac{\pi(6 \text{ ft})}{8}$$

$$= 2.36 \text{ ft}$$

From Eq. 19.79,

$$Fr = \frac{v}{\sqrt{gD_h}}$$

$$= \frac{10.67 \frac{\text{ft}}{\text{sec}}}{\sqrt{\left(32.2 \frac{\text{ft}}{\text{sec}^2}\right)(2.36 \text{ ft})}}$$

$$= \boxed{1.22}$$

Since $Fr > 1$, the flow is supercritical.

The answer is (C).

(j) From Eq. 19.92,

$$d_2 = -\tfrac{1}{2}d_1 + \sqrt{\frac{2v_1^2 d_1}{g} + \frac{d_1^2}{4}}$$

$$= -\left(\tfrac{1}{2}\right)(2 \text{ ft}) + \sqrt{\frac{(2)\left(15 \frac{\text{ft}}{\text{sec}}\right)^2 (2 \text{ ft})}{32.2 \frac{\text{ft}}{\text{sec}^2}} + \frac{(2 \text{ ft})^2}{4}}$$

$$= \boxed{4.38 \text{ ft} \quad (4.4 \text{ ft})}$$

The answer is (D).

9.

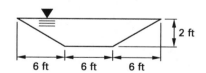

$$A = (6 \text{ ft})(2 \text{ ft}) + \frac{(2)(6 \text{ ft})(2 \text{ ft})}{2} = 24 \text{ ft}^2$$

$$P = 6 \text{ ft} + 2\sqrt{(6 \text{ ft})^2 + (2 \text{ ft})^2} = 18.65 \text{ ft}$$

$$R = \frac{A}{P} = \frac{24 \text{ ft}^2}{18.65 \text{ ft}} = 1.287 \text{ ft}$$

$$Q = (24 \text{ ft}^2)\left(\frac{1.49}{0.013}\right)(1.287 \text{ ft})^{2/3} \sqrt{0.002}$$

$$= \boxed{145.6 \text{ ft}^3/\text{sec} \quad (150 \text{ ft}^3/\text{sec})}$$

The answer is (B).

Alternate solution:

Solve using App. 19.E.

$$m = 3 \ (18.4°) \text{ column}$$

$$\frac{D}{b} = \frac{2 \text{ ft}}{6 \text{ ft}} = 0.33$$

Interpolating, $K \approx 6.69$.

$$Q = K\left(\frac{1}{n}\right)D^{8/3}\sqrt{S}$$

$$= (6.69)\left(\frac{1}{0.013}\right)(2 \text{ ft})^{8/3}\sqrt{0.002}$$

$$= 146.1 \text{ ft}^3/\text{sec}$$

The answer is (B).

10. The optimum trapezoidal channel has side slopes of 60° and a depth that is twice the hydraulic radius.

From App. 19.A, $n = 0.011$ for smooth concrete.

Rearranging Eq. 19.13,

$$AR^{2/3} = \frac{Qn}{\sqrt{S}}$$

$$= \frac{\left(17 \ \frac{\text{m}^3}{\text{s}}\right)(0.011)}{\sqrt{0.08}}$$

$$= 0.661$$

From Table 19.2,

$$A = \left(b + \frac{d}{\tan\theta}\right)d$$

$$R = \frac{bd\sin\theta + d^2\cos\theta}{b\sin\theta + 2d}$$

Since $R = d/2$ and $b = 2d/\sqrt{3}$,

$$AR^{2/3} = \left(\frac{2d}{\sqrt{3}} + \frac{d}{\tan\theta}\right)d\left(\frac{d}{2}\right)^{2/3} = 0.661$$

Simplifying, $d^{8/3} = 0.606$ with $\theta = 60°$, the optimal channel has a depth, d, of 0.828 m and a base, b, of 0.957 m. ⎡Use 0.8 m and 1.0 m.⎤

The answer is (A).

Alternate solution:

Use App. 19.F.

"Optimum" means $\alpha = 60°$ and $D/b = \sqrt{3}/2 = 0.866$.

Double interpolate $K' \approx 1.12$.

This is metric, so from App. 19.F footnote (c),

$$K' = \frac{1.12}{1.486} \approx 0.754$$

$$Q = K'\left(\frac{1}{n}\right)b^{8/3}\sqrt{S}$$

$$17 \ \frac{\text{m}^2}{\text{s}} = (0.754)\left(\frac{1}{0.011}\right)b^{8/3}\sqrt{0.08}$$

$$b = 0.952 \text{ m}$$

$$d = \frac{\sqrt{3}}{2}b = \left(\frac{\sqrt{3}}{2}\right)(0.952 \text{ m}) = 0.824 \text{ m}$$

The answer is (A).

11. From Eq. 19.53,

$$b = b_{\text{actual}} - 0.1NH$$

$$= 5 \text{ ft} - (0.1)(2)(0.43 \text{ ft})$$

$$= 4.914 \text{ ft}$$

From Eq. 19.50,

$$C_1 = \left(0.6035 + 0.0813\left(\frac{H}{Y}\right) + \frac{0.000295}{Y}\right)\left(1 + \frac{0.00361}{H}\right)^{3/2}$$

$$= \left(0.6035 + (0.0813)\left(\frac{0.43 \text{ ft}}{6 \text{ ft}}\right) + \frac{0.000295}{6 \text{ ft}}\right)$$

$$\times \left(1 + \frac{0.00361}{0.43 \text{ ft}}\right)^{3/2}$$

$$= 0.617$$

From Eq. 19.49,

$$Q = \tfrac{2}{3}C_1 b\sqrt{2g}H^{3/2}$$

$$= \left(\tfrac{2}{3}\right)(0.617)(4.914 \text{ ft})\sqrt{(2)\left(32.2 \ \frac{\text{ft}}{\text{sec}^2}\right)}(0.43 \text{ ft})^{3/2}$$

$$= \boxed{4.574 \text{ ft}^3/\text{sec} \quad (4.6 \text{ ft}^3/\text{sec})}$$

The answer is (C).

12. This is a Cipoletti weir. From Eq. 19.58,

$$Q = 3.367bH^{3/2}$$

$$= (3.367)\left(\frac{18 \text{ in}}{12 \ \frac{\text{in}}{\text{ft}}}\right)\left(\frac{9 \text{ in}}{12 \ \frac{\text{in}}{\text{ft}}}\right)^{3/2}$$

$$= \boxed{3.28 \text{ ft}^3/\text{sec} \quad (3.3 \text{ ft}^3/\text{sec})}$$

The answer is (B).

13.

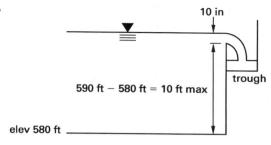

(a) Use Eq. 19.51.

$$C_1 \approx 0.602 + 0.083\left(\frac{H}{Y}\right)$$

$$= 0.602 + (0.083)\left(\frac{10 \text{ in}}{(590 \text{ ft} - 580 \text{ ft})\left(12\ \frac{\text{in}}{\text{ft}}\right)}\right)$$

$$= 0.609$$

Use Eq. 19.49 with $C_1 = 0.609$.

$$Q_{\text{MGD}} = \left(0.6463\ \frac{\text{MGD-sec}}{\text{ft}^3}\right)Q_{\text{ft}^3/\text{sec}}$$

$$= \left(0.6463\ \frac{\text{MGD-sec}}{\text{ft}^3}\right)(\tfrac{2}{3})(0.609)bH^{3/2}$$

$$\times \sqrt{(2)\left(32.2\ \frac{\text{ft}}{\text{sec}^2}\right)}$$

$$= 2.11bH^{3/2}$$

To satisfy the minimum head requirement, the minimum flow must be used. The minimum head requirement will then be automatically satisfied at maximum flow.

At minimum flow, $Q = 2$ MGD.

$$H = \frac{10 \text{ in}}{12\ \frac{\text{in}}{\text{ft}}} = 0.833 \text{ ft}$$

$$b_{\text{effective}} = \frac{2 \text{ MGD}}{(2.11)(0.833 \text{ ft})^{3/2}} = 1.25 \text{ ft}$$

From Eq. 19.53,

$$b_{\text{actual}} = 1.25 \text{ ft} + (0.1)(2)(0.833 \text{ ft})$$
$$= \boxed{1.42 \text{ ft} \quad (1.4 \text{ ft})}$$

The answer is (B).

(b) At maximum flow,

$$8 \text{ MGD} = (2.11)(1.25)H^{3/2}$$
$$H \approx 2.1 \text{ ft}$$

Recalculate C_1 from Eq. 19.52.

$$C_1 \approx 0.602 + 0.083\left(\frac{H}{Y}\right)$$

$$= 0.602 + (0.083)\left(\frac{2.1 \text{ ft}}{590 \text{ ft} - 580 \text{ ft}}\right)$$

$$= 0.619$$

From Eq. 19.49,

$$Q = \left(0.6463\ \frac{\text{MGD-sec}}{\text{ft}^3}\right)(\tfrac{2}{3})(0.619)bH^{3/2}$$

$$\times \sqrt{(2)\left(32.2\ \frac{\text{ft}}{\text{sec}^2}\right)}$$

$$= 2.14bH^{3/2}$$

Use Eq. 19.53 to calculate $b_{\text{effective}}$. Continue iterating until H stabilizes.

$$b_{\text{effective}} = 1.42 \text{ ft} - (0.1)(2)(2.1 \text{ ft}) = 1 \text{ ft}$$
$$8 \text{ MGD} = (2.14)(1)H^{3/2}$$
$$H = 2.41 \text{ ft}$$
$$b_{\text{effective}} = 1.42 \text{ ft} - (0.1)(2)(2.41 \text{ ft}) = 0.94 \text{ ft}$$
$$8 \text{ MGD} = (2.14)(0.94)H^{3/2}$$
$$H = 2.51 \text{ ft}$$
$$b_{\text{effective}} = 1.42 \text{ ft} - (0.1)(2)(2.51 \text{ ft}) = 0.918 \text{ ft}$$
$$8 \text{ MGD} = (2.14)(0.918)H^{3/2}$$
$$H = 2.55 \text{ ft}$$
$$b_{\text{effective}} = 1.42 \text{ ft} - (0.1)(2)(2.55 \text{ ft}) = 0.910 \text{ ft}$$
$$8 \text{ MGD} = (2.14)(0.910)H^{3/2}$$
$$H = 2.56 \text{ ft} \quad \text{[acceptable]}$$

Use a freeboard elevation of

$$580 \text{ ft} + 10 \text{ ft} + 2.56 \text{ ft} + \frac{4 \text{ in}}{12\ \frac{\text{in}}{\text{ft}}} = 592.89 \text{ ft}$$

The elevation at maximum flow is

$$592.89 \text{ ft} - \text{freeboard height} = 592.89 \text{ ft} - \frac{4 \text{ in}}{12\ \frac{\text{in}}{\text{ft}}}$$

$$= \boxed{592.56 \text{ ft} \quad (592.6 \text{ ft})}$$

The answer is (C).

(c) The channel elevation is

$$580 \text{ ft} + 10 \text{ ft} = 590 \text{ ft}$$

The elevation at minimum flow is

$$590 \text{ ft} + 0.833 \text{ ft} = \boxed{590.833 \text{ ft} \quad (590.8 \text{ ft})}$$

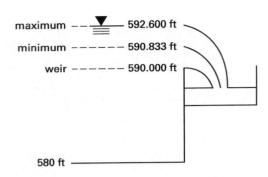

Flow of Fluids

The answer is (A).

14. From Eq. 19.94,

$$v_1 = \sqrt{\left(\frac{gd_2}{2d_1}\right)(d_1 + d_2)}$$

$$= \sqrt{\left(\frac{32.2 \ \frac{\text{ft}}{\text{sec}^2}}{2}\right)\left(\frac{2.4 \ \text{ft}}{1 \ \text{ft}}\right)(1 \ \text{ft} + 2.4 \ \text{ft})}$$

$$= 11.46 \ \text{ft/sec}$$

$$Q = Av = (5 \ \text{ft})(1 \ \text{ft})\left(11.46 \ \frac{\text{ft}}{\text{sec}}\right)$$

$$= \boxed{57.30 \ \text{ft}^3/\text{sec} \quad (57 \ \text{ft}^3/\text{sec})}$$

The answer is (D).

15. From Eq. 19.8,

$$A_1 v_1 = A_2 v_2$$

$$(0.2 \ \text{ft})(\text{width})\left(54.7 \ \frac{\text{ft}}{\text{sec}}\right) = (6 \ \text{ft})(\text{width})v_2$$

$$v_2 = 1.82 \ \text{ft/sec}$$

From Eq. 19.95,

$$\Delta E = \left(d_1 + \frac{v_1^2}{2g}\right) - \left(d_2 + \frac{v_2^2}{2g}\right)$$

$$= 0.2 \ \text{ft} + \frac{\left(54.7 \ \frac{\text{ft}}{\text{sec}}\right)^2}{(2)\left(32.2 \ \frac{\text{ft}}{\text{sec}^2}\right)}$$

$$- \left(6 \ \text{ft} + \frac{\left(1.82 \ \frac{\text{ft}}{\text{sec}}\right)^2}{(2)\left(32.2 \ \frac{\text{ft}}{\text{sec}^2}\right)}\right)$$

$$= \boxed{40.61 \ \text{ft} \quad (41 \ \text{ft-lbf/lbm})}$$

The answer is (A).

16. (a) From Eq. 19.94,

$$v_1 = \sqrt{\left(\frac{gd_2}{2d_1}\right)(d_1 + d_2)}$$

$$= \sqrt{\left(\frac{\left(32.2 \ \frac{\text{ft}}{\text{sec}^2}\right)(3.8 \ \text{ft})}{(2)(1.2 \ \text{ft})}\right)(1.2 \ \text{ft} + 3.8 \ \text{ft})}$$

$$= 15.97 \ \text{ft/sec}$$

$$Q = Av = (1.2 \ \text{ft})(6 \ \text{ft})\left(15.97 \ \frac{\text{ft}}{\text{sec}}\right)$$

$$= \boxed{115 \ \text{ft}^3/\text{sec}}$$

The answer is (B).

(b)

$$v_2 = \frac{Q}{A} = \frac{115 \ \frac{\text{ft}^3}{\text{sec}}}{(3.8 \ \text{ft})(6 \ \text{ft})} = 5.04 \ \text{ft/sec}$$

From Eq. 19.95,

$$\Delta E = \left(d_1 + \frac{v_1^2}{2g}\right) - \left(d_2 + \frac{v_2^2}{2g}\right)$$

$$= \left(1.2 \ \text{ft} + \frac{\left(15.97 \ \frac{\text{ft}}{\text{sec}}\right)^2}{(2)\left(32.2 \ \frac{\text{ft}}{\text{sec}^2}\right)}\right)$$

$$- \left(3.8 \ \text{ft} + \frac{\left(5.04 \ \frac{\text{ft}}{\text{sec}}\right)^2}{(2)\left(32.2 \ \frac{\text{ft}}{\text{sec}^2}\right)}\right)$$

$$= \boxed{0.97 \ \text{ft} \quad (1 \ \text{ft-lbf/lbm})}$$

The answer is (A).

17. (a) Disregard the velocity of approach. Use $C_s = 3.5$ in Eq. 19.61.

$$Q = C_s b H^{3/2}$$

$$= \left(3.5 \ \frac{\text{ft}^{1/2}}{\text{sec}}\right)(1 \ \text{ft})(2 \ \text{ft})^{3/2}$$

$$= \boxed{9.9 \ \text{ft}^3/\text{sec per ft of width}}$$

The answer is (A).

(b) The upstream energy is

$$E_1 = d = 2 \ \text{ft} \quad [v_1 \approx 0]$$

At the toe, from Eq. 19.73,

$$E_2 = E_1 + z_1 - z_2$$
$$= 2 \text{ ft} + 40 \text{ ft} - 0$$
$$= 42 \text{ ft}$$

$$E_2 = d_2 + \frac{v_2^2}{2g} = d_2 + \frac{Q^2}{2gA_2^2}$$

$$42 \text{ ft} = d_2 + \frac{\left(9.9 \ \frac{\text{ft}^3}{\text{sec}}\right)^2}{(2)\left(32.2 \ \frac{\text{ft}}{\text{sec}^2}\right)(1 \text{ ft}^2) d_2^2}$$

By trial and error, $d_2 = \boxed{0.19 \text{ ft.}}$

The answer is (A).

18.

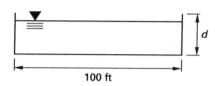

(a)
$$R = \frac{(100 \text{ ft}) d}{100 \text{ ft} + 2d} = \frac{50d}{50 + d}$$

From Eq. 19.13,

$$Q = \left(\frac{1.49}{n}\right) A R^{2/3} \sqrt{S}$$

$$10,000 \ \frac{\text{ft}^3}{\text{sec}} = \left(\frac{1.49}{0.012}\right)(100d)\left(\frac{50d}{50+d}\right)^{2/3}\sqrt{0.05}$$

$$0.26538 = \left(\frac{d^{5/2}}{50+d}\right)^{2/3}$$

$$0.1367 = \frac{d^{5/2}}{50+d}$$

By trial and error, $d = \boxed{2.2 \text{ ft.}}$

The answer is (B).

(b) From Eq. 19.75,

$$d_c = \sqrt[3]{\frac{Q^2}{gw^2}}$$

$$= \sqrt[3]{\frac{\left(10,000 \ \frac{\text{ft}^3}{\text{sec}}\right)^2}{\left(32.2 \ \frac{\text{ft}}{\text{sec}^2}\right)(100 \text{ ft})^2}}$$

$$= \boxed{6.77 \text{ ft} \quad (6.8 \text{ ft})}$$

The answer is (C).

(c) Since $d < d_c$, $\boxed{\text{flow is rapid (shooting).}}$
The answer is (C).

(d) $\quad v = \dfrac{Q}{A} = \dfrac{10,000 \ \frac{\text{ft}^3}{\text{sec}}}{(100 \text{ ft})(2.2 \text{ ft})} = 45.45 \text{ ft/sec}$

From Eq. 19.92,

$$d_2 = -\tfrac{1}{2}d_1 + \sqrt{\frac{2v_1^2 d_1}{g} + \frac{d_1^2}{4}}$$

$$= -\left(\tfrac{1}{2}\right)(2.2 \text{ ft})$$

$$+ \sqrt{\frac{(2)\left(45.45 \ \frac{\text{ft}}{\text{sec}}\right)^2 (2.2 \text{ ft})}{32.2 \ \frac{\text{ft}}{\text{sec}^2}} + \frac{(2.2 \text{ ft})^2}{4}}$$

$$= \boxed{15.74 \text{ ft} \quad (16 \text{ ft})}$$

The answer is (C).

19. (a) From Eq. 19.61 using $C_s = 3.7$,

$$H = 2425 \text{ ft} - 2420 \text{ ft} = 5 \text{ ft}$$
$$Q = C_3 b H^{3/2} = (3.7)(60 \text{ ft})(5 \text{ ft})^{3/2}$$
$$= \boxed{2482 \text{ ft}^3/\text{sec} \quad (2500 \text{ ft}^3/\text{sec})}$$

The answer is (C).

(b) Disregarding the velocity of approach, the initial energy is

$$E_1 = (\text{elev})_1 = 2425 \text{ ft}$$

At the toe (point 2),

$$E_1 = E_2 + h_f$$

$$2425 = (\text{elev})_2 + d_2 + \frac{Q^2}{2gw^2 d_2^2}$$
$$+ (0.20)(2425 \text{ ft} - 2345 \text{ ft})$$

$$(1 - 0.20)$$
$$\times (2425 \text{ ft} - 2345 \text{ ft}) = d_2 + \frac{\left(2482 \ \frac{\text{ft}^3}{\text{sec}}\right)^2}{(2)\left(32.2 \ \frac{\text{ft}}{\text{sec}^2}\right)(60 \text{ ft})^2 d_2^2}$$

$$64 \text{ ft} = d_2 + \frac{26.57}{d_2^2}$$

By trial and error, $d_2 = \boxed{0.647 \text{ ft } (0.65 \text{ ft}).}$

The answer is (A).

(c) $d_{toe} = d_1$ is implicitly a conjugate depth.

If there was a hydraulic jump, it would be between the conjugate depths, d_1 to d_2.

$$d_1 = 0.647 \text{ ft}$$

$$v_1 = \frac{Q}{A_1} = \frac{Q}{d_1 w} = \frac{2482 \frac{\text{ft}^3}{\text{sec}}}{(0.647 \text{ ft})(60 \text{ ft})}$$

$$= 63.94 \text{ ft/sec}$$

From Eq. 19.92,

$$d_2 = -\tfrac{1}{2} d_1 + \sqrt{\frac{2 v_1^2 d_1}{g} + \frac{d_1^2}{4}}$$

$$= -\left(\tfrac{1}{2}\right)(0.647 \text{ ft})$$

$$+ \sqrt{\frac{(2)\left(63.9 \frac{\text{ft}}{\text{sec}}\right)^2 (0.647 \text{ ft})}{32.2 \frac{\text{ft}}{\text{sec}^2}} + \frac{(0.647 \text{ ft})^2}{4}}$$

$$= 12.49 \text{ ft}$$

The answer is (D).

(d) $v_2 = \dfrac{Q}{A_2} = \dfrac{Q}{d_2 w} = \dfrac{2482 \frac{\text{ft}^3}{\text{sec}}}{(12.49 \text{ ft})(60 \text{ ft})} = 3.31 \text{ ft/sec}$

Use d_2 and v_2 in Eq. 19.95.

$$\Delta E = \left(d_1 + \frac{v_1^2}{2g}\right) - \left(d_2 + \frac{v_2^2}{2g}\right)$$

$$= \left(0.647 \text{ ft} + \frac{\left(63.9 \frac{\text{ft}}{\text{sec}}\right)^2}{(2)\left(32.2 \frac{\text{ft}}{\text{sec}^2}\right)}\right)$$

$$- \left(12.49 \text{ ft} + \frac{\left(3.31 \frac{\text{ft}}{\text{sec}}\right)^2}{(2)\left(32.2 \frac{\text{ft}}{\text{sec}^2}\right)}\right)$$

$$= \boxed{51.4 \text{ ft} \quad (50 \text{ ft-lbf/lbm})}$$

The answer is (B).

20. (a) From Eq. 19.65,

$$n = 1.522 b^{0.026}$$

$$= (1.522)(6 \text{ ft})^{0.026}$$

$$= 1.595$$

From Eq. 19.64,

$$Q = K b H_a^n$$

$$= (4)(6 \text{ ft})\left(\frac{18 \text{ in}}{12 \frac{\text{in}}{\text{ft}}}\right)^{1.595}$$

$$= \boxed{45.8 \text{ ft}^3/\text{sec} \quad (46 \text{ ft}^3/\text{sec})}$$

The answer is (B).

(b) The flumes are self-cleansing, have low head loss, and can be operated by unskilled personnel. Parshall flumes are usually permanent structures.

The answer is (D).

21. (a) The normal depth is d.

$$R = \frac{8d}{8 + 2d}$$

From Eq. 19.13,

$$Q = \left(\frac{1.49}{n}\right) A R^{2/3} \sqrt{S}$$

$$150 \frac{\text{ft}^3}{\text{sec}} = \left(\frac{1.49}{0.015}\right)(8d)\left(\frac{8d}{8 + 2d}\right)^{2/3} \sqrt{0.0015}$$

$$4.875 = d\left(\frac{8d}{8 + 2d}\right)^{2/3}$$

By trial and error, $d = \boxed{3.28 \text{ ft} \ (3.3 \text{ ft}).}$

The answer is (B).

(b) To get the backwater curve, choose depths and compute distances to those depths.

Preliminary parameters:

d	$A = 8d$	$v = \dfrac{Q}{A}$	$\dfrac{v^2}{2g}$	$E = d + \dfrac{v^2}{2g}$	$P = 2d + 8$	R
6	48	3.12	0.151	6.151	20	2.40
5	40	3.75	0.218	5.218	18	2.22
4	32	4.69	0.342	4.342	16	2.00
3.28	26.24	5.72	0.502	3.782	14.56	1.80

The average velocities and average hydraulic radii are

d	v	v_{ave}	R	R_{ave}
6.00	3.12		2.40	
		3.435		2.31
5.00	3.75		2.22	
		4.22		2.11
4.00	4.69		2.00	
		5.21		1.90
3.28	5.72		1.80	

From Eq. 19.88, the average energy gradient is

$$S_{\text{ave}} = \left(\frac{n\text{v}_{\text{ave}}}{1.49R_{\text{ave}}^{2/3}}\right)^2$$

d	S_{ave}	ΔE	$S_0 - S_{\text{ave}}$
6.00			
	0.000392	0.933	0.001108
5.00			
	0.000667	0.876	0.000833
4.00			
	0.001169	0.552	0.000331
3.28			

From Eq. 19.90,

$$L_{6\text{-}5} = \frac{\Delta E}{S_0 - S_{\text{ave}}} = \frac{0.933 \text{ ft}}{0.001108}$$
$$= \boxed{842 \text{ ft} \quad (840 \text{ ft})}$$

The answer is (B).

(c)
$$L_{5\text{-}4} = \frac{\Delta E}{S_0 - S_{\text{ave}}} = \frac{0.876 \text{ ft}}{0.000833}$$
$$= \boxed{1052 \text{ ft} \quad (1100 \text{ ft})}$$

The answer is (D).

22. For concrete, $n = 0.012$.

The pipe diameter is

$$D = \frac{42 \text{ in}}{12 \frac{\text{in}}{\text{ft}}} = 3.5 \text{ ft}$$

The hydraulic radius is

$$R = \frac{D}{4} = \frac{3.5 \text{ ft}}{4}$$
$$= 0.875 \text{ ft}$$

From a table of orifice coefficients, Table 17.5, $C_d = 0.82$.

$$h_1 = 5 \text{ ft} + 3.5 \text{ ft} + (250 \text{ ft})(0.006)$$
$$= 10 \text{ ft}$$

Determine the culvert flow type.

$$\frac{h_1 - z}{D} = \frac{10 \text{ ft} - (250 \text{ ft})(0.006)}{3.5 \text{ ft}}$$
$$= 2.4$$

The tailwater submerges the culvert outlet, so this is type-4 flow (outlet control as per Table 19.9).

$$h_1 = 10 \text{ ft}$$
$$h_4 = D = 3.5 \text{ ft}$$

$$Q = C_d A_0 \sqrt{2g\left(\frac{h_1 - h_4}{1 + \dfrac{29C_d^2 n^2 L}{R^{4/3}}}\right)}$$
$$= \left(\frac{(0.82)\pi(3.5 \text{ ft})^2}{4}\right)$$
$$\times \sqrt{(2)\left(32.2 \frac{\text{ft}}{\text{sec}^2}\right) \times \left(\frac{10 \text{ ft} - 3.5 \text{ ft}}{1 + \dfrac{(29)(0.82)^2(0.012)^2(250 \text{ ft})}{(0.875 \text{ ft})^{4/3}}}\right)}$$
$$= 119 \text{ ft}^3/\text{sec}$$

Check to see if the flow rate is limited by inlet geometry. Evaluate type-6 flow. From Eq. 19.108 with $h_3 = d$ and $h_f = 0$,

$$Q = C_d A_o \sqrt{2g(h_1 - h_3 - h_f)}$$
$$= (0.82)\left(\frac{\pi(3.5 \text{ ft})^2}{4}\right)$$
$$\times \sqrt{(2)\left(32.2 \frac{\text{ft}}{\text{sec}^2}\right)(10 \text{ ft} - 3.5 \text{ ft} - 0)}$$
$$= 161.4 \text{ ft}^3/\text{sec}$$

The culvert has the capacity.

$$Q = \boxed{119 \text{ ft}^3/\text{sec} \quad (120 \text{ ft}^3/\text{sec})}$$

The answer is (D).

23. (a) Rearrange Eq. 19.61.

$$H = \left(\frac{Q}{C_s b}\right)^{2/3} = \left(\frac{2000 \frac{\text{m}^3}{\text{s}}}{(2.2)(150 \text{ m})}\right)^{2/3}$$
$$= \boxed{3.32 \text{ m} \quad (3.3 \text{ m})}$$

The answer is (D).

(b) At the crest,

$$v_1 = \frac{Q}{A_1} = \frac{2000 \frac{m^3}{s}}{(150 \text{ m})(3.32 \text{ m})}$$
$$= 4.02 \text{ m/s}$$

From Eq. 19.97,

$$E_1 = y_{crest} + H + \frac{v_1^2}{2g}$$
$$= 30 \text{ m} + 3.32 \text{ m} + \frac{\left(4.02 \frac{m}{s}\right)^2}{(2)\left(9.81 \frac{m}{s^2}\right)}$$
$$= \boxed{34.1 \text{ m}}$$

The answer is (B).

(c) At the toe,

$$v_2 = \frac{Q}{A_2} = \frac{2000 \frac{m^3}{s}}{(150 \text{ m})(0.6 \text{ m})} = 22.22 \text{ m/s}$$

$$E_2 = d_2 + \frac{v_2^2}{2g}$$
$$= 0.6 \text{ m} + \frac{\left(22.22 \frac{m}{s}\right)^2}{(2)\left(9.81 \frac{m}{s^2}\right)}$$
$$= \boxed{25.8 \text{ m}}$$

The answer is (A).

(d) The friction loss is

$$E_1 - E_2 = 34.1 \text{ m} - 25.8 \text{ m}$$
$$= \boxed{8.3 \text{ m}}$$

The answer is (C).

(e) From Eq. 19.13,

$$Q = \left(\frac{1}{n}\right) A R^{2/3} \sqrt{S}$$

$$2000 \frac{m^3}{s} = \left(\frac{1}{0.016}\right)(150 \text{ m}) d_n \left(\frac{(150 \text{ m}) d_n}{2 d_n + 150 \text{ m}}\right)^{2/3}$$
$$\times \sqrt{0.002}$$

By trial and error, $d_n = \boxed{2.6 \text{ m.}}$

The answer is (B).

(f) From Eq. 19.75,

$$d_c = \left(\frac{Q^2}{gw^2}\right)^{1/3}$$
$$= \left(\frac{\left(2000 \frac{m^3}{s}\right)^2}{\left(9.81 \frac{m}{s^2}\right)(150 \text{ m})^2}\right)^{1/3}$$
$$= \boxed{2.6 \text{ m}}$$

The answer is (B).

(g) The normal depth of flow coincides with the critical depth. A hydraulic jump is an abrupt rise from below the critical depth to above the critical depth. In this case the water will rise gradually and no jump occurs.

The answer is (D).

(h) From Table 19.7, the maximum velocity (assuming the water is clear) is

$$v_{max} = \left(4.0 \frac{ft}{sec}\right)\left(0.3 \frac{m}{ft}\right) = \boxed{1.2 \text{ m/s}}$$

The answer is (A).

(i) From App. 19.A, for natural channels in good condition, $n = 0.025$.

The depth of flow is

$$d = \frac{Q}{vb} = \frac{2000 \frac{m^3}{sec}}{\left(1.2 \frac{m}{s}\right)(150 \text{ m})}$$
$$= 11.11 \text{ m}$$
$$R = \frac{bd}{2d + b}$$
$$= \frac{(150 \text{ m})(11.1 \text{ m})}{(2)(11.1 \text{ m}) + 150 \text{ m}}$$
$$= 9.67 \text{ m}$$

Rearranging Eq. 19.12,

$$S = \left(\frac{vn}{R^{2/3}}\right)^2$$
$$= \left(\frac{\left(1.2 \frac{m}{s}\right)(0.025)}{(9.67 \text{ m})^{2/3}}\right)^2$$
$$= \boxed{0.000044}$$

The answer is (B).

(j) From Eq. 19.29,

$$h_f = LS = (1000 \text{ m})(0.002) = \boxed{2.0 \text{ m}}$$

The answer is (D).

20 Meteorology, Climatology, and Hydrology

PRACTICE PROBLEMS

Hydrographs

1. A 2 h storm over a 111 km^2 area produces a total runoff volume of 4×10^6 m^3 with a peak discharge of 260 m^3/s.

(a) What is the total excess precipitation?

(A) 1.4 cm

(B) 2.6 cm

(C) 3.6 cm

(D) 4.0 cm

(b) What is the unit hydrograph peak discharge?

(A) 72 m^3/s·cm

(B) 120 m^3/s·cm

(C) 210 m^3/s·cm

(D) 260 m^3/s·cm

(c) If a 2 h storm producing 6.5 cm of runoff is to be used to design a culvert, what is the design flood hydrograph volume?

(A) 4.0×10^6 m^3

(B) 7.2×10^6 m^3

(C) 2.6×10^7 m^3

(D) 3.6×10^7 m^3

(d) What is the design discharge?

(A) 89 m^3/s

(B) 130 m^3/s

(C) 260 m^3/s

(D) 470 m^3/s

(e) The recurrence interval of the 6.5 cm storm is 50 yr, and the culvert is to be designed for a 30 yr life. What is the probability that the capacity will be exceeded during the design life?

(A) 0%

(B) 33%

(C) 45%

(D) 92%

(f) The unit hydrograph represents water flowing into a stream from

I.   base flow

II.   evapotranspiration

III. overland flow

IV.   surface flow

V.   interflow

(A) I only

(B) I and III

(C) II, III, and IV

(D) III, IV, and V

Use the following information for parts (g) through (j).

Stream discharges recorded during a 3 h storm on a 40 km^2 watershed are as follows.

t (h)	Q (m^3/s)
0	17
1	16
2	48
3	90
4	108
5	85
6	58
7	38
8	26
9	20
10	18
11	17
12	16

Flow of Fluids

(g) What is the unit hydrograph peak discharge from surface runoff for this 3 h storm?

(A) 5.2 m^3/s·cm

(B) 29 m^3/s·cm

(C) 65 m^3/s·cm

(D) 120 m^3/s·cm

(h) What is the peak discharge for a 6 h storm producing 5 cm of runoff in this watershed?

(A) 92 m^3/s

(B) 140 m^3/s

(C) 170 m^3/s

(D) 300 m^3/s

(i) What is the peak discharge for a 5 h storm producing 5 cm of runoff in this watershed?

(A) 63 m^3/s

(B) 90 m^3/s

(C) 110 m^3/s

(D) 260 m^3/s

(j) What is the design flood hydrograph volume for a 5 h storm producing 5 cm of runoff in this watershed?

(A) 1.3×10^4 m^3

(B) 4.0×10^5 m^3

(C) 8.3×10^5 m^3

(D) 2.0×10^6 m^3

2. (*Time limit: one hour*) A stream gaging station recorded the following discharges from a 1.2 mi^2 drainage area after a storm.

t (hr)	Q (ft^3/sec)
0	102
1	99
2	101
3	215
4	507
5	625
6	455
7	325
8	205
9	145
10	100
11	70
12	55
13	49
14	43
15	38

(a) Draw the actual hydrograph.

(b) Separate the groundwater and surface water.

(c) Draw the unit hydrograph.

(d) What is the length of the direct runoff recession limb?

(A) 5 hr

(B) 6 hr

(C) 10 hr

(D) 13 hr

Rational Equation

3. A 0.5 mi^2 drainage area has a runoff coefficient of 0.6 and a time of concentration of 60 min. The drainage area is in Steel region no. 3, and a 10 yr storm is to be used for design purposes. What is the peak runoff?

(A) 310 ft^3/sec

(B) 390 ft^3/sec

(C) 460 ft^3/sec

(D) 730 ft^3/sec

4. Four contiguous 5 ac watersheds are served by an adjacent 1200 ft storm drain ($n = 0.013$ and slope = 0.005). Inlets to the storm drain are placed every 300 ft along the storm drain. The inlet time for each area served by an inlet is 15 min, and the area's runoff coefficient is 0.55. A storm to be used for design purposes has the following characteristics (I is in in/hr, t is in min).

$$I = \frac{100}{t_c + 10}$$

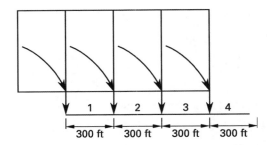

All flows are maximum, and all pipe sizes are available. n is constant. What is the diameter of the last section of storm drain?

(A) 28 in

(B) 32 in

(C) 36 in

(D) 42 in

5. (*Time limit: one hour*) A 75 ac urbanized section of land drains into a rectangular 5 ft × 7 ft channel that directs runoff through a round culvert under a roadway. The culvert is concrete and is 60 in in diameter. It is 36 ft long and placed on a 1% slope.

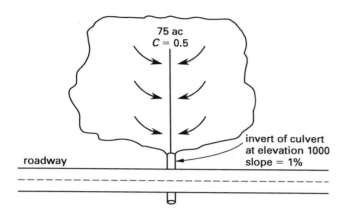

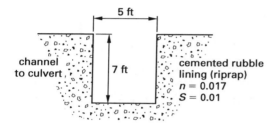

It is desired to evaluate the culvert design based on a 50 yr storm. It is known that the time of concentration to the entrance of the culvert is 30 min. The corresponding rainfall intensity is 5.3 in/hr.

(a) If the minimum road surface elevation is 1010 ft, will the road surface be flooded?

    (A) Yes; the culvert needs 10 ft^3/sec additional capacity.

    (B) Yes; the culvert needs 70 ft^3/sec additional capacity.

    (C) No; the culvert has 20 ft^3/sec excess capacity.

    (D) No; the culvert has 70 ft^3/sec excess capacity.

(b) What is the depth of flow elevation upstream at the culvert entrance after 30 min?

    (A) 2.7 ft

    (B) 3.1 ft

    (C) 3.6 ft

    (D) 4.1 ft

(c) If $n_{full} = 0.013$ and n varies with depth of flow, what is the depth of flow at the culvert exit after 30 min?

    (A) 2.7 ft

    (B) 3.1 ft

    (C) 3.6 ft

    (D) 4.1 ft

Hydrograph Synthesis

6. (*Time limit: one hour*) A standard 4 hr storm produces 2 in net of runoff. A stream gaging report is produced from successive sampling every few hours. Two weeks later, the first 4 hr of an 8 hr storm over the same watershed produces 1 in of runoff. The second 4 hr produces 2 in of runoff. Neglecting groundwater, draw a hydrograph of the 8 hr storm.

t (hr)	Q (ft^3/sec)
0	0
2	100
4	350
6	600
8	420
10	300
12	250
14	150
16	100
18	–
20	50
22	–
24	0

Reservoir Sizing

7. A class A evaporation pan located near a reservoir shows a 1 day evaporation loss of 0.8 in. If the pan coefficient is 0.7, what is the approximate evaporation loss in the reservoir?

    (A) 0.56 in

    (B) 0.63 in

    (C) 0.98 in

    (D) 1.1 in

8. A reservoir has a total capacity of 7 volume units. At the beginning of a study, the reservoir contains 5.5 units. The monthly demand on the reservoir from a nearby city is 0.7 units. The monthly inflow to the reservoir is normally distributed with a mean of 0.9 units and a standard deviation of 0.2 units. Simulate one year of reservoir operation with a 99.7% confidence level.

9. Repeat Prob. 8 assuming that the monthly demand on the reservoir is normally distributed with a mean of 0.7 units and a standard deviation of 0.2 units.

10. (*Time limit: one hour*) A reservoir is needed to provide 20 ac-ft of water each month. The inflow for each of 13 representative months is given. The reservoir starts full. Size the reservoir.

month	inflow (ac-ft)
February	30
March	60
April	20
May	10
June	5
July	10
August	5
September	10
October	20
November	90
December	85
January	75
February	50

(A) 40 ac-ft

(B) 60 ac-ft

(C) 90 ac-ft

(D) 120 ac-ft

11. (*Time limit: one hour*) A reservoir with a constant draft of 240 MG/mi^2-yr is being designed. The inflow distribution is known.

month	inflow (MG/mi^2)	inflow (m^3/m^2)
January	20	0.029
February	30	0.044
March	45	0.066
April	30	0.044
May	40	0.058
June	30	0.044
July	15	0.022
August	5	0.007
September	15	0.022
October	60	0.088
November	90	0.132
December	40	0.058

(a) What should be the minimum reservoir size?

(A) 25 MG/mi^2

(B) 35 MG/mi^2

(C) 45 MG/mi^2

(D) 65 MG/mi^2

(b) The reservoir will start to spill at the end of

(A) March

(B) June

(C) September

(D) March and October

Flow of Fluids

SOLUTIONS

1. (a) From Eq. 20.21,

$$P_{\text{ave}} = \frac{V}{A_d} = \frac{4 \times 10^6 \text{ m}^3}{(111 \text{ km}^2)\left(1000 \dfrac{\text{m}}{\text{km}}\right)^2}$$

$$= \boxed{0.036 \text{ m} \quad (3.6 \text{ cm})}$$

The answer is (C).

(b) The unit hydrograph discharge is the peak discharge divided by the average precipitation.

$$Q_{\text{unit hydrograph}} = \frac{Q_p}{P} = \frac{260 \dfrac{\text{m}^3}{\text{s}}}{3.6 \text{ cm}}$$

$$= \boxed{72.2 \text{ m}^3/\text{s·cm} \quad (72 \text{ m}^3/\text{s·cm})}$$

The answer is (A).

(c) The design flood hydrograph volume for a 6.5 cm storm is determined by multiplying the unit hydrograph volume by 6.5 cm. For the unit hydrograph,

$$V_{\text{hydrograph}} = \frac{V}{P} = \frac{4 \times 10^6 \text{ m}^3}{3.6 \text{ cm}}$$

$$= 1.11 \times 10^6 \text{ m}^3/\text{cm}$$

For the 6.5 cm storm,

$$V = \left(1.11 \times 10^6 \dfrac{\text{m}^3}{\text{cm}}\right)(6.5 \text{ cm}) = \boxed{7.2 \times 10^6 \text{ m}^3}$$

The answer is (B).

(d) The design discharge is determined by multiplying the unit hydrograph discharge by 6.5 cm.

$$Q_p = Q_{\text{hydrograph}}(6.5 \text{ cm}) = \left(72.2 \dfrac{\text{m}^3}{\text{s·cm}}\right)(6.5 \text{ cm})$$

$$= \boxed{469 \text{ m}^3/\text{s} \quad (470 \text{ m}^3/\text{s})}$$

The answer is (D).

(e) From Eq. 20.20,

$$p\{F \text{ event in } n \text{ years}\} = 1 - \left(1 - \frac{1}{F}\right)^n$$

$$p\{50 \text{ yr flood in 30 yr}\} = 1 - \left(1 - \frac{1}{50}\right)^{30}$$

$$= \boxed{0.45 \quad (45\%)}$$

The answer is (C).

(f) The unit hydrograph represents all discharge into a stream except for groundwater or base flow, which are separated out. Evapotranspiration refers to water that is returned to the atmosphere and, therefore, is not measured in the stream discharge. The unit hydrograph includes overland flow, surface flow, and interflow.

The answer is (D).

(g) The actual runoff is plotted. To separate base flow from overland flow, use the straight line method. Draw a horizontal line from the start of the rising limb to the falling limb. In the table, subtract the base flow from the overland flow.

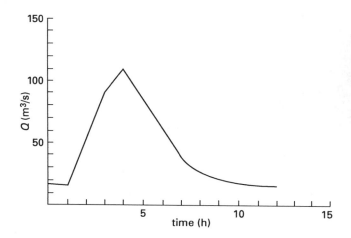

hour	runoff	ground water	surface water	surface water 3.14
0	17	16	1	0.32
1	16	16	0	0
2	48	16	32	10.19
3	90	16	74	23.56
4	108	16	92	29.29
5	85	16	69	21.97
6	58	16	42	13.37
7	38	16	22	7.00
8	26	16	10	3.18
9	20	16	4	1.27
10	18	16	2	0.64
11	17	16	1	0.32
12	16	16	0	0
		total	349 m³/s	

The average precipitation is given by Eq. 20.21.

$$P = \frac{V}{A_d}$$

The volume of runoff is the area under the separated hydrograph curve. Since data are given for each hour, the "width" of each histogram cell is 1 h.

$$V = \left(349 \dfrac{\text{m}^3}{\text{s}}\right)(1 \text{ h})\left(3600 \dfrac{\text{s}}{\text{h}}\right)$$

$$= 1\,256\,400 \text{ m}^3$$

$$P = \frac{1\,256\,400 \text{ m}^3}{(40 \text{ km}^2)\left(1000 \dfrac{\text{m}}{\text{km}}\right)^2}$$

$$= 0.0314 \text{ m} \quad (3.14 \text{ cm})$$

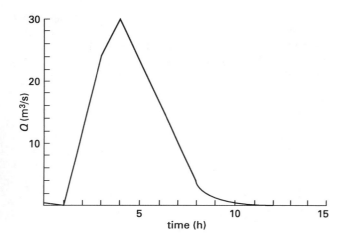

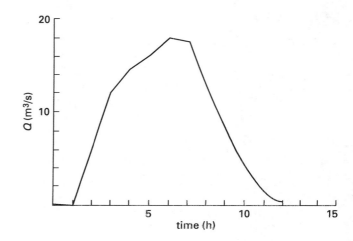

The unit hydrograph peak discharge is

$$\frac{108 \ \frac{m^3}{s} - 16 \ \frac{m^3}{s}}{3.14 \ cm} = \boxed{29.3 \ m^3/s \cdot cm \quad (29 \ m^3/s \cdot cm)}$$

The answer is (B).

(h) There are two methods for constructing hydrographs for a longer storm than that of the unit hydrograph. For storms that are whole multiples of the unit hydrograph duration, the lagging storm method is simplest.

In a table, add another unit hydrograph separated by $t_r = 3$ h. Then, add the ordinates for a hydrograph of $Nt_r = (2)(3 \ h) = 6$ h duration. The ordinates must then be divided by $N = 2$ to get the unit hydrograph ordinates.

hour	unit hydrograph 3 h storm	second storm	total	unit hydrograph 6 h storm
0	0.32	0	0.32	0.16
1	0	0	0	0
2	10.19	0	10.19	5.09
3	23.56	0.32	23.88	11.94
4	29.29	0	29.29	14.65
5	21.97	10.19	32.16	16.08
6	13.37	23.56	36.93	18.47
7	7.00	29.29	36.29	18.15
8	3.18	21.97	25.15	12.58
9	1.27	13.37	14.65	7.32
10	0.64	7.00	7.64	3.82
11	0.32	3.18	3.50	1.75
12	0	1.27	1.27	0.64
13	0	0.64	0.64	0.32
14	0	0.32	0.32	0.16
15	0	0	0	0

The peak discharge for a 5 cm storm is calculated from the peak unit hydrograph discharge.

$$Q_{p,\text{unit hydrograph}} = 18.47 \ m^3/s \cdot cm$$

$$Q_{p,5 \text{ cm storm}} = (5 \ cm)\left(18.47 \ \frac{m^3}{s \cdot cm}\right)$$

$$= \boxed{92.35 \ m^3/s \quad (92 \ m^3/s)}$$

The answer is (A).

(i) If a storm is not a multiple of the unit hydrograph duration, the S-curve method must be used to construct a hydrograph for that storm. In the table, add the ordinates of five 3 h unit hydrographs, each offset by 3 h, to produce an S-curve.

hour	3 h unit hydrograph	second storm	third storm	fourth storm	fifth storm	total (S-curve)
0	0.32	0	0	0	0	0.32
1	0	0	0	0	0	0
2	10.19	0	0	0	0	10.19
3	23.56	0.32	0	0	0	23.88
4	29.29	0	0	0	0	29.29
5	21.97	10.19	0	0	0	32.16
6	13.37	23.56	0.32	0	0	37.25
7	7.00	29.29	0	0	0	36.29
8	3.18	21.97	10.19	0	0	35.33
9	1.27	13.37	23.56	0.32	0	38.52
10	0.64	7.00	29.29	0	0	36.93
11	0.32	3.18	21.97	10.19	0	35.66
12	0	1.27	13.37	23.56	0.32	38.52
13	0	0.64	7.00	29.29	0	36.93
14	0	0.32	3.18	21.97	10.19	35.66
15	0	0	1.27	13.37	23.56	38.20

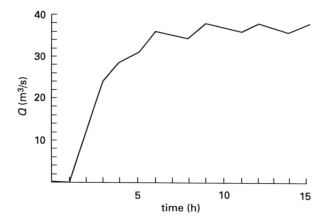

After approximately 5 h, the S-curve levels off to a steady value. (It actually alternates slightly between 35 m³/s and 39 m³/s.) The 5 h unit hydrograph is calculated by tabulating another S-curve lagging the first by 5 h, finding the difference, and scaling to a ratio of 3:5. (The negative values obtained reflect the approximate nature of the method.)

hour	S-curve	lagging S-curve	5 h unit hydrograph
0	0.32	0	0.19
1	0	0	0
2	10.19	0	6.11
3	23.88	0	14.32
4	29.29	0	17.57
5	32.16	0.32	19.10
6	37.25	0	22.35
7	36.29	10.19	15.66
8	35.33	23.88	6.88
9	38.52	29.29	5.54
10	36.93	32.16	2.87
11	35.66	37.25	−0.96
12	38.52	36.29	1.34
13	36.93	35.33	0.96
14	35.66	38.52	−1.72
15	38.20	36.93	0.96

total $\overline{111.17 \text{ m}^3/\text{s·cm}}$

The peak discharge for a 5 cm storm is calculated from the peak unit hydrograph discharge.

$$Q_{p,\text{unit hydrograph}} = 22.35 \text{ m}^3/\text{s·cm} \quad [\text{per cm of rainfall}]$$

$$Q_{p,5\text{ cm storm}} = (5 \text{ cm})\left(22.35 \frac{\text{m}^3}{\text{s·cm}}\right)$$

$$= \boxed{111.75 \text{ m}^3/\text{s} \quad (110 \text{ m}^3/\text{s})}$$

The answer is (C).

(j) The design flood volume is derived from the 5 h unit hydrograph.

$$V_{\text{unit hydrograph}} = \left(111.17 \frac{\text{m}^3}{\text{s·cm}}\right)(1 \text{ h})\left(3600 \frac{\text{s}}{\text{h}}\right)$$

$$= 400\,212 \text{ m}^3/\text{cm}$$

For a 5 cm storm,

$$V_{5\text{ cm storm}} = \left(400\,212 \frac{\text{m}^3}{\text{cm}}\right)(5 \text{ cm})$$

$$= \boxed{2.00 \times 10^6 \text{ m}^3 \quad (2.0 \times 10^6 \text{ m}^3)}$$

The answer is (D).

2. (a) The actual hydrograph is a plot of time versus flow quantity, both of which are given in the problem statement.

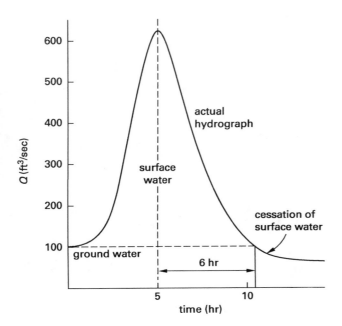

(b) Use the fixed-base method to separate the base flow from the overland flow. The base flow before the storm is projected to a point directly under the peak (in this case a more or less horizontal line), and then a straight line connects the projection to the falling limb. The connection point is between 5 hr and 7 hr later, so use 6 hr.

(c) The ordinates of the unit hydrograph are found by separating the base flow and then dividing the ordinates of the actual hydrograph by the average precipitation. The average precipitation is determined by tabulating the surface water flow and adding to find the total volume.

hour	runoff (ft^3/sec)	ground water (ft^3/sec)	surface water (ft^3/sec)	surface water/2.43 (ft^3/sec)
0	102	100	0	0
1	99	100	0	0
2	101	100	0	0
3	215	100	115	47.3
4	507	100	407	167.5
5	625	100	525	215.0
6	455	93	362	149.0
7	325	87	238	97.9
8	205	80	125	51.4
9	145	74	71	29.2
10	100	67	33	13.6
11	70	60	10	4.1
12	55	55	0	0
13	49	49	0	0
14	43	43	0	0
15	38	38	0	0
		total	$\overline{1886 \text{ ft}^3/\text{sec}}$	

The total volume of surface water is the total flow multiplied by the time interval (1 hr).

$$V = \left(1886 \ \frac{\text{ft}^3}{\text{sec}}\right)(1 \text{ hr})\left(3600 \ \frac{\text{sec}}{\text{hr}}\right)$$
$$= 6.79 \times 10^6 \text{ ft}^3$$

The watershed area is

$$A_d = (1.2 \text{ mi}^2)\left(5280 \ \frac{\text{ft}}{\text{mi}}\right)^2$$
$$= 3.35 \times 10^7 \text{ ft}^2$$

The average precipitation is the total volume divided by the drainage area.

$$P = \frac{(6.79 \times 10^6 \text{ ft}^3)\left(12 \ \frac{\text{in}}{\text{ft}}\right)}{3.35 \times 10^7 \text{ ft}^2}$$
$$= 2.43 \text{ in}$$

The ordinates for the unit hydrograph are tabulated in the last column of the table.

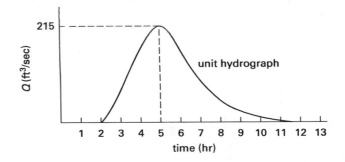

(d) The time base for direct runoff is approximately 11 hr − 2 hr = 9 hr. The time base includes all the time in which direct runoff is observed, that is, both the rising and falling limbs.

The actual hydrograph is as shown. From the graph, the recession limb starts at $t = 5$ and continues to $t = 11$. Therefore, the length of the direct runoff recession limb is $\boxed{6 \text{ hr.}}$

The answer is (B).

3. Use Table 20.2.

$$K = 170$$
$$b = 23$$

From Eq. 20.14, the intensity is

$$I = \frac{K}{t_c + b} = \frac{170 \ \frac{\text{in-min}}{\text{hr}}}{60 \text{ min} + 23 \text{ min}}$$

$$= 2.05 \text{ in/hr}$$

$$A_d = (0.5)(640 \text{ ac}) = 320 \text{ ac}$$

From Eq. 20.36,

$$Q_p = CIA_d = (0.6)\left(2.05 \ \frac{\text{in}}{\text{hr}}\right)(320 \text{ ac})$$

$$= \boxed{394 \text{ ft}^3/\text{sec} \quad (390 \text{ ft}^3/\text{sec})}$$

The answer is (B).

4.

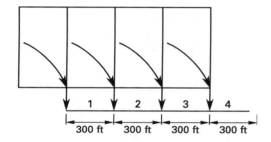

- First inlet:

Peak flow for the first inlet begins at $t_c = 15$ min.

$$I = \frac{K}{t_c + b} = \frac{100 \ \frac{\text{in-min}}{\text{hr}}}{15 \text{ min} + 10 \text{ min}}$$

$$= 4 \text{ in/hr}$$

From Eq. 20.36, the runoff is

$$Q_p = CIA_d = (0.55)\left(4 \ \frac{\text{in}}{\text{hr}}\right)(5 \text{ ac}) = 11 \text{ ft}^3/\text{sec}$$

Use Eq. 19.16.

$$D = 1.335\left(\frac{nQ}{\sqrt{S}}\right)^{3/8} = (1.335)\left(\frac{(0.013)\left(11 \ \frac{\text{ft}^3}{\text{sec}}\right)}{\sqrt{0.005}}\right)^{3/8}$$

$$= 1.74 \text{ ft} \quad (21 \text{ in})$$

$$v_{\text{full}} = \frac{Q}{A} = \frac{11 \ \frac{\text{ft}^3}{\text{sec}}}{\left(\frac{\pi}{4}\right)\left(\frac{21 \text{ in}}{12 \ \frac{\text{in}}{\text{ft}}}\right)^2}$$

$$= 4.6 \text{ ft/sec}$$

(An integer-inch pipe diameter must be used to calculate velocity, because 1.73 ft diameter pipes are not manufactured.)

- Second inlet:

The flow time from inlet 1 to inlet 2 is

$$t = \frac{L}{v} = \frac{300 \text{ ft}}{\left(4.6 \ \frac{\text{ft}}{\text{sec}}\right)\left(60 \ \frac{\text{sec}}{\text{min}}\right)} = 1.09 \text{ min}$$

The intensity at $t_c = 15.0$ min $+ 1.09$ min $= 16.09$ min is

$$I = \frac{K}{t_c + b} = \frac{100 \ \frac{\text{in-min}}{\text{hr}}}{16.09 \text{ min} + 10 \text{ min}}$$

$$= 3.83 \text{ in/hr}$$

From Eq. 20.36, the runoff is

$$Q = CIA_d = (0.55)\left(3.83 \ \frac{\text{in}}{\text{hr}}\right)(10 \text{ ac})$$

$$= 21.07 \text{ ft}^3/\text{sec}$$

Use Eq. 19.16.

$$D = 1.335\left(\frac{nQ}{\sqrt{S}}\right)^{3/8}$$

$$= (1.335)\left(\frac{(0.013)\left(21.07 \ \frac{\text{ft}^3}{\text{sec}}\right)}{\sqrt{0.005}}\right)^{3/8}$$

$$= 2.22 \text{ ft} \quad (27 \text{ in})$$

$$v_{\text{full}} = \frac{Q}{A} = \frac{21.07 \ \frac{\text{ft}^3}{\text{sec}}}{\left(\frac{\pi}{4}\right)\left(\frac{27 \text{ in}}{12 \ \frac{\text{in}}{\text{ft}}}\right)^2}$$

$$\approx 5.3 \text{ ft/sec}$$

- Third inlet:

The flow time from inlet 2 is

$$t = \frac{L}{v} = \frac{300 \text{ ft}}{\left(5.3 \ \frac{\text{ft}}{\text{sec}}\right)\left(60 \ \frac{\text{sec}}{\text{min}}\right)} = 0.94 \text{ min}$$

Flow of Fluids

The intensity at $t_c = 16.09$ min $+ 0.94$ min $= 17.03$ min is

$$I = \frac{K}{t_c + b} = \frac{100 \, \frac{\text{in-min}}{\text{hr}}}{17.03 \, \text{min} + 10 \, \text{min}}$$
$$= 3.70 \, \text{in/hr}$$

From Eq. 20.36, the runoff is

$$Q = CIA_d = (0.55)\left(3.70 \, \frac{\text{in}}{\text{hr}}\right)(15 \, \text{ac})$$
$$= 30.53 \, \text{ft}^3/\text{sec}$$

Use Eq. 19.16.

$$D = 1.335\left(\frac{nQ}{\sqrt{S}}\right)^{3/8}$$
$$= (1.335)\left(\frac{(0.013)\left(30.53 \, \frac{\text{ft}^3}{\text{sec}}\right)}{\sqrt{0.005}}\right)^{3/8}$$
$$= 2.55 \, \text{ft} \quad (31 \, \text{in})$$

$$v_{\text{full}} = \frac{Q}{A} = \frac{30.53 \, \frac{\text{ft}^3}{\text{sec}}}{\left(\frac{\pi}{4}\right)\left(\frac{31 \, \text{in}}{12 \, \frac{\text{in}}{\text{ft}}}\right)^2}$$
$$\approx 5.82 \, \text{ft/sec}$$

- Fourth inlet:

The flow time from inlet 3 is

$$t = \frac{L}{v} = \frac{300 \, \text{ft}}{\left(5.82 \, \frac{\text{ft}}{\text{sec}}\right)\left(60 \, \frac{\text{sec}}{\text{min}}\right)} = 0.86 \, \text{min}$$

The intensity at $t_c = 17.03$ min $+ 0.86$ min $= 17.89$ min is

$$I = \frac{K}{t_c + b} = \frac{100 \, \frac{\text{in-min}}{\text{hr}}}{17.89 \, \text{min} + 10 \, \text{min}}$$
$$= 3.59 \, \text{in/hr}$$

From Eq. 20.36, the runoff is

$$Q = CIA_d = (0.55)\left(3.59 \, \frac{\text{in}}{\text{hr}}\right)(20 \, \text{ac})$$
$$= 39.49 \, \text{ft}^3/\text{sec}$$

Use Eq. 19.16.

$$D = 1.335\left(\frac{nQ}{\sqrt{S}}\right)^{3/8}$$
$$= (1.335)\left(\frac{(0.013)\left(39.49 \, \frac{\text{ft}^3}{\text{sec}}\right)}{\sqrt{0.005}}\right)^{3/8}$$
$$= 2.81 \, \text{ft} \quad (33.7 \, \text{in})$$

Use $\boxed{36 \, \text{in}}$ (standard size) pipe.

The answer is (C).

5.
$$I = 5.3 \, \text{in/hr} \quad [\text{given}]$$
$$C = 0.5 \quad [\text{given}]$$
$$Q = CIA = (0.5)\left(5.3 \, \frac{\text{in}}{\text{hr}}\right)(75 \, \text{ac})$$
$$= 198.8 \, \text{ac-in/hr} \quad (200 \, \text{ft}^3/\text{sec})$$

(a) Take a worst-case approach. See if the culvert has a capacity equal to or greater than 200 ft³/sec when water is at the elevation of the roadway.

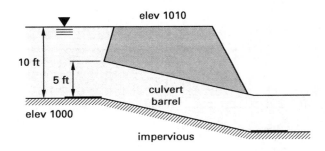

Disregard barrel friction. $K_e = 0.90$ for the projecting, square end. Assume type 6 flow.

$$H = 10 \, \text{ft} + (36 \, \text{ft})(0.01) - 5 \, \text{ft} = 5.36 \, \text{ft}$$

From Eq. 19.102,

$$v = \sqrt{\frac{H}{\frac{1 + k_e}{2g}}} = \sqrt{\frac{5.36 \, \text{ft}}{\frac{1 + 0.9}{(2)\left(32.2 \, \frac{\text{ft}}{\text{sec}^2}\right)}}}$$
$$= 13.48 \, \text{ft/sec}$$

$$Q = Av = \left(\frac{\pi}{4}\right)(5 \, \text{ft})^2\left(13.48 \, \frac{\text{ft}}{\text{sec}}\right) = 265 \, \text{ft}^3/\text{sec}$$

The excess capacity is

$$265 \; \frac{\text{ft}^3}{\text{sec}} - 200 \; \frac{\text{ft}^3}{\text{sec}} = 65 \; \text{ft}^3/\text{sec} \quad [\text{say } 70 \; \text{ft}^3/\text{sec}]$$

> There will be no flooding, as the culvert has excess capacity. The water cannot be maintained at elevation 1010.

The answer is (D).

- Alternate method:

  From Table 17.5,

  $$C_d = 0.72$$

  $$H = 5.36 \; \text{ft}$$

  $$Q = C_d A \sqrt{2gh} = (0.72)\left(\frac{\pi}{4}\right)(5 \; \text{ft})^2$$

  $$\times \sqrt{(2)\left(32.2 \; \frac{\text{ft}}{\text{sec}^2}\right)(5.36 \; \text{ft})}$$

  $$= 263 \; \text{ft}^3/\text{sec}$$

(b) For the rubble channel,

$$n = 0.017$$

$$S = 0.01$$

$$R = \frac{5d}{2d+5}$$

$$C = \left(\frac{1.49}{0.017}\right)\left(\frac{5d}{2d+5}\right)^{1/6}$$

$$Q = Av = (5d)\left(\frac{1.49}{0.017}\right)\left(\frac{5d}{2d+5}\right)^{2/3}\sqrt{0.01}$$

$$200 \; \frac{\text{ft}^3}{\text{sec}} = (43.8d)\left(\frac{5d}{2d+5}\right)^{2/3}$$

By trial and error,

$$d = \text{depth in channel} = \boxed{3.55 \; \text{ft} \quad (3.6 \; \text{ft})}$$

The answer is (C).

(c) The full capacity of the culvert is calculated from the Chezy-Manning equation.

$$R = \frac{D}{4} = \frac{5 \; \text{ft}}{4} = 1.25 \; \text{ft}$$

$$A = \frac{\pi}{4}D^2 = \left(\frac{\pi}{4}\right)(5 \; \text{ft})^2 = 19.63 \; \text{ft}^2$$

$$n = 0.013$$

From Eq. 19.13,

$$Q_{\text{full}} = \left(\frac{1.49}{n}\right)AR^{2/3}\sqrt{S}$$

$$= \left(\frac{1.49}{0.013}\right)(19.63 \; \text{ft}^2)(1.25 \; \text{ft})^{2/3}\sqrt{0.01}$$

$$= 261 \; \text{ft}^3/\text{sec}$$

This is close to (but not the same as) the 265 ft^3/sec calculated for the pressure flow.

$$\frac{Q}{Q_{\text{full}}} = \frac{200 \; \frac{\text{ft}^3}{\text{sec}}}{261 \; \frac{\text{ft}^3}{\text{sec}}} = 0.77$$

Use the table of circular channel ratios. From App. 19.C, the depth is

$$\frac{d}{D} \approx 0.72$$

$$d_{\text{barrel}} = (0.72)(5 \; \text{ft}) = \boxed{3.6 \; \text{ft}}$$

The answer is (C).

6. Since the first storm produces 1 in net, and the second storm produces 2 in, divide all runoffs by two. Offset the second storm by 4 hr.

hour	first storm	second storm	total
0	0		0
2	50		50
4	175	0	175
6	300	100	400
8	210	350	560
10	150	600	750
12	125	420	545
14	75	300	375
16	50	250	300
18	≈ 38	150	188
20	25	100	125
22	≈ 12	≈ 75	87
24	0	50	50
26		≈ 25	25
28		0	0

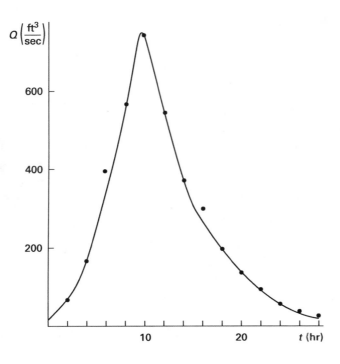

7. From Eq. 20.48,

$$E_R = K_p E_p = (0.7)(0.8 \text{ in}) = \boxed{0.56 \text{ in}}$$

The answer is (A).

8. For 99.7% (two-tailed) confidence, $z = \pm 3$. The inflow distribution is a normal distribution that extends $\pm 3\sigma$, or from 0.3 to 1.5. Take the cell width as $\frac{1}{2}\sigma$ or 0.1. Then, for the first cell,

  actual endpoints: 0.3 to 0.4
  midpoint: 0.35
  z limits: -3.0 to -2.5
  area under curve: $0.5 - 0.49 = 0.01$

The following inflow distribution is produced similarly.

endpoints	midpoint	z limits	area under curve	cum. × 100
0.3 to 0.4	0.35	−3.0 to −2.5	0.01	1
0.4 to 0.5	0.45	−2.5 to −2.0	0.02	3
0.5 to 0.6	0.55	−2.0 to −1.5	0.04	7
0.6 to 0.7	0.65	−1.5 to −1.0	0.09	16
0.7 to 0.8	0.75	−1.0 to −0.5	0.15	31
0.8 to 0.9	0.85	−0.5 to 0	0.19	50
0.9 to 1.0	0.95	0 to 0.5	0.19	69
1.0 to 1.1	1.05	0.5 to 1.0	0.15	84
1.1 to 1.2	1.15	1.0 to 1.5	0.09	93
1.2 to 1.3	1.25	1.5 to 2.0	0.04	97
1.3 to 1.4	1.35	2.0 to 2.5	0.02	99
1.4 to 1.5	1.45	2.5 to 3.0	0.01	100

Choose 12 random numbers that are less than 100 from App. 20.B. Use the third row, reading to the right.

- Inflow distribution:

month	random number	corresponding midpoint
1	06	0.55
2	40	0.85
3	18	0.75
4	73	1.05
5	97	1.25
6	72	1.05
7	89	1.15
8	83	1.05
9	24	0.75
10	41	0.85
11	88	1.15
12	86	1.15

Simulate the reservoir operation.

month	starting volume	+ inflow	− constant use	= ending volume	+ spill
1	5.5	0.55	0.7	5.35	
2	5.35	0.85	0.7	5.5	
3	5.5	0.75	0.7	5.55	
4	5.55	1.05	0.7	5.9	
5	5.9	1.25	0.7	6.45	
6	6.45	1.05	0.7	6.8	
7	6.8	1.15	0.7	7.0	0.25
8	7.0	1.05	0.7	7.0	0.35
9	7.0	0.75	0.7	7.0	0.05
10	7.0	0.85	0.7	7.0	0.15
11	7.0	1.15	0.7	7.0	0.45
12	7.0	1.15	0.7	7.0	0.45

9. Use the same simulation procedure for inflow as in Prob. 8. Proceed similarly.

- Demand distribution:

endpoints	midpoint	z limits	cum. × 100
0.1 to 0.2	0.15	−3.0 to −2.5	1
0.2 to 0.3	0.25	−2.5 to −2.0	3
0.3 to 0.4	0.35	−2.0 to −1.5	7
0.4 to 0.5	0.45	−1.5 to −1.0	16
0.5 to 0.6	0.55	−1.0 to −0.5	31
0.6 to 0.7	0.65	−0.5 to 0	50
0.7 to 0.8	0.75	0 to 0.5	69
0.8 to 0.9	0.85	0.5 to 1.0	84
0.9 to 1.0	0.95	1.0 to 1.5	93
1.0 to 1.1	1.05	1.5 to 2.0	97
1.1 to 1.2	1.15	2.0 to 2.5	99
1.2 to 1.3	1.25	2.5 to 3.0	100

Choose 12 random numbers. Use the fourth row, reading to the right.

• Demand distribution:

month	random number	use
1	04	0.35
2	75	0.85
3	41	0.65
4	44	0.65
5	89	0.95
6	39	0.65
7	42	0.65
8	09	0.45
9	42	0.65
10	11	0.45
11	58	0.75
12	04	0.35

Simulate the reservoir operation.

month	starting volume	+ inflow	− demand	= ending volume	+ spill
1	5.5	0.55	0.35	5.7	
2	5.7	0.85	0.85	5.7	
3	5.7	0.75	0.65	5.8	
4	5.8	1.05	0.65	6.2	
5	6.2	1.25	0.95	6.5	
6	6.5	1.05	0.65	6.9	
7	6.9	1.15	0.65	7.0	0.4
8	7.0	1.05	0.45	7.0	0.6
9	7.0	0.75	0.65	7.0	0.1
10	7.0	0.85	0.45	7.0	0.4
11	7.0	1.15	0.75	7.0	0.4
12	7.0	1.15	0.35	7.0	0.8

10. Note that all dates correspond to the end of the month. Draw the mass diagram.

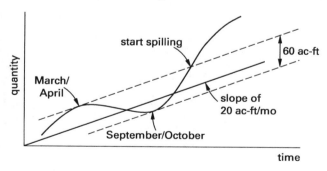

If the reservoir had been full in March or April, the maximum shortfall would have been $\boxed{60 \text{ ac-ft,}}$ and the reservoir would be essentially empty in September and October.

The answer is (B).

11. (a) The monthly draft is

$$\text{monthly draft} = \frac{240 \; \dfrac{\text{MG}}{\text{mi}^2\text{-yr}}}{12 \; \dfrac{\text{mo}}{\text{yr}}}$$

$$= 20 \text{ MG/mi}^2\text{-mo}$$

Assume the reservoir is initially empty. Draw the mass diagram.

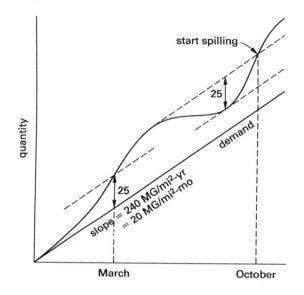

$$\text{capacity} = \boxed{25 \text{ MG/mi}^2}$$

The answer is (A).

(b) The reservoir would start spilling around the end of March. It would also begin spilling around the end of October.

The answer is (D).

21 Groundwater

PRACTICE PROBLEMS

Wells

1. A well extends from the ground surface at elevation 383 ft through a gravel bed to a layer of bedrock at elevation 289 ft. The screened well is 1500 ft from a river whose surface level is 363 ft. The well is pumped by a 10 in diameter schedule-40 steel pipe which draws 120,000 gal/day. The hydraulic conductivity of the well is 1600 gal/day-ft^2. The pump discharges into a piping network whose friction head is 100 ft. What net power is required for steady flow?

(A) 1.1 hp

(B) 2.6 hp

(C) 7.8 hp

(D) 15 hp

2. (*Time limit: one hour*) An aquifer consists of a homogeneous material that is 300 ft thick. The surface of the water table in this aquifer is 100 ft below ground surface. An 18 in diameter well extends through the top 100 ft and then 200 ft below the water table, for a total depth of 300 ft. The aquifer transmissivity is 10,000 gal/day-ft. The well's radius of influence is 900 ft with a 20 ft drawdown at the well.

(a) The geologic formation in which the well is installed is called

(A) an aquifuge

(B) an artesian well

(C) a connate aquifer

(D) an unconfined aquifer

(b) What is the hydraulic conductivity of the aquifer?

(A) 50 gal/day-ft^2

(B) 500 gal/day-ft^2

(C) 1300 gal/day-ft^2

(D) 10,000 gal/day-ft^2

(c) What steady discharge is possible?

(A) 0.26 ft^3/sec

(B) 0.52 ft^3/sec

(C) 1.4 ft^3/sec

(D) 1.9 ft^3/sec

(d) What is the drawdown 100 ft from the well?

(A) 1 ft

(B) 6 ft

(C) 13 ft

(D) 18 ft

(e) Assuming a reasonable pump efficiency, what horsepower motor should be selected to achieve the steady discharge?

(A) 3.6 hp

(B) 5.5 hp

(C) 7.5 hp

(D) 12.2 hp

The following applies to parts (f) through (j).

The same well conditions are found at an adjacent site where the aquifer extends 100 ft below a layer of low permeability that is 200 ft thick. The piezometric surface is 100 ft below the ground surface.

(f) The geologic formation in which the well is installed is called

(A) an aquifuge

(B) a confined aquifer

(C) a spring

(D) a vadose well

(g) The 18 in diameter well extends 300 ft below the ground surface to the bottom of the aquifer. The aquifer transmissivity is 10,000 gal/day-ft. What is the hydraulic conductivity of the aquifer?

(A) 50 gal/day-ft^2

(B) 100 gal/day-ft^2

(C) 500 gal/day-ft^2

(D) 800 gal/day-ft^2

(h) What is the steady discharge in the well if the radius of influence is 900 ft with a 20 ft drawdown at the well?

(A) 0.27 ft³/sec

(B) 0.62 ft³/sec

(C) 1.2 ft³/sec

(D) 3.5 ft³/sec

(i) At what distance from the well is the drawdown equal to 10 ft?

(A) 3 ft

(B) 26 ft

(C) 130 ft

(D) 450 ft

(j) After pumping is stopped, groundwater flows in one direction through the aquifer. If the aquifer is only 100 ft wide and the porosity is 0.4, what is the area in clear flow?

(A) 2000 ft²

(B) 4000 ft²

(C) 8000 ft²

(D) 20,000 ft²

Flow Nets

3. (*Time limit: one hour*) An impervious concrete dam is shown. The dam reduces seepage by using two impervious sheets extending 5 m below the dam bottom.

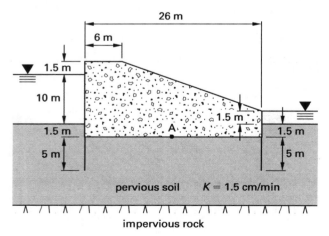

(a) Sketch the flow net.

(b) Determine the seepage per meter of width.

(A) 0.03 m³/min

(B) 0.1 m³/min

(C) 0.3 m³/min

(D) 0.9 m³/min

(c) What is the uplift pressure on the dam at point A, midway between the left and right edges?

(A) 17 kPa

(B) 33 kPa

(C) 71 kPa

(D) 91 kPa

4. (*Time limit: one hour*) A cofferdam in a river is shown. Sheet piles extend below the mud line, and the cofferdam floor is unlined. (The figure is drawn to scale.) Draw the initial flow net.

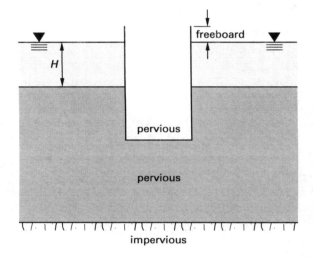

5. (*Time limit: one hour*) The concrete dam shown is drawn to scale. The silt has a hydraulic conductivity of 0.15 ft/hr.

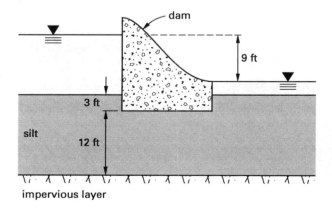

(a) Sketch the flow net for the dam.

(b) What is the approximate seepage rate per foot of width?

    (A) 0.05 ft^3/hr-ft width

    (B) 0.20 ft^3/hr-ft width

    (C) 0.40 ft^3/hr-ft width

    (D) 0.90 ft^3/hr-ft width

SOLUTIONS

1. The well is constructed as shown.

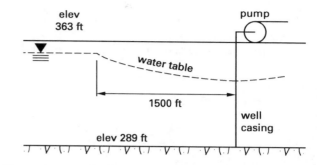

From App. 16.B, for schedule-40 pipe,

$$D_i = 0.835 \text{ ft}$$

$$A_i = 0.5476 \text{ ft}^2$$

$$Q = \left(120{,}000 \ \frac{\text{gal}}{\text{day}}\right)\left(1.547 \times 10^{-6} \ \frac{\text{ft}^3\text{-day}}{\text{sec-gal}}\right)$$

$$= 0.1856 \text{ ft}^3/\text{sec}$$

$$K_p = \left(1600 \ \frac{\text{gal}}{\text{day-ft}^2}\right)\left(0.1337 \ \frac{\text{ft}^3}{\text{gal}}\right)$$

$$= 213.9 \text{ ft}^3/\text{day-ft}^2$$

$$y_1 = 363 \text{ ft} - 289 \text{ ft} = 74 \text{ ft}$$

$$r_1 = 1500 \text{ ft}$$

$$r_2 = \frac{0.835 \text{ ft}}{2} = 0.4175 \text{ ft}$$

From Eq. 21.25,

$$Q = \frac{\pi K (y_1^2 - y_2^2)}{\ln \dfrac{r_1}{r_2}}$$

$$0.1856 \ \frac{\text{ft}^3}{\text{sec}} = \frac{\pi \left(213.9 \ \dfrac{\text{ft}^3}{\text{day-ft}^2}\right)\left((74 \text{ ft})^2 - y_2^2\right)}{\ln \left(\dfrac{1500 \text{ ft}}{0.4175 \text{ ft}}\right)\left(86{,}400 \ \dfrac{\text{sec}}{\text{day}}\right)}$$

$$195.36 \text{ ft}^2 = (74 \text{ ft})^2 - y_2^2$$

$$y_2 = 72.67 \text{ ft}$$

The drawdown is

$$d = 74 \text{ ft} - 72.67 \text{ ft} = 1.33 \text{ ft}$$

d is small compared to y.

The velocity of the water in the pipe is

$$v = \frac{Q}{A} = \frac{0.1856 \frac{ft^3}{sec}}{0.5476 \; ft^2} = 0.34 \; ft/sec$$

This velocity is too small to include the velocity head. The suction lift is

$$383 \; ft - 363 \; ft + 1.33 \; ft = 21.33 \; ft$$

$$H = 21.33 \; ft + 100 \; ft = 121.33 \; ft$$

The mass flow of the water is

$$\dot{m} = \left(0.1856 \frac{ft^3}{sec}\right)\left(62.4 \frac{lbm}{ft^3}\right) = 11.58 \; lbm/sec$$

From Table 18.5, the net water horsepower is

$$P = \frac{h_A \dot{m}}{550} \times \frac{g}{g_c}$$

$$= \frac{(121.33 \; ft)\left(11.58 \frac{lbm}{sec}\right)}{550 \frac{ft\text{-}lbf}{hp\text{-}sec}} \times \frac{32.2 \frac{ft}{sec^2}}{32.2 \frac{lbm\text{-}ft}{lbf\text{-}sec^2}}$$

$$= \boxed{2.55 \; hp \quad (2.6 \; hp)}$$

The answer is (B).

2. (a) The geologic formation is an unconfined aquifer. If the aquifer were overlain by an impermeable layer, it would be a confined, or artesian, aquifer.

The answer is (D).

(b) From Eq. 21.13,

$$K = \frac{T}{Y} = \frac{10,000 \frac{gal}{day\text{-}ft}}{200 \; ft}$$

$$= \boxed{50 \; gal/day\text{-}ft^2}$$

The answer is (A).

(c) $\quad y_1 = 200 \; ft$ at $r_1 = 900 \; ft$

$\quad y_2 = 200 \; ft - 20 \; ft = 180 \; ft$ at r_2

$$r_2 = \frac{\frac{18 \; in}{2}}{12 \frac{in}{ft}} = 0.75 \; ft$$

From Eq. 21.25,

$$Q = \frac{\pi K(y_1^2 - y_2^2)}{\ln \frac{r_1}{r_2}}$$

$$= \frac{\pi\left(50 \frac{gal}{day\text{-}ft^2}\right)\left(0.1337 \frac{ft^3}{gal}\right)}{\ln\left(\frac{900 \; ft}{0.75 \; ft}\right)\left(86,400 \frac{sec}{day}\right)}$$

$$\times \left((200 \; ft)^2 - (180 \; ft)^2\right)$$

$$= \boxed{0.261 \; ft^3/sec \quad (0.26 \; ft^3/sec)}$$

The answer is (A).

(d) $\quad y_1 = 200 \; ft$ at r_1

$\quad r_1 = 900 \; ft$

$\quad r_2 = 100 \; ft$

$\quad Q = 0.261 \; ft^3/sec$

Rearranging Eq. 21.25,

$$y_2^2 = y_1^2 - \frac{Q \ln \frac{r_1}{r_2}}{\pi K}$$

$$= (200 \; ft)^2$$

$$- \frac{\left(0.261 \frac{ft^3}{sec}\right)\ln\left(\frac{900 \; ft}{100 \; ft}\right)\left(86,400 \frac{sec}{day}\right)}{\pi\left(50 \frac{gal}{day\text{-}ft^2}\right)\left(0.1337 \frac{ft^3}{gal}\right)}$$

$$y_2 = 194 \; ft$$

The drawdown is

$$200 \; ft - 194 \; ft = \boxed{6 \; ft}$$

The answer is (B).

(e) This is a low-capacity pump, so the efficiency will be low. Therefore, assume a pump efficiency of 0.65. From Table 18.5, the hydraulic horsepower is

$$P = \frac{h_A \dot{V}(SG)}{8.814}$$

$$= \frac{(100 \; ft + 20 \; ft)\left(0.261 \frac{ft^3}{sec}\right)(1)}{\left(8.814 \frac{sec}{ft^4\text{-}hp}\right)(0.65)}$$

$$= 5.47 \; hp$$

Choose a standard motor size of $\boxed{7.5 \; hp.}$

The answer is (C).

(f) The geologic formation is a confined aquifer. The well is an artesian well.

The answer is (B).

(g) The aquifer depth, Y, is 100 ft. Y is the thickness of the aquifer, not the height of the water table or piezometric surface.

From Eq. 21.13,

$$K = \frac{T}{Y}$$

$$= \frac{10{,}000 \ \dfrac{\text{gal}}{\text{day-ft}}}{100 \ \text{ft}}$$

$$= \boxed{100 \ \text{gal/day-ft}^2}$$

The answer is (B).

(h) The well radius of influence is

$$r_1 = 900 \ \text{ft}$$

$$y_1 = 200 \ \text{ft at } r_1$$

Calculate the well casing radius.

$$r_2 = \frac{\dfrac{18 \ \text{in}}{2}}{12 \ \dfrac{\text{in}}{\text{ft}}} = 0.75 \ \text{ft}$$

$$y_2 = 200 \ \text{ft} - 20 \ \text{ft} = 180 \ \text{ft at } r_2$$

From Eq. 21.27,

$$Q = \frac{2\pi KY(y_1 - y_2)}{\ln \dfrac{r_1}{r_2}}$$

$$= \frac{2\pi(100 \ \text{ft})\left(100 \ \dfrac{\text{gal}}{\text{day-ft}^2}\right)(200 \ \text{ft} - 180 \ \text{ft})}{\ln\left(\dfrac{900 \ \text{ft}}{0.75 \ \text{ft}}\right)}$$

$$= 177{,}239 \ \text{gal/day}$$

$$Q = \frac{\left(177{,}239 \ \dfrac{\text{gal}}{\text{day}}\right)\left(0.002228 \ \dfrac{\text{ft}^3\text{-min}}{\text{gal-sec}}\right)}{\left(24 \ \dfrac{\text{hr}}{\text{day}}\right)\left(60 \ \dfrac{\text{min}}{\text{hr}}\right)}$$

$$= \boxed{0.274 \ \text{ft}^3/\text{sec} \quad (0.27 \ \text{ft}^3/\text{sec})}$$

The answer is (A).

(i) Rearrange Eq. 21.27 to find the drawdown distance, r_2.

$$y_2 = 200 \ \text{ft} - 10 \ \text{ft} = 190 \ \text{ft}$$

$$\ln \frac{r_1}{r_2} = 2\pi K(y_1 - y_2)\left(\frac{Y}{Q}\right)$$

$$= \frac{2\pi\left(100 \ \dfrac{\text{gal}}{\text{day-ft}^2}\right)(200 \ \text{ft} - 190 \ \text{ft})(100 \ \text{ft})}{177{,}239 \ \dfrac{\text{gal}}{\text{day}}}$$

$$= 3.545$$

$$r_2 = \frac{r_1}{e^{3.545}} = \frac{900 \ \text{ft}}{e^{3.545}}$$

$$= \boxed{26 \ \text{ft}}$$

The answer is (B).

(j) Darcy's law assumes the total cross-sectional area. To obtain the cross-sectional area of the pores (voids), multiply by the porosity.

$$A_{\text{clear flow}} = nA = nbY$$

$$= (0.4)(100 \ \text{ft})(100 \ \text{ft})$$

$$= \boxed{4000 \ \text{ft}^2}$$

The answer is (B).

3. (a) One possible flow net is shown.

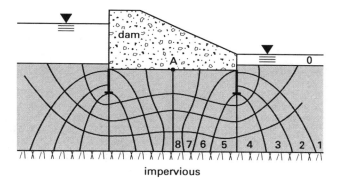

(b) Use Eq. 21.33 to solve for the flow rate.

$$Q = KH\left(\frac{N_f}{N_p}\right)$$

From the flow net as drawn, $N_f = 4$ and $N_p = 16$.

$$H = 10 \ \text{m} - 1.5 \ \text{m} = 8.5 \ \text{m}$$

$$Q = \frac{\left(1.5 \ \dfrac{\text{cm}}{\text{min}}\right)(8.5 \ \text{m})\left(\dfrac{4}{16}\right)\left(1 \ \dfrac{\text{m}}{\text{m}}\right)}{100 \ \dfrac{\text{cm}}{\text{m}}}$$

$$= 0.031875 \ \text{m}^3/\text{min}$$

Use $\boxed{0.03 \ \text{m}^3/\text{min}}$ per meter width of wall.

The answer is (A).

(c) Point A is located at an elevation 3 m lower than the downstream water level. From Eq. 21.36,

$$p_u = \left(\left(\frac{j}{N_p} \right) H + z \right) \rho_w g$$

$$= \left(\left(\frac{8}{16} \right) (8.5 \text{ m}) + 3.0 \text{ m} \right) \left(1000 \ \frac{\text{kg}}{\text{m}^3} \right) \left(9.81 \ \frac{\text{m}}{\text{s}^2} \right)$$

$$= \boxed{71\,122 \text{ Pa} \quad (71 \text{ kPa})}$$

The answer is (C).

4. The initial flow net is drawn as shown.

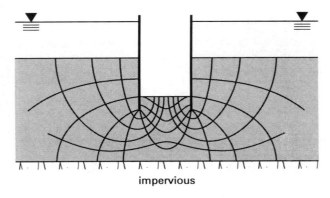

impervious

5. (a) The dam's flow net is drawn as shown.

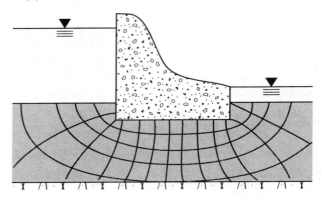

(b) From Eq. 21.33, with 4 flow paths and 13 equipotential drops,

$$Q = KH \left(\frac{N_f}{N_p} \right)$$

$$= \left(0.15 \ \frac{\text{ft}}{\text{hr}} \right) (9.0 \text{ ft}) \left(\frac{4}{13} \right)$$

$$= \boxed{0.42 \text{ ft}^3/\text{hr-ft width} \quad (0.40 \text{ ft}^3/\text{hr-ft width})}$$

The answer is (C).

22 Inorganic Chemistry

PRACTICE PROBLEMS

Empirical Formula Development

1. The gravimetric analysis of a compound is 40% carbon, 6.7% hydrogen, and 53.3% oxygen. The simplest formula for the compound is most nearly

(A) HCO

(B) HCO_2

(C) CH_2O

(D) CHO_2

Corrosion

2. When used together, which of the following metal pairs is LEAST likely to experience galvanic corrosion in sea water?

(A) zinc and platinum

(B) zinc and steel

(C) steel and lead

(D) brass and copper

3. When use of dissimilar metals is unavoidable, which of the following CANNOT be used to prevent or reduce the likelihood of galvanic corrosion?

(A) inert spacers between the dissimilar metals

(B) sacrificial anodes

(C) paints, coatings, and natural oxide buildups

(D) layers of cadmium, nickel, chromium, or zinc plating

4. A sacrificial galvanic protection system is proposed for a water storage tank made of low-carbon steel. Which of the following materials is most suitable as the sacrificial anode?

(A) brass

(B) nickel

(C) stainless steel

(D) zinc

5. A linear actuator mechanism incorporates a naval brass threaded rod that passes through a threaded metallic collar. The mechanism is exposed to both fresh and saltwater. To limit galvanic corrosion, which material would be best for the mechanism's collar?

(A) zinc-plated carbon steel

(B) nickel-plated carbon steel

(C) passivated 316 stainless steel

(D) non-passivated 316 stainless steel

Precipitation Softening

6. A municipal water supply has the following ionic concentrations.

Al^{+++}	0.5 mg/L
Ca^{++}	80.2 mg/L
Cl^-	85.9 mg/L
CO_2	19 mg/L
CO_3^{--}	(none)
Fe^{++}	1.0 mg/L
Fl^-	(none)
HCO_3^-	185 mg/L
Mg^{++}	24.3 mg/L
Na^+	46.0 mg/L
NO_3^-	(none)
SO_4^{--}	125 mg/L

(a) The total hardness is most nearly

(A) 160 mg/L as $CaCO_3$

(B) 200 mg/L as $CaCO_3$

(C) 260 mg/L as $CaCO_3$

(D) 300 mg/L as $CaCO_3$

(b) Approximately how much slaked lime is required to combine with the carbonate hardness?

(A) 45 mg/L as substance

(B) 90 mg/L as substance

(C) 130 mg/L as substance

(D) 150 mg/L as substance

(c) Approximately how much soda ash is required to react with the carbonate hardness?

(A) none

(B) 15 mg/L as substance

(C) 35 mg/L as substance

(D) 60 mg/L as substance

7. A municipal water supply contains the following ionic concentrations.

$Ca(HCO_3)_2$	137 mg/L as $CaCO_3$
$MgSO_4$	72 mg/L as $CaCO_3$
CO_2	(none)

(a) Approximately how much slaked lime is required to soften 1,000,000 gal of this water to a hardness of 100 mg/L if 30 mg/L of excess lime is used?

(A) 930 lbm

(B) 1200 lbm

(C) 1300 lbm

(D) 1700 lbm

(b) Approximately how much soda ash is required to soften 1,000,000 gal of this water to a hardness of 100 mg/L if 30 mg/L of excess lime is used?

(A) none

(B) 300 lbm

(C) 500 lbm

(D) 700 lbm

SOLUTIONS

1. Calculate the relative mole ratios of the atoms by assuming there are 100 g of sample.

For 100 g of sample,

substance	mass	$\frac{m}{AW}$ = no. moles	relative mole ratio
C	40 g	$\frac{40}{12}$ = 3.33	1
H	6.7 g	$\frac{6.7}{1}$ = 6.7	2
O	53.3 g	$\frac{53.3}{16}$ = 3.33	1

The empirical formula is $\boxed{CH_2O.}$

The answer is (C).

2. Galvanic corrosion (galvanic action) results from a difference in oxidation potentials of metallic ions. The greater the difference in oxidation potentials, the greater the likelihood of galvanic action. To determine which of the options are least likely to experience galvanic action, use a galvanic series chart, such as Table 22.7. The closer the pair of metals are, the less likely they are to experience galvanic corrosion. Therefore, of the options, $\boxed{\text{brass and copper}}$ are least likely to experience galvanic corrosion.

The answer is (D).

3. While $\boxed{\text{cadmium, nickel, chromium, and zinc}}$ are often used as protective deposits on steel, porosities in the surfaces can act as small galvanic cells, resulting in invisible subsurface corrosion.

The answer is (D).

4. From Table 22.7, brass, nickel, and stainless steel have lower potentials than low-carbon steel; therefore, they will be cathodic, not anodic. $\boxed{\text{Zinc}}$ has a higher potential than low-carbon steel, so it can be used as the sacrificial anode.

The answer is (D).

5. From Table 22.7, nickel is the closest to brass, so $\boxed{\text{nickel-plated carbon steel}}$ should be used to make the actuator's threaded collar.

The answer is (B).

6. (a) Hardness is the sum of the concentrations of all doubly and triply charged positive ions, expressed as $CaCO_3$. Find the total hardness by multiplying the mg/L as substance by the factor from App. 22.C.

	mg/L as substance		factor from App. 22.C		
Ca^{++}:	80.2	×	2.50	=	200.5 mg/L
Mg^{++}:	24.3	×	4.10	=	99.63 mg/L
Fe^{++}:	1	×	1.79	=	1.79 mg/L
Al^{+++}:	0.5	×	5.56	=	2.78 mg/L

$$\text{hardness} = \boxed{\begin{array}{c} 304.7 \text{ mg/L} \\ (300 \text{ mg/L}) \end{array}}$$

The answer is (D).

(b) Add lime to remove the carbonate hardness. It does not matter whether the HCO_3^- comes from Mg^{++}, Ca^{++}, or Fe^{++}; adding lime will remove it.

There may be Mg^{++}, Ca^{++}, or Fe^{++} ions left over in the form of noncarbonate hardness, but the problem asked for carbonate hardness. Converting from mg/L of substance to mg/L as $CaCO_3$,

$$CO_2: \quad \left(19 \ \frac{mg}{L}\right)(2.27) = 43.13 \text{ mg/L as } CaCO_3$$

$$HCO_3^-: \quad \left(185 \ \frac{mg}{L}\right)(0.82) = 151.7 \text{ mg/L as } CaCO_3$$

The total equivalents to be neutralized are

$$43.13 \ \frac{mg}{L} + 151.7 \ \frac{mg}{L} = 194.83 \text{ mg/L as } CaCO_3$$

Convert $Ca(OH)_2$ to substance using App. 22.C.

$$\frac{mg}{L} \text{ of } Ca(OH)_2 = \frac{194.83 \ \frac{mg}{L}}{1.35}$$

$$= \boxed{\begin{array}{c} 144.3 \text{ mg/L as substance} \\ (150 \text{ mg/L as substance}) \end{array}}$$

The answer is (D).

(c) $\boxed{\text{No soda ash is required since it is used to remove noncarbonate hardness.}}$

The answer is (A).

7. (a) $Ca(HCO_3)_2$ and $MgSO_4$ both contribute to hardness. Since 100 mg/L of hardness is the goal, leave all $MgSO_4$ in the water. Take out 137 mg/L + 72 mg/L − 100 mg/L = 109 mg/L of $Ca(HCO_3)_2$. From App. 22.C (including the excess even though the reaction is not complete),

$$\text{pure } Ca(OH)_2 = 30 \ \frac{mg}{L} + \frac{109 \ \frac{mg}{L}}{1.35}$$

$$= 110.74 \text{ mg/L}$$

$$\left(110.74 \ \frac{mg}{L}\right)\left(8.345 \ \frac{lbm\text{-}L}{mg\text{-}MG}\right) = \boxed{\begin{array}{c} 924 \text{ lbm/MG} \\ (930 \text{ lbm}) \end{array}}$$

The answer is (A).

(b) $\boxed{\text{No soda ash is required.}}$

The answer is (A).

Water Treatment

23 Water Supply Quality and Testing

PRACTICE PROBLEMS

Phosphorus

1. (*Time limit: one hour*) In a study of a small pond to determine phosphorus impact, the following information was collected.

> pond size: 12 ac-ft (14.8×10^6 L)
> watershed area: 4 ac (grassland)
> average annual rainfall: 5 in
> bioavailable P in runoff: 0.01 mg/L
> runoff coefficient: 0.1

Biological processes in the pond biota convert phosphorus to a nonbioavailable form at the rate of 22% per year. Recycling of sediment phosphorus by rooted plants and by anaerobic conditions in the hypolimnion converts 12% per year of the nonbioavailable phosphorus back to bioavailable forms.

Runoff into the pond evaporates during the year, so no change in pond volume occurs.

(a) Starting from an initial condition of 0.1 mg/L bioavailable P, what is the P concentration after five annual cycles?

(A) 0.03 mg/L

(B) 0.11 mg/L

(C) 0.13 mg/L

(D) 0.18 mg/L

(b) What can be done to reduce the P accumulation?

(A) Add chemicals to precipitate phosphorus in the pond.

(B) Add chemicals to combine with phosphorus in the pond.

(C) Reduce use of phosphorus-based fertilizers in the surrounding fields.

(D) Use natural-based soaps and detergents in the home.

Alkalinity

2. (*Time limit: one hour*) Groundwater is used for a water supply. It is taken from the ground at 25°C. The initial properties of the water are as follows.

CO_2	60 mg/L (as $CaCO_3$)
pH	7.1

The water is treated for CO_2 removal by spraying into the atmosphere through a nozzle. The final CO_2 concentration is 5.6 mg/L (as $CaCO_3$) at 25°C. The final alkalinity is 200 mg/L (as $CaCO_3$). All final alkalinity is in the form of bicarbonate. The first ionization constant of carbonic acid is 4.45×10^{-7}. What is the final pH of the water after spraying and recovery?

(A) 7.3

(B) 7.9

(C) 8.2

(D) 8.8

Solids

3. (*Time limit: one hour*) The solids concentration of a stream water sample is to be determined. The total solids concentration is determined by placing a portion of the sample into a porcelain evaporating dish, drying the sample at 105°C, and igniting the residue by placing the dried sample in a muffle furnace at 550°C. The following masses are recorded.

> mass of empty dish: 50.326 g
> mass of dish and sample: 118.400 g
> mass of dish and dry solids: 50.437 g
> mass of dish and ignited solids: 50.383 g

(a) The total solids concentration is

(A) 900 mg/L

(B) 1000 mg/L

(C) 1100 mg/L

(D) 1600 mg/L

(b) The total volatile solids concentration is

(A) 630 mg/L

(B) 710 mg/L

(C) 790 mg/L

(D) 830 mg/L

(c) The total fixed solids concentration is

(A) 270 mg/L

(B) 300 mg/L

(C) 420 mg/L

(D) 840 mg/L

The suspended solids concentration is determined by filtering a portion of the sample through a glass-fiber filter disk, drying the disk at 105°C, and igniting the residue by placing the dried sample in a muffle furnace at 550°C. The follow masses are recorded.

volume of sample: 30 mL
mass of filter disk: 0.1170 g
mass of disk and dry solids: 0.1278 g
mass of disk and ignited solids: 0.1248 g

(d) The total suspended solids concentration is

(A) 240 mg/L

(B) 360 mg/L

(C) 370 mg/L

(D) 820 mg/L

(e) The volatile suspended solids concentration is

(A) 100 mg/L

(B) 120 mg/L

(C) 230 mg/L

(D) 640 mg/L

(f) The fixed suspended solids concentration is

(A) 120 mg/L

(B) 180 mg/L

(C) 240 mg/L

(D) 260 mg/L

The dissolved solids concentration is determined by filtering a portion of the sample through a glass-fiber filter disk into a porcelain evaporating dish, drying the sample at 105°C, and igniting the residue by placing the dried sample in a muffle furnace at 550°C. The following masses are recorded.

volume of sample: 25 mL
mass of empty dish: 51.494 g
mass of dish and dry solids: 51.524 g
mass of dish and ignited solids: 51.506 g

(g) The total dissolved solids concentration is

(A) 1000 mg/L

(B) 1100 mg/L

(C) 1200 mg/L

(D) 1400 mg/L

(h) The volatile dissolved solids concentration is

(A) 680 mg/L

(B) 720 mg/L

(C) 810 mg/L

(D) 900 mg/L

(i) The fixed dissolved solids concentration is

(A) 230 mg/L

(B) 300 mg/L

(C) 410 mg/L

(D) 480 mg/L

(j) Assume the total solids determined in part (a) is equal to the sum of the total suspended solids determined in part (d) plus the total dissolved solids determined in part (g). Does the total solids concentration equal the sum of the suspended and dissolved solids?

(A) yes

(B) no; probably due to poor laboratory technique

(C) no; probably due to rounding of measured values

(D) no; probably because the samples were representative but not identical

Hardness

4. (*Time limit: one hour*) The laboratory analysis of a water sample is as follows. All concentrations are "as substance."

Ca^{++}	74.0 mg/L
Mg^{++}	18.3 mg/L
Na^{+}	27.6 mg/L
K^{+}	39.1 mg/L
pH	7.8
HCO_3^-	274.5 mg/L
SO_4^{--}	72.0 mg/L
Cl^-	49.7 mg/L

Water Treatment

(a) The hardness of the water in terms of mg/L of calcium carbonate equivalent is

(A) 1.3 mg/L as $CaCO_3$

(B) 4.7 mg/L as $CaCO_3$

(C) 66 mg/L as $CaCO_3$

(D) 260 mg/L as $CaCO_3$

(b) Based on the laboratory analysis, which of the following can be said?

(A) It is surprising that no carbonates were found in the sample.

(B) The cations in solution, when converted to milliequivalents, will equal the anions, when converted to milliequivalents.

(C) The large amount of bicarbonate in the solution tends to make the water acidic.

(D) None of the above can be said.

(c) Assuming that hypothetical compounds are formed proportionally to the relative concentrations of ions, the hypothetical concentration of calcium bicarbonate in the sample is

(A) 3.7 meq/L

(B) 4.5 meq/L

(C) 4.7 meq/L

(D) 74 meq/L

(d) The hypothetical concentration of magnesium bicarbonate is

(A) 0.8 meq/L

(B) 1.5 meq/L

(C) 1.6 meq/L

(D) 5.2 meq/L

(e) The hypothetical concentration of magnesium sulfate is

(A) 0.7 meq/L

(B) 1.5 meq/L

(C) 4.5 meq/L

(D) 5.2 meq/L

(f) The hypothetical concentration of sodium sulfate is

(A) 0.8 meq/L

(B) 1.2 meq/L

(C) 1.5 meq/L

(D) 6.4 meq/L

(g) The hypothetical concentration of sodium chloride is

(A) 0.2 meq/L

(B) 0.4 meq/L

(C) 1.2 meq/L

(D) 1.4 meq/L

(h) The amount of lime (CaO) necessary in a lime softening process to remove the hardness caused by calcium bicarbonate is

(A) 17 mg/L

(B) 28 mg/L

(C) 95 mg/L

(D) 100 mg/L

(i) To remove hardness caused by magnesium bicarbonate, it is necessary to raise the pH by adding 35 mg/L of CaO in excess of the stoichiometric requirements. The amount of lime (CaO) necessary to remove the carbonate hardness caused by magnesium bicarbonate is

(A) 12 mg/L

(B) 45 mg/L

(C) 56 mg/L

(D) 80 mg/L

(j) The water is subsequently recarbonated to reduce the pH. Assume that by this softening process calcium hardness can be reduced to 30 mg/L and magnesium hardness can be reduced to 10 mg/L, both measured in terms of equivalent calcium carbonate. The amount of hardness that remains in the water is

(A) 10 mg/L as $CaCO_3$

(B) 30 mg/L as $CaCO_3$

(C) 45 mg/L as $CaCO_3$

(D) 75 mg/L as $CaCO_3$

Water Treatment

SOLUTIONS

1. (a) annual rainfall $= (4 \text{ ac})(5 \text{ in}) = 20 \text{ ac-in}$

runoff to pond $= (0.1)(20 \text{ ac-in})$

$$\times \left(102{,}790 \ \frac{\text{L}}{\text{ac-in}}\right)$$

$$= 205{,}580 \text{ L}$$

$$\begin{aligned}\text{bioavailable P} \\ \text{reaching pond}\end{aligned} = (205{,}580 \text{ L})\left(0.01 \ \frac{\text{mg}}{\text{L}}\right)$$

$$= 2056 \text{ mg}$$

$$\begin{aligned}\text{initial bioavailable} \\ \text{P in pond}\end{aligned} = (14.8 \times 10^6 \text{ L})\left(0.1 \ \frac{\text{mg}}{\text{L}}\right)$$

$$= 14.8 \times 10^5 \text{ mg}$$

The total bioavailable P in the pond is

$$2056 \text{ mg} + 14.8 \times 10^5 \text{ mg} = 14.82056 \times 10^5 \text{ mg}$$

Of this, after biological processes and recycling, the following percentage remains.

$$100\% - 22\% + \frac{(12\%)(22\%)}{100\%} = 80.64\%$$

Therefore,

$$(0.8064)(14.82056 \times 10^5 \text{ mg}) = 11.95130 \times 10^5 \text{ mg}$$

The following table is prepared in a similar manner.

year	P at start of year	P from rainfall	total P available for activity	P at end of year
1	14.8×10^5	2056	14.82056×10^5	11.95130×10^5
2	11.95130×10^5	2056	11.97186×10^5	9.65411×10^5
3	9.65411×10^5	2056	9.67467×10^5	7.80165×10^5
4	7.80165×10^5	2056	7.82221×10^5	6.30783×10^5
5	6.30783×10^5	2056	6.32839×10^5	5.10322×10^5

The concentration after five years will be

$$\frac{5.10322 \times 10^5 \text{ mg}}{14.8 \times 10^6 \text{ L}} = \boxed{0.0345 \text{ mg/L} \quad (0.03 \text{ mg/L})}$$

The answer is (A).

(b) To reduce the phosphorus accumulation in the pond, the arrival of additional bioavailable phosphorous must be reduced. This entails watershed management to reduce the amount of phosphorus applied as fertilizer and released through other sources. As eutrophication is a natural process accelerated by the availability of plant nutrients (nitrogen and phosphorus especially), reducing phosphorus can slow the process.

While there are chemical means to alter the forms of phosphorus to nonbioavailable states, this usually is not practical on a large scale and is potentially harmful in itself.

It is unlikely that the pond receives untreated discharge from local homes. Use of phosphate-rich detergents in the home will not affect the pond.

The answer is (C).

2. Since pH < 8.3 (see Eq. 23.2 and Eq. 23.3), all alkalinity is in the bicarbonate (HCO_3^-) form. Therefore, the equilibrium expression for the first ionization of carbonic acid may be used.

$$CO_2 + H_2O \rightarrow H_2CO_3 \rightleftharpoons H^+ + HCO_3^-$$

$$K_1 = \frac{[H^+][HCO_3^-]}{[H_2CO_3]} = 4.45 \times 10^{-7}$$

The coefficients of CO_2 and H_2CO_3 are both 1. The number of moles of each compound is the same. One mole of CO_2 produces one mole of H_2CO_3.

The "as $CaCO_3$" values must be converted to substance for both CO_2 and HCO_3^-. Use the conversion factors in App. 22.C.

$$CO_2: \ \frac{5.6 \ \frac{\text{mg}}{\text{L}}}{2.27} = 2.47 \text{ mg/L}$$

$$HCO_3^-: \ \frac{200 \ \frac{\text{mg}}{\text{L}}}{0.82} = 244 \text{ mg/L}$$

Convert to moles per liter.

$$[H_2CO_3] = [CO_2] = \frac{2.47 \ \frac{\text{mg}}{\text{L}}}{\left(44 \ \frac{\text{g}}{\text{mol}}\right)\left(1000 \ \frac{\text{mg}}{\text{g}}\right)}$$

$$= 5.61 \times 10^{-5} \text{ mol/L}$$

$$[HCO_3^-] = \frac{244 \ \frac{\text{mg}}{\text{L}}}{\left(61 \ \frac{\text{g}}{\text{mol}}\right)\left(1000 \ \frac{\text{mg}}{\text{g}}\right)}$$

$$= 4.00 \times 10^{-3} \text{ mol/L}$$

Solve for the hydrogen ion concentration.

$$[H^+] = \frac{K_1[H_2CO_3]}{[HCO_3^-]} = \frac{(4.45 \times 10^{-7})(5.61 \times 10^{-5})}{4.00 \times 10^{-3}}$$

$$= 6.24 \times 10^{-9} \text{ mol/L}$$

The pH is

$$pH = -\log[H^+] = -\log(6.24 \times 10^{-9}) = \boxed{8.2}$$

The answer is (C).

3. (a) The density of water is 1 g/mL. The volume of the tested sample is

$$V = \frac{m}{\rho} = \frac{118.4 \text{ g} - 50.326 \text{ g}}{1 \dfrac{\text{g}}{\text{mL}}} = 68.1 \text{ mL}$$

The total solids concentration is

$$TS = \frac{(50.437 \text{ g} - 50.326 \text{ g})\left(1000 \dfrac{\text{mg}}{\text{g}}\right)\left(1000 \dfrac{\text{mL}}{\text{L}}\right)}{68.1 \text{ mL}}$$

$$= \boxed{1630 \text{ mg/L} \quad (1600 \text{ mg/L})}$$

The answer is (D).

(b) The total volatile solids concentration is

$$TVS = \frac{(50.437 \text{ g} - 50.383 \text{ g})\left(1000 \dfrac{\text{mg}}{\text{g}}\right)\left(1000 \dfrac{\text{mL}}{\text{L}}\right)}{68.1 \text{ mL}}$$

$$= \boxed{793 \text{ mg/L} \quad (790 \text{ mg/L})}$$

The answer is (C).

(c) The total fixed solids concentration is

$$TFS = 1630 \frac{\text{mg}}{\text{L}} - 793 \frac{\text{mg}}{\text{L}}$$

$$= \boxed{837 \text{ mg/L} \quad (840 \text{ mg/L})}$$

The answer is (D).

(d) The total suspended solids concentration is

$$TSS = \frac{(0.1278 \text{ g} - 0.1170 \text{ g}) \times \left(1000 \dfrac{\text{mg}}{\text{g}}\right)\left(1000 \dfrac{\text{mL}}{\text{L}}\right)}{30 \text{ mL}}$$

$$= \boxed{360 \text{ mg/L}}$$

The answer is (B).

(e) The volatile suspended solids concentration is

$$VSS = \frac{(0.1278 \text{ g} - 0.1248 \text{ g}) \times \left(1000 \dfrac{\text{mg}}{\text{g}}\right)\left(1000 \dfrac{\text{mL}}{\text{L}}\right)}{30 \text{ mL}}$$

$$= \boxed{100 \text{ mg/L}}$$

The answer is (A).

(f) The fixed suspended solids concentration is

$$FFS = 360 \frac{\text{mg}}{\text{L}} - 100 \frac{\text{mg}}{\text{L}} = \boxed{260 \text{ mg/L}}$$

The answer is (D).

(g) The total dissolved solids concentration is

$$TDS = \frac{(51.524 \text{ g} - 51.494 \text{ g}) \times \left(1000 \dfrac{\text{mg}}{\text{g}}\right)\left(1000 \dfrac{\text{mL}}{\text{L}}\right)}{25 \text{ mL}}$$

$$= \boxed{1200 \text{ mg/L}}$$

The answer is (C).

(h) The volatile dissolved solids concentration is

$$VDS = \frac{(51.524 \text{ g} - 51.506 \text{ g}) \times \left(1000 \dfrac{\text{mg}}{\text{g}}\right)\left(1000 \dfrac{\text{mL}}{\text{L}}\right)}{25 \text{ mL}}$$

$$= \boxed{720 \text{ mg/L}}$$

The answer is (B).

(i) The fixed dissolved solids concentration is

$$FDS = 1200 \frac{\text{mg}}{\text{L}} - 720 \frac{\text{mg}}{\text{L}} = \boxed{480 \text{ mg/L}}$$

The answer is (D).

(j) If only the first two procedures were performed, the dissolved solids would be calculated as

$$TDS = TS - TSS = 1630 \frac{\text{mg}}{\text{L}} - 360 \frac{\text{mg}}{\text{L}}$$

$$= 1270 \text{ mg/L} \quad (\text{vs. } 1200 \text{ mg/L})$$

$$VDS = TVS - VSS = 793 \frac{\text{mg}}{\text{L}} - 100 \frac{\text{mg}}{\text{L}}$$

$$= 693 \text{ mg/L} \quad (\text{vs. } 720 \text{ mg/L})$$

$$FDS = TFS - FSS = 837 \frac{\text{mg}}{\text{L}} - 260 \frac{\text{mg}}{\text{L}}$$

$$= 577 \text{ mg/L} \quad (\text{vs. } 480 \text{ mg/L})$$

The results of the three sets of tests are not the same. These differences are too great to be caused by improper rounding or faulty laboratory technique. This is probably the result of using samples that are not truly identical. For this reason, the suspended solids values are mathematically determined from the total solids and dissolved solids tests.

The answer is (D).

Water Treatment

4. (a) Hardness is caused by multivalent cations: Ca^{++} and Mg^{++} in this example. Calculate the milliequivalents by dividing the measured concentration by the milliequivalent weight.

$$Ca^{++}: \dfrac{74.0 \ \frac{mg}{L}}{20 \ \frac{mg}{meq}} = 3.7 \ meq/L$$

$$Mg^{++}: \dfrac{18.3 \ \frac{mg}{L}}{12.2 \ \frac{mg}{meq}} = 1.5 \ meq/L$$

$$\text{total hardness} = 3.7 \ \frac{meq}{L} + 1.5 \ \frac{meq}{L} = 5.2 \ meq/L$$

Determine the calcium carbonate equivalent by multiplying by the equivalent weight of calcium carbonate.

$$\text{hardness} = \left(5.2 \ \frac{meq}{L}\right)\left(50 \ \frac{mg}{meq}\right)$$

$$= \boxed{260 \ mg/L \ \text{as} \ CaCO_3}$$

The answer is (D).

(b) Alkalinity in the form of carbonate radical does not exist at a pH below 8.3. Bicarbonate is a form of alkalinity, and it neutralizes, not creates, acidity. To determine the cation/anion relationship, convert all of the concentrations to milliequivalents by dividing the measured concentrations by the milliequivalent weights.

$$Ca^{++}: \dfrac{74.0 \ \frac{mg}{L}}{20 \ \frac{mg}{meq}} = 3.7 \ meq/L$$

$$Mg^{++}: \dfrac{18.3 \ \frac{mg}{L}}{12.2 \ \frac{mg}{meq}} = 1.5 \ meq/L$$

$$Na^{+}: \dfrac{27.6 \ \frac{mg}{L}}{23 \ \frac{mg}{meq}} = 1.2 \ meq/L$$

$$K^{+}: \dfrac{39.1 \ \frac{mg}{L}}{39.1 \ \frac{mg}{meq}} = 1.0 \ meq/L$$

$$\text{total cations} = 3.7 \ \frac{meq}{L} + 1.5 \ \frac{meq}{L} + 1.2 \ \frac{meq}{L}$$
$$+ 1.0 \ \frac{meq}{L}$$
$$= 7.4 \ meq/L$$

$$HCO_3^{-}: \dfrac{274.5 \ \frac{mg}{L}}{61 \ \frac{mg}{meq}} = 4.5 \ meq/L$$

$$SO_4^{--}: \dfrac{72.0 \ \frac{mg}{L}}{48 \ \frac{mg}{meq}} = 1.5 \ meq/L$$

$$Cl^{-}: \dfrac{49.7 \ \frac{mg}{L}}{35.5 \ \frac{mg}{meq}} = 1.4 \ meq/L$$

$$\text{total anions} = 4.5 \ \frac{meq}{L} + 1.5 \ \frac{meq}{L} + 1.4 \ \frac{meq}{L}$$
$$= 7.4 \ meq/L$$

$\boxed{\text{The total cations equals the total anions.}}$

The answer is (B).

(c) To determine the hypothetical compounds, construct a milliequivalent per liter bar graph for cations and anions, as shown.

0		3.7		5.2		6.4		7.4
Ca^{++} 3.7		Mg^{++} 1.5		Na^{+} 1.2		K^{+} 1.0	
HCO_3^{-} 4.5			SO_4^{--} 1.5			Cl^{-} 1.4	
0			4.5			6.0		7.4

The hypothetical compounds are determined by moving from left to right.

hypothetical	compound concentration (meq/L)	remarks
$Ca(HCO_3)_2$	3.7	Ca^{++} exhausted
$Mg(HCO_3)_2$	0.8	HCO_3^{-} exhausted
$MgSO_4$	0.7	Mg^{++} exhausted
Na_2SO_4	0.8	SO_4^{-} exhausted
$NaCl$	0.4	Na^{+} exhausted
KCl	1.0	K^{+} and Cl^{-} exhausted

The hypothetical concentration of calcium bicarbonate is $\boxed{3.7 \ meq/L.}$

The answer is (A).

(d) From part (c), the hypothetical concentration of magnesium bicarbonate is $\boxed{0.8 \ meq/L.}$

The answer is (A).

(e) From part (c), the hypothetical concentration of magnesium sulfate is $\boxed{0.7 \ meq/L.}$

The answer is (A).

(f) From part (c), the hypothetical concentration of sodium sulfate is $\boxed{0.8 \text{ meq/L.}}$

The answer is (A).

(g) From part (c), the hypothetical concentration of sodium chloride is $\boxed{0.4 \text{ meq/L.}}$

The answer is (B).

(h) The reactions are

$$CaO + H_2O \rightarrow Ca(OH)_2$$

$$Ca(HCO_3)_2 + Ca(OH)_2 \rightarrow 2CaCO_3\downarrow + 2H_2O$$

One molecule of CaO forms one molecule of $Ca(OH)_2$, which in turn reacts with one molecule of $Ca(HCO_3)_2$. There are 3.7 meq/L of $Ca(HCO_3)_2$ in solution. The equivalent weight of CaO is $(40 + 16)/2 = 28$. Therefore, 3.7 meq/L of CaO are needed.

$$\left(3.7 \ \frac{\text{meq}}{\text{L}} \ CaO \right) \left(28 \ \frac{\text{mg}}{\text{meq}} \right)$$
$$= \boxed{103.6 \text{ mg/L CaO} \quad (100 \text{ mg/L CaO})}$$

The answer is (D).

(i) The reactions are

$$CaO + H_2O \rightarrow Ca(OH)_2$$

$$Mg(HCO_3)_2 + 2Ca(OH)_2 \rightarrow$$
$$2CaCO_3\downarrow + Mg(OH)_2\downarrow + 2H_2O$$

Two molecules of $Ca(OH)_2$ are required for each molecule of $Mg(HCO_3)_2$.

$$\left(0.8 \ \frac{\text{meq}}{\text{L}} \ Mg(HCO_3)_2 \right) (2) \left(28 \ \frac{\text{mg}}{\text{meq}} \right)$$
$$+ 35 \ \frac{\text{mg}}{\text{L}} \ \text{excess} = \boxed{79.8 \text{ mg/L} \quad (80 \text{ mg/L})}$$

The answer is (D).

(j) The original hardness was in the hypothetical forms of calcium bicarbonate, magnesium bicarbonate, and magnesium sulfate. The bicarbonates have been removed in parts (h) and (i), leaving the residuals of 30 mg/L calcium hardness and 10 mg/L magnesium hardness. However, no attempt was made to remove the noncarbonate magnesium hardness represented by magnesium sulfate (this would have required the addition of soda ash and additional lime).

The residual hardness is as follows.

$$\text{calcium hardness:} \quad \frac{30 \ \dfrac{\text{mg}}{\text{L}}}{50 \ \dfrac{\text{mg}}{\text{meq}}} = 0.6 \text{ meq/L}$$

$$\text{magnesium carbonate hardness:} \quad \frac{10 \ \dfrac{\text{mg}}{\text{L}}}{50 \ \dfrac{\text{mg}}{\text{meq}}} = 0.2 \text{ meq/L}$$

From part (c), the magnesium noncarbonate hardness is 0.7 meq/L.

$$\text{residual hardness} = 0.6 \ \frac{\text{meq}}{\text{L}} + 0.2 \ \frac{\text{meq}}{\text{L}} + 0.7 \ \frac{\text{meq}}{\text{L}}$$
$$= 1.5 \text{ meq/L}$$
$$\left(1.5 \ \frac{\text{meq}}{\text{L}} \right) \left(50 \ \frac{\text{mg}}{\text{meq}} \right) = \boxed{75 \text{ mg/L as } CaCO_3}$$

The answer is (D).

Water Treatment

24 Water Supply Treatment and Distribution

PRACTICE PROBLEMS

Plain Sedimentation

1. A settling tank's overflow rate is 100,000 gal/ft²-day. Water carrying sediment of various sizes enters the tank. The sediment has the following distribution of settling velocities.

settling velocity (ft/min)	mass fraction remaining
10.0	0.54
5.0	0.45
2.0	0.35
1.0	0.20
0.75	0.10
0.50	0.03

(a) What is the gravimetric percentage of sediment particles completely removed?

(A) 32%

(B) 39%

(C) 47%

(D) 55%

(b) What is the total gravimetric percentage of all sediment particles removed?

(A) 35%

(B) 40%

(C) 50%

(D) 60%

2. A spherical sand particle has a specific gravity of 2.6 and a diameter of 1 mm. What is the settling velocity?

(A) 0.2 ft/sec

(B) 0.7 ft/sec

(C) 1.1 ft/sec

(D) 1.6 ft/sec

3. A mechanically cleaned circular clarifier is to be designed with the following characteristics.

flow rate	2.8 MGD
detention period	2 hr
surface loading	700 gal/ft²-day

(a) What is the approximate diameter?

(A) 45 ft

(B) 60 ft

(C) 70 ft

(D) 90 ft

(b) What is the approximate depth?

(A) 6 ft

(B) 8 ft

(C) 12 ft

(D) 15 ft

(c) If the initial flow rate is reduced to 1.1 MGD, what is the surface loading?

(A) 190 gal/day-ft²

(B) 230 gal/day-ft²

(C) 250 gal/day-ft²

(D) 280 gal/day-ft²

(d) If the initial flow rate is reduced to 1.1 MGD, what is the average detention period?

(A) 4 hr

(B) 5 hr

(C) 6 hr

(D) 8 hr

4. (*Time limit: one hour*) A water treatment plant is designed to handle a total flow rate of 1.5 MGD. The current design includes two identical sedimentation basins that run in parallel to collectively handle the entire load and have the following characteristics.

plan area	90 ft × 16 ft
depth	12 ft
total weir length	48 ft per basin
three-month sustained average low	70% of the design average daily flow
three-month sustained average high	200% of the design average daily flow

Each of the basins must meet the following government standards.

minimum retention time	4.0 hr
maximum weir load	20,000 gpd/ft
maximum velocity	0.5 ft/min

Do the basins meet the standards?

(A) Yes, specifications are met at both peak and low flows.

(B) No, specifications are not met at low flow.

(C) No, specifications are not met at high flow.

(D) No, specifications are not met at either high or low flow.

Mixing Physics

5. (*Time limit: one hour*) A flocculator tank with a volume of 200,000 ft^3 uses a paddle wheel to mix the coagulant in 60°F water. The operating characteristics are as follows.

mean velocity gradient	45 sec^{-1}
paddle drag coefficient	1.75
paddle tip velocity	2 ft/sec
paddle velocity relative to the water	1.5 ft/sec

(a) What is the theoretical power required to drive the paddle?

(A) 12 hp

(B) 17 hp

(C) 60 hp

(D) 120 hp

(b) What is the drag force on the paddle?

(A) 450 lbf

(B) 910 lbf

(C) 5500 lbf

(D) 6400 lbf

(c) What is the required paddle area?

(A) 900 ft^2

(B) 1200 ft^2

(C) 1500 ft^2

(D) 1700 ft^2

Filtration

6. A water treatment plant has four square rapid sand filters. The flow rate is 4 gal/min-ft^2. Each filter has a treatment capacity of 600,000 gal/day. Each filter is backwashed once a day for 8 min. The rate of rise during washing is 24 in/min.

(a) What are the inside dimensions of each filter?

(A) 6 ft × 6 ft

(B) 8 ft × 8 ft

(C) 10 ft × 10 ft

(D) 12 ft × 12 ft

(b) What percentage of the filtered water is used for backwashing?

(A) 2%

(B) 4%

(C) 8%

(D) 10%

7. A water treatment plant has five identical rapid sand filters. Each is square in plan and has a capacity of 1.0 MGD. The application rate is 4 gal/min-ft^2, and the total wetted depth is 10 ft.

(a) What should be the cross-section dimensions of each filter?

(A) 7 ft × 7 ft

(B) 9 ft × 9 ft

(C) 10 ft × 10 ft

(D) 13 ft × 13 ft

(b) Each filter is backwashed every day for 5 min. The rate of rise of the backwash water is 2 ft/min. What percentage of the plant's filtered water will be used for backwashing?

(A) 1.3%

(B) 1.9%

(C) 2.4%

(D) 3.1%

Water Treatment

Precipitation Softening

8. A town's water supply has the following ionic concentrations.

Al^{+++}	0.5 mg/L
Ca^{++}	80.2 mg/L
Cl^-	85.9 mg/L
CO_2	19 mg/L
CO_3^{--}	0
Fe^{++}	1.0 mg/L
Fl^-	0
HCO_3^-	185 mg/L
Mg^{++}	24.3 mg/L
Na^+	46.0 mg/L
NO_3^-	0
SO_4^{--}	125 mg/L

(a) What is the total hardness (as $CaCO_3$)?

(A) 160 mg/L

(B) 200 mg/L

(C) 260 mg/L

(D) 300 mg/L

(b) How much slaked lime (as substance) is required to combine with the carbonate hardness?

(A) 45 mg/L

(B) 90 mg/L

(C) 130 mg/L

(D) 150 mg/L

(c) How much soda ash (as substance) is required to react with the carbonate hardness?

(A) 0 mg/L

(B) 15 mg/L

(C) 35 mg/L

(D) 60 mg/L

9. A city's water supply contains the following ionic concentrations.

$Ca(HCO_3)_2$	137 mg/L as $CaCO_3$
$MgSO_4$	72 mg/L as $CaCO_3$
CO_2	0

(a) How much slaked lime will be required to soften 1,000,000 gal of this water to a hardness of 100 mg/L (as $CaCO_3$) if 30 mg/L (as $CaCO_3$) of excess lime is used?

(A) 860 lbm

(B) 930 lbm

(C) 1170 lbm

(D) 1310 lbm

(b) The soda ash required to soften 1,000,000 gal of this water to a hardness of 100 mg/L (as $CaCO_3$) if there is 30 mg/L (as $CaCO_3$) of excess lime is most nearly

(A) 0 lbm

(B) 300 lbm

(C) 500 lbm

(D) 700 lbm

(c) How much soda ash (as $CaCO_3$) is required to soften 1,000,000 gal of this water to a hardness of 50 mg/L (as $CaCO_3$) assuming all carbonate hardness has been removed?

(A) 10 mg/L

(B) 20 mg/L

(C) 40 mg/L

(D) 50 mg/L

Zeolite Softening

10. The water described in Prob. 9 is to be softened using a zeolite process with the following characteristics: exchange capacity, 10,000 grains/ft^3; and salt requirement, 0.5 lbm per 1000 grains hardness removed. How much salt is required to soften the water to 100 mg/L hardness?

(A) 700 lbm/MG

(B) 1800 lbm/MG

(C) 2500 lbm/MG

(D) 3200 lbm/MG

11. The hardness of water from an underground aquifer is to be reduced from 245 mg/L (as $CaCO_3$) to 80 mg/L (as $CaCO_3$) using a zeolite process. The volumetric flow rate is 20,000 gal/day. The process has the following characteristics: resin exchange capacity, 20,000 grains/ft^3; and zeolite volume, 2 ft^3.

(a) What fraction of the water is bypassed around the process?

(A) 0.15

(B) 0.33

(C) 0.67

(D) 0.85

(b) What is the time between regenerations of the softener?

(A) 5 hr

(B) 16 hr

(C) 24 hr

(D) 30 hr

Demand

12. (*Time limit: one hour*) An area is expected to attain a population of 40,000 in 20 years. The cumulative per capita water demand for a peak day in the area is shown in the following diagram.

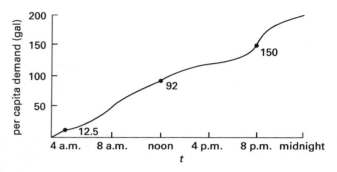

(a) What is the total per capita demand for a peak day?

(A) 160 gal

(B) 200 gal

(C) 240 gal

(D) 290 gal

(b) Assuming uniform operation over 24 hours, what storage volume is required in the treatment plant for all uses, including fire fighting demand?

(A) 1.1 MG

(B) 2.5 MG

(C) 3.2 MG

(D) 5.5 MG

(c) Assume that the pumping station runs uniformly from 4 a.m. until 8 a.m. to fill the storage tanks for the day. What storage is required to meet all uses, including fire fighting? (Use the American Insurance Association equation to calculate fire fighting demand.)

(A) 3.2 MG

(B) 3.8 MG

(C) 5.5 MG

(D) 8.0 MG

13. The water supply for a town of 15,000 people is taken from a river. The average consumption is 110 gpcd. The water has the following characteristics.

turbidity	20–100 NTU (varies)
total hardness	less than 60 mg/L as $CaCO_3$
coliform count	200 to 1000 per 100 mL (varies)

(a) What capacity should the distribution and treatment system have? (Use the American Insurance Association equation to calculate fire fighting demand.)

(A) 6000 gpm

(B) 7500 gpm

(C) 9500 gpm

(D) 12,000 gpm

(b) If the application rate is 4 gal/min-ft^2, what total filter area is required?

(A) 290 ft^2

(B) 580 ft^2

(C) 910 ft^2

(D) 1100 ft^2

(c) Is softening required?

(A) No, 60 mg/L is soft water.

(B) No, softening would interfere with turbidity removal.

(C) Yes, the turbidity and coliform counts would also benefit.

(D) Yes, all municipal water should be softened.

(d) If a chlorine dose of 2 mg/L is required to obtain the desired chlorine residual, how much chlorine is required every 24 hours? (Disregard fire fighting flow.)

(A) 15 lbm

(B) 22 lbm

(C) 28 lbm

(D) 32 lbm

SOLUTIONS

1. (a) Calculate the overflow rate.

$$v^* = \frac{100{,}000\ \dfrac{\text{gal}}{\text{ft}^2\text{-day}}}{\left(7.48\ \dfrac{\text{gal}}{\text{ft}^3}\right)\left(24\ \dfrac{\text{hr}}{\text{day}}\right)\left(60\ \dfrac{\text{min}}{\text{hr}}\right)}$$

$$= 9.28\ \text{ft/min}$$

Using interpolation, the mass fraction remaining is

$$x_c = 0.45 + \left(\frac{9.28\ \dfrac{\text{ft}}{\text{min}} - 5.0\ \dfrac{\text{ft}}{\text{min}}}{10.0\ \dfrac{\text{ft}}{\text{min}} - 5.0\ \dfrac{\text{ft}}{\text{min}}}\right)(0.54 - 0.45)$$

$$= 0.527\ \text{remains in flow}$$

$$1 - x_c = 1 - 0.527 = 0.473$$

$$\boxed{47\%\ \text{completely removed}}$$

The answer is (C).

(b) Plot the mass fraction remaining versus the settling velocity.

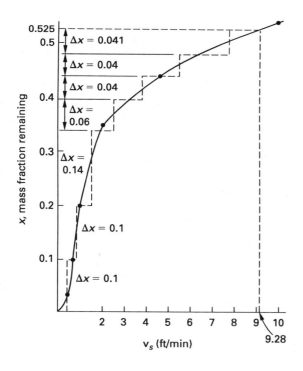

$$x_c \approx 0.525 \quad (52.5\%)$$

This corresponds to the value in part (a).

Determine $\Delta x\,v_t$ by graphical integration.

Δx	v_t	$\Delta x v_t$
0.1	0.5	0.05
0.1	0.8	0.08
0.14	1.4	0.196
0.06	2.5	0.15
0.04	3.75	0.15
0.04	5.5	0.22
0.041	7.75	0.318
	total	1.164

The overall removal efficiency is

$$f = 1 - x_c + \sum \frac{\Delta x v_t}{v^*}$$

$$= 1 - 0.525 + \frac{1.164\ \dfrac{\text{ft}}{\text{min}}}{9.28\ \dfrac{\text{ft}}{\text{min}}}$$

$$= \boxed{0.60 \quad (60\%)}$$

The answer is (D).

2. From Fig. 24.2, $v_s = \boxed{0.7\ \text{ft/sec.}}$

The answer is (B).

3. (a) The surface area is

$$A_{\text{surface}} = \frac{Q}{v^*} = \frac{2.8 \times 10^6\ \dfrac{\text{gal}}{\text{day}}}{700\ \dfrac{\text{gal}}{\text{ft}^2\text{-day}}} = 4000\ \text{ft}^2$$

Since $A = \dfrac{\pi}{4} D^2$,

$$D = \sqrt{\frac{(4)(4000\ \text{ft}^2)}{\pi}} = \boxed{71.4\ \text{ft}}$$

The answer is (C).

(b) From Eq. 24.17, the volume is

$$V = tQ = \frac{\left(2.8 \times 10^6\ \dfrac{\text{gal}}{\text{day}}\right)(2\ \text{hr})}{24\ \dfrac{\text{hr}}{\text{day}}}$$

$$= 2.333 \times 10^5\ \text{gal}$$

The depth is

$$d = \frac{V}{A_{\text{surface}}} = \frac{2.333 \times 10^5 \text{ gal}}{(4000 \text{ ft}^2)\left(7.48 \frac{\text{gal}}{\text{ft}^3}\right)}$$

$$= \boxed{7.8 \text{ ft} \quad (8 \text{ ft})}$$

The answer is (B).

(c) Use Eq. 24.7.

$$v^* = \frac{Q}{A} = \frac{1.1 \times 10^6 \frac{\text{gal}}{\text{day}}}{4000 \text{ ft}^2}$$

$$= \boxed{275 \text{ gal/day-ft}^2 \quad (280 \text{ gal/day-ft}^2)}$$

The answer is (D).

(d) Use Eq. 24.6.

$$t = \frac{V}{Q} = \frac{(2.333 \times 10^5 \text{ gal})\left(24 \frac{\text{hr}}{\text{day}}\right)}{1.1 \times 10^6 \frac{\text{gal}}{\text{day}}}$$

$$= \boxed{5.09 \text{ hr} \quad (5 \text{ hr})}$$

The answer is (B).

4. volume per basin $= (90 \text{ ft})(16 \text{ ft})(12 \text{ ft})$

$$= 17,280 \text{ ft}^3$$

(The freeboard is not given.)

The area per basin is

$$A = (90 \text{ ft})(16 \text{ ft}) = 1440 \text{ ft}^2$$

The three-month peak flow per basin is

$$(2)\left(\frac{1.5 \text{ MGD}}{2}\right) = 1.5 \text{ MGD}$$

$$(1.5 \text{ MGD})\left(1.547 \frac{\frac{\text{ft}^3}{\text{sec}}}{\text{MGD}}\right) = 2.32 \text{ ft}^3/\text{sec}$$

The detention time at peak flow is given by Eq. 24.6.

$$t = \frac{V}{Q} = \frac{17,280 \text{ ft}^3}{\left(2.32 \frac{\text{ft}^3}{\text{sec}}\right)\left(60 \frac{\text{min}}{\text{hr}}\right)\left(60 \frac{\text{sec}}{\text{min}}\right)} = 2.07 \text{ hr}$$

Since 2.07 hr < 4 hr, this is not acceptable.

The weir loading is

$$\frac{1.5 \times 10^6 \frac{\text{gal}}{\text{day}}}{48 \text{ ft}} = 31,250 \text{ gal/day-ft}$$

Since 31,250 gal/day-ft > 20,000 gal/day-ft, this is not acceptable either.

The overflow rate is

$$v^* = \frac{Q}{A} = \frac{\left(2.32 \frac{\text{ft}^3}{\text{sec}}\right)\left(60 \frac{\text{sec}}{\text{min}}\right)}{1440 \text{ ft}^2} = 0.0967 \text{ ft/min}$$

Since 0.0967 ft/min < 0.5 ft/min, this is acceptable.

At low flow,

$$\text{flow} = \frac{0.7}{2.0} = 0.35 \quad [35\% \text{ of high flow}]$$

$$t = \frac{2.07 \text{ hr}}{0.35} = 5.91 \text{ hr} \quad [\text{acceptable}]$$

$$\text{weir loading} = (0.35)\left(31,250 \frac{\text{gal}}{\text{day-ft}}\right)$$

$$= 10,938 \frac{\text{gal}}{\text{day-ft}} \quad [\text{acceptable}]$$

$$v^* = (0.35)\left(0.1 \frac{\text{ft}}{\text{min}}\right) = 0.035 \text{ ft/min}$$

$$[\text{acceptable}]$$

> The basins have been correctly designed for low flow but not for peak flow. One or more basins should be used.

The answer is (C).

5. For 60°F water, $\mu = 2.359 \times 10^{-5}$ lbf-sec/ft^2.
From Eq. 24.25,

$$P = \mu G^2 V_{\text{tank}}$$

$$= \left(2.359 \times 10^{-5} \frac{\text{lbf-sec}}{\text{ft}^2}\right)\left(45 \frac{1}{\text{sec}}\right)^2 (200,000 \text{ ft}^3)$$

$$= 9554 \text{ ft-lbf/sec}$$

(a) water horsepower $= \dfrac{9554 \frac{\text{ft-lbf}}{\text{sec}}}{550 \frac{\text{ft-lbf}}{\text{hp-sec}}}$

$$= \boxed{17.4 \text{ hp} \quad (17 \text{ hp})}$$

The answer is (B).

(b) Since work = force × distance, then power = force × velocity.

$$D = \frac{P}{v_{mixing}} = \frac{9554 \ \frac{\text{ft-lbf}}{\text{sec}}}{1.5 \ \frac{\text{ft}}{\text{sec}}} = \boxed{6369 \ \text{lbf} \quad (6400 \ \text{lbf})}$$

The answer is (D).

(c) Use Eq. 24.20.

$$A = \frac{2gF_D}{C_D \gamma v^2} = \frac{(2)\left(32.2 \ \frac{\text{ft}}{\text{sec}^2}\right)(6369 \ \text{lbf})}{(1.75)\left(62.4 \ \frac{\text{lbf}}{\text{ft}^3}\right)\left(1.5 \ \frac{\text{ft}}{\text{sec}}\right)^2}$$

$$= \boxed{1669 \ \text{ft}^2 \quad (1700 \ \text{ft}^2)}$$

The answer is (D).

6. (a) The required area is

$$A = \frac{Q}{v^*} = \frac{600,000 \ \frac{\text{gal}}{\text{day}}}{\left(4 \ \frac{\text{gal}}{\text{min-ft}^2}\right)\left(24 \ \frac{\text{hr}}{\text{day}}\right)\left(60 \ \frac{\text{min}}{\text{hr}}\right)}$$

$$= 104.2 \ \text{ft}^2$$

$$\boxed{\text{Use 10 ft} \times \text{10 ft.}}$$

The answer is (C).

(b) The required volume is

$$V = tAv = \left(8 \ \frac{\text{min}}{\text{day}}\right)(100 \ \text{ft}^2)\left(2 \ \frac{\text{ft}}{\text{min}}\right)\left(7.48 \ \frac{\text{gal}}{\text{ft}^3}\right)$$

$$= 11,968 \ \text{gal/day}$$

The backwash fraction is

$$\frac{11,968 \ \frac{\text{gal}}{\text{day}}}{600,000 \ \frac{\text{gal}}{\text{day}}} = \boxed{0.01995 \quad (2\%)}$$

The answer is (A).

7. (a) $A = \dfrac{Q}{v^*} = \dfrac{1 \times 10^6 \ \frac{\text{gal}}{\text{day}}}{\left(24 \ \frac{\text{hr}}{\text{day}}\right)\left(60 \ \frac{\text{min}}{\text{hr}}\right)\left(4 \ \frac{\text{gal}}{\text{min-ft}^2}\right)}$

$$= 173.6 \ \text{ft}^2$$

$$\text{width} = \sqrt{173.6 \ \text{ft}^2} = 13.2 \ \text{ft}$$

$$\boxed{\text{Use 13 ft} \times \text{13 ft.}}$$

The answer is (D).

(b) The water volume is

$$V = tAv$$

$$= \left(5 \ \frac{\text{min}}{\text{day}}\right)(173.6 \ \text{ft}^2)\left(2 \ \frac{\text{ft}}{\text{min}}\right)(5 \ \text{filters})\left(7.48 \ \frac{\text{gal}}{\text{ft}^3}\right)$$

$$= 64,926 \ \text{gal/day}$$

The backwash fraction is

$$\frac{64,926 \ \frac{\text{gal}}{\text{day}}}{5 \times 10^6 \ \frac{\text{gal}}{\text{day}}} = \boxed{0.013 \quad (1.3\%)}$$

The answer is (A).

8. (a) Using the factors from App. 22.C,

	mg/L as substance		factor		
Ca^{++}:	80.2	×	2.5	=	200.5 mg/L
Mg^{++}:	24.3	×	4.1	=	99.63 mg/L
Fe^{++}:	1	×	1.79	=	1.79 mg/L
Al^{+++}:	0.5	×	5.56	=	2.78 mg/L
			hardness =		304.7 mg/L

The total hardness is approximately $\boxed{300 \ \text{mg/L}}$ as $CaCO_3$.

The answer is (D).

(b) To remove the carbonate hardness, the carbon dioxide must first be removed. The CO_2 concentration in $CaCO_3$ equivalents is

$$CO_2\text{: } 19 \ \frac{\text{mg}}{\text{L}} \times 2.27 = 43.13 \ \text{mg/L as } CaCO_3$$

Add lime to remove the carbonate hardness. It does not matter whether the HCO_3^- comes from Mg^{++}, Ca^{++}, or Fe^{++}; adding lime will remove it.

There may be Mg^{++}, Ca^{++}, or Fe^{++} ions left over in the form of noncarbonate hardness, but the problem asked for carbonate hardness. Converting from mg/L of substance to mg/L as $CaCO_3$,

$$HCO_3^-\text{: } 185 \ \frac{\text{mg}}{\text{L}} \times 0.82 = 151.7 \ \text{mg/L}$$

The total equivalents to be neutralized are

$$43.13 \ \frac{\text{mg}}{\text{L}} + 151.7 \ \frac{\text{mg}}{\text{L}} = 194.83 \ \text{mg/L}$$

Water Treatment

Convert $Ca(OH)_2$ using App. 22.B.

$$\frac{mg}{L} \text{ of } Ca(OH)_2 = \frac{194.83 \ \frac{mg}{L}}{1.35}$$

$$= \boxed{144.3 \text{ mg/L} \quad (150 \text{ mg/L})}$$

The answer is (D).

(c) $\boxed{\begin{array}{l} \text{No soda ash is required since it is used to} \\ \text{remove noncarbonate hardness.} \end{array}}$

The answer is (A).

9. (a) $Ca(HCO_3)_2$ and $MgSO_4$ both contribute to hardness. Since 100 mg/L of hardness is the goal, leave all $MgSO_4$ in the water. Take out 137 mg/L + 72 mg/L − 100 mg/L = 109 mg/L of $Ca(HCO_3)_2$. From App. 22.B (including the excess even though the reaction is not complete),

$$\text{pure } Ca(OH)_2 = \frac{30 \ \frac{mg}{L} + 109 \ \frac{mg}{L}}{1.35}$$

$$= 102.96 \text{ mg/L}$$

$$\left(102.96 \ \frac{mg}{L}\right)\left(8.345 \ \frac{lbm\text{-}L}{mg\text{-}MG}\right) = 859 \text{ lbm/MG}$$

The amount of slaked time required is approximately $\boxed{860 \text{ lbm/MG.}}$

The answer is (A).

(b) Since the goal is a residual hardness of 100 mg/L and the magnesium sulfate only contributes 72 mg/L, it does not have to be removed. No soda ash is required.

The answer is (A).

(c) In order to reduce hardness to 50 mg/L, 72 mg/L − 50 mg/L = 22 mg/L of $MgSO_4$ must be removed. (This assumes that all of the carbonate hardness has already been removed.) This will require 22 mg/L (as $CaCO_3$) of soda ash.

The answer is (B).

10. There are 7000 grains in a pound. The hardness removed is

$$137 \ \frac{mg}{L} + 72 \ \frac{mg}{L} - 100 \ \frac{mg}{L} = 109 \text{ mg/L}$$

$$\left(109 \ \frac{mg}{L}\right)\left(8.345 \ \frac{lbm\text{-}L}{mg\text{-}MG}\right) = 909.6 \text{ lbm hardness/MG}$$

$$\left(\frac{0.5 \text{ lbm}}{1000 \text{ gr}}\right)\left(909.6 \ \frac{lbm}{MG}\right)\left(7000 \ \frac{gr}{lbm}\right)$$

$$= \boxed{3184 \text{ lbm/MG} \quad (3200 \text{ lbm/MG})}$$

The answer is (D).

11. (a) A bypass process is required.

$$\text{fraction bypassed: } \frac{80 \ \frac{mg}{L}}{245 \ \frac{mg}{L}} = \boxed{0.327 \quad (0.33)}$$

fraction processed: $1 - 0.327 = 0.673$

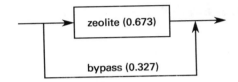

The answer is (B).

(b) The maximum hardness reduction is

$$\frac{(2 \text{ ft}^3)\left(20{,}000 \ \frac{gr}{ft^3}\right)}{7000 \ \frac{gr}{lbm}} = 5.71 \text{ lbm}$$

The hardness removal rate is

$$\frac{(0.673)\left(245 \ \frac{mg}{L}\right)\left(8.345 \times 10^{-6} \ \frac{lbm}{gal}\right)\left(20{,}000 \ \frac{gal}{day}\right)}{24 \ \frac{hr}{day}}$$

$$= 1.15 \text{ lbm/hr}$$

$$t = \frac{5.71 \text{ lbm}}{1.15 \ \frac{lbm}{hr}} = \boxed{4.97 \text{ hr} \quad (5 \text{ hr})}$$

The answer is (A).

12. (a) The given graph shows that by the end of the day, the cumulative demand has risen to 200 gal per person. The daily demand is $\boxed{200 \text{ gal.}}$

The answer is (B).

(b) The analysis is similar to that used with reservoir sizing.

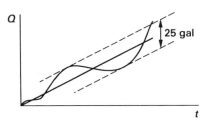

From the cumulative flow quantity, the storage requirement per capita-day is 25 gal.

The population use is

$$\left(25 \ \frac{\text{gal}}{\text{day-person}}\right)(40{,}000 \ \text{people}) = 1{,}000{,}000 \ \text{gal/day}$$

From Eq. 24.69, the fire fighting requirement is

$$\begin{aligned}
Q &= 1020\sqrt{P}(1 - 0.01\sqrt{P}) \\
&= 1020\sqrt{40}(1 - 0.01\sqrt{40}) \\
&= 6043 \ \text{gal/min}
\end{aligned}$$

Therefore, the flow rate is $6043/1000 = 6$ thousands of gallons. Maintain the flow for 4 hr (approximate ISO specifications).

$$\begin{aligned}
\text{capacity} &= 1{,}000{,}000 \ \text{gal} \\
&\quad + \left(6043 \ \frac{\text{gal}}{\text{min}}\right)\left(60 \ \frac{\text{min}}{\text{hr}}\right)(4 \ \text{hr}) \\
&= \boxed{2{,}450{,}000 \ \text{gal} \quad (2.5 \ \text{MG})}
\end{aligned}$$

The answer is (B).

(c) The pump supplies demand from 4 a.m. until 8 a.m. The storage supplies demand from 8 a.m. until 4 a.m.

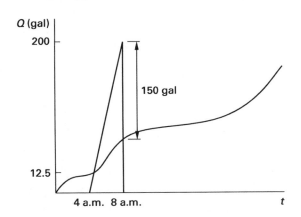

The population use is

$$(150 \ \text{gal} + 12.5 \ \text{gal})(40{,}000) = 6{,}500{,}000 \ \text{gal}$$

Add fire fighting.

$$\begin{aligned}
6{,}500{,}000 \ \text{gal} &+ \left(6043 \ \frac{\text{gal}}{\text{min}}\right)\left(60 \ \frac{\text{min}}{\text{hr}}\right)(4 \ \text{hr}) \\
&= \boxed{7{,}950{,}000 \ \text{gal} \quad (8.0 \ \text{MG})}
\end{aligned}$$

The answer is (D).

13. (a) From Table 24.7, use 3 as the peak multiplier.

$$\frac{\left(110 \ \dfrac{\text{gal}}{\text{day}}\right)(15{,}000)(3)}{\left(24 \ \dfrac{\text{hr}}{\text{day}}\right)\left(60 \ \dfrac{\text{min}}{\text{hr}}\right)} = 3437 \ \text{gal/min}$$

The fire fighting requirements are given by Eq. 24.69.

$$\begin{aligned}
Q &= 1020\sqrt{P}(1 - 0.01\sqrt{P}) \\
&= 1020\sqrt{15}(1 - 0.01\sqrt{15}) \\
&= 3797 \ \text{gal/min}
\end{aligned}$$

The total maximum demand for which the distribution system should be designed is

$$\begin{aligned}
3437 \ \frac{\text{gal}}{\text{min}} &+ 3797 \ \frac{\text{gal}}{\text{min}} \\
&= \boxed{7234 \ \text{gal/min} \quad (7500 \ \text{gal/min})}
\end{aligned}$$

The answer is (B).

(b) The filter area should not be based on the maximum hourly rate since some of the demand during the peak hours can come from storage (clearwell or tanks). Also, fire requirements can bypass the filters if necessary.

$$\left(110 \ \frac{\text{gal}}{\text{day-person}}\right)(15{,}000 \ \text{people}) = 1.65 \times 10^6 \ \text{gal/day}$$

Using a flow rate of 4 gpm/ft², the required filter area is

$$\begin{aligned}
A = \frac{Q}{\text{v}^*} &= \frac{1.65 \times 10^6 \ \dfrac{\text{gal}}{\text{day}}}{\left(4 \ \dfrac{\text{gal}}{\text{min-ft}^2}\right)\left(24 \ \dfrac{\text{hr}}{\text{day}}\right)\left(60 \ \dfrac{\text{min}}{\text{hr}}\right)} \\
&= \boxed{286.5 \ \text{ft}^2 \quad (290 \ \text{ft}^2)}
\end{aligned}$$

The answer is (A).

(c) No, 60 mg/L is soft water.

The answer is (A).

(d) Disregarding the fire flow, the required average (not peak) daily chlorine mass is given by Eq. 24.16.

$$F = \frac{DQ\left(8.345 \ \frac{\text{lbm-L}}{\text{mg-MG}}\right)}{PG}$$

$$= \frac{\left(110 \ \frac{\text{gal}}{\text{day-person}}\right)(15{,}000 \ \text{people})}{10^6 \ \frac{\text{gal}}{\text{MG}}}$$

$$\times \left(2 \ \frac{\text{mg}}{\text{L}}\right)\left(8.345 \ \frac{\text{lbm-L}}{\text{mg-MG}}\right)$$

$$= \boxed{27.54 \ \text{lbm/day}}$$

The answer is (C).

25 Wastewater Quantity and Quality

PRACTICE PROBLEMS

Sewer Velocities and Sizing

1. (*Time limit: one hour*) A town of 10,000 people (125 gpcd) has its own primary treatment plant.

(a) What mass of total solids should the treatment plant expect?

(A) 57 lbm/day

(B) 740 lbm/day

(C) 3800 lbm/day

(D) 8300 lbm/day

(b) If the town is 4 mi from the treatment plant and 400 ft above it in elevation, what minimum size pipe should be used between the town and the plant, assuming that the pipe flows full? Disregard infiltration.

(A) 8 in

(B) 12 in

(C) 14 in

(D) 18 in

2. (*Time limit: one hour*) Your client has just completed a subdivision, and his sewage lines hook up to a collector that goes into a trunk. A problem has developed in the first manhole up from the trunk, and the collector pipe overflows periodically. An industrial plant is hooked directly into the trunk, and it is this plant's flow that is making your client's line back up. The subdivision is in a flood plain, and the sewer lines are very flat and cannot be steepened. (a) Describe two possible solutions to this problem. (b) Sketch plan views of your solutions.

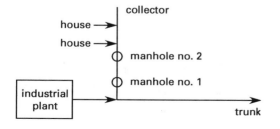

Dilution

3. Town A wants to discharge wastewater into a river 5.79 km upstream of town B. The state standard level for dissolved oxygen is 5 mg/L. Other data pertinent to this problem are given in the following table.

parameter	wastewater	river
flow (m³/s)	0.280	0.877
ultimate BOD at 28°C (mg/L)	6.44	7.00
DO (mg/L)	1.00	6.00
K_d at 28°C, d⁻¹ (base-10)	N/A	0.199
K_r at 28°C, d⁻¹ (base-10)	N/A	0.370
velocity (m/s)	N/A	0.650
temperature (°C)	28	28

(a) Will the dissolved oxygen level at town A be reduced below the state standard level?

(A) yes; $DO_{min} = 4.1$ mg/L

(B) yes; $DO_{min} = 4.8$ mg/L

(C) no; $DO_{min} = 5.3$ mg/L

(D) no; $DO_{min} = 5.9$ mg/L

(b) Will the dissolved oxygen level be below the state standard level at town B?

(A) yes; $DO_{min} = 4.1$ mg/L

(B) yes; $DO_{min} = 4.8$ mg/L

(C) no; $DO_{min} = 5.3$ mg/L

(D) no; $DO_{min} = 5.9$ mg/L

(c) At any point downstream, will the dissolved oxygen level be below the state standard level?

(A) yes; $DO_{min} = 4.1$ mg/L

(B) yes; $DO_{min} = 4.7$ mg/L

(C) no; $DO_{min} = 5.1$ mg/L

(D) no; $DO_{min} = 5.5$ mg/L

(d) Determine if the discharge will reduce the dissolved oxygen in the river at town B below the state standard level if the winter river temperature drops to 12°C (before mixing) and all other characteristics remain the same.

(A) yes; $DO_{min} = 4.5$ mg/L

(B) yes; $DO_{min} = 4.9$ mg/L

(C) no; $DO_{min} = 5.0$ mg/L

(D) no; $DO_{min} = 5.9$ mg/L

4. To protect aquatic life, the limit on increase in temperature in a certain stream is 2°C at seasonal low flow. In addition, the stream must not contain more than 0.002 mg/L of un-ionized ammonia.

A manufacturing facility proposes to draw cooling water from the stream and return it to the stream at a higher temperature. They will also discharge some un-ionized ammonia in the cooling water.

The critical low flow in the stream is 1200 m³/s at 20°C. The stream has no measurable un-ionized ammonia in its natural state. The facility intends to withdraw 60 m³/s of cooling water and return it to the stream without loss of mass.

(a) What is the maximum temperature for the discharge water if the stream is not to be damaged?

(A) 40°C

(B) 60°C

(C) 80°C

(D) 90°C

(b) What is the maximum ammonia concentration for the discharge water if the stream is not to be damaged?

(A) 0.04 mg/L

(B) 0.06 mg/L

(C) 0.09 mg/L

(D) 0.13 mg/L

5. A rural town of 10,000 people discharges partially treated wastewater into a stream. The stream has the following characteristics.

minimum flow rate	120 ft³/sec
velocity	3 mi/hr
minimum dissolved oxygen	7.5 mg/L at 15°C
temperature	15°C
BOD	0

The reoxygenation and deoxygenation coefficients (base-10) are

reoxygenation coefficient of stream and effluent mixture	0.2 d⁻¹ at 20°C
deoxygenation coefficient	0.15 d⁻¹ at 20°C

The town's effluent has the following characteristics.

source	volume	BOD at 20°C	temp.
domestic	122 gpcd	0.191 lbm/cd	64°F
infiltration	116,000 gpd		51°F
industrial no. 1	180,000 gpd	800 mg/L	95°F
industrial no. 2	76,000 gpd	1700 mg/L	84°F

(a) What is the domestic waste BOD in mg/L at 20°C?

(A) 140 mg/L

(B) 160 mg/L

(C) 190 mg/L

(D) 220 mg/L

(b) What is the approximate total effluent 20°C BOD in mg/L just before discharge into the stream?

(A) 245 mg/L

(B) 270 mg/L

(C) 295 mg/L

(D) 315 mg/L

(c) What is the average temperature of the effluent just before discharge into the stream?

(A) 20°C

(B) 30°C

(C) 50°C

(D) 70°C

(d) If the wastewater does not contribute to dissolved oxygen at all, how far downstream is the theoretical point of minimum dissolved oxygen concentration?

(A) 40 mi

(B) 70 mi

(C) 80 mi

(D) 120 mi

(e) What is the theoretical minimum dissolved oxygen concentration in the stream?

(A) 4.7 mg/L

(B) 5.5 mg/L

(C) 7.3 mg/L

(D) 8.3 mg/L

6. A sewage treatment plant is being designed to handle both domestic and industrial wastewaters. The city population is 20,000, and an excess capacity factor of 15% is to be used for future domestic expansion. The wastewaters have the following characteristics.

source	volume	BOD
domestic	100 gpcd	0.18 lbm/cd
industrial no. 1	1.3 MGD	1100 mg/L
industrial no. 2	1.0 MGD	500 mg/L

(a) What is the design population equivalent for the plant?

(A) 60,000

(B) 75,000

(C) 90,000

(D) 110,000

(b) What is the plant's organic loading?

(A) 2300 lbm/day

(B) 8100 lbm/day

(C) 14,000 lbm/day

(D) 20,000 lbm/day

(c) What is the plant's hydraulic loading?

(A) 4.6 MGD

(B) 5.8 MGD

(C) 6.3 MGD

(D) 7.5 MGD

SOLUTIONS

1. (a) Assume: 125 gpcd for average flow
800 mg/L total solids

$$\dot{m} = \frac{(10{,}000)\left(125\ \dfrac{\text{gal}}{\text{day}}\right)\left(800\ \dfrac{\text{mg}}{\text{L}}\right)\left(8.345\ \dfrac{\text{lbm-L}}{\text{mg-MG}}\right)}{1 \times 10^6\ \dfrac{\text{gal}}{\text{MG}}}$$

$$= \boxed{8345\ \text{lbm/day}}$$

(Minor variations in the assumptions should not affect the answer choice.)

The answer is (D).

(b) $$S = \frac{400\ \text{ft}}{(4\ \text{mi})\left(5280\ \dfrac{\text{ft}}{\text{mi}}\right)} = 0.01894$$

Use Eq. 25.1.

$$\frac{Q_{\text{peak}}}{Q_{\text{ave}}} = \frac{18 + \sqrt{P}}{4 + \sqrt{P}} = \frac{18 + \sqrt{10}}{4 + \sqrt{10}}$$

$$\approx 3.0$$

$$Q_{\text{peak}} = \frac{\left(125\ \dfrac{\text{gal}}{\text{day}}\right)(3)(10{,}000)\left(0.1337\ \dfrac{\text{ft}^3}{\text{gal}}\right)}{\left(24\ \dfrac{\text{hr}}{\text{day}}\right)\left(60\ \dfrac{\text{min}}{\text{hr}}\right)\left(60\ \dfrac{\text{sec}}{\text{min}}\right)}$$

$$= 5.80\ \text{ft}^3/\text{sec}$$

Assume $n = 0.013$. From Eq. 19.16,

$$D = 1.335\left(\frac{nQ}{\sqrt{S}}\right)^{3/8}$$

$$= (1.335)\left(\frac{(0.013)\left(5.80\ \dfrac{\text{ft}^3}{\text{sec}}\right)}{\sqrt{0.01894}}\right)^{3/8}\left(12\ \dfrac{\text{in}}{\text{ft}}\right)$$

$$= \boxed{12.8\ \text{in}\quad(14\ \text{in})}$$

(A 12 in pipe does not have the capacity of a 12.8 in pipe.)

The answer is (C).

2. (a) If the first manhole overflows, the piezometric head at the rim must be less than the head in the trunk. Further up at manhole no. 2, the increase in elevation is sufficient to raise rim no. 2, so the head in the trunk must be lowered.

Alternate solutions:

- relief storage (surge chambers) between trunk and manhole no. 1
- storage at plant and gradual release
- private trunk for plant
- private treatment and discharge for plant
- larger trunk capacity using larger or parallel pipes
- backflow preventors

(b)

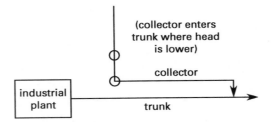

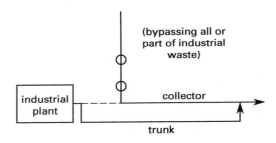

3. (a) Find the dissolved oxygen at town A using Eq. 25.35.

$$DO = \frac{Q_w DO_w + Q_r DO_r}{Q_w + Q_r}$$

$$= \frac{\left(0.280 \frac{m^3}{s}\right)\left(1 \frac{mg}{L}\right) + \left(0.877 \frac{m^3}{s}\right)\left(6 \frac{mg}{L}\right)}{0.280 \frac{m^3}{s} + 0.877 \frac{m^3}{s}}$$

$$= \boxed{4.79 \text{ mg/L}}$$

So, immediately after mixing, the water will be reduced below the state standard of 5 mg/L.

The answer is (B).

(b) The composite ultimate BOD is

$$BOD_{u,28°C} = \frac{Q_w BOD_w + Q_r BOD_r}{Q_w + Q_r}$$

$$= \frac{\left(0.280 \frac{m^3}{s}\right)\left(6.44 \frac{mg}{L}\right)}{0.280 \frac{m^3}{s} + 0.877 \frac{m^3}{s}}$$

$$= 6.8645 \text{ mg/L}$$

From App. 22.D, $DO_{sat} = 7.92$ mg/L at 28°C. The initial deficit is

$$D_0 = DO_{sat} - DO = 7.92 \frac{mg}{L} - 4.79 \frac{mg}{L}$$

$$= 3.13 \text{ mg/L}$$

Calculate travel time from town A to town B.

$$t = \frac{(5.79 \text{ km})\left(1000 \frac{m}{km}\right)}{\left(0.650 \frac{m}{s}\right)\left(86\,400 \frac{s}{d}\right)}$$

$$= 0.1031 \text{ d}$$

Calculate the deficit 5.79 km downstream. Use base-10 exponents because the K values are in base-10. Use Eq. 25.36.

$$D_t = \left(\frac{K_d BOD_u}{K_r - K_d}\right)(10^{-K_d t} - 10^{-K_r t}) + D_0(10^{-K_r t})$$

$$= \frac{(0.199 \text{ d}^{-1})\left(6.8645 \frac{mg}{L}\right)}{0.370 \frac{mg}{L} - 0.199 \frac{mg}{L}}$$

$$\times \left(\begin{matrix} 10^{-(0.199 \text{ d}^{-1})(0.1031 \text{ d})} \\ - 10^{-(0.370 \text{ d}^{-1})(0.1031 \text{ d})} \end{matrix}\right)$$

$$+ \left(3.13 \frac{mg}{L}\right)\left(10^{-(0.370 \text{ d}^{-1})(0.1031 \text{ d})}\right)$$

$$= 3.17 \text{ mg/L}$$

Calculate the dissolved oxygen downstream.

$$DO_{town B} = DO_{sat} - D$$

$$= 7.92 \frac{mg}{L} - 3.17 \frac{mg}{L}$$

$$= \boxed{4.75 \text{ mg/L}}$$

This is still below the state standard of 5 mg/L. More distance is required for reoxygenation to increase the oxygen concentration.

The answer is (B).

(c) Determine the critical time. Use Eq. 25.38.

$$t_c = \left(\frac{1}{K_r - K_d}\right)$$
$$\times \log_{10}\left(\left(\frac{K_d\text{BOD}_u - K_r D_0 + K_d D_0}{K_d\text{BOD}_u}\right)\left(\frac{K_r}{K_d}\right)\right)$$
$$= \left(\frac{1}{0.370 \text{ d}^{-1} - 0.199 \text{ d}^{-1}}\right)$$
$$\times \log_{10}\left(\begin{array}{c}\left(\begin{array}{c}(0.199 \text{ d}^{-1})\left(6.8645 \frac{\text{mg}}{\text{L}}\right) \\ - (0.370 \text{ d}^{-1})\left(3.13 \frac{\text{mg}}{\text{L}}\right) \\ + (0.199 \text{ d}^{-1})\left(3.13 \frac{\text{mg}}{\text{L}}\right) \\ \hline (0.199 \text{ d}^{-1})\left(6.8645 \frac{\text{mg}}{\text{L}}\right) \end{array}\right) \\ \times \left(\frac{0.370 \text{ d}^{-1}}{0.199 \text{ d}^{-1}}\right)\end{array}\right)$$
$$= 0.3122 \text{ d}$$

Calculate critical deficit and dissolved oxygen. Use Eq. 25.39.

$$D_c = \left(\frac{K_d\text{BOD}_u}{K_r}\right)10^{-K_d t_c}$$
$$= \left(\frac{(0.199 \text{ d}^{-1})\left(6.8645 \frac{\text{mg}}{\text{L}}\right)}{0.370 \text{ d}^{-1}}\right)$$
$$\times \left(10^{-(0.199 \text{ d}^{-1})(0.3122 \text{ d})}\right)$$
$$= 3.20 \text{ mg/L}$$
$$\text{DO} = \text{DO}_{\text{sat}} - D_c$$
$$= 7.92 \frac{\text{mg}}{\text{L}} - 3.20 \frac{\text{mg}}{\text{L}}$$
$$= \boxed{4.72 \text{ mg/L}}$$

This is below the state standard of 5 mg/L.

The answer is (B).

(d) From Eq. 25.35, the composite dissolved oxygen is

$$\text{DO} = \frac{Q_w\text{DO}_w + Q_r\text{DO}_r}{Q_w + Q_r}$$
$$= \frac{\left(0.280 \frac{\text{m}^3}{\text{s}}\right)\left(1 \frac{\text{mg}}{\text{L}}\right) + \left(0.877 \frac{\text{m}^3}{\text{s}}\right)\left(6 \frac{\text{mg}}{\text{L}}\right)}{0.280 \frac{\text{m}^3}{\text{s}} + 0.877 \frac{\text{m}^3}{\text{s}}}$$
$$= 4.79 \text{ mg/L} \quad [\text{no change}]$$

Calculate the temperature of the river water and wastewater mixture.

$$T_1 = \frac{Q_w T_w + Q_r T_r}{Q_w + Q_r}$$
$$= \frac{\left(0.280 \frac{\text{m}^3}{\text{s}}\right)(28°\text{C}) + \left(0.877 \frac{\text{m}^3}{\text{s}}\right)(12°\text{C})}{0.280 \frac{\text{m}^3}{\text{s}} + 0.877 \frac{\text{m}^3}{\text{s}}}$$
$$= 15.87°\text{C}$$

Calculate K_d at this temperature. Two different values of θ_d are used for the two temperature ranges. Use Eq. 25.29, written for base-10 constants.

$$K_{d,T_1} = K_{d,T_2}\theta^{T_1 - T_2}$$
$$K_{d,20°\text{C}} = (0.199 \text{ d}^{-1})(1.056)^{20°\text{C}-28°\text{C}}$$
$$= 0.1287 \text{ d}^{-1}$$
$$K_{d,T_1} = K_{d,T_2}\theta^{T_1 - T_2}$$
$$K_{d,15.87°\text{C}} = (0.1287 \text{ d}^{-1})(1.135)^{15.87°\text{C}-20°\text{C}}$$
$$= 0.07629 \text{ d}^{-1}$$

Similarly, use Eq. 25.25 to correct K_r.

$$K_{r,T_1} = K_{r,T_2}\theta^{T_1 - T_2}$$
$$K_{r,15.87°\text{C}} = K_{r,28°\text{C}}(1.024)^{15.87°\text{C}-28°\text{C}}$$
$$= (0.370 \text{ d}^{-1})(1.024)^{15.87°\text{C}-28°\text{C}}$$
$$= 0.2775 \text{ d}^{-1}$$

Correct BOD_u to the new temperature. Use Eq. 25.33 twice.

$$\text{BOD}_{u,20°\text{C}} = \frac{\text{BOD}_{u,28°\text{C}}}{(0.02)(28°\text{C}) + 0.6}$$
$$\text{BOD}_{u,15.87°\text{C}} = (\text{BOD}_{u,20°\text{C}})\big((0.02)(15.87°\text{C}) + 0.6\big)$$
$$= \frac{\left(6.8645 \frac{\text{mg}}{\text{L}}\right)\big((0.02)(15.87°\text{C}) + 0.6\big)}{(0.02)(28°\text{C}) + 0.6}$$
$$= 5.43 \text{ mg/L}$$

At 15.87°C, the saturation dissolved oxygen is found from App. 22.D. $DO_{sat,15.87°C} = 9.98$ mg/L.

Calculate the initial deficit from Eq. 25.17.

$$D_0 = DO_{sat} - DO = 9.98 \frac{mg}{L} - 4.79 \frac{mg}{L}$$
$$= 5.19 \text{ mg/L}$$

Calculate the deficit at town B. Use Eq. 25.36.

$$D_B = \left(\frac{K_d BOD_u}{K_r - K_d}\right)(10^{-K_d t} - 10^{-K_r t}) + D_0(10^{-K_r t})$$

$$= \left(\frac{(0.07629 \text{ d}^{-1})\left(5.43 \frac{mg}{L}\right)}{0.2775 \text{ d}^{-1} - 0.07629 \text{ d}^{-1}}\right)$$

$$\times \left(\begin{array}{c} 10^{-(0.07629 \text{ d}^{-1})(0.1031 \text{ d})} \\ - 10^{-(0.2775 \text{ d}^{-1})(0.1031 \text{ d})} \end{array}\right)$$

$$+ \left(5.19 \frac{mg}{L}\right)\left(10^{-(0.2775 \text{ d}^{-1})(0.1031 \text{ d})}\right)$$

$$= 4.95 \text{ mg/L}$$

The dissolved oxygen at town B is

$$DO_B = DO_{sat} - D_B$$
$$= 9.98 \frac{mg}{L} - 4.95 \frac{mg}{L}$$
$$= \boxed{5.03 \text{ mg/L} \quad (5.0 \text{ mg/L})}$$

This meets the state standard of 5 mg/L.

The answer is (C).

4. (a) $\quad T = 20°C + 2°C = 22°C$

$T_1 = 20°C$

$Q_1 + Q_2 = 1200$ m³/s

$Q_2 = 60$ m³/s

$Q_1 = 1200 \frac{m^3}{s} - 60 \frac{m^3}{s} = 1140$ m³/s

$T_f = \frac{T_1 Q_1 + T_2 Q_2}{Q_1 + Q_2}$

$$22°C = \frac{(20°C)\left(1140 \frac{m^3}{s}\right) + C_2\left(60 \frac{m^3}{s}\right)}{1200 \frac{m^3}{s}}$$

$$= \boxed{60°C}$$

The answer is (B).

(b)
$$C = 0 + 0.002 \frac{mg}{L}$$

$$C_1 = 0$$

$$C_f = \frac{C_1 Q_1 + C_2 Q_2}{Q_1 + Q_2}$$

$$0.002 \frac{mg}{L} = \frac{(0)\left(1140 \frac{m^3}{s}\right) + C_2\left(60 \frac{m^3}{s}\right)}{1200 \frac{m^3}{s}}$$

$$= \boxed{0.04 \text{ mg/L}}$$

The answer is (A).

5. (a) The domestic BOD concentration is

$$BOD = \frac{\left(0.191 \frac{lbm}{capita\text{-}day}\right)\left(10^6 \frac{gal}{MG}\right)}{122 \frac{gal}{capita\text{-}day}} \times \left(0.1198 \frac{mg\text{-}MG}{L\text{-}lbm}\right)$$

$$= \boxed{187.6 \text{ mg/L}}$$

The answer is (C).

(b) Use Eq. 25.35.

$$BOD = \frac{\sum Q_i BOD_i}{\sum Q_i}$$

$$= \frac{\begin{array}{c}\left(122 \frac{gal}{day}\right)(10{,}000)\left(187.6 \frac{mg}{L}\right) \\ + \left(116{,}000 \frac{gal}{day}\right)(0) \\ + \left(180{,}000 \frac{gal}{day}\right)\left(800 \frac{mg}{L}\right) \\ + \left(76{,}000 \frac{gal}{day}\right)\left(1700 \frac{mg}{L}\right)\end{array}}{\begin{array}{c}\left(122 \frac{gal}{day}\right)(10{,}000) + 116{,}000 \frac{gal}{day} \\ + 0 + 180{,}000 \frac{gal}{day} + 76{,}000 \frac{gal}{day}\end{array}}$$

$$= \boxed{315.4 \text{ mg/L}}$$

The answer is (D).

(c) Use Eq. 25.35.

$$T_{\circ F} = \frac{\sum Q_i T_i}{\sum Q_i}$$

$$= \frac{\begin{pmatrix} \left(122 \ \frac{gal}{day}\right)(10{,}000)(64°F) \\ + \left(116{,}000 \ \frac{gal}{day}\right)(51°F) \\ + \left(180{,}000 \ \frac{gal}{day}\right)(95°F) \\ + \left(76{,}000 \ \frac{gal}{day}\right)(84°F) \end{pmatrix}}{\begin{pmatrix} \left(122 \ \frac{gal}{day}\right)(10{,}000) + 116{,}000 \ \frac{gal}{day} \\ + 180{,}000 \ \frac{gal}{day} + 76{,}000 \ \frac{gal}{day} \end{pmatrix}}$$

$$= 67.5°F$$

$$T_{\circ C} = \tfrac{5}{9}(T_{\circ F} - 32°) = \left(\tfrac{5}{9}\right)(67.5°F - 32°F)$$

$$= \boxed{19.7°C}$$

The answer is (A).

(d) The total discharge into the river is

$$\begin{pmatrix} \left(122 \ \frac{gal}{day}\right)(10{,}000) \\ + 116{,}000 \ \frac{gal}{day} \\ + 180{,}000 \ \frac{gal}{day} \\ + 76{,}000 \ \frac{gal}{day} \end{pmatrix} \left(1.547 \times 10^{-6} \ \frac{ft^3\text{-}day}{sec\text{-}day}\right)$$

$$= 2.46 \ ft^3/sec$$

step 1: Find the stream conditions immediately after mixing.

$$BOD_{5,20°C} = \frac{\left(2.46 \ \frac{ft^3}{sec}\right)\left(315.4 \ \frac{mg}{L}\right) + \left(120 \ \frac{ft^3}{sec}\right)(0)}{2.46 \ \frac{ft^3}{sec} + 120 \ \frac{ft^3}{sec}}$$

$$= 6.34 \ mg/L$$

$$DO = \frac{\left(2.46 \ \frac{ft^3}{sec}\right)(0) + \left(120 \ \frac{ft^3}{sec}\right)\left(7.5 \ \frac{mg}{L}\right)}{2.46 \ \frac{ft^3}{sec} + 120 \ \frac{ft^3}{sec}}$$

$$= 7.35 \ mg/L$$

$$T = \frac{\left(2.46 \ \frac{ft^3}{sec}\right)(19.7°C) + \left(120 \ \frac{ft^3}{sec}\right)(15°C)}{2.46 \ \frac{ft^3}{sec} + 120 \ \frac{ft^3}{sec}}$$

$$= 15.1°C$$

step 2: Calculate the rate constants at 15.1°C. Use Eq. 25.28, written for base-10 constants.

$$K_{d,T} = K_{d,20°C} \ \theta^{T-20°C}$$

$$K_{d,15.1°C} = (0.15 \ day^{-1})(1.135)^{15.1°C-20°C}$$

$$= 0.0807 \ day^{-1}$$

$$K_{r,15.1°C} = (0.2 \ day^{-1})(1.024)^{15.1°C-20°C}$$

$$= 0.178 \ day^{-1}$$

step 3: Estimate BOD_u. Use Eq. 25.31.

$$BOD_{u,20°C} = \frac{BOD_t}{1 - 10^{-K_d t}}$$

$$= \frac{6.34 \ \frac{mg}{L}}{1 - 10^{-(0.15 \ day^{-1})(5 \ days)}}$$

$$= 7.71 \ mg/L$$

Use Eq. 25.33 to convert BOD_u to 15.1°C.

$$BOD_{u,15.1°C} = BOD_{u,20°C}(0.02 \ T_{\circ C} + 0.6)$$

$$= \left(7.71 \ \frac{mg}{L}\right)((0.02)(15.1°C) + 0.6)$$

$$= 6.95 \ mg/L$$

step 4: From App. 22.D at 15°C, saturated DO = 10.15 mg/L. Since the actual is 7.35 mg/L, the deficit is

$$D_0 = DO_{sat} - DO = 10.15 \ \frac{mg}{L} = 7.35 \ \frac{mg}{L}$$

$$= 2.8 \ mg/L$$

step 5: Calculate t_c. Use Eq. 25.38.

$$t_c = \left(\frac{1}{K_r - K_d}\right)$$
$$\times \log_{10}\left(\left(\frac{K_d BOD_u - K_r D_0 + K_d D_0}{K_d BOD_u}\right)\left(\frac{K_r}{K_d}\right)\right)$$

$$= \left(\frac{1}{0.178 \text{ day}^{-1} - 0.0807 \text{ day}^{-1}}\right)$$

$$\times \log_{10}\left(\begin{array}{c}\left(\dfrac{\begin{array}{c}(0.0807 \text{ day}^{-1})\left(6.95 \frac{mg}{L}\right)\\[4pt] - (0.178 \text{ day}^{-1})\left(2.8 \frac{mg}{L}\right)\\[4pt] + (0.0807 \text{ day}^{-1})\left(2.8 \frac{mg}{L}\right)\end{array}}{(0.0807 \text{ day}^{-1})\left(6.95 \frac{mg}{L}\right)}\right)\\[30pt] \times \left(\dfrac{0.178 \text{ day}^{-1}}{0.0807 \text{ day}^{-1}}\right)\end{array}\right)$$

$$= 0.562 \text{ day}$$

step 6: The distance downstream is

$$(0.562 \text{ day})\left(3 \frac{mi}{hr}\right)\left(24 \frac{hr}{day}\right) = \boxed{40.5 \text{ mi}}$$

The answer is (A).

step 7: Use Eq. 25.39.

$$D_c = \left(\frac{K_d BOD_u}{K_r}\right)10^{-K_d t_c}$$

$$= \left(\frac{(0.0807 \text{ day}^{-1})\left(6.95 \frac{mg}{L}\right)}{0.178 \text{ day}^{-1}}\right)$$
$$\times \left(10^{-(0.0807 \text{ day}^{-1})(0.562 \text{ day})}\right)$$
$$= 2.84 \text{ mg/L}$$

step 8:

$$DO_{min} = DO_{sat} - D_c$$
$$= 10.15 \frac{mg}{L} - 2.84 \frac{mg}{L}$$
$$= \boxed{7.31 \text{ mg/L}}$$

The answer is (C).

6. (a) Do not apply the population expansion factor to the industrial effluents. Use Eq. 25.3, modified for the given population equivalent (in thousands of people) for the domestic flow contribution.

$$P_{e,1000s} = P_{\text{domestic flow}} + P_{\text{industrial source 1}} + P_{\text{industrial source 2}}$$
$$= (20)(1.15)$$
$$+ \frac{\left(1100 \frac{mg}{L}\right)\left(1.3 \times 10^6 \frac{gal}{day}\right) \times \left(8.345 \times 10^{-9} \frac{lbm\text{-}L}{MG\text{-}mg}\right)}{0.18 \frac{lbm}{day\text{-}person}}$$
$$+ \frac{(500)\left(1.0 \times 10^6 \frac{gal}{day}\right) \times \left(8.345 \times 10^{-9} \frac{lbm\text{-}L}{MG\text{-}mg}\right)}{0.18 \frac{lbm}{day\text{-}person}}$$
$$= \boxed{112.5 \quad (112{,}500)}$$

The answer is (D).

(b) Since the plant loading is requested, the organic loading can be given in lbm/day. The population from part (a) is 112,500.

$$L_{BOD} = (112{,}500)\left(0.18 \frac{lbm}{day}\right) = \boxed{20{,}250 \text{ lbm/day}}$$

The answer is (D).

(c) $\quad L_H = (20{,}000)(1.15)\left(100 \frac{gal}{day}\right)$
$$+ 1.3 \times 10^6 \frac{gal}{day} + 1.0 \times 10^6 \frac{gal}{day}$$
$$= \boxed{4.6 \times 10^6 \text{ gal/day} \quad (4.6 \text{ MGD})}$$

The answer is (A).

26 Wastewater Treatment: Equipment and Processes

PRACTICE PROBLEMS

Lagoons

1. A cheese factory located in a normally warm state has liquid waste with the following characteristics.

> *waste no. 1*
> volume   10,000 gal/day
> BOD   1000 mg/L
> *waste no. 2*
> volume   25,000 gal/day
> BOD   250 mg/L

The factory will use a 4 ft deep, on-site, nonaerated lagoon to stabilize the waste.

(a) What is the total BOD loading?

(A) 60 lbm/day

(B) 85 lbm/day

(C) 110 lbm/day

(D) 140 lbm/day

(b) What lagoon size is required if the BOD loading is 20 lbm BOD/ac-day?

(A) 2.7 ac

(B) 6.8 ac

(C) 8.3 ac

(D) 10.9 ac

(c) What is the detention time?

(A) 5 wk

(B) 14 wk

(C) 36 wk

(D) 52 wk

Trickling Filters

2. It is estimated that the BOD of raw sewage received at a treatment plant serving a population of 20,000 will be 300 mg/L. It is estimated that the per capita BOD loading is 0.17 lbm/day. 30% of the influent BOD is removed by settling. One single-stage high-rate trickling filter is to be used to reduce the plant effluent to 50 mg/L. Recirculation is from the filter effluent to the primary settling influent. *Ten States' Standards* is in effect.

(a) What is the design flow rate?

(A) 0.9 MGD

(B) 1.1 MGD

(C) 1.4 MGD

(D) 2.0 MGD

(b) Assuming a design flow rate of 1.35 MGD, what is the total organic load on the filter?

(A) 1700 lbm/day

(B) 2100 lbm/day

(C) 2400 lbm/day

(D) 3400 lbm/day

(c) Assuming an incoming volume of 1.35 MGD and an organic loading of 60 lbm/day-1000 ft^3, what flow should be recirculated?

(A) 0.6 MGD

(B) 0.9 MGD

(C) 1.2 MGD

(D) 2.2 MGD

(d) What is the overall plant efficiency?

(A) 45%

(B) 76%

(C) 83%

(D) 91%

3. The average wastewater flow from a community of 20,000 is 125 gpcd. The 5-day, 20°C BOD is 250 mg/L. The suspended solids content is 300 mg/L. A final plant effluent of 50 mg/L of BOD is to be achieved through the use of two sets of identical settling tanks and trickling filters operating in parallel. The settling tanks are to be designed to a standard of 1000 gpd/ft^2. The trickling filters are to be 6 ft deep. There is no recirculation.

(a) What settling tank surface area is required?

(A) 2500 ft^2

(B) 3000 ft^2

(C) 3500 ft^2

(D) 4500 ft^2

(b) What settling tank diameter is required?

(A) 40 ft

(B) 50 ft

(C) 65 ft

(D) 80 ft

(c) Estimate the BOD removal in the settling tanks.

(A) 15%

(B) 30%

(C) 45%

(D) 60%

(d) What is the trickling filter diameter?

(A) 65 ft

(B) 75 ft

(C) 85 ft

(D) 95 ft

4. Wastewater from a city with a population of 40,000 has an average daily flow of 4.4 MGD. The sewage has the following characteristics.

BOD$_5$ at 20°C	160 mg/L
COD	800 mg/L
total solids	900 mg/L
suspended solids	180 mg/L
volatile solids	320 mg/L
settleable solids	8 mg/L
pH	7.8

The wastewater is to be treated with primary settling and secondary trickling filtration. The settling basins are to be circular, 8 ft deep, and designed to a standard overflow rate of 1000 gal/day-ft^2.

(a) Assuming two identical basins operating in parallel, what should be the diameter of each sedimentation basin in order to remove 30% of the BOD?

(A) 45 ft

(B) 53 ft

(C) 77 ft

(D) 86 ft

(b) What is the detention time?

(A) 1.4 hr

(B) 1.8 hr

(C) 2.3 hr

(D) 3.1 hr

(c) What is the weir loading?

(A) 8200 gpd/ft

(B) 11,000 gpd/ft

(C) 13,000 gpd/ft

(D) 15,000 gpd/ft

5. (*Time limit: one hour*) A small community has a projected average flow of 1 MGD, with a peaking factor of 2.0. Incoming wastewater has the following properties: BOD, 250 mg/L; grit specific gravity, 2.65; and total suspended solids, 400 mg/L. The community wants to have a wastewater treatment plant consisting of a single aerated grit chamber, a single primary clarifier, two identical circular trickling filters in parallel, and a single secondary clarifier. There will be no equalization basin. Recirculation from the second clarifier to the entrance of the trickling filters will be 100% of the average flow. The final effluent is to have a BOD of 30 mg/L.

(a) What is the peak design flow at the grit chamber?

(A) 1.0 MGD

(B) 1.5 MGD

(C) 2.0 MGD

(D) 2.5 MGD

(b) Determine the aerated grit chamber width assuming the following: a 3 min detention time, a 20 ft length, and a width:depth ratio of 1.25.

(A) 4 ft

(B) 6 ft

(C) 8 ft

(D) 10 ft

(c) Determine the approximate air requirements for the grit chamber in order to capture approximately 95% of the grit.

(A) 60 cfm

(B) 140 cfm

(C) 280 cfm

(D) 420 cfm

(d) If the clarifier is 12 ft deep, what should be its diameter?

(A) 50 ft

(B) 54 ft

(C) 63 ft

(D) 81 ft

(e) What is the peak flow entering the trickling filters?

(A) 1.5 MGD

(B) 2.0 MGD

(C) 3.0 MGD

(D) 4.0 MGD

(f) Assume the primary clarifier removes 30% of the incoming BOD, and the filters see the average flow only. Determine the diameter of the trickling filters assuming a 6 ft deep rock bed.

(A) 45 ft

(B) 55 ft

(C) 65 ft

(D) 85 ft

(g) Determine the diameter of the final clarifier.

(A) 35 ft

(B) 50 ft

(C) 65 ft

(D) 80 ft

Recirculating Biological Contactors

6. (*Time limit: one hour*) 1.5 MGD of wastewater with a BOD of 250 mg/L is processed by a high-rate rock trickling filter followed by recirculating biological contactor (RBC) processing. The trickling filter is 75 ft in diameter and 6 ft deep and was designed to NRC standards. The BOD removal efficiency of the RBC process is given by the following equation. (k is 2.45 gal/day-ft^2, and Q (in units of gal/day) does not include recirculation. A is the immersed area of the RBC in ft^2.)

$$\eta_{BOD} = 1 - \frac{1}{\left(1 + \dfrac{kA}{Q}\right)^3}$$

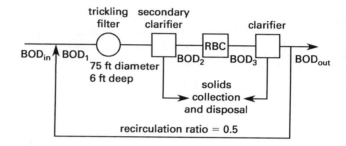

(a) What total RBC surface area is required to achieve an effluent BOD$_{out}$ of 30 mg/L?

(A) 4.2×10^4 ft^2

(B) 8.4×10^4 ft^2

(C) 1.3×10^5 ft^2

(D) 2.0×10^5 ft^2

(b) The recirculation pick-up point is relocated from after the final clarifier to after the trickling filter. The efficiency of the RBC process is 65%. Determine the recirculation ratio such that BOD$_{out}$ is 30 mg/L.

(A) 25%

(B) 50%

(C) 75%

(D) 100%

(c) If BOD$_2 = 85$ mg/L, BOD$_{out} = 30$ mg/L, the yield is 0.4 lbm/lbm BOD removed, and the sludge specific gravity is essentially 1.0, what is the approximate sludge volume produced from the clarifiers?

(A) 4 ft^3/day

(B) 12 ft^3/day

(C) 21 ft^3/day

(D) 35 ft^3/day

SOLUTIONS

1. (a) The total BOD is

$$\left(1000 \ \frac{\text{mg}}{\text{L}}\right)\left(8.345 \ \frac{\text{lbm-L}}{\text{mg-MG}}\right)\left(\frac{10{,}000 \ \dfrac{\text{gal}}{\text{day}}}{1{,}000{,}000 \ \dfrac{\text{gal}}{\text{MG}}}\right)$$

$$+ \left(250 \ \frac{\text{mg}}{\text{L}}\right)\left(8.345 \ \frac{\text{lbm-L}}{\text{mg-MG}}\right)\left(\frac{25{,}000 \ \dfrac{\text{gal}}{\text{day}}}{1{,}000{,}000 \ \dfrac{\text{gal}}{\text{MG}}}\right)$$

$$= \boxed{135.6 \ \text{lbm/day} \quad (140 \ \text{lbm/day})}$$

The answer is (D).

(b) The warm weather and depth contribute to a decrease in pond effectiveness. Assume 20 lbm BOD/ac-day for a nonaerated stabilization pond. From Eq. 26.4, the required area is

$$A = \frac{Q}{v^*} = \frac{135.6 \ \dfrac{\text{lbm}}{\text{day}}}{20 \ \dfrac{\text{lbm}}{\text{ac-day}}} = \boxed{6.78 \ \text{ac} \quad (6.8 \ \text{ac})}$$

The answer is (B).

(c) Use Eq. 26.5.

$$t_d = \frac{V}{Q}$$

$$= \frac{(6.78 \ \text{ac})\left(43{,}560 \ \dfrac{\text{ft}^2}{\text{ac}}\right)(4 \ \text{ft})\left(7.48 \ \dfrac{\text{gal}}{\text{ft}^3}\right)}{\left(35{,}000 \ \dfrac{\text{gal}}{\text{day}}\right)\left(7 \ \dfrac{\text{days}}{\text{wk}}\right)}$$

$$= \boxed{36.07 \ \text{wk} \quad (36 \ \text{wk})}$$

The answer is (C).

2. (a) The design flow rate is

$$Q = \frac{\dot{m}}{C} = \frac{\left(0.17 \ \dfrac{\text{lbm}}{\text{capita-day}}\right)(20{,}000 \ \text{people})}{\left(8.345 \times 10^{-6} \ \dfrac{\text{lbm-L}}{\text{gal-mg}}\right)\left(300 \ \dfrac{\text{mg}}{\text{L}}\right)}$$

$$= \boxed{1.358 \times 10^6 \ \text{gal/day} \quad (1.4 \ \text{MGD})}$$

(*Ten States' Standards* specifies 100 gpcd in the absence of other information. In such a case, $Q = (100 \ \text{gal/capita-day})(20{,}000 \ \text{people}) = 2 \times 10^6 \ \text{gal/day}$.)

The answer is (C).

(b) The total BOD load leaving the primary clarifier and entering the filter is

$$\text{BOD}_i = (1 - 0.30)\left(300 \ \frac{\text{mg}}{\text{L}}\right) = 210 \ \text{mg/L}$$

$$L_{\text{BOD}} = (1.35 \ \text{MGD})\left(210 \ \frac{\text{mg}}{\text{L}}\right)\left(8.345 \ \frac{\text{lbm-L}}{\text{mg-MG}}\right)$$

$$= \boxed{2366 \ \text{lbm/day} \quad (2400 \ \text{lbm/day})}$$

The answer is (C).

(c) The efficiency of the filter and secondary clarifier is found from Eq. 26.9.

$$\eta = \frac{S_{\text{ps}} - S_o}{S_{\text{ps}}} = \frac{210 \ \dfrac{\text{mg}}{\text{L}} - 50 \ \dfrac{\text{mg}}{\text{L}}}{210 \ \dfrac{\text{mg}}{\text{L}}}$$

$$= 0.762 \quad (76.2\%)$$

Use Eq. 26.13.

$$\eta = \frac{1}{1 + 0.0561\sqrt{\dfrac{L_{\text{BOD}}}{F}}}$$

$$0.762 = \frac{1}{1 + 0.0561\sqrt{\dfrac{60 \ \dfrac{\text{lbm}}{\text{day-1000 ft}^3}}{F}}}$$

$$F = 1.936$$

Use Eq. 26.15.

$$F = \frac{1 + R}{(1 + wR)^2}$$

$$1.936 = \frac{1 + R}{(1 + 0.1R)^2}$$

$$R = 1.6$$

Use Eq. 26.10.

$$R = \frac{Q_r}{Q_w}$$

$$= (1.6)(1.35 \ \text{MGD})$$

$$= \boxed{2.16 \ \text{MGD} \quad (2.2 \ \text{MGD})}$$

The answer is (D).

(d) Use Eq. 26.9.

$$\eta = \frac{S_{ps} - S_o}{S_{ps}}$$

$$= \frac{300 \, \frac{mg}{L} - 50 \, \frac{mg}{L}}{300 \, \frac{mg}{L}}$$

$$= \boxed{0.833 \quad (83\%)}$$

The answer is (C).

3. (a) Use Eq. 26.4. Disregarding variations in peak flow, the average design volume is

$$v^* = \frac{Q}{A} = \frac{\left(125 \, \frac{gal}{capita\text{-}day}\right)(20{,}000 \text{ people})}{1{,}000{,}000 \, \frac{gal}{MG}}$$

$$= 2.5 \text{ MGD}$$

The settling tank surface area is

$$\frac{2.5 \times 10^6 \, \frac{gal}{day}}{1000 \, \frac{gal}{day\text{-}ft^2}} = \boxed{2500 \text{ ft}^2}$$

The answer is (A).

(b) The required diameter when using two tanks in parallel is

$$D = \sqrt{\frac{(4)(2500 \text{ ft}^2)}{2\pi}} = \boxed{39.9 \text{ ft (use 40 ft) each}}$$

The answer is (A).

(c) A 30% removal is typical.

The answer is (B).

(d) BOD entering the filter is

$$(1 - 0.30)\left(250 \, \frac{mg}{L}\right) = 175 \text{ mg/L}$$

The filter efficiency is

$$\eta = \frac{175 \, \frac{mg}{L} - 50 \, \frac{mg}{L}}{175 \, \frac{mg}{L}} = 0.71$$

From Fig. 26.5 with 71% efficiency and $R = 0$, $L_{BOD} = 55$ lbm/day-1000 ft^3.

The total load is found from Eq. 26.12 (rearranged in terms of V_1).

$$V_1 = \frac{QS\left(8.345 \, \frac{lbm\text{-}L}{MG\text{-}mg}\right)(1000)}{L_{BOD}}$$

$$= \frac{(2.5 \text{ MGD})\left(175 \, \frac{mg}{L}\right)}{\left(55 \, \frac{lbm}{day\text{-}1000 \text{ ft}^3}\right)(2 \text{ filters})}$$

$$= 33{,}190 \text{ ft}^3/\text{filter}$$

With a depth of 6 ft, the total required surface area is

$$A = \frac{V}{Z} = \frac{33{,}190 \text{ ft}^3}{6 \text{ ft}} = 5532 \text{ ft}^2$$

The required diameter per filter is

$$D = \sqrt{\frac{4A}{\pi}} = \sqrt{\frac{(4)(5532 \text{ ft}^2)}{\pi}}$$

$$= \boxed{83.9 \text{ ft} \quad (85 \text{ ft})}$$

The answer is (C).

4. (a) There is nothing particularly special about a basin that removes 30% BOD. Choose two basins in parallel, each working with half of the total flow. Choose an overflow rate of 1000 gpd/ft^2. The area per basin is given by Eq. 26.4.

$$A = \frac{Q}{v^*} = \frac{(4.4 \text{ MGD})\left(10^6 \, \frac{gal}{MG}\right)}{(2 \text{ basins})\left(1000 \, \frac{gal}{day\text{-}ft^2}\right)}$$

$$= 2200 \text{ ft}^2$$

$$D = \sqrt{\frac{4}{\pi} A} = \sqrt{\left(\frac{4}{\pi}\right)(2200 \text{ ft}^2)}$$

$$= \boxed{52.9 \text{ ft} \quad (53 \text{ ft})}$$

The answer is (B).

Water Treatment

(b) The detention time is given by Eq. 26.5.

$$t_d = \frac{V}{Q}$$

$$= \frac{(2200 \text{ ft}^2)(8 \text{ ft})\left(7.48 \ \frac{\text{gal}}{\text{ft}^3}\right)\left(24 \ \frac{\text{hr}}{\text{day}}\right)}{\left(2.2 \ \frac{\text{MGD}}{\text{tank}}\right)\left(10^6 \ \frac{\text{gal}}{\text{MG}}\right)}$$

$$= \boxed{1.436 \text{ hr} \quad (1.4 \text{ hr})}$$

The answer is (A).

(c) Find the circumference of the basins.

$$\text{circumference} = \pi D = \pi(52.9 \text{ ft})$$
$$= 166.2 \text{ ft}$$

The weir loading is

$$\text{weir loading} = \frac{2.2 \times 10^6 \ \frac{\text{gal}}{\text{day}}}{166.2 \text{ ft}}$$

$$= \boxed{13{,}237 \text{ gal/day-ft} \quad (13{,}000 \text{ gpd/ft})}$$

The answer is (C).

5.

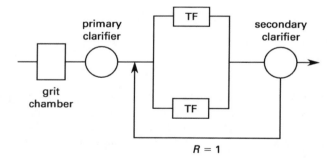

$$R = 1$$

(a) The peak flow is

$$(2)(1 \text{ MGD}) = \boxed{2 \text{ MGD}}$$

The answer is (C).

(b) The peak flow rate per second is

$$Q_{\text{peak}} = \frac{(2)(1 \text{ MGD})\left(10^6 \ \frac{\text{gal}}{\text{MG}}\right)\left(0.1337 \ \frac{\text{ft}^3}{\text{gal}}\right)}{\left(24 \ \frac{\text{hr}}{\text{day}}\right)\left(60 \ \frac{\text{min}}{\text{hr}}\right)\left(60 \ \frac{\text{sec}}{\text{min}}\right)}$$

$$= 3.095 \text{ ft}^3/\text{sec}$$

With a detention time of 3 min, the volume of the grit chamber would be

$$V = Qt = \left(3.095 \ \frac{\text{ft}^3}{\text{sec}}\right)\left(60 \ \frac{\text{sec}}{\text{min}}\right)(3 \text{ min}) = 557.1 \text{ ft}^3$$

$$557.1 \text{ ft}^3 = (\text{length})(\text{width})(\text{depth})$$
$$= (20 \text{ ft})(1.25)(\text{depth})^2$$
$$\text{water depth} = 4.72 \text{ ft} \quad [\text{round up to } 4\tfrac{3}{4} \text{ ft}]$$
$$\text{chamber width} = (1.25)(4.72 \text{ ft})$$
$$= \boxed{5.9 \text{ ft} \quad (6 \text{ ft})}$$

The answer is (C).

(c) This is a shallow chamber. Use 3 cfm/ft.

$$\left(3 \ \frac{\text{ft}^3}{\text{min-ft}}\right)(20 \text{ ft}) = \boxed{60 \text{ ft}^3/\text{min (cfm) of air}}$$

The answer is (A).

(d) (Note that the *Ten States' Standards* requires two basins.) Surface loading is the primary design parameter. Choose a surface loading of 1000 gal/day-ft². The diameter is

$$A = \frac{Q}{v^*} = \frac{(2 \text{ MGD})\left(10^6 \ \frac{\text{gal}}{\text{day-MGD}}\right)}{1000 \ \frac{\text{gal}}{\text{day-ft}^2}}$$

$$= 2000 \text{ ft}^2$$

$$D = \sqrt{\frac{4A}{\pi}} = \sqrt{\frac{(4)(2000 \text{ ft}^2)}{\pi}}$$

$$= \boxed{50.5 \text{ ft} \quad (50 \text{ ft})}$$

The answer is (A).

(e) The volume of the primary clarifier is fixed by the weir height. Therefore, the clarifier does not provide any storage (i.e., no damping of the flow rates). The fluctuations in flow will be passed on to the trickling filters.

$$Q_p + Q_r = 2 \text{ MGD} + 1 \text{ MGD} = 3 \text{ MGD}$$

The answer is (C).

(f) First, assume the primary sedimentation basin removes 30% of the BOD. Find the BOD incoming to the trickle filter to solve for the water flow quantity, Q_w.

$$\text{BOD}_{in} = (1 - 0.30)\left(250 \ \frac{mg}{L}\right) = 175 \ mg/L$$

$$Q_w = \left(175 \ \frac{mg}{L}\right)\left(8.345 \ \frac{lbm\text{-}L}{mg\text{-}MG}\right)(1 \ MGD)$$
$$= 1460 \ lbm/day$$

Second, find the required trickling filter-clarifier process efficiency.

$$\eta = \frac{\text{BOD}_{in} - \text{BOD}_{out}}{\text{BOD}_{in}} = \frac{175 \ \frac{mg}{L} - 30 \ \frac{mg}{L}}{175 \ \frac{mg}{L}}$$
$$= 0.829 \quad (82.9\%)$$

From Eq. 26.15 with $w = 0.1$ and $R = 1$,

$$F = \frac{1 + R}{(1 + wR)^2} = \frac{1 + 1}{\left(1 + (0.1)(1)\right)^2} = 1.65$$

From Eq. 26.13,

$$\eta = \frac{1}{1 + 0.0561\sqrt{\dfrac{L_{BOD}}{F}}}$$

$$0.83 = \frac{1}{1 + 0.0561\sqrt{\dfrac{L_{BOD}}{1.65}}}$$

$$L_{BOD} \approx 22 \ lbm/day\text{-}1000 \ ft^3$$

Third, find the filter volume.

$$V = \frac{\dot{m}_{BOD}}{L_{BOD}}$$
$$= \frac{\left(1460 \ \frac{lbm}{day}\right)\left(1000 \ \frac{ft^3}{1000 \ ft^3}\right)}{\left(22 \ \frac{lbm}{day\text{-}1000 \ ft^3}\right)(2 \ filters)}$$
$$= 33,182 \ ft^3$$

Finally, for 6 ft deep rock bed, the diameter for the trickling filters is

$$D = \sqrt{\frac{4}{\pi} \frac{V}{Z}} = \sqrt{\frac{\left(\frac{4}{\pi}\right)(33,182 \ ft^3)}{6 \ ft}}$$
$$= \boxed{83.9 \ ft \quad (85 \ ft)}$$

Use two 85 ft diameter, 6 ft deep filters.

The answer is (D).

(g) For the final clarifier, the maximum overflow rate is 1100 gpd/ft^2, and the minimum depth is 10 ft. Assume volumetric flow fluctuations will be damped out by previous processes.

$$A = \frac{1 \times 10^6 \ \dfrac{gal}{day}}{1100 \ \dfrac{gal}{day\text{-}ft^2}} = 909 \ ft^2$$

$$D = \sqrt{\frac{4}{\pi} A} = \sqrt{\left(\frac{4}{\pi}\right)(909 \ ft^2)}$$
$$= \boxed{34.0 \ ft \quad (35 \ ft)}$$

Use a 35 ft diameter, 10 ft deep basin.

The answer is (A).

6. (a) Assume the last clarifier removes only sloughed off biological material and removes no BOD.

$$\text{BOD}_3 = 30 \ mg/L$$

There is no recirculation that matches the NRC model. The NRC model "recirculation" is from the trickling filter discharge directly back to the entrance to the filter. This problem's recirculation is from several processes beyond the trickling filter. The recirculation increases the BOD loading and dilutes the influent.

The BOD loading, L_{BOD}, to the trickling filter must include recirculation, L_r. Use Eq. 26.12.

$$L_{BOD} + L_r = \frac{\begin{array}{c}(1.5 \ MGD)\left(250 \ \frac{mg}{L}\right) \\ \times \left(8.345 \ \frac{lbm\text{-}L}{MG\text{-}mg}\right)\left(1000 \ \frac{ft^3}{1000 \ ft^3}\right)\end{array}}{\pi\left(\frac{75 \ ft}{2}\right)^2(6 \ ft)}$$
$$+ \frac{\begin{array}{c}(0.5)(1.5 \ MGD)\left(30 \ \frac{mg}{L}\right) \\ \times \left(8.345 \ \frac{lbm\text{-}L}{mg\text{-}MG}\right)\left(1000 \ \frac{ft^3}{1000 \ ft^3}\right)\end{array}}{\pi\left(\frac{75 \ ft}{2}\right)^2(6 \ ft)}$$
$$= 118.1 \ \frac{lbm}{day\text{-}1000 \ ft^3} + 7.1 \ \frac{lbm}{day\text{-}1000 \ ft^3}$$
$$= 125.2 \ lbm/day\text{-}1000 \ ft^3$$

From Eq. 26.15, since $R = 0$, then $F = 1$.

The filter/clarifier efficiency is given by Eq. 26.13.

$$\eta = \frac{1}{1 + 0.0561 \sqrt{\dfrac{L_{BOD}}{F}}}$$

$$= \frac{1}{1 + 0.0561 \sqrt{\dfrac{125.2 \dfrac{\text{lbm}}{\text{day-1000 ft}^3}}{1.00}}}$$

$$= 0.61 \quad (61\%)$$

$$\text{BOD}_2 = (1 - 0.61)\left(250 \ \frac{\text{mg}}{\text{L}}\right) = 97.5 \ \text{mg/L}$$

The removal fraction in the RBC must be

$$\eta = \frac{\text{BOD}_2 - \text{BOD}_3}{\text{BOD}_2}$$

$$= \frac{97.5 \ \dfrac{\text{mg}}{\text{L}} - 30 \ \dfrac{\text{mg}}{\text{L}}}{97.5 \ \dfrac{\text{mg}}{\text{L}}}$$

$$= 0.69$$

Solving the given performance equation, the immersed area is

$$\eta_{BOD} = \frac{1}{\left(1 + \dfrac{kA}{Q}\right)^3}$$

$$0.69 = 1 - \frac{1}{\left(1 + \dfrac{2.45A}{1.5 \times 10^6}\right)^3}$$

$$A = 80{,}609 \ \text{ft}^2$$

From Table 26.11, only 40% of the total RBC area is immersed at one time. The total RBC area is

$$A_{\text{total}} = \frac{80{,}609 \ \text{ft}^2}{0.4} = \boxed{201{,}523 \ \text{ft}^2 \quad (2.0 \times 10^5 \ \text{ft}^2)}$$

The answer is (D).

(b) This is one of the recirculation modes to which the NRC model applies. Solve the problem backward to get $\text{BOD}_{\text{out}} = 30$ mg/L.

$$\text{BOD}_3 = 30 \ \text{mg/L}$$

$$\text{BOD}_2 = \frac{30 \ \dfrac{\text{mg}}{\text{L}}}{1 - 0.65} = 85.7 \ \text{mg/L}$$

The efficiency in the NRC model includes the effect of the clarifier, even though the recirculation occurs before the clarifier.

$$\eta_{\text{trickling filter and clarifier}} = \frac{\text{BOD}_1 - \text{BOD}_2}{\text{BOD}_1}$$

$$= \frac{250 \ \dfrac{\text{mg}}{\text{L}} - 85.7 \ \dfrac{\text{mg}}{\text{L}}}{250 \ \dfrac{\text{mg}}{\text{L}}}$$

$$= 0.657$$

In this configuration, the BOD loading does not include the effects of recirculation, as the NRC model places a higher emphasis on organic loading than on hydraulic loading. From part (a), $L_{BOD} = 118.1$ lbm/day-1000 ft^3.

Equation 26.13 could be solved for F, and then that value used in Eq. 26.15 to find R. It is easier to use Fig. 26.5 with $L_{BOD} = 118.1$ and $\eta = 65.7\%$.

$$R \approx \boxed{0.5 \quad (50\%)}$$

The answer is (B).

(c) Approximate the BOD removal of the secondary (first in-line) clarifier.

$$\text{BOD}_{\text{entering}} = \frac{\text{BOD}_2}{1 - \eta} = \frac{85 \ \dfrac{\text{mg}}{\text{L}}}{1 - 0.30} = 121.4 \ \text{mg/L}$$

$$[\text{assuming 30\% removal}]$$

The sludge production is

$$\left(0.4 \ \frac{\text{lbm}}{\text{lbm}}\right) \left(\begin{array}{c} \left(121.4 \ \dfrac{\text{mg}}{\text{L}} - 85 \ \dfrac{\text{mg}}{\text{L}}\right) \\ + \left(43 \ \dfrac{\text{mg}}{\text{L}} - 30 \ \dfrac{\text{mg}}{\text{L}}\right) \end{array} \right)$$

$$\times \left(8.345 \ \frac{\text{lbm-L}}{\text{mg-MG}}\right)(1.5 \ \text{MGD})$$

$$= 247.3 \ \text{lbm/day}$$

The sludge volume is

$$V = \frac{247.3 \ \dfrac{\text{lbm}}{\text{day}}}{62.4 \ \dfrac{\text{lbm}}{\text{ft}^3}}$$

$$= \boxed{3.96 \ \text{ft}^3/\text{day} \quad (4 \ \text{ft}^3/\text{day})}$$

The answer is (A).

27 Activated Sludge and Sludge Processing

PRACTICE PROBLEMS

Sludge Quantities

1. 33 m³/d of thickened sludge with a suspended solids content of 3.8% and 13 m³/d of anaerobic digester sludge with a suspended solids content of 7.8% are produced in a wastewater treatment plant.

(a) What would be the decrease in sludge volume per year by using a filter press to increase the solids content of the thickened sludge to 24%?

    (A) 4000 m³/yr

    (B) 8000 m³/yr

    (C) 10 000 m³/yr

    (D) 15 000 m³/yr

(b) What volume of digester sludge must be disposed of from sand drying beds that increase the solids concentration to 35%?

    (A) 500 m³/yr

    (B) 1000 m³/yr

    (C) 2000 m³/yr

    (D) 4000 m³/yr

2. An activated sludge plant processes 10 MGD of wastewater with 240 mg/L BOD and 225 mg/L suspended solids. 70% of the suspended solids are inorganic. The discharge from the final clarifier contains 15 mg/L BOD (all organic) and 20 mg/L suspended solids (all inorganic). Primary clarification removes 60% of the suspended solids and 35% of the BOD. The BOD reduction in the primary clarifier does not contribute to sludge production. The sludge produced has a specific gravity of 1.02 and a solids content of 6%. The cell yield (conversion of BOD reduction to biological solids) is 60%. The final clarifier does not reduce BOD.

(a) What is the daily mass of dry sludge solids produced?

    (A) 2500 lbm/day

    (B) 5000 lbm/day

    (C) 10,000 lbm/day

    (D) 20,000 lbm/day

(b) Assuming that the sludge is completely dried and compressed solid, what is the daily sludge volume?

    (A) 200 ft³/day

    (B) 300 ft³/day

    (C) 1000 ft³/day

    (D) 1600 ft³/day

(c) Assuming that the final gravimetric fraction of water in the sludge is 70% and air voids increase the volume by 10%, what is the daily sludge volume?

    (A) 200 ft³/day

    (B) 500 ft³/day

    (C) 1000 ft³/day

    (D) 1600 ft³/day

SOLUTIONS

1. (a) Sludge volume is inversely proportional to the solids content. The sludge volume from the filter press is

$$V_2 = V_1 \left(\frac{SS_1}{SS_2} \right) = \left(33 \ \frac{m^3}{d} \right) \left(\frac{0.038}{0.24} \right)$$
$$= 5.23 \ m^3/d$$

The yearly decrease in volume is

$$\left(33 \ \frac{m^3}{d} - 5.23 \ \frac{m^3}{d} \right) \left(365 \ \frac{d}{yr} \right)$$
$$= \boxed{10\,136 \ m^3/yr \quad (10\,000 \ m^3/yr)}$$

The answer is (C).

(b) The sludge volume from the sand drying bed is

$$V_2 = V_1 \left(\frac{SS_1}{SS_2} \right) = \left(13 \ \frac{m^3}{d} \right) \left(\frac{0.078}{0.35} \right)$$
$$= 2.90 \ m^3/d$$

The yearly disposal volume is

$$\left(2.9 \ \frac{m^3}{d} \right) \left(365 \ \frac{d}{yr} \right) = \boxed{1059 \ m^3/yr \quad (1000 \ m^3/yr)}$$

The answer is (B).

2. The BOD and suspended solids are not mutually exclusive. Some of the suspended solids are organic in nature and show up as BOD.

The inorganic suspended solids are

$$(0.70) \left(225 \ \frac{mg}{L} \right) = 157.5 \ mg/L$$

The inorganic suspended solids that survive primary settling are

$$(1 - 0.60) \left(157.5 \ \frac{mg}{L} \right) = 63 \ mg/L$$

The BOD that survives primary settling is

$$(1 - 0.35) \left(240 \ \frac{mg}{L} \right) = 156 \ mg/L$$

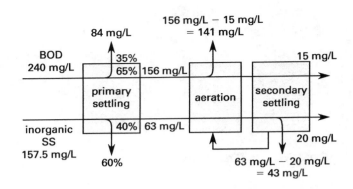

(a) The total dry weight of suspended solids removed in all processes is

$$\left(157.5 \ \frac{mg}{L} - 20 \ \frac{mg}{L} \right) \left(8.345 \ \frac{lbm\text{-}L}{MG\text{-}mg} \right) \left(10 \ \frac{mg}{day} \right)$$
$$= 11{,}474 \ lbm/day$$

Solids from BOD reduction in the secondary process are

$$Y(\Delta BOD)Q = (0.60) \left(156 \ \frac{mg}{L} - 15 \ \frac{mg}{L} \right)$$
$$\times \left(10 \ \frac{MG}{day} \right) \left(8.345 \ \frac{lbm\text{-}L}{mg\text{-}MG} \right)$$
$$= 7060 \ lbm/day$$

The total dry sludge mass is

$$11{,}474 \ \frac{lbm}{day} + 7060 \ \frac{lbm}{day}$$
$$= \boxed{18{,}534 \ lbm/day \quad (20{,}000 \ lbm/day)}$$

The answer is (D).

(b) The specific gravity of the sludge solids can be found from Eq. 27.47.

$$\frac{1}{SG} = \frac{1-s}{1} + \frac{s}{SG_{solids}}$$
$$\frac{1}{1.02} = \frac{1-0.06}{1} + \frac{0.06}{SG_{solids}}$$
$$SG_{solids} = 1.485$$

The density of the solids is

$$\rho_{solids} = SG_{solids}\rho_{water} = (1.485) \left(62.4 \ \frac{lbm}{ft^3} \right)$$
$$= 92.66 \ lbm/ft^3$$

The solid dry sludge volume is

$$V_{\text{solid}} = \frac{m}{\rho} = \frac{18{,}534 \ \dfrac{\text{lbm}}{\text{day}}}{92.66 \ \dfrac{\text{lbm}}{\text{ft}^3}}$$

$$= \boxed{200.0 \ \text{ft}^3/\text{day} \quad (200 \ \text{ft}^3/\text{day})}$$

The answer is (A).

(c) $1 - s$ (the moisture content) is given as 0.70. The disposal volume is

$$V_t = V_{\text{solid}} + V_{\text{water}}$$

$$V_{\text{solid}} = \frac{m_{\text{solid}}}{\rho}$$

$$V_{\text{water}} = \frac{m_{\text{water}}}{62.4 \ \dfrac{\text{lbm}}{\text{ft}^3}} = \frac{0.70 m_t}{62.4 \ \dfrac{\text{lbm}}{\text{ft}^3}}$$

$$= \frac{0.70 m_{\text{solids}}}{\left(62.4 \ \dfrac{\text{lbm}}{\text{ft}^3}\right)(1 - 0.70)}$$

$$V \approx (1.10)\left(\frac{18{,}534 \ \dfrac{\text{lbm}}{\text{day}}}{92.66 \ \dfrac{\text{lbm}}{\text{ft}^3}} + \frac{\left(18{,}534 \ \dfrac{\text{lbm}}{\text{day}}\right)(0.70)}{\left(62.4 \ \dfrac{\text{lbm}}{\text{ft}^3}\right)(1 - 0.70)}\right)$$

$$= \boxed{982.4 \ \text{ft}^3/\text{day} \quad (1000 \ \text{ft}^3/\text{day})}$$

Notice that the basis of moisture content, s, is the total sludge mass, whereas the traditional "water content" basis used in geotechnical calculations is the solids mass.

The answer is (C).

28 Protection of Wetlands

PRACTICE PROBLEMS

1. Which of the following best describe the main characteristics of wetlands?

I. Wetlands have water at the surface or in the root zone.

II. Wetlands support vegetation that is adapted to wet conditions.

III. Wetlands only have fresh water.

IV. Wetlands are flooded at least 6 months of the year.

V. Wetlands have unique soil conditions that usually differ from adjacent uplands.

  (A) I, III, and IV

  (B) II, III, and V

  (C) I, II, and V

  (D) III, IV, and V

2. The term "jurisdictional wetlands" means

  (A) lands that meet the U.S. Fish and Wildlife scientific definition of wetlands

  (B) lands that are regulated primarily by the EPA

  (C) lands that meet the legal definition for regulation under Section 404 of the Clean Water Act

  (D) lands that have met a court test

3. Palustrine wetlands do NOT include

  (A) tidal wetlands

  (B) nontidal wetlands

  (C) marshes

  (D) swamps

4. Riparian forested wetlands are

I. dry for portions of the growing season

II. periodically flooded by adjacent streams and rivers

III. usually of low productivity due to lack of nutrients

IV. referred to as bottomland hardwood forests in the southeast

V. have standing water for most of the growing season

  (A) I, II, and III

  (B) I, III, and V

  (C) II, III, and IV

  (D) I, II, and IV

5. The most important factor in determining the character, functions, and values of a wetland is

  (A) habitat

  (B) hydrology

  (C) salinity

  (D) wildlife

6. A hydroperiod is

  (A) the seasonal pattern of the rise and fall of the surface and subsurface water level

  (B) the depth of water at the start of the growing season

  (C) not affected by tides

  (D) always constant from year to year

7. Which of the following parameters is not included in the water "budget" for wetlands?

  (A) precipitation

  (B) groundwater inflow and outflow

  (C) evapotranspiration

  (D) average water depth

8. Which of the following is NOT true concerning a pulsing hydroperiod?

  (A) It is indicative of unproductive wetlands.

  (B) It is associated with input of nutrients.

  (C) It is typically associated with riparian wetlands.

  (D) It is a key factor in defining the character of a wetland.

9. Which of the following is NOT true about the biochemistry of wetlands?

(A) Wetlands can be sinks, sources, or transformers of chemicals.

(B) Wetlands are always high nutrient systems.

(C) Wetlands are considered open systems when there is substantial movement of nutrients across the ecosystem boundary.

(D) Eutrophic wetlands are rich in mineral and organic nutrients.

10. Which of the following changes would typically have the greatest impact on the character, functions, and values of wetlands?

(A) adverse change in hydrology

(B) oil and gas exploration

(C) pipeline construction

(D) marina construction

11. Which of the following is NOT applicable to wetlands protection?

(A) the Rivers and Harbors Act of 1899

(B) Section 404 of the Clean Water Act

(C) the National Environmental Protection Act

(D) the Safe Drinking Water Act

12. The "least environmentally damaging practicable alternative" refers to

(A) the alternative for which a permit can be issued by the Corps of Engineers for work in wetlands

(B) the least cost alternative that is proposed by the developer for a project in wetlands

(C) the alternative that must be selected to meet the requirements of the Swampbuster Act

(D) the alternative that the U.S. Fish and Wildlife Service must approve in the absence of a finding of no significant impact

SOLUTIONS

1. *The answer is (C).*

2. *The answer is (C).*

3. *The answer is (A).*

4. *The answer is (D).*

5. *The answer is (B).*

6. *The answer is (A).*

7. From Eq. 28.1,

$$\frac{\Delta V}{\Delta t} = P_n + S_i + G_i - \mathrm{ET} - S_o - G_o \pm T$$

The answer is (D).

8. *The answer is (A).*

9. *The answer is (B).*

10. *The answer is (A).*

11. *The answer is (D).*

12. *The answer is (A).*

29 Thermodynamic Properties of Substances

PRACTICE PROBLEMS

1. The molar enthalpy of 250°F (120°C) steam with a quality of 92% is most nearly

(A) 16,000 Btu/lbmol (37 MJ/kmol)

(B) 18,000 Btu/lbmol (41 MJ/kmol)

(C) 20,000 Btu/lbmol (46 MJ/kmol)

(D) 22,000 Btu/lbmol (51 MJ/kmol)

2. The ratio of specific heats for air at 600°F (300°C) is most nearly

(A) 1.33

(B) 1.38

(C) 1.41

(D) 1.67

3. The density of helium at 600°F (300°C) and one standard atmosphere is most nearly

(A) 0.0052 lbm/ft^3 (0.085 kg/m^3)

(B) 0.0061 lbm/ft^3 (0.098 kg/m^3)

(C) 0.0076 lbm/ft^3 (0.12 kg/m^3)

(D) 0.0095 lbm/ft^3 (0.15 kg/m^3)

4. What is the thermodynamic state of water at 600°F and 300 psia?

(A) subcooled liquid

(B) saturated vapor

(C) superheated vapor

(D) real or ideal gas

5. 106°F water flows in a closed feedwater heater. At the pressure in the feedwater heater, water has a saturation temperature of 263°F and a saturation enthalpy of 262 Btu/lbm. What is most nearly the enthalpy of the water?

(A) 105 Btu/lbm

(B) 111 Btu/lbm

(C) 119 Btu/lbm

(D) 132 Btu/lbm

6. What is most nearly the enthalpy of saturated HFC-134a vapor at 180°F?

(A) 77 Btu/lbm

(B) 83 Btu/lbm

(C) 115 Btu/lbm

(D) 120 Btu/lbm

7. A 25% mixture (by volume) of ethylene glycol and water is used in a solar heating application. The mixture is intended to operate at standard atmospheric pressure and an average temperature of 150°F. At 150°F, ethylene glycol has a specific gravity of 1.11 (based on 60°F water) and a specific heat of 0.63 Btu/lbm-°F. What is most nearly the specific heat of the mixture at 150°F?

(A) 0.892 Btu/lbm-°F

(B) 0.899 Btu/lbm-°F

(C) 0.908 Btu/lbm-°F

(D) 0.913 Btu/lbm-°F

SOLUTIONS

1. *Customary U.S. Solution*

From App. 29.A, for 250°F steam, the enthalpy of saturated liquid, h_f, is 218.6 Btu/lbm. The heat of vaporization, h_{fg}, is 945.4 Btu/lbm. The enthalpy is given by Eq. 29.53.

$$h = h_f + xh_{fg} = 218.6 \ \frac{\text{Btu}}{\text{lbm}} + (0.92)\left(945.4 \ \frac{\text{Btu}}{\text{lbm}}\right)$$
$$= 1088.4 \ \text{Btu/lbm}$$

The molecular weight of water is 18 lbm/lbmol. The molar enthalpy is given by Eq. 29.14.

$$H = \text{MW} \times h = \left(18 \ \frac{\text{lbm}}{\text{lbmol}}\right)\left(1088.4 \ \frac{\text{Btu}}{\text{lbm}}\right)$$
$$= \boxed{19{,}591 \ \text{Btu/lbmol} \quad (20{,}000 \ \text{Btu/lbmol})}$$

The answer is (C).

SI Solution

From App. 29.N, for 120°C steam, the enthalpy of saturated liquid, h_f, is 503.81 kJ/kg. The heat of vaporization, h_{fg}, is 2202.1 kJ/kg. The enthalpy is given by Eq. 29.53.

$$h = h_f + xh_{fg} = 503.81 \ \frac{\text{kJ}}{\text{kg}} + (0.92)\left(2202.1 \ \frac{\text{kJ}}{\text{kg}}\right)$$
$$= 2529.7 \ \text{kJ/kg}$$

The molecular weight of water is 18 kg/kmol. Molar enthalpy is given by Eq. 29.14.

$$H = \text{MW} \times h = \left(18 \ \frac{\text{kg}}{\text{kmol}}\right)\left(2529.7 \ \frac{\text{kJ}}{\text{kg}}\right)$$
$$= \boxed{45\,535 \ \text{kJ/kmol} \quad (46 \ \text{MJ/kmol})}$$

The answer is (C).

2. *Customary U.S. Solution*

The absolute temperature is

$$600°\text{F} + 460° = 1060°\text{R}$$

From Table 37.1, the specific heat at constant pressure for air at 1060°R is $c_p = 0.250$ Btu/lbm-°R.

From Eq. 29.61, the specific gas constant is

$$R = \frac{R^*}{\text{MW}} = \frac{1545.33 \ \frac{\text{ft-lbf}}{\text{lbmol-°R}}}{28.967 \ \frac{\text{lbm}}{\text{lbmol}}}$$
$$= 53.35 \ \text{ft-lbf/lbm-°R}$$

From Eq. 29.108(b),

$$c_v = c_p - \frac{R}{J} = 0.250 \ \frac{\text{Btu}}{\text{lbm-°R}} - \frac{53.35 \ \frac{\text{ft-lbf}}{\text{lbm-°R}}}{778 \ \frac{\text{ft-lbf}}{\text{Btu}}}$$
$$= 0.1814 \ \text{Btu/lbm-°R}$$

The ratio of specific heats is given by Eq. 29.28.

$$k = \frac{c_p}{c_v} = \frac{0.250 \ \frac{\text{Btu}}{\text{lbm-°R}}}{0.1814 \ \frac{\text{Btu}}{\text{lbm-°R}}}$$
$$= \boxed{1.378 \quad (1.38)}$$

The answer is (B).

SI Solution

From Table 37.1, the specific heat at constant pressure for air is 0.250 Btu/lbm-°R. From the table footnote, the SI specific heat at constant pressure for air is

$$c_p = \left(0.250 \ \frac{\text{Btu}}{\text{lbm·°R}}\right)\left(4.187 \ \frac{\frac{\text{kJ}}{\text{kg·K}}}{\frac{\text{Btu}}{\text{lbm·°R}}}\right)$$
$$= 1.047 \ \text{kJ/kg·K}$$

From Eq. 29.61, the specific gas constant is

$$R = \frac{R^*}{\text{MW}} = \frac{8314.3 \ \frac{\text{J}}{\text{kmol·K}}}{28.967 \ \frac{\text{kg}}{\text{kmol}}}$$
$$= 287.0 \ \text{J/kg·K}$$

From Eq. 29.108(a),

$$c_v = c_p - R$$
$$= \left(1.047 \ \frac{\text{kJ}}{\text{kg·K}}\right)\left(1000 \ \frac{\text{J}}{\text{kJ}}\right) - 287.0 \ \frac{\text{J}}{\text{kg·K}}$$
$$= 760 \ \text{J/kg·K}$$

The ratio of specific heats is given by Eq. 29.28.

$$k = \frac{c_p}{c_v} = \frac{\left(1.047 \ \frac{\text{kJ}}{\text{kg·K}}\right)\left(1000 \ \frac{\text{J}}{\text{kJ}}\right)}{760 \ \frac{\text{J}}{\text{kg·K}}}$$
$$= \boxed{1.377 \quad (1.38)}$$

The answer is (B).

3. *Customary U.S. Solution*

From Eq. 29.61, the specific gas constant is

$$R = \frac{R^*}{MW} = \frac{1545.33 \ \dfrac{\text{ft-lbf}}{\text{lbmol-}°\text{R}}}{4 \ \dfrac{\text{lbm}}{\text{lbmol}}}$$

$$= 386.3 \ \text{ft-lbf/lbm-}°\text{R}$$

The absolute temperature is

$$600°\text{F} + 460° = 1060°\text{R}$$

From Eq. 29.63, the density of helium is

$$\rho = \frac{p}{RT} = \frac{\left(14.7 \ \dfrac{\text{lbf}}{\text{in}^2}\right)\left(12 \ \dfrac{\text{in}}{\text{ft}}\right)^2}{\left(386.3 \ \dfrac{\text{ft-lbf}}{\text{lbm-}°\text{R}}\right)(1060°\text{R})}$$

$$= \boxed{0.00517 \ \text{lbm/ft}^3 \quad (0.0052 \ \text{lbm/ft}^3)}$$

The answer is (A).

SI Solution

From Eq. 29.61, the specific gas constant is

$$R = \frac{R^*}{MW} = \frac{8314.3 \ \dfrac{\text{J}}{\text{kmol·K}}}{4 \ \dfrac{\text{kg}}{\text{kmol}}}$$

$$= 2079 \ \text{J/kg·K}$$

The absolute temperature is

$$300°\text{C} + 273° = 573\text{K}$$

From Eq. 29.63, the density of helium is

$$\rho = \frac{p}{RT} = \frac{1.013 \times 10^5 \ \text{Pa}}{\left(2079 \ \dfrac{\text{J}}{\text{kg·K}}\right)(573\text{K})}$$

$$= \boxed{0.0850 \ \text{kg/m}^3 \quad (0.085 \ \text{kg/m}^3)}$$

The answer is (A).

4. The saturation temperature for 300 psia steam is 417°F, so since the water's temperature is higher than this, the water is either a superheated vapor or a gas. The critical temperature for water is 705°F, so the water can't be considered a gas. Therefore, it is a $\boxed{\text{superheated vapor.}}$

The answer is (C).

5. Since the pressure of the water in the feedwater heater is not given, a subcooled liquid table can't be used directly. (The saturation temperature could be used to find the pressure, however, if this approach was taken.) The specific heat of liquid water is approximately 1 Btu/lbm-°F, which is the reason that saturated liquid enthalpy and saturation temperature have essentially the same numerical values. Calculate the subcooled enthalpy by subtracting the sensible heat from 106°F to 263°F.

$$h_{106°\text{F}} = h_{\text{saturation}} - c_p(T_{\text{saturation}} - T)$$

$$= 262 \ \frac{\text{Btu}}{\text{lbm}} - \left(1 \ \frac{\text{Btu}}{\text{lbm-}°\text{F}}\right)(263°\text{F} - 106°\text{F})$$

$$= \boxed{105 \ \text{Btu/lbm}}$$

The answer is (A).

6. Use App. 29.M. From the right side of the vapor dome on the 180°F isotherm, and dropping straight down to the enthalpy scale, the enthalpy is approximately $\boxed{120 \ \text{Btu/lbm.}}$

The answer is (D).

7. Use a saturated steam table, such as App. 29.A, to get the properties of 150°F water. The specific volume is the reciprocal of the density. Since $\Delta h = c_p \Delta T$, the specific heat can be found from the change in enthalpy over a known temperature range. Use the saturation enthalpies at 140°F and 160°F.

$$\rho = \frac{1}{v_f}$$

$$= \frac{1}{0.01634 \ \dfrac{\text{ft}^3}{\text{lbm}}}$$

$$= 61.20 \ \text{lbm/ft}^3$$

$$c_{p,\text{water}} = \frac{h_{\text{sat},T_2} - h_{\text{sat},T_1}}{T_2 - T_1}$$

$$= \frac{128.00 \ \dfrac{\text{Btu}}{\text{lbm}} - 107.99 \ \dfrac{\text{Btu}}{\text{lbm}}}{160°\text{F} - 140°\text{F}}$$

$$= 1.00 \ \text{Btu/lbm-}°\text{F}$$

Specific heats of liquid mixtures are gravimetrically weighted.

Consider 1 ft^3 of mixture, containing 0.25 ft^3 of ethylene glycol and $1 - 0.25$ ft$^3 = 0.75$ ft^3 water. The gravimetric fraction of ethylene glycol, G_{glycol}, in the mixture is

$$
\begin{aligned}
G_{glycol} &= \frac{m_{glycol}}{m_{glycol} + m_{water}} \\[6pt]
&= \frac{SG_{glycol}\rho_{water,60°F}\,V_{glycol}}{SG_{glycol}\rho_{water,60°F}\,V_{glycol} + \rho_{water,150°F}\,V_{water}} \\[6pt]
&= \frac{(1.11)\left(62.4\ \dfrac{lbm}{ft^3}\right)(0.25\ ft^3)}{\begin{array}{l}(1.11)\left(62.4\ \dfrac{lbm}{ft^3}\right)(0.25\ ft^3) \\[6pt] + \left(61.2\ \dfrac{lbm}{ft^3}\right)(0.75\ ft^3)\end{array}} \\[6pt]
&= 0.274
\end{aligned}
$$

The gravimetric fraction of water in the mixture is

$$
G_{water} = 1 - G_{glycol} = 1 - 0.274 = 0.726
$$

The specific heat of the mixture is

$$
\begin{aligned}
c_{p,mixture} &= G_{glycol}\,c_{p,glycol} + G_{water}\,c_{p,water} \\[6pt]
&= (0.274)\left(0.63\ \frac{Btu}{lbm\text{-}°F}\right) \\[6pt]
&\quad + (0.726)\left(1.00\ \frac{Btu}{lbm\text{-}°F}\right) \\[6pt]
&= \boxed{0.899\ Btu/lbm\text{-}°F}
\end{aligned}
$$

The answer is (B).

30 Changes in Thermodynamic Properties

PRACTICE PROBLEMS

1. Which statement is FALSE?

(A) The availability of a system depends on the location of the system.

(B) In the absence of friction and other irreversibilities, a heat engine cycle can have a thermal efficiency of 100%.

(C) The gas temperature always decreases when the gas expands isentropically.

(D) Water in a liquid state cannot exist at any temperature if the pressure is less than the triple point pressure.

2. Which statement is true?

(A) Entropy does not change in an adiabatic process.

(B) The entropy of a closed system cannot decrease.

(C) Entropy increases when a refrigerant passes through a throttling valve.

(D) The entropy of air inside a closed room with a running, electrically driven fan will always increase over time.

3. Which process CANNOT be modeled as an isenthalpic process?

(A) viscous drag on an object moving through air

(B) an ideal gas accelerating to supersonic speed through a converging-diverging nozzle

(C) refrigerant passing through a pressure-reducing throttling valve

(D) high-pressure steam escaping through a spring-loaded pressure-relief (safety) valve

4. Air expands isentropically in a steady-flow process from 700°F and 400 psia to 50 psia.

(a) From an air table, what is most nearly the change in enthalpy?

(A) −680 Btu/lbm

(B) −130 Btu/lbm

(C) −90 Btu/lbm

(D) −14 Btu/lbm

(b) From the ideal gas laws for isentropic expansion, and without using any values for specific heat capacity, what is most nearly the change in enthalpy?

(A) −680 Btu/lbm

(B) −120 Btu/lbm

(C) −90 Btu/lbm

(D) −14 Btu/lbm

(c) Based on the specific heat capacity at room temperature, what is most nearly the change in enthalpy?

(A) −680 Btu/lbm

(B) −120 Btu/lbm

(C) −90 Btu/lbm

(D) −14 Btu/lbm

5. 0.60 kg of air ($c_p = 1.005$ kJ/kg·K; $c_v = 0.718$ kJ/kg·K) are contained in a perfectly insulated, rigid enclosure. The ambient conditions outside the enclosure are 95 kPa and 20°C. The air inside the enclosure is initially at 200 kPa and 20°C. Subsequently, an internal impeller within the enclosure raises the air's pressure to 230 kPa through a shaft from an external motor with a motor efficiency of 65%.

(a) What is most nearly the temperature inside the enclosure after the pressure is increased?

(A) 23°C

(B) 45°C

(C) 64°C

(D) 340°C

(b) What is most nearly the initial specific volume of the air within the enclosure?

(A) 0.42 m³/kg

(B) 0.74 m³/kg

(C) 1.8 m³/kg

(D) cannot be determined

(c) What is most nearly the increase in density within the enclosure after the pressure increase?

(A) 0.00 kg/m³

(B) 0.14 kg/m³

(C) 0.21 kg/m³

(D) 0.36 kg/m³

(d) Approximately how much shaft work is required to raise the pressure from 200 kPa to 230 kPa?

(A) 19 kJ

(B) 25 kJ

(C) 32 kJ

(D) 140 kJ

(e) What is most nearly the increase in the available work potential of the air due to the pressure increase?

(A) 1.3 kJ

(B) 2.2 kJ

(C) 8.9 kJ

(D) 15 kJ

6. Cast iron is heated from 80°F to 780°F (27°C to 416°C). The heat required per unit mass is most nearly

(A) 70 Btu/lbm (160 kJ/kg)

(B) 120 Btu/lbm (280 kJ/kg)

(C) 170 Btu/lbm (390 kJ/kg)

(D) 320 Btu/lbm (740 kJ/kg)

7. 8.0 ft³ (0.25 m³) of 180°F, 14.7 psia (82°C, 101.3 kPa) air are cooled to 100°F (38°C) in a constant-pressure process. The amount of work done is most nearly

(A) −2100 ft-lbf (−3.1 kJ)

(B) −1500 ft-lbf (−2.3 kJ)

(C) −1100 ft-lbf (−1.5 kJ)

(D) −900 ft-lbf (−1.3 kJ)

8. Most nearly, the availability of an isentropic process using steam with an initial quality of 95% and operating between 300 psia and 50 psia (2 MPa and 0.35 MPa) is

(A) 100 Btu/lbm (230 kJ/kg)

(B) 130 Btu/lbm (300 kJ/kg)

(C) 210 Btu/lbm (480 kJ/kg)

(D) 340 Btu/lbm (780 kJ/kg)

9. A closed air heater receives 540°F, 100 psia (280°C, 700 kPa) air and heats it to 1540°F (840°C). The outside temperature is 100°F (40°C). The pressure of the air drops 20 psi (150 kPa) as it passes through the heater.

(a) The absolute temperatures at the inlet, T_1, and outlet, T_2, of the air heater are most nearly

(A) $T_1 = 460°R$, $T_2 = 1500°R$
($T_1 = 260K$, $T_2 = 860K$)

(B) $T_1 = 540°R$, $T_2 = 1000°R$
($T_1 = 300K$, $T_2 = 550K$)

(C) $T_1 = 1000°R$, $T_2 = 2000°R$
($T_1 = 550K$, $T_2 = 1100K$)

(D) $T_1 = 1500°R$, $T_2 = 1000°R$
($T_1 = 860K$, $T_2 = 550K$)

(b) The maximum work is most nearly

(A) −240 Btu/lbm (−560 kJ/kg)

(B) −200 Btu/lbm (−460 kJ/kg)

(C) −170 Btu/lbm (−390 kJ/kg)

(D) −150 Btu/lbm (−360 kJ/kg)

(c) The percentage loss in available energy due to the pressure drop is most nearly

(A) 5.0%

(B) 12%

(C) 18%

(D) 34%

10. Xenon gas at 20 psia and 70°F (150 kPa and 21°C) is compressed to 3800 psia and 70°F (25 MPa and 21°C) by a compressor/heat exchanger combination. The compressed gas is stored at 70°F (21°C) in a 100 ft³ (3 m³) rigid tank initially charged with xenon gas at 20 psia (150 kPa).

(a) The mass of the xenon gas initially in the tank is most nearly

(A) 35 lbm (18 kg)

(B) 42 lbm (22 kg)

(C) 46 lbm (24 kg)

(D) 51 lbm (27 kg)

(b) The average mass flow rate of xenon gas into the tank if the compressor fills the tank in exactly one hour is most nearly

(A) 6300 lbm/hr (0.88 kg/s)

(B) 9700 lbm/hr (1.3 kg/s)

(C) 12,000 lbm/hr (1.6 kg/s)

(D) 14,000 lbm/hr (1.9 kg/s)

(c) If filling takes exactly one hour and electricity costs $0.045 per kW-hr, the cost of filling the tank is most nearly

(A) $8.00

(B) $14

(C) $27

(D) $35

11. The mass of an insulated 20 ft³ (0.6 m³) steel tank is 40 lbm (20 kg). The steel has a specific heat of 0.11 Btu/lbm-°R (0.46 kJ/kg·K). The tank is placed in a room where the surrounding air is 70°F and 14.7 psia (21°C and 101.3 kPa). After the tank is evacuated to 1 psia and 70°F (7 kPa and 21°C), a valve is suddenly opened, allowing the tank to fill with room air. The air enters the tank in a well-mixed, turbulent condition.

(a) The volume of room air entering the tank is most nearly

(A) 1.4 ft³ (0.04 m³)

(B) 13 ft³ (0.40 m³)

(C) 14 ft³ (0.44 m³)

(D) 20 ft³ (0.64 m³)

(b) The work performed by the room air entering the tank is most nearly

(A) −42 Btu (−49 000 J)

(B) −34 Btu (−40 000 J)

(C) −8 Btu (−9500 J)

(D) −4 Btu (−4700 J)

(c) The air temperature of the air inside the tank after filling, after the gas and tank have reached thermal equilibrium, but before any heat loss from the tank to the room occurs is most nearly

(A) 70°F (20°C)

(B) 80°F (30°C)

(C) 100°F (40°C)

(D) 190°F (90°C)

SOLUTIONS

1. Availability depends on the temperature of the environment, T_L. Option A is true.

A heat engine's maximum efficiency is that of a Carnot engine cycle, which is always less than 100%. Option B is false.

An expansion includes a drop in enthalpy, and since $h = u + pv$, both u (manifested as temperature) and p decrease. Option C is true.

Liquid water cannot exist below the triple point pressure. Option D is true.

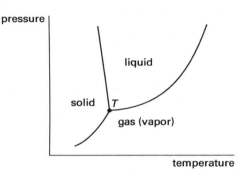

The answer is (B).

2. Entropy does not change in an isentropic (reversible adiabatic) process. However, not all adiabatic processes are reversible. Option A is false.

The entropy of a closed system can be decreased by decreasing the temperature. Option B is false.

When a refrigerant is throttled, the enthalpy remains constant, the pressure drops, and the entropy increases. Option C is true.

The fan and motor will certainly increase the entropy of the air inside the room. However, if the air temperature is decreased by heat loss to the outside, the entropy will decrease. Option D is false.

The answer is (C).

3. A throttling process (including flow through a control valve, safety relief valve, throttling valve, nozzle, or orifice) is modeled as an isenthalpic process. Consider steam expanding through a safety relief valve. To squeeze through the narrow restriction between the disk and the valve seat, steam has to accelerate to a high speed. It does so by converting enthalpy into kinetic energy. The process is not frictionless, but passage past the valve seat occurs so quickly as to be essentially so, and the process is considered to be isentropic. Once past the narrow restriction, the steam expands into the lower pressure region in the valve outlet. The steam decelerates as the flow area increases from the valve passageway to the downstream pipe. This decrease in velocity (kinetic energy) is manifested as an increase in temperature and enthalpy. The enthalpy drop associated with the initial increase in kinetic energy is reclaimed (except for a small portion lost due to the effects of friction). Therefore, the final process is essentially isenthalpic.

Viscous drag of an object in any fluid is an adiabatic process without work being performed on or by the fluid. The duration of the contact between fluid and object is short. Therefore, viscous drag is isenthalpic.

The increase in velocity in a supersonic nozzle comes entirely at the expense of enthalpy.

The answer is (B).

4. (a) From App. 29.F, for $T_1 = 700°F + 460° = 1160°R$ and 400 psia, $h_1 = 281.14$ Btu/lbm, and $p_{r,1} = 21.18$. Since the expansion is isentropic,

$$p_{r,2} = p_{r,1}\frac{p_2}{p_1} = (21.18)\left(\frac{50 \ \frac{\text{lbf}}{\text{in}^2}}{400 \ \frac{\text{lbf}}{\text{in}^2}}\right) = 2.6475$$

Interpolating from App. 29.F, the value of $p_{r,2}$ corresponds to $T_2 = 649.3°R$. For this temperature, $h_2 = 155.3$ Btu/lbm. The change in enthalpy is

$$\Delta h = h_2 - h_1 = 155.3 \ \frac{\text{Btu}}{\text{lbm}} - 281.14 \ \frac{\text{Btu}}{\text{lbm}}$$

$$= \boxed{-125.8 \ \text{Btu/lbm} \quad (-130 \ \text{Btu/lbm})}$$

The answer is (B).

(b) From Eq. 30.107 (applicable to both closed and open systems),

$$T_2 = T_1\left(\frac{p_2}{p_1}\right)^{\frac{k-1}{k}} = (1160°R)\left(\frac{50 \ \frac{\text{lbf}}{\text{in}^2}}{400 \ \frac{\text{lbf}}{\text{in}^2}}\right)^{\frac{1.4-1}{1.4}} = 640.4°R$$

The specific volumes are

$$v_1 = \frac{RT_1}{p_1} = \frac{\left(53.35 \ \frac{\text{ft-lbf}}{\text{lbm-°R}}\right)(1160°R)}{\left(400 \ \frac{\text{lbf}}{\text{in}^2}\right)\left(12 \ \frac{\text{in}}{\text{ft}}\right)^2}$$

$$= 1.074 \ \text{ft}^3/\text{lbm}$$

$$v_2 = \frac{RT_2}{p_2} = \frac{\left(53.35 \ \frac{\text{ft-lbf}}{\text{lbm-°R}}\right)(640.4°R)}{\left(50 \ \frac{\text{lbf}}{\text{in}^2}\right)\left(12 \ \frac{\text{in}}{\text{ft}}\right)^2}$$

$$= 4.745 \ \text{ft}^3/\text{lbm}$$

From Eq. 30.121,

$$h_2 - h_1 = \frac{k(p_2v_2 - p_1v_1)}{(k-1)J}$$

$$= \frac{(1.4)\left(\begin{array}{c}\left(50 \ \frac{\text{lbf}}{\text{in}^2}\right)\left(4.745 \ \frac{\text{ft}^3}{\text{lbm}}\right) \\ - \left(400 \ \frac{\text{lbf}}{\text{in}^2}\right)\left(1.074 \ \frac{\text{ft}^3}{\text{lbm}}\right)\end{array}\right) \times \left(12 \ \frac{\text{in}}{\text{ft}}\right)^2}{(1.4-1)\left(778.17 \ \frac{\text{ft-lbf}}{\text{Btu}}\right)}$$

$$= \boxed{-124.6 \ \text{Btu/lbm} \quad (-120 \ \text{Btu/lbm})}$$

The answer is (B).

(c) From part (b), $T_2 = 640.4°R$. From Table 29.7, $c_p = 0.240$ Btu/lbm-°R.

From Eq. 30.120,

$$\Delta h = c_p\Delta T = \left(0.240 \ \frac{\text{Btu}}{\text{lbm-°R}}\right)(640.4°R - 1160°R)$$

$$= \boxed{-124.7 \ \text{Btu/lbm} \quad (-120 \ \text{Btu/lbm})}$$

The answer is (B).

5. (a) Use Gay-Lussac's law.

$$T_2 = \frac{T_1p_2}{p_1} = \frac{(20°C + 273°)(230 \ \text{kPa})}{200 \ \text{kPa}} = 337\text{K}$$

$$T_{2,°C} = 337\text{K} - 273° = \boxed{64°C}$$

The answer is (C).

(b) The volume of the enclosure is not given, so specific volume cannot be calculated as $v = V/m$. Use the ideal gas law.

$$v_1 = \frac{RT_1}{p_1} = \frac{\left(287.03 \ \frac{\text{J}}{\text{kg·K}}\right)(20°C + 273°)}{(200 \ \text{Pa})\left(1000 \ \frac{\text{Pa}}{\text{kPa}}\right)}$$

$$= \boxed{0.42 \ \text{m}^3/\text{kg}}$$

The answer is (A).

(c) Although the pressure and temperature of the air inside the enclosure change, the total air mass and enclosure volume do not change. Therefore, the specific volume does not change, and $\mu_2 = \mu_1$.

The answer is (A).

(d) This is a closed system. The first law of thermodynamics is applicable.

$$Q = \Delta U + W$$

Since the system is insulated, it is adiabatic, and $Q = 0$.

$$W = -\Delta U$$

Work on a unit mass (not molar) basis. Use Eq. 30.51, valid for any process.

$$W = -\Delta U = -c_v(T_2 - T_1)$$

$$= (0.6 \text{ kg})\left(0.718 \frac{\text{kJ}}{\text{kg·K}}\right)(64°C - 20°C)$$

$$= \boxed{19 \text{ kJ}}$$

The answer is (A).

(e) From Eq. 30.81, with $\mu_2 = \mu_1$, the entropy change is

$$s_2 - s_1 = c_v \ln \frac{T_2}{T_1}$$

Use Eq. 30.179 to calculate the change in availability. Incorporate Eq. 30.50, valid for any process, and Eq. 30.81.

$$\Phi_2 - \Phi_1 = h_2 - h_1 - T_L(s_2 - s_1)$$

$$= c_p(T_2 - T_1) - T_L\left(c_v \ln \frac{T_2}{T_1}\right)$$

$$= \left(1.005 \frac{\text{kJ}}{\text{kg·K}}\right)(64°C - 20°C)$$

$$\quad - (20°C + 273°)$$

$$\quad \times \left(\left(0.718 \frac{\text{kJ}}{\text{kg·K}}\right)\ln\left(\frac{337\text{K}}{20°C + 273°}\right)\right)$$

$$= 14.79 \text{ kJ/kg}$$

For the air mass within the enclosure, the increase in availability is

$$m(\Phi_2 - \Phi_1) = (0.6 \text{ kg})\left(14.79 \frac{\text{kJ}}{\text{kg}}\right)$$

$$= \boxed{8.87 \text{ kJ} \quad (8.9 \text{ kJ})}$$

The answer is (C).

6. *Customary U.S. Solution*

From Table 29.2, the approximate value of specific heat for cast iron is $c_p = 0.10$ Btu/lbm-°F.

The heat required per unit mass is

$$q = c_p(T_2 - T_1) = \left(0.10 \frac{\text{Btu}}{\text{lbm-°F}}\right)(780°F - 80°F)$$

$$= \boxed{70 \text{ Btu/lbm}}$$

The answer is (A).

SI Solution

From Table 29.2, the approximate value of specific heat of cast iron is $c_p = 0.42$ kJ/kg·K.

The heat required per unit mass is

$$q = c_p(T_2 - T_1)$$

$$= \left(0.42 \frac{\text{kJ}}{\text{kg·K}}\right)(416°C - 27°C)$$

$$= \boxed{163.4 \text{ kJ/kg} \quad (160 \text{ kJ/kg})}$$

The answer is (A).

7. *Customary U.S. Solution*

The mass of air is

$$m = \frac{p_1 V_1}{RT_1} = \frac{\left(14.7 \frac{\text{lbf}}{\text{in}^2}\right)\left(12 \frac{\text{in}}{\text{ft}}\right)^2(8.0 \text{ ft}^3)}{\left(53.3 \frac{\text{ft-lbf}}{\text{lbm-°R}}\right)(180°F + 460°)}$$

$$= 0.4964 \text{ lbm}$$

For a constant pressure process from Eq. 30.65, on a per unit mass basis,

$$W = R(T_2 - T_1)$$

Since $\Delta T_{°R} = \Delta T_{°F}$, the total work for m in lbm is

$$W = mR(T_2 - T_1)$$

$$= (0.4964 \text{ lbm})\left(53.3 \frac{\text{ft-lbf}}{\text{lbm-°R}}\right)(100°F - 180°F)$$

$$= \boxed{-2116.6 \text{ ft-lbf} \quad (-2100 \text{ ft-lbf})}$$

This is negative because work is done on the system.

The answer is (A).

SI Solution

The mass of air is

$$m = \frac{p_1 V_1}{RT_1} = \frac{(101.3 \text{ kPa})\left(1000 \frac{\text{Pa}}{\text{kPa}}\right)(0.25 \text{ m}^3)}{\left(287 \frac{\text{J}}{\text{kg·K}}\right)(82°C + 273°)}$$

$$= 0.2486 \text{ kg}$$

For a constant pressure process from Eq. 30.65, on a per unit mass basis,

$$W = R(T_2 - T_1)$$

Ventilation

The total work for m in kg is

$$W = mR(T_2 - T_1)$$

$$= (0.2486 \text{ kg})\left(287 \; \frac{\text{J}}{\text{kg·K}}\right)(38°\text{C} - 82°\text{C})$$

$$= \boxed{-3139.3 \text{ J} \quad (-3.1 \text{ kJ})}$$

The answer is (A).

8. *Customary U.S. Solution*

From App. 29.B, for 300 psia, the enthalpy of saturated liquid, h_f, is 394.0 Btu/lbm. The heat of vaporization, h_{fg}, is 809.4 Btu/lbm. The enthalpy is given by Eq. 29.53.

$$h_1 = h_f + xh_{fg} = 394.0 \; \frac{\text{Btu}}{\text{lbm}} + (0.95)\left(809.4 \; \frac{\text{Btu}}{\text{lbm}}\right)$$

$$= 1162.9 \text{ Btu/lbm}$$

From the Mollier diagram, for an isentropic process from 300 psia to 50 psia, $h_2 = 1031$ Btu/lbm.

The availability is calculated from Eq. 30.178 using an isentropic process ($s_1 = s_2$).

$$\text{availability} = h_1 - h_2 = 1162.9 \; \frac{\text{Btu}}{\text{lbm}} - 1031 \; \frac{\text{Btu}}{\text{lbm}}$$

$$= \boxed{131.9 \text{ Btu/lbm} \quad (130 \text{ Btu/lbm})}$$

The answer is (B).

SI Solution

From App. 29.O, for 2 MPa, the enthalpy of saturated liquid, h_f, is 908.50 kJ/kg. The heat of vaporization, h_{fg}, is 1889.8 kJ/kg. The enthalpy is given by Eq. 29.53.

$$h_1 = h_f + xh_{fg} = 908.50 \; \frac{\text{kJ}}{\text{kg}} + (0.95)\left(1889.8 \; \frac{\text{kJ}}{\text{kg}}\right)$$

$$= 2703.8 \text{ kJ/kg}$$

From the Mollier diagram, for an isentropic process from 2 MPa to 0.35 MPa, $h_2 = 2405$ kJ/kg.

The availability is calculated from Eq. 30.178 using an isentropic process ($s_1 = s_2$).

$$\text{availability} = h_1 - h_2 = 2703.8 \; \frac{\text{kJ}}{\text{kg}} - 2405 \; \frac{\text{kJ}}{\text{kg}}$$

$$= \boxed{298.8 \text{ kJ/kg} \quad (300 \text{ kJ/kg})}$$

The answer is (B).

9. *Customary U.S. Solution*

(a) The absolute temperature at the inlet of the air heater is

$$T_1 = 540°\text{F} + 460° = \boxed{1000°\text{R}}$$

The absolute temperature at the outlet of the air heater is

$$T_2 = 1540°\text{F} + 460° = \boxed{2000°\text{R}}$$

The answer is (C).

(b) Since pressures are low and temperatures are high, use an air table.

From App. 29.F at 1000°R,

$$h_1 = 240.98 \text{ Btu/lbm}$$

$$\phi_1 = 0.75042 \text{ Btu/lbm-°R}$$

From App. 29.F at 2000°R,

$$h_2 = 504.71 \text{ Btu/lbm}$$

$$\phi_2 = 0.93205 \text{ Btu/lbm-°R}$$

The availability per unit mass is calculated from Eq. 30.178 using $T_L = 100°\text{F} + 460° = 560°\text{R}$.

$$W_{\max} = h_1 - h_2 + T_L(s_2 - s_1)$$

For no pressure drop,

$$s_2 - s_1 = \phi_2 - \phi_1$$

$$W_{\max} = h_1 - h_2 + T_L(\phi_2 - \phi_1)$$

$$= 240.98 \; \frac{\text{Btu}}{\text{lbm}} - 504.71 \; \frac{\text{Btu}}{\text{lbm}}$$

$$+ (560°\text{R})\left(0.93205 \; \frac{\text{Btu}}{\text{lbm-°R}} - 0.75042 \; \frac{\text{Btu}}{\text{lbm-°R}}\right)$$

$$= -162.02 \text{ Btu/lbm}$$

With a pressure drop from 100 psia to 80 psia, from Eq. 29.52,

$$s_2 - s_1 = \phi_2 - \phi_1 - \frac{R}{J}\ln\left(\frac{p_2}{p_1}\right)$$

Therefore, the maximum total work due to the pressure drop is

$$W_{\max,p \text{ loss}} = h_1 - h_2 + T_L\left(\phi_2 - \phi_1 - \frac{R}{J}\ln\left(\frac{p_2}{p_1}\right)\right)$$

$$= 240.98 \; \frac{\text{Btu}}{\text{lbm}} - 504.71 \; \frac{\text{Btu}}{\text{lbm}}$$

$$+ (560°\text{R})$$

$$\times \left(\begin{array}{c} 0.93205 \; \dfrac{\text{Btu}}{\text{lbm-°R}} - 0.75042 \; \dfrac{\text{Btu}}{\text{lbm-°R}} \\[2mm] - \left(\dfrac{53.3 \; \frac{\text{ft-lbf}}{\text{lbm-°R}}}{778 \; \frac{\text{ft-lbf}}{\text{Btu}}}\right)\ln\left(\dfrac{80 \text{ psia}}{100 \text{ psia}}\right) \end{array} \right)$$

$$= \boxed{-153.46 \text{ Btu/lbm} \quad (-150 \text{ Btu/lbm})}$$

The answer is (D).

(c) The percentage loss in available energy is

$$\frac{W_{\max} - W_{\max,p \text{ loss}}}{W_{\max}} \times 100\%$$

$$= \frac{-162.02 \ \frac{\text{Btu}}{\text{lbm}} - \left(-153.46 \ \frac{\text{Btu}}{\text{lbm}}\right)}{-162.02 \ \frac{\text{Btu}}{\text{lbm}}} \times 100\%$$

$$= \boxed{5.28\% \quad (5.0\%)}$$

The answer is (A).

SI Solution

(a) The absolute temperature at the inlet of the air heater is

$$T_1 = 280°C + 273° = \boxed{553K \quad (550K)}$$

The absolute temperature at the outlet of the air heater is

$$T_2 = 840°C + 273° = \boxed{1113K \quad (1100K)}$$

The answer is (C).

(b) Since pressures are low and temperatures are high, use an air table.

From App. 29.S at 553K,

$$h_1 = 557.9 \ \text{kJ/kg}$$

$$\phi_1 = 2.32372 \ \text{kJ/kg·K}$$

From App. 29.S at 1113K,

$$h_2 = 1176.2 \ \text{kJ/kg}$$

$$\phi_2 = 3.09092 \ \text{kJ/kg·K}$$

The availability per unit mass is calculated from Eq. 30.178 using $T_L = 40°C + 273° = 313K$.

$$W_{\max} = h_1 - h_2 + T_L(s_2 - s_1)$$

For no pressure drop,

$$s_2 - s_1 = \phi_2 - \phi_1$$

$$W_{\max} = h_1 - h_2 + T_L(\phi_2 - \phi_1)$$

$$= 557.9 \ \frac{\text{kJ}}{\text{kg}} - 1176.2 \ \frac{\text{kJ}}{\text{kg}}$$

$$\quad + (313K)\left(3.09092 \ \frac{\text{kJ}}{\text{kg·K}} - 2.32372 \ \frac{\text{kJ}}{\text{kg·K}}\right)$$

$$= -378.17 \ \text{kJ/kg}$$

With a pressure drop from 700 kPa to 550 kPa, from Eq. 29.52,

$$s_2 - s_1 = \phi_2 - \phi_1 - R\ln\left(\frac{p_2}{p_1}\right)$$

Therefore, the maximum total work due to the pressure drop is

$$W_{\max,p \text{ loss}} = h_1 - h_2 + T_L\left(\phi_2 - \phi_1 - R\ln\left(\frac{p_2}{p_1}\right)\right)$$

$$= 557.9 \ \frac{\text{kJ}}{\text{kg}} - 1176.2 \ \frac{\text{kJ}}{\text{kg}}$$

$$\quad + (313K)$$

$$\quad \times \left(\begin{array}{c} 3.09092 \ \frac{\text{kJ}}{\text{kg·K}} - 2.32372 \ \frac{\text{kJ}}{\text{kg·K}} \\ \\ - \left(\dfrac{287 \ \frac{\text{J}}{\text{kg·K}}}{1000 \ \frac{\text{J}}{\text{kJ}}}\right) \ln\left(\dfrac{550 \ \text{kPa}}{700 \ \text{kPa}}\right) \end{array} \right)$$

$$= \boxed{-356.50 \ \text{kJ/kg} \quad (-360 \ \text{kJ/kg})}$$

The answer is (D).

(c) The percentage loss in available energy is

$$\frac{W_{\max} - W_{\max,p \text{ loss}}}{W_{\max}} \times 100\%$$

$$= \frac{-378.17 \ \frac{\text{kJ}}{\text{kg}} - \left(-356.50 \ \frac{\text{kJ}}{\text{kg}}\right)}{-378.17 \ \frac{\text{kJ}}{\text{kg}}} \times 100\%$$

$$= \boxed{5.73\% \quad (5.0\%)}$$

The answer is (A).

10. *Customary U.S. Solution*

(a) The absolute temperature is

$$T = 70°F + 460° = 530°R$$

From Table 29.7, $R = 11.77$ ft-lbf/lbm-°R. From Eq. 29.60,

$$m = \frac{pV}{RT} = \frac{\left(20 \ \frac{\text{lbf}}{\text{in}^2}\right)\left(12 \ \frac{\text{in}}{\text{ft}}\right)^2 (100 \ \text{ft}^3)}{\left(11.77 \ \frac{\text{ft-lbf}}{\text{lbm-°R}}\right)(530°R)}$$

$$= \boxed{46.17 \ \text{lbm} \quad (46 \ \text{lbm})}$$

The answer is (C).

(b) From Table 29.4, the critical temperature and pressure of xenon are 521.9°R and 58.2 atm, respectively. The reduced variables are

$$T_r = \frac{T}{T_c} = \frac{530°R}{521.9°R} = 1.02$$

$$p_r = \frac{p}{p_c} = \frac{3800 \ \text{psia}}{(58.2 \ \text{atm})\left(14.7 \ \frac{\text{psia}}{\text{atm}}\right)} = 4.44$$

From App. 29.Z, Z is read as 0.61. Using Eq. 29.106,

$$m = \frac{pV}{ZRT} = \frac{\left(3800 \; \frac{\text{lbf}}{\text{in}^2}\right)\left(12 \; \frac{\text{in}}{\text{ft}}\right)^2 (100 \; \text{ft}^3)}{(0.61)\left(11.77 \; \frac{\text{ft-lbf}}{\text{lbm-}°\text{R}}\right)(530°\text{R})}$$

$$= 14{,}380 \; \text{lbm}$$

The average mass flow rate of xenon is

$$\dot{m} = \frac{14{,}380 \; \text{lbm} - 46.17 \; \text{lbm}}{1 \; \text{hr}}$$

$$= \boxed{14{,}334 \; \text{lbm/hr} \quad (14{,}000 \; \text{lbm/hr})}$$

The answer is (D).

(c) For isothermal compression, the work is calculated from Eq. 30.93.

$$W = mRT \ln\left(\frac{p_1}{p_2}\right)$$

$$= \frac{(14{,}334 \; \text{lbm})\left(11.77 \; \frac{\text{ft-lbf}}{\text{lbm-}°\text{R}}\right)}{\left(778 \; \frac{\text{ft-lbf}}{\text{Btu}}\right)\left(3413 \; \frac{\text{Btu}}{\text{kW-hr}}\right)}$$

$$= -176.7 \; \text{kW-hr} \quad (\text{for 1 hr})$$

The cost of electricity is

$$\left(\frac{\$0.045}{\text{kW-hr}}\right)(176.7 \; \text{kW-hr}) = \boxed{\$7.95 \quad (\$8.00)}$$

The answer is (A).

SI Solution

(a) The absolute temperature is

$$T = 21°\text{C} + 273° = 294\text{K}$$

From Table 29.7, $R = 63.32 \; \text{J/kg·K}$. From Eq. 30.60,

$$m = \frac{pV}{RT} = \frac{(150 \; \text{kPa})\left(1000 \; \frac{\text{Pa}}{\text{kPa}}\right)(3 \; \text{m}^3)}{\left(63.32 \; \frac{\text{J}}{\text{kg·K}}\right)(294\text{K})}$$

$$= \boxed{24.17 \; \text{kg} \quad (24 \; \text{kg})}$$

The answer is (C).

(b) From Table 29.4, the critical temperature and pressure of xenon are 289.9K and 58.2 atm, respectively. The reduced variables are

$$T_r = \frac{T}{T_c} = \frac{294\text{K}}{289.9\text{K}} = 1.01$$

$$p_r = \frac{p}{p_c} = \frac{25 \; \text{MPa}}{(58.2 \; \text{atm})\left(0.1013 \; \frac{\text{MPa}}{\text{atm}}\right)} = 4.24$$

From App. 29.Z, Z is read as 0.59. Using Eq. 29.106,

$$m = \frac{pV}{ZRT} = \frac{(25 \; \text{MPa})\left(10^6 \; \frac{\text{Pa}}{\text{MPa}}\right)(3 \; \text{m}^3)}{(0.59)\left(63.32 \; \frac{\text{J}}{\text{kg·K}}\right)(294\text{K})}$$

$$= 6828 \; \text{kg}$$

The average mass flow rate of xenon is

$$\dot{m} = \frac{6828 \; \text{kg} - 24.17 \; \text{kg}}{(1 \; \text{h})\left(3600 \; \frac{\text{s}}{\text{h}}\right)} = \boxed{1.89 \; \text{kg/s} \quad (1.9 \; \text{kg/s})}$$

The answer is (D).

(c) For isothermal compression, the work is calculated from Eq. 30.93.

$$W = mRT \ln\left(\frac{p_1}{p_2}\right)$$

$$= \frac{(6828 \; \text{kg} - 24.17 \; \text{kg})\left(63.32 \; \frac{\text{J}}{\text{kg·K}}\right)(294\text{K})}{\left(3600 \; \frac{\text{s}}{\text{h}}\right)\left(1000 \; \frac{\text{J}}{\text{kJ}}\right)}$$

$$\quad \times \ln\left(\frac{150 \; \text{kPa}}{(25 \; \text{MPa})\left(1000 \; \frac{\text{kPa}}{\text{MPa}}\right)}\right)$$

$$= -178.0 \; \text{kW·h}$$

The cost of electricity is

$$\left(\frac{\$0.045}{\text{kW·h}}\right)(178.0 \; \text{kW·h}) = \boxed{\$8.01 \quad (\$8.00)}$$

The answer is (A).

11. Choose the control volume to include the air outside the tank that is pushed into the tank (subscript "*e*" for "entering"), as well as the tank volume.

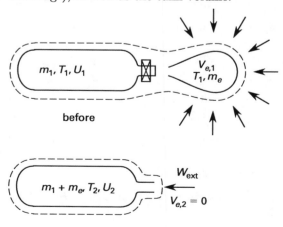

Ventilation

Customary U.S. Solution

(a) The absolute temperature of the air in the tank when evacuated is

$$T_1 = 70°F + 460° = 530°R$$

From Table 29.7, $R = 53.3$ ft-lbf/lbm-°R. From Eq. 30.60,

$$m = \frac{p_1 V_1}{RT_1} = \frac{\left(1\ \frac{lbf}{in^2}\right)\left(12\ \frac{in}{ft}\right)^2 (20\ ft^3)}{\left(53.3\ \frac{ft\text{-}lbf}{lbm\text{-}°R}\right)(530°R)}$$

$$= 0.102\ lbm$$

Assume $T_2 = 300°F$. The absolute temperature is

$$T_2 = 300°F + 460° = 760°R$$

From Eq. 30.60,

$$m_2 = m_1 + m_e = \frac{p_2 V_2}{RT_2}$$

$$= \frac{\left(14.7\ \frac{lbf}{in^2}\right)\left(12\ \frac{in}{ft}\right)^2 (20\ ft^3)}{\left(53.3\ \frac{ft\text{-}lbf}{lbm\text{-}°R}\right)(760°R)}$$

$$m_1 + m_e \approx 1.045\ lbm$$

$$m_e \approx 1.045\ lbm - m_1$$

$$= 1.045\ lbm - 0.102\ lbm$$

$$= 0.943\ lbm$$

From Eq. 30.60, the initial volume of the external air is

$$V_{e,1} = \frac{mRT}{p}$$

$$= \frac{(0.943\ lbm)\left(53.3\ \frac{ft\text{-}lbf}{lbm\text{-}°R}\right)(530°R)}{\left(14.7\ \frac{lbf}{in^2}\right)\left(12\ \frac{in}{ft}\right)^2}$$

$$= \boxed{12.58\ ft^3 \quad (13\ ft^3)}$$

The answer is (B).

(b) For a constant pressure, closed system, from Eq. 30.64, the total work is

$$W_{ext} = p(V_{e,2} - V_{e,1})$$

$$= \frac{\left(14.7\ \frac{lbf}{in^2}\right)\left(12\ \frac{in}{ft}\right)^2 (0\ ft - 12.58\ ft^3)}{778\ \frac{ft\text{-}lbf}{Btu}}$$

$$= \boxed{-34.23\ Btu \quad (-34\ Btu)} \quad \begin{bmatrix} \text{surroundings do} \\ \text{work on the system} \end{bmatrix}$$

The answer is (B).

(c) The energy in part (b) is used to raise the temperature of the air and tank. Consider air as an ideal gas. The process inside the tank is not constant-pressure, as was the compression of the external air mass. However, the process is adiabatic, so $\Delta U = -W$. From Eq. 29.61, $\Delta U = c_v \Delta T$.

$$W_{ext} = \big((m_1 + m_e)c_v\big)(T_2 - T_1)$$

$$34.23\ Btu = \left((1.045\ lbm)\left(0.171\ \frac{Btu}{lbm\text{-}°F}\right)\right)(T_2 - 70°F)$$

$$T_2 = 261.6°F$$

(Since $T_2 = 300°F$ was assumed in part (a), a second iteration may be required.)

Now, allow the 261.6°F air and the 70°F tank to reach thermal equilibrium.

$$q_{air} = q_{tank} \quad \text{[adiabatic]}$$

$$c_{p,air} m_{air}(T_{air} - T_{equilibrium})$$

$$= c_{p,tank} m_{tank}(T_{equilibrium} - T_1)$$

$$\left(0.24\ \frac{Btu}{lbm\text{-}°F}\right)(1.045\ lbm)(261.6°F - T_{equilibrium})$$

$$= \left(0.11\ \frac{Btu}{lbm\text{-}°F}\right)(40\ lbm)(T_{equilibrium} - 70°F)$$

$$T_{equilibrium} = \boxed{80.3°F \quad (80°F)}$$

If a second iteration is performed starting with $T_2 = 261.6°F$, the following values are obtained:

$$T_2 = 722°R$$

$$m_e = 0.998\ lbm$$

$$V_{e,1} = 13.32\ ft^3$$

$$W_{ext} = -36.24\ Btu$$

$$T_2 = 262.6°F$$

$$T_{equilibrium} = 80.9°F$$

The answer is (B).

SI Solution

(a) The absolute temperature of the air in the tank when evacuated is

$$T_1 = 21°C + 273° = 294K$$

From Table 29.7, $R = 287$ J/kg·K. From Eq. 30.60,

$$m = \frac{p_1 V_1}{RT_1} = \frac{(7\ kPa)\left(1000\ \frac{Pa}{kPa}\right)(0.6\ m^3)}{\left(287\ \frac{J}{kg\text{-}K}\right)(294K)} = 0.0498\ kg$$

Assume $T_2 = 127°C$. The absolute temperature is

$$T_2 = 127°C + 273° = 400K$$

From Eq. 30.60,

$$m_2 = m_1 + m_e = \frac{p_2 V_2}{R T_2}$$

$$= \frac{(101.3 \text{ kPa})\left(1000 \; \frac{\text{Pa}}{\text{kPa}}\right)(0.6 \text{ m}^3)}{\left(287 \; \frac{\text{J}}{\text{kg·K}}\right)(400\text{K})}$$

$$= 0.5294 \text{ kg}$$

$$m_e \approx 0.5294 \text{ kg} - m_1$$

$$= 0.5294 \text{ kg} - 0.0498 \text{ kg}$$

$$= 0.4796 \text{ kg}$$

From Eq. 30.60, the initial volume of the external air is

$$V_{e,1} = \frac{mRT}{p} = \frac{(0.4796 \text{ kg})\left(287 \; \frac{\text{J}}{\text{kg·K}}\right)(294\text{K})}{(101.3 \text{ kPa})\left(1000 \; \frac{\text{Pa}}{\text{kPa}}\right)}$$

$$= \boxed{0.3995 \text{ m}^3 \quad (0.40 \text{ m}^3)}$$

The answer is (B).

(b) For a constant pressure, closed system, from Eq. 30.64, the total work is

$$W_{\text{ext}} = p(V_{e,2} - V_{e,1})$$

$$= (101.3 \text{ kPa})\left(1000 \; \frac{\text{Pa}}{\text{kPa}}\right)(0 \text{ m}^3 - 0.3995 \text{ m}^3)$$

$$= \boxed{-40\,469 \text{ J} \quad (-40\,000 \text{ J})} \quad \begin{bmatrix} \text{surroundings do} \\ \text{work on the system} \end{bmatrix}$$

The answer is (B).

(c) The energy in part (b) is used to raise the temperature of the air and tank. Consider air as an ideal gas. The process inside the tank is not constant-pressure, as was the compression of the external air mass. However, the process is adiabatic, so $\Delta U = -W$. From Eq. 30.61, $\Delta U = c_v \Delta T$.

$$W_{\text{ext}} = ((m_1 + m_e)c_v)(T_2 - T_1)$$

$$40\,469 \text{ J} = \left((0.5294 \text{ kg})\left(718 \; \frac{\text{J}}{\text{kg·K}}\right)\right)(T_2 - 21°\text{C})$$

$$T_2 = 127.5°\text{C}$$

This is close enough to the assumed value of T_2 that a second iteration is not necessary.

Now, allow the 127.5°C air and the 21°C tank to reach thermal equilibrium.

$$q_{\text{air}} = q_{\text{tank}} \quad \text{[adiabatic]}$$

$$c_{p,\text{air}} m_{\text{air}} (T_{\text{air}} - T_{\text{equilibrium}})$$

$$= c_{p,\text{tank}} m_{\text{tank}} (T_{\text{equilibrium}} - T_1)$$

$$\left(1005 \; \frac{\text{J}}{\text{kg·K}}\right)(0.5294 \text{ kg})(127.5°\text{C} - T_{\text{equilibrium}})$$

$$= \left(460 \; \frac{\text{J}}{\text{kg·K}}\right)(20 \text{ kg})(T_{\text{equilibrium}} - 21°\text{C})$$

$$T_{\text{equilibrium}} = \boxed{26.8°\text{C} \quad (30°\text{C})}$$

The answer is (B).

31 HVAC: Psychrometrics

PRACTICE PROBLEMS

1. A room contains air at 80°F (27°C) dry-bulb and 67°F (19°C) wet-bulb. The total pressure is 1 atm.

(a) The humidity ratio is most nearly

   (A) 0.009 lbm/lbm (0.009 kg/kg)

   (B) 0.011 lbm/lbm (0.011 kg/kg)

   (C) 0.014 lbm/lbm (0.014 kg/kg)

   (D) 0.018 lbm/lbm (0.018 kg/kg)

(b) The enthalpy is most nearly

   (A) 30.2 Btu/lbm (51.3 kJ/kg)

   (B) 30.8 Btu/lbm (52.4 kJ/kg)

   (C) 31.5 Btu/lbm (53.9 kJ/kg)

   (D) 31.9 Btu/lbm (54.2 kJ/kg)

(c) The specific heat is most nearly

   (A) 0.234 Btu/lbm-°F (0.979 kJ/kg·K)

   (B) 0.237 Btu/lbm-°F (0.991 kJ/kg·K)

   (C) 0.239 Btu/lbm-°F (0.999 kJ/kg·K)

   (D) 0.242 Btu/lbm-°F (1.012 kJ/kg·K)

2. If one layer of cooling coils effectively bypasses one-third of the air passing through it, the theoretical bypass factor for four layers of identical cooling coils in series is most nearly

   (A) 0.01

   (B) 0.09

   (C) 0.33

   (D) 0.67

3. 1000 ft³/min (0.5 m³/s) of air at 50°F (10°C) dry-bulb and 95% relative humidity are mixed with 1500 ft³/min (0.75 m³/s) of air at 76°F (24°C) dry-bulb and 45% relative humidity.

(a) The dry-bulb temperature of the mixture is most nearly

   (A) 55°F (13°C)

   (B) 65°F (18°C)

   (C) 68°F (20°C)

   (D) 70°F (21°C)

(b) The specific humidity of the mixture is most nearly

   (A) 0.008 lbm/lbm (0.008 kg/kg)

   (B) 0.009 lbm/lbm (0.009 kg/kg)

   (C) 0.010 lbm/lbm (0.010 kg/kg)

   (D) 0.011 lbm/lbm (0.011 kg/kg)

(c) The dew point of the mixture is most nearly

   (A) 45°F (7.2°C)

   (B) 48°F (8.9°C)

   (C) 51°F (11°C)

   (D) 54°F (12°C)

4. Air at 60°F (16°C) dry-bulb and 45°F (7°C) wet-bulb passes through an air washer with a humidifying efficiency of 70%.

(a) The effective bypass factor of the system is most nearly

   (A) 0.30

   (B) 0.50

   (C) 0.67

   (D) 0.70

(b) The dry-bulb temperature of the air leaving the washer is most nearly

   (A) 45°F (7.2°C)

   (B) 50°F (9.7°C)

   (C) 54°F (12°C)

   (D) 57°F (14°C)

5. 95°F (35°C) dry-bulb, 75°F (24°C) wet-bulb air passes through a cooling tower and leaves at 85°F (29°C) dry-bulb and 90% relative humidity.

(a) The enthalpy change per cubic foot (meter) of air is most nearly

    (A) 0.57 Btu/ft³ (18 kJ/m³)

    (B) 1.3 Btu/ft³ (40 kJ/m³)

    (C) 3.2 Btu/ft³ (99 kJ/m³)

    (D) 7.6 Btu/ft³ (240 kJ/m³)

(b) The change in moisture content per cubic foot (meter) of air is most nearly

    (A) 1.8×10^{-4} lbm/ft³ (2.7×10^{-3} kg/m³)

    (B) 3.3×10^{-4} lbm/ft³ (4.5×10^{-3} kg/m³)

    (C) 6.7×10^{-4} lbm/ft³ (9.9×10^{-3} kg/m³)

    (D) 9.2×10^{-4} lbm/ft³ (14×10^{-3} kg/m³)

6. An air washer receives 1800 ft³/min (0.85 m³/s) of air at 70°F (21°C) and 40% relative humidity and discharges the air at 75% relative humidity. A recirculating water spray with a constant temperature of 50°F (10°C) is used.

(a) What are the conditions of the discharged air?

(b) What mass of makeup water is required per minute?

7. Repeat Prob. 6(a) and Prob. 6(b) using saturated steam at atmospheric pressure in place of the 50°F (10°C) water spray.

8. During performances, a theater experiences a sensible heat load of 500,000 Btu/hr (150 kW) and a moisture load of 175 lbm/hr (80 kg/h). Air enters the theater at 65°F (18°C) and 55% relative humidity and is removed when it reaches 75°F (24°C) or 60% relative humidity, whichever comes first.

(a) What is the ventilation rate in mass of air per hour?

(b) What are the conditions of the air leaving the theater?

9. 500 ft³/min (0.25 m³/s) of air at 80°F (27°C) dry-bulb and 70% relative humidity are removed from a room. 150 ft³/min (0.075 m³/s) pass through an air conditioner and leave saturated at 50°F (10°C). The remaining 350 ft³/min (0.175 m³/s) bypass the air conditioner and mix with the conditioned air at 1 atm.

(a) The mixture's temperature is most nearly

    (A) 55°F (13°C)

    (B) 66°F (19°C)

    (C) 71°F (22°C)

    (D) 74°F (23°C)

(b) The mixture's humidity ratio is most nearly

    (A) 0.010 lbm/lbm (1.0 g/kg)

    (B) 0.013 lbm/lbm (1.3 g/kg)

    (C) 0.017 lbm/lbm (1.7 g/kg)

    (D) 0.021 lbm/lbm (2.1 g/kg)

(c) The mixture's relative humidity is most nearly

    (A) 45%

    (B) 57%

    (C) 73%

    (D) 81%

(d) The heat load (in tons) of the air conditioner is most nearly

    (A) 0.90 ton

    (B) 1.3 ton

    (C) 2.4 tons

    (D) 2.9 tons

10. (*Time limit: one hour*) A dehumidifier takes 5000 ft³/min (2.36 m³/s) of air at 95°F (35°C) dry-bulb and 70% relative humidity and discharges it at 60°F (16°C) dry-bulb and 95% relative humidity. The dehumidifier uses a wet R-12 refrigeration cycle operating between 100°F (saturated) (38°C) and 50°F (10°C).

(a) Locate the air entering and leaving points on the psychrometric chart.

(b) Find the quantity of water removed from the air.

(c) Find the quantity of heat removed from the air.

(d) Draw the temperature-entropy and enthalpy-entropy diagrams for the refrigeration cycle.

(e) Find the temperature, pressure, enthalpy, entropy, and specific volume for each endpoint of the refrigeration cycle.

11. (*Time limit: one hour*) 1500 ft³/min (0.71 m³/s) of saturated 25 psia (170 kPa) air is heated from 200°F to 400°F (93°C to 204°C) in a constant pressure, constant moisture drying process.

(a) The final relative humidity is most nearly

    (A) 4.7%

    (B) 9.2%

    (C) 13%

    (D) 23%

Ventilation

(b) The final specific humidity is most nearly

(A) 0.41 lbm/lbm (0.41 kg/kg)

(B) 0.53 lbm/lbm (0.53 kg/kg)

(C) 0.66 lbm/lbm (0.66 kg/kg)

(D) 0.79 lbm/lbm (0.79 kg/kg)

(c) The heat required per unit mass of dry air is most nearly

(A) 31 Btu/lbm (71 kJ/kg)

(B) 57 Btu/lbm (130 kJ/kg)

(C) 99 Btu/lbm (230 kJ/kg)

(D) 120 Btu/lbm (280 kJ/kg)

(d) The final dew point is most nearly

(A) 180°F (82°C)

(B) 200°F (93°C)

(C) 220°F (100°C)

(D) 240°F (120°C)

12. (*Time limit: one hour*) 410 lbm/hr (0.052 kg/s) of dry 800°F (427°C) air pass through a scrubber to reduce particulate emissions. To protect the elastomeric seals in the scrubber, the air temperature is reduced to 350°F (177°C) by passing the air through a spray of 80°F (27°C) water. The pressure in the spray chamber is 20 psia (140 kPa).

(a) Approximately how much water is evaporated per hour?

(A) 18 lbm/hr (0.0023 kg/s)

(B) 27 lbm/hr (0.0035 kg/s)

(C) 31 lbm/hr (0.0040 kg/s)

(D) 39 lbm/hr (0.0050 kg/s)

(b) The relative humidity of the air leaving the spray chamber is most nearly

(A) 1.3%

(B) 2.0%

(C) 3.1%

(D) 4.4%

13. Combustion products leaving a gas turbine combustor are released at a temperature of 180°F into the atmosphere. The combustion products have a relative humidity of 20%. What is most nearly the dew point temperature of the combustion products?

(A) 110°F

(B) 120°F

(C) 130°F

(D) 140°F

14. Each hour, 100 lbm of methane are burned with excess air. The combustion products pass through a heat exchanger used to heat water. The combustion products enter the heat exchanger at 340°F, and they leave the heat exchanger at 110°F. The total pressure of the combustion products in the heat exchanger is 19 psia. The humidity ratio of the combustion products is 560 gr/lbm. The dew point temperature of the combustion products is most nearly

(A) 90°F

(B) 100°F

(C) 120°F

(D) 130°F

15. (*Time limit: one hour*) An evaporative counterflow air cooling tower removes 1×10^6 Btu/hr (290 kW) from a water flow. The temperature of the water is reduced from 120°F to 110°F (49°C to 43°C). Air enters the cooling tower at 91°F (33°C) and 60% relative humidity, and air leaves at 100°F (38°C) and 82% relative humidity.

(a) Calculate the air flow rate.

(b) Calculate the quantity of makeup water.

Ventilation

SOLUTIONS

1. *Customary U.S. Solution*

((a) and (b)) Locate the intersection of 80°F dry-bulb and 67°F wet-bulb on the psychrometric chart (see App. 31.A). Read the value of humidity and enthalpy.

$$\omega = 0.0112 \text{ lbm moisture/lbm dry air}$$
$$(0.011 \text{ lbm/lbm})$$
$$h = 31.5 \text{ Btu/lbm dry air}$$

The answer is (B) for Prob. 1(a).

The answer is (C) for Prob. 1(b).

(c) c_p is gravimetrically weighted. c_p for air is 0.240 Btu/lbm-°F, and c_p for steam is approximately 0.40 Btu/lbm-°F.

$$G_{air} = \frac{1}{1 + 0.0112} = 0.989$$
$$G_{steam} = \frac{0.0112}{1 + 0.0112} = 0.011$$
$$c_{p,mixture} = G_{air}c_{p,air} + G_{steam}c_{p,steam}$$
$$= (0.989)\left(0.240 \ \frac{\text{Btu}}{\text{lbm-°F}}\right)$$
$$+ (0.011)\left(0.40 \ \frac{\text{Btu}}{\text{lbm-°F}}\right)$$
$$= \boxed{0.242 \text{ Btu/lbm-°F}}$$

The answer is (D).

SI Solution

((a) and (b)) Locate the intersection of 27°C dry-bulb and 19°C wet-bulb on the psychrometric chart (see App. 31.B). Read the value of humidity and enthalpy.

$$\omega = \frac{10.5 \ \dfrac{\text{g}}{\text{kg dry air}}}{1000 \ \dfrac{\text{g}}{\text{kg}}} = 0.0105 \text{ kg/kg dry air}$$
$$(0.011 \text{ kg/kg})$$
$$h = 53.9 \text{ kJ/kg dry air}$$

The answer is (B) for Prob. 1(a).

The answer is (C) for Prob. 1(b).

(c) c_p is gravimetrically weighted. c_p for air is 1.0048 kJ/kg·K, and c_p for steam is approximately 1.675 kJ/kg·K.

$$G_{air} = \frac{1}{1 + 0.0105} = 0.990$$
$$G_{steam} = \frac{0.0105}{1 + 0.0105} = 0.010$$
$$c_{p,mixture} = G_{air}c_{p,air} + G_{air}c_{p,steam}$$
$$= (0.990)\left(1.0048 \ \frac{\text{kJ}}{\text{kg·K}}\right)$$
$$+ (0.010)\left(1.675 \ \frac{\text{kJ}}{\text{kg·K}}\right)$$
$$= \boxed{1.0115 \text{ kJ/kg·K} \quad (1.012 \text{ kJ/kg·K})}$$

The answer is (D).

2. $\text{BF}_{n\,\text{layers}} = (\text{BF}_{1\,\text{layer}})^n = \left(\frac{1}{3}\right)^4 = \boxed{0.0123 \quad (0.01)}$

Only 1% of the air will be untreated.

The answer is (A).

3. Use the illustration shown for both the customary U.S. and SI solutions.

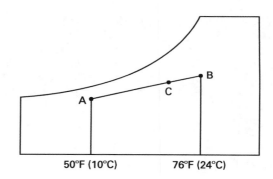

Customary U.S. Solution

(a) Locate the two points on the psychrometric chart and draw a line between them.

Reading from the chart (specific volumes),

$$v_A = 13.0 \text{ ft}^3/\text{lbm}$$
$$v_B = 13.7 \text{ ft}^3/\text{lbm}$$

The density at each point is

$$\rho_A = \frac{1}{v_A} = \frac{1}{13.0 \ \dfrac{\text{ft}^3}{\text{lbm}}} = 0.0769 \text{ lbm/ft}^3$$

$$\rho_B = \frac{1}{v_B} = \frac{1}{13.7 \ \dfrac{\text{ft}^3}{\text{lbm}}} = 0.0730 \text{ lbm/ft}^3$$

The mass flow at each point is

$$\dot{m}_A = \rho_A \dot{V}_A = \left(0.0769 \ \frac{\text{lbm}}{\text{ft}^3}\right)\left(1000 \ \frac{\text{ft}^3}{\text{min}}\right)$$

$$= 76.9 \ \text{lbm/min}$$

$$\dot{m}_B = \rho_B \dot{V}_B = \left(0.0730 \ \frac{\text{lbm}}{\text{ft}^3}\right)\left(1500 \ \frac{\text{ft}^3}{\text{min}}\right)$$

$$= 109.5 \ \text{lbm/min}$$

The gravimetric fraction of flow A is

$$\frac{76.9 \ \dfrac{\text{lbm}}{\text{min}}}{76.9 \ \dfrac{\text{lbm}}{\text{min}} + 109.5 \ \dfrac{\text{lbm}}{\text{min}}} = 0.413$$

Since the scales are all linear,

$$0.413 = \frac{T_B - T_C}{T_B - T_A}$$

$$T_C = T_B - (0.413)(T_B - T_A)$$

$$= 76°\text{F} - (0.413)(76°\text{F} - 50°\text{F})$$

$$= \boxed{65.3°\text{F} \quad (65°\text{F})}$$

The answer is (B).

(b) $\omega = \boxed{\begin{array}{c}0.0082 \ \text{lbm moisture/lbm dry air} \\ (0.008 \ \text{lbm/lbm})\end{array}}$

The answer is (A).

(c) $T_{dp} = \boxed{51°\text{F}}$

The answer is (C).

SI Solution

(a) Locate the two points on the psychrometric chart and draw a line between them.

Reading from the chart (specific volumes),

$$v_A = 0.813 \ \text{m}^3/\text{kg dry air}$$

$$v_B = 0.856 \ \text{m}^3/\text{kg dry air}$$

The density at each point is

$$\rho_A = \frac{1}{v_A} = \frac{1}{0.813 \ \dfrac{\text{m}^3}{\text{kg}}} = 1.23 \ \text{kg/m}^3$$

$$\rho_B = \frac{1}{v_B} = \frac{1}{0.856 \ \dfrac{\text{m}^3}{\text{kg}}} = 1.17 \ \text{kg/m}^3$$

The mass flow at each point is

$$\dot{m}_A = \rho_A \dot{V}_A = \left(1.23 \ \frac{\text{kg}}{\text{m}^3}\right)\left(0.5 \ \frac{\text{m}^3}{\text{s}}\right) = 0.615 \ \text{kg/s}$$

$$\dot{m}_B = \rho_B \dot{V}_B = \left(1.17 \ \frac{\text{kg}}{\text{m}^3}\right)\left(0.75 \ \frac{\text{m}^3}{\text{s}}\right) = 0.878 \ \text{kg/s}$$

The gravimetric fraction of flow A is

$$\frac{0.615 \ \dfrac{\text{kg}}{\text{s}}}{0.615 \ \dfrac{\text{kg}}{\text{s}} + 0.878 \ \dfrac{\text{kg}}{\text{s}}} = 0.412$$

Since the scales are linear,

$$0.412 = \frac{T_B - T_C}{T_B - T_A}$$

$$T_C = T_B - (0.412)(T_B - T_A)$$

$$= 24°\text{C} - (0.412)(24°\text{C} - 10°\text{C})$$

$$= \boxed{18.2°\text{C} \quad (18°\text{C})}$$

The answer is (B).

(b) $\omega = \dfrac{8.0 \ \dfrac{\text{g}}{\text{kg dry air}}}{1000 \ \dfrac{\text{g}}{\text{kg}}}$

$$= \boxed{\begin{array}{c}0.008 \ \text{kg moisture/kg dry air} \\ (0.008 \ \text{kg/kg})\end{array}}$$

The answer is (A).

(c) $T_{dp} = \boxed{10.6°\text{C} \quad (11°\text{C})}$

The answer is (C).

4. *Customary U.S. Solution*

(a) From Eq. 31.24, the bypass factor is

$$\text{BF} = 1 - \eta_{\text{sat}} = 1 - 0.70 = \boxed{0.30}$$

The answer is (A).

(b) From Eq. 31.31, the dry-bulb temperature of air leaving the washer can be determined.

$$\eta_{sat} = \frac{T_{db,in} - T_{db,out}}{T_{db,in} - T_{wb,in}}$$

$$0.70 = \frac{60°F - T_{db,out}}{60°F - 45°F}$$

$$T_{db,out} = \boxed{49.5°F \quad (50°F)}$$

The answer is (B).

SI Solution

(a) From Eq. 31.24, the bypass factor is

$$BF = 1 - \eta_{sat} = 1 - 0.70 = \boxed{0.30}$$

The answer is (A).

(b) From Eq. 31.31, the dry-bulb temperature of air leaving the washer can be determined.

$$\eta_{sat} = \frac{T_{db,in} - T_{db,out}}{T_{db,in} - T_{wb,in}}$$

$$0.70 = \frac{16°C - T_{db,out}}{16°C - 7°C}$$

$$T_{db,out} = \boxed{9.7°C}$$

The answer is (B).

5. *Customary U.S. Solution*

(a) Refer to the psychrometric chart (see App. 31.A).

At point 1, properties of air at $T_{db} = 95°F$ and $T_{wb} = 75°F$ are

$$\omega_1 = 0.0141 \text{ lbm moisture/lbm air}$$
$$h_1 = 38.4 \text{ Btu/lbm air}$$
$$v_1 = 14.3 \text{ ft}^3/\text{lbm air}$$

At point 2, properties of air at $T_{db} = 85°F$ and 90% relative humidity are

$$\omega_2 = 0.0237 \text{ lbm moisture/lbm air}$$
$$h_2 = 46.6 \text{ Btu/lbm air}$$

The enthalpy change is

$$\frac{h_2 - h_1}{v_1} = \frac{46.6 \frac{\text{Btu}}{\text{lbm air}} - 38.4 \frac{\text{Btu}}{\text{lbm air}}}{14.3 \frac{\text{ft}^3}{\text{lbm air}}}$$

$$= \boxed{0.573 \text{ Btu/ft}^3 \text{ air} \quad (0.57 \text{ Btu/ft}^3)}$$

The answer is (A).

(b) The moisture added is

$$\frac{\omega_2 - \omega_1}{v_1} = \frac{0.0237 \frac{\text{lbm moisture}}{\text{lbm air}} - 0.0141 \frac{\text{lbm moisture}}{\text{lbm air}}}{14.3 \frac{\text{ft}^3}{\text{lbm air}}}$$

$$= \boxed{\begin{array}{c} 6.71 \times 10^{-4} \text{ lbm/ft}^3 \text{ air} \\ (6.7 \times 10^{-4} \text{ lbm/ft}^3) \end{array}}$$

The answer is (C).

SI Solution

(a) Refer to the psychrometric chart (see App. 31.B).

At point 1, properties of air at $T_{db} = 35°C$ and $T_{wb} = 24°C$ are

$$\omega_1 = 14.3 \text{ g/kg air}$$
$$h_1 = 71.8 \text{ kJ/kg air}$$
$$v_1 = 0.8893 \text{ m}^3/\text{kg air}$$

At point 2, properties of air at $T_{db} = 29°C$ and 90% relative humidity are

$$\omega_2 = 23.1 \text{ g/kg air}$$
$$h_2 = 88 \text{ kJ/kg air}$$

The enthalpy change is

$$\frac{h_2 - h_1}{v_1} = \frac{88 \frac{\text{kJ}}{\text{kg air}} - 71.8 \frac{\text{kJ}}{\text{kg air}}}{0.8893 \frac{\text{m}^3}{\text{kg air}}}$$

$$= \boxed{18.2 \text{ kJ/m}^3 \text{ air} \quad (18 \text{ kJ/m}^3)}$$

The answer is (A).

Ventilation

(b) The moisture added is

$$\frac{\omega_2 - \omega_1}{v_1} = \frac{23.1 \frac{\text{g}}{\text{kg air}} - 14.3 \frac{\text{kg}}{\text{kg air}}}{\left(0.8893 \frac{\text{m}^3}{\text{kg air}}\right)\left(1000 \frac{\text{g}}{\text{kg}}\right)}$$

$$= \boxed{\begin{array}{c} 9.90 \times 10^{-3} \text{ kg/m}^3 \text{ air} \\ (9.9 \times 10^{-3} \text{ kg/m}^3) \end{array}}$$

The answer is (C).

6. Use the illustration shown for both the customary U.S. and SI solutions.

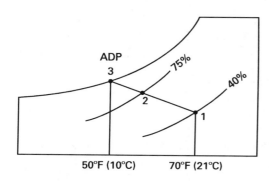

Customary U.S. Solution

(a) Refer to the psychrometric chart (see App. 31.A).

At point 1, properties of air at $T_{db} = 70°F$ and $\phi = 40\%$ are

$$h_1 = 23.6 \text{ Btu/lbm air}$$
$$\omega_1 = 0.00623 \text{ lbm moisture/lbm air}$$
$$v_1 = 13.48 \text{ ft}^3/\text{lbm air}$$

The mass flow rate of incoming air is

$$\dot{m}_{a,1} = \frac{\dot{V}_1}{v_1} = \frac{1800 \frac{\text{ft}^3}{\text{min}}}{13.48 \frac{\text{ft}^3}{\text{lbm air}}} = 133.53 \text{ lbm air/min}$$

Locate point 1 on the psychrometric chart.

Notice that the temperature of the recirculating water is constant but not equal to the air's entering wet-bulb temperature. Therefore, this is not an adiabatic process.

Locate point 3 as 50°F saturated condition (water being sprayed) on the psychrometric chart.

Draw a line from point 1 to point 3. The intersection of this line with 75% relative humidity defines point 2 as

$$\boxed{\begin{array}{l} h_2 = 21.4 \text{ Btu/lbm air} \\ \omega_2 = 0.0072 \text{ lbm moisture/lbm air} \\ T_{db,2} = 56°F \\ T_{wb,2} = 51.8°F \end{array}}$$

(b) The moisture (water) added is

$$\dot{m}_w = \dot{m}_{a,1}(\omega_2 - \omega_1)$$
$$= \left(133.53 \frac{\text{lbm air}}{\text{min}}\right)$$
$$\times \left(\begin{array}{c} 0.0072 \frac{\text{lbm moisture}}{\text{lbm air}} \\ - 0.00625 \frac{\text{lbm moisture}}{\text{lbm air}} \end{array}\right)$$
$$= \boxed{0.127 \text{ lbm/min}}$$

SI Solution

(a) Refer to the psychrometric chart (see App. 31.B).

At point 1, properties of air at $T_{db} = 21°C$ and $\phi = 40\%$ are

$$h_1 = 36.75 \text{ kJ/kg air}$$
$$\omega_1 = 6.2 \text{ g moisture/kg air}$$
$$v_1 = 0.842 \text{ m}^3/\text{kg air}$$

The mass flow rate of incoming air is

$$\dot{m}_{a,1} = \frac{\dot{V}_1}{v_1} = \frac{0.85 \frac{\text{m}^3}{\text{s}}}{0.842 \frac{\text{m}^3}{\text{kg air}}}$$
$$= 1.010 \text{ kg air/s}$$

Locate point 1 on the psychrometric chart.

Notice that the temperature of the recirculating water is constant but not equal to the air's entering wet-bulb temperature. Therefore, this is not an adiabatic process.

Locate point 3 as 10°C saturated condition (water being sprayed) on the psychrometric chart.

Ventilation

Draw a line from point 1 to point 3. The intersection of this line with 75% relative humidity defines point 2 as

$$h_2 = 34.1 \text{ kJ/kg air}$$
$$\omega_2 = 7.6 \text{ g moisture/kg air}$$
$$T_{db,2} = 14.7°C$$
$$T_{wb,2} = 12.0°C$$

(b) The water added is

$$\dot{m}_w = \dot{m}_{a,1}(\omega_2 - \omega_1)$$

$$= \frac{\left(1.010 \; \dfrac{\text{kg air}}{\text{s}}\right)\left(7.6 \; \dfrac{\text{g moisture}}{\text{kg air}} - 6.2 \; \dfrac{\text{g moisture}}{\text{kg air}}\right)}{1000 \; \dfrac{\text{g}}{\text{kg}}}$$

$$= \boxed{0.00141 \text{ kg/s}}$$

7. *Customary U.S. Solution*

((a) and (b)) From Prob. 6,

$$\omega_1 = 0.00623 \text{ lbm moisture/lbm air}$$
$$h_1 = 23.6 \text{ Btu/lbm air}$$
$$\dot{m}_{a,1} = 133.53 \text{ lbm air/min}$$

From the steam table (see App. 29.B) for 1 atm steam, $h_{steam} = 1150.3 \text{ Btu/lbm}$.

From the conservation of energy equation (see Eq. 31.33),

$$\dot{m}_{a,1}h_1 + \dot{m}_{steam}h_{steam} = \dot{m}_{a,1}h_2$$

$$\left(133.53 \; \frac{\text{lbm air}}{\text{min}}\right)$$
$$\times \left(23.6 \; \frac{\text{Btu}}{\text{lbm air}}\right)$$
$$+ \dot{m}_{steam}$$
$$\times \left(1150.3 \; \frac{\text{Btu}}{\text{lbm}}\right) = \left(133.53 \; \frac{\text{lbm air}}{\text{min}}\right)h_2 \quad \text{[Eq. 1]}$$

Solve for $\dot{m}_{steam}$.

$$\dot{m}_{steam} = 0.1161h_2 - 2.740 \quad \text{[Eq. 3]}$$

From a conservation of mass for the water (see Eq. 31.34),

$$\dot{m}_{a,1}\omega_1 + \dot{m}_{steam} = \dot{m}_{a,2}\omega_2$$

$$\left(133.53 \; \frac{\text{lbm air}}{\text{min}}\right)\left(0.00623 \; \frac{\text{lbm moisture}}{\text{lbm air}}\right)$$
$$+ \dot{m}_{steam} = \left(133.53 \; \frac{\text{lbm air}}{\text{min}}\right)\omega_2 \quad \text{[Eq. 2]}$$

Solve for $\dot{m}_{steam}$.

$$\dot{m}_{steam} = 133.53\omega_2 - 0.8319 \quad \text{[Eq. 4]}$$

Equate Eq. 3 and Eq. 4.

$$0.1161h_2 - 2.740 = 133.53\omega_2 - 0.8319$$
$$h_2 = 1150.1\omega_2 + 16.43$$

Plot this line on the psychrometric chart, and determine properties where it crosses the 75% relative humidity line.

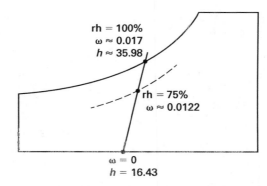

At 75% relative humidity, from the psychrometric chart,

$$\omega_2 = 0.0122 \text{ lbm moisture/lbm air}$$
$$h_2 = 30.5 \text{ Btu/lbm}$$
$$T_{db} = 72°F$$
$$T_{wb} = 66°F$$

The mass flow rate is

$$\dot{m}_{steam} = (0.1161)\left(30.5 \; \frac{\text{Btu}}{\text{lbm}}\right) - 2.740 = \boxed{0.801 \text{ lbm/min}}$$

(See the SI solution for a trial-and-error solution procedure.)

SI Solution

((a) and (b)) From Prob. 6,

$$\omega_1 = 6.2 \text{ g moisture/kg air}$$
$$h_1 = 36.75 \text{ kJ/kg air}$$
$$\dot{m}_{a,1} = 1.010 \text{ kg air/s}$$

From the steam table (see App. 29.O), for 1 atm steam, $h_{steam} = 2675.4 \text{ kJ/kg}$.

From the conservation of energy equation (see Eq. 31.33),

$$\dot{m}_{a,1}h_1 + \dot{m}_{steam}h_{steam} = \dot{m}_{a,1}h_2$$

$$\left(1.010 \ \frac{\text{kg air}}{\text{s}}\right)\left(36.75 \ \frac{\text{kJ}}{\text{kg air}}\right)$$

$$+ \dot{m}_{steam}\left(2675.4 \ \frac{\text{kJ}}{\text{kg}}\right) = \left(1.010 \ \frac{\text{kg air}}{\text{s}}\right)h_2$$

[Eq. 1]

From a conservation of mass for the water (see Eq. 31.34),

$$\dot{m}_{a,1}\omega_1 + \dot{m}_{steam} = \dot{m}_{a,2}\omega_2$$

$$\left(1.010 \ \frac{\text{kg air}}{\text{s}}\right)\left(6.2 \ \frac{\text{g moisture}}{\text{kg air}}\right)$$

$$+ \dot{m}_{steam} = \left(1.010 \ \frac{\text{kg air}}{\text{s}}\right)\omega_2$$

[Eq. 2]

Since no single relationship exists between ω_2, $\dot{m}_{steam}$, and h_2, a trial-and-error solution can be used. (See the customary U.S. solution for a graphical method.) Once $\dot{m}_{steam}$ is selected, ω_2 and h_2 can be found from Eq. 1 and Eq. 2 as

$$h_2 = 36.75 + 2648.9\dot{m}_{steam}$$

$$\omega_2 = 0.0062 + 0.99\dot{m}_{steam}$$

Once h_2 and ω_2 are known, the relative humidity can be determined from the psychrometric chart. Continue the process until a relative humidity of 75% is achieved.

$\dot{m}_{steam}$ $\left(\dfrac{\text{kg}}{\text{s}}\right)$	ω_2 $\left(\dfrac{\text{kg moisture}}{\text{kg air}}\right)$	h_2 $\left(\dfrac{\text{kJ}}{\text{kg air}}\right)$	ϕ_2 (%)
0.005	0.0112	49.99	69.5
0.0055	0.0116	51.32	74.5
0.0056	0.0117	51.58	75.0

$$\dot{m}_{steam} = 0.0056 \ \text{kg/s}$$

$$\omega_2 = \left(0.0117 \ \frac{\text{kg moisture}}{\text{kg air}}\right)\left(1000 \ \frac{\text{g}}{\text{kg}}\right)$$

$$= 11.7 \ \text{g moisture/kg air}$$

$$T_{db} = 21.3°C$$

$$T_{wb} = 18.2°C$$

8. *Customary U.S. Solution*

(a) From the psychrometric chart (see App. 31.A), for incoming air at 65°F and 55% relative humidity, $\omega_1 = 0.0072$ lbm moisture/lbm air.

With sensible heating as a limiting factor, calculate the mass flow rate of air entering the theater from Eq. 31.25 (ventilation rate).

$$\dot{q} = \dot{m}_a(c_{p,air} + \omega c_{p,moisture})(T_2 - T_1)$$

$$500,000 \ \frac{\text{Btu}}{\text{hr}} = \dot{m}_a$$

$$\times \left(\begin{array}{c} 0.240 \ \dfrac{\text{Btu}}{\text{lbm-°F}} \\ + \ 0.0072 \ \dfrac{\text{lbm moisture}}{\text{lbm air}} \\ \times \left(0.444 \ \dfrac{\text{Btu}}{\text{lbm-°F}}\right) \end{array} \right)$$

$$\times (75°F - 65°F)$$

$$\dot{m}_a = \boxed{2.056 \times 10^5 \ \text{lbm air/hr}}$$

(b) Assume that this air absorbs all the moisture. Then, the final humidity ratio is given by

$$\dot{m}_w = \dot{m}_a(\omega_2 - \omega_1)$$

$$\omega_2 = \left(\frac{\dot{m}_w}{\dot{m}_a}\right) + \omega_1$$

$$= \frac{175 \ \dfrac{\text{lbm moisture}}{\text{hr}}}{2.056 \times 10^5 \ \dfrac{\text{lbm air}}{\text{hr}}} + 0.0072 \ \frac{\text{lbm moisture}}{\text{lbm air}}$$

$$= 0.00805 \ \text{lbm moisture/lbm air}$$

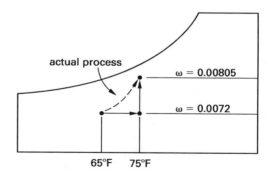

The final conditions are

$$\boxed{\begin{array}{l} T_{db} = 75°F \quad \text{[given]} \\ \omega_2 = 0.00805 \ \text{lbm moisture/lbm air} \end{array}}$$

From the psychrometric chart (see App. 31.A), the relative humidity is 44%. This is below 60%.

SI Solution

(a) From the psychrometric chart (see App. 31.B), for incoming air at 18°C and 55% relative humidity, $\omega_1 = 7.1$ g moisture/kg air.

With sensible heating as a limiting factor, calculate the mass flow rate of air entering the theater from Eq. 31.25 (ventilation rate).

$$\dot{q} = \dot{m}_a(c_{p,\text{air}} + \omega c_{p,\text{moisture}})(T_2 - T_1)$$

$$(150 \text{ kW}) \times \left(1000 \frac{\text{W}}{\text{kW}}\right) = \dot{m}_a \begin{pmatrix} \left(1.005 \frac{\text{kJ}}{\text{kg·°C}}\right)\left(1000 \frac{\text{J}}{\text{kJ}}\right) \\ + \left(1.805 \frac{\text{kJ}}{\text{kg·°C}}\right)\left(1000 \frac{\text{J}}{\text{kJ}}\right) \\ \times \left(\dfrac{7.1 \frac{\text{g moisture}}{\text{kg air}}}{1000 \frac{\text{g}}{\text{kg}}}\right) \end{pmatrix}$$
$$\times (24°C - 18°C)$$

$$\dot{m}_a = \boxed{24.56 \text{ kg/s}}$$

(b) Assume that this air absorbs all the moisture. Then, the final humidity ratio is given by

$$\dot{m}_w = \dot{m}_a(\omega_2 - \omega_1)$$

$$\omega_2 = \frac{\dot{m}_w}{\dot{m}_a} + \omega_1$$

$$= \frac{80 \frac{\text{kg}}{\text{h}}}{\left(24.56 \frac{\text{kg}}{\text{s}}\right)\left(3600 \frac{\text{s}}{\text{h}}\right)} + \frac{7.0 \frac{\text{g moisture}}{\text{kg air}}}{1000 \frac{\text{g}}{\text{kg}}}$$

$$= 0.00790 \text{ kg moisture/kg air}$$

The final conditions are

$$\boxed{\begin{array}{l} T_{\text{db}} = 24°C \\ \omega_2 = \left(0.00790 \frac{\text{kg moisture}}{\text{kg air}}\right)\left(1000 \frac{\text{g}}{\text{kg}}\right) \\ \quad = 7.9 \text{ g moisture/kg air} \end{array}}$$

From the psychrometric chart (see App. 31.B), the relative humidity is 44%. This is below 60%.

9. Use the illustration shown for both customary U.S. and SI solutions.

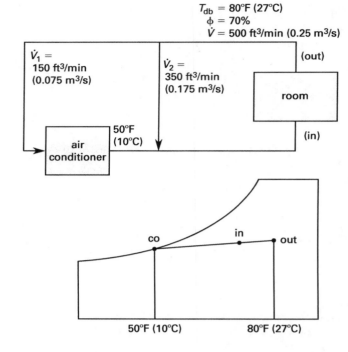

Customary U.S. Solution

Locate point "out" ($T_{\text{db}} = 80°F$ and $\phi = 70\%$) and point "co" (saturated at 50°F) on the psychrometric chart. At point "out" from App. 31.A,

$$v_{\text{out}} = 13.95 \text{ ft}^3/\text{lbm air}$$

$$h_{\text{out}} = 36.2 \text{ Btu/lbm air}$$

At point "co," $h_{\text{co}} = 20.3$ Btu/lbm air.

The air mass flow rate through the air conditioner is

$$\dot{m}_1 = \frac{\dot{V}_1}{v_1} = \frac{\dot{V}_1}{v_{\text{out}}} = \frac{150 \frac{\text{ft}^3}{\text{min}}}{13.95 \frac{\text{ft}^3}{\text{lbm air}}} = 10.75 \text{ lbm air/min}$$

The mass flow rate of the bypass air is

$$\dot{m}_2 = \frac{\dot{V}_2}{v} = \frac{350 \frac{\text{ft}^3}{\text{min}}}{13.95 \frac{\text{ft}^3}{\text{lbm air}}} = 25.09 \text{ lbm air/min}$$

The percentage of bypass air is

$$x = \frac{25.09 \frac{\text{lbm air}}{\text{min}}}{10.75 \frac{\text{lbm air}}{\text{min}} + 25.09 \frac{\text{lbm air}}{\text{min}}} = 0.70 \quad (70\%)$$

Using the lever rule and the fact that all of the temperature scales are linear,

$$T_{\text{db,in}} = T_{\text{co}} + (0.70)(T_{\text{out}} - T_{\text{co}})$$
$$= 50°F + (0.70)(80°F - 50°F)$$
$$= 71°F$$

At that point,

(a) $\boxed{T_{\text{db,in}} = 71°F}$

The answer is (C).

(b) $\boxed{\begin{array}{c} \omega_{\text{in}} = 0.0132 \text{ lbm moisture/lbm air} \\ (0.013 \text{ lbm/lbm}) \end{array}}$

The answer is (B).

(c) $\boxed{\phi_{\text{in}} = 81\%}$

The answer is (D).

(d) The air conditioner capacity is given by

$$\dot{Q} = \dot{m}_{\text{air}}(h_{t,2} - h_{t,1}) = \dot{m}_1(h_{\text{out}} - h_{\text{co}})$$
$$= \left(10.75 \ \frac{\text{lbm air}}{\text{min}}\right)\left(36.2 \ \frac{\text{Btu}}{\text{lbm air}} - 20.3 \ \frac{\text{Btu}}{\text{lbm air}}\right)$$
$$\times \left(\frac{1 \text{ ton}}{200 \ \frac{\text{Btu}}{\text{min}}}\right)$$
$$= \boxed{0.85 \text{ ton} \quad (0.90 \text{ ton})}$$

The answer is (A).

SI Solution

Locate point "out" ($T_{\text{db}} = 27°C$, $\phi = 70\%$) and point "co" (saturated at 10°C) on the psychrometric chart. At point "out" from App. 31.B,

$$v_{\text{out}} = 0.872 \text{ m}^3/\text{kg air}$$
$$h_{\text{out}} = 67.3 \text{ kJ/kg air}$$

At point "co" from App. 31.B, $h_{\text{co}} = 29.26$ kJ/kg air.

At mass flow rate through the air conditioner,

$$\dot{m}_1 = \frac{\dot{V}_1}{v} = \frac{0.075 \ \frac{\text{m}^3}{\text{s}}}{0.872 \ \frac{\text{m}^3}{\text{kg air}}} = 0.0860 \text{ kg air/s}$$

The flow rate of bypass air is

$$\dot{m}_2 = \frac{\dot{V}_2}{v} = \frac{0.175 \ \frac{\text{m}^3}{\text{s}}}{0.872 \ \frac{\text{m}^3}{\text{kg air}}} = 0.2007 \text{ kg air/s}$$

The percentage bypass air is

$$x = \frac{0.2007 \ \frac{\text{kg air}}{\text{s}}}{0.0860 \ \frac{\text{kg air}}{\text{s}} + 0.2007 \ \frac{\text{kg air}}{\text{s}}} = 0.70 \quad (70\%)$$

Using the lever rule and the fact that all of the temperature scales are linear,

$$T_{\text{db,in}} = T_{\text{co}} + (0.70)(T_{\text{out}} - T_{\text{co}})$$
$$= 10°C + (0.70)(27°C - 10°C)$$
$$= 21.9°C$$

At that point,

(a) $\boxed{T_{\text{db,in}} = 21.9°C \quad (22°C)}$

The answer is (C).

(b) $\boxed{\omega_{\text{in}} = 1.34 \text{ g moisture/kg air} \quad (1.3 \text{ g/kg})}$

The answer is (B).

(c) $\boxed{\phi_{\text{in}} = 81\%}$

The answer is (D).

(d) The air conditioner capacity is given by

$$\dot{Q} = \dot{m}_{\text{air}}(h_{t,2} - h_{t,1}) = \dot{m}_1(h_{\text{out}} - h_{\text{co}})$$
$$= \left(0.0860 \ \frac{\text{kg air}}{\text{s}}\right)\left(67.3 \ \frac{\text{kJ}}{\text{kg air}} - 29.26 \ \frac{\text{kJ}}{\text{kg air}}\right)$$
$$\times \left(0.2843 \ \frac{\text{ton}}{\text{kW}}\right)$$
$$= \boxed{0.93 \text{ ton} \quad (0.90 \text{ ton})}$$

The answer is (A).

10. Use the illustration shown for both customary U.S. and SI solutions.

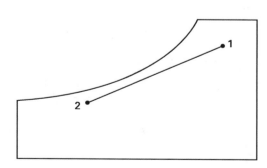

Customary U.S. Solution

(a) At point 1, from the psychrometric chart (see App. 31.A) at $T_{db} = 95°F$ and $\phi = 70\%$,

$$\boxed{\begin{array}{l} h_1 = 50.7 \text{ Btu/lbm air} \\ v_1 = 14.56 \text{ ft}^3/\text{lbm air} \\ \omega_1 = 0.0253 \text{ lbm water/lbm air} \end{array}}$$

At point 2, from the psychrometric chart (see App. 31.A) at $T_{db} = 60°F$ and $\phi = 95\%$,

$$\boxed{\begin{array}{l} h_2 = 25.8 \text{ Btu/lbm air} \\ \omega_2 = 0.0105 \text{ lbm water/lbm air} \end{array}}$$

(b) The air mass flow rate is

$$\dot{m}_a = \frac{\dot{V}}{v_1} = \frac{5000 \; \dfrac{\text{ft}^3}{\text{min}}}{14.56 \; \dfrac{\text{ft}^3}{\text{lbm air}}} = 343.4 \text{ lbm air/min}$$

From Eq. 31.26, the water removed is

$$\dot{m}_w = \dot{m}_a(\omega_1 - \omega_2)$$
$$= \left(343.4 \; \frac{\text{lbm air}}{\text{min}}\right)$$
$$\quad \times \left(0.0253 \; \frac{\text{lbm water}}{\text{lbm air}} - 0.0105 \; \frac{\text{lbm water}}{\text{lbm air}}\right)$$
$$= \boxed{5.08 \text{ lbm water/min}}$$

(c) From Eq. 31.27, the quantity of heat removed is

$$\dot{q} = \dot{m}_a(h_1 - h_2)$$
$$= \left(343.4 \; \frac{\text{lbm air}}{\text{min}}\right)\left(50.7 \; \frac{\text{Btu}}{\text{lbm air}} - 25.8 \; \frac{\text{Btu}}{\text{lbm air}}\right)$$
$$= \boxed{8551 \text{ Btu/min}}$$

(d) Considering an R-12 refrigeration cycle operating at saturated conditions at 100°F, the *T-s* and *h-s* diagrams are shown.

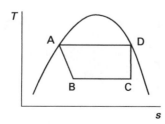 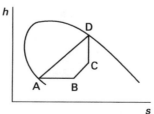

(e) Use App. 29.G for saturated conditions. At point A,

$$T = \boxed{100°F} \quad \text{[given]}$$
$$p = p_{sat} \text{ at } 100°F = \boxed{131.6 \text{ psia}}$$
$$h_A = h_f \text{ at } 100°F = \boxed{31.16 \text{ Btu/lbm}}$$
$$s_A = s_f \text{ at } 100°F = \boxed{0.06316 \text{ Btu/lbm-°R}}$$
$$v_A = v_f \text{ at } 100°F = \boxed{0.0127 \text{ ft}^3/\text{lbm}}$$

At point B,

$$T = \boxed{50°F} \quad \text{[given]}$$
$$p = p_{sat} \text{ at } 50°F = \boxed{61.39 \text{ psia}}$$
$$h_B = h_A = \boxed{31.16 \text{ Btu/lbm}}$$
$$h_{f,B} = 19.27 \text{ Btu/lbm}$$
$$h_{fg,B} = 64.51 \text{ Btu/lbm}$$
$$x = \frac{h_B - h_{f,B}}{h_{fg,B}} = \frac{31.16 \; \dfrac{\text{Btu}}{\text{lbm}} - 19.27 \; \dfrac{\text{Btu}}{\text{lbm}}}{64.51 \; \dfrac{\text{Btu}}{\text{lbm}}} = 0.184$$
$$s_B = s_{f,B} + x s_{fg,B}$$
$$= 0.04126 \; \frac{\text{Btu}}{\text{lbm-°R}} + (0.184)\left(0.12659 \; \frac{\text{Btu}}{\text{lbm-°R}}\right)$$
$$= \boxed{0.06455 \text{ Btu/lbm-°R}}$$
$$v_B = v_{f,B} + x(v_g - v_f)$$
$$= 0.0118 \; \frac{\text{ft}^3}{\text{lbm}} + (0.184)\left(0.673 \; \frac{\text{ft}^3}{\text{lbm}} - 0.0118 \; \frac{\text{ft}^3}{\text{lbm}}\right)$$
$$= \boxed{0.1335 \text{ ft}^3/\text{lbm}}$$

At point D,

$$T = \boxed{100°F} \quad \text{[given]}$$
$$p = p_{sat} \text{ at } 100°F = \boxed{131.6 \text{ psia}}$$
$$h_D = h_g \text{ at } 100°F = \boxed{88.62 \text{ Btu/lbm}}$$
$$s_D = s_g \text{ at } 100°F = \boxed{0.16584 \text{ Btu/lbm-°R}}$$
$$v_D = v_g \text{ at } 100°F = \boxed{0.319 \text{ ft}^3/\text{lbm}}$$

At point C,

$$T = \boxed{50°F} \quad \text{[given]}$$
$$p = p_{sat} \text{ at } 50°F = \boxed{61.39 \text{ psia}}$$
$$s_C = s_D = \boxed{0.16584 \text{ Btu/lbm-°R}}$$
$$s_{f,C} = 0.04126 \text{ Btu/lbm-°R}$$
$$s_{fg,C} = 0.12659 \text{ Btu/lbm-°R}$$

Ventilation

$$x = \frac{s_C - s_{f,C}}{s_{fg,C}} = \frac{0.16584 \, \frac{Btu}{lbm\text{-}°R} - 0.04126 \, \frac{Btu}{lbm\text{-}°R}}{0.12659 \, \frac{Btu}{lbm\text{-}°R}}$$

$$= 0.984$$

$$h_C = h_{f,C} + x h_{fg,C} = 19.27 \, \frac{Btu}{lbm} + (0.984)\left(64.51 \, \frac{Btu}{lbm}\right)$$

$$= \boxed{82.75 \text{ Btu/lbm}}$$

$$v_C = v_{f,C} + x(v_{g,C} - v_{f,C})$$

$$= 0.0118 \, \frac{ft^3}{lbm}$$

$$+ (0.984)\left(0.673 \, \frac{ft^3}{lbm} - 0.0118 \, \frac{ft^3}{lbm}\right)$$

$$= \boxed{0.662 \text{ ft}^3/\text{lbm}}$$

SI Solution

(a) At point 1, from the psychrometric chart (see App. 31.B) at $T_{db} = 35°C$ and $\phi = 70\%$,

$$\boxed{\begin{array}{l} h_1 = 99.9 \text{ kJ/kg air} \\[4pt] v_1 = 0.91 \text{ m}^3/\text{kg air} \\[4pt] \omega_1 = \dfrac{25.3 \, \frac{\text{g moisture}}{\text{kg air}}}{1000 \, \frac{\text{g}}{\text{kg}}} = 0.0253 \text{ kg moisture/kg air} \end{array}}$$

At point 2, from the psychrometric chart (see App. 31.B) at $T_{db} = 16°C$ and $\phi = 95\%$,

$$\boxed{\begin{array}{l} h_2 = 43.4 \text{ kJ/kg air} \\[4pt] \omega_2 = \dfrac{10.8 \, \frac{\text{g moisture}}{\text{kg air}}}{1000 \, \frac{\text{g}}{\text{kg}}} = 0.0108 \text{ kg moisture/kg air} \end{array}}$$

(b) The air mass flow rate is

$$\dot{m}_a = \frac{\dot{V}_1}{v_1} = \frac{\dot{V}_1}{v_{out}} = \frac{2.36 \, \frac{m^3}{s}}{0.91 \, \frac{m^3}{\text{kg air}}} = 2.593 \text{ kg air/s}$$

From Eq. 31.26, the water removed is

$$\dot{m}_w = \dot{m}_a(\omega_1 - \omega_2)$$

$$= \left(2.593 \, \frac{\text{kg air}}{s}\right)\left(\begin{array}{l} 0.0253 \, \dfrac{\text{kg moisture}}{\text{kg air}} \\[8pt] - 0.0108 \, \dfrac{\text{kg moisture}}{\text{kg air}} \end{array}\right)$$

$$= \boxed{0.0376 \text{ kg moisture/s}}$$

(c) From Eq. 31.27, the quantity of heat removed is

$$\dot{q} = \dot{m}_a(h_1 - h_2)$$

$$= \left(2.593 \, \frac{\text{kg air}}{s}\right)\left(99.9 \, \frac{\text{kJ}}{\text{kg air}} - 43.4 \, \frac{\text{kJ}}{\text{kg air}}\right)$$

$$= \boxed{146.5 \text{ kW}}$$

(d) The T-s and h-s diagrams are shown in the customary U.S. solution.

(e) Use App. 29.T for saturated conditions. At point A,

$$T = \boxed{38°C} \quad \text{[given]}$$

$$p = p_{sat} \text{ at } 38°C = \boxed{0.91324 \text{ MPa}}$$

$$h_A = h_f \text{ at } 38°C = \boxed{237.23 \text{ kJ/kg}}$$

$$s_A = s_f \text{ at } 38°C = \boxed{1.1259 \text{ kJ/kg·°C}}$$

$$v_A = v_f \text{ at } 38°C = \frac{1}{\rho_f} = \frac{1}{1261.9 \, \frac{\text{kg}}{\text{m}^3}}$$

$$= \boxed{0.0007925 \text{ m}^3/\text{kg}}$$

At point B,

$$T = \boxed{10°C} \quad \text{[given]}$$

$$p = p_{sat} \text{ at } 10°C = \boxed{0.42356 \text{ MPa}}$$

$$h_B = h_A = \boxed{237.23 \text{ kJ/kg}}$$

$$h_{f,B} = 209.48 \text{ kJ/kg}$$

$$h_{g,B} = 356.79 \text{ kJ/kg}$$

$$x = \frac{h_B - h_{f,B}}{h_{g,B} - h_{f,B}} = \frac{237.23 \, \frac{\text{kJ}}{\text{kg}} - 209.48 \, \frac{\text{kJ}}{\text{kg}}}{356.79 \, \frac{\text{kJ}}{\text{kg}} - 209.48 \, \frac{\text{kJ}}{\text{kg}}}$$

$$= 0.188$$

$$s_B = s_{f,B} + x(s_{g,B} - s_{f,B})$$

$$= 1.0338 \, \frac{\text{kJ}}{\text{kg·K}}$$

$$+ (0.188)\left(1.5541 \, \frac{\text{kJ}}{\text{kg·K}} - 1.0338 \, \frac{\text{kJ}}{\text{kg·K}}\right)$$

$$= \boxed{1.1316 \text{ kJ/kg·K}}$$

$$v_B = v_{f,B} + x(v_{g,B} - v_{f,B})$$

$$v_{f,B} = \frac{1}{\rho_{f,B}} = \frac{1}{1363.0 \, \frac{\text{kg}}{\text{m}^3}} = 0.0007337 \text{ m}^3/\text{kg}$$

$$v_{g,B} = 0.04119 \text{ m}^3/\text{kg}$$

Ventilation

$$v_B = 0.0007337 \ \frac{m^3}{kg}$$
$$+ (0.188)\left(0.04119 \ \frac{m^3}{kg} - 0.0007337 \ \frac{m^3}{kg}\right)$$
$$= \boxed{0.008339 \ m^3/kg}$$

At point D,

$$T = \boxed{38°C} \quad \text{[given]}$$
$$p = p_{sat} \text{ at } 38°C = \boxed{0.91324 \text{ MPa}}$$
$$h_D = h_g \text{ at } 38°C = \boxed{367.95 \text{ kJ/kg}}$$
$$s_D = s_g \text{ at } 38°C = \boxed{1.5461 \text{ kJ/kg·K}}$$
$$v_D = v_g \text{ at } 38°C = \boxed{0.01931 \text{ m}^3/\text{kg}}$$

At point C,

$$T = \boxed{10°C}$$
$$p = p_{sat} \text{ at } 10°C = \boxed{0.42356 \text{ MPa}}$$
$$s_C = s_D = \boxed{1.5461 \text{ kJ/kg·K}}$$
$$s_{f,C} = 1.0338 \text{ kJ/kg·K}$$
$$s_{g,C} = 1.5541 \text{ kJ/kg·K}$$

$$x = \frac{s_C - s_{f,C}}{s_{g,C} - s_{f,C}} = \frac{1.5461 \ \frac{kJ}{kg·K} - 1.0338 \ \frac{kJ}{kg·K}}{1.5541 \ \frac{kJ}{kg·K} - 1.0338 \ \frac{kJ}{kg·K}} = 0.985$$

$$h_C = h_{f,C} + x(h_{g,C} - h_{f,C})$$
$$= 209.48 \ \frac{kJ}{kg} + (0.985)\left(356.79 \ \frac{kJ}{kg} - 209.48 \ \frac{kJ}{kg}\right)$$
$$= \boxed{354.58 \text{ kJ/kg}}$$

$$v_C = v_{f,C} + x(v_{g,C} - v_{f,C})$$
$$v_{g,C} = 0.04119 \text{ m}^3/\text{kg}$$

$$v_C = 0.0007337 \ \frac{m^3}{kg}$$
$$+ (0.985)\left(0.04119 \ \frac{m^3}{kg} - 0.0007337 \ \frac{m^3}{kg}\right)$$
$$= \boxed{0.04058 \ m^3/kg}$$

11. *Customary U.S. Solution*

(a) The saturation pressure at 200°F from App. 29.A is $p_{sat,1} = 11.538$ psia.

Since air is saturated (100% relative humidity), the water vapor pressure is equal to the saturation pressure.

$$p_{w,1} = p_{sat,1} = 11.538 \text{ psia}$$

The partial pressure of the air is

$$p_{a,1} = p_1 - p_{w,1} = 25 \text{ psia} - 11.538 \text{ psia}$$
$$= 13.462 \text{ psia}$$

Use Eq. 31.7 to determine the humidity ratio.

$$\omega = 0.622\left(\frac{p_{w,1}}{p_{a,1}}\right) = (0.622)\left(\frac{11.538 \text{ psia}}{13.462 \text{ psia}}\right) = 0.533$$

Since it is a constant pressure, constant moisture drying process, mole fractions and partial pressures do not change.

$$p_{w,2} = p_{w,1} = 11.538 \text{ psia}$$

The saturation pressure at 400°F from App. 29.A is $p_{sat,2} = 247.3$ psia.

The relative humidity at state 2 is

$$\phi_2 = \frac{p_{w,2}}{p_{sat,2}} = \frac{11.538 \text{ psia}}{247.3 \text{ psia}} = \boxed{0.0467 \quad (4.7\%)}$$

The answer is (A).

(b) Although the volume may change, the mass does not. The specific humidity remains constant.

$$\omega_2 = \omega_1 = \boxed{0.533 \text{ lbm water/lbm air} \quad (0.53 \text{ lbm/lbm})}$$

The answer is (B).

(c) The heat required consists of two parts.

Obtain enthalpy for air from App. 29.F.

The absolute temperatures are

$$T_1 = 200°F + 460° = 660°R$$
$$T_2 = 400°F + 460° = 860°R$$
$$h_1 = 157.92 \text{ Btu/lbm}$$
$$h_2 = 206.46 \text{ Btu/lbm}$$

The heat absorbed by the air is

$$q_1 = h_2 - h_1 = 206.46 \ \frac{\text{Btu}}{\text{lbm}} - 157.92 \ \frac{\text{Btu}}{\text{lbm}}$$
$$= 48.54 \text{ Btu/lbm dry air}$$

(There will be a small error if constant specific heat is used instead.)

For water, use the Mollier diagram. From App. 29.E, h_1 at 200°F and 11.529 psia is 1146 Btu/lbm (almost saturated).

Follow a constant 11.529 psia pressure curve up to 400°F.

$$h_2 = 1240 \text{ Btu/lbm}$$

(There will be a small error if Eq. 31.19(b) is used instead.)

The heat absorbed by the steam is

$$q_2 = \omega(h_2 - h_1)$$
$$= \left(0.532 \, \frac{\text{lbm water}}{\text{lbm air}}\right)\left(1240 \, \frac{\text{Btu}}{\text{lbm}} - 1146 \, \frac{\text{Btu}}{\text{lbm}}\right)$$
$$= 50.01 \text{ Btu/lbm air}$$

The total heat absorbed is

$$q_{\text{total}} = q_1 + q_2 = 48.54 \, \frac{\text{Btu}}{\text{lbm air}} + 50.01 \, \frac{\text{Btu}}{\text{lbm air}}$$
$$= \boxed{98.55 \text{ Btu/lbm air} \quad (99 \text{ Btu/lbm})}$$

(d) The dew point is the temperature at which water starts to condense out in a constant pressure process. Following the constant 11.538 psia pressure line back to the saturation line, $\boxed{T_{\text{dp}} = 200°\text{F}.}$

The answer is (B).

SI Solution

(a) From App. 29.N, the saturation pressure at 93°C is

$$p_{\text{sat},1} = (0.7884 \text{ bar})\left(100 \, \frac{\text{kPa}}{\text{bar}}\right) = 78.84 \text{ kPa}$$

Since air is saturated (100% relative humidity), water vapor pressure is equal to saturation pressure.

$$p_{w,1} = p_{\text{sat},1} = 78.84 \text{ kPa}$$

The partial pressure of air is

$$p_{a,1} = p_1 - p_{w,1} = 170 \text{ kPa} - 78.84 \text{ kPa}$$
$$= 91.16 \text{ kPa}$$

Use Eq. 31.7 to determine the humidity ratio.

$$\omega = 0.622\left(\frac{p_{w,1}}{p_{a,1}}\right) = (0.622)\left(\frac{78.84 \text{ kPa}}{91.16 \text{ kPa}}\right)$$
$$= 0.538 \text{ kg water/kg air}$$

Since it is a constant pressure, constant moisture drying process, mole fractions and partial pressure do not change.

$$p_{w,2} = p_{w,1} = 78.84 \text{ kPa}$$

The saturation pressure at 204°C from App. 29.N is

$$p_{\text{sat},2} = (16.90 \text{ bar})\left(100 \, \frac{\text{kPa}}{\text{bar}}\right) = 1690 \text{ kPa}$$

The relative humidity at state 2 is

$$\phi_2 = \frac{p_{w,2}}{p_{\text{sat},2}} = \frac{78.84 \text{ kPa}}{1690 \text{ kPa}} = \boxed{0.0467 \quad (4.7\%)}$$

The answer is (A).

(b) Although the volume may change, the mass does not. The specific humidity remains constant.

$$\omega_2 = \omega_1 = \boxed{0.538 \text{ kg water/kg air} \quad (0.53 \text{ kg/kg})}$$

The answer is (B).

(c) The heat required consists of two parts.

Obtain enthalpy for air from App. 29.S.

The absolute temperatures are

$$T_1 = 93°\text{C} + 273° = 366\text{K}$$
$$T_2 = 204°\text{C} + 273° = 477\text{K}$$
$$h_1 = 366.63 \text{ kJ/kg}$$
$$h_2 = 479.42 \text{ kJ/kg}$$

The heat absorbed by air is

$$q_1 = h_2 - h_1 = 479.42 \, \frac{\text{kJ}}{\text{kg}} - 366.63 \, \frac{\text{kJ}}{\text{kg}}$$
$$= 112.79 \text{ kJ/kg air}$$

(There will be a small error if constant specific heat is used instead.)

For water, use the Mollier diagram. From App. 29.R, h_1 at 93°C and 78.79 kPa is 2670 kJ/kg (almost saturated).

Follow a constant 78.79 kPa pressure curve up to 204°C.

$$h_2 = 2890 \text{ kJ/kg}$$

(There will be a small error if Eq. 31.19(a) is used instead.)

The heat absorbed by steam is

$$q_2 = \omega(h_2 - h_1)$$
$$= \left(0.537 \, \frac{\text{kg water}}{\text{kg air}}\right)\left(2890 \, \frac{\text{kJ}}{\text{kg}} - 2670 \, \frac{\text{kJ}}{\text{kg}}\right)$$
$$= 118.14 \text{ kJ/kg air}$$

The total heat absorbed is

$$q_{\text{total}} = q_1 + q_2 = 112.79 \ \frac{\text{kJ}}{\text{kg air}} + 118.14 \ \frac{\text{kJ}}{\text{kg air}}$$

$$= \boxed{230.93 \ \text{kJ/kg air} \quad (230 \ \text{kJ/kg})}$$

The answer is (C).

(d) The dew point is the temperature at which water starts to condense out in a constant pressure process. Following the constant 78.84 kPa pressure line back to the saturation line, $\boxed{T_{\text{dp}} \approx 93°C.}$

The answer is (B).

12. *Customary U.S. Solution*

(a) The absolute air temperatures are

$$T_1 = 800°F + 460° = 1260°R$$

$$T_2 = 350°F + 460° = 810°R$$

At low pressures, use air tables. From App. 29.F,

$$h_1 = 306.65 \ \text{Btu/lbm}$$

$$h_2 = 194.25 \ \text{Btu/lbm}$$

From App. 29.A, the enthalpy of water at 80°F is $h_{w,1} = 48.07$ Btu/lbm (48 Btu/lbm).

From Eq. 31.19(b), the enthalpy of steam at 350°F is

$$h_{w,2} \approx \left(0.444 \ \frac{\text{Btu}}{\text{lbm-°F}} \right)(350°F) + 1061 \ \frac{\text{Btu}}{\text{lbm}}$$

$$= 1216.4 \ \text{Btu/lbm}$$

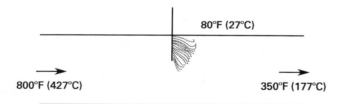

Air temperature is reduced from 800°F to 350°F, and this energy is used to change water at 80°F to steam at 350°F. From the energy balance equation,

$$\dot{m}_w(h_{w,2} - h_{w,1}) = \dot{m}_a(h_1 - h_2)$$

$$\dot{m}_w = \frac{\dot{m}_a(h_1 - h_2)}{h_{w,2} - h_{w,1}}$$

$$= \frac{\left(410 \ \frac{\text{lbm}}{\text{hr}} \right) \left(\begin{array}{c} 306.65 \ \dfrac{\text{Btu}}{\text{lbm}} \\ - 194.25 \ \dfrac{\text{Btu}}{\text{lbm}} \end{array} \right)}{1216.4 \ \dfrac{\text{Btu}}{\text{lbm}} - 48.07 \ \dfrac{\text{Btu}}{\text{lbm}}}$$

$$= \boxed{39.4 \ \text{lbm/hr water} \quad (39 \ \text{lbm/hr})}$$

The answer is (D).

(b) The number of moles of water evaporated (in the air mixture) is

$$\dot{n}_w = \frac{39.4 \ \dfrac{\text{lbm}}{\text{hr}}}{18.016 \ \dfrac{\text{lbm}}{\text{lbmol}}} = 2.19 \ \text{lbmol/hr}$$

The number of moles of air in the mixture at the exit is

$$\dot{n}_a = \frac{410 \ \dfrac{\text{lbm}}{\text{hr}}}{28.967 \ \dfrac{\text{lbm}}{\text{lbmol}}} = 14.15 \ \text{lbmol/hr}$$

The mole fraction of water in the mixture is

$$x_w = \frac{\dot{n}_w}{\dot{n}_a + \dot{n}_w} = \frac{2.19 \ \dfrac{\text{lbmol}}{\text{hr}}}{14.15 \ \dfrac{\text{lbmol}}{\text{hr}} + 2.19 \ \dfrac{\text{lbmol}}{\text{hr}}}$$

$$= 0.134$$

The partial pressure of water vapor is

$$p_w = x p_{\text{chamber}} = (0.134)(20 \ \text{psia}) = 2.68 \ \text{psia}$$

From App. 29.A, the saturation pressure at 350°F is $p_{\text{sat}} = 134.63$ psia.

The relative humidity is

$$\phi = \frac{p_w}{p_{\text{sat}}} = \frac{2.68 \ \text{psia}}{134.63 \ \text{psia}} = \boxed{0.020 \quad (2.0\%)}$$

The answer is (B).

SI Solution

(a) The absolute temperatures are

$$T_1 = 427°C + 273° = 700\text{K}$$

$$T_2 = 177°C + 273° = 450\text{K}$$

Air tables can be used at low pressures. From App. 29.S,

$$h_1 = 713.27 \ \text{kJ/kg}$$

$$h_2 = 451.80 \ \text{kJ/kg}$$

From App. 29.N, the enthalpy of water at 27°C is $h_{w,1} = 113.19$ kJ/kg (113 kJ/kg).

From Eq. 31.19(a), the enthalpy of steam at 177°C is

$$h_{w,2} = \left(1.805 \ \frac{\text{kJ}}{\text{kg·°C}} \right)(177°C) + 2501 \ \frac{\text{kJ}}{\text{kg}}$$

$$= 2820.5 \ \text{kJ/kg}$$

Air temperature is reduced from 427°C to 177°C, and this energy is used to change water at 27°C to steam at 177°C. From the energy balance equation,

$$\dot{m}_w = \frac{\dot{m}_a(h_1 - h_2)}{h_{w,2} - h_{w,1}}$$

$$= \frac{\left(0.052 \, \frac{kg}{s}\right)\left(713.27 \, \frac{kJ}{kg} - 451.80 \, \frac{kJ}{kg}\right)}{2820.5 \, \frac{kJ}{kg} - 113.19 \, \frac{kJ}{kg}}$$

$$= \boxed{0.00502 \, \text{kg/s} \quad (0.0050 \, \text{kg/s})}$$

The answer is (D).

(b) The number of moles of water evaporated (in the air mixture) is

$$\dot{n}_w = \frac{0.00502 \, \frac{kg}{s}}{18.016 \, \frac{kg}{kmol}} = 2.79 \times 10^{-4} \, \text{kmol/s}$$

The number of moles of air in the mixture at the exit is

$$\dot{n}_a = \frac{0.052 \, \frac{kg}{s}}{28.967 \, \frac{kg}{kmol}} = 1.80 \times 10^{-3} \, \text{kmol/s}$$

The mole fraction of water in the mixture is

$$x_w = \frac{\dot{n}_w}{\dot{n}_a + \dot{n}_w}$$

$$= \frac{2.79 \times 10^{-4} \, \frac{kmol}{s}}{1.80 \times 10^{-3} \, \frac{kmol}{s} + 2.79 \times 10^{-4} \, \frac{kmol}{s}}$$

$$= 0.134$$

The partial pressure of water vapor is

$$p_w = x p_{\text{chamber}} = (0.134)(140 \, \text{kPa})$$

$$= 18.76 \, \text{kPa}$$

From App. 29.N, the saturation pressure at 177°C is

$$p_{\text{sat}} = (9.368 \, \text{bar})\left(100 \, \frac{\text{kPa}}{\text{bar}}\right) = 936.8 \, \text{kPa}$$

The relative humidity is

$$\phi = \frac{p_w}{p_{\text{sat}}} = \frac{18.76 \, \text{kPa}}{936.8 \, \text{kPa}} = \boxed{0.020 \quad (2.0\%)}$$

The answer is (B).

13. *Method 1*: Use a high temperature psychrometric chart (such as App. 31.D). Locate the point where the 180°F line intersects the 20% relative humidity line. From that point, move horizontally to the left to intersect the saturation curve. Read a dew point temperature of $\boxed{116°\text{F} \, (120°\text{F})}$.

Method 2: Interpolating from App. 29.B, the saturation pressure, p_{sat}, of 180°F water vapor is 7.515 psia. Use Eq. 31.9 to find the partial pressure of the water vapor, p_w. The relative humidity, ϕ, is given as 20%.

$$\phi = \frac{p_w}{p_{\text{sat}}}$$

$$p_w = \phi p_{\text{sat}} = (0.20)\left(7.515 \, \frac{\text{lbf}}{\text{in}^2}\right)$$

$$= 1.5 \, \text{lbf/in}^2$$

From App. 29.B, the dew point temperature corresponding to 1.5 lbf/in² is $\boxed{115.64°\text{F} \, (120°\text{F})}$.

The answer is (B).

14. With a pressure of 19 psia, standard conditions do not apply, so a high-temperature psychrometric chart cannot not be used.

There are 7000 grains per pound. The humidity ratio expressed in lbm/lbm is

$$\omega = \frac{560 \, \frac{\text{gr}}{\text{lbm}}}{7000 \, \frac{\text{gr}}{\text{lbm}}} = 0.08 \, \frac{\text{lbm}}{\text{lbm}}$$

The total pressure, p_t, is 19 psia. From Eq. 31.16, the partial pressure of the water vapor in the combustion products is

$$p_w = \frac{p_t \, \omega}{0.622 + \omega} = \frac{(19 \, \text{psia})\left(0.08 \, \frac{\text{lbm}}{\text{lbm}}\right)}{0.622 + 0.08 \, \frac{\text{lbm}}{\text{lbm}}}$$

$$= 2.17 \, \text{psia}$$

The dew point is found from steam tables (see App. 29.B) as the saturation temperature corresponding to the partial pressure of the water vapor. Use 2.0 psia for convenience. The saturation temperature of 2.0 psia water vapor is $\boxed{126.03°\text{F} \, (130°\text{F})}$.

The answer is (D).

15. *Customary U.S. Solution*

(a) The cooled water flow rate is given by

$$Q = \dot{m}_w c_p \Delta T$$

$$1 \times 10^6 \text{ Btu} = \dot{m}_w \left(1.0 \ \frac{\text{Btu}}{\text{lbm-}^\circ\text{F}}\right)(120^\circ\text{F} - 110^\circ\text{F})$$

$$\dot{m}_w = 1.0 \times 10^5 \text{ lbm/hr}$$

From the psychrometric chart (see App. 31.A), for air in at $T_{db} = 91^\circ\text{F}$ and $\phi = 60\%$,

$$h_{in} \approx 42.7 \text{ Btu/lbm}$$

$$\omega_{in} = 0.0190 \text{ lbm moisture/lbm air}$$

For air out, the normal psychrometric chart (offscale) cannot be used. Use Eq. 31.11 to find the humidity ratio and Eq. 31.17, Eq. 31.18(b), and Eq. 31.19(b) to calculate the enthalpy of air. (Appendix 31.D could also be used as a simpler solution.)

From App. 29.A, the saturated steam pressure at 100°F is 0.9505 psia.

$$p_w = \phi p_{sat} = (0.82)(0.9505 \text{ psia})$$

$$= 0.7794 \text{ psia}$$

From Eq. 31.1,

$$p_a = p - p_w = 14.696 \text{ psia} - 0.7794 \text{ psia}$$

$$= 13.9166 \text{ psia}$$

From Eq. 31.11, the humidity ratio for air out is

$$\phi = 1.608 \omega_{out} \left(\frac{p_a}{p_{sat}}\right)$$

$$0.82 = 1.608 \omega_{out} \left(\frac{13.9166 \text{ psia}}{0.9505 \text{ psia}}\right)$$

$$\omega_{out} = 0.0348 \text{ lbm moisture/lbm air}$$

From Eq. 31.17, Eq. 31.18(b), and Eq. 31.19(b), the enthalpy of air out is

$$h_2 = h_a + \omega_2 h_w$$

$$= \left(0.240 \ \frac{\text{Btu}}{\text{lbm-}^\circ\text{F}}\right) T_{2,^\circ\text{F}} + \omega_2$$

$$\times \left(\left(0.444 \ \frac{\text{Btu}}{\text{lbm-}^\circ\text{F}}\right) T_{2,^\circ\text{F}} + 1061 \ \frac{\text{Btu}}{\text{lbm}}\right)$$

$$= \left(0.240 \ \frac{\text{Btu}}{\text{lbm-}^\circ\text{F}}\right)(100^\circ\text{F})$$

$$+ \left(0.0348 \ \frac{\text{lbm moisture}}{\text{lbm air}}\right) \begin{pmatrix} \left(0.444 \ \frac{\text{Btu}}{\text{lbm-}^\circ\text{F}}\right) \\ \times (100^\circ\text{F}) \\ + 1061 \ \frac{\text{Btu}}{\text{lbm}} \end{pmatrix}$$

$$= 62.47 \text{ Btu/lbm air}$$

The mass flow rate of air can be determined from Eq. 31.25.

$$q = \dot{m}_a(h_2 - h_1)$$

$$\dot{m}_{air} = \frac{q}{h_2 - h_1} = \frac{1 \times 10^6 \ \frac{\text{Btu}}{\text{hr}}}{62.47 \ \frac{\text{Btu}}{\text{lbm air}} - 42.7 \ \frac{\text{Btu}}{\text{lbm air}}}$$

$$= \boxed{5.058 \times 10^4 \text{ lbm air/hr}}$$

(b) From conservation of water vapor,

$$\omega_1 \dot{m}_{air} + \dot{m}_{make\text{-}up} = \omega_2 \dot{m}_{air}$$

$$\dot{m}_{make\text{-}up} = \dot{m}_{air}(\omega_{out} - \omega_{in})$$

$$= \left(5.058 \times 10^4 \ \frac{\text{lbm air}}{\text{hr}}\right)$$

$$\times \begin{pmatrix} 0.0348 \ \frac{\text{lbm moisture}}{\text{lbm air}} \\ - 0.0191 \ \frac{\text{lbm moisture}}{\text{lbm air}} \end{pmatrix}$$

$$= \boxed{794 \text{ lbm water/hr}}$$

SI Solution

(a) The cooled water flow rate is given by

$$Q = \dot{m}_w c_p \Delta T$$

$$290 \text{ kW} = \dot{m}_w \left(4.187 \ \frac{\text{kJ}}{\text{kg-}^\circ\text{C}}\right)(49^\circ\text{C} - 43^\circ\text{C})$$

$$\dot{m}_w = 11.54 \text{ kg/s}$$

From the psychrometric chart (see App. 31.B), for air in at $T_{db} = 33^\circ\text{C}$ and $\phi = 60\%$,

$$h_{in} = 82.3 \text{ kJ/kg air}$$

$$\omega_{in} = \frac{19.2 \ \frac{\text{g moisture}}{\text{kg air}}}{1000 \ \frac{\text{g}}{\text{kg}}}$$

$$= 0.0192 \text{ kg moisture/kg air}$$

For air out, the psychrometric chart (off scale) cannot be used. Use Eq. 31.11 to find the humidity ratio and Eq. 31.17, Eq. 31.18(a), and Eq. 31.19(a) to calculate enthalpy of air.

From App. 29.N, the saturated steam pressure at 38°C is 0.06633 bars.

$$p_w = \phi p_{sat} = (0.82)(0.06633 \text{ bar}) = 0.05439 \text{ bar}$$

From Eq. 31.1,

$$p_a = p - p_w = 1 \text{ bar} - 0.05439 \text{ bar} = 0.94561 \text{ bar}$$

From Eq. 31.11, the humidity ratio for air out is

$$\phi = 1.608\omega_{\text{out}}\left(\frac{p_a}{p_{\text{sat}}}\right)$$

$$0.82 = 1.608\omega_{\text{out}}\left(\frac{0.94561 \text{ bar}}{0.06633 \text{ bar}}\right)$$

$$\omega_{\text{out}} = 0.0358 \text{ kg moisture/kg air}$$

From Eq. 31.17, Eq. 31.18(a), and Eq. 31.19(a), the enthalpy of air out is

$$h_2 = h_a + \omega_2 h_w$$
$$= \left(1.005 \ \frac{\text{kJ}}{\text{kg·°C}}\right) T_{°C}$$
$$\quad + \omega_{\text{out}}\left(\left(1.805 \ \frac{\text{kJ}}{\text{kg·°C}}\right) T_{°C} + 2501 \ \frac{\text{kJ}}{\text{kg}}\right)$$
$$= \left(1.005 \ \frac{\text{kJ}}{\text{kg·°C}}\right)(38°C) + \left(0.0358 \ \frac{\text{kg moisture}}{\text{kg air}}\right)$$
$$\quad \times \left(\left(1.805 \ \frac{\text{kJ}}{\text{kg·°C}}\right)(38°C) + 2501 \ \frac{\text{kJ}}{\text{kg}}\right)$$
$$= 130.2 \text{ kJ/kg air}$$

The mass flow rate of air can be determined from Eq. 31.25.

$$q = \dot{m}_a(h_2 - h_1)$$
$$\dot{m}_{\text{air}} = \frac{q}{h_2 - h_1}$$
$$= \frac{290 \text{ kW}}{130.2 \ \dfrac{\text{kJ}}{\text{kg air}} - 82.3 \ \dfrac{\text{kJ}}{\text{kg air}}}$$
$$= \boxed{6.054 \text{ kg air/s}}$$

(b) From conservation of water vapor,

$$\omega_1 \dot{m}_{\text{air}} + \dot{m}_{\text{make-up}} = \omega_2 \dot{m}_{\text{air}}$$
$$\dot{m}_{\text{make-up}} = \dot{m}_{\text{air}}(\omega_{\text{out}} - \omega_{\text{in}})$$
$$= \left(6.054 \ \frac{\text{kg air}}{\text{s}}\right)$$
$$\quad \times \left(\begin{array}{c} 0.0358 \ \dfrac{\text{kg moisture}}{\text{kg air}} \\[2mm] - 0.0192 \ \dfrac{\text{kg moisture}}{\text{kg air}} \end{array}\right)$$
$$= \boxed{0.100 \text{ kg water/s}}$$

Ventilation

32 HVAC: Ventilation and Humidification Requirements

PRACTICE PROBLEMS

1. An office room has floor dimensions of 60 ft by 95 ft (18 m by 29 m) and a ceiling height of 10 ft (3 m). Cool air is supplied from ceiling diffusers. 45 people occupy the office.

(a) What is most nearly the ventilation rate based on six air changes per hour?

- (A) 1300 ft³/min (35 m³/min)
- (B) 5700 ft³/min (160 m³/min)
- (C) 13,000 ft³/min (350 m³/min)
- (D) 34,000 ft³/min (900 m³/min)

(b) What is most nearly the ventilation rate at the breathing zone based on ASHRAE Standard 62.1?

- (A) 520 ft³/min (250 L/s)
- (B) 570 ft³/min (270 L/s)
- (C) 700 ft³/min (330 L/s)
- (D) 850 ft³/min (400 L/s)

2. 150 ppm of methanol (TLV = 200 ppm; MW = 32.04; SG = 0.792) and 285 ppm of methylene chloride (TLV = 500 ppm; MW = 84.94; SG = 1.336) are found in the air in a plating booth. Two pints (1 L) of each are evaporated per hour. Use a mixing safety factor (i.e., a K-value) of 6. The minimum ventilation rate required for dilution is most nearly

- (A) 2500 ft³/min (1200 L/s)
- (B) 5200 ft³/min (2500 L/s)
- (C) 10,000 ft³/min (4800 L/s)
- (D) 12,000 ft³/min (6200 L/s)

3. A room is maintained at design conditions of 75°F (23.9°C) dry-bulb and 50% relative humidity. The air outside is at 95°F (35°C) dry-bulb and 75°F (23.9°C) wet-bulb. The outside air is conditioned and mixed with some room exhaust air. The mixed, conditioned air enters the room and increases 20°F (11.1°C) in temperature before being removed from the room. The sensible and latent loads are 200,000 Btu/hr (60 kW) and

50,000 Btu/hr (15 kW), respectively. Air leaves the coil at 50.8°F (10°C). The volume of air flowing through the coil is most nearly

- (A) 4200 ft³/min (120 m³/min)
- (B) 5800 ft³/min (160 m³/min)
- (C) 7600 ft³/min (220 m³/min)
- (D) 9300 ft³/min (260 m³/min)

4. According to ASHRAE Standard 62.1, the occupant diversity as it applies to the calculation of outdoor air in multi-zone systems accounts for the

- (A) variation in total system occupancy over time
- (B) variation in zonal occupancy over time
- (C) fractional mixture of women and children in the system population
- (D) likelihood of simultaneous peak zone populations

5. Modern ventilation standards consider both population and floor area when calculating outdoor air requirements because

- (A) exact occupancy numbers are difficult to predict
- (B) contaminants originate from both the population and the floor
- (C) the floor area component ensures a minimum ventilation rate
- (D) modern buildings are more "tight," and infiltration cannot be counted on

6. How should the pressurization (relative to the surroundings) in clean rooms and laboratories be maintained?

- (A) Clean rooms and laboratories should both be at positive pressures.
- (B) Clean rooms and laboratories should both be at negative pressures.
- (C) Clean rooms should be at positive pressures, while laboratories should be at negative pressures.
- (D) Clean rooms should be at negative pressures, while laboratories should be at positive pressures.

7. Assuming atmospheric air is a mixture of oxygen and nitrogen only, above what approximate ambient volumetric nitrogen concentration should supplemental oxygen be provided by an employer to employees?

    (A) 77% nitrogen

    (B) 79% nitrogen

    (C) 81% nitrogen

    (D) 88% nitrogen

8. The volumetric fraction of carbon dioxide in a submarine is measured to be 0.06%. What is most nearly the concentration measured in parts per million (ppm)?

    (A) 0.6 ppm

    (B) 60 ppm

    (C) 600 ppm

    (D) 6000 ppm

SOLUTIONS

1. *Customary U.S. Solution*

(a) The office volume is

$$V = (60 \text{ ft})(95 \text{ ft})(10 \text{ ft}) = 57{,}000 \text{ ft}^3$$

Based on six air changes per hour, the flow rate is

$$\dot{V} = \frac{\left(57{,}000 \; \dfrac{\text{ft}^3}{\text{air change}}\right)\left(6 \; \dfrac{\text{air changes}}{\text{hr}}\right)}{60 \; \dfrac{\text{min}}{\text{hr}}}$$

$$= \boxed{5700 \text{ ft}^3/\text{min}}$$

The answer is (B).

(b) The floor area is

$$A = (60 \text{ ft})(95 \text{ ft}) = 5700 \text{ ft}^2$$

From Table 32.1 (office building: offices), $R_p = 5$ cfm/person, and $R_a = 0.06$ cfm/ft^2. From Eq. 32.1,

$$\dot{V}_{\text{bz}} = R_p P_z + R_a A_z$$

$$= \left(5 \; \frac{\text{ft}^3}{\text{min-person}}\right)(45 \text{ people})$$

$$+ \left(0.06 \; \frac{\text{ft}^3}{\text{min-ft}^2}\right)(5700 \text{ ft}^2)$$

$$= 567 \text{ ft}^3/\text{min}$$

Since cool air is supplied from the ceiling, from Table 32.2, $E_z = 1.0$. Using Eq. 32.2, the total outdoor air to the zone is

$$\dot{V}_{\text{oz}} = \frac{\dot{V}_{\text{bz}}}{E_z} = \frac{567 \; \dfrac{\text{ft}^3}{\text{min}}}{1.0} = \boxed{567 \text{ ft}^3/\text{min} \quad (570 \text{ ft}^3/\text{min})}$$

The answer is (B).

SI Solution

(a) The office volume is

$$V = (18 \text{ m})(29 \text{ m})(3 \text{ m}) = 1566 \text{ m}^3$$

Based on six air changes per hour, the flow rate is

$$\dot{V} = \frac{\left(1566 \; \dfrac{\text{m}^3}{\text{air change}}\right)\left(6 \; \dfrac{\text{air changes}}{\text{h}}\right)}{60 \; \dfrac{\text{min}}{\text{h}}}$$

$$= \boxed{156.6 \text{ m}^3/\text{min} \quad (160 \text{ m}^3/\text{min})}$$

The answer is (B).

Ventilation

(b) The floor area is

$$A = (18 \text{ m})(29 \text{ m}) = 522 \text{ m}^2$$

From Table 32.1 (office building: offices), $R_p = 2.5$ L/s·person, and $R_a = 0.3$ L/s·m². From Eq. 32.1,

$$\dot{V}_{bz} = R_p P_z + R_a A_z$$

$$= \left(2.5 \; \frac{\text{L}}{\text{s·person}}\right)(45 \text{ people})$$

$$\quad + \left(0.3 \; \frac{\text{L}}{\text{s·m}^2}\right)(522 \text{ m}^2)$$

$$= 269.1 \text{ L/s}$$

Since cool air is supplied from the ceiling, from Table 32.2, $E_z = 1.0$. Using Eq. 32.2, the total outdoor air to the zone is

$$\dot{V}_{oz} = \frac{\dot{V}_{bz}}{E_z} = \frac{269.1 \; \frac{\text{L}}{\text{s}}}{1.0} = \boxed{269.1 \text{ L/s} \quad (270 \text{ L/s})}$$

The answer is (B).

2. *Customary U.S. Solution*

Use Eq. 32.25(b) to find the required dilution ventilation rate. The maximum concentration, C, is equal to the threshold limit value, TLV.

$$\dot{V}_{\text{ft}^3/\text{min}} = \frac{(4.03 \times 10^8)K(\text{SG})R_{\text{pints/min}}}{(\text{MW})\text{TLV}_{\text{ppm}}}$$

For the methanol,

$$\dot{V}_{\text{ft}^3/\text{min}} = \frac{(4.03 \times 10^8)(6)(0.792)\left(2 \; \frac{\text{pints}}{\text{hr}}\right)}{(32.04)(200 \text{ ppm})\left(60 \; \frac{\text{min}}{\text{hr}}\right)}$$

$$= 9962 \text{ ft}^3/\text{min}$$

For the methylene chloride,

$$\dot{V}_{\text{ft}^3/\text{min}} = \frac{(4.03 \times 10^8)(6)(1.336)\left(2 \; \frac{\text{pints}}{\text{hr}}\right)}{(84.94)(500 \text{ ppm})\left(60 \; \frac{\text{min}}{\text{hr}}\right)}$$

$$= 2535 \text{ ft}^3/\text{min}$$

The total dilution ventilation rate is

$$9962 \; \frac{\text{ft}^3}{\text{min}} + 2535 \; \frac{\text{ft}^3}{\text{min}}$$

$$= \boxed{12{,}497 \text{ ft}^3/\text{min} \quad (12{,}000 \text{ ft}^3/\text{min})}$$

The answer is (D).

Use Eq. 32.25(a) to find the required dilution ventilation rate. The maximum concentration, C, is equal to the threshold limit value, TLV.

$$\dot{V}_{\text{L/s}} = \frac{(4.02 \times 10^8)K(\text{SG})R_{\text{L/min}}}{(\text{MW})C_{\max}}$$

For the methanol,

$$\dot{V}_{\text{L/s}} = \frac{(4.02 \times 10^8)(6)(0.792)\left(1 \; \frac{\text{L}}{\text{h}}\right)}{(32.04)(200 \text{ ppm})\left(60 \; \frac{\text{min}}{\text{h}}\right)}$$

$$= 4969 \text{ L/s}$$

For the methylene chloride,

$$\dot{V}_{\text{L/s}} = \frac{(4.02 \times 10^8)(6)(1.336)\left(1 \; \frac{\text{L}}{\text{h}}\right)}{(84.94)(500 \text{ ppm})\left(60 \; \frac{\text{min}}{\text{h}}\right)}$$

$$= 1265 \text{ L/s}$$

The total dilution ventilation rate is

$$4969 \; \frac{\text{L}}{\text{s}} + 1265 \; \frac{\text{L}}{\text{s}} = \boxed{6234 \text{ L/s} \quad (6200 \text{ L/s})}$$

The answer is (D).

3. *Customary U.S. Solution*

The mixed conditioned air enters the room at

$$T_{\text{db,in}} = 75°\text{F} - 20°\text{F} = 55°\text{F}$$

From Eq. 32.20(b), the volumetric flow rate of air entering the room is

$$\dot{V}_{\text{in,ft}^3/\text{min}} = \frac{\dot{q}_{s,\text{Btu/hr}}}{\left(1.08 \; \frac{\text{Btu-min}}{\text{ft}^3\text{-hr-}°\text{F}}\right)(T_{\text{id,}°\text{F}} - T_{\text{in,}°\text{F}})}$$

$$= \frac{200{,}000 \; \frac{\text{Btu}}{\text{hr}}}{\left(1.08 \; \frac{\text{Btu-min}}{\text{ft}^3\text{-hr-}°\text{F}}\right)(75°\text{F} - 55°\text{F})}$$

$$= 9259 \text{ ft}^3/\text{min}$$

This is a mixing problem.

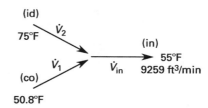

Using the lever rule, and since the temperature scales are all linear, the fraction of air passing through the coil is

$$\frac{T_{id} - T_{in}}{T_{id} - T_{co}} = \frac{75°F - 55°F}{75°F - 50.8°F}$$
$$= 0.826$$

$$\dot{V}_1 = (0.826)\left(9259 \ \frac{ft^3}{min}\right)$$

$$= \boxed{7648 \ ft^3/min \quad (7600 \ ft^3/min)}$$

The answer is (C).

SI Solution

The mixed conditioned air enters the room at

$$T_{db,in} = 23.9°C - 11.1°C = 12.8°C$$

From Eq. 32.20(a),

$$\dot{V}_{in,m^3/min} = \frac{\dot{q}_{s,kW}}{\left(0.02 \ \frac{kJ \cdot min}{m^3 \cdot s \cdot °C}\right)(T_{id,°C} - T_{in,°C})}$$

$$= \frac{60 \ kW}{\left(0.02 \ \frac{kJ \cdot min}{m^3 \cdot s \cdot °C}\right)(23.9°C - 12.8°C)}$$

$$= 270.3 \ m^3/min$$

This is a mixing problem.

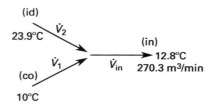

Using the lever rule, and since the temperature scales are all linear, the fraction of air passing through the coil is

$$\frac{T_{id} - T_{in}}{T_{id} - T_{co}} = \frac{23.9°C - 12.8°C}{23.9°C - 10°C}$$
$$= 0.799$$

$$\dot{V}_1 = (0.799)\left(270.3 \ \frac{m^3}{min}\right)$$

$$= \boxed{216.0 \ m^3/min \quad (220 \ m^3/min)}$$

The answer is (C).

4. The occupancy diversity, D, is the ratio of the total system population to the sum of the zonal peak populations. Diversity will have a value of 1.0 if all zones simultaneously operate at peak occupancy, in which case, the total system population will be the same as the sum of the zonal peak populations. In most cases, the zones will experience peak occupancies at different times, and the diversity will be less than 1.0.

The answer is (D).

5. The occupants (i.e., the population) and the building both contribute contaminants independently to the occupied space. Occupant contaminants include moisture, heat, odors, carbon dioxide, and in some cases, tobacco smoke. Buildings produce odors, dust, organic vapors, and mold and fungi spores.

The answer is (B).

6. Clean rooms should be maintained at positive pressures with filtered forced air in order to prevent the incursion of airborne particles from the atmosphere. The intake air is filtered and cleaned, and the interior of the clean room is protected. Most laboratories should be maintained at negative pressures in order to keep airborne hazardous materials in the air distribution system. The exhaust is filtered and cleaned, and the environment is protected.

The answer is (C).

7. OSHA defines an oxygen-deficient atmosphere as one having less than 19.5% oxygen. In this scenario, the maximum nitrogen content would be

$$100\% - 19.5\% = \boxed{80.5\% \quad (81\%)}$$

The answer is (C).

8. Air contaminants are measured by volume, hence "ppm" are parts per million by volume (ppmv). In this problem, there are 0.06 volumes of carbon dioxide for every 100 volumes of air (i.e., 0.06 parts per 100 parts). In one million volumes of air, there would be $10^6/10^2 = 10^4$ more volumes of carbon dioxide.

$$(0.06 \ pphv)\left(10^4 \ \frac{ppmv}{pphv}\right) = \boxed{600 \ ppmv \quad (600 \ ppm)}$$

The answer is (C).

33 HVAC: Heating Load

PRACTICE PROBLEMS

1. A conditioned room contains 12,000 W of fluorescent lights (with older coil ballasts) and twelve 90% efficient, 10 hp motors operating at 80% of their rated capacities. The lights are pendant-mounted on chains from the ceiling. The motors drive various pieces of machinery located in the conditioned space. The internal heat gain is most nearly

(A) 1.9×10^5 Btu/hr (55 kW)

(B) 3.2×10^5 Btu/hr (94 kW)

(C) 3.9×10^5 Btu/hr (110 kW)

(D) 4.6×10^5 Btu/hr (130 kW)

2. A building is located in the city of New York. There are 4772 (2651 in SI) degree-days (basis of 65°F (18°C)) during the October 15 to May 15 period for this area. The building is heated by fuel oil whose heating value is 153,600 Btu/gal (42 800 MJ/m³) and which costs $0.15/gal ($0.04/L). The calculated design heat loss is 3.5×10^6 Btu/hr (1 MW) based on 70°F (21.1°C) inside and 0°F (−17.8°C) outside design temperatures. The furnace has an efficiency of 70%. The cost of heating this building during the winter (October 15 through May 15) is most nearly

(A) $7000

(B) $8000

(C) $9000

(D) $10,000

3. A 12 ft × 12 ft (3.6 m × 3.6 m) floor is constructed as a concrete slab with two exposed edges. The slab edge heat loss coefficient for these two edges is 0.55 Btu/hr-ft-°F (0.95 W/m·°C). The other two edges form part of a basement wall exposed to 70°F air. The inside design temperature is 70°F (21.1°C). The outdoor design temperature is −10°F (−23.3°C). Using the slab edge method, the heat loss from the slab is most nearly

(A) 500 Btu/hr (140 W)

(B) 800 Btu/hr (240 W)

(C) 1100 Btu/hr (300 W)

(D) 4200 Btu/hr (1300 W)

4. A first-floor office in a remodeled historic building has floor dimensions of 100 ft × 40 ft (30 m × 12 m) and a ceiling height of 10 ft (3 m). One of the 40 ft (12 m) walls is shared with an adjacent heated space. The three remaining walls have one 4 ft (1.2 m) wide × 6 ft (1.8 m) high, double glass with ¼ in (6.4 mm) air space, weather-stripped, double-hung window per 10 ft (3 m). The crack coefficient for the windows is 32 ft³/hr-ft (3.0 m³/h·m). One wall of 10 windows is exfiltrating. The basement and second floor are heated to 70°F (21.1°C). The wall coefficient (exclusive of film resistance) is 0.2 Btu/hr-ft²-°F (1.1 W/m²·°C), and the outside wind velocity is 15 mi/hr (24 km/h). The inside design temperature is 70°F (21.1°C); the outside design temperature is −10°F (−23.3°C). Including infiltration but disregarding ventilation air, the heating load is most nearly

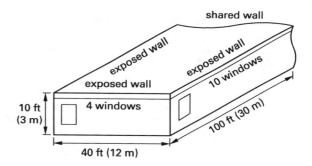

(A) 53,000 Btu/hr (15 kW)

(B) 66,000 Btu/hr (19 kW)

(C) 71,000 Btu/hr (20 kW)

(D) 79,000 Btu/hr (22 kW)

5. A building is located in Newark, New Jersey. The heating season lasts for 245 days. There are 5252 (2918 in SI) degree-days (basis of 65°F (18°C)), and the outside design temperature is 0°F (−17.8°C). The temperature in the building is maintained at 70°F (21.1°C) between the hours of 8:30 a.m. and 5:30 p.m. During the rest of the day and the night, the temperature is allowed to drop to 50°F (10°C). The building is heated with coal that has a heating value of 13,000 Btu/lbm (30.2 MJ/kg). A heat loss of 650,000 Btu/hr (0.19 MW) has been calculated based on 70°F (21.1°C) inside and

0°F (−17.8°C) outside temperatures. The furnace has an efficiency of 70%. The mass of coal required each year is most nearly

(A) 18,000 lbm/yr (4000 kg/yr)

(B) 43,000 lbm/yr (9500 kg/yr)

(C) 69,000 lbm/yr (15 000 kg/yr)

(D) 84,000 lbm/yr (40 000 kg/yr)

6. The design heat loss of a building is 200,000 Btu/hr (60 kW) originally based on 70°F (21.1°C) inside and 0°F (−17.8°C) outside design temperatures. At that location, there are 4200 (2333 in SI) degree-days (basis of 65°F (18°C)) over the 210-day heating season. The building is occupied 24 hr/day. If the thermostat is lowered from 70°F to 68°F (21.1°C to 20°C), the percentage reduction in heating fuel is most nearly

(A) 2.0%

(B) 4.0%

(C) 8.0%

(D) 14%

7. A building with the characteristics listed is located where the annual heating season lasts 21 weeks. The inside design temperature is 70°F (21.1°C). The average outside air temperature is 30°F (−1°C). The gas furnace has an efficiency of 75%. Fuel costs $0.25 per therm. The building is occupied only from 8 a.m. until 6 p.m., Monday through Friday. In the past, the building was maintained at 70°F (21.1°C) at all times and ventilated with one air change per hour (based on inside conditions). The thermostat is now being set back 12°F (6.7°C), and the ventilation is being reduced by 50% when the building is unoccupied. Infiltration through cracks and humidity changes are disregarded.

internal volume	801,000 ft³ (22 700 m³)
wall area	11,040 ft² (993 m²)
wall overall heat transfer coefficient	0.15 Btu/hr-ft²-°F (0.85 W/m²·°C)
window area	2760 ft² (260 m²)
window overall heat transfer coefficient	1.13 Btu/hr-ft²-°F (6.42 W/m²·°C)
roof area	26,700 ft² (2480 m²)
roof overall heat transfer coefficient	0.05 Btu/hr-ft²-°F (0.3 W/m²·°C)
slab on grade	690 lineal ft (210 m) of exposed slab edge
slab edge coefficient	1.5 Btu/hr-ft-°F (2.6 W/m·°C)

(a) The unoccupied time for a 21-week period over a year is most nearly

(A) 1500 hr/yr

(B) 2100 hr/yr

(C) 2500 hr/yr

(D) 3600 hr/yr

(b) At 70°F (21.1°C), the energy savings are most nearly

(A) 5.6×10^8 Btu/hr (5.9×10^8 kJ/yr)

(B) 7.9×10^8 Btu/hr (8.3×10^8 kJ/yr)

(C) 9.3×10^8 Btu/hr (9.8×10^8 kJ/yr)

(D) 14×10^9 Btu/hr (15×10^9 kJ/yr)

(c) The savings due to reduced heat losses from the roof, floor, and walls is most nearly

(A) 4.1×10^7 Btu/yr (4.4×10^7 kJ/yr)

(B) 9.2×10^7 Btu/yr (9.8×10^7 kJ/yr)

(C) 1.8×10^8 Btu/yr (1.9×10^8 kJ/yr)

(D) 2.1×10^8 Btu/yr (2.3×10^8 kJ/yr)

(d) The energy saved per year is most nearly

(A) 11,000 therm/yr

(B) 13,000 therm/yr

(C) 15,000 therm/yr

(D) 18,000 therm/yr

(e) The annual savings is most nearly

(A) $3100/yr

(B) $3800/yr

(C) $4600/yr

(D) $6200/yr

SOLUTIONS

1. *Customary U.S. Solution*

Based on Sec. 33.10, for fluorescent lights, the rated wattage should be increased by 20–25% to account for ballast heating. Since the lights are pendant-mounted on chains from the ceiling, most of this heat enters the conditioned space. Assume rated wattage increases by 20%. η is given as 1. From Eq. 33.8(b), the internal heat gain due to lights (SF = 1.2) is

$$\dot{q}_{lights} = \left(3412 \ \frac{Btu}{kW\text{-}hr}\right)(SF)\left(\frac{P_{kW}}{\eta}\right)$$

$$= \left(3412 \ \frac{Btu}{kW\text{-}hr}\right)(1.2)\left(\frac{12{,}000 \ W}{(1)\left(1000 \ \frac{W}{kW}\right)}\right)$$

$$= 4.9133 \times 10^4 \ Btu/hr$$

Use Eq. 33.8(b) to find the internal heat gain due to the twelve motors (SF = 0.8). η is given as 0.90.

$$\dot{q}_{motors} = \left(2545 \ \frac{Btu}{hr\text{-}hp}\right)(SF)\left(\frac{P_{hp}}{\eta}\right)$$

$$= (12 \ motors)\left(2545 \ \frac{Btu}{hp\text{-}hr}\right)(0.8)\left(\frac{10 \ hp}{0.90}\right)$$

$$= 2.7147 \times 10^5 \ Btu/hr$$

$$\dot{q}_{total} = \dot{q}_{lights} + \dot{q}_{motors}$$

$$= 4.9147 \times 10^4 \ \frac{Btu}{hr} + 2.7147 \times 10^5 \ \frac{Btu}{hr}$$

$$= \boxed{3.206 \times 10^5 \ Btu/hr \quad (3.2 \times 10^5 \ Btu/hr)}$$

The answer is (B).

SI Solution

Based on the customary U.S. solution, SF = 1.2 for lights. η is given as 1. From Eq. 33.8(a), the internal heat gain due to lights is

$$\dot{q}_{lights} = (SF)\left(\frac{P_{kW}}{\eta}\right)$$

$$= (1.2)\left(\frac{12\,000 \ W}{(1)\left(1000 \ \frac{W}{kW}\right)}\right)$$

$$= 14.4 \ kW$$

Use Eq. 33.8(a), to find the internal heat gain due to the twelve motors (SF = 0.8). η is given as 0.90.

$$\dot{q}_{motors} = \left(0.7457 \ \frac{kW}{hp}\right)(SF)\left(\frac{P_{hp}}{\eta}\right)$$

$$= (12 \ motors)\left(0.7457 \ \frac{kW}{hp}\right)(0.8)\left(\frac{10 \ hp}{0.90}\right)$$

$$= 79.5 \ kW$$

$$\dot{q}_{total} = \dot{q}_{lights} + \dot{q}_{motors} = 14.4 \ kW + 79.5 \ kW$$

$$= \boxed{93.9 \ kW \quad (94 \ kW)}$$

The answer is (B).

2. *Customary U.S. Solution*

From Eq. 33.13(b), the fuel consumption (in gal/heating season) is

$$\frac{\left(24 \ \frac{hr}{day}\right)\dot{q}_{Btu/hr}(HDD)}{(T_i - T_o)(HV_{Btu/gal})\eta_{furnace}}$$

$$= \frac{\left(24 \ \frac{hr}{day}\right)\left(3.5 \times 10^6 \ \frac{Btu}{hr}\right)(4772°F\text{-days})}{(70°F - 0°F)\left(153{,}600 \ \frac{Btu}{gal}\right)(0.70)}$$

$$= 53{,}260 \ gal$$

The total cost of 53,260 gal fuel at \$0.15/gal is

$$(53{,}260 \ gal)\left(\frac{\$0.15}{gal}\right) = \boxed{\$7989 \quad (\$8000)}$$

The answer is (B).

SI Solution

$$HV_{kJ/L} = \frac{\left(42\,800 \ \frac{MJ}{m^3}\right)\left(1000 \ \frac{kJ}{MJ}\right)}{1000 \ \frac{L}{m^3}} = 42\,800 \ kJ/L$$

From Eq. 33.13(a), fuel consumption (in L/heating season) is

$$\frac{\left(86\,400 \ \frac{s}{d}\right)\dot{q}_{kW}(HDD)}{(T_i - T_o)(HV_{kJ/L})\eta_{furnace}}$$

$$= \frac{\left(86\,400 \ \frac{s}{d}\right)(1 \ MW)\left(1000 \ \frac{kW}{MW}\right)}{(21.1°C - (-17.8°C))\left(42\,800 \ \frac{kJ}{L}\right)(0.70)}$$

$$\times (2651°C\cdot d)$$

$$= 196\,531 \ L$$

The total cost at $0.04/L is

$$(196\,531 \text{ L})\left(\frac{\$0.04}{\text{L}}\right) = \boxed{\$7861 \quad (\$8000)}$$

The answer is (B).

3. *Customary U.S. Solution*

The slab edge coefficient is given as $F = 0.55$ Btu/hr-ft-°F.

The perimeter length is

$$p = 12 \text{ ft} + 12 \text{ ft} \quad \text{[2 edges only]}$$
$$= 24 \text{ ft}$$

From Eq. 33.5, the heat loss from the slab is

$$\dot{q} = pF(T_i - T_o)$$
$$= (24 \text{ ft})\left(0.55 \frac{\text{Btu}}{\text{hr-ft-°F}}\right)(70°\text{F} - (-10°\text{F}))$$
$$= \boxed{1056 \text{ Btu/hr} \quad (1100 \text{ Btu/hr})}$$

The answer is (C).

SI Solution

The slab edge coefficient is given as $F = 0.95$ W/m·°C.

The perimeter length is

$$p = 3.6 \text{ m} + 3.6 \text{ m} \quad \text{[2 edges only]}$$
$$= 7.2 \text{ m}$$

From Eq. 33.5, the heat loss from the slab is

$$\dot{q} = pF(T_i - T_o)$$
$$= (7.2 \text{ m})\left(0.95 \frac{\text{W}}{\text{m·°C}}\right)(21.1°\text{C} - (-23.3°\text{C}))$$
$$= \boxed{303.7 \text{ W} \quad (300 \text{ W})}$$

The answer is (C).

4. *Customary U.S. Solution*

For three unshared walls, the total exposed (walls + windows) area is

$$(10 \text{ ft})(40 \text{ ft} + 100 \text{ ft} + 100 \text{ ft}) = 2400 \text{ ft}^2$$

- There are two windows per 20 ft.
- The number of windows for a 40 ft wall is 4.

- The number of windows for each 100 ft wall is 10.
- The total number of windows is

$$4 + 10 + 10 = 24$$

- The total window area is

$$(24)(4 \text{ ft})(6 \text{ ft}) = 576 \text{ ft}^2$$

- The total wall area is

$$2400 \text{ ft}^2 - 576 \text{ ft}^2 = 1824 \text{ ft}^2$$

From Table 33.2, the outside film coefficient for a vertical wall in a 15 mph wind is $h_o = 6.00$ Btu/hr-ft^2-°F.

From Table 33.2, the inside film coefficient for still air (vertical, heat flow horizontal) is $h_i = 1.46$ Btu/hr-ft^2-°F.

Notice that L/k was given in the problem statement, not k. From Eq. 33.3, the overall coefficient of heat transfer for the wall is

$$U_{\text{wall}} = \frac{1}{\dfrac{1}{h_i} + \dfrac{L_{\text{wall}}}{k_{\text{wall}}} + \dfrac{1}{h_o}}$$

$$= \frac{1}{\dfrac{1}{1.46 \frac{\text{Btu}}{\text{hr-ft}^2\text{-°F}}} + \dfrac{1}{0.2 \frac{\text{Btu}}{\text{hr-ft}^2\text{-°F}}} + \dfrac{1}{6.00 \frac{\text{Btu}}{\text{hr-ft}^2\text{-°F}}}}$$

$$= 0.1709 \text{ Btu/hr-ft}^2\text{-°F}$$

From Eq. 33.2, the heat loss from the wall is

$$\dot{q}_{\text{wall}} = U_{\text{wall}} A_{\text{wall}} \Delta T$$
$$= \left(0.1709 \frac{\text{Btu}}{\text{hr-ft}^2\text{-°F}}\right)(1824 \text{ ft}^2)(70°\text{F} - (-10°\text{F}))$$
$$= 24{,}938 \text{ Btu/hr}$$

From App. 33.A, for double, vertical ($^1/_4$ in air space) glass windows, the thermal resistance is 1.63 hr-ft^2-°F/Btu. From the footnote, this includes the inside and outside film coefficients.

$$U_{\text{windows}} = \frac{1}{1.63 \frac{\text{hr-ft}^2\text{-°F}}{\text{Btu}}} = 0.613 \text{ Btu/hr-ft}^2\text{-°F}$$

The heat loss from the windows is

$$\dot{q}_{\text{windows}} = U_{\text{windows}} A_{\text{windows}} \Delta T$$
$$= \left(0.613 \frac{\text{Btu}}{\text{hr-ft}^2\text{-°F}}\right)(576 \text{ ft}^2)(70°\text{F} - (-10°\text{F}))$$
$$= 28{,}247 \text{ Btu/hr}$$

Ventilation

From Eq. 32.9, the infiltration flow rate is

$$\dot{V} = BL$$

The crack coefficient is given as $B = 32$ ft^3/hr-ft.

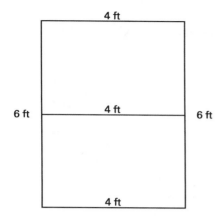

For double-hung windows, the crack length is

$$L = 4 \text{ ft} + 4 \text{ ft} + 4 \text{ ft} + 6 \text{ ft} + 6 \text{ ft} = 24 \text{ ft}$$

$$\dot{V} = BL = \left(32 \; \frac{\text{ft}^3}{\text{hr-ft}}\right)(24 \text{ ft}) = 768 \text{ ft}^3/\text{hr per window}$$

Although there are 24 windows, one wall of 10 is exfiltrating.

$$\dot{V} = (24 - 10)\left(768 \; \frac{\text{ft}^3}{\text{hr}}\right) = 10{,}752 \text{ ft}^3/\text{hr}$$

For atmospheric air, $p = 14.7$ psia and $c_p = 0.24$ Btu/lbm-°F.

The air density is

$$\rho = \frac{p}{RT} = \frac{\left(14.7 \; \frac{\text{lbf}}{\text{in}^2}\right)\left(12 \; \frac{\text{in}}{\text{ft}}\right)^2}{\left(53.35 \; \frac{\text{ft-lbf}}{\text{lbm-°R}}\right)(-10°\text{F} + 460°)}$$

$$= 0.088 \text{ lbm/ft}^3$$

The mass flow rate of infiltrated air is

$$\dot{m} = \rho\dot{V} = \left(0.088 \; \frac{\text{lbm}}{\text{ft}^3}\right)\left(10{,}752 \; \frac{\text{ft}^3}{\text{hr}}\right)$$

$$= 946.2 \text{ lbm/hr}$$

The heat loss due to infiltration is

$$\dot{q}_{\text{infiltration}} = \dot{m}c_p\Delta T$$

$$= \left(946.2 \; \frac{\text{lbm}}{\text{hr}}\right)\left(0.24 \; \frac{\text{Btu}}{\text{lbm-°F}}\right)$$

$$\times \left(70°\text{F} - (-10°\text{F})\right)$$

$$= 18{,}167 \text{ Btu/hr}$$

The total heat loss is

$$\dot{q}_{\text{total}} = \dot{q}_{\text{wall}} + \dot{q}_{\text{windows}} + \dot{q}_{\text{infiltration}}$$

$$= 24{,}938 \; \frac{\text{Btu}}{\text{hr}} + 28{,}247 \; \frac{\text{Btu}}{\text{hr}} + 18{,}167 \; \frac{\text{Btu}}{\text{hr}}$$

$$= \boxed{71{,}352 \text{ Btu/hr} \quad (71{,}000 \text{ Btu/hr})}$$

The answer is (C).

SI Solution

For three unshared walls, the total exposed (walls + windows) area is

$$(3 \text{ m})(12 \text{ m} + 30 \text{ m} + 30 \text{ m}) = 216 \text{ m}^2$$

For 24 windows (from the customary U.S. solution), the total window area is

$$(24)(1.2 \text{ m})(1.8 \text{ m}) = 51.8 \text{ m}^2$$

The total wall area is

$$216 \text{ m}^2 - 51.8 \text{ m}^2 = 164.2 \text{ m}^2$$

From Table 33.2, the outside film coefficient for a vertical wall in a 15 mph wind is $h_o = 34.1$ W/m^2·°C.

From Table 33.2, the inside film coefficient for still air (vertical, heat flow horizontal) is $h_i = 8.29$ W/m^2·°C.

Notice that L/k was given in the problem statement, not k. From Eq. 33.3, the overall coefficient of heat transfer for the wall is

$$U_{\text{wall}} = \frac{1}{\dfrac{1}{h_i} + \dfrac{L_{\text{wall}}}{k_{\text{wall}}} + \dfrac{1}{h_o}}$$

$$= \frac{1}{\dfrac{1}{8.29 \; \dfrac{\text{W}}{\text{m}^2\cdot°\text{C}}} + \dfrac{1}{1.1 \; \dfrac{\text{W}}{\text{m}^2\cdot°\text{C}}} + \dfrac{1}{34.1 \; \dfrac{\text{W}}{\text{m}^2\cdot°\text{C}}}}$$

$$= 0.944 \text{ W/m}^2\cdot°\text{C}$$

From Eq. 33.2, the heat loss from the wall is

$$\dot{q}_{\text{wall}} = U_{\text{wall}}A_{\text{wall}}\Delta T$$

$$= \left(0.944 \; \frac{\text{W}}{\text{m}^2\cdot°\text{C}}\right)(164.2 \text{ m}^2)(21.1°\text{C} - (-23.3°\text{C}))$$

$$= 6882 \text{ W}$$

From the customary U.S. solution,

$$U_{\text{windows}} = \left(0.613 \; \frac{\text{Btu}}{\text{hr-ft}^2\text{-}^\circ\text{F}}\right)\left(5.68 \; \frac{\frac{\text{W}}{\text{m}^2\cdot^\circ\text{C}}}{\frac{\text{Btu}}{\text{hr-ft}^2\text{-}^\circ\text{F}}}\right)$$

$$= 3.48 \; \text{W/m}^2\cdot^\circ\text{C}$$

The heat loss from the windows is

$$\dot{q}_{\text{windows}} = U_{\text{windows}} A_{\text{windows}} \Delta T$$
$$= \left(3.48 \; \frac{\text{W}}{\text{m}^2\cdot^\circ\text{C}}\right)(51.8 \; \text{m}^2)(21.1^\circ\text{C} - (-23.3^\circ\text{C}))$$
$$= 8004 \; \text{W}$$

From Eq. 32.9, the infiltration flow rate is

$$\dot{V} = BL$$

The crack coefficient is given as $B = 3.0 \; \text{m}^3/\text{h}\cdot\text{m}$.

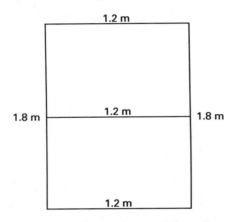

For double-hung windows, the crack length is

$$L = 1.2 \; \text{m} + 1.2 \; \text{m} + 1.2 \; \text{m} + 1.8 \; \text{m} + 1.8 \; \text{m} = 7.2 \; \text{m}$$

$$\dot{V} = BL = \left(3.0 \; \frac{\text{m}^3}{\text{h}\cdot\text{m}}\right)(7.2 \; \text{m}) = 21.6 \; \text{m}^3/\text{h per window}$$

Although there are 24 windows, one wall of 10 is exfiltrating.

$$\dot{V} = (24 - 10)\left(21.6 \; \frac{\text{m}^3}{\text{h}}\right) = 302.4 \; \text{m}^3/\text{h}$$

For atmospheric air, $p = 100\,000$ Pa and $c_p = 1.005$ kJ/kg·°C.

The air density is

$$\rho = \frac{p}{RT} = \frac{100\,000 \; \text{Pa}}{\left(287.03 \; \frac{\text{J}}{\text{kg}\cdot\text{K}}\right)(-23.3^\circ\text{C} + 273^\circ)}$$

$$= 1.4 \; \text{kg/m}^3$$

The mass flow rate of infiltrated air is

$$\dot{m} = \rho\dot{V} = \left(1.4 \; \frac{\text{kg}}{\text{m}^3}\right)\left(302.4 \; \frac{\text{m}^3}{\text{h}}\right) = 423.4 \; \text{kg/h}$$

The heat loss due to infiltration is

$$\dot{q}_{\text{infiltration}} = \dot{m}c_p\Delta T$$

$$= \frac{\left(423.4 \; \frac{\text{kg}}{\text{h}}\right)\left(1.005 \; \frac{\text{kJ}}{\text{kg}\cdot^\circ\text{C}}\right)\left(1000 \; \frac{\text{J}}{\text{kJ}}\right)}{3600 \; \frac{\text{s}}{\text{h}}}$$
$$\times \left(21.1^\circ\text{C} - (-23.3^\circ\text{C})\right)$$

$$= 5248 \; \text{W}$$

The total heat loss is

$$\dot{q}_{\text{total}} = \dot{q}_{\text{wall}} + \dot{q}_{\text{windows}} + \dot{q}_{\text{infiltration}}$$
$$= 6882 \; \text{W} + 8004 \; \text{W} + 5248 \; \text{W}$$
$$= \boxed{20\,134 \; \text{W} \quad (20 \; \text{kW})}$$

The answer is (C).

5. *Customary U.S. Solution*

Use Hitchin's empirical formula to calculate the approximate average outside temperature from the degree-days. The standard base temperature for degree-days is 65°F. The empirical constant, k, for Fahrenheit temperatures is approximately 0.39.

$$\frac{\text{HDD}}{N} = \frac{T_b - \overline{T}}{1 - e^{-k(T_b - \overline{T})}}$$

$$\frac{5252 \; \text{degree-days}}{245 \; \text{days}} = \frac{65^\circ\text{F} - \overline{T}}{1 - e^{-(0.39)(65^\circ\text{F} - \overline{T})}}$$

By trial and error or any other method, $\overline{T} \approx 43.56^\circ\text{F}$.

Since $\overline{T}$ is so much lower than T_b, the denominator is essentially 1.0, and the answer derived from Hitchin's formula is essentially what would have been derived directly from Eq. 33.12. However, Eq. 33.12 is less accurate when $\overline{T}$ and T_b are closer.

The heat loss per degree of temperature difference is

$$\frac{\dot{q}}{\Delta T} = \frac{650,000 \; \frac{\text{Btu}}{\text{hr}}}{70^\circ\text{F} - 0^\circ\text{F}} = 9286 \; \text{Btu/hr-}^\circ\text{F}$$

The period between 8:30 a.m. and 5:30 p.m. is 9 hr. For this period, the temperature in the building is 70°F. For the remaining 15 hr, the temperature in the building is 50°F. The total winter heat loss is

$$\dot{q}_{\text{total,winter}} = (245 \text{ days})\left(9286 \; \frac{\text{Btu}}{\text{hr-°F}}\right)$$
$$\times \left(\begin{array}{c} \left(9 \; \dfrac{\text{hr}}{\text{day}}\right)(70°F - 43.56°F) \\ + \left(15 \; \dfrac{\text{hr}}{\text{day}}\right)(50°F - 43.56°F) \end{array} \right)$$
$$= 7.61 \times 10^8 \text{ Btu}$$

The fuel consumption (per year) is

$$\text{fuel consumption} = \frac{\dot{q}_{\text{total,winter}}}{(\text{HV})\eta_{\text{furnace}}}$$
$$= \frac{7.61 \times 10^8 \text{ Btu}}{\left(13{,}000 \; \dfrac{\text{Btu}}{\text{lbm}}\right)(0.70)}$$
$$= \boxed{83{,}626 \text{ lbm/yr} \quad (84{,}000 \text{ lbm/yr})}$$

The answer is (D).

SI Solution

Use Hitchin's empirical formula to calculate the approximate average outside temperature from the degree-days. The standard base temperature for degree-days is 18°C. The empirical constant, k, for Celsius temperatures is approximately 0.71.

$$\frac{\text{HDD}}{N} = \frac{T_b - \overline{T}}{1 - e^{-k(T_b - \overline{T})}}$$
$$\frac{2918 \text{ degree-days}}{245 \text{ days}} = \frac{18°C - \overline{T}}{1 - e^{-(0.71)(18°C - \overline{T})}}$$

By trial and error or any other method, $\overline{T} \approx 6.09°C$.

Since $\overline{T}$ is so much lower than T_b, the denominator is essentially 1.0, and the answer derived from Hitchin's formula is essentially what would have been derived directly from Eq. 33.12. However, Eq. 33.12 is less accurate when $\overline{T}$ and T_b are closer.

The heat loss per degree of temperature difference is

$$\frac{\dot{q}}{\Delta T} = \frac{(0.19 \text{ MW})\left(10^6 \; \dfrac{\text{W}}{\text{MW}}\right)}{21.1°C - (-17.8°C)} = 4884 \text{ W/°C}$$

The period between 8:30 a.m. and 5:30 p.m. is 9 h. For this period, the temperature in the building is 21.1°C. For the remaining 15 h, the temperature in the building is 10°C. The total winter heat loss is

$$\dot{q}_{\text{total,winter}} = (245 \text{ d})\left(4884 \; \frac{\text{W}}{°C}\right)$$
$$\times \left(\begin{array}{c} \left(\left(9 \; \dfrac{\text{h}}{\text{d}}\right)\left(3600 \; \dfrac{\text{s}}{\text{h}}\right) \right. \\ \left. \times (21.1°C - 6.09°C)\right) \\ + \left(\left(15 \; \dfrac{\text{h}}{\text{d}}\right)\left(3600 \; \dfrac{\text{s}}{\text{h}}\right) \right. \\ \left. \times (10°C - 6.09°C)\right) \end{array} \right)$$
$$= 8.345 \times 10^5 \text{ MJ}$$

The fuel consumption (per year) is

$$\text{fuel consumption} = \frac{\dot{q}_{\text{total,winter}}}{(\text{HV})\eta_{\text{furnace}}} = \frac{8.345 \times 10^5 \text{ MJ}}{\left(30.2 \; \dfrac{\text{MJ}}{\text{kg}}\right)(0.70)}$$
$$= \boxed{39\,475 \text{ kg/yr} \quad (40\,000 \text{ kg/yr})}$$

The answer is (D).

6. *Customary U.S. Solution*

The heat loss per degree of temperature difference is

$$\frac{\dot{q}}{\Delta T} = \frac{200{,}000 \; \dfrac{\text{Btu}}{\text{hr}}}{70°F - 0°F} = 2857 \text{ Btu/hr-°F}$$

From Eq. 33.12, the average temperature, $\overline{T}$, over the entire heating season can be estimated.

$$\frac{\text{HDD}}{N} = T_b - \overline{T}$$
$$\text{HDD} = N(T_b - \overline{T}) = N(65°F - \overline{T})$$

The heating degree-days, HDD, are 4200.

The number of days in the heating season, N, is 210.

$$4200 \text{ days} = (210 \text{ days})(65°F - \overline{T})$$
$$\overline{T} = 45°F$$

The total original winter heat loss based on 70°F inside is

$$(210 \text{ days})\left(24 \; \frac{\text{hr}}{\text{day}}\right)\left(2857 \; \frac{\text{Btu}}{\text{hr-°F}}\right)$$
$$\times (70°F - 45°F) = 3.60 \times 10^8 \text{ Btu}$$

The reduced heat loss based on 68°F inside is

$$(210 \text{ days})\left(24 \ \frac{\text{hr}}{\text{day}}\right)\left(2857 \ \frac{\text{Btu}}{\text{hr-}°\text{F}}\right)$$
$$\times (68°\text{F} - 45°\text{F}) = 3.31 \times 10^8 \text{ Btu}$$

The reduction is

$$\frac{3.60 \times 10^8 \text{ Btu} - 3.31 \times 10^8 \text{ Btu}}{3.60 \times 10^8 \text{ Btu}} = \boxed{0.081 \quad (8.0\%)}$$

The answer is (C).

SI Solution

Since the problem only concerns the percentage reduction in heat loss, determine only the average temperature, $\overline{T}$, as all other parameters except inside temperature remain constant.

From Eq. 33.12, the average temperature, $\overline{T}$, over the entire heating season can be calculated.

$$\frac{\text{HDD}}{N} = T_b - \overline{T}$$
$$\text{HDD} = N(T_b - \overline{T}) = N(18°\text{C} - \overline{T})$$

The heating Kelvin degree-days, HDD, are 2333 K·d.

The number of days in the entire heating season, N, is 210.

$$2333 \text{ d} = (210 \text{ d})(18°\text{C} - \overline{T})$$
$$\overline{T} = 6.89°\text{C}$$

The reduction is

$$\frac{\Delta T_{\text{original}} - \Delta T_{\text{reduced}}}{\Delta T_{\text{original}}}$$
$$\Delta T_{\text{original}} = 21.1°\text{C} - 6.89°\text{C} = 14.21°\text{C}$$
$$\Delta T_{\text{reduced}} = 20°\text{C} - 6.89°\text{C} = 13.11°\text{C}$$

The reduction is

$$\frac{14.21°\text{C} - 13.11°\text{C}}{14.21°\text{C}} = \boxed{0.077 \quad (8.0\%)}$$

The answer is (C).

7. *Customary U.S. Solution*

(a) Prior to the thermostat change, the energy requirement for ventilation air is

$$q_{\text{air},1} = m_1 c_p (T_{i,1} - T_o) = N_1 V t \rho c_p (T_{i,1} - T_o)$$

After the change, the energy requirement is

$$q_{\text{air},2} = m_2 c_p (T_{i,2} - T_o)$$
$$= N_2 V t \rho c_p (T_{i,2} - T_o)$$

The energy savings are

$$\Delta q_{\text{air}} = q_{\text{air},1} - q_{\text{air},2}$$
$$= N_1 V t \rho c_p (T_{i,1} - T_o) - N_2 V t \rho c_p (T_{i,2} - T_o)$$
$$= V t \rho c_p \big(N_1(T_{i,1} - T_o) - N_2(T_{i,2} - T_o)\big)$$

During an entire week, the building is occupied from 8:00 a.m. to 6:00 p.m. (for 10 hr) and is unoccupied 14 hr each day for 5 days, and it is unoccupied 24 hr each day for 2 days (over the weekend).

The unoccupied time for a 21-week period over a year is

$$t = \left(14 \ \frac{\text{hr}}{\text{day}}\right)\left(5 \ \frac{\text{days}}{\text{wk}}\right)\left(21 \ \frac{\text{wk}}{\text{yr}}\right)$$
$$+ \left(24 \ \frac{\text{hr}}{\text{day}}\right)\left(2 \ \frac{\text{days}}{\text{wk}}\right)\left(21 \ \frac{\text{wk}}{\text{yr}}\right)$$
$$= \boxed{2478 \text{ hr/yr} \quad (2500 \text{ hr/yr})}$$

The answer is (C).

(b) With air at 70°F, $c_p \approx 0.24$ Btu/lbm-°F and $\rho \approx 0.075$ lbm/ft³. From part (a), the energy savings are

$$\Delta q_{\text{air}} = V t \rho c_p \big(N_1(T_{i,1} - T_o) - N_2(T_{i,2} - T_o)\big)$$
$$= \left(801{,}000 \ \frac{\text{ft}^3}{\text{air change}}\right)\left(2478 \ \frac{\text{hr}}{\text{yr}}\right)\left(0.075 \ \frac{\text{lbm}}{\text{ft}^3}\right)$$
$$\times \left(0.24 \ \frac{\text{Btu}}{\text{lbm-}°\text{F}}\right)$$
$$\times \left(\begin{array}{l} \left(1 \ \dfrac{\text{air change}}{\text{hr}}\right) \\ \times (70°\text{F} - 30°\text{F}) - \left(0.5 \ \dfrac{\text{air change}}{\text{hr}}\right) \\ \times ((70°\text{F} - 12°\text{F}) - 30°\text{F}) \end{array}\right)$$
$$= \boxed{9.289 \times 10^8 \text{ Btu/hr} \quad (9.3 \times 10^8 \text{ Btu/hr})}$$

The answer is (C).

Ventilation

(c) The savings due to reduced heat losses from the roof, floor, and walls (for a 12°F setback) are

$$\dot{q} = UA\Delta T$$

$$
= \left(\begin{array}{c} \left(0.15 \ \dfrac{\text{Btu}}{\text{hr-ft}^2\text{-}°\text{F}}\right)(11{,}040 \ \text{ft}^2) \\[2mm] + \left(1.13 \ \dfrac{\text{Btu}}{\text{hr-ft}^2\text{-}°\text{F}}\right)(2760 \ \text{ft}^2) \\[2mm] + \left(0.05 \ \dfrac{\text{Btu}}{\text{hr-ft}^2\text{-}°\text{F}}\right)(26{,}700 \ \text{ft}^2) \\[2mm] + \left(1.5 \ \dfrac{\text{Btu}}{\text{hr-ft-}°\text{F}}\right)(690 \ \text{ft}) \end{array} \right)(12°\text{F})
$$

$$= 85{,}738 \ \text{Btu/hr}$$

$$q = \dot{q}t$$

$$= \left(85{,}738 \ \frac{\text{Btu}}{\text{hr}}\right)\left(2478 \ \frac{\text{hr}}{\text{yr}}\right)$$

$$= \boxed{2.125 \times 10^8 \ \text{Btu/yr} \quad (2.1 \times 10^8 \ \text{Btu/yr})}$$

The answer is (D).

(d) The energy saved per year is

$$q_{\text{actual}} = \frac{q_{\text{ideal}}}{\eta_{\text{furnace}}}$$

$$= \frac{\Delta q_{\text{air}} + q}{\eta_{\text{furnace}}}$$

$$= \frac{9.289 \times 10^8 \ \dfrac{\text{Btu}}{\text{yr}} + 2.125 \times 10^8 \ \dfrac{\text{Btu}}{\text{yr}}}{(0.75)\left(100{,}000 \ \dfrac{\text{Btu}}{\text{therm}}\right)}$$

$$= \boxed{15{,}218 \ \text{therm/yr} \quad (15{,}000 \ \text{therm/yr})}$$

The answer is (C).

(e) The annual cost savings are

$$\left(15{,}218 \ \frac{\text{therm}}{\text{yr}}\right)\left(\frac{\$0.25}{\text{therm}}\right) = \boxed{\$3805/\text{yr} \quad (\$3800/\text{yr})}$$

The answer is (B).

SI Solution

(a) Prior to the thermostat change, the energy requirement for ventilation air is

$$q_{\text{air},1} = m_1 c_p (T_{i,1} - T_o)$$

$$= N_1 V t \rho c_p (T_{i,1} - T_o)$$

After the change, the energy requirement is

$$q_{\text{air},2} = m_2 c_p (T_{i,2} - T_o)$$

$$= N_2 V t \rho c_p (T_{i,2} - T_o)$$

The energy savings are

$$\Delta q_{\text{air}} = q_{\text{air},1} - q_{\text{air},2}$$

$$= N_1 V t \rho c_p (T_{i,1} - T_o) - N_2 V t \rho c_p (T_{i,2} - T_o)$$

$$= V t \rho c_p \big(N_1 (T_{i,1} - T_o) - N_2 (T_{i,2} - T_o) \big)$$

During an entire week, the building is occupied from 8:00 a.m. to 6:00 p.m. (for 10 h) and is unoccupied 14 h each day for 5 d, and it is unoccupied 24 h each day for 2 d (over the weekend).

The unoccupied time for a 21-week period over a year is

$$t = \left(14 \ \frac{\text{h}}{\text{d}}\right)\left(5 \ \frac{\text{d}}{\text{wk}}\right)\left(21 \ \frac{\text{wk}}{\text{yr}}\right)$$

$$+ \left(24 \ \frac{\text{h}}{\text{d}}\right)\left(2 \ \frac{\text{d}}{\text{wk}}\right)\left(21 \ \frac{\text{wk}}{\text{yr}}\right)$$

$$= \boxed{2478 \ \text{h/yr} \quad (2500 \ \text{h/yr})}$$

The answer is (C).

(b) With air at 21°C, $c_p \approx 1.005$ kJ/kg·°C and $\rho \approx 1.2$ kg/m³.

From part (a), the energy savings are

$$\Delta q_{\text{air}} = V t \rho c_p \big(N_1 (T_{i,1} - T_o) - N_2 (T_{i,2} - T_o) \big)$$

$$= \left(22\,700 \ \frac{\text{m}^3}{\text{air change}}\right)\left(2478 \ \frac{\text{h}}{\text{yr}}\right)\left(1.2 \ \frac{\text{kg}}{\text{m}^3}\right)$$

$$\times \left(1.005 \ \frac{\text{kJ}}{\text{kg·°C}}\right)$$

$$\times \left(\begin{array}{c} \left(1 \ \dfrac{\text{air change}}{\text{h}}\right)\left(\begin{array}{c}21.1°\text{C} \\ - (-1°\text{C})\end{array}\right) \\[3mm] - \left(0.5 \ \dfrac{\text{air change}}{\text{h}}\right)\left(\begin{array}{c}21.1°\text{C} - 6.7°\text{C} \\ - (-1°\text{C})\end{array}\right) \end{array} \right)$$

$$= \boxed{9.769 \times 10^8 \ \text{kJ/yr} \quad (9.8 \times 10^8 \ \text{kJ/yr})}$$

The answer is (C).

(c) The savings due to reduced heat losses from the roof, floor, and walls for 6.7°C setback are

$$\dot{q} = UA\Delta T$$

$$= \begin{pmatrix} \left(0.85 \; \dfrac{\text{W}}{\text{m}^2 \cdot {}^\circ\text{C}}\right)(993 \text{ m}^2) \\[2mm] + \left(6.42 \; \dfrac{\text{W}}{\text{m}^2 \cdot {}^\circ\text{C}}\right)(260 \text{ m}^2) \\[2mm] + \left(0.3 \; \dfrac{\text{W}}{\text{m}^2 \cdot {}^\circ\text{C}}\right)(2480 \text{ m}^2) \\[2mm] + \left(2.6 \; \dfrac{\text{W}}{\text{m} \cdot {}^\circ\text{C}}\right)(210 \text{ m}) \end{pmatrix} (6.7{}^\circ\text{C})$$

$$= 25\,482 \text{ W}$$

$$q = \dot{q}t$$

$$= \frac{(25\,482 \text{ W}) \left(2478 \; \dfrac{\text{h}}{\text{yr}}\right) \left(3600 \; \dfrac{\text{s}}{\text{h}}\right)}{1000 \; \dfrac{\text{J}}{\text{kJ}}}$$

$$= \boxed{2.273 \times 10^8 \text{ kJ/yr} \quad (2.3 \times 10^8 \text{ kJ/yr})}$$

The answer is (D).

(d) The energy saved per year is

$$q_{\text{actual}} = \frac{q_{\text{ideal}}}{\eta_{\text{furnace}}} = \frac{\Delta q_{\text{air}} + q}{\eta_{\text{furnace}}}$$

$$= \frac{9.769 \times 10^8 \; \dfrac{\text{kJ}}{\text{yr}} + 2.273 \times 10^8 \; \dfrac{\text{kJ}}{\text{yr}}}{(0.75) \left(1000 \; \dfrac{\text{kJ}}{\text{MJ}}\right) \left(105.506 \; \dfrac{\text{MJ}}{\text{therm}}\right)}$$

$$= \boxed{15\,218 \text{ therm/yr} \quad (15\,000 \text{ therm/yr})}$$

The answer is (C).

(e) The annual cost savings are

$$\left(15\,218 \; \frac{\text{therm}}{\text{yr}}\right) \left(\frac{\$0.25}{\text{therm}}\right) = \boxed{\$3805/\text{yr} \quad (\$3800/\text{yr})}$$

The answer is (B).

Ventilation

34 HVAC: Cooling Load

PRACTICE PROBLEMS

1. An air conditioning unit has a SEER-13 rating and a cooling load of 8000 Btu/hr. The unit operates eight hours per day for 140 days each year. The average cost of electricity is $0.25/kW-hr.

(a) What is most nearly the air conditioning unit's average power usage?

(A) 450,000 W-hr/yr

(B) 690,000 W-hr/yr

(C) 9,000,000 W-hr/yr

(D) 120,000,000 W-hr/yr

(b) What is most nearly the hourly cost of operating the air conditioning unit?

(A) $0.15/hr

(B) $0.47/hr

(C) $1.10/hr

(D) $2.00/hr

2. A small, one-story building in Austin, TX, consists of a single room. The room is 40 ft wide × 50 ft long and has an 8 ft ceiling. A blower door has been used to determine the CFM50 as 1800 ft³/min. The energy climate factor is 17.

(a) What is most nearly ACH50?

(A) 0.11

(B) 0.40

(C) 6.8

(D) 9.1

(b) What is most nearly ACHnat?

(A) 0.40 air changes/hr

(B) 0.52 air changes/hr

(C) 0.67 air changes/hr

(D) 0.79 air changes/hr

3. Multiple blower door leakage tests performed on a building are used to develop a CFM50 building leakage curve correlation with coefficient $C = 110.2$ and exponent $n = 0.702$. What airflow is needed to create a 5 Pa pressure difference in the building?

(A) 210 ft³/min

(B) 340 ft³/min

(C) 390 ft³/min

(D) 430 ft³/min

4. The bypass air conditioning system shown has the following operating characteristics.

total air pressure	14.7 psia (101.3 kPa)
supply temperature	58°F (14.4°C) dry-bulb
sensible load	200,000 Btu/hr (58.6 kW)
latent load	450,000 grains/hr (29 kg/h)
outside air	90°F (32.2°C) dry-bulb, 76°F (24.4°C) wet-bulb
make-up outside air	2000 ft³/min (940 L/s)
condition of air leaving air handler	saturated

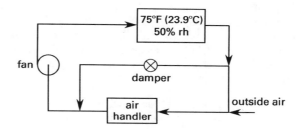

(a) The temperature of the air leaving the air handler is most nearly

(A) 34°F (1°C)

(B) 38°F (3°C)

(C) 49°F (9°C)

(D) 55°F (13°C)

(b) The rate of supply air is most nearly

(A) 7400 ft³/min (3400 L/s)

(B) 11,000 ft³/min (5100 L/s)

(C) 14,000 ft³/min (6400 L/s)

(D) 19,000 ft³/min (8700 L/s)

(c) The moisture content of the supply air is most nearly

(A) 0.0081 lbm/lbm (8.1 g/kg)

(B) 0.0097 lbm/lbm (9.7 g/kg)

(C) 0.0110 lbm/lbm (11 g/kg)

(D) 0.0130 lbm/lbm (13 g/kg)

(d) What is most nearly the efficiency of the process experienced by the air passing through the air handler's coils?

(A) 0%

(B) 35%

(C) 65%

(D) 100%

(e) The bypass factor of the system is most nearly

(A) 0

(B) 0.35

(C) 0.65

(D) 1.0

5. A building is located at 32°N latitude. The inside design temperature is 78°F (25.6°C). The outside design temperature is 95°F (35°C). The daily temperature range is 22°F (12.2°C). Energy gains through the floor, from lights, and from occupants are insignificant. The building's construction is as follows.

walls	1600 ft^2 (144 m^2) facing north
	1400 ft^2 (126 m^2) facing south
	1500 ft^2 (135 m^2) facing east
	1400 ft^2 (126 m^2) facing west
	4 in (100 mm) brick facing
	3 in (75 mm) concrete block
	1 in (25 mm) mineral wool
	(emissivity = 0.96)
	2 in (50 mm) furring
	$^3/_8$ in (9.5 mm) drywall gypsum
	(emissivity = 0.95)
	$^1/_2$ in (12 mm) plaster
roof	6000 ft^2 (540 m^2)
	4 in concrete (100 mm)
	2 in insulation (50 mm)
	insulation conductivity
	2.28 hr-ft^2-°F/Btu-in
	single felt layer
	1 in (25 mm) air gap ($\epsilon = 0.90$)
	$^1/_2$ in (12 mm) acoustical ceiling tile
windows	100 ft^2 (9 m^2) facing east
	$^1/_4$ in (6.4 mm) thick, single glazing
	cream-colored Venetian shades (shading
	factor = 0.75)
	no exterior shading

The maximum temperature differences occur between 4 p.m. and 6 p.m. The equivalent temperature difference for the concrete roof is 74°F at 4 p.m. and 68°F at 6 p.m. The equivalent temperature differences for 4 in brick face walls are as follows.

time	facing direction	temperature difference
4 p.m.	N	17°F
	E	20°F
	S	28°F
	W	20°F
6 p.m.	N	20°F
	E	22°F
	S	28°F
	W	35°F

(a) What is the cooling load for mid-July at 4:00 p.m. sun time? (Solve using customary U.S. units.)

(A) 68,400 Btu/hr

(B) 68,800 Btu/hr

(C) 69,000 Btu/hr

(D) 71,000 Btu/hr

(b) Is this the peak cooling load? Explain. (Solve using customary U.S. units.)

(A) Yes. The instantaneous heat absorption by the building peaks around 12:00 p.m. (noon) local sun time; the cooling load peaks around 4:00 p.m. local sun time.

(B) Yes. The instantaneous heat absorption by the building peaks around 2:00 p.m. local sun time; the cooling load peaks around 4:00 p.m. local sun time.

(C) No. The instantaneous heat absorption by the building peaks around 12:00 p.m. (noon) local sun time; the cooling load peaks around 2:00 p.m. local sun time.

(D) No. The instantaneous heat absorption by the building peaks around 4:00 p.m. local sun time; the cooling load peaks around 6:00 p.m. local sun time.

SOLUTIONS

1. (a) The total annual cooling load is

$$\dot{q}_{c,\text{annual}} = \left(8000 \ \frac{\text{Btu}}{\text{hr}}\right)\left(140 \ \frac{\text{days}}{\text{yr}}\right)\left(8 \ \frac{\text{hr}}{\text{day}}\right)$$
$$= 8{,}960{,}000 \ \text{Btu/yr}$$

With a SEER-13 rating, the annual electrical energy usage is

$$\text{annual energy} = \frac{\dot{q}_{c,\text{annual}}}{\text{SEER}_{\text{Btu/W-hr}}}$$

$$= \frac{8{,}960{,}000 \ \dfrac{\text{Btu}}{\text{yr}}}{13 \ \dfrac{\text{Btu}}{\text{W-hr}}}$$

$$= \boxed{\begin{array}{c} 689{,}231 \ \text{W-hr/yr} \\ (690{,}000 \ \text{W-hr/yr}) \end{array}}$$

The answer is (B).

(b) The power usage while the air conditioning unit is operating is

$$P_{\text{kW}} = \frac{\dot{q}_{c,\text{hourly}}}{\text{SEER}_{\text{Btu/kW-hr}}}$$

$$= \frac{8000 \ \dfrac{\text{Btu}}{\text{hr}}}{\left(13 \ \dfrac{\text{Btu}}{\text{W-hr}}\right)\left(1000 \ \dfrac{\text{W}}{\text{kW}}\right)}$$

$$= 0.615 \ \text{kW}$$

The hourly cost is

$$\text{cost}_{\text{hourly}} = P_{\text{kW}}(\text{cost}_{\text{avg}})$$
$$= (0.615 \ \text{kW})\left(0.25 \ \frac{\$}{\text{kW-hr}}\right)$$
$$= \boxed{\$0.154/\text{hr} \quad (\$0.15/\text{hr})}$$

The answer is (A).

2. (a) The building's volume is

$$V_{\text{structure,ft}^3} = (40 \ \text{ft})(50 \ \text{ft})(8 \ \text{ft})$$
$$= 16{,}000 \ \text{ft}^3$$

Rearranging Eq. 34.12, ACH50 is

$$\text{CFM50} = \frac{(\text{ACH50}) \, V_{\text{structure,ft}^3}}{60 \ \dfrac{\text{min}}{\text{hr}}}$$

$$\text{ACH50} = \frac{(\text{CFM50})\left(60 \ \dfrac{\text{min}}{\text{hr}}\right)}{V_{\text{structure,ft}^3}}$$

$$= \frac{\left(1800 \ \dfrac{\text{ft}^3}{\text{min}}\right)\left(60 \ \dfrac{\text{min}}{\text{hr}}\right)}{16{,}000 \ \text{ft}^3}$$

$$= \boxed{6.75 \quad (6.8)}$$

The answer is (C).

(b) Use Eq. 34.16. The energy climate factor is the same as the LBL factor or the N factor.

$$\text{ACHnat} = \frac{\text{ACH50}}{\text{LBL}} = \frac{6.75}{17}$$

$$= \boxed{\begin{array}{c} 0.397 \ \text{air changes/hr} \\ (0.40 \ \text{air changes/hr}) \end{array}}$$

The answer is (A).

3. Using Eq. 34.11,

$$Q_{\text{ft}^3/\text{min}} = C\Delta p_{\text{Pa}}^n = (110.2)(5 \ \text{Pa})^{0.702}$$
$$= \boxed{341.1 \ \text{ft}^3/\text{min} \quad (340 \ \text{ft}^3/\text{min})}$$

The answer is (B).

4. *Customary U.S. Solution*

(a) This is a standard bypass problem.

The indoor conditions are

$$T_i = 75°\text{F}$$
$$\phi_i = 50\%$$

The outdoor conditions are

$$T_{o,\text{db}} = 90°\text{F}$$
$$T_{o,\text{wb}} = 76°\text{F}$$

The ventilation rate is 2000 ft^3/min.

The loads are

$$q_s = 200{,}000 \text{ Btu/hr}$$
$$q_l = 450{,}000 \text{ gr/hr}$$

To find the sensible heat ratio, q_l must be expressed in Btu/hr.

From the psychrometric chart (see App. 31.A), for the room conditions, $\omega = 0.0095$ lbm moisture/lbm dry air.

From Eq. 31.7,

$$\omega = 0.622 \left(\frac{p_w}{p_a} \right)$$
$$= 0.622 \left(\frac{p_w}{p - p_w} \right)$$
$$0.0095 \ \frac{\text{lbm moisture}}{\text{lbm dry air}} = (0.622) \left(\frac{p_{w,\text{psia}}}{14.7 \text{ psia} - p_{w,\text{psia}}} \right)$$
$$p_w = 0.22 \text{ psia}$$

From steam tables (see App. 29.A), for 0.22 psia,

$$h_{fg} = 1061.7 \text{ Btu/lbm}$$
$$q_l = \frac{\left(450{,}000 \ \frac{\text{gr}}{\text{hr}} \right) \left(1061.7 \ \frac{\text{Btu}}{\text{lbm}} \right)}{7000 \ \frac{\text{gr}}{\text{lbm}}}$$
$$= 68{,}252 \text{ Btu/hr}$$

From Eq. 34.21, the room sensible heat ratio is

$$\text{RSHR} = \frac{q_s}{q_s + q_l} = \frac{200{,}000 \ \frac{\text{Btu}}{\text{hr}}}{200{,}000 \ \frac{\text{Btu}}{\text{hr}} + 68{,}252 \ \frac{\text{Btu}}{\text{hr}}}$$
$$= 0.75$$

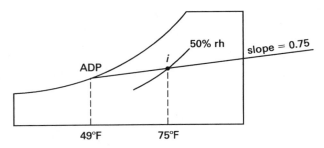

Locate 75°F and 50% relative humidity on the psychrometric chart (see App. 31.A). Draw a line with a slope of 0.75 through this point.

The left-hand intersection shows $T_{co} = 49°\text{F}$. Since the air leaves the conditioner saturated,

$$\text{BF}_{\text{coil}} = 0$$
$$T_{co} = \boxed{49°\text{F}}$$

The answer is (C).

(b) Calculate the airflow through the room from Eq. 34.23(b).

$$\dot{V}_{\text{in,ft}^3/\text{min}} = \frac{\dot{q}_{s,\text{Btu/hr}}}{\left(1.08 \ \frac{\text{Btu-min}}{\text{hr-ft}^3\text{-}°\text{F}} \right)(T_i - T_{\text{in}})}$$
$$= \frac{200{,}000 \ \frac{\text{Btu}}{\text{hr}}}{\left(1.08 \ \frac{\text{Btu-min}}{\text{hr-ft}^3\text{-}°\text{F}} \right)(75°\text{F} - 58°\text{F})}$$
$$= \boxed{10{,}893 \text{ ft}^3/\text{min} \quad (11{,}000 \text{ ft}^3/\text{min})}$$

The answer is (B).

(c) Since $T_{\text{in}} = 58°\text{F}$, locate this dry-bulb temperature on the condition line. Reading from the chart (see App. 31.A),

$$\omega_{\text{in}} = \boxed{0.0081 \text{ lbm/lbm}}$$

The answer is (A).

(d) Since the air leaves the coils saturated, the efficiency of the air handler is $\boxed{100\%.}$ No air escapes being saturated.

The answer is (D).

(e) The bypass factor of the system can be calculated from the dry-bulb temperatures.

$$\text{BF} = \frac{T_{\text{supply}} - T_{\text{air handler}}}{T_{\text{bypass}} - T_{\text{air hander}}} = \frac{58°\text{F} - 49°\text{F}}{75°\text{F} - 49°\text{F}}$$
$$= \boxed{0.35}$$

The answer is (B).

SI Solution

(a) The indoor conditions are

$$T_i = 23.9°\text{C}$$
$$\phi_i = 50\%$$

The outdoor conditions are

$$T_{o,db} = 32.2°C$$
$$T_{o,wb} = 24.4°C$$

The ventilation rate is 940 L/s.

The loads are

$$q_s = 58.6 \text{ kW}$$
$$q_l = 29 \text{ kg/h}$$

To find the sensible heat ratio, q_l must be expressed in kilowatts.

From the psychrometric chart (see App. 31.B), for the room conditions, $\omega \approx 9.3$ g moisture/kg dry air (9.3×10^{-3} kg moisture/kg dry air).

From Eq. 31.7,

$$\omega = 0.622\left(\frac{p_w}{p_a}\right)$$
$$= 0.622\left(\frac{p_w}{p - p_w}\right)$$
$$9.3 \times 10^{-3} \frac{\text{kg moisture}}{\text{kg dry air}} = (0.622)\left(\frac{p_w}{101.3 \text{ kPa} - p_w}\right)$$
$$p_w = 1.49 \text{ kPa}$$

From steam tables (see App. 29.N), for 1.49 kPa,

$$h_{fg} = 2470.3 \text{ kJ/kg}$$
$$q_l = \frac{\left(29 \frac{\text{kg}}{\text{h}}\right)\left(2470.3 \frac{\text{kJ}}{\text{kg}}\right)}{3600 \frac{\text{s}}{\text{h}}}$$
$$= 19.9 \text{ kW}$$

From Eq. 31.22, the sensible heat ratio is

$$\text{SHR} = \frac{q_s}{q_s + q_l}$$
$$= \frac{58.6 \text{ kW}}{58.6 \text{ kW} + 19.9 \text{ kW}}$$
$$= 0.75$$

Locate 23.9°C and 50% relative humidity on the psychrometric chart (see App. 31.B). Draw a line with a slope of 0.75 through this point.

The left-hand intersection shows ADP = 9°C. Since the air leaves the conditioner saturated,

$$\text{BF}_{coil} = 0$$
$$T_{co} = \boxed{9°C}$$

The answer is (C).

(b) Calculate the airflow through the room from Eq. 34.23(a).

$$\dot{V}_{in,L/s} = \frac{\dot{q}_{s,W}}{\left(1.20 \frac{\text{J}}{\text{L·°C}}\right)(T_i - T_{in})}$$
$$= \frac{(58.6 \text{ kW})\left(1000 \frac{\text{W}}{\text{kW}}\right)}{\left(1.20 \frac{\text{J}}{\text{L·°C}}\right)(23.9°C - 14.4°C)}$$
$$= \boxed{5140 \text{ L/s} \quad (5100 \text{ L/s})}$$

The answer is (B).

(c) Since $T_{in} = 14.4°C$, locate this dry-bulb temperature on the condition line. Reading from the chart (see App. 31.B),

$$\omega_{in} = \boxed{8.1 \text{ g/kg dry air}}$$

The answer is (A).

(d) Since the air leaves the coils saturated, the efficiency of the air handler is $\boxed{100\%.}$ No air escapes being saturated.

The answer is (D).

(e) The bypass factor of the system can be calculated from the dry-bulb temperatures.

$$\text{BF} = \frac{T_{supply} - T_{air\,handler}}{T_{bypass} - T_{air\,handler}} = \frac{58°C - 49°C}{75°C - 49°C}$$
$$= \boxed{0.35}$$

The answer is (B).

5. (a) *step 1:* Determine thermal resistances for the walls, roof, and windows.

For the walls:

Assume 2 in furring to be 2 in air space.

From Eq. 33.4, the effective space emissivity is

$$\frac{1}{E} = \frac{1}{\epsilon_1} + \frac{1}{\epsilon_2} - 1 = \frac{1}{0.96} + \frac{1}{0.95} - 1$$
$$E = 0.91$$

From Table 33.3, for vertical air space orientation, the thermal conductance (by extrapolating for $E = 0.91$) is

$$\frac{1}{C} = \frac{1}{1.07 \; \dfrac{\text{Btu}}{\text{hr-ft}^2\text{-}°\text{F}}} = 0.93 \; \text{hr-ft}^2\text{-}°\text{F/Btu}$$

Use App. 33.A.

walls

type of construction	$R, \dfrac{\text{hr-ft}^2\text{-}°\text{F}}{\text{Btu}}$	notes
4 in brick spacing	0.44	
3 in concrete block	0.40	
1 in mineral wool	3.85	value given per in
2 in furring	0.93	see calculation above
3/8 in drywall	0.34	0.45 for 1/2 is given
gypsum		
1/2 in plaster (sand aggregate)	0.09	
surface outside (Table 33.3)	0.25	assume 7 1/2 mph air speed
surface inside (Table 33.3)	0.68	still air, $U = 1.46 \; \dfrac{\text{Btu}}{\text{hr-ft}^2\text{-}°\text{F}}$

roof

type of construction	$R, \dfrac{\text{hr-ft}^2\text{-}°\text{F}}{\text{Btu}}$	notes
4 in concrete	0.32	sand aggregate
2 in insulation	5.56	2.78/in
felt	0.06	
1 in air gap	0.87	$\epsilon = 0.90$ for buildup material
1/2 in accoustical ceiling tile	1.19	
surface outside (Table 33.3)	0.25	assume 7 1/2 mph air speed and downward heat flow ~ upward heat flow
surface inside (Table 33.3)	0.93	still air, $U = 1.08 \; \dfrac{\text{Btu}}{\text{hr-ft}^2\text{-}°\text{F}}$

windows

type of construction	$R, \dfrac{\text{hr-ft}^2\text{-}°\text{F}}{\text{Btu}}$	notes
1/4 in thick, single glazing	0.66	without shades, $R = 0.88 \; \dfrac{\text{hr-ft}^2\text{-}°\text{F}}{\text{Btu}}$
cream colored venetian shades, no exterior shades		shading factor = 0.75

step 2: Determine the overall heat transfer coefficient for the walls, roof, and windows using Eq. 33.3.

$$U = \frac{1}{\sum R_i}$$

For the walls,

$$U = \cfrac{1}{\begin{array}{l} 0.25 \; \dfrac{\text{hr-ft}^2\text{-}°\text{F}}{\text{Btu}} + 0.44 \; \dfrac{\text{hr-ft}^2\text{-}°\text{F}}{\text{Btu}} \\[2mm] + 0.40 \; \dfrac{\text{hr-ft}^2\text{-}°\text{F}}{\text{Btu}} + 3.85 \; \dfrac{\text{hr-ft}^2\text{-}°\text{F}}{\text{Btu}} \\[2mm] + 0.93 \; \dfrac{\text{hr-ft}^2\text{-}°\text{F}}{\text{Btu}} + 0.34 \; \dfrac{\text{hr-ft}^2\text{-}°\text{F}}{\text{Btu}} \\[2mm] + 0.09 \; \dfrac{\text{hr-ft}^2\text{-}°\text{F}}{\text{Btu}} + 0.68 \; \dfrac{\text{hr-ft}^2\text{-}°\text{F}}{\text{Btu}} \end{array}}$$

$$= 0.143 \; \text{Btu/hr-ft}^2\text{-}°\text{F}$$

The cooling load is calculated from the equivalent temperature differences for the walls and roof. Tables of these values are not common; they are now often incorporated directly into HVAC computer application databases. Results from the remainder of this problem will vary with the temperature differences used.

For 4 in brick walls facing at 4 p.m,

$$Q_{\text{walls}} = \left(0.143 \; \frac{\text{Btu}}{\text{hr-ft}^2\text{-}°\text{F}} \right)$$
$$\times \left(\begin{array}{l} (1600 \; \text{ft}^2)(17°\text{F}) + (1400 \; \text{ft}^2)(28°\text{F}) \\ + (1500 \; \text{ft}^2)(20°\text{F}) + (1400 \; \text{ft}^2)(20°\text{F}) \end{array} \right)$$
$$= 17{,}789 \; \text{Btu/hr}$$

For the roof,

$$U = \cfrac{1}{\begin{array}{l} 0.25 \; \dfrac{\text{hr-ft}^2\text{-}°\text{F}}{\text{Btu}} + 0.32 \; \dfrac{\text{hr-ft}^2\text{-}°\text{F}}{\text{Btu}} \\[2mm] + 5.56 \; \dfrac{\text{hr-ft}^2\text{-}°\text{F}}{\text{Btu}} + 0.06 \; \dfrac{\text{hr-ft}^2\text{-}°\text{F}}{\text{Btu}} \\[2mm] + 0.87 \; \dfrac{\text{hr-ft}^2\text{-}°\text{F}}{\text{Btu}} + 1.19 \; \dfrac{\text{hr-ft}^2\text{-}°\text{F}}{\text{Btu}} \\[2mm] + 0.93 \; \dfrac{\text{hr-ft}^2\text{-}°\text{F}}{\text{Btu}} \end{array}}$$

$$= 0.109 \; \text{Btu/hr-ft}^2\text{-}°\text{F}$$

For a 4 in concrete roof at 4 p.m.,

$$Q_{roof} = \left(0.109 \ \frac{Btu}{hr\text{-}ft^2\text{-}°F}\right)(6000 \ ft^2)(74°F)$$
$$= 48{,}396 \ Btu/hr$$

For the windows:

Since all windows facing east will receive no direct sunlight at 4 p.m., consider $\Delta T = 95°F - 78°F$.

$$Q_{windows} = \left(\cfrac{1}{0.66 \ \cfrac{hr\text{-}ft^2\text{-}°F}{Btu}}\right)(100 \ ft^2)(95°F - 78°F)$$
$$= 2576 \ Btu/hr$$

The total sensible transmission load (at 4 p.m.) is

$$Q_{total} = 17{,}789 \ \frac{Btu}{hr} + 48{,}396 \ \frac{Btu}{hr} + 2576 \ \frac{Btu}{hr}$$
$$= \boxed{68{,}761 \ Btu/hr \quad (69{,}000 \ Btu/hr)}$$

The answer is (C).

(b) This $\boxed{\text{may not}}$ be the peak cooling load. Since Q_{walls} and Q_{roof} are the major contributors and ΔT equivalent temperature differences are maximum between 4 p.m. and 6 p.m., the peak cooling load will be maximum somewhere between $\boxed{\text{4 p.m. and 6 p.m.}}$ Consider Q_{walls} and Q_{roof} for 6 p.m.

For 4 in brick walls facing at 6 p.m.,

$$Q_{walls} = \left(0.143 \ \frac{Btu}{hr\text{-}ft^2\text{-}°F}\right)$$
$$\times \left(\begin{array}{c} (1600 \ ft^2)(20°F) + (1400 \ ft^2)(28°F) \\ + (1500 \ ft^2)(22°F) + (1400 \ ft^2)(35°F) \end{array}\right)$$
$$= 21{,}908 \ Btu/hr$$

For the roof,

$$Q_{roof} = \left(0.109 \ \frac{Btu}{hr\text{-}ft^2\text{-}°F}\right)(6000 \ ft^2)(68°F)$$
$$= 44{,}472 \ Btu/hr$$

The total sensible transmission load at 6 p.m. is

$$Q_{total} = 21{,}908 \ \frac{Btu}{hr} + 44{,}472 \ \frac{Btu}{hr} + 2576 \ \frac{Btu}{hr}$$
$$= 68{,}956 \ Btu/hr$$

This is very close to the 4 p.m. load.

The answer is (D).

Ventilation

35 Air Conditioning Equipment and Controls

PRACTICE PROBLEMS

1. What is the typical approximate supply air pressure in a pneumatic HVAC control system?

(A) 7 psig

(B) 18 psig

(C) 35 psig

(D) 140 psig

2. The control signal in a pneumatic control system is air pressure that falls within which approximate range?

(A) 0–4 psig

(B) 3–15 psig

(C) 12–45 psig

(D) 30–120 psig

3. What kind of device is a pneumatic thermostat?

(A) reverse acting

(B) normally closed

(C) normally open

(D) direct acting

4. What is the term for a device that converts a pneumatic signal of 15 psig to a control pressure of 20 psig?

(A) a pneumatic transformer

(B) a pneumatic amplifier

(C) a pneumatic relay

(D) a pneumatic booster

5. What kind of device is a pneumatic hot water valve?

(A) reverse acting

(B) normally closed

(C) normally open

(D) direct acting

6. What kind of device is a pneumatic steam valve?

(A) reverse acting

(B) normally closed

(C) normally open

(D) direct acting

Relay Logic Diagrams

7. The control relay diagram shown is missing a rung. What are the most likely elements of that rung?

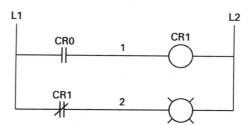

(A) a switch

(B) a relay coil

(C) a switch and a relay coil

(D) a switch and a relay contact

8. What is the most likely description of element M1?

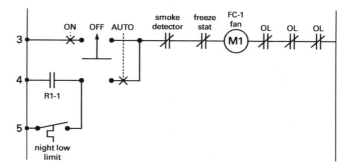

(A) a start/run enable solenoid or exciter circuit

(B) a fan motor

(C) a relay coil

(D) a fan interrupt circuit

9. Which Boolean truth table describes the control of motor M in the relay logic diagram shown?

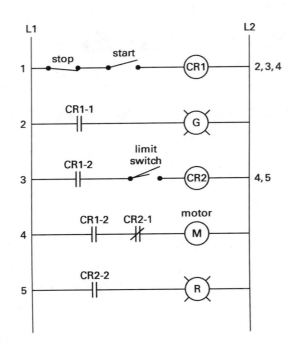

(A)

CR1	CR2	M
0	0	1
0	1	0
1	0	0
1	1	0

(B)

CR1	CR2	M
0	0	0
0	1	1
1	0	0
1	1	0

(C)

CR1	CR2	M
0	0	0
0	1	0
1	0	1
1	1	0

(D)

CR1	CR2	M
0	0	0
0	1	0
1	0	0
1	1	1

10. Given that the stop switch is closed as shown and motor M is not running, what are the functions of switches PB1 and PB2 in the relay logic diagram?

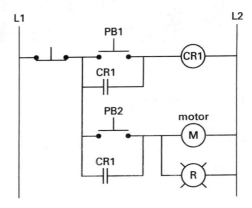

(A) PB1 starts the motor running continuously, and PB2 stops the motor.

(B) PB1 starts the motor for as long as PB1 is pushed, and PB2 starts the motor running continuously.

(C) PB1 stops the motor, and PB2 starts the motor running continuously.

(D) PB1 starts the motor running continuously, and PB2 starts the motor for as long as PB2 is pushed.

11. What are the functions of the (initially) NC M2 and M1 relay contacts in the two main rungs?

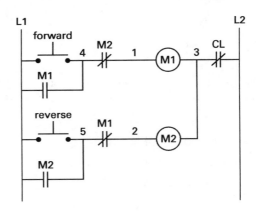

(A) When running forward, NC M2 latches and prevents the device from stopping.

(B) When running forward, NC M2 prevents the device from being placed in reverse.

(C) When running forward, NC M1 prevents the device from being placed in reverse.

(D) When running forward, NC M1 latches and prevents the device from stopping.

12. What is most likely the purpose of the transformer in the control circuit shown?

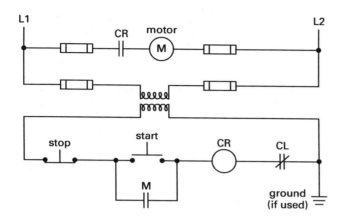

(A) isolate the motor M from start/stop transients

(B) step up the control voltage to motor M level

(C) step up the line voltage to control relay CR level

(D) step down the line voltage for safety

SOLUTIONS

1. Pneumatic air pressure is typically supplied at 18 psig to 20 psig, but may be as high as 25 psig.

The answer is (B).

2. A device's pneumatic signal cannot be less than the supply signal, and it must be sufficient for detection. 3–15 psig is a typical range of output pressures.

The answer is (B).

3. Thermostats are direct acting (DA) because, as the room cools, an increase in control pressure is needed.

The answer is (D).

4. A pneumatic relay has three ports: a signal input, a supply input, and a control output. When the full input signal is received, the control output valve is opened to the full supply input.

The answer is (C).

5. A hot water valve is normally open so that the valve will provide uncontrolled heat in the event of a power failure or air pressure outage.

The answer is (C).

6. A steam valve is normally closed so that uncontrolled steam will not be introduced into the HVAC system in the event of a power failure or air pressure outage.

The answer is (B).

7. Relay contact CR0 does not have any way to change state. The missing rung must have the CR0 relay coil, and the coil itself needs a control element, probably a switch of some type.

The answer is (C).

8. The smoke detector, freeze stat, and overload relay contacts are all NC, which means power is supplied through them to M1 during normal operation in order to keep fan FC-1 running.

M1 is probably a start/run enable solenoid or fan motor exciter circuit.

The answer is (A).

9. When the start switch in rung 1 is pushed, the coil in control relay CR1 is energized. The coil controls contacts in rungs 2, 3, and 4. The NO contacts CR1-2 in rung 3 close, making the rung live, but power is not supplied to coil CR2 until the limit switch is closed. In rung 4, the NO contacts CR1-2 also close, and power is supplied through the NC contacts CR2-1 to motor M. If the limit switch is closed by overtravel, relay coil CR2 will be energized, opening the NC CR2-1 contacts in rung 4, and stopping the motor, M. Therefore, in order for the motor to run, CR1 must be energized, and CR2 must not be energized.

The answer is (C).

10. If the motor is not running, when PB1 is pushed, (circled) relay coil CR1 is energized, and both NO CR1 contacts are closed. This supplies power through the lowest sub-rung to motor M, turning on the motor ⌐continuously⌐ (because contacts CR1 are latched on). If the motor is not running, pushing PB2 will supply power to the motor ⌐for as long as PB2 is pushed.⌐ PB2 does not control a relay coil and has no effect on the CR1 contacts.

The answer is (D).

11. When the forward switch is pressed, power is supplied through the switch and NC M2 to coil M1, opening the NC contacts M1 in rung 2.

⌐This prevents the device from being placed in reverse.⌐

The answer is (C).

12. The lower circuit does not contain a battery or connection to a power source. The voltage in the circuit is provided by the transformer. The transformer is probably a ⌐step-down⌐ transformer that allows the start/stop control circuit to operate at a lower (and, therefore, safer) voltage than the controlled fused upper circuit.

The answer is (D).

PRACTICE PROBLEMS

(Note: Round all duct dimensions to the next larger whole inch or multiples of 25 mm.)

1. A round 18 in (457 mm) duct is to be replaced by a rectangular duct with an aspect ratio of 4:1. The dimensions of the rectangular duct are most nearly

(A) 8 in × 30 in (200 mm × 760 mm)

(B) 9 in × 35 in (220 mm × 880 mm)

(C) 12 in × 18 in (300 mm × 460 mm)

(D) 18 in × 72 in (460 mm × 1800 mm)

2. A fan moves 10,000 SCFM (5700 L/s) through ductwork that has a dynamic friction pressure of 4 in wg (1 kPa). If the fan speed is decreased so that the flow rate becomes 8000 SCFM (4700 L/s), the total pressure in the duct after the fan will most nearly be

(A) 2.2 in wg (0.53 kPa)

(B) 2.4 in wg (0.58 kPa)

(C) 2.6 in wg (0.62 kPa)

(D) 2.8 in wg (0.68 kPa)

3. A $^1/_8$ scale model fan is tested at 300 rpm. At standard air conditions and at that speed, the model fan moves 40 ft³/min (19 L/s) against 0.5 in wg (125 Pa). If a full-sized fan operates at 5000 ft (1500 m) altitude at half the model's speed, the air power is most nearly

(A) 11 hp (8.6 kW)

(B) 14 hp (11 kW)

(C) 19 hp (15 kW)

(D) 24 hp (19 kW)

4. A duct system consists of 750 ft (230 m) of galvanized 20 in (508 mm) diameter duct with four round elbows (radius-to-diameter ratio of 1.5). A 2 in (51 mm) diameter pipe passes perpendicularly through the duct at two locations. The flow rate is 6000 SCFM (4200 L/s). The friction loss is most nearly

(A) 1.7 in wg (1.1 kPa)

(B) 2.9 in wg (2.0 kPa)

(C) 3.7 in wg (2.4 kPa)

(D) 4.8 in wg (3.2 kPa)

5. 1500 SCFM (700 L/s) of air flows in an 18 in (457 mm) diameter round duct. After a branch reduction of 300 SCFM (140 L/s), the fitting reduces to 14 in (356 mm) in the through direction. The static regain in the through direction is most nearly

(A) −0.037 in wg loss (−8.7 Pa loss)

(B) −0.073 in wg loss (−17 Pa loss)

(C) −0.11 in wg loss (−26 Pa loss)

(D) −0.19 in wg loss (−45 Pa loss)

6. A fan in a small theater delivers 1500 ft³/min (700 L/s) through the system shown. The duct is rectangular with an aspect ratio of 1.5:1 and the shorter side vertical. All elbows have a radius-to-width ratio of 1.5. The terminal pressure at each outlet must be 0.25 in wg (63 Pa) or higher. Takeoff fitting losses are to be disregarded. Use the equal-friction method to size the system.

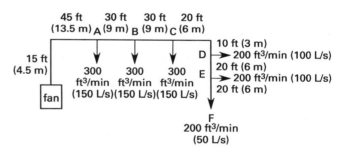

(a) The static pressure at the fan is most nearly

(A) 0.5 in wg (110 Pa)

(B) 0.7 in wg (160 Pa)

(C) 0.8 in wg (200 Pa)

(D) 1.0 in wg (260 Pa)

(b) The total pressure at the fan is most nearly

(A) 1.0 in wg (230 Pa)

(B) 1.2 in wg (280 Pa)

(C) 1.6 in wg (380 Pa)

(D) 2.0 in wg (500 Pa)

7. A fan in a theater delivers 300 ft³/min (140 L/s) of air through round ducts to each of the twelve outlets shown. The minimum terminal pressure at the outlets is

Ventilation

0.15 in wg (38 Pa). All elbows have a radius-to-diameter ratio of 1.25. Disregard branch takeoff fitting losses. Use the equal-friction method to size the system.

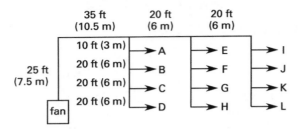

(a) The static pressure at the fan is most nearly

(A) 0.5 in wg (130 Pa)

(B) 0.7 in wg (160 Pa)

(C) 0.8 in wg (210 Pa)

(D) 1.0 in wg (260 Pa)

(b) The total pressure at the fan is most nearly

(A) 0.5 in wg (110 Pa)

(B) 0.7 in wg (170 Pa)

(C) 0.8 in wg (210 Pa)

(D) 1.0 in wg (260 Pa)

8. Size the longest run in Prob. 7 using the static regain method with a regain coefficient of 0.75. (Customary U.S. solution only required.)

(a) The diameter of the main duct is most nearly

(A) 10 in

(B) 20 in

(C) 25 in

(D) 35 in

(b) The equivalent length of the main duct from the fan to the first takeoff and bend is most nearly

(A) 75 ft

(B) 85 ft

(C) 95 ft

(D) 110 ft

(c) The friction loss in the main duct up to the first takeoff is most nearly

(A) 0.12 in wg

(B) 0.14 in wg

(C) 0.15 in wg

(D) 0.17 in wg

(d) The total pressure supplied by the fan is most nearly

(A) 0.14 in wg

(B) 0.27 in wg

(C) 0.29 in wg

(D) 0.32 in wg

9. 3000 ft³/min (1400 L/s) of air with a density of 0.075 lbm/ft³ (1.2 kg/m³) enters a 12 in (305 mm) diameter duct at section A. The four-piece 90° elbows have a radius-to-diameter ratio of 1.5 and an equivalent length of 14 diameters. The Darcy friction factor is 0.02 everywhere in the system. Use the static regain method with a static regain coefficient of 0.65.

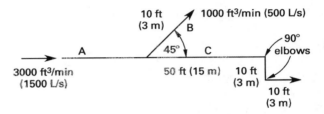

(a) Assuming a 10 in (250 mm) diameter duct, the total equivalent length of run C is most nearly

(A) 75 ft (21 m)

(B) 85 ft (24 m)

(C) 95 ft (29 m)

(D) 110 ft (31 m)

(b) The friction loss in section C is most nearly

(A) 8.0 ft (2.0 m)

(B) 26 ft (9.0 m)

(C) 52 ft (16 m)

(D) 88 ft (27 m)

(c) The diameter at section B is most nearly

(A) 11 in (280 mm)

(B) 14 in (340 mm)

(C) 18 in (460 mm)

(D) 21 in (530 mm)

(d) The regain (in feet of air) in sections B and C is most nearly

(A) 26 ft (8.0 m)

(B) 31 ft (9.0 m)

(C) 34 ft (10 m)

(D) 40 ft (12 m)

Ventilation

(e) Which of these sections will require dampers?

  (A) section A only

  (B) section B only

  (C) sections B and C only

  (D) sections A, B, and C

10. A motor rated at 5 hp delivers 4.3 hp to a fan. The fan rotates at 920 rpm and moves 160°F air at standard pressure into a drying process. The static pressure across the fan is 2.3 in wg. The fan efficiency is 62%. The fan flow rate at standard conditions is most nearly

  (A) 5100 ft³/min

  (B) 5800 ft³/min

  (C) 6300 ft³/min

  (D) 7400 ft³/min

11. Which of the following would NOT normally be used to vary the flow rate of a fan-driven ventilation system?

  (A) system dampers

  (B) variable pitch blades

  (C) inlet vanes

  (D) variable speed motors

12. Which of the following parameters is normally sensed in order to control a variable frequency drive (VFD) in a variable air volume (VAV) system?

  (A) duct velocity pressure

  (B) supply air fan rate of flow

  (C) fan motor current

  (D) VAV box airflow

13. Fan A and fan B are identical variable-speed fans operating in parallel. They are of the same rating, and they have the same efficiency. Both fans draw from the same source. Fan A always turns at 2000 rev/min and has a volumetric flow rate of 4000 ft³/min. Fan B's speed is varied according to demand. When fan B turns at 1800 rev/min, its volumetric flow rate is 5000 ft³/min. At a particular moment, the total flow rate supplied by fans A and B is 10,000 ft³/min. At that moment, when operating in parallel with fan A, the speed of fan B is most nearly

  (A) 1500 rev/min

  (B) 2100 rev/min

  (C) 2200 rev/min

  (D) 4200 rev/min

SOLUTIONS

1. *Customary U.S. Solution*

With an aspect ratio of $R = 4$, the short side is given by Eq. 36.36.

$$\text{short side} = \frac{D(1+R)^{1/4}}{1.3R^{5/8}} = \frac{(18 \text{ in})(1+4)^{1/4}}{(1.3)(4)^{5/8}}$$
$$= 8.7 \text{ in} \quad (9 \text{ in})$$

From Eq. 36.37, the long side is

$$\text{long side} = R(\text{short side}) = (4)(8.7 \text{ in})$$
$$= 34.8 \text{ in} \quad (35 \text{ in})$$

The dimensions of the rectangular duct are $\boxed{9 \text{ in} \times 35 \text{ in.}}$

The answer is (B).

SI Solution

With an aspect ratio of $R = 4$, the short side is given by Eq. 36.36.

$$\text{short side} = \frac{D(1+R)^{1/4}}{1.3R^{5/8}} = \frac{(457 \text{ mm})(1+4)^{1/4}}{(1.3)(4)^{5/8}}$$
$$= 221 \text{ mm} \quad (220 \text{ mm})$$

From Eq. 36.37, the long side is

$$\text{long side} = R(\text{short side}) = (4)(221 \text{ mm})$$
$$= 884 \text{ mm} \quad (880 \text{ mm})$$

The dimensions of the rectangular duct are $\boxed{220 \text{ mm} \times 880 \text{ mm.}}$

The answer is (B).

2. *Customary U.S. Solution*

Use Eq. 36.21 to calculate the friction pressure at the reduced flow rate.

$$\text{FP}_2 = \text{FP}_1\left(\frac{Q_2}{Q_1}\right)^2 = (4 \text{ in wg})\left(\frac{8000 \text{ SCFM}}{10{,}000 \text{ SCFM}}\right)^2$$
$$= 2.56 \text{ in wg}$$

The dynamic loss in the duct at the reduced flow rate is 2.56 in wg. However, the pressure in the duct after the fan includes the terminal pressure of 0.1–0.3 in wg. The duct pressure will be 2.7–2.9 in wg. Therefore, the total pressure in the duct after the fan will be $\boxed{2.8 \text{ in wg.}}$

The answer is (D).

SI Solution

Use Eq. 36.21 to calculate the friction pressure at the reduced flow rate.

$$FP_2 = FP_1\left(\frac{Q_2}{Q_1}\right)^2 = (1 \text{ kPa})\left(\frac{3700 \frac{\text{L}}{\text{s}}}{4700 \frac{\text{L}}{\text{s}}}\right)^2 = 0.619 \text{ kPa}$$

The dynamic loss in the duct at the reduced flow rate is 0.619 kPa. However, the pressure in the duct after the fan includes the terminal pressure of 25–75 Pa. The duct pressure will be 0.64–0.69 kPa. Therefore, the total pressure in the duct after the fan will be $\boxed{0.68 \text{ kPa.}}$

The answer is (D).

3. *Customary U.S. Solution*

The air horsepower is given by Eq. 36.13(b).

$$AHP_1 = \frac{Q_{\text{ft}^3/\text{min}}(TP_{\text{in wg}})}{6356} = \frac{\left(40 \frac{\text{ft}^3}{\text{min}}\right)(0.5 \text{ in wg})}{6356 \frac{\text{in-ft}^3}{\text{hp-min}}}$$

$$= 3.147 \times 10^{-3} \text{ hp}$$

In order to predict the performance of a dynamically similar fan, use Eq. 36.28.

$$\frac{AHP_1}{AHP_2} = \left(\frac{D_1}{D_2}\right)^5\left(\frac{n_1}{n_2}\right)^3\left(\frac{\gamma_1}{\gamma_2}\right)$$

$$AHP_2 = AHP_1\left(\frac{D_2}{D_1}\right)^5\left(\frac{n_2}{n_1}\right)^3\left(\frac{\gamma_2}{\gamma_1}\right)$$

$$\frac{D_2}{D_1} = \frac{1}{\frac{1}{8}} = 8$$

$$\frac{n_2}{n_1} = \frac{1}{2}$$

At standard air conditions, $\rho_1 = 0.075 \text{ lbm/ft}^3$. Use standard atmospheric data from App. 18.A for 5000 ft altitude.

$$\rho_{5000\,\text{ft}} = \rho_2 = \frac{\rho}{RT} = \frac{\left(12.225 \frac{\text{lbf}}{\text{in}^2}\right)\left(12 \frac{\text{in}}{\text{ft}}\right)^2}{\left(53.3 \frac{\text{ft-lbf}}{\text{lbm-}^\circ\text{R}}\right)(500.9^\circ\text{R})}$$

$$= 0.06594 \text{ lbf/ft}^3$$

$$\frac{\gamma_2}{\gamma_1} = \frac{\frac{\rho_2 g}{g_c}}{\frac{\rho_1 g}{g_c}} = \frac{\rho_2}{\rho_1} = \frac{0.06594 \frac{\text{lbm}}{\text{ft}^3}}{0.075 \frac{\text{lbm}}{\text{ft}^3}} = 0.8792$$

$$AHP_2 = (3.147 \times 10^{-3} \text{ hp})(8)^5\left(\tfrac{1}{2}\right)^3(0.8792)$$

$$= \boxed{11.33 \text{ hp} \quad (11 \text{ hp})}$$

The answer is (A).

SI Solution

The air power is given by Eq. 36.13(a).

$$AkW_1 = \frac{Q_{\text{L/s}}(TP_{\text{Pa}})}{10^6}$$

$$= \frac{\left(19 \frac{\text{L}}{\text{s}}\right)(125 \text{ Pa})}{10^6 \frac{\text{L·W}}{\text{m}^3\text{·kW}}}$$

$$= 2.375 \times 10^{-3} \text{ kW}$$

In order to predict the performance of a dynamically similar fan, use Eq. 36.28.

$$\frac{AkW_1}{AkW_2} = \left(\frac{D_1}{D_2}\right)^5\left(\frac{n_1}{n_2}\right)^3\left(\frac{\gamma_1}{\gamma_2}\right)$$

$$AkW_2 = AkW_1\left(\frac{D_2}{D_1}\right)^5\left(\frac{n_2}{n_1}\right)^3\left(\frac{\gamma_2}{\gamma_1}\right)$$

$$\frac{D_2}{D_1} = \frac{1}{\frac{1}{8}} = 8$$

$$\frac{n_2}{n_1} = \frac{1}{2}$$

At standard air conditions, $\rho = 1.2 \text{ kg/m}^3$. Use standard atmospheric data from App. 18.A for 1500 m altitude.

$$\rho_{1500\,\text{m}} = \rho_2 = \frac{p}{RT} = \frac{(0.8456 \text{ bar})\left(10^5 \frac{\text{Pa}}{\text{bar}}\right)}{\left(287 \frac{\text{J}}{\text{kg·K}}\right)(278.4\text{K})}$$

$$= 1.058 \text{ kg/m}^3$$

$$\frac{\gamma_2}{\gamma_1} = \frac{\rho_2 g}{\rho_1 g} = \frac{\rho_2}{\rho_1} = \frac{1.058 \frac{\text{kg}}{\text{m}^3}}{1.2 \frac{\text{kg}}{\text{m}^3}}$$

$$= 0.882$$

$$AkW_2 = (2.375 \times 10^{-3} \text{ kW})(8)^5\left(\tfrac{1}{2}\right)^3(0.882)$$

$$= \boxed{8.58 \text{ kW} \quad (8.6 \text{ kW})}$$

The answer is (A).

4. *Customary U.S. Solution*

From the standard friction loss chart (see Fig. 36.5), the friction loss is 0.42 in of water per 100 ft of duct and v = 2700 ft/min. The friction loss in 750 ft of duct is

$$\text{FP}_{f,1} = \left(0.42 \ \frac{\text{in wg}}{100 \ \text{ft}}\right)\left(\frac{750 \ \text{ft}}{100 \ \text{ft}}\right) = 3.15 \ \text{in wg}$$

The equivalent length of each round elbow (radius-to-diameter ratio of 1.5) can be found from Table 36.5.

$$L_e = 12D$$

For four round elbows,

$$L_e = (4)(12D) = \frac{(4)(12)(20 \ \text{in})}{12 \ \frac{\text{in}}{\text{ft}}} = 80 \ \text{ft}$$

The friction loss is

$$\text{FP}_{f,2} = \left(0.42 \ \frac{\text{in wg}}{100 \ \text{ft}}\right)\left(\frac{80 \ \text{ft}}{100 \ \text{ft}}\right) = 0.336 \ \text{in wg}$$

For a cross duct,

$$\frac{D_1}{D_2} = \frac{2 \ \text{in}}{20 \ \text{in}} = 0.1$$

From Table 36.4, the loss coefficient is $K_{up} = 0.2$. For two cross pipes, the friction loss is given by Eq. 36.39(b).

$$\text{FP}_{f,3} = 2K_{up}\left(\frac{v_{\text{ft/min}}}{4005}\right)^2 = (2)(0.2)\left(\frac{2700 \ \frac{\text{ft}}{\text{min}}}{4005 \ \frac{\text{ft}}{\text{min}}}\right)^2$$

$$= 0.182 \ \text{in wg}$$

The total loss is

$$\text{FP}_{f,1} + \text{FP}_{f,2} + \text{FP}_{f,3} = 3.15 \ \text{in wg} + 0.336 \ \text{in wg}$$

$$+ \ 0.182 \ \text{in wg}$$

$$= \boxed{3.668 \ \text{in wg} \quad (3.7 \ \text{in wg})}$$

The answer is (C).

SI Solution

From the standard friction loss chart (see Fig. 36.6), the friction loss is 9 Pa/m and v = 21.5 m/s. The friction loss in 230 m of duct is

$$\text{FP}_{f,1} = \left(9 \ \frac{\text{Pa}}{\text{m}}\right)(230 \ \text{m}) = 2070 \ \text{Pa}$$

The equivalent length of each round elbow (radius-to-diameter ratio of 1.5) can be found in Table 36.5.

$$L_e = 12D$$

For four round elbows,

$$L_e = (4)(12D) = \frac{(4)(12)(508 \ \text{mm})}{1000 \ \frac{\text{mm}}{\text{m}}} = 24.38 \ \text{m}$$

The friction loss is

$$\text{FP}_{f,2} = \left(9 \ \frac{\text{Pa}}{\text{m}}\right)(24.38 \ \text{m}) = 219.4 \ \text{Pa}$$

For a cross duct,

$$\frac{D_1}{D_2} = \frac{51 \ \text{mm}}{508 \ \text{mm}} = 0.1$$

From Table 36.4, the loss coefficient K_{up} is 0.2. For two cross pipes, the friction loss is given by Eq. 36.39(a).

$$\text{FP}_{f,3} = (2)0.6K_{up}v_{\text{m/s}}^2$$

$$= (2)\left(0.6 \ \frac{\text{Pa} \cdot \text{s}^2}{\text{m}^2}\right)(0.2)\left(21.5 \ \frac{\text{m}}{\text{s}}\right)^2$$

$$= 110.9 \ \text{Pa}$$

The total loss is

$$\text{FP}_{f,1} + \text{FP}_{f,2} + \text{FP}_{f,3} = 2070 \ \text{Pa} + 219.4 \ \text{Pa} + 110.9 \ \text{Pa}$$

$$= \boxed{2400.3 \ \text{Pa} \quad (2.4 \ \text{kPa})}$$

The answer is (C).

5. *Customary U.S. Solution*

For the 18 in duct,

$$A = \frac{\pi}{4}D^2 = \left(\frac{\pi}{4}\right)\left(\frac{18 \ \text{in}}{12 \ \frac{\text{in}}{\text{ft}}}\right)^2 = 1.767 \ \text{ft}^2$$

$$v_1 = \frac{Q}{A} = \frac{1500 \ \frac{\text{ft}^3}{\text{min}}}{1.767 \ \text{ft}^2} = 848.9 \ \text{ft/min}$$

For the 14 in duct,

$$A = \frac{\pi}{4}D^2 = \left(\frac{\pi}{4}\right)\left(\frac{14 \text{ in}}{12 \frac{\text{in}}{\text{ft}}}\right)^2 = 1.069 \text{ ft}^2$$

$$v_2 = \frac{Q}{A} = \frac{1500 \frac{\text{ft}^3}{\text{min}} - 300 \frac{\text{ft}^3}{\text{min}}}{1.069 \text{ ft}^2} = 1122.5 \text{ ft/min}$$

Since $v_2 > v_1$, there will not be an increase in static pressure. There will be a static pressure loss. Use $R = 1.1$. From Eq. 36.45(b), the static pressure loss is

$$SR_{\text{actual}} = R\left(\frac{v_{\text{up}}^2 - v_{\text{down}}^2}{(4005)^2}\right)$$

$$= (1.1)\left(\frac{\left(848.9 \frac{\text{ft}}{\text{min}}\right)^2 - \left(1122.5 \frac{\text{ft}}{\text{min}}\right)^2}{\left(4005 \frac{\text{ft}}{\text{min}}\right)^2}\right)$$

$$= \boxed{-0.037 \text{ in wg loss}}$$

The answer is (A).

SI Solution

For the 457 mm duct,

$$A = \frac{\pi}{4}D^2 = \left(\frac{\pi}{4}\right)\left(\frac{457 \text{ mm}}{1000 \frac{\text{mm}}{\text{m}}}\right)^2 = 0.164 \text{ m}^2$$

$$v_1 = \frac{Q}{A} = \frac{700 \frac{\text{L}}{\text{s}}}{(0.164 \text{ m}^2)\left(1000 \frac{\text{L}}{\text{m}^3}\right)} = 4.27 \text{ m/s}$$

For the 356 mm duct,

$$A = \frac{\pi}{4}D^2 = \left(\frac{\pi}{4}\right)\left(\frac{356 \text{ mm}}{1000 \frac{\text{mm}}{\text{m}}}\right)^2 = 0.10 \text{ m}^2$$

$$v_2 = \frac{Q}{A} = \frac{700 \frac{\text{L}}{\text{s}} - 140 \frac{\text{L}}{\text{s}}}{(0.10 \text{ m}^2)\left(1000 \frac{\text{L}}{\text{m}^3}\right)} = 5.60 \text{ m/s}$$

Since $v_2 > v_1$, there will not be an increase in static pressure. There will be a static pressure loss. Use $R = 1.1$. From Eq. 36.45(a), the static pressure loss is

$$SR = R0.6(v_{\text{up}}^2 - v_{\text{down}}^2)$$

$$= (1.1)\left(0.6 \frac{\text{Pa·s}^2}{\text{m}^2}\right)\left(\left(4.27 \frac{\text{m}}{\text{s}}\right)^2 - \left(5.60 \frac{\text{m}}{\text{s}}\right)^2\right)$$

$$= \boxed{-8.66 \text{ Pa loss} \quad (-8.7 \text{ Pa loss})}$$

The answer is (A).

6. *Customary U.S. Solution*

Use the equal-friction method to size the system.

step 1: From Table 36.8, choose the main duct velocity as approximately 1600 ft/min.

step 2: The total air flow from the fan is 1500 ft³/min. From Fig. 36.5, the main duct diameter is 13 in. The friction loss is 0.27 in wg per 100 ft.

step 3: After the first takeoff at A, the flow rate in section A–B is

$$1500 \frac{\text{ft}^3}{\text{min}} - 300 \frac{\text{ft}^3}{\text{min}} = 1200 \text{ ft}^3/\text{min}$$

From Fig. 36.5, for 1200 ft³/min and 0.27 in wg per 100 ft, the diameter is 11.8 in (say 12 in). The velocity from the chart, 1500 ft/min, cannot be used because the duct is not circular. Similarly, the diameters for the other sections are obtained and listed in the following table.

section	Q (ft³/min)	D (in)
fan–A	1500	13.0
A–B	1200	12.0
B–C	900	11.0
C–D	600	9.5
D–E	400	8.0
E–F	200	6.3

These diameters are for round duct. The equivalent rectangular duct sides with an aspect ratio of 1.5 are found as follows.

$$a = \frac{D(1.5 + 1)^{0.25}}{(1.3)(1.5)^{5/8}}$$

$$= 0.75D$$

$$b = Ra = (1.5)(0.75D)$$

$$= 1.125D$$

Ventilation

Convert the diameters to sides a and b.

$$a_{\text{fan–A}} = (0.75)(13 \text{ in}) = 9.75 \text{ in}$$
$$b_{\text{fan–A}} = (1.125)(13 \text{ in}) = 14.63 \text{ in}$$

Use a 10 in × 15 in duct.

The following table is prepared similarly, rounding up as appropriate.

section	D (in)	a (in)	b (in)	v (ft/min)
fan–A	13.0	10	15	1440
A–B	12.0	9	14	1370
B–C	11.0	9	13	1110
C–D	9.5	8	11	980
D–E	8.0	6	9	1070
E–F	6.3	5	8	720

The velocity in section fan–A is

$$v_{\text{fan–A}} = \frac{\left(1500 \ \frac{\text{ft}^3}{\text{min}}\right)\left(12 \ \frac{\text{in}}{\text{ft}}\right)^2}{(10 \text{ in})(15 \text{ in})} = 1440 \text{ ft/min}$$

The velocities in the other sections are found similarly.

(a) *step 4:* By inspection, the longest run is fan–F.

Use Eq. 36.41 to estimate the equivalent lengths of the bends. Since both bends have the same radius-width and height-width ratios, the equivalent lengths can be combined.

$$L_e = (W_{\text{fan–A}} + W_{\text{C–D}})\left(0.33\frac{r}{W}\right)^{-2.13(H/W)^{0.126}}$$
$$= \left(\frac{15 \text{ in} + 11 \text{ in}}{12 \ \frac{\text{in}}{\text{ft}}}\right)((0.33)(1.5))^{-(2.13)(1.5)^{0.126}}$$
$$= 10.5 \text{ ft}$$

The equivalent length of the entire run is

15 ft	(fan to first bend)
45 ft	(first bend to point A)
30 ft	(section A–B)
30 ft	(section B–C)
20 ft	(point A to second bend)
10.5 ft	(equivalent length of two bends)
10 ft	(second bend to point D)
20 ft	(section D–E)
20 ft	(section E–F)
total: 200.5 ft	

The straight-through friction loss in the longest run is

$$\left(\frac{200.5 \text{ ft}}{100 \text{ ft}}\right)\left(0.27 \ \frac{\text{in wg}}{100 \text{ ft}}\right) = 0.54 \text{ in wg}$$

The fan must be able to supply a static pressure of

$$SP_{\text{fan}} = 0.54 \text{ in wg} + 0.25 \text{ in wg}$$
$$= \boxed{0.79 \text{ in wg} \quad (0.8 \text{ in wg})}$$

The answer is (C).

(b) The total pressure supplied by the fan is

$$TP_{\text{fan}} = 0.79 \text{ in wg} + \left(\frac{1440 \ \frac{\text{ft}}{\text{min}}}{4005 \ \frac{\text{ft}}{\text{min}}}\right)^2$$
$$= \boxed{0.92 \text{ in wg} \quad (1.0 \text{ in wg})}$$

The answer is (A).

SI Solution

Use the equal-friction method to size the system.

step 1: From Table 36.8, choose the main duct velocity as approximately 1600 ft/min. From the table footnote, the SI velocity is

$$v_{\text{main}} = \left(1600 \ \frac{\text{ft}}{\text{min}}\right)\left(0.00508 \ \frac{\text{m·min}}{\text{s·ft}}\right) = 8.1 \text{ m/s}$$

step 2: The total air flow from the fan is 700 L/s. From Fig. 36.6, the main duct diameter is approximately 340 mm. The friction loss is 2.3 Pa/m.

step 3: After the first takeoff at A, the flow rate in section A–B is

$$700 \ \frac{\text{L}}{\text{s}} - 150 \ \frac{\text{L}}{\text{s}} = 550 \text{ L/s}$$

From Fig. 36.6 for 550 L/s and 2.3 Pa/m, the diameter is 300 mm. The velocity from the chart, 7.7 m/s, cannot be used because the duct is not circular. Similarly, the diameters for the other sections are obtained and listed in the following table.

section	Q (L/s)	D (mm)
fan–A	700	340
A–B	550	300
B–C	400	275
C–D	250	225
D–E	150	180
E–F	50	125

Ventilation

These diameters are for round duct. The equivalent rectangular duct sides with an aspect ratio of 1.5 are found as follows.

$$a = \frac{D(1.5+1)^{0.25}}{(1.3)(1.5)^{5/8}} = 0.75D$$

$$b = Ra = (1.5)(0.75D) = 1.125D$$

Convert the diameters to sides a and b.

$$a_{\text{fan}-A} = (0.75)(340 \text{ mm}) = 255 \text{ mm}$$

$$b_{\text{fan}-A} = (1.125)(340 \text{ mm}) = 383 \text{ mm}$$

Use a 275 mm × 400 mm duct.

The following table is prepared similarly, rounding up as appropriate.

section	D (mm)	a (mm)	b (mm)	v (m/s)
fan–A	340	275	400	6.4
A–B	300	225	350	7.0
B–C	275	225	325	5.5
C–D	225	175	275	5.2
D–E	180	150	225	4.4
E–F	125	100	150	3.3

The velocity in section fan–A is

$$v_{\text{fan}-A} = \frac{700 \dfrac{\text{L}}{\text{s}}}{\left(\dfrac{(275 \text{ mm})(400 \text{ mm})}{\left(1000 \dfrac{\text{mm}}{\text{m}}\right)^2}\right)\left(1000 \dfrac{\text{L}}{\text{m}^3}\right)}$$

$$= 6.4 \text{ m/s}$$

The velocities in the other sections are found similarly.

(a) *step 4:* By inspection, the longest run is fan–F.

Use Eq. 36.41 to estimate the equivalent lengths of the bends. Since both bends have the same radius-width and height-width ratios, the equivalent lengths can be combined.

$$L_e = (W_{\text{fan}-A} + W_{C-D})\left(0.33\frac{r}{W}\right)^{-2.13(H/W)^{0.126}}$$

$$= \left(\frac{400 \text{ mm} + 275 \text{ mm}}{1000 \dfrac{\text{mm}}{\text{m}}}\right)\left((0.33)(1.5)\right)^{-(2.13)(1.5)^{0.126}}$$

$$= 3.27 \text{ m} (3.3 \text{ m})$$

The equivalent length of the entire run is

4.5 m	(fan to first bend)
13.5 m	(first bend to point A)
9 m	(section A–B)
9 m	(section B–C)
6 m	(point C to second bend)
3.3 m	(equivalent length of two bends)
3 m	(second bend to point D)
6 m	(section D–E)
6 m	(section E–F)
total:	60.3 m

The straight-through friction loss in the longest run is

$$(60.3 \text{ m})\left(2.3 \frac{\text{Pa}}{\text{m}}\right) = 139 \text{ Pa}$$

The fan must be able to supply a static pressure of

$$SP_{\text{fan}} = 139 \text{ Pa} + 63 \text{ Pa} = \boxed{202 \text{ Pa} (200 \text{ Pa})}$$

The answer is (C).

(b) The total pressure supplied by the fan is

$$TP_{\text{fan}} = 202 \text{ Pa} + \left(0.6 \frac{\text{Pa·s}^2}{\text{m}^2}\right)\left(6.4 \frac{\text{m}}{\text{s}}\right)^2$$

$$= \boxed{227 \text{ Pa} (230 \text{ Pa})}$$

The answer is (A).

7. *Customary U.S. Solution*

Use the equal-friction method to size the system.

step 1: From Table 36.8, choose the main duct velocity as 1600 ft/min.

step 2: The total air flow from the fan is

$$(12)\left(300 \frac{\text{ft}^3}{\text{min}}\right) = 3600 \text{ ft}^3/\text{min}$$

From Fig. 36.5, the main duct diameter is 20 in. The friction loss is 0.16 in wg per 100 ft.

step 3: After the first takeoff, the flow rate in the main duct is

$$3600 \frac{\text{ft}^3}{\text{min}} - (4)\left(300 \frac{\text{ft}^3}{\text{min}}\right) = 2400 \text{ ft}^3/\text{min}$$

From Fig. 36.5, for 2400 ft³/min and 0.16 in wg per 100 ft, the diameter is 17.2 in and the velocity is 1440 ft/min. Similarly, the diameters and the velocities for other sections are obtained and listed in the following table.

section	Q (ft³/min)	D (in)	v (ft/min)
fan–first takeoff	3600	20.0	1600
first–second takeoff	2400	17.2	1440
second–third takeoff	1200	13.2	1210
first–A	1200	13.2	1210
A–B	900	12.0	1150
B–C	600	10.2	1030
C–D	300	7.9	860
second–E	1200	13.2	1210
E–F	900	12.0	1150
F–G	600	10.2	1030
G–H	300	7.9	860
I–J	900	12.0	1150
J–K	600	10.2	1030
K–L	300	7.9	860

(a) *step 4:* By inspection, the longest run is fan–L. From Table 36.5, the equivalent length of each bend is $14.5D$ (interpolated). For two elbows,

$$L_{e,\text{bend}} = 14.5D$$

$$= (14.5)\left(\frac{20 \text{ in} + 13.2 \text{ in}}{12 \frac{\text{in}}{\text{ft}}}\right)$$

$$= 40.1 \text{ ft}$$

The equivalent length of the entire run is

25 ft	(fan to first bend)
35 ft	(first bend to first section)
20 ft	(first section to second section)
20 ft	(second section to third section)
10 ft	(third section to point I)
40.1 ft	(equivalent length of two bends)
20 ft	(point I to point J)
20 ft	(point J to point K)
20 ft	(point K to point L)
total: 210.1 ft	

The straight-through friction loss in the longest run is

$$\left(\frac{210.1 \text{ ft}}{100 \text{ ft}}\right)\left(0.16 \frac{\text{in wg}}{100 \text{ ft}}\right) = 0.34 \text{ in wg}$$

The fan must be able to supply a static pressure of

$$SP_\text{fan} = 0.34 \text{ in wg} + 0.15 \text{ in wg}$$

$$= \boxed{0.49 \text{ in wg} \quad (0.5 \text{ in wg})}$$

The answer is (A).

(b) The total pressure supplied by the fan is

$$TP_\text{fan} = 0.49 \text{ in wg} + \left(\frac{1600 \frac{\text{ft}}{\text{min}}}{4005 \frac{\text{ft}}{\text{min}}}\right)^2$$

$$= \boxed{0.65 \text{ in wg} \quad (0.7 \text{ in wg})}$$

The answer is (B).

SI Solution

Use the equal-friction method to size the system.

step 1: From Table 36.8, choose the main duct velocity as 1600 ft/min. From the table footnote, the SI velocity is

$$v_\text{main} = \left(1600 \frac{\text{ft}}{\text{min}}\right)\left(0.00508 \frac{\text{m·min}}{\text{s·ft}}\right)$$

$$= 8.1 \text{ m/s}$$

step 2: The total air flow from the fan is

$$(12)\left(140 \frac{\text{L}}{\text{s}}\right) = 1680 \text{ L/s}$$

From Fig. 36.6, the main duct diameter is 500 mm. The friction loss is 1.5 Pa/m.

step 3: After the first takeoff, the flow rate in the main duct is

$$1680 \frac{\text{L}}{\text{s}} - (4)\left(140 \frac{\text{L}}{\text{s}}\right) = 1120 \text{ L/s}$$

From Fig. 36.6, for 1120 L/s and 1.5 Pa/m, the diameter is 440 mm and the velocity is 7.5 m/s. Similarly, the diameters and the velocities for other sections are obtained and listed in the following table.

section	Q (L/s)	D (mm)	v (m/s)
fan–first takeoff	1680	500	8.1
first–second takeoff	1120	440	7.5
second–third takeoff	560	335	6.4
first–A	560	335	6.4
A–B	420	305	5.9
B–C	280	260	5.4
C–D	140	195	4.5
second–E	560	335	6.4
E–F	420	305	5.9
F–G	280	260	5.4
G–H	140	195	4.5
I–J	420	305	5.9
J–K	280	260	5.4
K–L	140	195	4.5

(a) *step 4:* By inspection, the longest run is fan–L. From Table 36.5, the equivalent length of each bend is $14.5D$ (interpolated). For two elbows,

$$L_{e,bend} = 14.5D = (14.5)\left(\frac{500 \text{ mm} + 335 \text{ mm}}{1000 \frac{\text{mm}}{\text{m}}}\right)$$

$$= 12.1 \text{ m}$$

The equivalent length of the entire run is

7.5 m	(fan to first bend)
10.5 m	(first bend to first section)
6 m	(first section to second section)
6 m	(second section to third section)
3 m	(third section to point I)
12.1 m	(equivalent length of two bends)
6 m	(point I to point J)
6 m	(point J to point K)
6 m	(point K to point L)
total:	63.1 m

The straight-through friction loss in the longest run is

$$(63.1 \text{ m})\left(1.5 \frac{\text{Pa}}{\text{m}}\right) = 95 \text{ Pa}$$

The fan must be able to supply a static pressure of

$$SP_{fan} = 95 \text{ Pa} + 38 \text{ Pa} = \boxed{133 \text{ Pa} \quad (130 \text{ Pa})}$$

The answer is (A).

(b) The total pressure supplied by the fan is

$$TP_{fan} = 133 \text{ Pa} + (0.6)\left(8.1 \frac{\text{m}}{\text{s}}\right)^2 = \boxed{172 \text{ Pa} \quad (170 \text{ Pa})}$$

The answer is (B).

8. (a) *step 1:* From Table 36.8, choose the main duct velocity as 1600 ft/min.

step 2: The diameter of the main duct is

$$A = \frac{Q}{v} = \frac{3600 \frac{\text{ft}^3}{\text{min}}}{1600 \frac{\text{ft}}{\text{min}}} = 2.25 \text{ ft}^2$$

$$D = \sqrt{\frac{4A}{\pi}} = \sqrt{\frac{(4)(2.25 \text{ ft}^2)}{\pi}}\left(12 \frac{\text{in}}{\text{ft}}\right)$$

$$= \boxed{20.3 \text{ in} \quad (20 \text{ in})}$$

The answer is (B).

(b) *step 3:* From Table 36.5, the equivalent length of each bend is $14.5D$ (interpolated). For the first elbow,

$$L_{e,bend} = 14.5D = (14.5)\left(\frac{20.3 \text{ in}}{12 \frac{\text{in}}{\text{ft}}}\right) = 24.5 \text{ ft}$$

The equivalent length of the main duct from the fan to the first takeoff and bend is

$$L = 25 \text{ ft} + 24.5 \text{ ft} + 35 \text{ ft} = \boxed{84.5 \text{ ft} \quad (85 \text{ ft})}$$

The answer is (B).

(c) *step 4:* From Fig. 36.5, the friction loss in the main run up to the branch takeoff is approximately 0.16 in wg per 100 ft. The actual friction loss is

$$FP_{main} = \left(0.16 \frac{\text{in wg}}{100 \text{ ft}}\right)\left(\frac{84.5 \text{ ft}}{100 \text{ ft}}\right)$$

$$= \boxed{0.135 \text{ in wg} \quad (0.14 \text{ in wg})}$$

The answer is (B).

(d) *step 5:* After the first takeoff,

$$Q = 3600 \frac{\text{ft}^3}{\text{min}} - 1200 \frac{\text{ft}^3}{\text{min}} = 2400 \text{ ft}^3/\text{min}$$

$$L = 20 \text{ ft}$$

$$\frac{L}{Q^{0.61}} = \frac{20 \text{ ft}}{\left(2400 \frac{\text{ft}^3}{\text{min}}\right)^{0.61}} = 0.173$$

(This equation is dimensionally inconsistent.)

From Fig. 36.10, the velocity, v_2, is 1390 ft/min.

step 6: Solve for the duct size from $A = Q/v$.

$$D_2 = \sqrt{\frac{4A_2}{\pi}} = \sqrt{\frac{4Q_2}{\pi v_2}} = \sqrt{\frac{(4)\left(2400 \frac{\text{ft}^3}{\text{min}}\right)}{\pi\left(1390 \frac{\text{ft}}{\text{min}}\right)}}\left(12 \frac{\text{in}}{\text{ft}}\right)$$

$$= 17.8 \text{ in} \quad (18 \text{ in})$$

step 7: After the second takeoff,

$$Q = 2400 \frac{\text{ft}^3}{\text{min}} - 1200 \frac{\text{ft}^3}{\text{min}} = 1200 \text{ ft}^3/\text{min}$$

$$L = 20 \text{ ft} + 14.5D + 10 \text{ ft}$$

$$= 30 \text{ ft} + 14.5D$$

Since L contains the unknown diameter, D, this will require an iterative procedure. Using velocity $v_1 = 1390$ ft/min before the takeoff and using an iterative procedure, $D = 15$ in and $v = 980$ ft/min.

step 8: Proceeding similarly, the following table is developed for the remaining sizes.

section	L (ft)	Q (ft^3/min)	$\dfrac{L}{Q^{0.61}}$	v (ft/min)
I–J	20	900	0.32	830
J–K	20	600	0.40	660
K–L	20	300	0.62	500

step 9: Solve for the duct size from $A = Q/\mathrm{v}$.

$$D = \sqrt{\frac{4A}{\pi}} = \sqrt{\frac{4Q}{\pi \mathrm{v}}}$$

$$D_{\text{I–J}} = \sqrt{\frac{(4)\left(900 \ \dfrac{\text{ft}^3}{\text{min}}\right)}{\pi\left(830 \ \dfrac{\text{ft}}{\text{min}}\right)}}\left(12 \ \frac{\text{in}}{\text{ft}}\right)$$

$$= 14.1 \text{ in} \quad (14 \text{ in})$$

$$D_{\text{J–K}} = \sqrt{\frac{(4)\left(600 \ \dfrac{\text{ft}^3}{\text{min}}\right)}{\pi\left(660 \ \dfrac{\text{ft}}{\text{min}}\right)}}\left(12 \ \frac{\text{in}}{\text{ft}}\right)$$

$$= 12.9 \text{ in} \quad (13 \text{ in})$$

$$D_{\text{K–L}} = \sqrt{\frac{(4)\left(300 \ \dfrac{\text{ft}^3}{\text{min}}\right)}{\pi\left(500 \ \dfrac{\text{ft}}{\text{min}}\right)}}\left(12 \ \frac{\text{in}}{\text{ft}}\right)$$

$$= 10.5 \text{ in} \quad (11 \text{ in})$$

step 10: The total pressure supplied by the fan is

$$0.135 \text{ in wg} + 0.15 \text{ in wg} = \boxed{0.285 \text{ in wg} \quad (0.29 \text{ in wg})}$$

The answer is (C).

9. *Customary U.S. Solution*

(a) No dampers are needed in duct A. The area of section A is

$$A = \frac{\pi}{4}D^2 = \left(\frac{\pi}{4}\right)\left(\frac{12 \text{ in}}{12 \ \dfrac{\text{in}}{\text{ft}}}\right)^2 = 0.7854 \text{ ft}^2$$

The velocity in section A is

$$\mathrm{v_A} = \frac{Q}{A} = \frac{3000 \ \dfrac{\text{ft}^3}{\text{min}}}{(0.7854 \text{ ft}^2)\left(60 \ \dfrac{\text{sec}}{\text{min}}\right)} = 63.66 \text{ ft/sec}$$

For a four-piece elbow with a radius-to-diameter ratio of 1.5, $L_e = 15D$. The total equivalent length of run C is

$$L_e = 50 \text{ ft} + 10 \text{ ft} + 10 \text{ ft} + (2)\left((15)\left(\frac{10 \text{ in}}{12 \ \dfrac{\text{in}}{\text{ft}}}\right)\right)$$

$$= \boxed{95 \text{ ft}}$$

The answer is (C).

(b) For any diameter, D, in inches of section C, the velocity will be

$$\mathrm{v_C} = \frac{Q}{A} = \frac{2000 \ \dfrac{\text{ft}^3}{\text{min}}}{\left(\dfrac{\pi}{4}\right)\left(\dfrac{D}{12 \ \dfrac{\text{in}}{\text{ft}}}\right)^2\left(60 \ \dfrac{\text{sec}}{\text{min}}\right)}$$

$$= 6111.5/D^2$$

From Eq. 17.27, the friction loss in section C will be

$$h_{f,\text{C}} = \frac{fL\mathrm{v}^2}{2Dg} = \frac{(0.02)(93.3 \text{ ft})\left(\dfrac{6111.5}{D^2}\right)^2}{(2)\left(\dfrac{D}{12 \ \dfrac{\text{in}}{\text{ft}}}\right)\left(32.2 \ \dfrac{\text{ft}}{\text{sec}^2}\right)}$$

$$= \frac{1.3 \times 10^7}{D^5} \quad [\text{ft of air}]$$

The regain between A and C will be

$$h_{\text{regain}} = R\left(\frac{\mathrm{v_A^2} - \mathrm{v_G^2}}{2g}\right)$$

$$= (0.65)\left(\frac{\left(63.66 \ \dfrac{\text{ft}}{\text{sec}}\right)^2 - \left(\dfrac{6111.5}{D^2}\right)^2}{(2)\left(32.2 \ \dfrac{\text{ft}}{\text{sec}^2}\right)}\right)$$

$$= 40.9 - \frac{3.77 \times 10^5}{D^4} \quad [\text{ft of air}]$$

The principle of static regain is that

$$h_{f,\text{C}} = h_{\text{regain}}$$

$$\frac{1.3 \times 10^7}{D^5} = 40.9 - \frac{3.77 \times 10^5}{D^4}$$

By trial and error, $D \approx 13.5$ in. Since the assumed value of D is different from the calculated value, this process should be repeated.

$$L_e = 50 \text{ ft} + 10 \text{ ft} + 10 \text{ ft} + (2)\left((14)\left(\frac{13.5 \text{ in}}{12 \frac{\text{in}}{\text{ft}}}\right)\right)$$

$$= 101.5 \text{ ft}$$

From Eq. 17.27, the friction loss in section C will be

$$h_{f,\text{C}} = \frac{fLv^2}{2Dg} = \frac{(0.02)(101.5 \text{ ft})\left(\frac{6111.5}{D^2}\right)^2}{(2)\left(\frac{D}{12 \frac{\text{in}}{\text{ft}}}\right)\left(32.2 \frac{\text{ft}}{\text{sec}^2}\right)}$$

$$= \frac{1.41 \times 10^7}{D^5} \quad \text{[ft of air]}$$

The principle of static regain is that

$$h_{f,\text{C}} = h_{\text{regain}}$$

$$\frac{1.41 \times 10^7}{D^5} = 40.9 - \frac{3.77 \times 10^5}{D^4}$$

By trial and error, $D_\text{C} = 13.63$ in (14 in). This results in a friction loss of

$$h_{f,\text{C}} = \frac{1.41 \times 10^7}{(14 \text{ in})^5} = \boxed{26.2 \text{ ft } (26 \text{ ft}) \text{ of air}}$$

Because the regain cancels friction loss, the pressure loss from A to C is zero. A damper is not needed in duct C. A damper is needed in section B, however.

The answer is (B).

(c) For any diameter, D, in inches in section B, the velocity will be

$$v_\text{B} = \frac{Q}{A} = \frac{1000 \frac{\text{ft}^3}{\text{min}}}{\left(\frac{\pi}{4}\right)\left(\frac{D}{12 \frac{\text{in}}{\text{ft}}}\right)^2\left(60 \frac{\text{sec}}{\text{min}}\right)}$$

$$= \frac{3055.8}{D^2} \quad \text{[ft/sec]}$$

From Eq. 17.27, the friction loss in section B will be

$$h_{f,\text{B}} = \frac{fLv^2}{2Dg} = \frac{(0.02)(10 \text{ ft})\left(\frac{3055.8}{D^2}\right)^2}{(2)\left(\frac{D}{12 \frac{\text{in}}{\text{ft}}}\right)\left(32.2 \frac{\text{ft}}{\text{sec}^2}\right)}$$

$$= \frac{3.48 \times 10^5}{D^5} \quad \text{[ft of air]}$$

From Eq. 36.48, the friction loss in the branch takeoff between sections A and B will be

$$\text{TP}_\text{A} - \text{TP}_\text{B} = K_{\text{br}}(\text{VP}_{\text{up}})$$

At this point, assume $v_\text{B}/v_\text{A} = 1.0$. From Table 36.7, for a 45° angle of takeoff, $K_{\text{br}} = 0.5$.

$$\text{TP}_\text{A} - \text{TP}_\text{B} = (0.5)\left(\frac{3820 \frac{\text{ft}}{\text{min}}}{4005 \frac{\text{ft}}{\text{min}}}\right)^2 = 0.455 \text{ in wg}$$

From Eq. 36.6,

$$p_{\text{psig}} = (0.455 \text{ in wg})\left(0.0361 \frac{\text{lbf}}{\text{in}^3}\right) = 0.01643 \text{ lbf/in}^2$$

From Eq. 36.5,

$$h_f = \frac{p_{\text{psig}}}{\gamma} = \frac{\left(0.01643 \frac{\text{lbf}}{\text{in}^2}\right)\left(12 \frac{\text{in}}{\text{ft}}\right)^2}{0.075 \frac{\text{lbf}}{\text{ft}^3}} = 31.5 \text{ ft of air}$$

The regain between A and B will be

$$h_{\text{regain}} = R\left(\frac{v_\text{A}^2 - v_\text{B}^2}{2g}\right)$$

$$= (0.65)\left(\frac{\left(63.66 \frac{\text{ft}}{\text{sec}}\right)^2 - \left(\frac{3055.8}{D^2}\right)^2}{(2)\left(32.2 \frac{\text{ft}}{\text{sec}^2}\right)}\right)$$

$$= 40.9 \text{ ft} - \frac{9.42 \times 10^4}{D^4} \quad \text{[ft of air]}$$

Set the regain equal to the loss.

$$h_{\text{regain}} = h_{f,\text{B}}$$

$$40.9 \text{ ft} - \frac{9.42 \times 10^4}{D^4} = \frac{3.48 \times 10^5}{D^5} + 31.5 \text{ ft}$$

By trial and error,

$$D_B = \boxed{10.77 \text{ in} \quad (11 \text{ in})}$$

The answer is (A).

(d) Calculate the regain in sections B and C.

$$v_B = \frac{3055.8}{D^2} = \frac{3055.8}{(11 \text{ in})^2} = 25.25 \text{ ft/sec}$$

$$\frac{v_B}{v_A} = \frac{25.25 \frac{\text{ft}}{\text{sec}}}{63.6 \frac{\text{ft}}{\text{sec}}} \approx 0.4$$

From Table 36.7, for a 45° angle of takeoff, $K_{br} = 0.5$. Since the value of K_{br} remains the same, there is no need to repeat the preceding procedure. For section B, the friction canceled by the regain is

$$h_{f,B} = 31.5 \text{ ft} + \frac{3.48 \times 10^5}{D^5} = 31.5 \text{ ft} + \frac{3.48 \times 10^5}{(11 \text{ in})^5}$$

$$= \boxed{33.7 \text{ ft} \ (34 \text{ ft}) \text{ of air}}$$

The answer is (C).

(e) From part (a) and part (b), a damper is required in $\boxed{\text{section B only.}}$

The answer is (B).

SI Solution

(a) No dampers are needed in duct A. The area of section A is

$$A = \frac{\pi}{4}D^2 = \left(\frac{\pi}{4}\right)\left(\frac{305 \text{ mm}}{1000 \frac{\text{mm}}{\text{m}}}\right)^2$$

$$= 0.0731 \text{ m}^2$$

The velocity in section A is

$$v_A = \frac{Q}{A} = \frac{1400 \frac{\text{L}}{\text{s}}}{(0.0731 \text{ m}^2)\left(1000 \frac{\text{L}}{\text{m}^3}\right)}$$

$$= 19.15 \text{ m/s}$$

For a four-piece elbow with a radius-to-diameter ratio of 1.5, $L_e = 15D$. The total equivalent length of run C is

$$L_e = 15 \text{ m} + 3 \text{ m} + 3 \text{ m} + (2)\left((15)\left(\frac{250 \text{ mm}}{1000 \frac{\text{mm}}{\text{m}}}\right)\right)$$

$$= \boxed{28.5 \text{ m} \quad (29 \text{ m})}$$

The answer is (C).

(b) For any diameter, D, in mm of section C, the velocity will be

$$v_C = \frac{Q}{A} = \frac{900 \frac{\text{L}}{\text{s}}}{\left(\frac{\pi}{4}\right)\left(\frac{D}{1000 \frac{\text{mm}}{\text{m}}}\right)^2\left(1000 \frac{\text{L}}{\text{m}^3}\right)}$$

$$= \frac{1.146 \times 10^6}{D^2} \quad [\text{m/s}]$$

From Eq. 17.27, the friction loss in section C will be

$$h_{f,C} = \frac{fLv^2}{2Dg} = \frac{(0.02)(28 \text{ m})\left(\frac{1.146 \times 10^6}{D^2}\right)^2}{(2)\left(\frac{D}{1000 \frac{\text{mm}}{\text{m}}}\right)\left(9.81 \frac{\text{m}}{\text{s}^2}\right)}$$

$$= \frac{3.75 \times 10^{13}}{D^5} \quad [\text{m of air}]$$

The regain between A and C will be

$$h_{regain} = R\left(\frac{v_A^2 - v_C^2}{2g}\right)$$

$$= (0.65)\left(\frac{\left(19.15 \frac{\text{m}}{\text{s}}\right)^2 - \left(\frac{1.146 \times 10^6}{D^2}\right)^2}{(2)\left(9.81 \frac{\text{m}}{\text{s}^2}\right)}\right)$$

$$= 12.15 \text{ m} - \frac{4.35 \times 10^{10}}{D^4} \quad [\text{m of air}]$$

The principle of static regain is that

$$h_{f,C} = h_{regain}$$

$$\frac{3.75 \times 10^{13}}{D^5} = 12.15 \text{ m} - \frac{4.35 \times 10^{10}}{D^4}$$

By trial and error, $D \approx 335$ mm. Since the assumed value of D is different from the calculated value, this process should be repeated.

$$L_e = 15 \text{ m} + 3 \text{ m} + 3 \text{ m} + (2)\left((14)\left(\frac{335 \text{ mm}}{1000 \frac{\text{mm}}{\text{m}}}\right)\right)$$

$$= 30.4 \text{ m}$$

From Eq. 17.27, the friction loss in section C will be

$$h_{f,C} = \frac{fLv^2}{2Dg} = \frac{(0.02)(30.4 \text{ m})\left(\frac{1.146 \times 10^6}{D^2}\right)^2}{(2)\left(\frac{D}{1000 \frac{\text{mm}}{\text{m}}}\right)\left(9.81 \frac{\text{m}}{\text{s}^2}\right)}$$

$$= \frac{4.07 \times 10^{13}}{D^5} \quad \text{[m of air]}$$

The principle of static regain is that

$$h_{f,C} = h_{\text{regain}}$$

$$\frac{4.07 \times 10^{13}}{D^5} = 12.15 - \frac{4.35 \times 10^{10}}{D^4}$$

By trial and error, $D = 340$ mm. This results in a friction loss of

$$h_{f,C} = \frac{4.07 \times 10^{13}}{(340 \text{ mm})^5} = \boxed{8.96 \text{ m} \ (9.0 \text{ m}) \text{ of air}}$$

Because the regain cancels this friction loss, the pressure loss from A to C is zero. A damper is not needed in duct C. A damper is needed at B, however.

The answer is (B).

(c) For any diameter, D, in mm in section B, the velocity will be

$$v_B = \frac{Q}{A} = \frac{500 \frac{\text{L}}{\text{s}}}{\left(\frac{\pi}{4}\right)\left(\frac{D}{1000 \frac{\text{mm}}{\text{m}}}\right)^2 \left(1000 \frac{\text{L}}{\text{m}^3}\right)}$$

$$= \frac{6.366 \times 10^5}{D^2} \quad \text{[m/s]}$$

From Eq. 17.27, the friction loss in section B will be

$$h_{f,B} = \frac{fLv^2}{2Dg} = \frac{(0.02)(3 \text{ m})\left(\frac{6.366 \times 10^5}{D^2}\right)^2}{(2)\left(\frac{D}{1000 \frac{\text{mm}}{\text{m}}}\right)\left(9.81 \frac{\text{m}}{\text{s}^2}\right)}$$

$$= \frac{1.239 \times 10^{12}}{D^5} \quad \text{[m of air]}$$

From Eq. 36.48, the friction loss in the branch takeoff between sections A and B will be

$$TP_A - TP_B = K_{\text{br}}(VP_{\text{up}})$$

At this point, assume $v_B/v_A = 1.0$. From Table 36.7, for a 45° angle of takeoff, $K_{\text{br}} = 0.5$.

$$TP_A - TP_B = (0.5)(0.6)\left(19.15 \frac{\text{m}}{\text{s}}\right)^2$$

$$= 110 \text{ Pa}$$

From Eq. 36.5,

$$h_f = \frac{p}{\rho g} = \frac{110 \text{ Pa}}{\left(1.2 \frac{\text{kg}}{\text{m}^3}\right)\left(9.81 \frac{\text{m}}{\text{s}^2}\right)}$$

$$= 9.34 \text{ m of air}$$

The regain between A and B will be

$$h_{\text{regain}} = R\left(\frac{v_A^2 - v_B^2}{2g}\right)$$

$$= (0.65)\left(\frac{\left(19.15 \frac{\text{m}}{\text{s}}\right)^2 - \left(\frac{6.366 \times 10^5}{D^2}\right)^2}{(2)\left(9.81 \frac{\text{m}}{\text{s}^2}\right)}\right)$$

$$= 12.15 \text{ m} - \frac{1.343 \times 10^{10}}{D^4} \quad \text{[m of air]}$$

Set the regain equal to the loss.

$$12.15 \text{ m} - \frac{1.343 \times 10^{10}}{D^4} = \frac{1.239 \times 10^{12}}{D^5} + 9.34 \text{ m}$$

By trial and error,

$$D_B = \boxed{280 \text{ mm}}$$

The answer is (A).

Ventilation

(d) Calculate the regain in sections B and C. By trial and error,

$$v_B = 8.12 \text{ m/s}$$

Then,

$$\frac{v_B}{v_A} = \frac{8.12 \frac{\text{m}}{\text{s}}}{19.15 \frac{\text{m}}{\text{s}}} \approx 0.4$$

From Table 36.7, for a 45° angle of takeoff, $K_{br} = 0.5$. Since the value of K_{br} remains the same, there is no need to repeat the preceding procedure. For section B, the friction canceled by the regain is

$$h_{f,B} = 9.34 \text{ m} + \frac{1.239 \times 10^{12}}{D^5}$$

$$= 9.34 \text{ m} + \frac{1.239 \times 10^{12}}{(280 \text{ mm})^5}$$

$$= \boxed{10.06 \text{ m } (10 \text{ m}) \text{ of air}}$$

The answer is (C).

(e) From part (a) and part (b), a damper is required in $\boxed{\text{section B only.}}$

The answer is (B).

10. Rearrange Eq. 36.13(b) to find the volumetric flow rate.

$$Q_{\text{ft}^3/\text{min}} = \frac{6356(\text{AHP})}{\text{TP}_{\text{in wg}}}$$

$$= \frac{\left(6356 \frac{\text{in-ft}^3}{\text{hp-min}}\right)(4.3 \text{ hp})}{2.3 \text{ in wg}}$$

$$= 11{,}883 \text{ ft}^3/\text{min}$$

At 62% efficiency, the actual cubic feet per minute, ACFM, is

$$\text{ACFM} = \eta Q$$

$$= (0.62)\left(11{,}883 \frac{\text{ft}^3}{\text{min}}\right)$$

$$= 7367 \text{ ft}^3/\text{min}$$

Use Eq. 36.1 and Eq. 36.2 to find the standard flow rate, SCFM. The pressure is standard.

$$\text{SCFM} = \frac{\text{ACFM}}{K_d} = \frac{\text{ACFM}}{\dfrac{T_{\text{actual}}}{T_{\text{std}}}} = \frac{7367 \frac{\text{ft}^3}{\text{min}}}{\dfrac{160°\text{F} + 460°}{70°\text{F} + 460°}}$$

$$= \boxed{6298 \text{ ft}^3/\text{min} \quad (6300 \text{ ft}^3/\text{min})}$$

The answer is (C).

11. Variable flow rates are achieved with variable blade pitches, inlet vanes in the supply line, and variable speed motors. $\boxed{\text{System dampers}}$ in the discharge lines should not be used because they increase friction loss, are noisy, and are nonlinear in their response.

The answer is (A).

12. Velocity pressure can be used to determine velocity and is an indication of total system flow. At the entrance to the VAV box, a pitot tube senses total pressure, and a static port senses static pressure. Velocity pressure is calculated and a signal sent to the controller by a differential pressure transmitter. The $\boxed{\text{duct velocity pressure}}$ can be used to determine the fan speed needed.

The answer is (A).

13. The volumetric flow rate of fan A is 4000 ft³/min. Therefore, the volumetric flow rate of fan B is

$$Q_B = Q_{\text{total}} - Q_A = 10{,}000 \frac{\text{ft}^3}{\text{min}} - 4000 \frac{\text{ft}^3}{\text{min}}$$

$$= 6000 \text{ ft}^3/\text{min}$$

Since the fans are of the same size, have the same efficiency, and move air with the same density, Eq. 36.22 may be used.

$$\frac{Q_B}{Q_A} = \frac{n_B}{n_A}$$

$$n_B = n_A\left(\frac{Q_B}{Q_A}\right) = (1800 \frac{\text{rev}}{\text{min}})\left(\frac{6000 \frac{\text{ft}^3}{\text{min}}}{5000 \frac{\text{ft}^3}{\text{min}}}\right)$$

$$= \boxed{2160 \text{ rev/min} \quad (2200 \text{ rev/min})}$$

The answer is (C).

Ventilation

37 Fuels and Combustion

PRACTICE PROBLEMS

1. Methane (MW = 16.043) with a heating value of 24,000 Btu/lbm (55.8 MJ/kg) is burned with a 50% efficiency. If the heat of vaporization of any water vapor formed is recovered, approximately how much water (specific heat of 1 Btu/lbm-°F (4.1868 kJ/kg·°C)) can be heated from 60°F to 200°F (15°C to 95°C) when 7 ft³ (200 L) of methane at 60°F and 14.73 psia (288.9K and 101.51 kPa) are burned?

(A) 25 lbm (11 kg)

(B) 35 lbm (16 kg)

(C) 50 lbm (23 kg)

(D) 95 lbm (43 kg)

2. 15 lbm/hr (6.8 kg/h) of propane (C_3H_8, MW = 44.097) is burned stoichiometrically in air. Approximately what volume of dry carbon dioxide (CO_2, MW = 44.011) is formed after cooling to 70°F (21°C) and 14.7 psia (101 kPa)?

(A) 180 ft³/hr (5.0 m³/h)

(B) 270 ft³/hr (7.6 m³/h)

(C) 390 ft³/hr (11 m³/h)

(D) 450 ft³/hr (13 m³/h)

3. In a particular installation, 30% excess air at 15 psia (103 kPa) and 100°F (40°C) is needed for the combustion of methane. Approximately how much nitrogen (MW = 28.016) passes through the furnace if methane is burned at the rate of 4000 ft³/hr (31 L/s)?

(A) 270 lbm/hr (0.033 kg/s)

(B) 930 lbm/hr (0.11 kg/s)

(C) 1800 lbm/hr (0.22 kg/s)

(D) 2700 lbm/hr (0.34 kg/s)

4. Approximately how much air is required to completely burn one unit mass of a fuel that is 84% carbon, 15.3% hydrogen, 0.3% sulfur, and 0.4% nitrogen by weight?

(A) 9 lbm air/lbm fuel (9 kg air/kg fuel)

(B) 12 lbm air/lbm fuel (12 kg air/kg fuel)

(C) 15 lbm air/lbm fuel (15 kg air/kg fuel)

(D) 18 lbm air/lbm fuel (18 kg air/kg fuel)

5. Propane (C_3H_8) is burned with 20% excess air. The gravimetric percentage of carbon dioxide in the flue gas is most nearly

(A) 8%

(B) 12%

(C) 15%

(D) 22%

6. The ultimate analysis of a coal is 80% carbon, 4% hydrogen, 2% oxygen, and the rest ash. The flue gases are 60°F and 14.7 psia (15.6°C and 101.3 kPa) when sampled, and are 80% nitrogen, 12% carbon dioxide, 7% oxygen, and 1% carbon monoxide by volume. The air required to burn 1 lbm (1 kg) of coal under these conditions is most nearly

(A) 11 lbm (11 kg)

(B) 15 lbm (15 kg)

(C) 19 lbm (19 kg)

(D) 23 lbm (23 kg)

7. What is the approximate heating value of an oil with a specific gravity of 40° API?

(A) 20,000 Btu/lbm (46 MJ/kg)

(B) 25,000 Btu/lbm (58 MJ/kg)

(C) 30,000 Btu/lbm (69 MJ/kg)

(D) 35,000 Btu/lbm (81 MJ/kg)

8. The ultimate analysis of a coal is 75% carbon, 5% hydrogen, 3% oxygen, 2% nitrogen, and the rest ash. Atmospheric air is 60°F (16°C) and at standard pressure.

(a) The theoretical temperature of the combustion products is most nearly

(A) 3500°F (1900°C)

(B) 4000°F (2200°C)

(C) 4500°F (2500°C)

(D) 5000°F (2800°C)

(b) Estimate the actual temperature of the combustion products at the boiler outlet. Neglect dissociation. Assume 40% excess air is used and 75% of the heat is transferred to the boiler.

(A) 650°F (340°C)

(B) 880°F (470°C)

(C) 970°F (520°C)

(D) 1300°F (700°C)

9. A fuel oil has the following ultimate analysis: 85.43% carbon, 11.31% hydrogen, 2.7% oxygen, 0.34% sulfur, and 0.22% nitrogen. The oil is burned with 60% excess air. Evaluate flue gas volumes at 600°F (320°C).

(a) The volume of wet flue gases that will be produced is most nearly

(A) 450 ft³ (28 m³)

(B) 500 ft³ (32 m³)

(C) 550 ft³ (35 m³)

(D) 600 ft³ (38 m³)

(b) The volume of dry flue gases that will be produced is most nearly

(A) 480 ft³ (30 m³)

(B) 560 ft³ (35 m³)

(C) 630 ft³ (40 m³)

(D) 850 ft³ (79 m³)

(c) The volumetric fraction of carbon dioxide is most nearly

(A) 6%

(B) 8%

(C) 10%

(D) 13%

10. The ultimate analysis of a coal is 51.45% carbon, 16.69% ash, 15.71% moisture, 7.28% oxygen, 4.02% hydrogen, 3.92% sulfur, and 0.93% nitrogen. 15,395 lbm (6923 kg) of the coal are burned, and 2816 lbm (1267 kg) of ash containing 20.9% carbon (by weight) are recovered. 13.3 lbm (kg) of dry gases are produced per pound (kilogram) of fuel burned. The air used per unit mass of fuel is most nearly

(A) 13 lbm air/lbm fuel (13 kg air/kg fuel)

(B) 14 lbm air/lbm fuel (14 kg air/kg fuel)

(C) 15 lbm air/lbm fuel (15 kg air/kg fuel)

(D) 17 lbm air/lbm fuel (17 kg air/kg fuel)

11. A coal is 65% carbon by weight. During combustion, 3% of the coal is lost in the ash pit. Combustion uses 9.87 lbm (kg) of air per pound (kg) of fuel. The flue gas analysis is 81.5% nitrogen, 9.5% carbon dioxide, and 9% oxygen. The percentage of excess air by mass is most nearly

(A) 10%

(B) 30%

(C) 70%

(D) 140%

12. A coal has an ultimate analysis of 67.34% carbon, 4.91% oxygen, 4.43% hydrogen, 4.28% sulfur, 1.08% nitrogen, and the rest ash. 3% of the carbon is lost during combustion. The flue gases are 81.9% nitrogen, 15.5% carbon dioxide, 1.6% carbon monoxide, and 1% oxygen by volume. The heat loss due to the formation of carbon monoxide is most nearly

(A) 600 Btu/lbm (1.4 MJ/kg)

(B) 800 Btu/lbm (1.9 MJ/kg)

(C) 1000 Btu/lbm (2.3 MJ/kg)

(D) 1200 Btu/lbm (2.8 MJ/kg)

13. A natural gas is 93% methane, 3.73% nitrogen, 1.82% hydrogen, 0.45% carbon monoxide, 0.35% oxygen, 0.25% ethylene, 0.22% carbon dioxide, and 0.18% hydrogen sulfide by volume. The gas is burned with 40% excess air. Atmospheric air is 60°F and at standard atmospheric pressure.

(a) The gas density is most nearly

(A) 0.017 lbm/ft³

(B) 0.043 lbm/ft³

(C) 0.069 lbm/ft³

(D) 0.110 lbm/ft³

(b) The theoretical air requirements are most nearly

(A) 5 ft³ air/ft³ fuel

(B) 7 ft³ air/ft³ fuel

(C) 9 ft³ air/ft³ fuel

(D) 13 ft³ air/ft³ fuel

(c) The percentage of CO_2 in the flue gas (wet basis) is most nearly

(A) 6.9%

(B) 7.7%

(C) 8.1%

(D) 11%

(d) The percentage of CO_2 in the flue gas (dry basis) is most nearly

(A) 6.9%

(B) 7.7%

(C) 8.1%

(D) 11%

14. A coal enters a steam generator at 73°F (23°C). The coal has an ultimate analysis of 78.42% carbon, 8.25% oxygen, 5.68% ash, 5.56% hydrogen, 1.09% nitrogen, and 1.0% sulfur. The coal's heating value is 14,000 Btu/lbm (32.6 MJ/kg). 7.03% of the coal's weight is lost in the ash pit. The ash contains 31.5% carbon. Air at 67°F (19°C) wet bulb and 73°F (23°C) dry bulb is supplied. The flue gases consist of 80.08% nitrogen, 14.0% carbon dioxide, 5.5% oxygen, and 0.42% carbon monoxide. The flue gases are at a temperature of 575°F (300°C). Saturated water at 212°F (100°C) and 1.0 atm enters the steam generator. 11.12 lbm (kg) of water are evaporated in the boiler per pound (kilogram) of dry coal consumed. From a complete heat balance, combustion losses in the furnace are most nearly

(A) 400 Btu/lbm (1040 kJ/kg)

(B) 650 Btu/lbm (1690 kJ/kg)

(C) 960 Btu/lbm (2500 kJ/kg)

(D) 1500 Btu/lbm (3900 kJ/kg)

15. A coal has a gravimetric analysis of 83% carbon, 5% hydrogen, 5% oxygen, and 7% noncombustible matter. 10% of the fired coal mass, including all of the noncombustible matter, is recovered in the ash pit. 26 lbm (26 kg) of air at 70°F (21°C) and 1 atmosphere are used per lbm (kg) of coal burned. When loaded into the furnace, the coal temperature is 60°F (15.6°C). The stack gas from coal combustion has a temperature of 550°F (290°C). The percentage of the combustion heat carried away by the stack gases, based on the coal's as-delivered properties, is most nearly

(A) 18%

(B) 24%

(C) 37%

(D) 49%

16. A utility boiler burns coal with an ultimate analysis of 76.56% carbon, 7.7% oxygen, 6.1% silicon, 5.5% hydrogen, 2.44% sulfur, and 1.7% nitrogen. 410 lbm/hr of refuse are removed with a composition of 30% carbon and 0% sulfur. All the sulfur and the remaining carbon are burned. The power plant has the following characteristics.

- coal feed rate: 15,300 lbm/hr

- electric power rating: 17 MW

- generator efficiency: 95%

- steam generator efficiency: 86%

- cooling water rate: 225 ft³/sec

(a) The emission rate of solid particulates in lbm/hr is most nearly

(A) 23 lbm/hr

(B) 150 lbm/hr

(C) 810 lbm/hr

(D) 1700 lbm/hr

(b) The sulfur dioxide produced per hour is most nearly

(A) 220 lbm/hr

(B) 340 lbm/hr

(C) 750 lbm/hr

(D) 1100 lbm/hr

(c) The temperature rise of the cooling water is most nearly

(A) 2.4°F

(B) 6.5°F

(C) 9.8°F

(D) 13°F

(d) The efficiency the flue gas particulate collectors must have in order to meet a limit of 0.1 lbm of particulates per million Btus per hour (0.155 kg/MW) is most nearly

(A) 93.1%

(B) 97.4%

(C) 98.8%

(D) 99.1%

Combustion

17. 250 SCFM (118 L/s) of propane are mixed with an oxidizer consisting of 60% oxygen and 40% nitrogen by volume in a proportion resulting in 40% excess oxygen by weight. The maximum velocity for the two reactants when combined is 400 ft/min (2 m/s) at 14.7 psia (101 kPa) and 80°F (27°C). Maximum velocity for the products is 800 ft/min (4 m/s) at 8 psia (55 kPa) and 460°F (240°C).

(a) The actual flow of oxygen is most nearly

(A) 150 lbm/min (1.1 kg/s)

(B) 180 lbm/min (1.3 kg/s)

(C) 220 lbm/min (1.6 kg/s)

(D) 270 lbm/min (2.0 kg/s)

(b) The minimum size of the inlet pipe is most nearly

(A) 8 ft^2 (0.8 m^2)

(B) 11 ft^2 (1.1 m^2)

(C) 14 ft^2 (1.3 m^2)

(D) 17 ft^2 (1.6 m^2)

(c) The volume of flue gases is most nearly

(A) 4000 ft^3/min (1.9 m^3/s)

(B) 7000 ft^3/min (3.3 m^3/s)

(C) 9000 ft^3/min (4.3 m^3/s)

(D) 11,000 ft^3/min (5.3 m^3/s)

(d) The minimum area of the stack is most nearly

(A) 8 ft^2 (0.8 m^2)

(B) 11 ft^2 (1.1 m^2)

(C) 14 ft^2 (1.3 m^2)

(D) 17 ft^2 (1.6 m^2)

(e) The dew point of the flue gases is most nearly

(A) 40°F (4°C)

(B) 100°F (38°C)

(C) 130°F (55°C)

(D) 180°F (80°C)

18. An industrial process uses hot gas at 3600°R (1980°C) and 14.7 psia (101 kPa). It is proposed that propane be burned stoichiometrically in a mixture of nitrogen and oxygen. After passing through the process, gas will be exhausted through a duct, being cooled slowly to 100°F (38°C) and 14.7 psia (101 kPa) before discharge. The following data are available.

- The enthalpies of formation (at the standard reference temperature) are

$C_3H_8(g)$: $\Delta H_f = +28,800$ Btu/lbmol (+67.0 GJ/kmol)
$CO_2(g)$: $\Delta H_f = -169,300$ Btu/lbmol (−393.8 GJ/kmol)
$H_2O(g)$: $\Delta H_f = -104,040$ Btu/lbmol (−242 GJ/kmol)

- The enthalpy increases from the standard reference temperature to 3600°R (1980°C) are

CO_2: 39,791 Btu/lbmol (92.6 GJ/kmol)
H_2O: 31,658 Btu/lbmol (73.6 GJ/kmol)
N_2: 24,471 Btu/lbmol (56.9 GJ/kmol)

(a) What are most nearly the combining weights of the oxygen and nitrogen, respectively, per pound-mole of propane?

(A) 760 lbm/lbmol, 128 lbm/lbmol (760 kg/kmol, 128 kg/kmol)

(B) 810 lbm/lbmol, 160 lbm/lbmol (810 kg/kmol, 160 kg/kmol)

(C) 845 lbm/lbmol, 160 lbm/lbmol (845 kg/kmol, 160 kg/kmol)

(D) 850 lbm/lbmol, 160 lbm/lbmol (850 kg/kmol, 160 kg/kmol)

(b) The amount of water vapor, if any, present in the stack gas immediately after combustion is most nearly

(A) 0 lbmol/lbmol C_3H_8 (0 kmol H_2O/kmol C_3H_8)

(B) 1.2 lbmol/lbmol C_3H_8 (1.2 kmol H_2O/kmol C_3H_8)

(C) 2.3 lbmol/lbmol C_3H_8 (2.3 kmol H_2O/kmol C_3H_8)

(D) 3.5 lbmol/lbmol C_3H_8 (3.5 kmol H_2O/kmol C_3H_8)

(c) The amount of water removed from the stack gas is most nearly

(A) 0 lbm H_2O/lbmol C_3H_8 (0 kg H_2O/kmol C_3H_8)

(B) 9 lbm H_2O/lbmol C_3H_8 (9 kg H_2O/kmol C_3H_8)

(C) 30 lbm H_2O/lbmol C_3H_8 (30 kg H_2O/kmol C_3H_8)

(D) 50 lbm H_2O/lbmol C_3H_8 (50 kg H_2O/kmol C_3H_8)

Combustion

19. An electrical power-generating plant burns refuse-derived fuel (RDF). After sorting, incoming refuse is shredded and compressed before being fed into the combustor. The raw refuse averages 7% by weight incombustible solids. 5000 lbm/hr of processed RDF produces 20,070 lbm/hr of saturated steam at 200 lbf/in². The combustion products are used to heat incoming feedwater to a saturated temperature of 160°F before entering the combustor. 2000 lbm/hr of water vapor condense in the feedwater heater at a partial pressure of 4 lbf/in² and are removed. All thermal losses are to be disregarded. What is most nearly the higher heating value of the RDF?

(A) 4300 Btu/lbm

(B) 4700 Btu/lbm

(C) 4900 Btu/lbm

(D) 5100 Btu/lbm

SOLUTIONS

1. *Customary U.S. Solution*

$$T = 60°F + 460° = 520°R$$

$$p = 14.73 \text{ psia}$$

The specific gas constant, R, is calculated from the universal gas constant, R^*, and the molecular weight.

$$R = \frac{R^*}{\text{MW}} = \frac{1545.35 \frac{\text{ft-lbf}}{\text{lbmol-°R}}}{16.043 \frac{\text{lbm}}{\text{lbmol}}}$$

$$= 96.33 \text{ ft-lbf/lbm-°R}$$

$$m = \frac{pV}{RT} = \frac{\left(14.73 \frac{\text{lbf}}{\text{in}^2}\right)\left(12 \frac{\text{in}}{\text{ft}}\right)^2 (7 \text{ ft}^3)}{\left(96.33 \frac{\text{ft-lbf}}{\text{lbm-°R}}\right)(520°R)}$$

$$= 0.296 \text{ lbm}$$

The combustion energy available from methane is

$$Q = \eta m(\text{HHV}) = (0.5)(0.296 \text{ lbm})\left(24{,}000 \frac{\text{Btu}}{\text{lbm}}\right)$$

$$= 3552 \text{ Btu}$$

This energy is used to heat water from 60°F to 200°F.

$$Q = m_{\text{water}} c_p(T_2 - T_1)$$

$$m_{\text{water}} = \frac{3552 \text{ Btu}}{\left(1 \frac{\text{Btu}}{\text{lbm-°F}}\right)(200°F - 60°F)}$$

$$= \boxed{25.37 \text{ lbm} \quad (25 \text{ lbm})}$$

The answer is (A).

SI Solution

The specific gas constant, R, is calculated from the universal gas constant, R^*, and the molecular weight.

$$R = \frac{R^*}{\text{MW}} = \frac{8314.5 \frac{\text{J}}{\text{kmol·K}}}{16.043 \frac{\text{kg}}{\text{kmol}}}$$

$$= 518.26 \text{ J/kg·K}$$

$$m = \frac{pV}{RT}$$

$$= \frac{(101.51 \text{ kPa})\left(1000 \frac{\text{Pa}}{\text{kPa}}\right)(200 \text{ L})}{\left(518.26 \frac{\text{J}}{\text{kg·K}}\right)(288.9\text{K})\left(1000 \frac{\text{L}}{\text{m}^3}\right)}$$

$$= 0.136 \text{ kg}$$

The combustion energy available from methane is

$$Q = \eta m (\text{HHV})$$

$$= (0.5)(0.136 \text{ kg})\left(55.8 \frac{\text{MJ}}{\text{kg}}\right)\left(1000 \frac{\text{kJ}}{\text{MJ}}\right)$$

$$= 3794 \text{ kJ}$$

This energy is used to heat water from 15°C to 95°C.

$$Q = m_{\text{water}} c_p (T_2 - T_1)$$

$$m_{\text{water}} = \frac{3794 \text{ kJ}}{\left(4.1868 \frac{\text{kJ}}{\text{kg·C}}\right)(95°C - 15°C)}$$

$$= \boxed{11.33 \text{ kg} \quad (11 \text{ kg})}$$

The answer is (A).

2. *Customary U.S. Solution*

From Table 37.7,

	C_3H_8	+	$5O_2$	→	$3CO_2$	+	$4H_2O$
MW	44.097		(5)(32)		(3)(44.011)		
	44.097		160		132.033		

The amount of carbon dioxide produced is 132.033 lbm/ 44.097 lbm propane. For 15 lbm/hr of propane, the amount of carbon dioxide produced is

$$\left(\frac{132.033 \text{ lbm}}{44.097 \text{ lbm}}\right)\left(15 \frac{\text{lbm}}{\text{hr}}\right) = 44.91 \text{ lbm/hr}$$

$$R = \frac{R^*}{\text{MW}} = \frac{1545.35 \frac{\text{ft-lbf}}{\text{lbmol-°R}}}{44.011 \frac{\text{lbm}}{\text{lbmol}}}$$

$$= 35.11 \text{ ft-lbf/lbm-°R}$$

$$T = 70°F + 460° = 530°R$$

$$\dot{V} = \frac{\dot{m}RT}{p}$$

$$= \frac{\left(44.91 \frac{\text{lbm}}{\text{hr}}\right)\left(35.11 \frac{\text{ft-lbf}}{\text{lbm-°R}}\right)(530°R)}{\left(14.7 \frac{\text{lbf}}{\text{in}^2}\right)\left(12 \frac{\text{in}}{\text{ft}}\right)^2}$$

$$= \boxed{394.8 \text{ ft}^3/\text{hr} \quad (390 \text{ ft}^3/\text{hr})}$$

The answer is (C).

SI Solution

From Table 37.7,

	C_3H_8	+	$5O_2$	→	$3CO_2$	+	$4H_2O$
MW	44.097		(5)(32)		(3)(44.011)		
	44.097		160		132.033		

The amount of carbon dioxide produced is 132.033 kg/ 44.097 kg propane. For 6.8 kg/h of propane, the amount of carbon dioxide produced is

$$\left(\frac{132.033 \text{ kg}}{44.097 \text{ kg}}\right)\left(6.8 \frac{\text{kg}}{\text{h}}\right) = 20.36 \text{ kg/h}$$

$$R = \frac{R^*}{\text{MW}} = \frac{8314.5 \frac{\text{J}}{\text{kmol·K}}}{44.01 \frac{\text{kg}}{\text{kmol}}} = 188.92 \text{ J/kg·K}$$

$$T = 21°C + 273° = 294\text{K}$$

$$V = \frac{mRT}{p}$$

$$= \frac{\left(20.36 \frac{\text{kg}}{\text{h}}\right)\left(188.92 \frac{\text{J}}{\text{kg·K}}\right)(294\text{K})}{(101 \text{ kPa})\left(1000 \frac{\text{Pa}}{\text{kPa}}\right)}$$

$$= \boxed{11.20 \text{ m}^3/\text{h} \quad (11 \text{ m}^3/\text{h})}$$

The answer is (C).

3. Use the balanced chemical reaction equation from Table 37.7.

$$CH_4 + 2O_2 \rightarrow CO_2 + 2H_2O$$

Use Table 37.6 and Table 37.7. With 30% excess air and considering that there are 3.773 volumes of nitrogen for every volume of oxygen, the reaction equation is

$$CH_4 + (1.3)(2)O_2 + (1.3)(2)(3.773)N_2$$
$$\rightarrow CO_2 + 2H_2O + (1.3)(2)(3.773)N_2 + 0.6O_2$$

$$CH_4 + 2.6O_2 + 9.81N_2$$
$$\rightarrow CO_2 + 2H_2O + 9.81N_2 + 0.6O_2$$

Customary U.S. Solution

The volume of nitrogen that accompanies 4000 ft³/hr of entering methane is

$$V_{N_2} = \left(9.81 \frac{\text{ft}^3 \text{ N}_2}{\text{ft}^3 \text{ CH}_4}\right)\left(4000 \frac{\text{ft}^3 \text{ CH}_4}{\text{hr}}\right)$$

$$= 39{,}240 \text{ ft}^3 \text{ N}_2/\text{hr}$$

This is the "partial volume" of nitrogen in the input stream.

$$R = \frac{R^*}{MW} = \frac{1545.35 \ \frac{\text{ft-lbf}}{\text{lbmol-°R}}}{28.016 \ \frac{\text{lbm}}{\text{lbmol}}}$$

$$= 55.16 \ \text{ft-lbf/lbm-°R}$$

The absolute temperature is

$$T = 100°F + 460° = 560°R$$

$$\dot{m}_{N_2} = \frac{p_{N_2} V_{N_2}}{RT} = \frac{\left(15 \ \frac{\text{lbf}}{\text{in}^2}\right)\left(12 \ \frac{\text{in}}{\text{ft}}\right)^2 \left(39{,}240 \ \frac{\text{ft}^3}{\text{hr}}\right)}{\left(55.16 \ \frac{\text{ft-lbf}}{\text{lbm-°R}}\right)(560°R)}$$

$$= \boxed{2744 \ \text{lbm/hr} \quad (2700 \ \text{lbm/hr})}$$

The answer is (D).

SI Solution

The volume of nitrogen that accompanies 31 L/s of entering methane is

$$\frac{\left(9.81 \ \frac{\text{m}^3 \ N_2}{\text{m}^3 \ CH_4}\right)\left(31 \ \frac{\text{L CH}_4}{\text{s}}\right)}{1000 \ \frac{\text{L}}{\text{m}^3}} = 0.3041 \ \text{m}^3/\text{s}$$

This is the "partial volume" of nitrogen in the input stream.

$$R = \frac{R^*}{MW} = \frac{8314.5 \ \frac{\text{J}}{\text{kmol·K}}}{28.016 \ \frac{\text{kg}}{\text{kmol}}} = 296.8 \ \text{J/kg·K}$$

The absolute temperature is

$$T = 40°C + 273° = 313K$$

$$\dot{m}_{N_2} = \frac{p_{N_2} V_{N_2}}{RT} = \frac{(103 \ \text{kPa})\left(1000 \ \frac{\text{Pa}}{\text{kPa}}\right)\left(0.3041 \ \frac{\text{m}^3}{\text{s}}\right)}{\left(296.8 \ \frac{\text{J}}{\text{kg·K}}\right)(313K)}$$

$$= \boxed{0.337 \ \text{kg/s} \quad (0.34 \ \text{kg/s})}$$

The answer is (D).

4. From Table 37.7, combustion reactions are

$$\begin{array}{ccccc} & C & + & O_2 & \rightarrow & CO_2 \\ MW & 12 & & 32 & & \end{array}$$

The mass of oxygen required per unit mass of carbon is

$$\frac{32}{12} = 2.67$$

$$\begin{array}{ccccc} & 2H_2 & + & O_2 & \rightarrow & 2H_2O \\ MW & (2)(2) & & 32 & & \end{array}$$

The mass of oxygen required per unit mass of hydrogen is

$$\frac{32}{(2)(2)} = 8.0$$

$$\begin{array}{ccccc} & S & + & O_2 & \rightarrow & SO_2 \\ MW & 32.1 & & 32 & & \end{array}$$

The mass of oxygen required per unit mass of sulfur is

$$\frac{32}{32.1} = 1.0$$

Nitrogen does not burn.

The mass of oxygen required per unit mass of fuel is

$$(0.84)(2.67) + (0.153)(8) + (0.003)(1)$$
$$= 3.47 \ \text{units of mass of O}_2/\text{unit mass fuel}$$

From Table 37.6, air is 0.2315 O_2/unit mass.

Customary U.S. Solution

The air required is

$$\frac{3.47}{0.2315} = \boxed{14.99 \quad (15 \ \text{lbm air/lbm fuel})}$$

SI Solution

The air required is

$$\frac{3.47}{0.2315} = \boxed{14.99 \quad (15 \ \text{kg air/kg fuel})}$$

(Equation 37.6 could also have been used to solve this problem.)

The answer is (C).

5. From Table 37.7, the balanced chemical reaction equation is

$$C_3H_8 + 5O_2 \rightarrow 3CO_2 + 4H_2O$$

With 20% excess air, the oxygen volume is $(1.2)(5) = 6$.

$$C_3H_8 + 6O_2 \rightarrow 3CO_2 + 4H_2O + O_2$$

From Table 37.6, there are 3.773 volumes of nitrogen for every volume of oxygen.

$$(6)(3.773) = 22.6$$

$$C_3H_8 + 6O_2 + 22.6N_2 \rightarrow 3CO_2 + 4H_2O + O_2 + 22.6N_2$$

The percentage of carbon dioxide by weight in flue gas is

$$G_{CO_2} = \frac{m_{CO_2}}{m_{total}} = \frac{B_{CO_2}(MW_{CO_2})}{\sum B_i(MW_i)}$$

$$= \frac{(3)(44.011)}{(3)(44.011) + (4)(18.016)}$$

$$+ 32 + (22.6)(28.016)$$

$$= \boxed{0.152 \quad (15\%)}$$

The answer is (C).

6. The actual air/fuel ratio can be estimated from the flue gas analysis and the fraction of carbon in fuel.

From Eq. 37.9,

$$\frac{m_{air}}{m_{fuel}} = \frac{3.04 B_{N_2} G_C}{B_{CO_2} + B_{CO}} = \frac{(3.04)(0.80)(0.80)}{0.12 + 0.01}$$

$$= 14.97$$

Customary U.S. Solution

The air required to burn 1 lbm of coal is $\boxed{14.97 \text{ lbm } (15 \text{ lbm}).}$

SI Solution

The air required to burn 1 kg of coal is $\boxed{14.97 \text{ kg } (15 \text{ kg}).}$

The answer is (B).

Alternative Solution

The use of Eq. 37.9 obscures the process of finding the air/fuel ratio. (The SI solution is similar but is not presented here.)

step 1: Find the mass of oxygen in the stack gases.

$$R_{CO_2} = 35.11 \text{ ft-lbf/lbm-}°R$$
$$R_{CO} = 55.17 \text{ ft-lbf/lbm-}°R$$
$$R_{O_2} = 48.29 \text{ ft-lbf/lbm-}°R$$

The partial densities are

$$\rho_{CO_2} = \frac{p}{RT} = \frac{(0.12)\left(14.7 \frac{lbf}{in^2}\right)\left(12 \frac{in}{ft}\right)^2}{\left(35.11 \frac{ft\text{-}lbf}{lbm\text{-}°R}\right)(60°F + 460°)}$$

$$= 1.391 \times 10^{-2} \text{ lbm/ft}^3$$

$$\rho_{CO} = \frac{(0.01)\left(14.7 \frac{lbf}{in^2}\right)\left(12 \frac{in}{ft}\right)^2}{\left(55.11 \frac{ft\text{-}lbf}{lbm\text{-}°R}\right)(60°F + 460°)}$$

$$= 7.387 \times 10^{-4} \text{ lbm/ft}^3$$

$$\rho_{CO_2} = \frac{(0.07)\left(14.7 \frac{lbf}{in^2}\right)\left(12 \frac{in}{ft}\right)^2}{\left(48.29 \frac{ft\text{-}lbf}{lbm\text{-}°R}\right)(60°F + 460°)}$$

$$= 5.901 \times 10^{-3} \text{ lbm/ft}^3$$

The fraction of oxygen in the three components is

$$CO_2: \frac{32.0}{44} = 0.7273$$

$$CO: \frac{16}{28} = 0.5714$$

$$O_2: 1.00$$

In 100 ft^3 of stack gases, the total oxygen mass will be

$$(100 \text{ ft}^3)\left(\begin{array}{c} (0.7273)\left(1.391 \times 10^{-2} \frac{lbm}{ft^3}\right) \\ + (0.5714)\left(7.387 \times 10^{-4} \frac{lbm}{ft^3}\right) \\ + (1.00)\left(5.901 \times 10^{-3} \frac{lbm}{ft^3}\right) \end{array} \right)$$

$$= 1.644 \text{ lbm}$$

step 2: Since air is 23.15% oxygen by weight, the mass of air per 100 ft^3 of stack gases is

$$\frac{1.644 \text{ lbm}}{0.2315} = 7.102 \text{ lbm/100 ft}^3$$

step 3: Find the mass of carbon in the stack gases by a similar process.

$$CO_2: \frac{12}{44} = 0.2727$$

$$CO: \frac{12}{28} = 0.4286$$

$$(100 \text{ ft}^3)\left(\begin{array}{c} (0.2727)\left(1.391 \times 10^{-2} \frac{lbm}{ft^3}\right) \\ + (0.4286)\left(7.387 \times 10^{-4} \frac{lbm}{ft^3}\right) \end{array} \right)$$

$$= 0.411 \text{ lbm}$$

step 4: The coal is 80% carbon, so the air per lbm of coal for combustion of the carbon is

$$\left(\frac{0.80 \ \frac{\text{lbm carbon}}{\text{lbm coal}}}{0.411 \ \frac{\text{lbm carbon}}{100 \ \text{ft}^3}}\right)\left(7.102 \ \frac{\text{lbm}}{100 \ \text{ft}^3}\right)$$

$$= 13.824 \ \text{lbm air/lbm coal}$$

step 5: This does not include air to burn hydrogen, since Orsat is a dry analysis.

The theoretical air for the hydrogen is given by Eq. 37.6.

$$R_{a/f,\text{H}} = \left(34.5 \ \frac{\text{lbm}}{\text{lbm}}\right)\left(G_\text{H} - \frac{G_\text{O}}{8}\right)$$

$$= \left(34.5 \ \frac{\text{lbm}}{\text{lbm}}\right)\left(0.04 - \frac{0.02}{8}\right)$$

$$= 1.294 \ \text{lbm air/lbm fuel}$$

Ignoring any excess air for the hydrogen, the total air per pound of coal is

$$13.824 \ \frac{\text{lbm air}}{\text{lbm coal}} + 1.294 \ \frac{\text{lbm air}}{\text{lbm coal}}$$

$$= \boxed{15.1 \ \text{lbm air/lbm coal} \quad (15 \ \text{lbm})}$$

The answer is (B).

7. From Eq. 14.10,

$$\text{SG} = \frac{141.5}{°\text{API} + 131.5} = \frac{141.5}{40 + 131.5} = 0.825$$

Customary U.S. Solution

From Eq. 37.18(b),

$$\text{HHV} = 22{,}320 - 3780(\text{SG})^2$$

$$= 22{,}320 - (3780)(0.825)^2$$

$$= \boxed{19{,}747 \ \text{Btu/lbm} \quad (20{,}000 \ \text{Btu/lbm})}$$

The answer is (A).

SI Solution

From Eq. 37.18(a),

$$\text{HHV} = 51.92 - 8.792(\text{SG})^2$$

$$= 51.92 - (8.792)(0.825)^2$$

$$= \boxed{45.94 \ \text{MJ/kg} \quad (46 \ \text{MJ/kg})}$$

The answer is (A).

8. *Customary U.S. Solution*

(a) *step 1:* From Eq. 37.16(b), substituting the lower heating value of hydrogen from App. 37.A, the lower heating value of coal is

$$\text{LHV} = 14{,}093 G_\text{C} + (51{,}623)\left(G_\text{H} - \frac{G_\text{O}}{8}\right) + 3983 G_\text{S}$$

$$= (14{,}093)(0.75)$$

$$+ (51{,}623)\left(0.05 - \frac{0.03}{8}\right) + (3983)(0)$$

$$= 12{,}957 \ \text{Btu/lbm}$$

step 2: The gravimetric analysis of 1 lbm of coal is

carbon: 0.75 lbm

free hydrogen: $G_{\text{H,total}} - \frac{G_\text{O}}{8} = 0.05 - \frac{0.03}{8}$

$$= 0.0463$$

The ratio of the molecular weight of water (18) to the molecular weight of hydrogen (2) is 9.

water: $(9)(0.05 - 0.0463) = 0.0333$

nitrogen: 0.02

step 3: From Table 37.8, the theoretical stack gases per lbm coal for 0.75 lbm of carbon are

$$\text{CO}_2 = (0.75)(3.667 \ \text{lbm}) = 2.750 \ \text{lbm}$$

$$\text{N}_2 = (0.75)(8.883 \ \text{lbm}) = 6.662 \ \text{lbm}$$

All products are calculated similarly (as the following table summarizes). All values are per pound of fuel.

	CO_2	N_2	H_2O
from C:	2.750 lbm	6.662 lbm	
from H_2:		1.217 lbm	0.414 lbm
from H_2O:			0.0333 lbm
from O_2:	shows up in	CO_2 and H_2	
from N_2:		0.02 lbm	
total:	2.750 lbm	7.899 lbm	0.4473 lbm

step 4: Assume the combustion gases leave at 1000°F.

$$T_\text{ave} = \left(\tfrac{1}{2}\right)(60°\text{F} + 1000°\text{F}) = 530°\text{F}$$

$$T_\text{ave} = 530°\text{F} + 460° = 990°\text{R}$$

The specific heat values are given in Table 37.1.

$$c_{p,\text{CO}_2} = 0.251 \ \text{Btu/lbm-°F}$$

$$c_{p,\text{N}_2} = 0.255 \ \text{Btu/lbm-°F}$$

$$c_{p,\text{H}_2\text{O}} = 0.475 \ \text{Btu/lbm-°F}$$

Combustion

The energy required to raise the combustion products (from 1 lbm of coal) 1°F is

$$m_{CO_2} c_{p,CO_2} + m_{N_2} c_{p,N_2} + m_{H_2O} c_{p,H_2O}$$

$$= (2.750 \text{ lbm}) \left(0.251 \frac{\text{Btu}}{\text{lbm-°F}} \right)$$

$$+ (7.899 \text{ lbm}) \left(0.255 \frac{\text{Btu}}{\text{lbm-°F}} \right)$$

$$+ (0.4473 \text{ lbm}) \left(0.475 \frac{\text{Btu}}{\text{lbm-°F}} \right)$$

$$= 2.92 \text{ Btu/°F}$$

step 5: Assuming all combustion heat goes into the stack gases, the temperature is given by Eq. 37.19.

$$T_{\max} = T_i + \frac{\text{lower heat of combustion}}{\text{energy required}}$$

$$= 60°\text{F} + \frac{12{,}957 \frac{\text{Btu}}{\text{lbm}}}{2.92 \frac{\text{Btu}}{\text{lbm-°F}}}$$

$$= \boxed{4497°\text{F} \quad (4500°\text{F})} \quad \text{[unreasonable]}$$

The answer is (C).

(b) *step 6:* Nitrogen in the coal does not contribute to excess air. With 40% excess air and 75% of heat absorbed by the boiler, the excess air (based on 76.85% N_2 by weight) per pound of fuel is

$$(0.40) \left(\frac{7.899 \frac{\text{lbm}}{\text{lbm}} - 0.02 \frac{\text{lbm}}{\text{lbm}}}{0.7685} \right)$$

$$= 4.101 \text{ lbm/lbm}$$

From Table 37.1, $c_{p,\text{air}} = 0.249$ Btu/lbm-°F.

Therefore,

$$T_{\max} = 60°\text{F} + \frac{\left(12{,}957 \frac{\text{Btu}}{\text{lbm}} \right)(1 - 0.75)}{\left(\begin{array}{c} 2.92 \frac{\text{Btu}}{\text{lbm-°F}} + \left(4.101 \frac{\text{lbm}}{\text{lbm}} \right) \\ \times \left(0.249 \frac{\text{Btu}}{\text{lbm-°F}} \right) \end{array} \right)}$$

$$= \boxed{881.9°\text{F} \quad (880°\text{F})}$$

The answer is (B).

SI Solution

(a) *step 1:* From Eq. 37.16(a), and substituting the lower heating value of hydrogen from App. 37.A, the lower heating value of coal is

$$\text{LHV} = 32.78\,G_C + \frac{\left(51{,}623 \frac{\text{Btu}}{\text{lbm}} \right) \left(2.326 \frac{\text{kJ-lbm}}{\text{kg-Btu}} \right)}{1000 \frac{\text{kJ}}{\text{MJ}}}$$

$$\times \left(G_H - \frac{G_O}{8} \right) + 9.264\,G_S$$

$$= (32.78)(0.75) + (120.1) \left(0.05 - \frac{0.03}{8} \right)$$

$$+ (9.264)(0)$$

$$= 30.14 \text{ MJ/kg}$$

Steps 2 and 3 are the same as for the customary U.S. solution except that all masses are in kg.

step 4: Assume the combustion gases leave at 550°C.

$$T_{\text{ave}} = \left(\tfrac{1}{2} \right)(16°\text{C} + 550°\text{C}) + 273° = 556\text{K}$$

Specific heat values are given in Table 37.1. Using the footnote for SI units,

$$c_{p,CO_2} = 1.051 \text{ kJ/kg·K}$$

$$c_{p,N_2} = 1.068 \text{ kJ/kg·K}$$

$$c_{p,H_2O} = 1.989 \text{ kJ/kg·K}$$

The energy required to raise the combustion products (from 1 kg of coal) 1°C is

$$m_{CO_2} c_{p,CO_2} + m_{N_2} c_{p,N_2} + m_{H_2O} c_{p,H_2O}$$

$$= (2.750 \text{ kg}) \left(1.051 \frac{\text{kJ}}{\text{kg·K}} \right)$$

$$+ (7.899 \text{ kg}) \left(1.068 \frac{\text{kJ}}{\text{kg·K}} \right)$$

$$+ (0.4473 \text{ kg}) \left(1.989 \frac{\text{kJ}}{\text{kg·K}} \right)$$

$$= 12.22 \text{ kJ/K}$$

step 5: Assuming all combustion heat goes into the stack gases, the temperature is given by Eq. 37.19.

$$T_{\max} = T_i + \frac{\text{lower heat of combustion}}{\text{energy required}}$$

$$= 16°\text{C} + \frac{\left(30.14 \frac{\text{MJ}}{\text{kg}} \right) \left(1000 \frac{\text{kJ}}{\text{MJ}} \right)}{12.22 \frac{\text{kJ}}{\text{K}}}$$

$$= \boxed{2482°\text{C} \quad (2500°\text{C})} \quad \text{[unreasonable]}$$

The answer is (C).

Combustion

(b) *step 6:* Nitrogen in the coal does not contribute to excess air. With 40% excess air and 75% of heat absorbed by the boiler, the excess air (based on 76.85% N_2 by weight) per kilogram of fuel is

$$(0.40)\left(\dfrac{7.899\ \frac{kg}{kg} - 0.02\ \frac{kg}{kg}}{0.7685}\right) = 4.101\ kg/kg$$

From Table 37.1, using the table footnote, $c_{p,air} = 1.043$ kJ/kg·K.

Therefore,

$$T_{max} = 16°C + \dfrac{\left(30.14\ \frac{MJ}{kg}\right)\left(1000\ \frac{kJ}{MJ}\right)(1-0.75)}{12.22\ \frac{kJ}{K} + (4.101\ kg)\left(1.043\ \frac{kJ}{kg·K}\right)}$$

$$= \boxed{472.7°C \quad (470°C)}$$

The answer is (B).

9. Assume the oxygen is in the form of moisture in the fuel.

The available hydrogen is

$$G_{H,free} = G_H - \dfrac{G_O}{8} = 0.1131 - \dfrac{0.027}{8} = 0.1097$$

Customary U.S. Solution

step 1: From Table 37.8, find the stoichiometric oxygen required per pound of fuel oil.

$$C \to CO_2: O_2 \text{ required} = (0.8543)(2.667\ lbm)$$
$$= 2.2784\ lbm$$
$$H_2 \to H_2O: O_2 \text{ required} = (0.1097)(7.936\ lbm)$$
$$= 0.8706\ lbm$$
$$S \to SO_2: O_2 \text{ required} = (0.0034)(0.998\ lbm)$$
$$= 0.0034\ lbm$$

The total amount of oxygen required per pound of fuel oil is

$$2.2784\ lbm + 0.8706\ lbm + 0.0034\ lbm = 3.1524\ lbm$$

step 2: With 60% excess air, the excess oxygen per pound of fuel oil is

$$(0.6)(3.1524\ lbm) = 1.8914\ lbm\ O_2$$

From Eq. 29.60, this oxygen occupies a volume of

$$V = \dfrac{mRT}{p}$$

At standard conditions,

$$p = 14.7\ psia$$
$$T = 60°F + 460° = 520°R$$

From Table 29.7, for oxygen, $R = 48.29$ ft-lbf/lbm-°R.

$$V = \dfrac{(1.8914\ lbm)\left(48.29\ \frac{ft\text{-}lbf}{lbm\text{-}°R}\right)(520°R)}{\left(14.7\ \frac{lbf}{in^2}\right)\left(12\ \frac{in}{ft}\right)^2} = 22.44\ ft^3$$

step 3: The theoretical nitrogen per pound of fuel based on Table 37.6 is

$$\left(\dfrac{3.1524\ lbm}{0.2315}\right)(0.7685) = 10.465\ lbm\ N_2$$

The actual nitrogen per pound of fuel with 60% excess air and nitrogen in the fuel is

$$(10.465\ lbm)(1.6) + 0.0022\ lbm = 16.746\ lbm\ N_2$$

From Eq. 29.60, this nitrogen occupies a volume of

$$V = \dfrac{mRT}{p}$$

From Table 29.7, R for nitrogen $= 55.16$ ft-lbf/lbm-°R.

$$V = \dfrac{(16.746\ lbm)\left(55.16\ \frac{ft\text{-}lbf}{lbm\text{-}°R}\right)(520°R)}{\left(14.7\ \frac{lbf}{in^2}\right)\left(12\ \frac{in}{ft}\right)^2}$$
$$= 226.91\ ft^3$$

step 4: From Table 37.8, the 60°F combustion product volumes per pound of fuel will be

CO_2:	$(0.8543)(31.63\ ft^3)$	=	$27.02\ ft^3$
H_2O:	$(0.1131)(188.25\ ft^3)$	=	$21.29\ ft^3$
SO_2:	$(0.0034)(11.84\ ft^3)$	=	$0.04\ ft^3$
N_2:	from step 3	=	$226.91\ ft^3$
O_2:	from step 2	=	$22.44\ ft^3$
		total =	$297.7\ ft^3$

(a) At 60°F, the wet volume per pound of fuel will be 297.7 ft³.

At 600°F, the wet volume per pound of fuel is

$$V_{wet,600°F} = (297.7\ ft^3)\left(\dfrac{600°F + 460°}{60°F + 460°}\right)$$
$$= \boxed{606.9\ ft^3 \quad (600\ ft^3)}$$

The answer is (D).

(b) At 60°F, the dry volume per pound of fuel will be

$$297.7 \text{ ft}^3 - 21.29 \text{ ft}^3 = 276.4 \text{ ft}^3$$

At 600°F, the dry volume per pound of fuel is

$$V_{\text{dry,600°F}} = (276.4 \text{ ft}^3)\left(\frac{600°F + 460°}{60°F + 460°}\right)$$

$$= \boxed{563.4 \text{ ft}^3 \quad (560 \text{ ft}^3)}$$

The answer is (B).

(c) The volumetric fraction of dry carbon dioxide is

$$\frac{27.02 \text{ ft}^3}{276.41 \text{ ft}^3} = \boxed{0.098 \quad (10\%)}$$

The answer is (C).

SI Solution

step 1: From Table 37.8, find the stoichiometric oxygen required per kilogram of fuel oil.

$$C \rightarrow CO_2: O_2 \text{ required} = (0.8543)(2.667 \text{ kg})$$
$$= 2.2784 \text{ kg}$$

$$H_2 \rightarrow H_2O: O_2 \text{ required} = (0.1097)(7.936 \text{ kg})$$
$$= 0.8706 \text{ kg}$$

$$S \rightarrow SO_2: O_2 \text{ required} = (0.0034)(0.998 \text{ kg})$$
$$= 0.0034 \text{ kg}$$

The total amount of oxygen required per kilogram of fuel oil is

$$2.2784 \text{ kg} + 0.8706 \text{ kg} + 0.0034 \text{ kg} = 3.1524 \text{ kg}$$

step 2: With 60% excess air, the excess oxygen per kilogram of fuel oil is

$$(0.6)(3.1524 \text{ kg}) = 1.8914 \text{ kg } O_2$$

From Eq. 29.60, this oxygen occupies a volume of

$$V = \frac{mRT}{p}$$

At standard conditions,

$$p = 101.3 \text{ kPa}$$
$$T = 16°C + 273° = 289K$$

From Table 29.7, for oxygen, $R = 259.82$ J/kg·K.

$$V = \frac{(1.8914 \text{ kg})\left(259.82 \dfrac{J}{\text{kg·K}}\right)(289K)}{(101.3 \text{ kPa})\left(1000 \dfrac{Pa}{\text{kPa}}\right)}$$

$$= 1.402 \text{ m}^3$$

step 3: The theoretical nitrogen per kilogram of fuel based on Table 37.6 is

$$\left(\frac{3.1524 \text{ kg}}{0.2315}\right)(0.7685) = 10.465 \text{ kg } N_2$$

The actual nitrogen per kilogram of fuel with 60% excess air and nitrogen in the fuel is

$$(10.465 \text{ kg})(1.6) + 0.0022 \text{ kg} = 16.746 \text{ kg } N_2$$

From Eq. 29.60, this nitrogen occupies a volume of

$$V = \frac{mRT}{p}$$

From Table 29.7, R for nitrogen $= 296.77$ J/kg·K.

$$V = \frac{(16.746 \text{ kg})\left(296.77 \dfrac{J}{\text{kg·K}}\right)(289K)}{(101.3 \text{ kPa})\left(1000 \dfrac{Pa}{\text{kPa}}\right)}$$

$$= 14.178 \text{ m}^3$$

step 4: From Table 37.8, the 16°C combustion product volumes per kilogram of fuel will be

CO_2:	$(0.8543)(31.63 \text{ ft}^3)(0.06243)$	=	1.687 m^3
H_2O:	$(0.1131)(188.25 \text{ ft}^3)(0.06243)$	=	1.329 m^3
SO_2:	$(0.0034)(11.84 \text{ ft}^3)(0.06243)$	=	0.003 m^3
N_2:	from step 3	=	14.178 m^3
O_2:	from step 2	=	1.402 m^3
		total =	18.599 m^3

(a) At 16°C, the wet volume per kilogram of fuel will be 18.599 m³.

At 320°C, the wet volume per kilogram of fuel is

$$V_{\text{wet,320°C}} = (18.599 \text{ m}^3)\left(\frac{320°C + 273°}{16°C + 273°}\right)$$

$$= \boxed{38.16 \text{ m}^3 \quad (38 \text{ m}^3)}$$

The answer is (D).

(b) At 16°C, the dry volume per kilogram of fuel will be

$$18.599 \text{ m}^3 - 1.329 \text{ m}^3 = 17.27 \text{ m}^3$$

At 320°C, the dry volume per kilogram of fuel is

$$V_{\text{dry,320°C}} = (17.27 \text{ m}^3)\left(\frac{320°\text{C} + 273°}{16°\text{C} + 273°}\right)$$

$$= \boxed{35.44 \text{ m}^3 \quad (35 \text{ m}^3)}$$

The answer is (B).

(c) The volumetric fraction of dry carbon dioxide is

$$\frac{1.687 \text{ m}^3}{17.27 \text{ m}^3} = \boxed{0.098 \quad (10\%)}$$

The answer is (C).

10. *Customary U.S. Solution*

step 1: Based on 15,395 lbm of coal burned producing 2816 lbm of ash containing 20.9% carbon by weight, the usable percentage of carbon per pound of fuel is

$$0.5145 - \frac{(2816 \text{ lbm})(0.209)}{15,395 \text{ lbm}} = 0.4763$$

step 2: Since moisture is reported separately, assume all of the oxygen and hydrogen are free. (This is not ordinarily the case.) From Table 37.8, find the stoichiometric oxygen required per pound of fuel.

$$\text{C} \rightarrow \text{CO}_2: \text{O}_2 \text{ required} = (0.4763)(2.667 \text{ lbm})$$
$$= 1.2703 \text{ lbm}$$
$$\text{H}_2 \rightarrow \text{H}_2\text{O}: \text{O}_2 \text{ required} = (0.0402)(7.936 \text{ lbm})$$
$$= 0.3190 \text{ lbm}$$
$$\text{S} \rightarrow \text{SO}_2: \text{O}_2 \text{ required} = (0.0392)(0.998 \text{ lbm})$$
$$= 0.0391 \text{ lbm}$$

The total amount of O_2 required per pound of fuel is

$$1.2703 \text{ lbm} + 0.3190 \text{ lbm}$$
$$+ 0.0391 \text{ lbm} - 0.0728 \text{ lbm} = 1.5556 \text{ lbm}$$

step 3: The theoretical air per pound of fuel based on Table 37.6 is

$$\frac{1.5556 \text{ lbm}}{0.2315} = 6.720 \text{ lbm air}$$

step 4: Ignoring fly ash, the theoretical dry products per pound of fuel are given from Table 37.8.

CO$_2$:	(0.4763)(3.667 lbm)	= 1.7466 lbm
SO$_2$:	(0.0392)(1.998 lbm)	= 0.0783 lbm
N$_2$:	0.0093 lbm + (6.720 lbm)(0.7685)	= 5.1736 lbm
	total =	6.999 lbm

step 5: The excess air per pound of fuel is

$$13.3 \text{ lbm} - 6.999 \text{ lbm} = 6.301 \text{ lbm}$$

step 6: The total air supplied per pound of fuel is

$$6.301 \text{ lbm} + 6.720 \text{ lbm} = \boxed{13.02 \text{ lbm} \quad (13 \text{ lbm})}$$

The answer is (A).

SI Solution

step 1: Based on 6923 kg of coal burned producing 1267 kg of ash containing 20.9% carbon by weight, the usable percentage of carbon per kilogram of fuel is

$$0.5145 - \frac{(1267 \text{ kg})(0.209)}{6923 \text{ kg}} = 0.4763$$

step 2: From Table 37.8, find the stoichiometric oxygen required per kilogram of fuel.

$$\text{C} \rightarrow \text{CO}_2: \text{O}_2 \text{ required} = (0.4763)(2.667 \text{ kg})$$
$$= 1.2703 \text{ kg}$$
$$\text{H}_2 \rightarrow \text{H}_2\text{O}: \text{O}_2 \text{ required} = (0.0402)(7.936 \text{ kg})$$
$$= 0.3190 \text{ kg}$$
$$\text{S} \rightarrow \text{SO}_2: \text{O}_2 \text{ required} = (0.0392)(0.998 \text{ kg})$$
$$= 0.0391 \text{ kg}$$

The total amount of O_2 required per kilogram of fuel is

$$1.2703 \text{ kg} + 0.3190 \text{ kg}$$
$$+ 0.0391 \text{ kg} - 0.0728 \text{ kg} = 1.5556 \text{ kg}$$

step 3: The theoretical air per kilogram of fuel based on Table 37.6 is

$$\frac{1.5556 \text{ kg}}{0.2315} = 6.720 \text{ kg air}$$

step 4: Ignoring fly ash, the theoretical dry products per kilogram of fuel are given from Table 37.8.

CO$_2$:	(0.4763)(3.667 kg)	= 1.7466 kg
SO$_2$:	(0.0392)(1.998 kg)	= 0.0783 kg
N$_2$:	0.0093 kg + (6.720 kg)(0.7685)	= 5.1736 kg
	total =	6.999 kg

Combustion

step 5: The excess air per kilogram of fuel is

$$13.3 \text{ kg} - 6.999 \text{ kg} = 6.301 \text{ kg}$$

step 6: The total air supplied per kilogram of fuel is

$$6.301 \text{ kg} + 6.720 \text{ kg} = \boxed{13.02 \text{ kg} \quad (13 \text{ kg})}$$

The answer is (A).

11. (The customary U.S. and SI solutions are essentially identical.)

From Eq. 37.9, the actual air-fuel ratio can be estimated as

$$R_{a/f,\text{actual}} = \frac{3.04 B_{N_2} G_C}{B_{CO_2} + B_{CO}}$$

A fraction of carbon is reduced due to the percentage of coal lost in the ash pit.

$$G_C = (1 - 0.03)(0.65) = 0.6305$$

$$R_{a/f,\text{actual}} = \frac{(3.04)(0.815)(0.6305)}{0.095 + 0}$$
$$= 16.44 \text{ lbm air/lbm fuel} \quad (\text{kg air/kg fuel})$$

Combustion uses 9.87 lbm of air/lbm of fuel.

$$(9.87 \text{ lbm})(1 - 0.03) = 9.57 \text{ lbm of air/lbm of fuel}$$

$$\% \text{ of excess air} = \left(\frac{16.44 \dfrac{\text{lbm}}{\text{lbm}} - 9.57 \dfrac{\text{lbm}}{\text{lbm}}}{9.57 \dfrac{\text{lbm}}{\text{lbm}}} \right)$$
$$\times 100\%$$
$$= \boxed{71.8\% \quad (70\%)}$$

The answer is (C).

12. From Eq. 37.23, the heat loss due to the formation of carbon monoxide is

$$q = \frac{(\text{HHV}_C - \text{HHV}_{CO}) G_C B_{CO}}{B_{CO_2} + B_{CO}}$$

From App. 37.A, the difference in the two heating values is

$$\text{HHV}_C - \text{HHV}_{CO} = 14{,}093 \frac{\text{Btu}}{\text{lbm}} - 4347 \frac{\text{Btu}}{\text{lbm}}$$
$$= 9746 \text{ Btu/lbm} \quad (22.67 \text{ MJ/kg})$$
$$G_C = (1 - 0.03)(0.6734) = 0.6532$$

Customary U.S. Solution

$$q = \frac{\left(9746 \dfrac{\text{Btu}}{\text{lbm}}\right)(0.6532)(0.016)}{0.155 + 0.016}$$
$$= \boxed{596 \text{ Btu/lbm} \quad (600 \text{ Btu/lbm})}$$

The answer is (A).

SI Solution

$$q = \frac{\left(22.67 \dfrac{\text{MJ}}{\text{kg}}\right)(0.6532)(0.016)}{0.155 + 0.016}$$
$$= \boxed{1.39 \text{ MJ/kg} \quad (1.4 \text{ MJ/kg})}$$

The answer is (A).

13. (a) For methane, $B = 0.93$.

From ideal gas laws (R for methane $= 96.32$ ft-lbf/lbm-°R),

$$\rho = \frac{p}{RT} = \frac{\left(14.7 \dfrac{\text{lbf}}{\text{in}^2}\right)\left(12 \dfrac{\text{in}}{\text{ft}}\right)^2}{\left(96.32 \dfrac{\text{ft-lbf}}{\text{lbm-°R}}\right)(60°\text{F} + 460°)}$$
$$= 0.0423 \text{ lbm/ft}^3$$

From Table 37.9, $K = 9.55$ ft^3 air/ft^3 fuel.

From Table 37.8,

$$\text{products:} \quad 1 \text{ ft}^3 \text{ CO}_2, 2 \text{ ft}^3 \text{ H}_2\text{O}$$

From App. 37.A, HHV $= 1013$ Btu/lbm.

Similar results for all the other fuel components are tabulated in the table.

Density is volumetrically weighted. The composite density is

$$\rho = \sum B_i \rho_i$$
$$= (0.93)\left(0.0423 \frac{\text{lbm}}{\text{ft}^3}\right) + (0.0373)\left(0.0738 \frac{\text{lbm}}{\text{ft}^3}\right)$$
$$+ (0.0045)\left(0.0738 \frac{\text{lbm}}{\text{ft}^3}\right) + (0.0182)\left(0.0053 \frac{\text{lbm}}{\text{ft}^3}\right)$$
$$+ (0.0025)\left(0.0739 \frac{\text{lbm}}{\text{ft}^3}\right) + (0.0018)\left(0.0900 \frac{\text{lbm}}{\text{ft}^3}\right)$$
$$+ (0.0035)\left(0.0843 \frac{\text{lbm}}{\text{ft}^3}\right) + (0.0022)\left(0.1160 \frac{\text{lbm}}{\text{ft}^3}\right)$$
$$= \boxed{0.0434 \text{ lbm/ft}^3 \quad (0.043 \text{ lbm/ft}^3)}$$

The answer is (B).

(sidebar) Combustion

Reaction Products for Prob. 13

gas	B	$\dfrac{\text{lbm}}{\text{ft}^3}$	ft^3 air	$\dfrac{\text{Btu}}{\text{lbm}}$	volumes of products		
					CO_2	H_2O	other
CH_4	0.93	0.0422	9.556	1013	1	2	–
N_2	0.0373	0.0738	–	–	–	–	$1\ N_2$
CO	0.0045	0.0738	2.389	322	1	–	–
H_2	0.0182	0.0053	2.389	325	–	1	–
C_2H_4	0.0025	0.0739	14.33	1614	2	2	–
H_2S	0.0018	0.0900	7.167	647	–	1	$1\ SO_2$
O_2	0.0035	0.0843	–	–	–	–	–
CO_2	0.0022	0.1160	–	–	1	–	–

(b) The air is 20.9% oxygen by volume. The theoretical air requirements are

$$\sum B_i V_{\text{air},i} - \frac{O_2 \text{ in fuel}}{0.209}$$
$$= (0.93)(9.556 \text{ ft}^3) + (0.0373)(0 \text{ ft}^3)$$
$$+ (0.0045)(2.389 \text{ ft}^3) + (0.0182)(2.389 \text{ ft}^3)$$
$$+ (0.0025)(14.33 \text{ ft}^3) + (0.0018)(7.167 \text{ ft}^3)$$
$$+ (0.0035)(0 \text{ ft}^3) + (0.0022)(0 \text{ ft}^3) - \frac{0.0035}{0.209}$$
$$= \boxed{8.9733 \text{ ft}^3 \text{ air/ft}^3 \text{ fuel} \quad (9 \text{ ft}^3 \text{ air/ft}^3 \text{ fuel})}$$

The answer is (C).

((c) and (d)) The theoretical oxygen will be

$$\left(8.9733 \ \frac{\text{ft}^3}{\text{ft}^3}\right)(0.209) = 1.875 \text{ ft}^3/\text{ft}^3$$

The excess oxygen will be

$$\left(1.875 \ \frac{\text{ft}^3}{\text{ft}^3}\right)(0.4) = 0.75 \text{ ft}^3/\text{ft}^3$$

Similarly, with 40% excess air, the total nitrogen in the stack gases is

$$(1.4)(0.791)\left(8.9733 \ \frac{\text{ft}^3}{\text{ft}^3}\right) + 0.0373 \ \frac{\text{ft}^3}{\text{ft}^3}$$
$$= 9.974 \text{ ft}^3/\text{ft}^3 \text{ fuel}$$

The wet stack gas volumes per ft^3 of fuel are

excess O_2:	$=$	0.7500 ft^3
excess N_2:	$=$	9.974 ft^3
excess SO_2:	$=$	0.0018 ft^3
excess CO_2: $(0.93)(1) + (0.0045)(1)$		
$+ (0.0025)(2) + (0.0022)(1)$	$=$	0.942 ft^3
excess H_2O: $(0.93)(2) + (0.0182)(1)$		
$+ (0.0025)(2) + (0.0018)(1)$	$=$	1.885 ft^3
	total $=$	13.55 ft^3

The total wet volume is $13.55 \text{ ft}^3/\text{ft}^3$ fuel.

The total dry volume is $11.67 \text{ ft}^3/\text{ft}^3$ fuel.

The volumetric analyses are

	O_2	N_2	SO_2	CO_2	H_2O
wet:	$\dfrac{0.7500 \text{ ft}^3}{13.55 \text{ ft}^3}$	$\dfrac{9.974 \text{ ft}^3}{13.55 \text{ ft}^3}$	$\dfrac{0.0018 \text{ ft}^3}{13.55 \text{ ft}^3}$	$\dfrac{0.942 \text{ ft}^3}{13.55 \text{ ft}^3}$	$\dfrac{1.885 \text{ ft}^3}{13.55 \text{ ft}^3}$
	$= 0.0553$	0.736	–	0.069	0.139
dry:	$\dfrac{0.7500 \text{ ft}^3}{11.67 \text{ ft}^3}$	$\dfrac{9.974 \text{ ft}^3}{11.67 \text{ ft}^3}$	$\dfrac{0.0018 \text{ ft}^3}{11.67 \text{ ft}^3}$	$\dfrac{0.942 \text{ ft}^3}{11.67 \text{ ft}^3}$	
	$= 0.0643$	0.855	–	0.081	–

The percentage of CO_2 in the flue gas (wet basis) is

$$B_{CO_2,\text{wet}} = \boxed{6.9\%}$$

The answer is (A).

The percentage of CO_2 in the flue gas (dry basis) is

$$B_{CO_2,\text{dry}} = \boxed{8.1\%}$$

The answer is (C).

14. *Customary U.S. Solution*

step 1: From App. 29.B, the heat of vaporization at 212°F is $h_{fg} = 970.1$ Btu/lbm.

The heat absorbed in the boiler is

$$m_{H_2O}h_{fg} = (11.12 \text{ lbm } H_2O)\left(970.1 \ \frac{\text{Btu}}{\text{lbm}}\right)$$
$$= 10,787.5 \text{ Btu/lbm fuel}$$

step 2: The losses for heating stack gases can be found as follows.

The burned carbon per lbm of fuel is

$$0.7842 - (0.315)(0.0703) = 0.7621 \text{ lbm/lbm fuel}$$

The mass ratio of dry flue gases to solid fuel is given by Eq. 37.12.

$$\frac{\text{mass of flue gas}}{\text{mass of solid fuel}}$$

$$= \frac{\left(\begin{array}{c} 11B_{CO_2} + 8B_{O_2} \\ + (7)(B_{CO} + B_{N_2}) \end{array}\right)\left(G_C + \dfrac{G_S}{1.833}\right)}{(3)(B_{CO_2} + B_{CO})}$$

$$= \frac{\left(\begin{array}{c} (11)(14.0) + (8)(5.5) \\ + (7)(0.42 + 80.08) \end{array}\right)\left(0.7621 + \dfrac{0.01}{1.833}\right)}{(3)(14.0 + 0.42)}$$

$$= 13.51 \text{ lbm stack gases/lbm fuel}$$

Properties of nitrogen are commonly assumed for dry flue gas. From Table 37.1, for nitrogen at an average temperature of $(575°F + 73°F)/2 = 324°F$ $(784°R)$, $c_p \approx 0.252$ Btu/lbm-°F.

The losses for heating stack gases are given by Eq. 37.20.

$$q_1 = m_{\text{flue gas}} c_p (T_{\text{flue gas}} - T_{\text{incoming air}})$$

$$= (13.51 \text{ lbm})\left(0.252 \frac{\text{Btu}}{\text{lbm-°F}}\right)(575°F - 73°F)$$

$$= 1709.1 \text{ Btu/lbm fuel}$$

The heat loss in the vapor formed during the combustion of hydrogen is given by Eq. 37.21.

$$q_2 = 8.94 G_H (h_g - h_f)$$

Assume that the partial pressure of the water vapor is below 1 psia (the lowest App. 29.C goes).

h_g at 575°F and 1 psia can be found from the superheat tables, App. 29.C.

$$h_g \approx 1324.3 \text{ Btu/lbm}$$

h_f at 73°F can be found from App. 29.A.

$$h_f = 41.07 \text{ Btu/lbm}$$

$$G_{H,\text{available}} = G_{H,\text{total}} - \frac{G_O}{8} = 0.0556 - \frac{0.0825}{8}$$

$$= 0.0453$$

$$q_2 = (8.94)(0.0453)\left(1324.3 \frac{\text{Btu}}{\text{lbm}} - 41.07 \frac{\text{Btu}}{\text{lbm}}\right)$$

$$= 519.7 \text{ Btu/lbm fuel}$$

Heat is also lost when it is absorbed by the moisture that was originally in the combustion air.

$$q_3 = \omega m_{\text{combustion air}}(h_g - h'_g)$$

Assume that the partial pressure of the water vapor is below 1 psia (the lowest App. 29.C goes).

From App. 29.B, at 73°F and 0.4 psia, $h'_g \approx 1093$ Btu/lbm. From the psychrometric chart,

$$\omega = 90 \text{ grains/lbm air} = 0.0129 \text{ lbm water/lbm air}$$

Considering the sulfur content, find the air/fuel ratio.

$$\frac{\text{lbm air}}{\text{lbm fuel}} = \frac{3.04 B_{N_2}\left(G_C + \dfrac{G_S}{1.833}\right)}{B_{CO_2} + B_{CO}}$$

$$= \frac{(3.04)(0.8008)\left(0.7621 + \dfrac{0.01}{1.833}\right)}{0.14 + 0.0042}$$

$$= 12.96 \text{ lbm air/lbm fuel}$$

$$q_3 = \left(0.0129 \frac{\text{lbm water}}{\text{lbm air}}\right)\left(12.96 \frac{\text{lbm air}}{\text{lbm fuel}}\right)$$

$$\times \left(1324.3 \frac{\text{Btu}}{\text{lbm}} - 1093 \frac{\text{Btu}}{\text{lbm}}\right)$$

$$= 38.7 \text{ Btu/lbm fuel}$$

The energy lost in incomplete combustion is given by Eq. 37.23.

$$q_4 = \frac{(\text{HHV}_C - \text{HHV}_{CO}) G_C B_{CO}}{B_{CO_2} + B_{CO}}$$

$$= \frac{\left(9746 \frac{\text{Btu}}{\text{lbm}}\right)(0.7621)(0.0042)}{0.14 + 0.0042}$$

$$= 216.3 \text{ Btu/lbm fuel}$$

The energy lost in unburned carbon is given by Eq. 37.24.

$$q_5 = \left(14{,}093 \frac{\text{Btu}}{\text{lbm}}\right) m_{\text{ash}} G_{C,\text{ash}}$$

$$= \left(14{,}093 \frac{\text{Btu}}{\text{lbm}}\right)(0.0703)(0.315)$$

$$= 312.1 \text{ Btu/lbm fuel}$$

The energy lost in radiation and unaccounted losses per pound of fuel is

$$14{,}000 \frac{\text{Btu}}{\text{lbm}} - 10{,}787.5 \frac{\text{Btu}}{\text{lbm}} - 1709.1 \frac{\text{Btu}}{\text{lbm}} - 519.7 \frac{\text{Btu}}{\text{lbm}}$$

$$- 38.7 \frac{\text{Btu}}{\text{lbm}} - 216.3 \frac{\text{Btu}}{\text{lbm}} - 312.1 \frac{\text{Btu}}{\text{lbm}}$$

$$= \boxed{416.6 \text{ Btu/lbm} \quad (400 \text{ Btu/lbm})}$$

The answer is (A).

SI Solution

step 1: From App. 29.N, the heat of vaporization at 100°C is $h_{fg} = 2256.4$ kJ/kg.

The heat absorbed in the boiler is

$$m_{H_2O} h_{fg} = (11.12 \text{ kg}) \left(2256.4 \frac{\text{kJ}}{\text{kg}} \right) = 25\,091 \text{ kJ/kg fuel}$$

step 2: From step 2 of the U.S. solution,

$$\frac{\text{mass of flue gas}}{\text{mass of solid fuel}} = 13.51 \text{ kg stack gases/kg fuel}$$

Properties of nitrogen are commonly assumed for dry flue gas. From Table 37.1, for nitrogen at an average temperature of $(1/2)(300°C + 23°C) + 273° = 434.5K$ (782°R), $c_p \approx 0.252$ Btu/lbm-°R.

c_p from Table 37.1 can be found (using the table footnote) as

$$c_p = \left(0.252 \frac{\text{Btu}}{\text{lbm-°R}} \right) \left(4.187 \frac{\frac{\text{kJ}}{\text{kg·K}}}{\frac{\text{Btu}}{\text{lbm-°R}}} \right) = 1.055 \text{ kJ/kg·K}$$

The losses for heating stack gases are given by Eq. 37.20.

$$q_1 = m_{\text{flue gas}} c_p (T_{\text{flue gas}} - T_{\text{incoming air}})$$

$$= (13.51 \text{ kg}) \left(1.055 \frac{\text{kJ}}{\text{kg·K}} \right) (300°C - 23°C)$$

$$= 3948.1 \text{ kJ/kg fuel}$$

The heat loss in the vapor formed during the combustion of hydrogen is given by Eq. 37.21.

$$q_2 = 8.94 G_H (h_g - h_f)$$

Assume that the partial pressure of the water vapor is below 10 kPa (the lowest App. 29.P goes).

h_g at 300°C and 10 kPa can be found from the superheat tables, App. 29.P.

$$h_g = 3076.7 \text{ kJ/kg}$$

h_f at 23°C can be found from App. 29.N.

$$h_f = 96.47 \text{ kJ/kg}$$

$$G_{H,\text{available}} = G_{H,\text{total}} - \frac{G_O}{8} = 0.0556 - \frac{0.0825}{8} = 0.0453$$

$$q_2 = (8.94)(0.0453) \left(3076.7 \frac{\text{kJ}}{\text{kg}} - 96.47 \frac{\text{kJ}}{\text{kg}} \right)$$

$$= 1206.9 \text{ kJ/kg fuel}$$

Heat is also lost when it is absorbed by the moisture originally in the combustion air.

$$q_3 = \omega m_{\text{combustion air}} (h_g - h'_g)$$

Assume that the partial pressure of the water vapor is low. At 23°C, from App. 29.N, $h'_g \approx 2542.9$ kJ/kg.

From the psychrometric chart for 19°C wet bulb and 23°C dry bulb, $\omega = 12.2$ g/kg dry air.

The air/fuel ratio from the customary U.S. solution is

$$\frac{\text{kg air}}{\text{kg fuel}} = 12.96 \text{ kg air/kg fuel}$$

$$q_3 = \frac{\left(12.2 \frac{\text{g}}{\text{kg}} \right) \left(12.96 \frac{\text{kg}}{\text{kg}} \right) \times \left(3076.7 \frac{\text{kJ}}{\text{kg}} - 2542.9 \frac{\text{kJ}}{\text{kg}} \right)}{1000 \frac{\text{g}}{\text{kg}}}$$

$$= 84.40 \text{ kJ/kg}$$

Energy lost in incomplete combustion is given by Eq. 37.23.

$$q_4 = \frac{(HHV_C - HHV_{CO}) G_C B_{CO}}{B_{CO_2} + B_{CO}}$$

$$= \left(\frac{\left(22.67 \frac{\text{MJ}}{\text{kg}} \right)(0.7621)(0.42)}{14 + 0.42} \right) \left(1000 \frac{\text{kJ}}{\text{MJ}} \right)$$

$$= \left(0.5032 \frac{\text{MJ}}{\text{kg}} \right) \left(1000 \frac{\text{kJ}}{\text{MJ}} \right)$$

$$= 503.2 \text{ kJ/kg fuel}$$

The energy lost in unburned carbon is given by Eq. 37.24.

$$q_5 = \left(32.8 \frac{\text{MJ}}{\text{kg}} \right) m_{\text{ash}} G_{C,\text{ash}}$$

$$= \left(32.8 \frac{\text{MJ}}{\text{kg}} \right) \left(1000 \frac{\text{kJ}}{\text{MJ}} \right)(0.0703)(0.315)$$

$$= 726.3 \text{ kJ/kg fuel}$$

The energy lost in radiation and unaccounted losses per kilogram of fuel is

$$\left(32.6 \frac{\text{MJ}}{\text{kg}} \right) \left(1000 \frac{\text{kJ}}{\text{MJ}} \right) - 25\,091 \frac{\text{kJ}}{\text{kg}}$$

$$- 3948.1 \frac{\text{kJ}}{\text{kg}} - 1206.9 \frac{\text{kJ}}{\text{kg}} - 84.40 \frac{\text{kJ}}{\text{kg}}$$

$$- 503.2 \frac{\text{kJ}}{\text{kg}} - 726.3 \frac{\text{kJ}}{\text{kg}}$$

$$= \boxed{1040.1 \text{ kJ/kg} \quad (1040 \text{ kJ/kg})}$$

The answer is (A).

15. *Customary U.S. Solution*

step 1: The incoming reactants on a per-pound basis are

$$0.07 \text{ lbm ash}$$

$$0.05 \text{ lbm hydrogen}$$

$$0.05 \text{ lbm oxygen}$$

$$0.83 \text{ lbm carbon}$$

This is an ultimate analysis. Assume that only the hydrogen that is not locked up with oxygen in the form of water is combustible. From Eq. 37.15, the available hydrogen is

$$G_{\text{H,available}} = G_{\text{H,total}} - \frac{G_O}{8} = 0.05 \text{ lbm} - \frac{0.05 \text{ lbm}}{8}$$

$$= 0.04375 \text{ lbm}$$

The mass of water produced is the hydrogen mass plus eight times as much oxygen. The locked hydrogen is

$$0.05 \text{ lbm} - 0.04375 \text{ lbm} = 0.00625 \text{ lbm}$$

$$\text{lbm of moisture} = G_H + G_O = G_H + 8G_H$$

$$= 0.00625 \text{ lbm} + (8)(0.00625 \text{ lbm})$$

$$= 0.05625 \text{ lbm}$$

The air is 23.15% oxygen by weight (see Table 37.6), so other reactants for 26 lbm of air are

$$(0.2315)(26 \text{ lbm}) = 6.019 \text{ lbm O}_2$$

$$(0.7685)(26 \text{ lbm}) = 19.981 \text{ lbm N}_2$$

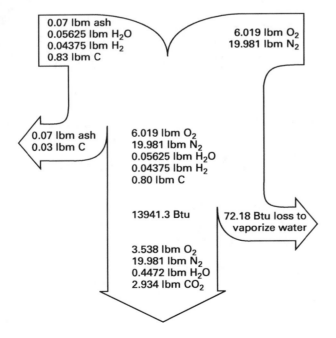

step 2: Ash pit material losses are 10%, or 0.1 lbm, which includes all of the ash.

$$0.07 \text{ lbm ash (noncombustible matter)}$$

$$0.03 \text{ lbm unburned carbon}$$

step 3: Determine what remains.

$$6.019 \text{ lbm oxygen}$$

$$19.981 \text{ lbm nitrogen}$$

$$0.05625 \text{ lbm water}$$

$$0.04375 \text{ lbm hydrogen}$$

$$0.80 \text{ lbm carbon}$$

step 4: Determine the energy loss in vaporizing the moisture.

$$q = (\text{moisture})(h_g - h_f)$$

From App. 29.C, h_g at 550°F and 14.7 psia is 1311.4 Btu/lbm.

From App. 29.A, h_f at 60°F is 28.08 Btu/lbm.

$$q = (0.05625 \text{ lbm})\left(1311.4 \frac{\text{Btu}}{\text{lbm}} - 28.08 \frac{\text{Btu}}{\text{lbm}}\right)$$

$$= 72.19 \text{ Btu}$$

step 5: Calculate the heating value of the remaining fuel components using App. 37.A.

$$\text{HV}_C = (0.80 \text{ lbm})\left(14{,}093 \frac{\text{Btu}}{\text{lbm}}\right) = 11{,}274.4 \text{ Btu}$$

$$\text{HV}_H = (0.04375 \text{ lbm})\left(60{,}958 \frac{\text{Btu}}{\text{lbm}}\right) = 2666.9 \text{ Btu}$$

The heating value after the coal moisture is evaporated (i.e., the as-delivered heating value) is

$$11{,}274.4 \text{ Btu} + 2666.9 \text{ Btu} - 72.19 \text{ Btu}$$

$$= 13{,}869 \text{ Btu}$$

step 6: Using Table 37.8, determine the combustion products.

$$\text{oxygen required by carbon} = (0.80)(2.667 \text{ lbm})$$

$$= 2.134 \text{ lbm}$$

$$\text{oxygen required by hydrogen} = (0.04375)(7.936 \text{ lbm})$$

$$= 0.3472 \text{ lbm}$$

$$\begin{aligned} \text{carbon dioxide} \\ \text{produced by carbon} &= (0.8)(3.667 \text{ lbm}) \end{aligned}$$

$$= 2.934 \text{ lbm}$$

$$\text{water produced by hydrogen} = (0.04375)(8.936 \text{ lbm})$$

$$= 0.3910 \text{ lbm}$$

The remaining oxygen is

$$6.019 \text{ lbm} - 2.134 \text{ lbm} - 0.3472 \text{ lbm} = 3.538 \text{ lbm}$$

step 7: The gaseous products must be heated from 70°F to 550°F. The average temperature is

$$\left(\tfrac{1}{2}\right)(70°F + 550°F) + 460° = 770°R$$

From Table 37.1, the specific heat of gaseous products is

gas	$\dfrac{\text{Btu}}{\text{lbm-°R}}$
oxygen	0.228
nitrogen	0.252
water	0.460
carbon dioxide	0.225

$$Q_{\text{heating}} = \begin{pmatrix} (3.538 \text{ lbm})\left(0.228 \dfrac{\text{Btu}}{\text{lbm-°R}}\right) \\ + (19.981 \text{ lbm})\left(0.252 \dfrac{\text{Btu}}{\text{lbm-°R}}\right) \\ + (0.3910 \text{ lbm})\left(0.460 \dfrac{\text{Btu}}{\text{lbm-°R}}\right) \\ + (2.934 \text{ lbm})\left(0.225 \dfrac{\text{Btu}}{\text{lbm-°R}}\right) \end{pmatrix} \\ \times (550°F - 70°F) \\ = 3207.3 \text{ Btu}$$

step 8: The percentage loss based on the coal's as-delivered heating value is

$$\frac{3207.3 \text{ Btu} + 72.19 \text{ Btu}}{13{,}869 \text{ Btu}} = \boxed{0.236 \quad (24\%)}$$

The answer is (B).

SI Solution

Steps 1 through 3 are the same as for the customary U.S. solution except that everything is based on kg.

step 4: Determine the energy loss in the vaporizing moisture.

$$q = (\text{moisture})(h_g - h_f)$$

From App. 29.P, h_g at 290°C and 101.3 kPa is 3054.5 kJ/kg.

h_f at 15.6°C from App. 29.N is 65.49 kJ/kg.

$$q = (0.05625 \text{ kg})\left(3054.5 \frac{\text{kJ}}{\text{kg}} - 65.49 \frac{\text{kJ}}{\text{kg}}\right) \\ = 168.1 \text{ kJ}$$

step 5: Calculate the heating value of the remaining fuel components using App. 37.A and the table footnote.

$$HV_C = (0.80 \text{ kg})\left(14\,093 \frac{\text{Btu}}{\text{lbm}}\right)\left(2.326 \frac{\frac{\text{kJ}}{\text{kg}}}{\frac{\text{Btu}}{\text{lbm}}}\right) \\ = 26\,224 \text{ kJ}$$

$$HV_H = (0.04375 \text{ kg})\left(60\,958 \frac{\text{Btu}}{\text{lbm}}\right)\left(2.326 \frac{\frac{\text{kJ}}{\text{kg}}}{\frac{\text{Btu}}{\text{lbm}}}\right) \\ = 6203 \text{ kJ}$$

The heating value after the coal moisture is evaporated is

$$26\,224 \text{ kJ} + 6203 \text{ kJ} - 168.1 \text{ kJ} = 32\,259 \text{ kJ}$$

step 6: This step is the same as for the customary U.S. solution except that all quantities are in kg.

step 7: The gaseous products must be heated from 21°C to 290°C. The average temperature is

$$\left(\tfrac{1}{2}\right)(21°C + 290°C) + 273° = 428.5\text{K}$$
$$(428.5\text{K})\left(1.8 \frac{°R}{K}\right) = 771°R$$

From Table 37.1, the specific heat of gaseous products is calculated using the table footnote.

gas	$\dfrac{\text{kJ}}{\text{kg·K}}$
oxygen	0.955
nitrogen	1.055
water	1.926
carbon dioxide	0.942

$$Q_{\text{heating}} = \begin{pmatrix} (3.538 \text{ kg})\left(0.955 \dfrac{\text{kJ}}{\text{kg·K}}\right) \\ + (19.981 \text{ kg})\left(1.055 \dfrac{\text{kJ}}{\text{kg·K}}\right) \\ + (0.3910 \text{ kg})\left(1.926 \dfrac{\text{kJ}}{\text{kg·K}}\right) \\ + (2.934 \text{ kg})\left(0.942 \dfrac{\text{kJ}}{\text{kg·K}}\right) \end{pmatrix} \\ \times (290°C - 21°C) \\ = 7525.4 \text{ kJ}$$

step 8: The percentage loss is

$$\frac{7525.4 \text{ kJ} + 168.1 \text{ kJ}}{32\,259 \text{ kJ}} = \boxed{0.238 \quad (24\%)}$$

The answer is (B).

16. (a) Silicon in ash is SiO_2 with a molecular weight of

$$28.09 \ \frac{\text{lbm}}{\text{lbmol}} + (2)\left(16 \ \frac{\text{lbm}}{\text{lbmol}}\right) = 60.09 \ \text{lbm/lbmol}$$

The oxygen used with 6.1% by mass silicon is

$$\left(\frac{(2)(16 \ \text{lbm})}{28.09 \ \text{lbm}}\right)\left(0.061 \ \frac{\text{lbm}}{\text{lbm coal}}\right)$$
$$= 0.0695 \ \text{lbm/lbm coal}$$

Silicon ash produced per lbm of coal is

$$0.061 \ \frac{\text{lbm}}{\text{lbm coal}} + 0.0695 \ \frac{\text{lbm}}{\text{lbm coal}}$$
$$= 0.1305 \ \text{lbm/lbm coal}$$

Silicon ash produced per hour is

$$\left(0.1305 \ \frac{\text{lbm}}{\text{lbm coal}}\right)\left(15,300 \ \frac{\text{lbm coal}}{\text{hr}}\right)$$
$$= 1996.7 \ \text{lbm/hr}$$

The silicon in 410 lbm/hr refuse is

$$\left(410 \ \frac{\text{lbm}}{\text{hr}}\right)(1 - 0.3) = 287 \ \text{lbm/hr}$$

The emission rate is

$$1996.7 \ \frac{\text{lbm}}{\text{hr}} - 287 \ \frac{\text{lbm}}{\text{hr}}$$
$$= \boxed{1709.7 \ \text{lbm/hr} \quad (1700 \ \text{lbm/hr})}$$

The answer is (D).

(b) From Table 37.7, the stoichiometric reaction for sulfur is

$$S + O_2 \rightarrow SO_2$$
$$\text{MW} \quad 32 \quad 32 \quad 64$$

Sulfur dioxide produced for 15,300 lbm/hr of coal feed is

$$\left(15,300 \ \frac{\text{lbm fuel}}{\text{hr}}\right)\left(\frac{0.0244 \ \text{lbm S}}{\text{lbm fuel}}\right)\left(\frac{64 \ \frac{\text{lbm } SO_2}{\text{lbmol}}}{32 \ \frac{\text{lbm S}}{\text{lbmol}}}\right)$$
$$= \boxed{746.6 \ \text{lbm } SO_2/\text{hr} \quad (750 \ \text{lbm/hr})}$$

The answer is (C).

(c) From Eq. 37.16(b), the heating value of the fuel is

$$\text{HHV} = 14{,}093G_C + (60{,}958)\left(G_H - \frac{G_O}{8}\right) + 3983G_S$$
$$= \left(14{,}093 \ \frac{\text{Btu}}{\text{lbm}}\right)(0.7656)$$
$$+ \left(60{,}958 \ \frac{\text{Btu}}{\text{lbm}}\right)\left(0.055 - \frac{0.077}{8}\right)$$
$$+ \left(3983 \ \frac{\text{Btu}}{\text{lbm}}\right)(0.0244)$$
$$= 13{,}653 \ \text{Btu/lbm}$$

The gross available combustion power is

$$\dot{m}_f(\text{HV}) = \left(15{,}300 \ \frac{\text{lbm}}{\text{hr}}\right)\left(13{,}653 \ \frac{\text{Btu}}{\text{lbm}}\right)$$
$$= 2.089 \times 10^8 \ \text{Btu/hr}$$

The carbon in 410 lbm/hr refuse is

$$\left(410 \ \frac{\text{lbm}}{\text{hr}}\right)(0.3) = 123 \ \text{lbm/hr}$$

Power lost in unburned carbon in refuse is $\dot{m}_c(\text{HV})$.

From App. 37.A, the gross heat of combustion for carbon is 14,093 Btu/lbm.

$$\left(123 \ \frac{\text{lbm}}{\text{hr}}\right)\left(14{,}093 \ \frac{\text{Btu}}{\text{lbm}}\right) = 1.733 \times 10^6 \ \text{Btu/hr}$$

The remaining combustion power is

$$2.089 \times 10^8 \ \frac{\text{Btu}}{\text{hr}} - 1.733 \times 10^6 \ \frac{\text{Btu}}{\text{hr}}$$
$$= 2.072 \times 10^8 \ \text{Btu/hr}$$

Losses in the steam generator and electrical generator will further reduce this to

$$(0.86)\left(2.072 \times 10^8 \ \frac{\text{Btu}}{\text{hr}}\right) = 1.782 \times 10^8 \ \text{Btu/hr}$$

With an electrical output of 17 MW, thermal energy removed by cooling water is

$$Q = 1.782 \times 10^8 \ \frac{\text{Btu}}{\text{hr}}$$
$$- (17 \ \text{MW})\left(1000 \ \frac{\text{kW}}{\text{MW}}\right)\left(3413 \ \frac{\frac{\text{Btu}}{\text{hr}}}{\text{kW}}\right)$$
$$= 1.202 \times 10^8 \ \text{Btu/hr}$$

At 60°F, the specific heat of water is $c_p = 1$ Btu/lbm-°F. The temperature rise of the cooling water is

$$\Delta T = \frac{Q}{\dot{m}c_p}$$

$$= \frac{1.202 \times 10^8 \ \frac{\text{Btu}}{\text{hr}}}{\left(225 \ \frac{\text{ft}^3}{\text{sec}}\right)\left(62.4 \ \frac{\text{lbm}}{\text{ft}^3}\right)\left(3600 \ \frac{\text{sec}}{\text{hr}}\right)\left(1 \ \frac{\text{Btu}}{\text{lbm-°F}}\right)}$$

$$= \boxed{2.38°\text{F} \quad (2.4°\text{F})}$$

The electrical generation is not cooled by the cooling water. Therefore, it is not correct to include the generation efficiency in the calculation of losses.

The answer is (A).

(d) Limiting 0.1 lbm of particulates per million Btu per hour, the allowable emission rate is

$$\frac{\left(0.1 \ \frac{\text{lbm}}{\text{MBtu}}\right)\left(15,300 \ \frac{\text{lbm}}{\text{hr}}\right)\left(13,653 \ \frac{\text{Btu}}{\text{lbm}}\right)}{10^6 \ \frac{\text{MBtu}}{\text{Btu}}} = 20.89 \ \text{lbm/hr}$$

The efficiency of the flue gas particulate collectors is

$$\eta = \frac{\text{actual emission rate} - \text{allowable emission rate}}{\text{actual emission rate}}$$

$$= \frac{1709.7 \ \frac{\text{lbm}}{\text{hr}} - 20.89 \ \frac{\text{lbm}}{\text{hr}}}{1709.7 \ \frac{\text{lbm}}{\text{hr}}}$$

$$= \boxed{0.988 \quad (98.8\%)}$$

The answer is (C).

17. *Customary U.S. Solution*

(a) The stoichiometric reaction for propane is given in Table 37.7.

$$\text{C}_3\text{H}_8 \ + \quad 5\text{O}_2 \quad \rightarrow \quad 3\text{CO}_2 \quad + \quad 4\text{H}_2\text{O}$$

MW 44.097   (5)(32.000)   (3)(44.011)   (4)(18.016)

With 40% excess O_2 by weight,

$$\text{C}_3\text{H}_8 \qquad (1.4)(5)\text{O}_2$$

MW 44.097 + (7)(32.000)

44.097   224

$$3\text{CO}_2 \qquad\qquad 4\text{H}_2\text{O} \qquad\qquad 2\text{O}_2$$

$$\rightarrow (3)(44.011) + (4)(18.016) + (2)(32.000)$$

132.033   72.064   64

The excess oxygen is

$$(2)(32) = 64 \ \text{lbm/lbmol C}_3\text{H}_8$$

The mass ratio of nitrogen to oxygen is

$$\frac{G_\text{N}}{G_\text{O}} = \left(\frac{B_\text{N}}{R_\text{N}}\right)\left(\frac{R_{\text{O}_2}}{B_{\text{O}_2}}\right)$$

$$= \left(\frac{0.40}{55.16 \ \frac{\text{ft-lbf}}{\text{lbm-°R}}}\right)\left(\frac{48.29 \ \frac{\text{ft-lbf}}{\text{lbm-°R}}}{0.60}\right)$$

$$= 0.584$$

(Values of R_N and R_{O_2} are taken from Table 29.7.)

The nitrogen accompanying the oxygen is

$$(7)(32)(0.584) = 130.8 \ \text{lbm}$$

The mass balance per mole of propane is

mass per mole

$$\text{C}_3\text{H}_8 + \ \text{O}_2 + \ \text{N}_2 \ \rightarrow \ \text{CO}_2 \ + \ \text{H}_2\text{O} \ + \text{O}_2 + \text{N}_2$$

$$44.097 + 224 + 130.8 \rightarrow 132.033 + 72.064 + 64 \ + 130.8$$

Standard conditions are 60°F and 1 atm. The propane density is given by Eq. 29.63.

$$\rho = \frac{p}{RT}$$

The absolute temperature, T, is

$$60°\text{F} + 460° = 520°\text{R}$$

R for propane from Table 29.7 is 35.04 ft-lbf/lbm-°R.

$$\rho = \frac{\left(14.7 \ \frac{\text{lbf}}{\text{in}^2}\right)\left(12 \ \frac{\text{in}}{\text{ft}}\right)^2}{\left(35.04 \ \frac{\text{ft-lbf}}{\text{lbm-°R}}\right)(520°\text{R})} = 0.1162 \ \text{lbm/ft}^3$$

The mass flow rate of propane is

$$\left(250 \ \frac{\text{ft}^3}{\text{min}}\right)\left(0.1162 \ \frac{\text{lbm}}{\text{ft}^3}\right) = 29.05 \ \text{lbm/min}$$

Scaling the other mass balance factors down by

$$\frac{29.05 \ \frac{\text{lbm}}{\text{min}}}{44.097 \ \frac{\text{lbm}}{\text{lbmol}}} = 0.6588 \ \text{lbmol/min}$$

$\frac{\text{lbm}}{\text{min}}$

$$\text{C}_3\text{H}_8 + \ \text{O}_2 \ + \ \text{N}_2 \ \rightarrow \ \text{CO}_2 + \text{H}_2\text{O} + \ \text{O}_2 \ + \ \text{N}_2$$

$$29.05 + 147.57 + 86.17 \rightarrow 86.98 + 47.48 + 42.16 + 86.17$$

The oxygen flow rate is $\boxed{147.57 \ \text{lbm/min} \ (150 \ \text{lbm/min}).}$

The answer is (A).

(b) Using R from Table 29.7, the specific volumes of the reactants are given by Eq. 29.63.

$$v_{C_3H_8} = \frac{RT}{p} = \frac{\left(35.04 \ \frac{\text{ft-lbf}}{\text{lbm-}°\text{R}}\right)(80°\text{F} + 460°)}{\left(14.7 \ \frac{\text{lbf}}{\text{in}^2}\right)\left(12 \ \frac{\text{in}}{\text{ft}}\right)^2}$$

$$= 8.939 \ \text{ft}^3/\text{lbm}$$

$$v_{O_2} = \frac{\left(48.29 \ \frac{\text{ft-lbf}}{\text{lbm-}°\text{R}}\right)(80°\text{F} + 460°)}{\left(14.7 \ \frac{\text{lbf}}{\text{in}^2}\right)\left(12 \ \frac{\text{in}}{\text{ft}}\right)^2}$$

$$= 12.319 \ \text{ft}^3/\text{lbm}$$

$$v_{N_2} = \frac{\left(55.16 \ \frac{\text{ft-lbf}}{\text{lbm-}°\text{R}}\right)(80°\text{F} + 460°)}{\left(14.7 \ \frac{\text{lbf}}{\text{in}^2}\right)\left(12 \ \frac{\text{in}}{\text{ft}}\right)^2}$$

$$= 14.071 \ \text{ft}^3/\text{lbm}$$

The total incoming volume is

$$\dot{V} = \left(29.05 \ \frac{\text{lbm}}{\text{min}}\right)\left(8.939 \ \frac{\text{ft}^3}{\text{lbm}}\right)$$

$$+ \left(147.57 \ \frac{\text{lbm}}{\text{min}}\right)\left(12.319 \ \frac{\text{ft}^3}{\text{lbm}}\right)$$

$$+ \left(86.17 \ \frac{\text{lbm}}{\text{min}}\right)\left(14.071 \ \frac{\text{ft}^3}{\text{lbm}}\right)$$

$$= 3290 \ \text{ft}^3/\text{min}$$

Since the velocity must be kept below 400 ft/min, the minimum area of inlet pipe is

$$A_{\text{inlet}} = \frac{\dot{V}}{v} = \frac{3290 \ \frac{\text{ft}^3}{\text{min}}}{400 \ \frac{\text{ft}}{\text{min}}} = \boxed{8.23 \ \text{ft}^2 \quad (8 \ \text{ft}^2)}$$

The answer is (A).

(c) Similarly, the specific volumes of the products are

$$v = \frac{RT}{p}$$

$$v_{CO_2} = \frac{\left(35.11 \ \frac{\text{ft-lbf}}{\text{lbm-}°\text{R}}\right)(460°\text{F} + 460°)}{\left(8 \ \frac{\text{lbf}}{\text{in}^2}\right)\left(12 \ \frac{\text{in}}{\text{ft}}\right)^2}$$

$$= 28.04 \ \text{ft}^3/\text{lbm}$$

$$v_{H_2O} = \frac{\left(85.78 \ \frac{\text{ft-lbf}}{\text{lbm-}°\text{R}}\right)(460°\text{F} + 460°)}{\left(8 \ \frac{\text{lbf}}{\text{in}^2}\right)\left(12 \ \frac{\text{in}}{\text{ft}}\right)^2}$$

$$= 68.50 \ \text{ft}^3/\text{lbm}$$

$$v_{O_2} = \frac{\left(48.29 \ \frac{\text{ft-lbf}}{\text{lbm-}°\text{R}}\right)(460°\text{F} + 460°)}{\left(8 \ \frac{\text{lbf}}{\text{in}^2}\right)\left(12 \ \frac{\text{in}}{\text{ft}}\right)^2}$$

$$= 38.56 \ \text{ft}^3/\text{lbm}$$

$$v_{N_2} = \frac{\left(55.16 \ \frac{\text{ft-lbf}}{\text{lbm-}°\text{R}}\right)(460°\text{F} + 460°)}{\left(8 \ \frac{\text{lbf}}{\text{in}^2}\right)\left(12 \ \frac{\text{in}}{\text{ft}}\right)^2}$$

$$= 44.05 \ \text{ft}^3/\text{lbm}$$

The total exhaust volume is

$$\dot{V} = \left(86.98 \ \frac{\text{lbm}}{\text{min}}\right)\left(28.04 \ \frac{\text{ft}^3}{\text{lbm}}\right)$$

$$+ \left(47.48 \ \frac{\text{lbm}}{\text{min}}\right)\left(68.50 \ \frac{\text{ft}^3}{\text{lbm}}\right)$$

$$+ \left(42.16 \ \frac{\text{lbm}}{\text{min}}\right)\left(38.56 \ \frac{\text{ft}^3}{\text{lbm}}\right)$$

$$+ \left(86.17 \ \frac{\text{lbm}}{\text{min}}\right)\left(44.05 \ \frac{\text{ft}^3}{\text{lbm}}\right)$$

$$= \boxed{11{,}112.8 \ \text{ft}^3/\text{min} \quad (11{,}000 \ \text{ft}^3/\text{min})}$$

The answer is (D).

(d) Since the velocity of the products must be kept below 800 ft/min, the minimum area of the stack is

$$A_{\text{stack}} = \frac{Q}{v} = \frac{11{,}112.8 \ \frac{\text{ft}^3}{\text{min}}}{800 \ \frac{\text{ft}}{\text{min}}} = \boxed{13.89 \ \text{ft}^2 \quad (14 \ \text{ft}^2)}$$

The answer is (C).

(e) For ideal gases, the partial pressure is volumetrically weighted. The water vapor partial pressure in the stack is

$$(8 \ \text{psia})\left(\frac{\left(47.48 \ \frac{\text{lbm}}{\text{min}}\right)\left(68.50 \ \frac{\text{ft}^3}{\text{lbm}}\right)}{11{,}112.8 \ \frac{\text{ft}^3}{\text{min}}}\right) = 2.34 \ \text{psia}$$

The saturation temperature corresponding to 2.34 psia is $T_{\text{dp}} = \boxed{131°\text{F} \ (130°\text{F}).}$

The answer is (C).

SI Solution

(a) Following the procedure for the customary U.S. solution, the mass balance per mole of propane is

$$\text{C}_3\text{H}_8 + \text{O}_2 + \text{N}_2 \rightarrow \text{CO}_2 + \text{H}_2\text{O} + \text{O}_2 + \text{N}_2$$

kg
per   44.097 + 224 + 130.8 → 132.033 + 72.064 + 64 + 130.8
mole

Standard conditions are 16°C and 101.3 kPa. The propane density is given by Eq. 29.63.

$$\rho = \frac{p}{RT}$$

The absolute temperature is

$$T = 16°C + 273° = 289K$$

From Table 29.7, R for propane is 188.55 J/kg·K.

$$\rho = \frac{(101.3 \text{ kPa})\left(1000 \frac{\text{Pa}}{\text{kPa}}\right)}{\left(188.55 \frac{\text{J}}{\text{kg·K}}\right)(289K)} = 1.86 \text{ kg/m}^3$$

The mass flow rate of propane is

$$\frac{\left(118 \frac{\text{L}}{\text{s}}\right)\left(1.86 \frac{\text{kg}}{\text{m}^3}\right)}{1000 \frac{\text{L}}{\text{m}^3}} = 0.2195 \text{ kg/s}$$

Scale the other mass balance factors down.

$$\frac{0.2195 \frac{\text{kg}}{\text{s}}}{44.097 \frac{\text{kg}}{\text{kmol}}} = 0.004978 \text{ kmol/s}$$

$$
\begin{array}{ccccccccc}
\text{C}_3\text{H}_8 & + & \text{O}_2 & + & \text{N}_2 & \rightarrow & \text{CO}_2 & + & \text{H}_2\text{O} & + & \text{O}_2 & + & \text{N}_2 \\
\end{array}
$$
$$\frac{\text{kg}}{\text{s}} \quad 0.2195 + 1.115 + 0.6511 \rightarrow 0.6572 + 0.3587 + 0.3186 + 0.6511$$

The oxygen flow rate is $\boxed{1.115 \text{ kg/s } (1.1 \text{ kg/s})}$.

The answer is (A).

(b) Using R from Table 29.7, the specific volumes of the reactants are given by Eq. 29.63.

$$v_{C_3H_8} = \frac{RT}{p} = \frac{\left(188.55 \frac{\text{J}}{\text{kg·K}}\right)(27°C + 273°)}{(101 \text{ kPa})\left(1000 \frac{\text{Pa}}{\text{kPa}}\right)}$$

$$= 0.5600 \text{ m}^3/\text{kg}$$

$$v_{O_2} = \frac{\left(259.82 \frac{\text{J}}{\text{kg·K}}\right)(27°C + 273°)}{(101 \text{ kPa})\left(1000 \frac{\text{Pa}}{\text{kPa}}\right)}$$

$$= 0.7717 \text{ m}^3/\text{kg}$$

$$v_{N_2} = \frac{\left(296.77 \frac{\text{J}}{\text{kg·K}}\right)(27°C + 273°)}{(101 \text{ kPa})\left(1000 \frac{\text{Pa}}{\text{kPa}}\right)}$$

$$= 0.8815 \text{ m}^3/\text{kg}$$

The total incoming volume, $\dot{V}$, is

$$\dot{V} = \left(0.2195 \frac{\text{kg}}{\text{s}}\right)\left(0.5600 \frac{\text{m}^3}{\text{kg}}\right)$$

$$+ \left(1.115 \frac{\text{kg}}{\text{s}}\right)\left(0.7717 \frac{\text{m}^3}{\text{kg}}\right)$$

$$+ \left(0.6511 \frac{\text{kg}}{\text{s}}\right)\left(0.8815 \frac{\text{m}^3}{\text{kg}}\right)$$

$$= 1.557 \text{ m}^3/\text{s}$$

Since the velocity for the reactants must be kept below 2 m/s, the minimum area of inlet pipe is

$$A_{\text{inlet}} = \frac{\dot{V}}{\text{v}} = \frac{1.557 \frac{\text{m}^3}{\text{s}}}{2 \frac{\text{m}}{\text{s}}} = \boxed{0.779 \text{ m}^2 \quad (0.8 \text{ m}^2)}$$

The answer is (A).

(c) Similarly, the specific volumes of the products are

$$v = \frac{RT}{p}$$

$$v_{CO_2} = \frac{\left(188.92 \frac{\text{J}}{\text{kg·K}}\right)(240°C + 273°)}{(55 \text{ kPa})\left(1000 \frac{\text{Pa}}{\text{kPa}}\right)} = 1.762 \text{ m}^3/\text{kg}$$

$$v_{H_2O} = \frac{\left(461.5 \frac{\text{J}}{\text{kg·K}}\right)(240°C + 273°)}{(55 \text{ kPa})\left(1000 \frac{\text{Pa}}{\text{kPa}}\right)} = 4.305 \text{ m}^3/\text{kg}$$

$$v_{O_2} = \frac{\left(259.82 \frac{\text{J}}{\text{kg·K}}\right)(240°C + 273°)}{(55 \text{ kPa})\left(1000 \frac{\text{Pa}}{\text{kPa}}\right)} = 2.423 \text{ m}^3/\text{kg}$$

$$v_{N_2} = \frac{\left(296.77 \frac{\text{J}}{\text{kg·K}}\right)(240°C + 273°)}{(55 \text{ kPa})\left(1000 \frac{\text{Pa}}{\text{kPa}}\right)} = 2.768 \text{ m}^3/\text{kg}$$

The total exhaust volume is

$$\dot{V} = \left(0.6572 \frac{\text{kg}}{\text{s}}\right)\left(1.762 \frac{\text{m}^3}{\text{kg}}\right)$$

$$+ \left(0.3587 \frac{\text{kg}}{\text{s}}\right)\left(4.305 \frac{\text{m}^3}{\text{kg}}\right)$$

$$+ \left(0.3186 \frac{\text{kg}}{\text{s}}\right)\left(2.423 \frac{\text{m}^3}{\text{kg}}\right)$$

$$+ \left(0.6511 \frac{\text{kg}}{\text{s}}\right)\left(2.768 \frac{\text{m}^3}{\text{kg}}\right)$$

$$= \boxed{5.276 \text{ m}^3/\text{s} \quad (5.3 \text{ m}^3/\text{s})}$$

The answer is (D).

Combustion

(d) Since the velocity of products must be kept below 4 m/s, the minimum area of the stack is

$$A_{\text{stack}} = \frac{\dot{V}}{\text{v}} = \frac{5.276 \ \frac{m^3}{s}}{4 \ \frac{m}{s}}$$

$$= \boxed{1.319 \ m^2 \quad (1.3 \ m^2)}$$

The answer is (C).

(e) mFor ideal gases, the partial pressure is volumetrically weighted. The water vapor partial pressure in the stack is

$$(55 \ \text{kPa}) \left(\frac{\left(0.3587 \ \frac{kg}{s} \right) \left(4.305 \ \frac{m^3}{kg} \right)}{5.276 \ \frac{m^3}{s}} \right) = 16.098 \ \text{kPa}$$

The saturation temperature corresponding to 16.098 kPa is found from App. 29.O to be $T_{\text{dp}} = \boxed{54.5°C \ (55°C).}$

The answer is (C).

18. *Customary U.S. Solution*

(a) Since atmospheric air is not used, the nitrogen and oxygen can be varied independently. Furthermore, since enthalpy increase information is not given for oxygen, a 0% excess of oxygen can be assumed.

From Table 37.7,

$$C_3H_8 \ + \ 5O_2 \ \rightarrow \ 3CO_2 \ + \ 4H_2O$$
$$\text{moles} \quad (1) \qquad (5) \qquad (3) \qquad (4)$$

Subtract the reactant enthalpies from the product enthalpies to calculate the heat of reaction. The enthalpy of formation of oxygen is zero since it is an element in its natural state. The heat of reaction is

$$n_{CO_2}(\Delta H_f)_{CO_2} + n_{H_2O}(\Delta H_f)_{H_2O}$$
$$\quad - n_{C_3H_8}(\Delta H_f)_{C_3H_8} - n_{O_2}(\Delta H_f)_{O_2}$$
$$= \left(3 \ \frac{\text{lbmol}}{\text{lbmol}} \right) \left(-169{,}300 \ \frac{\text{Btu}}{\text{lbmol}} \right)$$
$$\quad + \left(4 \ \frac{\text{lbmol}}{\text{lbmol}} \right) \left(-104{,}040 \ \frac{\text{Btu}}{\text{lbmol}} \right)$$
$$\quad - \left(1 \ \frac{\text{lbmol}}{\text{lbmol}} \right) \left(28{,}800 \ \frac{\text{Btu}}{\text{lbmol}} \right)$$
$$\quad - \left(5 \ \frac{\text{lbmol}}{\text{lbmol}} \right) \left(0 \ \frac{\text{Btu}}{\text{lbmol}} \right)$$
$$= -952{,}860 \ \text{Btu/lbmol of propane}$$

The negative sign indicates an exothermal reaction.

Let x be the number of moles of nitrogen per mole of propane. Use the nitrogen to cool the combustion. The heat of reaction will increase the enthalpy of products from the standard reference temperature to 3600°R.

$$952{,}860 \ \frac{\text{Btu}}{\text{lbmol}} = \left(3 \ \frac{\text{lbmol}}{\text{lbmol}} \right) \left(39{,}791 \ \frac{\text{Btu}}{\text{lbmol}} \right)$$
$$\quad + \left(4 \ \frac{\text{lbmol}}{\text{lbmol}} \right) \left(31{,}658 \ \frac{\text{Btu}}{\text{lbmol}} \right)$$
$$\quad + x \left(24{,}471 \ \frac{\text{Btu}}{\text{lbmol}} \right)$$
$$x = 28.89 \ \text{lbmol/lbmol propane}$$

The mass of nitrogen per pound-mole of propane is

$$m_{N_2} = \left(28.89 \ \frac{\text{lbmol}}{\text{lbmol propane}} \right) \left(28.016 \ \frac{\text{lbm}}{\text{lbmol}} \right)$$
$$= \boxed{\begin{array}{c} 809.4 \ \text{lbm/lbmol propane} \\ (810 \ \text{lbm/lbmol propane}) \end{array}}$$

The mass of oxygen per pound-mole of propane is

$$m_{O_2} = \left(5 \ \frac{\text{lbmol}}{\text{lbmol}} \right) \left(32 \ \frac{\text{lbm}}{\text{lbmol}} \right)$$
$$= \boxed{160 \ \text{lbm/lbmol propane}}$$

The answer is (B).

(b) The partial pressure is volumetrically weighted. This is the same as molar (mole fraction) weighting.

product	lbmol	mole fraction
CO_2	3	$3/35.89 = 0.0836$
H_2O	4	$4/35.89 = 0.1115$
N_2	28.89	$28.89/35.89 = 0.8049$
O_2	0	$0/35.89 = 0$
	35.89 lbmol	1.000

The partial pressure of water vapor is

$$p_{H_2O} = \left(\frac{n_{H_2O}}{n} \right) p = (0.1115)(14.7 \ \text{psia}) = 1.64 \ \text{psia}$$

From App. 29.B, this pressure corresponds to approximately 118°F. Since the stack temperature is 100°F, some of the water will condense. From App. 29.A, the maximum vapor pressure of water at the stack temperature is 0.9505 psia. Let n be the number of moles of water per mole of propane in the stack gas.

$$n_{H_2O} = \left(\frac{p_{H_2O}}{p} \right) n = \left(\frac{0.9505 \ \text{psia}}{14.7 \ \text{psia}} \right) \left(35.89 \ \frac{\text{lbmol}}{\text{lbmol}} \right)$$
$$= \boxed{\begin{array}{c} 2.321 \ \text{lbmol/lbmol } C_3H_8 \\ (2.3 \ \text{lbmol/lbmol } C_3H_8) \end{array}}$$

The answer is (C).

Combustion

(c) The water removed is

$$4 - n_{H_2O} = 4 \; \frac{\text{lbmol}}{\text{lbmol}} - 2.321 \; \frac{\text{lbmol}}{\text{lbmol}}$$

$$= 1.679 \text{ lbmol of } H_2O/\text{lbmol of } C_3H_8$$

$$m = \left(1.679 \; \frac{\text{lbmol}}{\text{lbmol}}\right)\left(18.016 \; \frac{\text{lbm}}{\text{lbmol}}\right)$$

$$\boxed{\begin{array}{c} = \; 30.25 \text{ lbm } H_2O/\text{lbmol } C_3H_8 \\ (30 \text{ lbm } H_2O/\text{lbmol } C_3H_8) \end{array}}$$

The answer is (C).

SI Solution

(a) From the customary U.S. solution, the heat of reaction is

$$n_{CO_2}(\Delta H_f)_{CO_2} + n_{H_2O}(\Delta H_f)_{H_2O}$$
$$- n_{C_3H_8}(\Delta H_f)_{C_3H_8} - n_{O_2}(\Delta H_f)_{O_2}$$

$$= \left(3 \; \frac{\text{kmol}}{\text{kmol}}\right)\left(-393.8 \; \frac{\text{GJ}}{\text{kmol}}\right)$$
$$+ \left(4 \; \frac{\text{kmol}}{\text{kmol}}\right)\left(-242 \; \frac{\text{GJ}}{\text{kmol}}\right)$$
$$+ \left(1 \; \frac{\text{kmol}}{\text{kmol}}\right)\left(-67.0 \; \frac{\text{GJ}}{\text{kmol}}\right)$$
$$- \left(5 \; \frac{\text{kmol}}{\text{kmol}}\right)\left(0 \; \frac{\text{GJ}}{\text{kmol}}\right)$$

$$= -2216.4 \text{ GJ/kmol propane}$$

The negative sign indicates an exothermal reaction.

Let x be the number of moles of nitrogen per mole of propane. Use the nitrogen to cool the combustion. The heat of reaction will increase the enthalpy of products from the standard reference temperature to 1980°C.

$$2216.4 \; \frac{\text{GJ}}{\text{mol}} = \left(3 \; \frac{\text{kmol}}{\text{kmol}}\right)\left(92.6 \; \frac{\text{GJ}}{\text{kmol}}\right)$$
$$+ \left(4 \; \frac{\text{kmol}}{\text{kmol}}\right)\left(73.6 \; \frac{\text{GJ}}{\text{kmol}}\right)$$
$$+ x\left(56.9 \; \frac{\text{GJ}}{\text{kmol}}\right)$$

$$x = 28.90 \text{ kmol/kmol propane}$$

The mass of nitrogen per mole of propane is

$$m_{N_2} = \left(28.90 \; \frac{\text{kmol}}{\text{kmol}}\right)\left(28.016 \; \frac{\text{kg}}{\text{kmol}}\right)$$

$$= \boxed{809.7 \text{ kg/kmol} \quad (810 \text{ kg/kmol})}$$

The mass of oxygen per kmol of propane is

$$m_{O_2} = \left(5 \; \frac{\text{kmol}}{\text{kmol}}\right)\left(32 \; \frac{\text{kg}}{\text{kmol}}\right) = \boxed{160 \text{ kg/kmol}}$$

The answer is (B).

(b) The partial pressure is volumetrically weighted. This is the same as molar (mole fraction) weighting.

product	kmol	mole fraction
CO_2	3	$3/35.90 = 0.0836$
H_2O	4	$4/35.90 = 0.1114$
N_2	28.90	$28.89/35.90 = 0.8050$
O_2	0	$0/35.90 = 0$
	35.90 lbmol	1.000

The partial pressure of water vapor is

$$p_{H_2O} = \left(\frac{n_{H_2O}}{n}\right)p = (0.1114)(101 \text{ kPa})$$
$$= 11.25 \text{ kPa}$$

From App. 29.O, this pressure corresponds to approximately 47.6°C. Since the stack temperature is 38°C, some of the water will condense. From App. 29.N, the maximum vapor pressure of water at the stack temperature is 6.633 kPa. Let n be the number of moles of water per mole of propane in the stack gas.

$$n_{H_2O} = \left(\frac{p_{H_2O}}{p}\right)n = \left(\frac{6.633 \text{ kPa}}{101 \text{ kPa}}\right)\left(35.90 \; \frac{\text{kmol}}{\text{kmol}}\right)$$

$$\boxed{\begin{array}{c} = \; 2.358 \text{ kmol } H_2O/\text{kmol } C_3H_8 \\ (2.3 \text{ kmol } H_2O/\text{kmol } C_3H_8) \end{array}}$$

The answer is (C).

(c) The liquid water removed is

$$4 - n_{H_2O} = 4 \; \frac{\text{kmol}}{\text{kmol}} - 2.358 \; \frac{\text{kmol}}{\text{kmol}}$$

$$= 1.642 \text{ kmol of } H_2O/\text{kmol propane}$$

$$m = \left(1.642 \; \frac{\text{kmol}}{\text{kmol}}\right)\left(18.016 \; \frac{\text{kg}}{\text{kmol}}\right)$$

$$\boxed{\begin{array}{c} = \; 29.58 \text{ kg } H_2O/\text{kmol } C_3H_8 \\ (30 \text{ kg } H_2O/\text{kmol } C_3H_8) \end{array}}$$

The answer is (C).

19. The heat rate, Q, required to generate the steam is found from the generation rate of steam, $\dot{m}$, the enthalpy of the saturated vapor, h_g, and the enthalpy of the incoming saturated feedwater, h_f. From App. 29.B, for an absolute pressure of 200 lbf/in^2, h_g is 1198.8 Btu/lbm. From App. 29.A, for a temperature of 160°F, h_f is 128.0 Btu/lbm.

$$Q = \dot{m}(h_g - h_f)$$
$$= \left(20{,}070 \; \frac{\text{lbm}}{\text{hr}}\right)\left(1198.8 \; \frac{\text{Btu}}{\text{lbm}} - 128.0 \; \frac{\text{Btu}}{\text{lbm}}\right)$$
$$= 21.5 \times 10^6 \text{ Btu/hr}$$

Combustion

The lower heating value of the fuel, LHV, is

$$\text{LHV} = \frac{Q}{\dot{m}_{\text{fuel}}} = \frac{21.5 \times 10^6 \ \dfrac{\text{Btu}}{\text{hr}}}{5000 \ \dfrac{\text{lbm}}{\text{hr}}} = 4300 \ \text{Btu/lbm}$$

The higher heating value, HHV, includes the heat of vaporization (same as the heat of condensation) of the water in the fuel and generated by the combustion of the fuel. The mass of water in the combustion products produced per point of fuel is

$$m_{\text{water}} = \frac{2000 \ \dfrac{\text{lbm water}}{\text{hr}}}{5000 \ \dfrac{\text{lbm fuel}}{\text{hr}}} = 0.4 \ \text{lbm water/lbm fuel}$$

From App. 29.B, for a partial pressure of 4 lbf/in^2, the enthalpy of vaporization, h_{fg}, is 1006.0 Btu/lbm. From Eq. 37.14, the higher heating value is

$$\begin{aligned}
\text{HHV} &= \text{LHV} + m_{\text{water}} h_{fg} \\
&= 4300 \ \frac{\text{Btu}}{\text{lbm}} + \left(0.4 \ \frac{\text{lbm}}{\text{lbm}}\right)\left(1006.0 \ \frac{\text{Btu}}{\text{lbm}}\right) \\
&= \boxed{4702.4 \ \text{Btu/lbm} \quad (4700 \ \text{Btu/lbm})}
\end{aligned}$$

The answer is (B).

38 Air Quality

PRACTICE PROBLEMS

1. The following questions pertain to air pollution regulations of the Clean Air Act (CAA) and the Resource Conservation and Recovery Act (RCRA).

(a) What regulatory standards define acceptable levels for the criteria pollutants?

    (A) National Ambient Air Quality Standards (NAAQS)

    (B) New Source Performance Standards (NSPS)

    (C) National Emission Standards for Hazardous Air Pollutants (NESHAP)

    (D) Lowest Achievable Emission Rate (LAER) for Non-Attainment Areas (NA)

(b) Which of the following is NOT included under the CAA?

    (A) regional classification as part of Prevention of Significant Deterioration (PSD)

    (B) State Implementation Plans (SIP) to define control strategies for criteria pollutants

    (C) hazardous air pollutants such as asbestos, mercury, beryllium, and radionuclides

    (D) performance standards for principal organic hazardous constituent (POHC) emissions from hazardous waste incinerators

(c) What option is NOT available under emissions trading?

    (A) bubble

    (B) emission reduction credit

    (C) interregional averaging

    (D) offset

(d) What does destruction and removal efficiency (DRE) encompass?

    (A) destruction and removal in the incinerator of the principal organic hazardous constituent (POHC) prior to the application of any air pollution control equipment

    (B) destruction and removal in the incinerator of all pollutants regulated under the National Emission Standards for Hazardous Air Pollutants (NESHAP) prior to the application of any air pollution control equipment

    (C) destruction in the incinerator and removal by any air pollution control equipment of the principal organic hazardous constituent (POHC)

    (D) destruction in the incinerator and removal by any air pollution control equipment of all pollutants regulated under the National Emission Standards for Hazardous Air Pollutants (NESHAP)

(e) What is the corrected combustion emission particulate concentration for a particulate measured as 347 mg/m^3 with 4% O_2?

    (A) 58 mg/m^3

    (B) 286 mg/m^3

    (C) 347 mg/m^3

    (D) 421 mg/m^3

2. The following questions address issues related to classification of air pollutants.

(a) What average particle size defines the upper limit of respirable particulate matter?

    (A) 0.5 μm

    (B) 1 μm

    (C) 2.5 μm

    (D) 10 μm

(b) What is the settling Reynolds number for a particle with a 5 μm diameter and 1000 kg/m^3 density in air at 60°C and 1 atm?

(A) 7.5×10^{-5}

(B) 1.9×10^{-4}

(C) 3.7×10^{-4}

(D) 1.4×10^{-3}

(c) Historically, what has been the primary source of airborne lead pollution?

(A) automobile exhaust emissions

(B) lead smelting emissions

(C) lead-based paint weathering and flaking

(D) electronic equipment soldering emissions

(d) What are the most common primary forms of nitrogen oxides (NOx) and sulfur oxides (SOx) emitted?

(A) nitric oxide (NO) and sulfur dioxide (SO_2)

(B) nitric oxide (NO) and sulfur trioxide (SO_3)

(C) nitrogen dioxide (NO_2) and sulfur dioxide (SO_2)

(D) nitrogen dioxide (NO_2) and sulfur trioxide (SO_3)

(e) What condensed chemical formula represents an alkane radical?

(A) $CH_3-CH=CH_2$

(B) $CH_2=O$

(C) CH_3-CH_3

(D) $CH_2-CH_2-CH_3$

3. A mixture of 20% methane, 30% oxygen, and 50% nitrogen is stored in a pressurized container at 1.5 atm and 15°C. What is the change in the density of the gas mixture if the temperature is increased to 25°C at constant pressure?

4. An air quality test procedure has produced the following data for particulate matter.

particle size collected	10 μm
clean filter mass	10.1453 g
filter mass at 24 h	10.5818 g
initial air flow	1.6 m^3/min
final air flow	1.58 m^3/min

(a) What is the PM10 concentration?

(A) 174 μg/m^3

(B) 191 μg/m^3

(C) 275 μg/m^3

(D) 395 μg/m^3

(b) Considering only 10 μm particulates, what is the Air Quality Index (AQI)?

(A) 112

(B) 119

(C) 163

(D) 264

(c) What is the air quality descriptor corresponding to the AQI based on the PM10 concentration?

(A) moderate

(B) unhealthy

(C) very unhealthy

(D) hazardous

(d) If ozone (O_3) had been present at a 1 h concentration of 0.42 ppm along with the PM10, what would be the AQI?

(A) 120

(B) 163

(C) 302

(D) 316

(e) What is the air quality descriptor corresponding to the AQI when ozone is included?

(A) moderate

(B) unhealthy

(C) very unhealthy

(D) hazardous

5. An air pollutant is emitted from a stack with an effective height of 185 m at 27 kg/s. The wind speed at 10 m above grade is 5.3 m/s under a clear night sky. (a) What is the ground-level concentration at a point 2000 m downwind and 300 m crosswind of the stack? (b) If the plume temperature at the stack is 30°C and the ambient lapse rate and ground level air temperature are −0.0088°C/m and 18°C, respectively, will the pollutant reach the ground level?

6. An electrostatic precipitator has been selected to treat a particulate-laden air stream characterized as follows.

particulate concentration	11 g/m³
particulate temperature	175°C
air flow	4 m³/s
average particle diameter	6 μm
required removal efficiency	95%
applied voltage	80 kV
distance between collectors	36 cm

(a) What is the spacing between the discharge electrodes and the collectors?

(A) 9 cm

(B) 18 cm

(C) 36 cm

(D) 72 cm

(b) What is the average electric field?

(A) 1.0×10^5 N/C

(B) 2.2×10^5 N/C

(C) 4.4×10^5 N/C

(D) 8.9×10^5 N/C

(c) What is the drift velocity?

(A) 0.0071 m/s

(B) 0.034 m/s

(C) 0.14 m/s

(D) 0.56 m/s

(d) What is the required collector area?

(A) 22 m²

(B) 86 m²

(C) 350 m²

(D) 1700 m²

(e) What is the specific collection area?

(A) 6 m²·s/m³

(B) 21 m²·s/m³

(C) 88 m²·s/m³

(D) 430 m²·s/m³

7. The following questions relate to mobile source air pollution.

(a) Which of the following would NOT contribute to overall improved air quality in urban areas?

(A) Increase access to public transportation.

(B) Increase the proportion of new cars to old cars.

(C) Increase the proportion of cars with diesel engines to those with gasoline engines.

(D) Require routine maintenance of automobile air pollution control equipment.

(b) What strategy does the EPA employ to regulate mobile source emissions?

(A) pre-consumer certification and enforcement

(B) pre-consumer certification and post-manufacture enforcement

(C) pre-consumer enforcement and post-manufacture certification

(D) post-manufacture certification and enforcement

(c) What is the EPA average fuel economy for a car rated at 20 mpg city and 34 mpg highway?

(A) 24.3 mpg

(B) 24.5 mpg

(C) 25.9 mpg

(D) 27.0 mpg

(d) Why are fuel economy ratings important to air quality?

(A) Lower fuel consumption results in lower emissions per mile driven.

(B) Lower fuel consumption reduces fuel production demand, thereby reducing refinery emissions.

(C) Lower fuel consumption reduces fuel tank fill-ups, thereby reducing hydrocarbon emissions.

(D) Lower fuel consumption entices consumers to purchase less-polluting, smaller cars.

Combustion

(e) What are the primary sources of emissions from automobiles to the air?

(A) ventilation from the fuel tank, exhaust from the tailpipe, and particulates from tire wear and roadway surface abrasion

(B) ventilation from the crankcase, ventilation from the fuel tank, exhaust from the tailpipe, and particulates from tire wear and roadway surface abrasion

(C) volatilization from the carburetor or throttle body, exhaust from the tailpipe, and particulates from tire wear and roadway surface abrasion

(D) ventilation from the crankcase, ventilation from the fuel tank, volatilization from the carburetor or throttle body, and exhaust from the tailpipe

8. Radon gas enters a 1000 m^3 basement room at 2600 pCi/min.

(a) What ventilation flow rate is required to keep the radon concentration in the room below 4 pCi/L?

(A) 0.65 m^3/min

(B) 6.5 m^3/min

(C) 10 m^3/min

(D) 15 m^3/min

(b) At the calculated ventilation rate, how long is required to change out one room-volume of air?

(A) 1.2 h

(B) 1.7 h

(C) 2.5 h

(D) 26 h

(c) What is the primary source of radon in indoor air?

(A) carpeting and furnishings

(B) concrete and masonry

(C) electronic equipment

(D) soil

(d) How does environmental tobacco smoke (ETS) influence radon exposure?

(A) There is no influence.

(B) Radon adsorbs to the ETS particulate, reducing exposure.

(C) Radon adsorbs to the ETS particulate, increasing exposure.

(D) Radon absorbs ETS gases, increasing exposure.

(e) What is the primary health risk associated with radon exposure?

(A) lung cancer

(B) genetic defects

(C) bone decalcification

(D) permanent memory loss

9. Nitrogen dioxide (NO_2) and sulfur trioxide (SO_3) concentrations in the air in an urban area are 254 μg/m^3 and 427 μg/m^3, respectively. Assume that 1 L of rainwater will scrub 10 m^3 of air and that the natural pH of rainwater is 5.5. What will be the rainwater pH after scrubbing the NO_2 and SO_3?

SOLUTIONS

1. (a) The National Ambient Air Quality Standards (NAAQS) define acceptable levels for the criteria pollutants.

The answer is (A).

(b) The CAA does not include performance standards for the principal organic hazardous constituent (POHC) emissions from hazardous waste incinerators. The POHC emissions from hazardous waste incinerators are regulated under RCRA.

The answer is (D).

(c) Interregional averaging is not an option available under emissions trading.

The answer is (C).

(d) Destruction and removal efficiency (DRE) encompasses destruction in the incinerator and removal by any air pollution control equipment of the principal organic hazardous constituent (POHC).

The answer is (C).

(e) From Eq. 38.2,

$$P_c = P_m\left(\frac{14}{21 - Y}\right)$$
$$= \left(347 \ \frac{\text{mg}}{\text{m}^3}\right)\left(\frac{14}{21 - 4}\right)$$
$$= \boxed{286 \ \text{mg/m}^3}$$

The answer is (B).

2. (a) The average particle size that defines the upper limit of respirable particulate matter is 2.5 μm.

The answer is (C).

(b) From App. 38.C, $\mu_g = 1.97 \times 10^{-5}$ kg/m·s, and $\rho_g = 1.06$ kg/m³ at 60°C.

Use Eq. 38.3.

$$\text{v}_s = \frac{d_p^2(\rho_p - \rho_g)g}{18\mu_g}$$

$$= \frac{\begin{array}{c}(5 \ \mu\text{m})^2\left(10^{-6} \ \frac{\text{m}}{\mu\text{m}}\right)^2\left(1000 \ \frac{\text{kg}}{\text{m}^3} - 1.06 \ \frac{\text{kg}}{\text{m}^3}\right)\\ \times \left(9.81 \ \frac{\text{m}}{\text{s}^2}\right)\end{array}}{(18)\left(1.97 \times 10^{-5} \ \frac{\text{kg}}{\text{m·s}}\right)}$$

$$= 6.9 \times 10^{-4} \ \text{m/s}$$

Use Eq. 38.4.

$$\text{Re} = \frac{\text{v}_s d_p \rho_g}{\mu_g}$$

$$= \frac{\left(6.9 \times 10^{-4} \ \frac{\text{m}}{\text{s}}\right)(5 \ \mu\text{m})\left(10^{-6} \ \frac{\text{m}}{\mu\text{m}}\right)\left(1.06 \ \frac{\text{kg}}{\text{m}^3}\right)}{1.97 \times 10^{-5} \ \frac{\text{kg}}{\text{m·s}}}$$

$$= \boxed{1.9 \times 10^{-4}}$$

The answer is (B).

(c) Historically, the primary source of airborne lead pollution has been automobile exhaust emissions.

The answer is (A).

(d) The most common primary forms of nitrogen oxides (NOx) and sulfur oxides (SOx) emitted are nitric oxide (NO) and sulfur dioxide (SO_2).

The answer is (A).

(e) The formula representing an alkane radical is CH_2–CH_2–CH_3 (propyl).

The answer is (D).

3. The molecular weight of CH_4 is

$$\left(12 \ \frac{\text{g}}{\text{mol}}\right) + (4)\left(1 \ \frac{\text{g}}{\text{mol}}\right) = 16 \ \text{g/mol}$$

The molecular weight of O_2 is

$$(2)\left(16 \ \frac{\text{g}}{\text{mol}}\right) = 32 \ \text{g/mol}$$

The molecular weight of N_2 is

$$(2)\left(14 \ \frac{\text{g}}{\text{mol}}\right) = 28 \ \text{g/mol}$$

From Eq. 38.15, the average molecular weight of the gas mixture is

$$\begin{aligned}\text{average} & \text{ molecular weight}\\ &= f_{v,\text{CH}_4}(\text{MW})_{\text{CH}_4} + f_{v,\text{O}_2}(\text{MW})_{\text{O}_2}\\ &\quad + f_{v,\text{N}_2}(\text{MW})_{\text{N}_2}\\ &= (0.20)\left(16 \ \frac{\text{g}}{\text{mol}}\right) + (0.30)\left(32 \ \frac{\text{g}}{\text{mol}}\right)\\ &\quad + (0.50)\left(28 \ \frac{\text{g}}{\text{mol}}\right)\\ &= 26.8 \ \text{g/mol}\end{aligned}$$

Combustion

Use Eq. 38.12.

ρ_g at 1.5 atm and 15°C is

$$\rho_g = \frac{p(MW)}{R^*T}$$

$$= \frac{(1.5 \text{ atm})\left(26.8 \ \frac{\text{g}}{\text{mol}}\right)\left(\frac{1 \text{ kg}}{1000 \text{ g}}\right)}{\left(8.2 \times 10^{-5} \ \frac{\text{atm·m}^3}{\text{mol·K}}\right)(15°C + 273°)}$$

$$= 1.702 \text{ kg/m}^3$$

ρ_g at 1.5 atm and 25°C is

$$\rho_g = \frac{(1.5 \text{ atm})\left(26.8 \ \frac{\text{g}}{\text{mol}}\right)\left(\frac{1 \text{ kg}}{1000 \text{ g}}\right)}{\left(8.2 \times 10^{-5} \ \frac{\text{atm·m}^3}{\text{mol·K}}\right)(25°C + 273°)}$$

$$= 1.645 \text{ kg/m}^3$$

$$\Delta\rho_g = 1.645 \ \frac{\text{kg}}{\text{m}^3} - 1.702 \ \frac{\text{kg}}{\text{m}^3}$$

$$= \boxed{-0.057 \text{ kg/m}^3}$$

4. (a) The air flow for 24 h PM10 is

$$\frac{1.6 \ \frac{\text{m}^3}{\text{min}} + 1.58 \ \frac{\text{m}^3}{\text{min}}}{2} = 1.59 \text{ m}^3/\text{min}$$

$$24 \text{ h PM10} = \frac{(10.5818 \text{ g} - 10.1453 \text{ g})\left(10^6 \ \frac{\mu\text{g}}{\text{g}}\right)}{(24 \text{ h})\left(60 \ \frac{\text{min}}{\text{h}}\right)\left(1.59 \ \frac{\text{m}^3}{\text{min}}\right)}$$

$$= \boxed{191 \ \mu\text{g/m}^3}$$

The answer is (B).

(b) The 24-hour PM10 concentration is 191 μg/m^3. From Table 38.6, this value is within the C_p range of 155–254 μg/m^3, corresponding to the subindex range of 101–150. From Eq. 38.29, the PM10 subindex is

$$I_{p,\text{PM10}} = I_{\text{low}} + \left(\frac{C_p - \text{BP}_{\text{low}}}{\text{BP}_{\text{high}} - \text{BP}_{\text{low}}}\right)(I_{\text{high}} - I_{\text{low}})$$

$$= 101 + \left(\frac{191 - 155}{254 - 155}\right)(150 - 101)$$

$$= \boxed{119}$$

The answer is (B).

(c) From Table 38.5, the air quality descriptor corresponding to an AQI of 119 is "unhealthy."

The answer is (B).

(d) The 1-hour O_3 concentration is 0.42 ppm. From Table 38.6, this value is within the C_p range of 0.405–0.604 ppm, corresponding to the subindex range of 301–500. From Eq. 38.29, the O_3 subindex is

$$I_{p,O_3} = I_{\text{low}} + \left(\frac{C_p - \text{BP}_{\text{low}}}{\text{BP}_{\text{high}} - \text{BP}_{\text{low}}}\right)(I_{\text{high}} - I_{\text{low}})$$

$$= 301 + \left(\frac{0.42 - 0.405}{0.604 - 0.405}\right)(500 - 301)$$

$$= \boxed{316}$$

Since the subindex for O_3 (316) is higher than the subindex for PM10 (119), the AQI is 316.

The answer is (D).

(e) From Table 38.5, the air quality descriptor corresponding to an AQI of 316 is "hazardous."

The answer is (D).

5. (a) From Table 38.7, the stability condition is D.

From Fig. 38.11, $\sigma_y = 150$ m when $x = 2000$ m for stability condition D.

From Fig. 38.12, $\sigma_z = 50$ m when $x = 2000$ m for stability condition D.

Use Eq. 38.30.

$$C_{x,y,0} = \frac{Q_g \exp\left((-0.5)\left(\frac{y}{\sigma_y}\right)^2\right) \times \exp\left((-0.5)\left(\frac{H}{\sigma_z}\right)^2\right)}{\pi\mu\sigma_y\sigma_z}$$

$$C_{2000,300,0} = \frac{\left(27 \ \frac{\text{kg}}{\text{s}}\right)\exp\left((-0.5)\left(\frac{300 \text{ m}}{150 \text{ m}}\right)^2\right) \times \exp\left((-0.5)\left(\frac{185 \text{ m}}{50 \text{ m}}\right)^2\right)}{\pi\left(5.3 \ \frac{\text{m}}{\text{s}}\right)(150 \text{ m})(50 \text{ m})\left(\frac{1 \text{ kg}}{10^9 \ \mu\text{g}}\right)}$$

$$= \boxed{30 \ \mu\text{g/m}^3}$$

(b) The ambient lapse rate is only slightly less than the adiabatic lapse rate, so this is a basically neutral condition. The resulting plume will be coning, and the pollutant will reach ground level.

6. (a) The spacing between the discharge electrodes and the collectors is

$$Z_e = \frac{36 \text{ cm}}{2}$$
$$= \boxed{18 \text{ cm}}$$

The answer is (B).

(b) Use Eq. 38.46.

$$E_f = \frac{V_d}{Z_e}$$
$$= \frac{80\,000 \text{ V}}{(18 \text{ cm})\left(\dfrac{1 \text{ m}}{100 \text{ cm}}\right)}$$
$$= \boxed{4.4 \times 10^5 \text{ V/m} \quad (4.4 \times 10^5 \text{ N/C})}$$

The answer is (C).

(c) From App. 38.C, $\mu_g = 2.50 \times 10^{-5}$ kg/m·s at 175°C.

Use Eq. 38.45.

$$w = \frac{8 i_0 E_f^2 d_p}{24 \mu_g}$$
$$= \frac{(8)\left(8.85 \times 10^{-12} \dfrac{C^2}{N \cdot m^2}\right)\left(4.4 \times 10^5 \dfrac{N}{C}\right)^2}{\times (6 \ \mu m)\left(10^{-6} \dfrac{m}{\mu m}\right)}{(24)\left(2.50 \times 10^{-5} \dfrac{kg}{m \cdot s}\right)\left(1 \dfrac{N \cdot s^2}{kg \cdot m}\right)}$$
$$= \boxed{0.14 \text{ m/s}}$$

The answer is (C).

(d) Use Eq. 38.47.

$$A = \frac{-\ln\left(1 - \dfrac{E}{100\%}\right) Q_g}{w}$$
$$= \frac{-\ln\left(1 - \dfrac{95\%}{100\%}\right)\left(4 \dfrac{m^3}{s}\right)}{0.14 \dfrac{m}{s}}$$
$$= \boxed{86 \text{ m}^2}$$

The answer is (B).

(e) Use Eq. 38.48.

$$\text{SCA} = \frac{A}{Q_g} = \frac{86 \text{ m}^2}{4 \dfrac{m^3}{s}}$$
$$= \boxed{21 \text{ m}^2 \cdot \text{s/m}^3}$$

The answer is (B).

7. (a) Increasing the proportion of cars with diesel engines to those with gasoline engines would not contribute to overall improved air quality in urban areas.

The answer is (C).

(b) The EPA employs a pre-consumer certification and enforcement strategy to regulate mobile source emissions.

The answer is (A).

(c) The EPA average fuel economy is

$$\text{average fuel economy} = \frac{1 \text{ mi}}{\dfrac{(1 \text{ mi})(0.55)}{20 \dfrac{\text{mi}}{\text{gal}}} + \dfrac{(1 \text{ mi})(0.45)}{34 \dfrac{\text{mi}}{\text{gal}}}}$$
$$= \boxed{24.5 \text{ mpg}}$$

The answer is (B).

(d) Fuel economy ratings are important to air quality because lower fuel consumption results in lower emissions per mile driven.

The answer is (A).

(e) The primary sources of emissions from automobiles to the air are ventilation from the crankcase, ventilation from the fuel tank, volatilization from the carburetor or throttle body, and exhaust from the tailpipe.

The answer is (D).

8. (a) Perform a mass balance with the room as the boundary. Ignore decay, and assume steady state.

$$\text{mass radon in} = \text{mass radon out}$$

The desired radon concentration in the room is 4 pCi/L.

$$2600 \text{ pCi/min} = \left(4 \dfrac{\text{pCi}}{\text{L}}\right)\left(1000 \dfrac{\text{L}}{\text{m}^3}\right)\left(Q_g \dfrac{\text{m}^3}{\text{min}}\right)$$
$$Q_g = \boxed{0.65 \text{ m}^3/\text{min}}$$

The answer is (A).

(b) The time required to change out 1000 m³ room-volume of air is

$$\frac{1000 \text{ m}^3}{\left(0.65 \dfrac{\text{m}^3}{\text{min}}\right)\left(60 \dfrac{\text{min}}{\text{h}}\right)} = \boxed{26 \text{ h}}$$

The answer is (D).

(c) The primary source of radon in indoor air is from soil.

The answer is (D).

Combustion

(d) Environmental tobacco smoke (ETS) increases radon exposure by adsorption of radon to ETS particulate.

The answer is (C).

(e) The primary health risk associated with radon exposure is lung cancer.

The answer is (A).

9. Since the pH is 5.5, $[H^+] = 10^{-5.5} = 3.2 \times 10^{-6}$ M.

From Eq. 38.25,

$$NO_2 + \cdot OH \rightarrow HNO_3 \text{ in the atmosphere}$$

$$HNO_3(aq) \rightarrow H^+ + NO_3^- \text{ in the solution}$$

1 mole NO_2 yields 1 mole H^+.

$$MW \ NO_2 = 14 \ \frac{g}{mol} + (2)\left(16 \ \frac{g}{mol}\right) = 46 \text{ g/mol}$$

$$[H^+] \text{ from } NO_2 = \frac{\left(254 \ \frac{\mu g}{m^3 \ air}\right)\left(\frac{1 \ g}{10^6 \ \mu g}\right)\left(1 \ \frac{mol \ H^+}{mol \ NO_2}\right)}{\left(\frac{1 \ L \ water}{10 \ m^3 \ air}\right)\left(46 \ \frac{g}{mol \ NO_2}\right)}$$

$$= 5.5 \times 10^{-5} \text{ M}$$

From Eq. 38.27 and dissociation of H_2SO_4,

$$SO_3 + H_2O \rightarrow H_2SO_4(aq) \rightarrow 2H^+ + SO_4^{-2}$$

1 mole SO_3 yields 2 mole H^+.

Since the concentration of SO_3 (not H_2SO_4) is given, determine the molecular weight of SO_3.

$$MW \ SO_3 = 32 \ \frac{g}{mol} + (3)\left(16 \ \frac{g}{mol}\right) = 80 \text{ g/mol}$$

$$[H^+] \text{ from } SO_3 = \frac{\left(427 \ \frac{\mu g}{m^3 \ air}\right)\left(\frac{1 \ g}{10^6 \ \mu g}\right)\left(2 \ \frac{mol \ H^+}{mol \ SO_3}\right)}{\left(\frac{1 \ L \ water}{10 \ m^3 \ air}\right)\left(80 \ \frac{g}{mol \ SO_3}\right)}$$

$$= 1.1 \times 10^{-4} \text{ M}$$

$$\text{total } [H^+] = 5.5 \times 10^{-5} \text{ M} + 1.1 \times 10^{-4} \text{ M}$$
$$+ 3.2 \times 10^{-6} \text{ M}$$
$$= 1.7 \times 10^{-4} \text{ M}$$

$$pH = -\log[H^+] = -\log(1.7 \times 10^{-4})$$
$$= \boxed{3.78}$$

39 Municipal Solid Waste

PRACTICE PROBLEMS

Landfill Capacity

1. A town has a current population of 10,000, which is expected to double in 15 yr. The town intends to dispose of its municipal solid waste in a 30 ac landfill that will be converted to a park in 20 yr. Solid waste is generated at the rate of 5 lbm/capita-day. The average compacted density in the landfill will be 1000 lbm/yd^3. Disregarding any soil addition for cover and cell construction, how long will it take to fill the landfill to a uniform height of 6 ft?

(A) 2100 days

(B) 4200 days

(C) 6300 days

(D) 9500 days

2. (*Time limit: one hour*) A town of 10,000 people has selected a 50 ac square landfill site to deposit its solid waste. The minimum unused side borders are 50 ft. The landfill currently consists of a square depression with an average depth of 20 ft below the surrounding grade. When the landfill is at final capacity, it will be covered with 10 ft of earth cover. The maximum height of the covered landfill is 20 ft above the surrounding grade. Solid waste is generated at the rate of 5 lbm/capita-day. The average compacted density in the landfill will be 1000 lbm/yd^3.

(a) What is the volumetric capacity of the landfill site?

(A) 1.1×10^6 ft^3

(B) 6.5×10^6 ft^3

(C) 3.1×10^7 ft^3

(D) 5.7×10^7 ft^3

(b) Using a loading factor of 1.25, what is the volume of landfill used each day?

(A) 60 yd^3/day

(B) 120 yd^3/day

(C) 180 yd^3/day

(D) 240 yd^3/day

(c) What is the service life of the landfill site?

(A) 30 yr

(B) 45 yr

(C) 60 yr

(D) 90 yr

Leachate

3. The pressure and temperature within a closed landfill cell are 1 atm and 45°C, respectively. The volumetric fraction of carbon dioxide in the landfill is 40%. The solubility, K_s, of carbon dioxide in leachate at 45°C is 0.8 g/kg. What is the concentration of carbon dioxide in the leachate in units of mg CO_2/L water?

4. Dissolved hydrogen sulfide gas, H_2S, in landfill leachate has a concentration of 270 mg/L. What is this concentration expressed in parts per million (ppm)?

SOLUTIONS

1. Find the rate of increase of waste production.

The mass of waste deposited in the first day will be

$$(10,000 \text{ people})\left(5 \ \frac{\text{lbm}}{\text{person-day}}\right) = 50,000 \text{ lbm/day}$$

The mass of waste deposited on the last day will be

$$(20,000 \text{ people})\left(5 \ \frac{\text{lbm}}{\text{person-day}}\right) = 100,000 \text{ lbm/day}$$

The increase in rate is

$$\frac{\Delta m}{\Delta t} = \frac{100,000 \ \frac{\text{lbm}}{\text{day}} - 50,000 \ \frac{\text{lbm}}{\text{day}}}{(15 \text{ yr})\left(365 \ \frac{\text{days}}{\text{yr}}\right)} = 9.132 \text{ lbm/day}^2$$

The mass deposited on day t is

$$m_D = 50,000 + 9.132(t - 1)$$
$$\approx 50,000 + 9.132t$$

The cumulative mass deposited is

$$m_t = \int_0^t m_D dt = 50,000t + \frac{9.132t^2}{2}$$

With a compacted density of 1000 lbm/yd³ and a loading factor of 1.00 (no soil cover), the capacity of the site with a 6 ft lift is

$$m_{\text{max}} = \frac{(30 \text{ ac})\left(43,560 \ \frac{\text{ft}^2}{\text{ac}}\right)(6 \text{ ft})\left(1000 \ \frac{\text{lbm}}{\text{yd}^3}\right)}{27 \ \frac{\text{ft}^3}{\text{yd}^3}}$$
$$= 2.9 \times 10^8 \text{ lbm}$$

The time to fill is found by solving the quadratic equation.

$$2.9 \times 10^8 = 50,000t + \frac{9.132t^2}{2}$$
$$t^2 + 10,951t = 6.351 \times 10^7$$
$$t = \boxed{4193 \text{ days} \quad (4200 \text{ days})}$$

The answer is (B).

2. (a) The side length of the square disposal site is

$$\text{length} = \sqrt{(50 \text{ ac})\left(43,560 \ \frac{\text{ft}^2}{\text{ac}}\right)} = 1476 \text{ ft}$$

With 50 ft borders, the usable area is

$$A = (1476 \text{ ft} - (2)(50 \text{ ft}))^2 = 1.893 \times 10^6 \text{ ft}^2$$

If the site is excavated 20 ft, 10 ft of soil is used as cover, and the maximum above-ground height is 20 ft, the service capacity of compacted waste is

$$(1.893 \times 10^6 \text{ ft}^2)(30 \text{ ft}) = \boxed{5.68 \times 10^7 \text{ ft}^3}$$

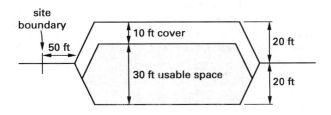

(not to scale or representative
of actual construction)

The answer is (D).

(b) The volume of landfill used per day is

$$\frac{(10,000 \text{ people})\left(5 \ \frac{\text{lbm}}{\text{day-person}}\right)(1.25)}{1000 \ \frac{\text{lbm}}{\text{yd}^3}}$$
$$= \boxed{62.5 \text{ yd}^3/\text{day}}$$

The answer is (A).

(c) The service life is

$$\frac{5.68 \times 10^7 \text{ ft}^3}{\left(27 \ \frac{\text{ft}^3}{\text{yd}^3}\right)\left(62.5 \ \frac{\text{yd}^3}{\text{day}}\right)\left(365 \ \frac{\text{days}}{\text{yr}}\right)} = \boxed{92.2 \text{ yr}}$$

The answer is (D).

Solid Waste

3. Use the ideal gas law to find the density of carbon dioxide at the landfill conditions.

$$\rho_{CO_2, g/L} = \frac{p}{RT} = \frac{B_{CO_2} p_t (MW)}{R^* T}$$

$$= \frac{(0.4)(1 \text{ atm})\left(44 \ \frac{g}{mol}\right)}{\left(0.08206 \ \frac{atm \cdot L}{mol \cdot K}\right)(45°C + 273°)}$$

$$= 0.674 \text{ g/L}$$

Use the ideal gas law to determine the volume of 0.8 g of carbon dioxide.

$$V_{CO_2} = \frac{m}{\rho} = \frac{0.8 \text{ g}}{0.674 \ \frac{g}{L}}$$

$$= 1.19 \text{ L}$$

The density of water at 45°C is approximately 990 kg/m^3. The volume of 1 kg of water is

$$V_{water} = \frac{m_{water}}{\rho_{water}} = \frac{(1 \text{ kg})\left(1000 \ \frac{L}{m^3}\right)}{990 \ \frac{kg}{m^3}}$$

$$= 1.01 \text{ L}$$

The solubility of carbon dioxide is

$$K_s = \frac{V_{CO_2}}{V_{water}} = \frac{1.19 \text{ L}}{1.01 \text{ L}}$$

$$= 1.18 \text{ L/L}$$

Find the concentration of carbon dioxide from Eq. 39.23.

$$C_{CO_2} = \rho K_s = \left(0.674 \ \frac{g}{L}\right)\left(1.18 \ \frac{L}{L}\right)\left(1000 \ \frac{mg}{L}\right)$$

$$= 795 \text{ mg/L}$$

4. ppm is a concentration based on a mass ratio; that is, milligrams of H_2S per million milligrams of leachate. The quantity of dissolved gas is already expressed in milligrams, so express the leachate quantity in milligrams. For dilute solutions, the leachate can be assumed to be water. Since the density of water is approximately 1000 kg/m^3, and there are 1000 liters per cubic meter, a liter of water has a mass of 10^6 mg (i.e., 1 kg). So, a concentration of 270 mg/L is equivalent to 270 ppm.

40 Environmental Pollutants

PRACTICE PROBLEMS

Dust and Particles

1. A PM-10 sample is collected with a high-volume sampler over a sampling interval of 24 hours and 10 minutes. The sampler draws an average flow of 41 ft^3/min and collects particles on a glass fiber filter. The initial and final masses of the filter are 4.4546 g and 4.4979 g, respectively. The PM-10 concentration is most nearly

(A) 11 μg/m^3

(B) 26 μg/m^3

(C) 42 μg/m^3

(D) 55 μg/m^3

Smog

2. Which statement related to smog is INCORRECT?

(A) Smog is produced when ozone precursors such as nitrogen dioxide, hydrocarbons, and volatile organic compounds (VOCs) react with sunlight.

(B) Peroxyacyl nitrates contribute to the formation of smog.

(C) Volatile organic compounds (VOCs) contributing to smog are the products of incomplete combustion reactions in automobiles, refineries, and hazardous waste incinerators.

(D) Ground-level ozone is a secondary pollutant that contributes to the formation of smog.

3. Smog generation generally increases with all of the following conditions EXCEPT

(A) temperature inversions

(B) absence of wind

(C) increased commuter traffic

(D) early morning clouds and fog

4. According to air pollutant criteria defined in the Clean Air Act, smog is a

(A) hazardous air pollutant (HAP)

(B) criteria air pollutant

(C) volatile organic compound (VOC)

(D) primary air toxic

5. The concentration, C, of nitrogen oxides in a city over a 24 hour period is shown.

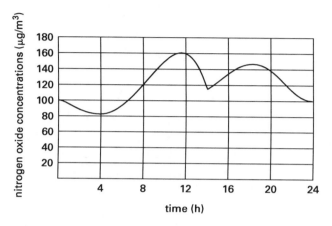

The concentration from $t = 0$ h to $t = 14$ h is described by

$$C_{\mu g/m^3} = 100 - 3.7848t - 2.2609t^2 + 0.6381t^3 - 0.0324t^4$$

The concentration from $t = 14$ h to $t = 24$ h is described by

$$C_{\mu g/m^3} = 5.1885 \times 10^3 - 1.1899 \times 10^3 t + 101.8740t^2 - 3.7638t^3 + 0.0507t^4$$

What is most nearly the average nitrogen oxide concentration over the 24 hour period?

(A) 118 μg/m^3

(B) 121 μg/m^3

(C) 123 μg/m^3

(D) 125 μg/m^3

SOLUTIONS

1. The duration of sampling is

$$t = (24 \text{ hr})\left(60 \frac{\text{min}}{\text{hr}}\right) + 10 \text{ min} = 1450 \text{ min}$$

The volume of air sampled is

$$V = Q_{\text{ave}} t = \left(\frac{41 \frac{\text{ft}^3}{\text{min}}}{\left(3.281 \frac{\text{ft}}{\text{m}}\right)^3}\right)(1450 \text{ min})$$

$$= 1683 \text{ m}^3$$

The PM-10 concentration is

$$C_{\text{PM-10}} = \frac{m_{\text{particles}}}{V} = \frac{m_{\text{final}} - m_{\text{initial}}}{V}$$

$$= \frac{(4.4979 \text{ g} - 4.4546 \text{ g})\left(10^6 \frac{\mu\text{g}}{\text{g}}\right)}{1683 \text{ m}^3}$$

$$= \boxed{25.7 \ \mu\text{g/m}^3 \quad (26 \ \mu\text{g/m}^3)}$$

The answer is (B).

2. Volatile organic compounds (VOCs) are emitted by manufacturing and refining processes, dry cleaners, gasoline stations, print shops, painting operations, municipal wastewater treatment plants, and many combustion sources. However, hazardous waste incinerators (which are used to treat VOCs) are operated at high enough temperatures to destroy VOCs.

The answer is (C).

3. Temperature inversions are conducive to smog generation because they keep warm air near the ground. An absence of wind causes smog to increase because smog precursors are not dispersed. Smog also increases with traffic, as automobile emissions contain nitrogen oxides and hydrocarbons, which are precursors to smog. Smog is primarily a sunlight-induced reaction, and smog production usually peaks in early afternoon when there is the most sunlight.

The answer is (D).

4. The Environmental Protection Agency (EPA) has identified six air pollutants common in the United States. These common air pollutants, known as *criteria air pollutants*, include ground-level ozone, carbon monoxide, sulfur oxides, nitrogen oxides, particle pollution, and lead. Since smog is composed primarily of ground-level ozone, it would be classified by the EPA as a $\boxed{\text{criteria air pollutant.}}$

The answer is (B).

5. Use integration to calculate the average, C_{ave}.

$$C_{\text{ave},\mu\text{g/m}^3} = \frac{\displaystyle\int_{0 \text{ hr}}^{14 \text{ hr}} \left(\begin{array}{c} 100 - 3.7848t - 2.2609t^2 \\ + 0.6381t^3 - 0.0324t^4 \end{array}\right) dt}{24 \text{ hr}}$$

$$+ \frac{\displaystyle\int_{14 \text{ hr}}^{24 \text{ hr}} \left(\begin{array}{c} 5.1885 \times 10^3 \\ - 1.1899 \times 10^3 t \\ + 101.8740t^2 \\ - 3.7638t^3 \\ + 0.0507t^4 \end{array}\right) dt}{24 \text{ hr}}$$

$$= \frac{\left(\begin{array}{c} 100t - 1.8924t^2 \\ - 7.5363 \times 10^{-1}t^3 \\ - 0.1595t^4 + 6.4800 \times 10^{-3}t^5 \end{array}\right)\Big|_{0 \text{ hr}}^{14 \text{ hr}}}{24 \text{ hr}}$$

$$+ \frac{\left(\begin{array}{c} 5.1885 \times 10^3 t \\ - 5.9495 \times 10^2 t^2 \\ + 33.9580t^3 - 0.9410t^4 \\ + 1.0140 \times 10^{-2}t^5 \end{array}\right)\Big|_{14 \text{ hr}}^{24 \text{ hr}}}{24 \text{ hr}}$$

$$= \boxed{121 \ \mu\text{g/m}^3}$$

The answer is (B).

41 Disposition of Hazardous Materials

PRACTICE PROBLEMS

1. The operator of an underground oil storage tank wishes to permanently convert to above-ground storage. To satisfy federal regulations, what must the operator do?

(A) Drain the existing tank.

(B) Drain, clean, and pressurize the existing tank to 150% of atmospheric pressure.

(C) Drain, clean, and fill the existing tank with sand.

(D) Remove the existing tank, remove and replace soil equal to 100% of the tank volume, and install test wells as per code.

2. According to the Environmental Protection Agency (EPA), leaking tanks and pipes may be repaired as long as the repairs meet "industry codes and standards." Which organization publishes such codes and standards?

(A) American Welding Society (AWS)

(B) National Institute of Standards and Technology (NIST)

(C) International Code Council (ICC)

(D) National Fire Protection Association (NFPA)

3. Contaminants migrate through soils and aquifers at a velocity equal to the

(A) pore velocity

(B) effective velocity

(C) superficial velocity

(D) Darcy velocity

4. Common paint solvents accumulated by a painting contractor are processed in a properly licensed and operating hazardous waste incineration facility. A small amount of mineral ash remains after incineration. Which statement regarding the mineral ash is true?

(A) The ash is considered nonhazardous solid waste and may be disposed of in a municipal solid waste facility that accepts ashes.

(B) The ash is considered an ignitable solid waste and must be buried in a buried (RCRA) municipal solid waste facility.

(C) The waste is considered a "derived from" (D) hazardous waste and must be disposed of in a RCRA hazardous waste facility.

(D) The ash is considered a "K" waste and must be disposed of in a RCRA hazardous waste facility.

5. A large section of town contains vacant factory buildings that, over the years, manufactured numerous unknown products. A developer now wishes to purchase the property from the current owners and convert it to noncommercial use. Public concern centers on possible site contamination that the developer cannot afford to clean up. The property can be referred to as a

(A) Superfund site

(B) designated protection zone

(C) brownfield

(D) undesignated watch site

SOLUTIONS

1. When an underground tank is decommissioned, (1) the regulatory authority must be notified at least 30 days before closing; (2) any contamination must be remediated; (3) the tank must be drained and cleaned by removing all liquids, dangerous vapor levels, and accumulated sludge; and (4) the tank must be removed from the ground or filled with a harmless, chemically inactive solid, such as sand.

The answer is (C).

2. The EPA identifies the following organizations as having relevant codes and standards: API (American Petroleum Institute); ASTM International (formerly American Society for Testing and Materials); KWA (Ken Wilcox Associates, Inc.); NACE International (formerly the National Association of Corrosion Engineers); NFPA (National Fire Protection Association); NLPA (National Leak Prevention Association); PEI (Petroleum Equipment Institute); STI (Steel Tank Institute); and UL (Underwriters Laboratories Inc.).

The answer is (D).

3. Contaminants migrate at the pore velocity, also known as seepage velocity or flow front velocity. Darcy velocity (also known as effective velocity and superficial velocity) does not take the porosity into consideration.

The answer is (A).

4. Although the source material was ignitable, the residual ash is not itself a "specially listed" waste due to its ignitability, corrosivity, reactivity, and toxicity characteristics. Painting is not designated as a specific source industry (such as wood preserving, petroleum refining, and organic chemical manufacturing), so the ash is not a "K" waste. The ash is derived from a hazardous waste, so it is a "derived from" (D) hazardous waste. It must be disposed of in a RCRA hazardous waste facility.

The answer is (C).

5. The current owner, not the developer, would be required to clean up any contamination found. Brownfields are abandoned, idled, or underused industrial and commercial properties where expansion or redevelopment is complicated by actual or suspected environmental contamination.

The answer is (C).

Solid Waste

42 Environmental Remediation

PRACTICE PROBLEMS

Oil Spill Remediation

1. Which statement about petroleum spills and spill remediation is INCORRECT?

   (A) As long as the roots of vegetation are kept moist by soil moisture, in situ burning of surface petroleum contaminants has limited long-term environmental effects.

   (B) When properly applied in large bodies of water, oil dispersants have little environmental impact.

   (C) Washing oil-contaminated sands, soils, and rocks with steam and hot water also removes organisms and nutrients that would otherwise contribute to bioremediation.

   (D) Magnetic particle technology helps recover spilled petroleum products for reuse.

2. The treatment of oil-contaminated soil in a large plastic-covered tank is known as

   (A) bioremediation

   (B) biofiltration

   (C) bioreaction

   (D) bioventing

Baghouses

3. The performance of a pilot study baghouse follows the filter drag model, with $K_e = 0.6$ in wg-min/ft and $K_s = 0.07$ in wg-ft^2-min/gr. After 20 minutes of operation of the scaled-up baghouse, the dust loading is 27 gr/ft^3, and the air-to-cloth ratio is 3.0 ft/min. Most nearly, what is the pressure drop for the full scale baghouse after 20 minutes?

   (A) 2.5 in wg

   (B) 4.1 in wg

   (C) 5.2 in wg

   (D) 7.5 in wg

4. 2500 ft^3/min of air flow into a baghouse. The air has a particulate concentration of 0.8 g/m^3. The instantaneous collection efficiency of the baghouse increases with time as the baghouse's pores fill with particles, as shown. The collection efficiency is consistent with a rate constant of -1.26 1/hr. Most nearly, what mass of particulate matter is removed in the first five hours of operation?

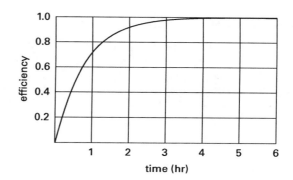

   (A) 32 lbm

   (B) 36 lbm

   (C) 42 lbm

   (D) 63 lbm

Incineration of Municipal Solid Waste

5. Which statement about incineration of municipal solid waste material is INCORRECT?

   (A) The primary source of lead and cadmium in the fly ash is the combustion of plastics.

   (B) Fly ash typically contains lead in concentrations 5000 times higher than natural soils and cadmium in concentrations 20,000 times higher than natural soils.

   (C) The level of carcinogenic metals found in the fly ash is typically high enough to classify the fly ash as extremely hazardous waste under state and federal regulations.

   (D) Incineration releases metals that are mobilized in air emissions or concentrated in ash residues, causing them to be highly bioavailable.

Incineration of Hazardous Waste

6. Which statement concerning hazardous waste incinerators is INCORRECT?

(A) The majority of incinerated radioactive waste is high-level waste (HLW) from nuclear power plants.

(B) Open pit incineration results in excessive smoke and high particulate emission.

(C) Single chamber incinerators generally do not meet federal air emission standards.

(D) Multiple chamber incinerators have low particulate emissions and generally meet federal air emission standards without additional air pollution control equipment.

Air Scrubbers

7. A 6 ft high spray tower has a collection efficiency of 80% when removing 10 μm diameter particles. If the tower's height is increased to 8 ft, and if all other characteristics are unchanged, the collection efficiency when removing 7 μm diameter particles will be most nearly

(A) 85%

(B) 91%

(C) 95%

(D) 98%

Electrostatic Precipitators

8. An electrostatic precipitator is being designed for 110,000 ACFM of 400°F flue gas containing incinerator fly ash at a concentration of 3.2 gr/SCF. For this installation, the effective drift velocity is 0.31 ft/sec. The required collection efficiency is 99.9%. The maximum pressure loss is 5 in wg. Most nearly, what should the total collection area be?

(A) 26,000 ft^2

(B) 31,000 ft^2

(C) 34,000 ft^2

(D) 41,000 ft^2

9. An electrostatic precipitator (ESP) has an efficiency of 95%. By increasing the collection area, the ESP's efficiency processing the same gas is increased to 99%. The ratio of the new collection area to the original collection area is most nearly

(A) 1.5

(B) 1.7

(C) 2.1

(D) 2.5

10. A two-chamber electrostatic precipitator (ESP) has an overall efficiency of 98% when the flow is split equally between the two chambers. If 80% of the same total flow goes through one chamber and 20% goes through the other, what is most nearly the new efficiency of the ESP?

(A) 84%

(B) 89%

(C) 92%

(D) 93%

SOLUTIONS

1. Dispersants are ultimately damaging to the environment. Rather than removing the oil, dispersants break up and distribute the oil, essentially hiding it and making it impossible remove it from the environment.

The answer is (B).

2. *Bioremediation* is the use of microorganisms to remove pollutants. *Biofiltration* is the use of composting and soil beds to remove pollutants. *Bioreactors* are open or closed tanks that contain dozens or hundreds of slowly rotating disks covered with a biological film of microorganisms used to remove pollutants. *Bioventing* is the treatment of contaminated soil in a large plastic-covered tank. Clean air, water, and nutrients are continuously supplied to the tank while off-gases are suctioned off. The off-gases are cleaned with activated carbon adsorption or with thermal or catalytic oxidation prior to discharge.

The answer is (D).

3. The filter drag is

$$
\begin{aligned}
S &= K_e + K_s W \\
&= 0.6 \ \frac{\text{in wg-min}}{\text{ft}} + \left(0.07 \ \frac{\text{in wg-ft}^2\text{-min}}{\text{gr}}\right)\left(27 \ \frac{\text{gr}}{\text{ft}^3}\right) \\
&= 2.49 \ \text{in wg-min/ft}
\end{aligned}
$$

The air-to-cloth ratio is the same as the filtering velocity. The pressure drop is

$$
\Delta p = v_{\text{filtering}} S = \left(3.0 \ \frac{\text{ft}}{\text{min}}\right)\left(2.49 \ \frac{\text{in wg-min}}{\text{ft}}\right)
$$

$$
= \boxed{7.47 \ \text{in wg} \quad (7.5 \ \text{in wg})}
$$

The answer is (D).

4. Instantaneous efficiency (given by the graph) is not average efficiency over time. From Eq. 42.13, the average efficiency of the baghouse, η, over five hours is

$$
\eta_{\text{ave}} = \frac{\displaystyle\int_{0 \ \text{hr}}^{5 \ \text{hr}} (1 - e^{-1.26t})\,dt}{5 \ \text{hr}} = \frac{\left(t + \dfrac{e^{-1.26t}}{1.26}\right)\Big|_{0 \ \text{hr}}^{5 \ \text{hr}}}{5 \ \text{hr}}
$$

$$
= \frac{\left(5 \ \text{hr} + \dfrac{e^{\left(-1.26\frac{1}{\text{hr}}\right)(5 \ \text{hr})}}{1.26}\right) - \left(0 \ \text{hr} + \dfrac{e^{\left(-1.26\frac{1}{\text{hr}}\right)(0 \ \text{hr})}}{1.26}\right)}{5 \ \text{hr}}
$$

$$
= 0.8416
$$

The total mass of particulate matter entering the baghouse in five hours is

$$
\begin{aligned}
m_{\text{in}} &= cQt \\
&= \frac{\left(0.8 \ \dfrac{\text{g}}{\text{m}^3}\right)\left(2500 \ \dfrac{\text{ft}^3}{\text{min}}\right)(5 \ \text{hr})\left(60 \ \dfrac{\text{min}}{\text{hr}}\right)}{\left(453.6 \ \dfrac{\text{g}}{\text{lbm}}\right)\left(3.281 \ \dfrac{\text{ft}}{\text{m}}\right)^3} \\
&= 37.45 \ \text{lbm}
\end{aligned}
$$

The mass of particulate matter removed by the filters, m_r, is

$$
m_r = \eta_{\text{ave}} m_{\text{in}} = (0.8416)(37.45 \ \text{lbm})
$$

$$
= \boxed{31.52 \ \text{lbm} \quad (32 \ \text{lbm})}
$$

The answer is (A).

5. The primary sources of lead and cadmium in incinerator fly ash are lead acid batteries and nickel cadmium batteries, respectively.

The answer is (A).

6. The majority of incinerated radioactive waste is low-level waste (LLW).

The answer is (A).

7. Use Eq. 42.40 with the initial conditions to calculate the scrubber constant.

$$
\begin{aligned}
K' &= \frac{\ln(1 - \eta)d_p}{L} \\
&= \frac{\ln(1 - 0.8)(10 \ \mu\text{m})\left(10^{-6} \ \dfrac{\text{m}}{\mu\text{m}}\right)\left(3.281 \ \dfrac{\text{ft}}{\text{m}}\right)}{6 \ \text{ft}} \\
&= -8.80 \times 10^{-6}
\end{aligned}
$$

Using the constant, the scrubber efficiency under the new conditions is

$$
\begin{aligned}
\eta &= 1 - \exp\left(\frac{K'L}{d_p}\right) \\
&= 1 - \exp\left(\frac{(-8.80 \times 10^{-6})(8 \ \text{ft})}{(7 \ \mu\text{m})\left(10^{-6} \ \dfrac{\text{m}}{\mu\text{m}}\right)\left(3.281 \ \dfrac{\text{ft}}{\text{m}}\right)}\right) \\
&= \boxed{0.953 \quad (95\%)}
\end{aligned}
$$

The answer is (C).

Solid Waste

8. Solve Eq. 42.27 for the specific collection area.

$$\eta = 1 - \exp\left(-0.06 w_{e,\text{ft/sec}} \text{SCA}_{\text{ft}^2/1000\,\text{ACFM}}\right)$$

$$\text{SCA} = \frac{-\ln(1-\eta)}{0.06 w_e}$$

$$= \frac{-\ln(1-0.999)}{(0.06)\left(0.31\,\dfrac{\text{ft}}{\text{sec}}\right)}$$

$$= 371.4\ \text{ft}^2/1000\ \text{ACFM}$$

Since the actual volumetric flow rate was given, it does not have to be calculated from the fly ash concentration. The total required plate area is

$$\text{SCA} = \frac{A_p}{Q}$$

$$A_p = Q(\text{SCA})$$

$$= \left(\frac{110{,}000\,\dfrac{\text{ft}^3}{\text{min}}}{1000\,\dfrac{\text{ft}^3}{1000\,\text{ft}^3}}\right)\left(371.4\,\frac{\text{ft}^2}{1000\,\dfrac{\text{ft}^3}{\text{min}}}\right)$$

$$= \boxed{40{,}854\ \text{ft}^2 \quad (41{,}000\ \text{ft}^2)}$$

The answer is (D).

9. Rearrange Eq. 42.22.

$$A = \frac{-\ln(1-\eta)Q}{w}$$

The flow and drift velocity are unchanged. The ratio of the final area to the original area is

$$\frac{A_2}{A_1} = \frac{-\ln(1-\eta_2)}{-\ln(1-\eta_1)}$$

$$= \frac{-\ln(1-0.99)}{-\ln(1-0.95)}$$

$$= \boxed{1.5}$$

The answer is (A).

10. Use Eq. 42.22. Let $Q_t = Q_1 + Q_2$. Use x_1 as the fraction of flow through chamber 1 and x_2 as the fraction of flow through chamber 2.

$$\eta = x_1\left(1 - e^{\frac{-wA_p}{Q_1}}\right) + x_2\left(1 - e^{\frac{-wA_p}{Q_2}}\right)$$

The efficiency is 0.98 when x_1 and x_2 are both 0.5. Rearranging and combining,

$$\eta_{50\text{-}50} = x_1\left(1 - e^{\frac{-wA_p}{Q_1}}\right) + x_2\left(1 - e^{\frac{-wA_p}{Q_2}}\right)$$

$$= (0.50)\left(1 - e^{\frac{-wA_p}{0.5Q_t}}\right) + (0.50)\left(1 - e^{\frac{-wA_p}{0.5Q_t}}\right)$$

$$= (0.50)\left(2 - 2e^{\frac{-wA_p}{0.5Q_t}}\right)$$

$$= 1 - e^{\frac{-wA_p}{0.5Q_t}}$$

$$wA_p = -\ln(1-\eta_{50\text{-}50})(0.5)Q_t$$

$$= -\ln(1-0.98)(0.5)Q_t$$

$$= 1.956Q_t$$

Let $Q_t = Q_1 + Q_2$. With an 80%–20% split, $x_1 = 0.80$, $x_2 = 0.20$, $Q_1 = 0.8Q_t$, and $Q_2 = 0.2Q_t$. The total efficiency is

$$\eta_{80\text{-}20} = x_1\left(1 - e^{\frac{-wA_p}{Q_1}}\right) + x_2\left(1 - e^{\frac{-wA_p}{Q_2}}\right)$$

$$= (0.80)\left(1 - e^{-\frac{1.956Q_t}{0.8Q_t}}\right) + (0.20)\left(1 - e^{-\frac{1.956Q_t}{0.2Q_t}}\right)$$

$$= (0.80)(1 - e^{-2.445}) + (0.20)(1 - e^{-9.780})$$

$$= \boxed{0.9306 \quad (93\%)}$$

The answer is (D).

43 Organic Chemistry

PRACTICE PROBLEMS

1. What is the name of the basic functional group that contains only carbon-carbon double bonds?

(A) alkene

(B) alkane

(C) alkyne

(D) alkyl halide

2. The vitamin niacin, C_5H_4NCOOH, is derived from chemical reactions with a reactant from one of the organic molecule families. What family is the reactant from?

(A) carboxylic acid

(B) amino acid

(C) sugar

(D) nitrile

3. What do C_6H_6 and C_8H_{10} have in common?

(A) They are alcohols.

(B) They are acids.

(C) They are aromatics.

(D) They are sugars.

4. How many double carbon bonds are present in the molecular structure shown?

(A) 1

(B) 2

(C) 5

(D) 10

5. Which of the following structures represents the 3-methylpentane isomer?

(A)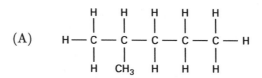

(B)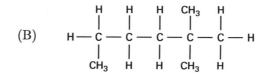

(C)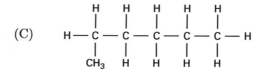

(D)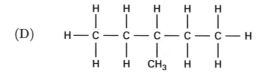

6. How many atoms or functional groups are attached to the nitrogen in *n*-ethyl-2-propanamine?

(A) 1

(B) 2

(C) 3

(D) 4

SOLUTIONS

1. Alkenes (such as ethylene, C_2H_4) have structures of C_nH_{2n} with carbon-carbon double bonds.

The answer is (A).

2. The compound contains a carboxyl (COOH) group, so it could be an organic acid (if combined with alkyl or an aryl), or it could be an amino acid (if combined with an amine). Since C_5H_4N appears to have an amine structure, this compound has been derived from an amino acid. (Niacin is derived from the amino acid tryptophan.)

The answer is (B).

3. Alcohols and sugars contain oxygen, and organic acids usually do also. By elimination, these are aromatics. The chemical formulas represent benzene and xylene, both of which are volatile organic compounds containing a benzene ring. They are members of the arene family, also known as aromatics.

The answer is (C).

4. This is naphthalene. The double lines represent double carbon bonds. There are five double carbon bonds.

![Naphthalene structure showing two fused six-membered carbon rings with hydrogen atoms attached]

The answer is (C).

5. In an isomer, the prefix number designates the carbon (counting from the left in a chain) to which the functional group (methyl in this case) is attached. In option A, the methyl group is attached to the second group, so this is 2-methylpentane. Option B is dimethylhexane. Option C is hexane. Option D is 3-methylpentane.

The answer is (D).

6. From the "-propan-," this is based on the propane molecule and has the form CH_3-CH_2-CH_3, visualized as

![Structural diagram of propane showing three carbon atoms each bonded to hydrogen atoms: H-C-C-C-H with H above and below each carbon]

From the "-amine," one of the hydrogen atoms has been replaced with an amine group, NH. From the "2-," the amine is attached to the second (middle) carbon atom.

From the "-ethyl-," the molecule contains an ethyl (CH_2-CH_3) group. From the "n-," the molecule retains its chain structure and doesn't have any side branches. So, the "-ethyl" group is attached to the nitrogen.

The nitrogen connects directly to the amine hydrogen, the ethyl carbon, and the propyl carbon.

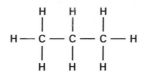

The answer is (C).

44 Biology and Bacteriology

PRACTICE PROBLEMS

1. (*Time limit: one hour*) A fresh wastewater sample containing nitrate ions, sulfate ions, and dissolved oxygen is placed in a sealed jar absent of air.

(a) What is the correct sequence of oxidation of the compounds?

  (A) nitrate, dissolved oxygen, and then sulfate

  (B) sulfate, nitrate, and then dissolved oxygen

  (C) dissolved oxygen, nitrate, and then sulfate

  (D) none of the above

(b) Obnoxious odors will

  (A) appear in the sample when the dissolved oxygen is exhausted

  (B) appear in the sample when the dissolved oxygen and nitrate are exhausted

  (C) appear in the sample when the dissolved oxygen, nitrate, and sulfate are exhausted

  (D) not appear

(c) Bacteria will convert ammonia to nitrate if the bacteria are

  (A) phototropic

  (B) autotrophic

  (C) thermophilic

  (D) obligate anaerobes

(d) Bacteria will generally reduce nitrate to nitrogen gas only if the bacteria are

  (A) photosynthetic

  (B) obligate aerobic

  (C) facultative heterotrophic

  (D) aerobic phototropic

(e) A bacteriophage is a

  (A) bacterial enzyme

  (B) virus that infects bacteria

  (C) mesophilic organism

  (D) virus that stimulates bacterial growth

(f) Algal growth in the wastewater

  (A) will be inhibited when the dissolved oxygen is increased

  (B) will be augmented when nitrifying bacteria are also found in the solution

  (C) will be inhibited when toxins are found in the solution or when the dissolved oxygen is exhausted

  (D) cannot be inhibited by chemical means

(g) In the presence of nitrifying bacteria, nontoxic inorganic compounds, and sunlight, algal growth in the wastewater sample will be

  (A) prevented

  (B) inhibited

  (C) unaffected

  (D) enhanced

(h) The addition of protozoa to the wastewater will

  (A) not change the biochemical composition of the wastewater

  (B) increase the growth of algae in wastewater

  (C) increase the growth of bacteria in wastewater

  (D) decrease the growth of algae and bacteria in the wastewater

(i) Coliform bacteria in wastewater from human, animal, or soil sources

(A) can be distinguished in the multiple-tube fermentation test

(B) can be categorized into only two groups: human/animal and soil

(C) can be distinguished if multiple-tube fermentation and Eschericheiae coli (EC) tests are both used

(D) cannot be distinguished

(j) Under what specific condition would a presence-absence (P-A) coliform test be used on this sample?

(A) if the sample contains known pathogens

(B) when multiple-tube fermentation indicates positive dilution in a presumptive test

(C) when multiple-tube fermentation indicates negative dilution in a presumptive test

(D) if the sample will be used as drinking water

2. A waste stabilization pond will be used in a municipal wastewater treatment system.

(a) Explain the function of bacteria and algae in the stabilization pond's application.

(b) Explain why algae would be a problem if it were present in the discharge from the pond.

(c) Define mechanisms that could be utilized to control algae and that would prove helpful in the development and use of this wastewater treatment system.

(d) Explain what is meant by *facultative pond*.

SOLUTIONS

1. (a) The sequence of oxygen usage reduced by bacteria is dissolved oxygen, nitrate, and then sulfate.

The answer is (C).

(b) Following the sequence of oxidation, obnoxious odors (e.g., hydrogen sulfide) will occur when dissolved oxygen and nitrate are exhausted.

The answer is (B).

(c) Nitrification is performed by autotrophic bacteria to gain energy for growth by synthesis of carbon dioxide in an aerobic environment.

The answer is (B).

(d) Facultative heterotropic bacteria can decompose organic matter to gain energy under anaerobic conditions by removing the oxygen from nitrate, releasing nitrogen gas.

The answer is (C).

(e) A bacteriophage is a virus that infects bacteria.

The answer is (B).

(f) Algae are photosynthetic (gaining energy from light), releasing oxygen during metabolism. They thrive in aerobic environments. The presence of nitrifying bacteria and toxins and the depletion of dissolved oxygen would inhibit the growth process.

The answer is (C).

(g) Nitrifying bacteria are nonphotosynthetic, gaining energy by taking in oxygen to oxidize reduced inorganic nitrogen. The presence of inorganic nutrients would maintain the nitrification process, thereby limiting algal growth.

The answer is (B).

(h) Protozoa consume bacteria and algae in wastewater treatment and in the aquatic food chain.

The answer is (D).

(i) Fecal coliforms from humans and other warm-blooded animals are the same bacterial species. Coliforms originating from the soil can be separated by a confirmatory procedure using EC medium broth incubated at the elevated temperature of 44.5°C (112°F).

The answer is (B).

(j) The presence-absence technique is used for the testing of drinking water.

The answer is (D).

2. (a) A waste stabilization pond's operation is dependent on the reaction of bacteria and algae. Organic matter is metabolized by bacteria to produce the principle products of carbon dioxide, water, and a small amount of ammonia nitrogen. Algae convert sunlight into energy through photosynthesis. They utilize the end products of cell synthesis and other nutrients to synthesize new cells and produce oxygen. The most important role of the algae is in the production of oxygen in the pond for use by aerobic bacteria. In the absence of sunlight, the algae will consume oxygen in the same manner as bacteria. Algae removal is important in producing a high-quality effluent from the pond.

(b) The discharge of algae increases suspended solids in the discharge and may present a problem in meeting water quality criteria. The algae exert an oxygen demand when they settle to the bottom of the stream and undergo respiration.

(c) The following methods have been suggested for control of algae: (1) multiple ponds in series, (2) drawing off of effluent from below the surface by use of a good baffling arrangement to avoid algae concentrations, (3) sand filter or rock filter for algae removal, (4) alum addition and flocculation, (5) microscreening, and (6) chlorination to kill algae. Chlorination may increase BOD loading due to dead algae cells releasing stored organic material.

(d) Facultative ponds have two zones of treatment: an aerobic surface layer in which oxygen is used by aerobic bacteria for waste stabilization and an anaerobic bottom zone in which sludge decomposition occurs. No artificially induced aeration is used.

45 Toxicology

PRACTICE PROBLEMS

1. An environmental engineer has been assigned to manage a laboratory that analyzes hazardous waste samples. She decides to check the workplace air for compliance with the ACGIH values because the ventilation system has been working poorly. She obtained data on possible toxicants in the breathing zone at four workstations.

station	toxicant	concentration for indicated duration (ppm)			
		2 h	4 h	2 h	peak <15 min
1	butane	100	80	120	200
2	pentane	500	70	900	1000
3	hexane	100	200	200	1050
4	mixture of all of above				

The environmental engineer reviewed the ACGIH PELs (permissible exposure limits) for the toxicants identified.

toxicant	TWA (ppm)	STEL (ppm)	ceiling (ppm)
butane	800	–	–
pentane	600	–	–
hexane	500	1000	–

(a) The TWA for butane at workstation 1 is most nearly

(A) 80 ppm

(B) 85 ppm

(C) 90 ppm

(D) 95 ppm

(b) The TWA for pentane at workstation 2 is most nearly

(A) 280 ppm

(B) 325 ppm

(C) 385 ppm

(D) 425 ppm

(c) The TWA for hexane at workstation 3 is most nearly

(A) 165 ppm

(B) 175 ppm

(C) 185 ppm

(D) 195 ppm

(d) Assuming the effects of the toxicants are additive, the TWA of the mixture is most nearly

(A) 0.77

(B) 0.94

(C) 1.1

(D) 1.4

(e) Which of the following statements regarding the laboratory are true?

I. Because the concentration of pentane of 900 ppm exceeds the PEL of 600 ppm, workstation 2 exceeds the ACGIH values.

II. Because the concentration of hexane exceeds the STEL, workstation 3 exceeds ACGIH values.

III. The mixture of butane, pentane, and hexane at workstation 4 exceeds ACGIH values.

IV. Even if hexane were not present, workstation 4 would still exceed ACGIH values.

V. If the effects of each toxicant were independent, workstation 4 would not exceed ACGIH values.

(A) I, III, IV, and V

(B) I, II, and IV

(C) II, IV, and V

(D) III and V

2. Two people work side by side at a chemical plant with poor ventilation. The average plant conditions have been 1 atm and 18°C. They have been working in an environment with air containing 6 ppmv of trichloroethylene (TCE) for 20 years. Standard factors for the two are as follows.

parameter	worker A	worker B
body mass (BW)	100 kg	70 kg
respiration rate (CR)	1.5 m³/h	0.9 m³/h
absorption factor	100%	90%

The molecular weight of TCE is 131.40 g/mol, and the slope factor-inhalation for TCE is 1.3×10^{-2} $(\text{mg/kg·d})^{-1}$. Their workstations are at normal temperature and pressure, and they both work 50 weeks a year, five days a week, and eight hours daily.

(a) The chronic daily intake for worker A is most nearly

(A) 2.4×10^{-4} mg/kg·d

(B) 3.8×10^{-3} mg/kg·d

(C) 1.4×10^{-2} mg/kg·d

(D) 7.8×10^{-1} mg/kg·d

(b) The chronic daily intake for worker B is most nearly

(A) 6.8×10^{-3} mg/kg·d

(B) 2.5×10^{-2} mg/kg·d

(C) 3.6×10^{-2} mg/kg·d

(D) 6.0×10^{-1} mg/kg·d

(c) The excess cancer risk for worker A is most nearly

(A) 1.3×10^{-6}

(B) 2.8×10^{-5}

(C) 1.0×10^{-2}

(D) 4.6×10^{-2}

(d) The excess cancer risk for worker B is most nearly

(A) 4.3×10^{-6}

(B) 3.4×10^{-5}

(C) 7.8×10^{-3}

(D) 4.3×10^{-2}

3. Drainage from a former pesticide manufacturing site enters Bear River at river kilometer 240. The drainage has occurred for the past 30 years. The drainage contains DDT, and the concentration of DDT after complete mixing is 5 μg/L. The town of Bear Valley, adult population 15 000, has used the river at river kilometer 40 for its water supply for the past 30 years. Chlorination is the only pre-distribution treatment. The river velocity is 5 cm/s. Assume the half-life of DDT is 70 days. The slope factor for DDT is 0.34 $(\text{mg/kg·d})^{-1}$.

(a) The concentration of DDT in the town's drinking water is most nearly

(A) 1.2 μg/L

(B) 2.3 μg/L

(C) 3.2 μg/L

(D) 3.9 μg/L

(b) The lifetime chronic daily intake of DDT for an adult living in Bear Valley is most nearly

(A) 4×10^{-5} mg/kg·d

(B) 8×10^{-4} mg/kg·d

(C) 6×10^{-3} mg/kg·d

(D) 2×10^{-2} mg/kg·d

(c) The number of excess lifetime cancers that can be expected in the adult population of Bear Valley is most nearly

(A) 0.05

(B) 0.2

(C) 4

(D) 10

4. (a) The major human pathways of exposure to toxic agents are

I.   inhalation

II. ingestion

III. whole body

IV. skin absorption

V.   eyes and hands

(A) I, II, and V

(B) I, III, and IV

(C) III, IV, and V

(D) I, II, and IV

(b) The absorption of toxicants into the bloodstream through the lungs depends upon the

I.   gradient between the concentration of the toxicant in the blood and its concentration in air

II. solubility of the toxicant in the blood

III. size of the toxicant molecule

IV. respiration rate

V.   condition of the lungs

(A) I, II, and IV

(B) I, II, IV, and V

(C) II, III, IV, and V

(D) I, II, III, IV, and V

(c) Toxicity effects are exerted on target organs or tissue if the rate of accumulation exceeds the

I.   rate of biotransformation
II. rate of elimination, transformation, and storage
III. slope of the dose-response curve
IV. metabolic rate of the exposed individual
V.   slope factor

   (A) I and II
   (B) II only
   (C) III and IV
   (D) V only

(d) LD_{50} means the dose in

   (A) mg/d at which 50 of the test animals died
   (B) mg/d at which 50% of the test animals died
   (C) mg/kg body mass at which 50% of the test animals died
   (D) mg/kg body mass at which an adverse effect was observed in 50% of the test animals

(e) Which of the following statements are true?

I.   Carcinogens have a dose threshold below which no adverse health response can be measured.
II. Exposure to a carcinogen below its threshold has no associated risk.
III. Carcinogens can produce malignant tumors that spread to other parts of the body through the bloodstream.
IV. Carcinogens cause cell mutation through damage to DNA beyond the ability of the body to make repairs.

   (A) I, III, and IV
   (B) II and III
   (C) I, II, and IV
   (D) III and IV

(f) What is the difference between acute and chronic exposure to toxic chemicals?

   (A) Acute exposure occurs through the inhalation pathway; chronic exposure occurs through skin absorption.
   (B) Acute exposure is a short duration, typically less than 24 hours; chronic exposure is for a long duration, typically 30 days or more.
   (C) Acute exposure results in death to 50% of test animals; chronic exposure results in 100% survival of test animals.
   (D) Acute exposure occurs above the threshold of the dose-response curve.

(g) Chemical A has an LD_{50} of 3 mg/kg. Chemical B has an LD_{50} of 9 mg/kg. Chemical A is

   (A) one-third as toxic as chemical B
   (B) three times as toxic as chemical B
   (C) not comparable to chemical B because the pathways are not specified
   (D) not comparable to chemical B because the target organs are not specified

(h) The law of additive effects

   (A) applies to all mixtures of gases and vapors
   (B) applies only when the effects from the component gases and vapors occur within the same organs
   (C) applies to noncarcinogens less than the TLV
   (D) applies to carcinogens with a high slope factor

(i) Biological exposure indices are

   (A) intended to replace the ACGIH-TLV values
   (B) the levels of contaminants that would be observed in specimens from a healthy worker exposed to the same level of TLVs
   (C) not suitable for use in calculating the total body burden
   (D) limited to use for the dermal pathway

(j) EPA reference doses (RfD) for noncarcinogens are

   (A) developed from precise chemical studies and are very accurate
   (B) independent of the pathway of exposure
   (C) typically based on the dose that results in no observed adverse effect, but the dose associated with the lowest observed adverse effect can be used if necessary
   (D) based on the most sensitive end point when there is more than one toxicological effect

SOLUTIONS

1. (a) From Eq. 45.6,

$$E = \tfrac{1}{8}\sum_{i=1}^{n} C_i T_i$$

$$E_{\text{butane}} = \left(\frac{1}{8\ \text{h}}\right)\left(\begin{array}{c}(100\ \text{ppm})(2\ \text{h}) + (80\ \text{ppm})(4\ \text{h}) \\ + (120\ \text{ppm})(2\ \text{h})\end{array}\right)$$

$$= \boxed{95\ \text{ppm}}$$

The answer is (D).

(b) From Eq. 45.6,

$$E = \tfrac{1}{8}\sum_{i=1}^{n} C_i T_i$$

$$E_{\text{pentane}} = \left(\frac{1}{8\ \text{h}}\right)\left(\begin{array}{c}(500\ \text{ppm})(2\ \text{h}) + (70\ \text{ppm})(4\ \text{h}) \\ + (900\ \text{ppm})(2\ \text{h})\end{array}\right)$$

$$= \boxed{385\ \text{ppm}}$$

The answer is (C).

(c) From Eq. 45.6,

$$E = \tfrac{1}{8}\sum_{i=1}^{n} C_i T_i$$

$$E_{\text{hexane}} = \left(\frac{1}{8\ \text{h}}\right)\left(\begin{array}{c}(100\ \text{ppm})(2\ \text{h}) + (200\ \text{ppm})(4\ \text{h}) \\ + (200\ \text{ppm})(2\ \text{h})\end{array}\right)$$

$$= \boxed{175\ \text{ppm}}$$

The answer is (B).

(d) From Eq. 45.7,

$$E_m = \sum_{i=1}^{n} \frac{C_i}{L_i}$$

$$= \frac{C_{\text{butane}}}{L_{\text{butane}}} + \frac{C_{\text{pentane}}}{L_{\text{pentane}}} + \frac{C_{\text{hexane}}}{L_{\text{hexane}}}$$

$$= \frac{95\ \text{ppm}}{800\ \text{ppm}} + \frac{385\ \text{ppm}}{600\ \text{ppm}} + \frac{175\ \text{ppm}}{500\ \text{ppm}}$$

$$= \boxed{1.11}$$

The answer is (C).

(e)

I. False. The TWA concentration of pentane is 385 ppm, and the PEL-TWA is 800 ppm, so the PEL-TWA is not exceeded.

II. False.

III. True.

IV. False. If hexane was not present at workstation 4, the PEL of the mixture would be 95/800 (butane) plus 385/600 (pentane) = 0.760, which is less than 1.0 and would not exceed the ACGIH limit.

V. True. The independent effects are each less than 1.0.

The answer is (D).

2. Since ppmv is a concentration ratio by volume, calculate the volume of TCE. Work with 1 m³ of air.

$$V_{\text{TCE}} = \frac{C_{\text{ppmv}} V_{\text{air}}}{10^6}$$

$$= \frac{(6\ \text{parts})(1\ \text{m}^3)}{10^6\ \text{parts}}$$

$$= 6 \times 10^{-6}\ \text{m}^3$$

Calculate the mass of TCE in 1 m³ of air. Use the ideal gas law.

$$m = \frac{pV}{RT} = \frac{pV}{\dfrac{R^* T}{\text{MW}}}$$

$$= \frac{(1\ \text{atm})\left(1.013 \times 10^5\ \dfrac{\text{Pa}}{\text{atm}}\right)(6 \times 10^{-6}\ \text{m}^3)}{\dfrac{\left(8314\ \dfrac{\text{J}}{\text{kmol·K}}\right)(18°\text{C} + 273°)}{\left(131.40\ \dfrac{\text{g}}{\text{mol}}\right)\left(1000\ \dfrac{\text{mol}}{\text{kmol}}\right)\left(1000\ \dfrac{\text{mg}}{\text{g}}\right)}}$$

$$= 33.0\ \text{mg}$$

Since this is the mass of TCE in 1 m³, it is also the value of the concentration in mg/m³.

(a) From Eq. 11.94, the CDI for worker A is

$$\text{CDI} = \frac{C_{\text{TCE/air}}(\text{CR})(\text{EF})(\text{ED})}{(\text{BW})(\text{AT})}$$

$$= \frac{\left(33.0\ \dfrac{\text{mg}}{\text{m}^3}\right)\left(1.5\ \dfrac{\text{m}^3}{\text{h}}\right)\left(8\ \dfrac{\text{h}}{\text{d}}\right)\left(5\ \dfrac{\text{d}}{\text{wk}}\right)}{(100\ \text{kg})(70\ \text{yr})\left(365\ \dfrac{\text{d}}{\text{yr}}\right)} \times \left(50\ \dfrac{\text{wk}}{\text{yr}}\right)(20\ \text{yr})(1)$$

$$= \boxed{0.775\ \text{mg/kg·d}}$$

The answer is (D).

(b) From Eq. 11.94, the CDI for worker B given an absorption of 0.9 is

$$\text{CDI} = \frac{C_{\text{TCE/air}}(\text{CR})(\text{EF})(\text{ED})}{(\text{BW})(\text{AT})}$$

$$= \frac{\left(33.0 \ \frac{\text{mg}}{\text{m}^3}\right)\left(0.9 \ \frac{\text{m}^3}{\text{h}}\right)\left(8 \ \frac{\text{h}}{\text{d}}\right)\left(5 \ \frac{\text{d}}{\text{wk}}\right)}{\times \left(50 \ \frac{\text{wk}}{\text{yr}}\right)(20 \ \text{yr})(0.9)}$$

$$= \boxed{0.598 \ \text{mg/kg·d}}$$

The answer is (D).

(c) From Eq. 45.4, the risk for worker A is

$$R = (\text{SF})(\text{CDI})$$

$$= \left(1.3 \times 10^{-2} \ \frac{\text{kg·d}}{\text{mg}}\right)\left(7.75 \times 10^{-1} \ \frac{\text{mg}}{\text{kg·d}}\right)$$

$$= \boxed{1.0 \times 10^{-2}}$$

The answer is (C).

(d) From Eq. 45.4, the risk for worker B is

$$R = (\text{SF})(\text{CDI})$$

$$= \left(1.3 \times 10^{-2} \ \frac{\text{kg·d}}{\text{mg}}\right)\left(5.98 \times 10^{-1} \ \frac{\text{mg}}{\text{kg·d}}\right)$$

$$= \boxed{7.8 \times 10^{-3}}$$

The answer is (C).

3. (a) travel distance $= 240 \ \text{km} - 40 \ \text{km} = 200 \ \text{km}$

$$\text{river velocity} = 5 \ \text{cm/s} \quad (5 \times 10^{-2} \ \text{m/s})$$

$$\text{travel time} = \frac{s}{\text{v}}$$

$$= \frac{(200 \ \text{km})\left(1000 \ \frac{\text{m}}{\text{km}}\right)}{\left(5 \times 10^{-2} \ \frac{\text{m}}{\text{s}}\right)\left(86\,400 \ \frac{\text{s}}{\text{d}}\right)}$$

$$= 46.3 \ \text{d}$$

Assuming a first-order reaction, $C_t = C_o e^{-kt}$

At half-life,

$$C_t = \frac{C_o}{2}$$

$$t = 70 \ \text{d}$$

$$0.5 = e^{-k(70 \ \text{d})}$$

$$k = \frac{-\ln 0.5}{70 \ \text{d}} = 9.902 \times 10^{-3} \ \text{d}^{-1}$$

$$C_t = \left(5 \ \frac{\mu\text{g}}{\text{L}}\right)e^{-\left(0.009902 \ \frac{1}{\text{d}}\right)(46.3 \ \text{d})}$$

$$= \boxed{3.16 \ \mu\text{g/L}}$$

The answer is (C).

(b) From Table 45.3, the standard factors for adults in Bear Valley are

$$C_{\text{DDT/water}} = 3 \ \mu\text{g/L}$$

$$\text{CR} = 2 \ \text{L/d}$$

$$\text{BW} = 71.8 \ \text{kg}$$

$$\text{ED} = 30 \ \text{yr}$$

$$\text{AT} = 70 \ \text{yr}$$

From Eq. 11.94, the CDI is

$$\text{CDI} = \frac{C_{\text{DDT/water}}(\text{CR})(\text{EF})(\text{ED})}{(\text{BW})(\text{AT})}$$

$$= \frac{\left(3 \times 10^{-6} \ \frac{\text{g}}{\text{L}}\right)\left(2 \ \frac{\text{L}}{\text{d}}\right)(30 \ \text{yr})\left(1000 \ \frac{\text{mg}}{\text{g}}\right)(1)}{(71.8 \ \text{kg})(70 \ \text{yr})}$$

$$= \boxed{3.58 \times 10^{-5} \ \text{mg/kg·d}}$$

The answer is (A).

(c) From Eq. 45.4,

$$R = (\text{SF})(\text{CDI})$$

$$= \left(0.34 \ \frac{\text{kg·d}}{\text{mg}}\right)\left(3.58 \times 10^{-5} \ \frac{\text{mg}}{\text{kg·d}}\right)$$

$$= 1.22 \times 10^{-5}$$

From Eq. 45.3,

$$\text{EC} = (\text{EP})R$$

$$= (15\,000)(1.22 \times 10^{-5})$$

$$= \boxed{0.183 \quad (0.2)}$$

The answer is (B).

4. (a) The major pathways are inhalation, ingestion, and skin absorption.

The answer is (D).

(b) All statements are true.

The answer is (D).

(c) Only II is true.

The answer is (B).

(d) Only C is true.

The answer is (C).

(e) III and IV are true.

The answer is (D).

(f) Only B is true.

The answer is (B).

(g) Only B is true.

The answer is (B).

(h) Only B is true.

The answer is (B).

(i) Only B is true.

The answer is (B).

(j) Only C is true.

The answer is (C).

46 Industrial Hygiene

PRACTICE PROBLEMS

1. A concentration of 2.2 ppmv of methane (CH_4) was measured at a landfill at an air temperature of 25°C. What is the concentration in mg/m^3?

2. A 25 L sample of landfill gas was bubbled through 50 mL of a solution that has an 85% collection efficiency for methane. The sample was analyzed at 25°C and 750 mm Hg pressure. The laboratory reported the result of 25 $\mu g/mL$. What is the concentration of methane in the landfill gas in ppmw?

3. During each shift, workers in a sewage treatment plant are exposed to noise from a sludge pump, a blower, and a ventilation fan as indicated.

source	sound pressure level (dB) (reference: 20 μPa)	period of operation (h)
sludge pump	96	0800–1200
		1600–1700
blower	94	0800–1000
		1500–1700
ventilation fan	98	0800–1000
		1300–1500

All noise is at 1000 Hz and referenced to the dBA scale.

(a) What is the sound pressure level the workers are exposed to during the period from 0800 to 1000 hours?

    (A) 99 dB

    (B) 100 dB

    (C) 101 dB

    (D) 103 dB

(b) What is the sound pressure level the workers are exposed to during the period from 1600 to 1700 hours?

    (A) 97 dB

    (B) 98 dB

    (C) 99 dB

    (D) 100 dB

(c) What is the time-weighted sound pressure level exposure for one shift?

    (A) 95.2 dB

    (B) 96.4 dB

    (C) 97.8 dB

    (D) 98.3 dB

(d) Relative to OSHA noise exposure standards, which of the following statements is true for this situation?

    (A) Standards are exceeded because the TWA exposure exceeds 85 dBA.

    (B) Standards are not exceeded because none of the permissible noise levels are exceeded for any duration allowed per day.

    (C) Standards are exceeded because the sum of the fractional allowable exposure duration for each sound pressure level exceeds 1.0 (100%).

    (D) Standards are not exceeded because the TWA exposure does not exceed the permissible TWA exposures.

4. Which of the following statements about biological hazards is/are FALSE?

I.   A bacterium is a prokaryote whose nucleus is not physically separated from its cytoplasm.

II.   Viruses depend on their host cells for reproduction and metabolism.

III.   Exogenous infections are caused by microorganisms normally present in a part of the body that have reached another site in the body.

IV.   Zoonoses are microorganisms capable of infecting both animals and humans.

    (A) I and III

    (B) I, II, and III

    (C) III only

    (D) IV only

5. Which of the following diseases should be of special concern to the environmental engineer when designing, operating, or managing water supply projects?

I. acquired immunodeficiency syndrome (AIDS)

II. Rocky Mountain Spotted Fever

III. botulism

IV. tuberculosis

V. Legionnaires' disease

VI. hepatitis A

    (A) I, III, and VI

    (B) II, IV, and VI

    (C) V only

    (D) V and VI

6. Which of the following statements concerning heat and cold stress are true?

I. Metabolic rate can affect heat gain by one or two orders of magnitude compared to radiation and convection, but affects heat loss to about the same extent as radiation and convection.

II. The NIOSH ceiling limit (CL), recommended exposure limit (REL), and recommended alert level (RAL) are functions of the time-weighted average of the metabolic rate.

III. Heat stress on unacclimatized workers should not exceed the recommended alert level (RAL).

IV. Two physiological functions that can be readily measured to confirm that heat stress is being controlled to allowable limits are core temperature and heart rate.

V. The main hazards from cold stress are hypothermia, tissue damage, and reduced blood circulation.

    (A) I, III, IV, and V

    (B) II, III, IV, and V

    (C) I, II, and V

    (D) I, II, III, and IV

7. An environmental engineer is the project manager for a project to upgrade two pumps that handle cooling water at a nuclear power plant. The pumps are in a restricted area with ionizing radiation at a rate of 1530 μSv/d. The upgrade will require two workers at all times and is expected to take a total of 6.5 h. The allowable radiation exposure limit for an 8 h workday is 140 μSv/d/person. How should the engineer schedule the work?

8. An environmental engineer must prepare some wastewater samples for analysis using isopropyl alcohol (CH_3CH_3CHOH). Vapor levels over an 8 h workday will be 420 ppm (2 h), 350 ppm (2.5 h), and 390 ppm (3.5 h). The temperature of the lab is 30°C because the air conditioner is broken. Will the engineer's exposure level be in compliance if the threshold limit value (TLV) for isopropyl alcohol is 400 ppm at 25°C?

9. Environmental engineers are testing some air monitoring equipment and need to make a 20 ppmv concentration of carbon tetrachloride (CCl_4) in a 200 L drum. How many milliliters of liquid CCl_4 at a liquid density of 1.5 g/mL need to be evaporated at 28°C and 750 mm Hg (0.987 atm, 100 kPa) to make the test drum?

10. Which of the following statements are true about cumulative trauma disorders (CTDs)?

I. Risk of CTDs is associated with highly repetitive use of the same body part.

II. Damage from CTDs is usually noticed by workers within 24 h of the activity.

III. CTDs may affect tendons, tendon sheaths, soft tissue, nerves, bones, and spinal vertebrae.

IV. High repetitiveness is defined as a cycle time of less than one minute.

V. Leaning over a workbench and resting the forearm on a hard edge or surface may lead to carpal tunnel syndrome.

    (A) I, III, and V

    (B) II, III, and V

    (C) I and III

    (D) I, IV, and V

SOLUTIONS

1. At STP (0°C and 1 atm), 1 mol of any gas occupies 22.4 L. Calculate its volume at 25°C.

$$T_1 = 0°C + 273° = 273K$$

$$T_2 = 25°C + 273° = 298K$$

$$V_2 = V_1\left(\frac{T_2}{T_1}\right) = \left(22.4\ \frac{L}{mol}\right)\left(\frac{298K}{273K}\right)$$

$$= 24.45\ L/mol$$

The molecular weight of methane is

$$MW = (1)\left(12\ \frac{g}{mol}\right) + (4)\left(1\ \frac{g}{mol}\right) = 16\ g/mol$$

Given a methane concentration of 2.2 ppmv (2.2×10^{-6} liters of methane per liter of air), calculate the number of moles of methane per unit volume of air.

$$\frac{\text{moles methane}}{\text{volume air}} = \frac{\text{methane concentration}}{\text{methane volume}}$$

$$= \frac{2.2 \times 10^{-6}\ \dfrac{L}{L}}{24.45\ \dfrac{L}{mol}}$$

$$= 9.0 \times 10^{-8}\ mol/L$$

To calculate the concentration in mg/m^3, use the calculation of the number of moles.

$$n = \frac{m}{MW}$$

$$\frac{m}{V_{air}} = \left(\frac{n}{V_{air}}\right)(MW)$$

$$= \left(9.0 \times 10^{-8}\ \frac{mol}{L}\right)\left(16\ \frac{g}{mol}\right)$$

$$\times \left(10^3\ \frac{mg}{g}\right)\left(10^3\ \frac{L}{m^3}\right)$$

$$= \boxed{1.44\ mg/m^3}$$

Alternatively, the concentration of methane in mg/m^3 can be calculated as

$$C_{m/V} = C_{ppm}\left(\frac{MW}{V_2}\right)$$

$$= \left(2.2 \times 10^{-6}\ \frac{L}{L}\right)\left(\frac{16\ \dfrac{g}{mol}}{24.45\ \dfrac{L}{mol}}\right)$$

$$\times \left(10^3\ \frac{mg}{g}\right)\left(10^3\ \frac{L}{m^3}\right)$$

$$= \boxed{1.44\ mg/m^3}$$

2. The total mass of methane in the 25 L sample is

$$m = \frac{\left(\begin{array}{c}\text{concentration of}\\ \text{sample in solution}\end{array}\right)\left(\begin{array}{c}\text{volume of}\\ \text{solution}\end{array}\right)}{\text{collection efficiency}}$$

$$= \frac{\left(25\ \dfrac{\mu g}{mL}\right)(50\ mL)}{0.85}$$

$$= 1470.6\ \mu g$$

At STP (0°C and 1 atm), 1 mol of any gas occupies 22.4 L. Calculate its volume at 25°C.

$$T_1 = 0°C + 273° = 273K$$

$$T_2 = 25°C + 273° = 298K$$

$$p_1 = 760\ mm\ Hg$$

$$p_2 = 750\ mm\ Hg$$

$$\frac{p_1 V_1}{T_1} = \frac{p_2 V_2}{T_2}$$

$$V_2 = V_1\left(\frac{T_2}{T_1}\right)\left(\frac{p_1}{p_2}\right)$$

$$= \left(22.4\ \frac{L}{mol}\right)\left(\frac{298K}{273K}\right)\left(\frac{760\ mm\ Hg}{750\ mm\ Hg}\right)$$

$$= 24.8\ L/mol$$

The molecular weight of methane is 16 g/mol. The concentration of methane in ppm is

$$C_{ppm} = C_{m/V}\left(\frac{V_2}{MW}\right)$$

$$= \left(\frac{1470.6\ \mu g}{25\ L}\right)\left(\frac{24.8\ \dfrac{L}{mol}}{16\ \dfrac{g}{mol}}\right)$$

$$= \boxed{91.2\ \mu g/g\quad(91.2\ ppmw)}$$

3. (a) Use Eq. 46.8.

$$L_{p,total} = 10\log\sum_i^3 10^{L_i/10}$$

$$= 10\log(10^{96/10} + 10^{94/10} + 10^{98/10})$$

$$= \boxed{101.1\ dB}$$

The answer is (C).

(b) Use Eq. 46.8.

$$L_{p,\text{total}} = 10 \log \sum_i^2 10^{L_i/10}$$

$$= 10 \log(10^{96/10} + 10^{94/10})$$

$$= \boxed{98.1 \text{ dB}}$$

The answer is (B).

(c) Calculate TWA exposure using Eq. 46.8.

period beginning hour	sludge pump (dB)	blower (dB)	ventilation fan (dB)	total exposure (dB)
0800	96	94	98	101.1
0900	96	94	98	101.1
1000	96			96
1100	96			96
1300			98	98
1400			98	98
1500		94		94
1600	96	94		98.1
total TWA				97.8

The answer is (C).

(d) Relative to OSHA noise exposure standards, the statements are as follows:

(A) False; the standards allow 85 dBA to be exceeded for specific durations at various sound pressure levels.

(B) False; the standards allow permissible noise levels to be exceeded for specific durations.

(C) True; from Eq. 46.8, Eq. 46.9, and Table 46.4,

total exposure (dB)	exposure duration, C (h)	permissible exposure duration, t (h)	C/t
101.1	2	1.5	$2/1.5 = 1.33$
96	2	3	$2/3 = 0.67$
98	2	2	$2/2 = 1.00$
94	1	4	$1/4 = 0.25$
98.1	1	2	$1/2 = 0.50$
Σ	8		3.75

Because the summation of 3.75 is greater than 1.0, the OSHA permissible noise exposure is exceeded. A hearing conservation program is required.

(D) False; the standards are based on an allowable duration at each exposure level, not on the TWA of the permissible exposures.

The answer is (C).

4. Statements I, II, and IV are true. Statement III is false; exogenous infections are caused by microorganisms that enter the body from the environment. Endogenous infections are caused by indigenous microorganisms.

The answer is (C).

5. Statements I through IV are not concerns: AIDS is not a waterborne disease. Rocky Mountain spotted fever is passed to humans by ticks. Botulism is foodborne or can enter the body through contaminated soil or needles. Tuberculosis is spread by inhalation of infectious droplets. Legionnaires' disease, on the other hand, is associated with heat-transfer systems, warm-temperature water, and stagnant water. Hepatitis A can also be spread through contaminated water.

The answer is (D).

6. Statements I, II, III, and IV are true. Statement V is false: Reduced blood circulation is one of the body's mechanisms to reduce heat loss but is not a main hazard.

The answer is (D).

7. Hourly exposure is

$$\frac{1530 \, \dfrac{\mu\text{Sv}}{\text{d}}}{24 \, \dfrac{\text{h}}{\text{d}}} = 63.75 \, \mu\text{Sv/h}$$

Maximum individual allowable exposure time is

$$\frac{140 \, \dfrac{\dfrac{\mu\text{Sv}}{\text{d}}}{\text{person}}}{63.75 \, \dfrac{\mu\text{Sv}}{\text{h}}} = 2.2 \text{ h/d/person}$$

> Use a total of four teams, three in 2 h shifts and one in a $1/2$ h shift. Other combinations are possible.

8. At STP (0°C and 1 atm), 1 mol of any gas occupies 22.4 L. Calculate its volume at 25°C and 30°C.

$$T_1 = 0°\text{C} + 273° = 273\text{K}$$

$$T_2 = 25°\text{C} + 273° = 298\text{K}$$

$$T_3 = 30°\text{C} + 273° = 303\text{K}$$

At constant pressure (1 atm),

$$\frac{V_1}{T_1} = \frac{V_2}{T_2} = \frac{V_3}{T_3}$$

$$V_2 = V_1\left(\frac{T_2}{T_1}\right)$$
$$= \left(22.4\ \frac{L}{mol}\right)\left(\frac{298K}{273K}\right)$$
$$= 24.45\ L/mol$$
$$V_3 = V_1\left(\frac{T_3}{T_1}\right)$$
$$= \left(22.4\ \frac{L}{mol}\right)\left(\frac{303K}{273K}\right)$$
$$= 24.86\ L/mol$$

The molecular weight of isopropyl alcohol is

$$MW = (3)\left(12\ \frac{g}{mol}\right) + (8)\left(1\ \frac{g}{mol}\right) + (1)\left(16\ \frac{g}{mol}\right)$$
$$= 60\ g/mol$$

Calculate the concentration of isopropyl alcohol at 25°C for each exposure period. Note that as the temperature increases, the volume increases, but the number of moles remains constant, so the concentration decreases.

$$C_2 = C_3\left(\frac{V_2}{V_3}\right) = (420\ ppm)\left(\frac{24.45\ \frac{L}{mol}}{24.86\ \frac{L}{mol}}\right)$$
$$= 413\ ppm$$

Similarly,

$$350\ ppm\ at\ 30°C = 344\ ppm\ at\ 25°C$$
$$390\ ppm\ at\ 30°C = 384\ ppm\ at\ 25°C$$

Calculate TLV-TWA for the 8 h day.

$$TLV\text{-}TWA = \frac{\binom{(2\ h)(413\ ppm) + (2.5\ h)(344\ ppm)}{+ (3.5\ h)(384\ ppm)}}{8\ h}$$
$$= 379\ ppm$$

The TLV-TWA of 400 ppm will not be exceeded.

9. At STP (0°C and 1 atm), 1 mol of any gas occupies 22.4 L. Calculate its volume at 28°C.

$$T_1 = 0°C + 273° = 273K$$
$$T_2 = 28°C + 273° = 301K$$
$$p_1 = 760\ mm\ Hg$$
$$p_2 = 750\ mm\ Hg$$

$$\frac{p_1 V_1}{T_1} = \frac{p_2 V_2}{T_2}$$
$$V_2 = V_1\left(\frac{T_2}{T_1}\right)\left(\frac{p_1}{p_2}\right)$$
$$= \left(22.4\ \frac{L}{mol}\right)\left(\frac{301K}{273K}\right)\left(\frac{760\ mm\ Hg}{750\ mm\ Hg}\right)$$
$$= 25\ L/mol$$

The molecular weight of carbon tetrachloride is

$$MW = (1)\left(12\ \frac{g}{mol}\right) + (4)\left(35.45\ \frac{g}{mol}\right)$$
$$= 154\ g/mol$$

Given the drum volume and a carbon tetrachloride concentration of 20 ppm (20×10^{-6} liters CCl_4 per liter of air), calculate the required volume of CCl_4 gas.

$$V_{gas} = C V_{drum}$$
$$= \left(20 \times 10^{-6}\ \frac{L}{L}\right)(200\ L)$$
$$= 4 \times 10^{-3}\ L$$

The required number of carbon tetrachloride moles is

$$n = \frac{V_{gas}}{V_2} = \frac{4 \times 10^{-3}\ L}{25\ \frac{L}{mol}}$$
$$= 1.6 \times 10^{-4}\ mol$$

$$m = n(MW)$$
$$= (1.6 \times 10^{-4}\ mol)\left(154\ \frac{g}{mol}\right)$$
$$= 2.46 \times 10^{-2}\ g$$

The volume of liquid carbon tetrachloride is

$$V_{liq} = \frac{m}{\rho} = \frac{2.46 \times 10^{-2}\ g}{1.5\ \frac{g}{mL}}$$
$$= \boxed{1.64 \times 10^{-2}\ mL}$$

10. Statements I and III are true. Statements II, IV, and V are false: CTDs may not show symptoms for months or years. High repetitiveness is defined as a cycle time of less than thiry seconds. Statement V should refer to cubital tunnel syndrome.

The answer is (C).

47 Health and Safety

PRACTICE PROBLEMS

1. Workers at a cleanup site have determined that benzene (C_6H_6) is a contaminant. The equilibrium vapor pressure of benzene in air is 7×10^{-3} mm Hg at normal temperature and pressure (NTP).

(a) The concentration of benzene in air at NTP in ppm is most nearly

(A) 4 ppm

(B) 6 ppm

(C) 9 ppm

(D) 12 ppm

(b) The concentration of benzene in air at NTP in mg/m^3 is most nearly

(A) 16 mg/m^3

(B) 20 mg/m^3

(C) 26 mg/m^3

(D) 29 mg/m^3

2. Toluene was evaporated into an airtight room filled with clean air at a temperature of 40°C and a pressure of 760 mm Hg. The specific gravity of toluene is 0.86. The room measures 10 m by 20 m by 10 m.

(a) If the volume of toluene liquid prior to vaporization is 500 mL, the concentration of toluene in the room is most nearly

(A) 60 ppmv

(B) 100 ppmv

(C) 190 ppmv

(D) 240 ppmv

(b) If the pressure in the room is increased to 780 mm Hg, the concentration of toluene in the room is most nearly

(A) 35 ppmv

(B) 60 ppmv

(C) 80 ppmv

(D) 100 ppmv

3. A supply chamber contains 0.06% ethylene oxide (C_2H_4O) at 10°C and 1 atm. A pump calibrated at 0.4 L/min is to deliver a mixture of clean air and ethylene oxide at a mixed concentration of 100 ppm to a test chamber that is maintained at 45°C and 1 atm.

(a) The concentration of ethylene oxide in the supply chamber is most nearly

(A) 200 ppmv

(B) 300 ppmv

(C) 500 ppmv

(D) 600 ppmv

(b) The flow rate of ethylene oxide needed is most nearly

(A) 0.01 L/min

(B) 0.07 L/min

(C) 0.10 L/min

(D) 0.15 L/min

(c) The flow rate of dilution air is most nearly

(A) 0.1 L/min

(B) 0.2 L/min

(C) 0.3 L/min

(D) 0.4 L/min

(d) If the test chamber measures 2 m by 1 m by 1 m, the volume of supply ethylene oxide needed to collect 2 g of ethylene oxide in the test chamber is most nearly

(A) 500 L

(B) 1800 L

(C) 2900 L

(D) 4500 L

4. Workers at a shipyard are exposed to lead while scrapping old navy warships. To determine lead exposure, the air sampling time required to produce sufficient mass for a reliable exposure result must be calculated. The OSHA PEL for lead is 0.050 mg/m³. The ACGIH TLV is 0.05. NIOSH test method 7082 for inorganic lead metals gives the following sampling information for lead.

range studied	0.13 to 0.4 mg/m³ or 10 to 200 μg/sample
estimated LOD	2.6 μg/sample
air sampling flow rate	1 to 4 L/min
PEL	0.05 mg/m³
sampling media	0.8 μm cellulose ester membrane

(a) The sampling technician wants to collect a lead sample mass of 10 times the LOD. He wants to try an exposure of 25% of the PEL for the first round of preliminary sampling. The maximum air sample he should collect at the field conditions of 30°C and 650 mm Hg is most nearly

(A) 500 L

(B) 1000 L

(C) 2000 L

(D) 2500 L

(b) A sampling pump calibrated at field conditions will deliver 4 L/min. The time to collect the sample is most nearly

(A) 300 min

(B) 400 min

(C) 500 min

(D) 600 min

(c) If the technician collects the minimum sample mass, how will the sampling protocol change?

(A) 2 samples at 4 h each

(B) 1 sample at 8 h each

(C) 4 samples at 2 h each

(D) 2 samples at 3 h each plus 1 sample at 2 h

SOLUTIONS

1. (a) At NTP,

$$T = 25°C + 273° = 298K$$

$$p = (760 \text{ mm Hg})\left(\frac{1 \text{ atm}}{760 \text{ mm Hg}}\right) = 1 \text{ atm}$$

Use Eq. 47.17.

$$C_{\text{ppm,benzene}} = \left(\frac{p_{\text{PVP,benzene}}}{p_{\text{total}}}\right)(10^6)$$

$$= \left(\frac{7 \times 10^{-3} \text{ mm Hg}}{760 \text{ mm Hg}}\right)(10^6)$$

$$= \boxed{9.21 \text{ ppm at NTP}}$$

The answer is (C).

(b) The molecular weight of benzene is

$$\text{MW}_{\text{benzene}} = (6)\left(12 \frac{\text{g}}{\text{mol}}\right) + (6)\left(1 \frac{\text{g}}{\text{mol}}\right)$$

$$= 78 \text{ g/mol}$$

Use Eq. 47.11.

$$C_{m/V,\text{benzene}} = \left(\frac{C_{\text{ppm,benzene}}(\text{MW}_{\text{benzene}})p}{R^*T}\right)(10^{-6})$$

$$= \frac{(9.21 \text{ ppm})\left(78 \frac{\text{g}}{\text{mol}}\right)(1 \text{ atm})}{\times (10^{-6})\left(10^3 \frac{\text{mg}}{\text{g}}\right)\left(10^3 \frac{\text{L}}{\text{m}^3}\right)}{\left(0.08205 \frac{\text{L·atm}}{\text{mol·K}}\right)(298K)}$$

$$= \boxed{29.4 \text{ mg/m}^3}$$

The answer is (D).

2. (a) $T = 40°C + 273° = 313K$

$$p = (760 \text{ mm Hg})\left(\frac{1 \text{ atm}}{760 \text{ mm Hg}}\right) = 1 \text{ atm}$$

From App. 14.B, at 1 atm and 40°C,

$$\rho_{\text{water}} = 992.25 \text{ kg/m}^3$$

$$\rho_{\text{toluene}} = 0.86\rho_{\text{water}}$$

$$= (0.86)\left(992.25 \frac{\text{kg}}{\text{m}^3}\right)$$

$$= 853.34 \text{ kg/m}^3$$

From Eq. 47.5, the mass of dissolved toluene is

$$m_{toluene} = \rho_{toluene} V_{toluene}$$
$$= \left(853.34 \ \frac{kg}{m^3}\right)(0.5 \ L)\left(\frac{1 \ m^3}{10^3 \ L}\right)$$
$$= 0.427 \ kg$$

The molecular weight of toluene (C_7H_8) is

$$MW_{toluene} = (7)\left(12 \ \frac{g}{mol}\right) + (8)\left(1 \ \frac{g}{mol}\right)$$
$$= 92 \ g/mol$$

Solve for the volume of the vaporized toluene. From Eq. 47.3 and Eq. 47.4,

$$V_{toluene} = \frac{m_{toluene}R^*T}{(MW_{toluene})p}$$
$$= \frac{(427 \ g)\left(0.08205 \ \frac{L \cdot atm}{mol \cdot K}\right)(313K)}{\left(92 \ \frac{g}{mol}\right)(1 \ atm)\left(10^3 \ \frac{L}{m^3}\right)}$$
$$= 0.119 \ m^3$$

Solve for the volume of air in the room.

$$V_{air} = (10 \ m)(20 \ m)(10 \ m) = 2000 \ m^3$$

Solve for the concentration of toluene in the room. From Eq. 47.8,

$$C_{ppm,toluene} = \left(\frac{V_{toluene}}{V_{air}}\right)(10^6)$$
$$= \left(\frac{0.119 \ m^3}{2000 \ m^3}\right)(10^6)$$
$$= \boxed{59.6 \ ppmv}$$

The answer is (A).

(b) The concentration in ppmv is unchanged by changes in temperature or pressure.

The answer is (B).

3. (a) The concentration of ethylene oxide is given as 0.06%.

$$C_{ppm} = (0.06\%)\left(\frac{10^6 \ ppm}{100\%}\right) = \boxed{600 \ ppmv}$$

The answer is (D).

(b) From the conservation of mass,

$$Q_i = \frac{Q_f C_f}{C_i} \quad [C_d = 0]$$
$$= \frac{\left(0.4 \ \frac{L}{min}\right)(100 \ ppm)}{600 \ ppm}$$
$$= \boxed{0.0667 \ L/min}$$

The answer is (B).

(c) From the continuity equation,

$$Q_d = Q_f - Q_i = 0.4 \ \frac{L}{min} - 0.0667 \ \frac{L}{min}$$
$$= \boxed{0.333 \ L/min}$$

The answer is (C).

(d) Find the mass concentration of the ethylene oxide (C_2H_4O) in the supply chamber.

$$T = 10°C + 273° = 283K$$
$$p = 1 \ atm$$

The molecular weight of ethylene oxide is

$$MW_{C_2H_4O} = (2)\left(12 \ \frac{g}{mol}\right) + (4)\left(1 \ \frac{g}{mol}\right)$$
$$+ (1)\left(16 \ \frac{g}{mol}\right)$$
$$= 44 \ g/mol$$

From Eq. 47.11,

$$C_{m/V,C_2H_4O} = \left(\frac{C_{ppm,C_2H_4O}MW_{C_2H_4O}p}{R^*T}\right)(10^{-6})$$
$$= \frac{(600 \ ppm)\left(44 \ \frac{g}{mol}\right)(1 \ atm)(10^{-6})}{\left(0.08205 \ \frac{L \cdot atm}{mol \cdot K}\right)(283K)}$$
$$= 0.00114 \ g/L$$

From Eq. 47.6, to deliver 2 g,

$$V_{supply,C_2H_4O} = \frac{m_{C_2H_4O}}{C_{m/V,C_2H_4O}} = \frac{2 \ g}{0.00114 \ \frac{g}{L}}$$
$$= \boxed{1760 \ L}$$

The answer is (B).

4. (a) At NTP,

$$T_S = 25°C + 273° = 298K$$

$$p_S = (760 \text{ mm Hg})\left(\frac{1 \text{ atm}}{760 \text{ mm Hg}}\right) = 1 \text{ atm}$$

At field conditions,

$$T_F = 30°C + 273° = 303K$$

$$p_F = (650 \text{ mm Hg})\left(\frac{1 \text{ atm}}{760 \text{ mm Hg}}\right) = 0.855 \text{ atm}$$

The mass of lead to be collected is 10 times the LOD.

$$m_{\text{lead}} = (10)(\text{LOD}) = (10)(2.6 \ \mu g) = 26 \ \mu g$$

The exposure is 25% of the PEL at NTP.

$$C_{m/V,\text{lead}} = (0.25)(\text{PEL}) = (0.25)\left(0.05 \ \frac{\text{mg}}{\text{m}^3}\right)$$

$$= 0.0125 \text{ mg/m}^3$$

From Eq. 47.6,

$$V_{S,\text{air}} = \frac{m_{\text{lead}}}{C_{m/V,\text{lead}}}$$

$$= \frac{(26 \ \mu g)\left(\frac{1 \text{ mg}}{10^3 \ \mu g}\right)\left(10^3 \ \frac{\text{L}}{\text{m}^3}\right)}{0.0125 \ \frac{\text{mg}}{\text{m}^3}}$$

$$= 2080 \text{ L}$$

Correct the sample volume to field conditions. From Eq. 47.1,

$$V_{F,\text{air}} = V_{S,\text{air}}\left(\frac{p_S}{p_F}\right)\left(\frac{T_F}{T_S}\right)$$

$$= (2080 \text{ L})\left(\frac{1 \text{ atm}}{0.855 \text{ atm}}\right)\left(\frac{303K}{298K}\right)$$

$$= \boxed{2470 \text{ L}}$$

The answer is (D).

(b)

$$t = \frac{V}{Q} = \frac{2470 \text{ L}}{4 \ \frac{\text{L}}{\text{min}}}$$

$$= \boxed{618 \text{ min}}$$

The answer is (D).

(c) Assume that the technician collects at least 10 μg/sample.

$$\frac{10 \ \mu g}{2.6 \ \mu g} = 3.85 \text{ times the LOD}$$

From Eq. 47.6,

$$V_{S,\text{air}} = \frac{m_{\text{lead}}}{C_{m/V,\text{lead}}}$$

$$= \frac{(10 \ \mu g)\left(\frac{1 \text{ mg}}{10^3 \ \mu g}\right)\left(10^3 \ \frac{\text{L}}{\text{m}^3}\right)}{0.0125 \ \frac{\text{mg}}{\text{m}^3}}$$

$$= 800 \text{ L} \quad [\text{at NTP}]$$

Correct the sample volume to field conditions. From Eq. 47.1,

$$V_{F,\text{air}} = V_{S,\text{air}}\left(\frac{p_S}{p_F}\right)\left(\frac{T_F}{T_S}\right)$$

$$= (800 \text{ L})\left(\frac{1 \text{ atm}}{0.855 \text{ atm}}\right)\left(\frac{303K}{298K}\right)$$

$$= 950 \text{ L}$$

$$t = \frac{V}{Q} = \frac{(950 \text{ L})\left(\frac{1 \text{ h}}{60 \text{ min}}\right)}{4 \ \frac{\text{L}}{\text{min}}}$$

$$= \boxed{3.96 \text{ h}}$$

All of the answer choices cover an 8 h period. Collect 2 samples at 4 h each.

The answer is (A).

48 Biological Effects of Radiation

PRACTICE PROBLEMS

1. Which is (are) the most appropriate unit(s) of measurement for use when discussing the biological effects of radiation on a human being?

(A) rad or gray

(B) rem or sievert

(C) roentgen

(D) curie or becquerel

2. Most nearly, what is the annual average effective dose equivalent received from background sources?

(A) 20 mrem

(B) 180 mrem

(C) 360 mrem

(D) 1000 mrem

3. What is the maximum whole body radiation dose allowed radiation workers per year?

(A) 5 mrem

(B) 50 mrem

(C) 5000 mrem

(D) 50 rem

4. Which of the following is FALSE?

(A) Increasing distance from a radiation field decreases radiation exposure.

(B) Increasing shielding from a radiation field decreases radiation exposure.

(C) Increasing walking speed in a radiation field decreases radiation exposure.

(D) Increasing time spent in a radiation field decreases radiation exposure.

5. A radiation source undergoes 87 039 256 decays per hour. Express this in (a) becquerels and (b) curies.

(A) 2.4×10^4 Bq; 6.5×10^{-7} Ci

(B) 6.5×10^{-7} Bq; 2.4×10^4 Ci

(C) 8.9×10^{14} Bq; 6.5×10^{-7} Ci

(D) 6.5×10^{-7} Bq; 8.9×10^{14} Ci

(handwritten: $3.7 \cdot 10^{10}/_s$)

6. How long can a new 15.5 mCi Cs-137 reference source be used if it is to be taken out of service when the activity reaches 10.0 mCi (for Cs-137, half-life $= 30.0$ yr)?

(A) 1.9 yr

(B) 5.0 yr

(C) 14 yr^{-1}

(D) 19 yr

(handwritten:
$t_{1/2} = 30 yr$
$\lambda = \ln 2/30 = 0.023$
$15.5 \to 10\ mCi$
$10 = 15.5 e^{-0.023t}$ *)*

7. (a) What is the decay constant for a radionuclide that has an initial activity of 225 mCi, and 110 days later has an activity of 81.8 mCi? (b) What is the half-life?

(A) 2.44×10^{-3} d^{-1}; 111 d

(B) 6.38×10^{-3} d^{-1}; 108 d

(C) 9.20×10^{-3} yr^{-1}; 75.3 yr

(D) 9.20×10^{-3} d^{-1}; 75.3 d

(handwritten:
$225 mCi$
$110 days - t$
$81.8 mCi$
$81.8 = 225 e^{-\lambda(110)}$
$\lambda = 0.009$ *)*

8. How long could a worker have 2 cm^2 of skin contaminated with 1 MBq carbon-14 and not exceed the SDE occupational exposure limit?

(A) 2.9 h

(B) 2.9 yr

(C) 29 h

(D) 0.5 yr

(handwritten:
$2 cm^2 \quad 1 MBq$
$1 MBq / 2 cm^2$ *)*

9. How far would a member of the public have to be kept away from an unshielded 750 mCi chromium-51 source to NOT exceed the NRC limits?

(A) 18 cm

(B) 180 cm

(C) 400 cm

(D) 4000 cm

(handwritten: $750 mCi$)

10. A worker accidentally ingests 100 μCi of tritiated water. Calculate the committed effective dose equivalent ($H_{E,50}$).

(A) 6.3×10^{-3} rad

(B) 6.3×10^{-3} rem

(C) 4.5×10^{-5} R

(D) 6.3×10^{-3} Sv

(handwritten:
$100 \mu Ci = 100\ rem$
H_{E50} *)*

11. What is the total effective dose equivalent to a person who inhales $^{11}CO_2$ at an average concentration of 5×10^{-5} μCi/mL for one year?

(A) 2.8 rem

(B) 28 mrem

(C) 1.1×10^{10} mrem

(D) 1.1×10^{10} mSv

12. What is the committed effective dose equivalent to a person whose drinking water was supplied by a facility that releases calcium-45 with an average monthly activity of 3×10^{-8} μCi/mL?

(A) 7.5×10^{-7} mrem

(B) 7.5×10^{-7} rem

(C) 7.5×10^{-10} mSv

(D) 7.5×10^{-7} Sv

13. How much farther away would one need to move an unshielded gamma source to reduce the radiation intensity to 1% of the original intensity?

(A) 9 times

(B) 10 times

(C) 100 times

(D) 200 times

SOLUTIONS

1. *The answer is (B).*

2. *The answer is (C).*

3. *The answer is (C).*

4. *The answer is (D).*

5. (a)

$$\frac{\left(87\,039\,259 \; \frac{\text{decay}}{\text{h}}\right)\left(\frac{\text{Bq}}{\frac{1}{\text{s}}}\right)}{\left(60 \; \frac{\text{min}}{\text{h}}\right)\left(60 \; \frac{\text{s}}{\text{min}}\right)} = \boxed{2.42 \times 10^4 \; \text{Bq}}$$

(b)

$$\frac{2.42 \times 10^4 \; \text{Bq}}{3.7 \times 10^{10} \; \frac{\text{Bq}}{\text{Ci}}} = \boxed{6.54 \times 10^{-7} \; \text{Ci}}$$

The answer is (A).

6. From Eq. 48.2,

$$\lambda = \frac{\ln 2}{t_{1/2}} = \frac{0.693}{30.0 \; \text{yr}} = 0.0231 \; \text{yr}^{-1}$$

From Eq. 48.1,

$$t = \frac{-\ln\left(\frac{A_t}{A_0}\right)}{\lambda} = \frac{-\ln\left(\frac{10.0 \; \text{mCi}}{15.5 \; \text{mCi}}\right)}{0.0231 \; \text{yr}^{-1}}$$

$$= \boxed{19 \; \text{yr}}$$

The answer is (D).

7. (a) From Eq. 48.1,

$$\lambda = \frac{-\ln\left(\frac{A_t}{A_0}\right)}{t} = \frac{-\ln\left(\frac{81.8 \; \text{mCi}}{225 \; \text{mCi}}\right)}{110 \; \text{d}}$$

$$= \boxed{9.20 \times 10^{-3} \; \text{d}^{-1}}$$

(b) From Eq. 48.2,

$$t_{1/2} = \frac{\ln 2}{\lambda} = \frac{0.693}{9.2 \times 10^{-3} \; \text{d}^{-1}}$$

$$= \boxed{75.3 \; \text{d}}$$

The answer is (D).

8.
$$A = 1 \text{ MBq} = 1 \times 10^6 \text{ Bq}$$
$$A_s = 2 \text{ cm}^2$$

From Table 48.4 for C-14 at 0.007 cm (where the SDE must be determined),

$$DF = 2.9 \times 10^{-3} \; \frac{\frac{\text{Sv}}{\text{yr}}}{\frac{\text{Bq}}{\text{cm}^2}}$$

Use Eq. 48.5.

$$X_b = DF\left(\frac{A}{A_s}\right)$$
$$= \left(2.9 \times 10^{-3} \; \frac{\text{Sv} \cdot \text{cm}^2}{\text{Bq} \cdot \text{yr}}\right)\left(\frac{1 \times 10^6 \text{ Bq}}{2 \text{ cm}^2}\right)$$
$$= 1.45 \times 10^3 \text{ Sv/yr}$$

The SDE limit for the skin is 50 rem (0.5 Sv) in a year.

$$\frac{0.5 \text{ Sv}}{1.45 \times 10^3 \; \frac{\text{Sv}}{\text{yr}}} = (3.3 \times 10^{-4} \text{ yr})\left(365 \; \frac{\text{d}}{\text{yr}}\right)\left(24 \; \frac{\text{h}}{\text{d}}\right)$$
$$= \boxed{3.0 \text{ h}}$$

The answer is (A).

9. Calculate the annual dose equivalent rate for Cr-51.

$$A = \frac{(750 \times 10^{-3} \text{ Ci})\left(3.7 \times 10^{10} \; \frac{\text{Bq}}{\text{Ci}}\right)}{10^6 \; \frac{\text{Bq}}{\text{MBq}}}$$
$$= 2.78 \times 10^3 \text{ MBq}$$

From Table 48.5, for Cr-51 at a distance, d_0, of 1 m,

$$\Gamma = 6.320 \times 10^{-6} \; \frac{\frac{\text{mSv}}{\text{h}}}{\text{MBq}}$$

From Eq. 48.6,

$$X_g = \Gamma A$$
$$= \left(6.320 \times 10^{-6} \; \frac{\text{mSv}}{\text{MBq} \cdot \text{h}}\right)(2.78 \times 10^4 \text{ MBq})$$
$$\times \left(24 \; \frac{\text{h}}{\text{d}}\right)\left(365 \; \frac{\text{d}}{\text{yr}}\right)$$
$$= 1.54 \times 10^3 \text{ mSv/yr} \quad [\text{at 1 m}]$$

Find the distance at which the dose equivalent rate is reduced to 1 mSv/yr, the total effective dose equivalent limit for individual members of the public.

From Eq. 48.15,

$$I_0 d_0^2 = I_f d_f^2$$
$$d_f = d_0\sqrt{\frac{I_0}{I_f}}$$
$$= (1 \text{ m})\sqrt{\frac{1.6 \times 10^3 \; \frac{\text{mSv}}{\text{yr}}}{1 \; \frac{\text{mSv}}{\text{yr}}}}$$
$$= \boxed{40 \text{ m}}$$

The answer is (D).

10. From Table 48.8, part 1, column 1, for oral ingestion of H-3 water, the ALI is 8×10^4 μCi. This would produce 5 rem $H_{E,50}$.

$$5 \text{ rem} = 8 \times 10^4 \; \mu\text{Ci}$$

The $H_{E,50}$ produced by 100 μCi is

$$H_{E,50} = 100 \; \mu\text{Ci}$$

Therefore,

$$\frac{H_{E,50}}{100 \; \mu\text{Ci}} = \frac{5 \text{ rem}}{8 \times 10^4 \; \mu\text{Ci}}$$

$$H_{E,50} = \frac{(5 \text{ rem})(100 \; \mu\text{Ci})}{8 \times 10^4 \; \mu\text{Ci}}$$
$$= \boxed{6.25 \times 10^{-3} \text{ rem}}$$

The answer is (B).

11. From Table 48.8, part 2, column 1, for C-11, 9×10^{-7} μCi/mL produces 0.05 rem TEDE, so

$$0.05 \text{ rem} = 9 \times 10^{-7} \; \mu\text{Ci/mL}$$

The TEDE produced by 5×10^{-5} μCi/mL is

$$\text{TEDE} = 5 \times 10^5 \; \mu\text{Ci/mL}$$

Health, Safety, Welfare

Therefore,

$$\frac{\text{TEDE}}{5 \times 10^{-5} \ \frac{\mu\text{Ci}}{\text{mL}}} = \frac{0.05 \text{ rem}}{9 \times 10^{-7} \ \frac{\mu\text{Ci}}{\text{mL}}}$$

$$\text{TEDE} = \frac{\left(5 \times 10^{-5} \ \frac{\mu\text{Ci}}{\text{mL}}\right)(0.05 \text{ rem})}{9 \times 10^{-7} \ \frac{\mu\text{Ci}}{\text{mL}}}$$

$$= \boxed{2.77 \text{ rem}}$$

The answer is (A).

12. From Table 48.8, part 3, for Ca-45, $2 \times 10^{-4} \ \mu\text{Ci/mL}$ produces 0.5 rem $H_{E,50}$, so

$$0.5 \text{ rem} = 2 \times 10^{-4} \ \mu\text{Ci/mL}$$

The $H_{E,50}$ produced by $3 \times 10^{-8} \ \mu\text{Ci/mL}$ is

$$H_{E,50} = 3 \times 10^{-8} \ \mu\text{Ci/mL}$$

Therefore,

$$\frac{H_{E,50}}{3 \times 10^{-8} \ \frac{\mu\text{Ci}}{\text{mL}}} = \frac{0.5 \text{ rem}}{2 \times 10^{-4} \ \frac{\mu\text{Ci}}{\text{mL}}}$$

$$H_{E,50} = \frac{\left(3 \times 10^{-8} \ \frac{\mu\text{Ci}}{\text{mL}}\right)(0.5 \text{ rem})}{2 \times 10^{-4} \ \frac{\mu\text{Ci}}{\text{mL}}}$$

$$= \boxed{7.5 \times 10^{-5} \text{ rem}}$$

The answer is (D).

13. For the intensity to be 1% of the original,

$$\frac{I_2}{I_1} = 0.01$$

From Eq. 48.15,

$$\frac{d_1}{d_2} = \sqrt{\frac{I_2}{I_1}} = \sqrt{0.01} = 0.1$$

$$d_2 = 10d_1 = d_1 + \boxed{9d_1}$$

The answer is (A).

49 Shielding

PRACTICE PROBLEMS

1. Using Einstein's $E = mc^2$ equation, calculate (a) the energy equivalent of a neutron's mass, and (b) the wavelength of a photon that has the equivalent energy.

(A) 3.2×10^{-10} J; 2.6×10^{-20} m

(B) 1.5×10^{-10} J; 1.3×10^{-20} m

(C) 3.2×10^{-10} J; 1.3×10^{-15} m

(D) 1.5×10^{-10} J; 1.3×10^{-15} m

2. What thickness of tin ($\rho = 6.5$ g/cm^3, $A_{Sn} = 118.7$) is needed to completely shield an isotope that emits a 7.0 MeV ($E_\alpha = 7.0$ MeV) alpha particle?

(A) 9.4×10^{-3} cm

(B) 2.6×10^{-3} cm

(C) 1.7×10^{-2} cm

(D) 5.66 cm

3. What is the density of a 5.0 mm thick shield that shields 100% of the beta particles from indium-118 ($E_{\beta,max} = 4.250$ MeV)?

(A) 0.31 g/cm^3

(B) 2.9 g/cm^3

(C) 3.4 g/cm^3

(D) 4.3 g/cm^3

4. (a) What is the mass attenuation factor for a shield that is 0.5 in thick, has a density of 10.3 g/cm^3, and reduces the exposure rate of 1.25 MeV gammas from 3.75 R/h to 1.44 mR/h? (b) What is the relaxation length?

(A) 0.5 cm^2/g; 5 cm

(B) 0.6 cm^2/g; 0.16 cm

(C) 1.9 cm^2/g; 5.87 cm

(D) 15 cm^2/g; 0.69 cm

5. A 4 cm thick large iron plate ($\rho_{Fe} = 7.86$ g/cm^3) is used to shield a high-activity 500 keV gamma ray source. The gamma intensity is measured to be 9.4 mR/h behind the shield. What is the unshielded intensity in front of the shield?

(A) 32 mR/h

(B) 84 mR/h

(C) 140 mR/h

(D) 670 mR/h

6. What percent of thermal neutrons will pass through a shield of 1 in of tungsten ($\rho_{a,W} = 19.3$ g/cm^3, $A_W = 183.9$)? The absorption cross section, σ_a, for thermal neutrons in tungsten is 19.2 barns. The total cross section is 24.2 barns.

(A) 0.01%

(B) 2.5%

(C) 3.7%

(D) 4.6%

7. What is the relaxation length for 500 keV gamma rays in aluminum ($\rho_{a,Al} = 2.7$ g/cm^3)?

(A) 4.4 μm

(B) 2.7 cm

(C) 4.4 cm

(D) 4.4 m

SOLUTIONS

1. (a)

$$m = (1.008665 \text{ amu})\left(1.661 \times 10^{-27} \; \frac{\text{kg}}{\text{amu}}\right)$$

$$= 1.675 \times 10^{-27} \text{ kg}$$

$$E = mc^2$$

$$= (1.675 \times 10^{-27} \text{ kg})\left(3 \times 10^8 \; \frac{\text{m}}{\text{s}}\right)^2$$

$$= \boxed{1.51 \times 10^{-10} \text{ J}}$$

(b) From Eq. 49.4,

$$\lambda = \frac{hc}{E}$$

$$= \frac{(6.63 \times 10^{-34} \text{ J·s})\left(3 \times 10^8 \; \frac{\text{m}}{\text{s}}\right)}{1.51 \times 10^{-10} \text{ J}}$$

$$= \boxed{1.32 \times 10^{-15} \text{ m}}$$

The answer is (D).

2. From Eq. 49.7,

$$R_{\alpha,\text{air}} = k_2 E_\alpha - k_3$$

$$= \left(1.24 \; \frac{\text{cm}}{\text{MeV}}\right)(7.0 \text{ MeV}) - 2.62 \text{ cm}$$

$$= 6.06 \text{ cm}$$

From Eq. 49.8,

$$R_{\alpha,\text{Sn}} = \frac{k_4 (A_{\text{Sn}})^{1/3} R_{\alpha,\text{air}}}{\rho_{\text{Sn}}}$$

$$= \frac{\left(0.56 \times 10^{-3} \; \frac{\text{g}}{\text{cm}^3}\right)(118.7)^{1/3}(6.06 \text{ cm})}{6.5 \; \frac{\text{g}}{\text{cm}^3}}$$

$$= \boxed{2.57 \times 10^{-3} \text{ cm}}$$

The answer is (B).

3. From Eq. 49.10,

$$\rho_{\text{shield}} = \frac{k_8 E_{\beta,\text{max}} - k_9}{R_{\beta,\text{shield}}}$$

$$= \frac{\left(0.530 \; \frac{\text{g}}{\text{cm}^2 \cdot \text{MeV}}\right)(4.250 \text{ MeV}) - 0.106 \; \frac{\text{g}}{\text{cm}^2}}{(5.0 \text{ mm})\left(\frac{1 \text{ cm}}{10 \text{ mm}}\right)}$$

$$= \boxed{4.29 \text{ g/cm}^3}$$

The answer is (D).

4. (a)

$$x = (0.5 \text{ in})\left(2.54 \; \frac{\text{cm}}{\text{in}}\right)$$

$$= 1.27 \text{ cm}$$

From Eq. 49.11,

$$\mu = \frac{-\ln\left(\frac{I_x}{I_0}\right)}{\rho x} = \frac{-\ln\left(\dfrac{1.44 \; \frac{\text{mR}}{\text{h}}}{3750 \; \frac{\text{mR}}{\text{h}}}\right)}{\left(10.3 \; \frac{\text{g}}{\text{cm}^3}\right)(1.27 \text{ cm})}$$

$$= \boxed{0.601 \text{ cm}^2/\text{g}}$$

(b) From Eq. 49.17,

$$\lambda = \frac{1}{\mu \rho}$$

$$= \frac{1}{\left(0.601 \; \frac{\text{cm}^2}{\text{g}}\right)\left(10.3 \; \frac{\text{g}}{\text{cm}^3}\right)}$$

$$= \boxed{0.16 \text{ cm}}$$

The answer is (B).

5. From Table 49.3, for 500 keV (0.5 MeV) gamma radiation in iron, $\mu = 0.0828 \text{ cm}^2/\text{g}$.

$$\mu \rho x = \left(0.0828 \; \frac{\text{cm}^2}{\text{g}}\right)\left(7.86 \; \frac{\text{g}}{\text{cm}^3}\right)(4 \text{ cm}) = 2.6$$

Interpolate for 0.5 MeV. From Table 49.4, $B \approx 4.00$.

From Eq. 49.12,

$$I_0 = \frac{I_x}{Be^{-\mu \rho x}} = \frac{9.4 \; \frac{\text{mR}}{\text{h}}}{4.0e^{-2.6}}$$

$$= \boxed{31.7 \text{ mR/h} \quad (32 \text{ mR/h})}$$

The answer is (A).

6. Calculate the number of lead nuclei per cubic centimeter.

From Eq. 49.15,

$$N = \frac{N_A \rho_W}{A_W}$$

$$= \frac{\left(6.02 \times 10^{23} \; \frac{\text{nuclei}}{\text{mol}}\right)\left(19.3 \; \frac{\text{g}}{\text{cm}^3}\right)}{183.9 \; \frac{\text{g}}{\text{mol}}}$$

$$= 6.32 \times 10^{22} \text{ nuclei/cm}^3$$

A neutron that is scattered by a tungsten nucleus is not necessarily removed. Only absorbed neutrons can be considered removed. The absorption cross section should be used. The percentage of neutrons transmitted through the shield (I_x/I_0) is given by Eq. 49.16.

$$\frac{I_x}{I_0} = e^{-N\sigma x}$$

$$= e^{-\left(6.32\times10^{22}\ \frac{\text{nuclei}}{\text{cm}^3}\right)\left(19.2\times10^{-24}\ \frac{\text{cm}^2}{\text{nucleus}}\right)(2.54\ \text{cm})}$$

$$= \boxed{0.0459 \quad (4.6\%)}$$

The answer is (D).

7. From Table 49.3, for 500 keV gammas in aluminum, $\mu = 0.0840\ \text{cm}^2/\text{g}$.

From Eq. 49.17,

$$\lambda = \frac{1}{\mu\rho_{\text{shield}}}$$

$$= \frac{1}{\left(0.0840\ \dfrac{\text{cm}^2}{\text{g}}\right)\left(2.7\ \dfrac{\text{g}}{\text{cm}^3}\right)}$$

$$= \boxed{4.41\ \text{cm}}$$

The answer is (C).

Health, Safety, Welfare

50 Illumination and Sound

PRACTICE PROBLEMS

Illumination

1. A 60 m × 24 m product assembly area is illuminated by a gridded arrangement of pendant lamps, each producing 18 000 lm. The minimum illumination required at the work surface level is 200 lux (lm/m^2). The lamps are well maintained and have a maintenance factor of 0.80. The utilization factor is 0.4. Approximately how many lamps are required?

(A) 34

(B) 42

(C) 50

(D) 66

Noise and Sound

2. A pipe is covered with acoustic pipe insulation to reduce noise. What is the best predictor of the perceived sound reduction?

(A) insertion loss (IL)

(B) transmission loss (TL)

(C) noise reduction rating (NRR)

(D) noise reduction coefficient (NRC)

3. A gear-driven electric power generation system has a prime mover noise level of 88 dBA, a gear system noise level of 82 dBA, and a generator noise level of 95 dBA. The overall noise level is most nearly

(A) 75 dBA

(B) 86 dBA

(C) 96 dBA

(D) 99 dBA

4. A 40 dB source is placed adjacent to a 35 dB source. The combined sound pressure level is most nearly

(A) 17 dB

(B) 28 dB

(C) 37 dB

(D) 41 dB

5. With no machinery operating, the background noise in a room has a sound pressure level of 43 dB. With the machinery operating, the sound pressure level is 45 dB. The sound pressure level due to the machinery alone is most nearly

(A) 2.0 dB

(B) 41 dB

(C) 47 dB

(D) 49 dB

6. An unenclosed source produces a sound pressure level of 100 dB. An enclosure is constructed from a material having a transmission loss of 30 dB. The sound pressure level inside the enclosure from the enclosed source increases to 110 dB. The reduction in sound pressure level outside the enclosure is most nearly

(A) 20 dB

(B) 30 dB

(C) 80 dB

(D) 90 dB

7. 4 ft (1.2 m) from an isotropic sound source, the sound pressure level is 92 dB. The sound pressure level 12 ft (3.6 m) from the source is most nearly

(A) 62 dB

(B) 73 dB

(C) 83 dB

(D) 87 dB

8. Octave band measurements of a noise source were made. The measurements were 85 dB, 90 dB, 92 dB, 87 dB, 82 dB, 78 dB, 65 dB, and 54 dB at frequencies of 63 Hz, 125 Hz, 250 Hz, 500 Hz, 1000 Hz, 2000 Hz, 4000 Hz, and 8000 Hz, respectively. The overall A-weighted sound pressure level is most nearly

(A) 83 dBA

(B) 86 dBA

(C) 89 dBA

(D) 93 dBA

9. If the number of sabins is 50% of the total room area, the maximum possible reduction in sound pressure level is most nearly a

(A) 1.2 dB decrease

(B) 3.0 dB decrease

(C) 4.6 dB decrease

(D) 6.1 dB decrease

10. A storage room has dimensions of 100 ft × 400 ft × 20 ft (30 m × 120 m × 6 m). All surfaces are plain concrete. 40% of the walls are treated acoustically with a material having a sound absorption coefficient of 0.8. The reduction in sound pressure level is most nearly a

(A) 1.2 dB decrease

(B) 3.0 dB decrease

(C) 4.6 dB decrease

(D) 6.1 dB decrease

11. A meeting room has dimensions of 20 ft × 50 ft × 10 ft (6 m × 15 m × 3 m). The floor is covered with roll vinyl. The walls and ceiling are sheetrock. 20% of the walls are glass windows. There are 15 seats, lightly upholstered, with 15 occupants, and 5 miscellaneous sabins. After complaints from the occupants, the ceiling is treated with a sound absorbing material with a sound absorption coefficient of 0.7. The reduction in sound level is most nearly a

(A) 3.5 dB decrease

(B) 5.7 dB decrease

(C) 6.3 dB decrease

(D) 12 dB decrease

12. A room has dimensions of 15 ft × 20 ft × 10 ft (4.5 m × 6 m × 3 m). The sound absorption coefficients are 0.03, 0.5, and 0.06 for the floor, ceiling, and walls, respectively. A machine with a sound power level of 65 dB is located at the intersection of the floor and wall, 7.5 ft (2.25 m) from the perpendicular walls. The ambient sound pressure level is 50 dB everywhere in the room. The sound pressure level 5 ft (1.5 m) from the machine is most nearly

(A) 45 dBA

(B) 61 dBA

(C) 73 dBA

(D) 81 dBA

SOLUTIONS

1. Use Eq. 50.8. The number of lamps required is

$$N = \frac{EA}{\Phi(\text{CU})(\text{LLF})} = \frac{\left(200 \, \frac{\text{lm}}{\text{m}^2}\right)(60 \text{ m})(24 \text{ m})}{(18\,000 \text{ lm})(0.4)(0.8)}$$

$$= \boxed{50}$$

The answer is (C).

2. The noise reduction coefficient (NRC) relates to sound absorption by features and furnishings within a space. The noise reduction rating (NRR) relates to hearing protection devices. The transmission loss (TL) is the amount of sound energy lost when sound passes through a barrier, such as acoustic pipe insulation. While the insulation may be quite dense, it is in direct contact with the pipe, and a high degree of structural vibration (i.e., noises) will transfer directly to the outer surface of the insulation. The insertion loss (IL) is the difference in sound pressure levels before and after insulation is applied. Therefore, the $\boxed{\text{insertion loss}}$ is the best predictor of the perceived sound reduction.

The answer is (A).

3. From Eq. 50.22, the governing equation for combining multiple noise sources is

$$L_p = 10 \log \sum 10^{L_i/10} = 10 \log \left(\begin{array}{l} 10^{88 \text{ dBA}/10} \\ + 10^{82 \text{ dBA}/10} \\ + 10^{95 \text{ dBA}/10} \end{array} \right)$$

$$= \boxed{96 \text{ dBA}}$$

The answer is (C).

4. Use Eq. 50.22.

$$L_p = 10 \log \sum 10^{L_i/10} = 10 \log\left(10^{40 \text{ dB}/10} + 10^{35 \text{ dB}/10}\right)$$

$$= \boxed{41.2 \text{ dB} \quad (41 \text{ dB})}$$

The answer is (D).

5. Use Eq. 50.22.

$$L_p = 10 \log \sum 10^{L_i/10} = 10 \log\left(10^{43 \text{ dB}/10} + 10^{L/10}\right)$$

$$= 45 \text{ dB}$$

Solve for the unknown machinery sound pressure level, L.

$$10^{L/10} = 10^{45 \text{ dB}/10} - 10^{43 \text{ dB}/10}$$

$$L = 10 \log\left(10^{45 \text{ dB}/10} - 10^{43 \text{ dB}/10}\right)$$

$$= \boxed{40.7 \text{ dB} \quad (41 \text{ dB})}$$

The answer is (B).

6. Define $L_{p,1}$ as the sound pressure level inside the enclosure, and define $L_{p,2}$ as the sound pressure level outside the enclosure. Use Eq. 50.29.

$$L_{p,2} = L_{p,1} - \text{TL} = 110 \text{ dB} - 30 \text{ dB} = 80 \text{ dB}$$

Define $L_{p,1}$ as the sound pressure level for the unenclosed source, and define $L_{p,2}$ as the sound pressure level for the enclosed source.

Use Eq. 50.28 to solve for the insertion loss.

$$\text{IL} = L_{p,1} - L_{p,2} = 100 \text{ dB} - 80 \text{ dB} = \boxed{20 \text{ dB}}$$

The answer is (A).

7. The free-field sound pressure is inversely proportional to the square of the distance from the source.

$$\frac{p_2}{p_1} = \left(\frac{r_1}{r_2}\right)^2$$

From Eq. 50.19,

$$L_{p,2} = L_{p,1} + 10 \log\left(\frac{p_2}{p_1}\right)^2 = L_{p,1} + 20 \log\frac{p_2}{p_1}$$

$$= L_{p,1} + 20 \log\left(\frac{r_1}{r_2}\right)^2$$

Customary U.S. Solution

$$L_{p,2} = L_{p,1} + 20 \log\left(\frac{r_1}{r_2}\right)^2 = 92 \text{ dB} + 20 \log\left(\frac{4 \text{ ft}}{12 \text{ ft}}\right)^2$$

$$= 92 \text{ dB} - 19.1 \text{ dB}$$

$$= \boxed{72.9 \text{ dB} \quad (73 \text{ dB})}$$

The answer is (B).

SI Solution

$$L_{p,2} = L_{p,1} + 20 \log\left(\frac{r_1}{r_2}\right)^2 = 92 \text{ dB} + 20 \log\left(\frac{1.2 \text{ m}}{3.6 \text{ m}}\right)^2$$

$$= 92 \text{ dB} - 19.1 \text{ dB}$$

$$= \boxed{72.9 \text{ dB} \quad (73 \text{ dB})}$$

The answer is (B).

8. Add the corrections from Table 50.5 to the measurements.

frequency (Hz)	measurement (dB)	correction (dB)	corrected value (dB)
63	85	−26.2	58.8
125	90	−16.1	73.9
250	92	−8.6	83.4
500	87	−3.2	83.8
1000	82	0	82.0
2000	78	+1.2	79.2
4000	65	+1.0	66.0
8000	54	−1.1	52.9

Use Eq. 50.22.

$$L_p = 10 \log \sum 10^{L_i/10}$$

$$= 10 \log \left(\begin{array}{l} 10^{58.8 \text{ dB}/10} + 10^{73.9 \text{ dB}/10} \\ \quad + 10^{83.4 \text{ dB}/10} + 10^{83.8 \text{ dB}/10} \\ \quad + 10^{82.0 \text{ dB}/10} + 10^{79.2 \text{ dB}/10} \\ \quad + 10^{66.0 \text{ dB}/10} + 10^{52.9 \text{ dB}/10} \end{array} \right)$$

$$= \boxed{88.6 \text{ dB} \quad (89 \text{ dBA})}$$

The answer is (C).

9. Define A as the total room area.

$$\sum S_1 = 0.50A$$

The maximum number of sabins is equal to the room area.

$$\sum S_2 = A$$

Use Eq. 50.27.

$$\text{NR} = 10 \log \frac{\sum S_1}{\sum S_2} = 10 \log \frac{0.50A}{A} = 10 \log 0.50$$

$$= \boxed{-3.0 \text{ dB} \quad (3.0 \text{ dB decrease})}$$

The answer is (B).

10. *Customary U.S. Solution*

The surface area of the room walls is

$$A_1 = \left((2)(100\text{ ft}) + (2)(400\text{ ft})\right)(20\text{ ft}) = 20{,}000\text{ ft}^2$$

The surface area of the room floor and ceiling is

$$A_2 = (2)(100\text{ ft})(400\text{ ft}) = 80{,}000\text{ ft}^2$$

The sound absorption coefficient of concrete is the NRC value of 0.02 from App. 50.A.

Define the sound absorption coefficient of precast concrete as α_{concrete}.

The sabin area of the room with all precast concrete is

$$\sum S_2 = \alpha_{\text{concrete}}(A_1 + A_2)$$
$$= (0.02)(20{,}000\text{ ft}^2 + 80{,}000\text{ ft}^2)$$
$$= 2000\text{ ft}^2$$

Define the sound absorption coefficient of the wall acoustical treatment as α_{wall}.

The sabin area of the room with 40% of the walls treated with $\alpha_{\text{wall}} = 0.8$ is

$$\sum S_1 = \alpha_{\text{concrete}}A_2 + \alpha_{\text{concrete}}(0.6A_1) + \alpha_{\text{wall}}(0.4A_1)$$
$$= (0.02)(80{,}000\text{ ft}^2) + (0.02)(0.6)(20{,}000\text{ ft}^2)$$
$$+ (0.8)(0.4)(20{,}000\text{ ft}^2)$$
$$= 8240\text{ ft}^2$$

Use Eq. 50.27.

$$\text{NR} = 10\log\frac{\sum S_1}{\sum S_2} = 10\log\frac{8240\text{ ft}^2}{2000\text{ ft}^2}$$
$$= \boxed{6.1\text{ dB decrease}}$$

The answer is (D).

SI Solution

The surface area of the room walls is

$$A_1 = \left((2)(30\text{ m}) + (2)(120\text{ m})\right)(6\text{ m}) = 1800\text{ m}^2$$

The surface area of the room floor and ceiling is

$$A_2 = (2)(30\text{ m})(120\text{ m}) = 7200\text{ m}^2$$

The sound absorption coefficient of precast concrete is the NRC value of 0.02 from App. 50.A.

Define the sound absorption coefficient of precast concrete as α_{concrete}.

The sabin area of the room with all precast concrete is

$$\sum S_2 = \alpha_{\text{concrete}}(A_1 + A_2)$$
$$= (0.02)(1800\text{ m}^2 + 7200\text{ m}^2)$$
$$= 180\text{ m}^2$$

Define the sound absorption coefficient of the wall acoustical treatment as α_{wall}.

The sabin area of the room with 40% of the walls treated with $\alpha_{\text{wall}} = 0.8$ is

$$\sum S_1 = \alpha_{\text{concrete}}A_2 + \alpha_{\text{concrete}}(0.6A_1) + \alpha_{\text{wall}}(0.4A_1)$$
$$= (0.02)(7200\text{ m}^2) + (0.02)(0.6)(1800\text{ m}^2)$$
$$+ (0.8)(0.4)(1800\text{ m}^2)$$
$$= 741.6\text{ m}^2$$

Use Eq. 50.27.

$$\text{NR} = 10\log\frac{\sum S_1}{\sum S_2} = 10\log\frac{741.6\text{ m}^2}{180\text{ m}^2}$$
$$= \boxed{6.1\text{ dB decrease}}$$

The answer is (D).

11. *Customary U.S Solution*

The gross area of the walls is

$$A_1 = \left((2)(20\text{ ft}) + (2)(50\text{ ft})\right)(10\text{ ft}) = 1400\text{ ft}^2$$
$$A_{\text{walls}} = (1 - 0.20)(1400\text{ ft}^2) = 1120\text{ ft}^2$$
$$A_{\text{glass}} = (0.20)(1400\text{ ft}^2) = 280\text{ ft}^2$$

From App. 50.A, the sound absorption coefficient of glass is $\alpha_1 = 0.03$.

The area of the floor is

$$A_2 = (20\text{ ft})(50\text{ ft}) = 1000\text{ ft}^2$$

From App. 50.A, the sound absorption coefficient of roll vinyl is $\alpha_2 = 0.03$.

The area of the ceiling is

$$A_3 = (20\text{ ft})(50\text{ ft}) = 1000\text{ ft}^2$$

The sound absorption coefficient of the sheetrock is $\alpha_3 = 0.05$.

From App. 50.A, the seats have approximately 1.5 sabins each, and the occupants have approximately 5.0 sabins each.

The total sabin area of the untreated room is

$$\sum S_2 = \alpha_{glass}A_{glass} + \alpha_{walls}A_{walls} + \alpha_{floor}A_{floor}$$
$$+ \alpha_{ceiling}A_{ceiling} + seats + occupants$$
$$+ miscellaneous$$
$$= (0.03)(280\ ft^2) + (0.05)(1120\ ft^2)$$
$$+ (0.03)(1000\ ft^2) + (0.05)(1000\ ft^2)$$
$$+ (15)(1.5\ ft^2) + (15)(5.0\ ft^2) + 5.0\ ft^2$$
$$= 246.9\ ft^2$$

The total sabin area excluding the ceiling is

$$246.9\ ft^2 - \alpha_3 A_3 = 246.9\ ft^2 - (0.03)(1000\ ft^2)$$
$$= 216.9\ ft^2$$

The total sabin area of the room with the ceiling treated with $\alpha_3 = 0.7$ sound absorption is

$$\sum S_1 = 216.9\ ft^2 + \alpha_3 A_3$$
$$= 216.9\ ft^2 + (0.7)(1000\ ft^2)$$
$$= 916.9\ ft^2$$

Use Eq. 50.27.

$$NR = 10\log\frac{\sum S_1}{\sum S_2} = 10\log\frac{916.9\ ft^2}{246.9\ ft^2}$$
$$= \boxed{5.7\ dB\ decrease}$$

The answer is (B).

SI Solution

The gross area of the walls is

$$A_1 = \Big((2)(6\ m) + (2)(15\ m)\Big)(3\ m) = 126\ m^2$$
$$A_{walls} = (1 - 0.20)(126\ m^2) = 100.8\ m^2$$
$$A_{glass} = (0.20)(126\ m^2) = 25.2\ m^2$$

From App. 50.A, the sound absorption coefficient of glass is $\alpha_1 = 0.03$.

The area of the floor is

$$A_2 = (6\ m)(15\ m) = 90\ m^2$$

From App. 50.A, the sound absorption coefficient of roll vinyl is $\alpha_2 = 0.03$.

The area of the ceiling is

$$A_3 = (6\ m)(15\ m) = 90\ m^2$$

The sound absorption coefficient of the sheetrock is $\alpha_3 = 0.05$.

From App. 50.A, the seats have approximately 1.5 sabins each, and the occupants have approximately 5.0 sabins each.

The total sabin area of the untreated room is

$$\sum S_2 = \alpha_{glass}A_{glass} + \alpha_{walls}A_{walls} + \alpha_{floor}A_{floor}$$
$$+ \alpha_{ceiling}A_{ceiling} + seats + occupants$$
$$+ miscellaneous$$
$$= (0.03)(25.2\ m^2) + (0.05)(100.8\ m^2)$$
$$+ (0.03)(90\ m^2) + (0.05)(90\ m^2)$$
$$+ (15)(1.5\ ft^2)\left(0.3048\ \frac{m}{ft}\right)^2 + (15)(5.0\ ft^2)$$
$$\times \left(0.3048\ \frac{m}{ft}\right)^2 + (5.0\ ft^2)\left(0.3048\ \frac{m}{ft}\right)^2$$
$$= 22.5\ m^2$$

The total sabin area excluding the ceiling is

$$22.5\ m^2 - \alpha_3 A_3 = 22.5\ m^2 - (0.03)(90\ m^2)$$
$$= 19.8\ m^2$$

The total sabin area of the room with the ceiling treated with $\alpha_3 = 0.7$ sound absorption is

$$\sum S_1 = 19.8\ m^2 + \alpha_3 A_3 = 19.8\ m^2 + (0.7)(90\ m^2)$$
$$= 82.8\ m^2$$

Use Eq. 50.27.

$$NR = 10\log\frac{\sum S_1}{\sum S_2} = 10\log\frac{82.8\ m^2}{22.5\ m^2}$$
$$= \boxed{5.7\ dB\ decrease}$$

The answer is (B).

12. *Customary U.S. Solution*

Define the sound absorption coefficients of the floor, ceiling, and walls as α_1, α_2, and α_3, respectively.

The floor area is

$$A_1 = (15\ ft)(20\ ft) = 300\ ft^2$$

The ceiling area is

$$A_2 = (15 \text{ ft})(20 \text{ ft}) = 300 \text{ ft}^2$$

The area of the walls is

$$A_3 = \Big((2)(15 \text{ ft}) + (2)(20 \text{ ft})\Big)(10 \text{ ft}) = 700 \text{ ft}^2$$

The total surface area of the room is

$$A = \sum A_i = A_1 + A_2 + A_3 = 300 \text{ ft}^2 + 300 \text{ ft}^2 + 700 \text{ ft}^2$$
$$= 1300 \text{ ft}^2$$

From Eq. 50.24, the average sound absorption coefficient of the room is

$$\bar{\alpha} = \frac{\sum S_i}{\sum A_i} = \frac{\alpha_1 A_1 + \alpha_2 A_2 + \alpha_3 A_3}{A}$$
$$= \frac{(0.03)(300 \text{ ft}^2) + (0.5)(300 \text{ ft}^2) + (0.06)(700 \text{ ft}^2)}{1300 \text{ ft}^2}$$
$$= 0.155$$

From Eq. 50.25, the room constant is

$$R = \frac{\bar{\alpha} A}{1 - \bar{\alpha}} = \frac{(0.155)(1300 \text{ ft}^2)}{1 - 0.155} = 238.5 \text{ ft}^2$$

From Eq. 50.21(b), the sound pressure level due to the machine is

$$L_{p,\text{dBA}} = 10.5 + L_W + 10 \log\left(\frac{Q}{4\pi r^2} + \frac{4}{R}\right)$$
$$= 10.5 \text{ dB} + 65 \text{ dB}$$
$$\quad + 10 \log\left(\frac{4}{(4\pi)(5 \text{ ft})^2} + \frac{4}{238.5 \text{ ft}^2}\right)$$
$$= 60.2 \text{ dBA}$$

Use Eq. 50.22 to combine the machine sound pressure level with the ambient sound pressure level.

$$L_p = 10 \log \sum 10^{L_i/10}$$
$$= 10 \log\left(10^{60.2 \text{ dBA}/10} + 10^{50 \text{ dBA}/10}\right)$$
$$= \boxed{60.6 \text{ dBA} \quad (61 \text{ dBA})}$$

The answer is (B).

SI Solution

Define the sound absorption coefficients of the floor, ceiling, and walls as α_1, α_2, and α_3, respectively.

The floor area is

$$A_1 = (4.5 \text{ m})(6 \text{ m}) = 27 \text{ m}^2$$

The ceiling area is

$$A_2 = (4.5 \text{ m})(6 \text{ m}) = 27 \text{ m}^2$$

The area of the walls is

$$A_3 = \Big((2)(4.5 \text{ m}) + (2)(6 \text{ m})\Big)(3 \text{ m}) = 63 \text{ m}^2$$

The total surface area of the room is

$$A = \sum A_i = A_1 + A_2 + A_3 = 27 \text{ m}^2 + 27 \text{ m}^2 + 63 \text{ m}^2$$
$$= 117 \text{ m}^2$$

From Eq. 50.24, the average sound absorption coefficient of the room is

$$\bar{\alpha} = \frac{\sum S_i}{\sum A_i} = \frac{\alpha_1 A_1 + \alpha_2 A_2 + \alpha_3 A_3}{A}$$
$$= \frac{(0.03)(27 \text{ m}^2) + (0.5)(27 \text{ m}^2) + (0.06)(63 \text{ m}^2)}{117 \text{ m}^2}$$
$$= 0.155$$

From Eq. 50.25, the room constant is

$$R = \frac{\bar{\alpha} A}{1 - \bar{\alpha}} = \frac{(0.155)(117 \text{ m}^2)}{1 - 0.155} = 21.5 \text{ m}^2$$

From Eq. 50.21(a), the sound pressure level due to the machine is

$$L_{p,\text{dBA}} = 0.2 + L_W + 10 \log\left(\frac{Q}{4\pi r^2} + \frac{4}{R}\right)$$
$$= 0.2 \text{ dB} + 65 \text{ dB} + 10 \log\left(\frac{4}{(4\pi)(1.5 \text{ m})^2} + \frac{4}{21.5 \text{ m}^2}\right)$$
$$= 60.4 \text{ dBA}$$

Use Eq. 50.22 to combine the machine sound pressure level with the ambient sound pressure level.

$$L_p = 10 \log \sum 10^{L_i/10}$$
$$= 10 \log\left(10^{60.4 \text{ dBA}/10} + 10^{50 \text{ dBA}/10}\right)$$
$$= \boxed{60.8 \text{ dBA} \quad (61 \text{ dBA})}$$

The answer is (B).

51 Emergency Management

PRACTICE PROBLEMS

1. A spill of DDT from a chemical tank truck accident has permeated a 100 m² area of soil next to a stream. The emergency response team is evaluating the risk to a town that uses the stream as a source of water 100 km downstream from the spill site. The instream concentration of DDT just below the spill site is 0.6 μg/L, and the stream velocity is 0.2 m/s. The town's water supply is treated only through disinfection.

(a) If the half-life of DDT ranges from 56 days to 110 days and its decay is a first-order reaction, the instream concentration of DDT at the point of use will be most nearly

- (A) 0.5 μg/L
- (B) 1.0 μg/L
- (C) 1.5 μg/L
- (D) 2.0 μg/L

(b) If the potency factor for DDT by the oral route is 0.34 $(\text{mg/kg·d})^{-1}$, the lifetime risk to adults in the community will be most nearly

- (A) 1 per 10,000
- (B) 2.5 per 100,000
- (C) 3 per 1,000,000
- (D) 5.5 per 1,000,000

2. Workers on a road construction project were exposed to vapors from an area contaminated with an herbicide identified as X49 (a noncarcinogen) for 30 working days (8 hr shifts) before it was recognized that individual protection might be needed. The mean concentration of vapors of X49 in air was 2 μg/m³ during this period.

(a) The standard factors for workers at the site are as follows.

air breathed	5.2 m³/h
retention rate (inhaled air)	100%
absorption rate	100%
adult body weight	70 kg

The chronic daily intake of X49 for each worker is most nearly

- (A) 1×10^{-2} mg/kg·d
- (B) 2×10^{-2} mg/kg·d
- (C) 1×10^{-3} mg/kg·d
- (D) 2×10^{-3} mg/kg·d

(b) If the reference dose for X49 for the inhalation route is 1.6×10^{-3} mg/kg·d and the acceptable risk is 1 in 1,000,000, how can exposure of the workers most accurately be characterized?

- (A) The exposure was acceptable because the hazard ratio is less than 1.0.
- (B) The exposure was not acceptable because the hazard ratio is greater than 1.0.
- (C) The exposure was acceptable because the risk is less than 1 in 1,000,000.
- (D) The exposure was not acceptable because the risk is greater than 1 in 1,000,000.

3. A rural community water supply was contaminated by wastes resulting from a railroad tank car derailment. The wastes contain the following noncarcinogenic pollutants and have the indicated reference doses.

pollutant	concentration	reference dose
chemical A	10 μg/L	5×10^{-3} mg/kg·d
metal B	25 μg/L	1.0×10^{-3} mg/kg·d
chemical C	15 μg/L	3.6×10^{-2} mg/kg·d

Standard factors for the community are as follows.

drinking water use, adult	2.0 L/d
drinking water use, child	1.2 L/d
body weight (mass), adult	70 kg
body weight (mass), child	30 kg

Assume retention and absorption rates of 100% each.

(a) The hazard ratio for adults is most nearly

- (A) 0.4
- (B) 0.8
- (C) 1.0
- (D) 1.4

(b) The hazard ratio for children is most nearly

- (A) 0.2
- (B) 0.5
- (C) 1.0
- (D) 1.5

(c) How should the exposure to the community be characterized?

(A) There is no significant risk to the community.

(B) There is an unacceptable risk to adults but not to children.

(C) There is an unacceptable risk to children but not to adults.

(D) There is an unacceptable risk to both adults and children.

4. Emergency responders have identified the following flammable liquids stored separately in a warehouse 50 m from a house fire. The atmosphere is at normal conditions.

liquid	auto-ignition temp.	vapor pressure at 20°C	ratio of specific heats	flammable range volume % LFL	UFL
ethane	472°C	3850 kPa	1.206	3.0	12.5
benzene	500°C	10 kPa	1.100	1.2	8.0
n-heptane	285°C	4.6 kPa	1.400	1.1	6.7

(a) If a spill occurs and equilibrium is reached, which chemicals would form a flammable mixture in air?

(A) ethane and n-heptane

(B) benzene only

(C) n-heptane only

(D) ethane and benzene

(b) If the drums are instantly crushed from a collapsing structure and the vapors mix with air and are compressed to 100 atm, which of the vapors would autoignite?

(A) ethane and benzene

(B) benzene

(C) n-heptane

(D) benzene and n-heptane

(c) If the vapors from the three liquids are mixed in air in the amounts given, what would the LFL of the mixture most nearly be?

	no. of moles
ethane	2
benzene	4
n-heptane	3

(A) 0.5%

(B) 1.5%

(C) 2.5%

(D) 3.5%

SOLUTIONS

1. (a) The travel time, t, from the spill to the town is

$$t = \frac{l}{v}$$

$$= \left(\frac{100 \text{ km}}{0.2 \frac{\text{m}}{\text{s}}}\right) \left(10^3 \frac{\text{m}}{\text{km}}\right) \left(\frac{1 \text{ min}}{60 \text{ s}}\right) \left(\frac{1 \text{ h}}{60 \text{ min}}\right) \left(\frac{1 \text{ d}}{24 \text{ h}}\right)$$

$$= 5.79 \text{ d}$$

The half-life, $t_{1/2}$, of DDT ranges from 56 days to 110 days. Use the average of 83 days.

The concentration can be found from the first-order reaction equation.

$$C_t = C_o e^{-kt}$$

Find k from the half-concentration and the half-life of DDT.

$$C_t = \frac{C_o}{2} = C_o e^{-k(83 \text{ d})}$$

$$0.5 = e^{-k(83 \text{ d})}$$

$$\ln 0.5 = -k(83 \text{ d})$$

$$k = 8.351 \times 10^{-3} \text{ d}^{-1}$$

The concentration of DDT at the point of use is

$$C_t = C_o e^{-kt}$$

$$= \left(0.6 \frac{\mu\text{g}}{\text{L}}\right) e^{-(8.351 \times 10^{-3} \text{ d}^{-1})(5.79 \text{ d})}$$

$$= \boxed{0.572 \ \mu\text{g/L}}$$

The answer is (A).

(b) Find the quantity of water ingested daily. Assume a standard adult has a body weight of 70 kg and ingests 2 L of water daily.

Find the chronic daily intake, CDI.

$$\text{CDI} = \frac{CQ}{m} = \frac{\left(0.572 \frac{\mu\text{g}}{\text{L}}\right) \left(2 \frac{\text{L}}{\text{d}}\right) \left(10^{-3} \frac{\text{mg}}{\mu\text{g}}\right)}{70 \text{ kg}}$$

$$= 1.634 \times 10^{-5} \text{ mg/kg·d}$$

The potency factor, PF, (also known as slope factor, SF) for DDT by the oral route is 0.34 $(\text{mg/kg·d})^{-1}$.

The lifetime risk is

$$R = (\text{CDI})(\text{PF})$$

$$= \left(1.634 \times 10^{-5} \frac{\text{mg}}{\text{kg·d}}\right) \left(0.34 \left(\frac{\text{mg}}{\text{kg·d}}\right)^{-1}\right)$$

$$= \boxed{5.554 \times 10^{-6} \quad \begin{array}{c} \text{(5.6 additional cancer cases} \\ \text{per million population)} \end{array}}$$

The answer is (D).

2. (a) The daily intake can be found as follows.

$$I_N = \frac{C(\text{CR})(\text{EF})(\text{ED})(\text{RR})(\text{ABS})}{(\text{BW})(\text{AT})}$$

$$C = \frac{2\,\frac{\mu g}{m^3}}{1000\,\frac{\mu g}{mg}} = 2 \times 10^{-3}\text{ mg/m}^3$$

$$\text{CR} = \left(5.2\,\frac{m^3}{h}\right)\left(8\,\frac{h}{d}\right) = 41.6\text{ m}^3/\text{d}$$

$$\text{EF} = 30\text{ d/yr}$$
$$\text{ED} = 1\text{ yr}$$
$$\text{RR} = 1.0$$
$$\text{ABS} = 1.0$$
$$\text{BW} = 70\text{ kg}$$
$$\text{AT} = 30\text{ d}$$

$$I_N = \frac{\left(2 \times 10^{-3}\,\frac{mg}{m^3}\right)\left(41.6\,\frac{m^3}{d}\right)\left(30\,\frac{d}{yr}\right)}{\times (1\text{ yr})(1.0)(1.0)}$$
$$= \boxed{1.19 \times 10^{-3}\text{ mg/kg·d}}$$

The answer is (C).

(b) For noncarcinogens, use the hazard ratio method.

$$\text{HR} = \frac{I_N}{\text{RfD}} = \frac{1.19 \times 10^{-3}\,\frac{mg}{kg\cdot d}}{1.6 \times 10^{-3}\,\frac{mg}{kg\cdot d}}$$
$$= 0.74$$

The HR is less than 1.0, so the exposure was acceptable.

The answer is (A).

3. (a)

$$\text{EF} = 1\text{ d/yr}$$
$$\text{ED} = 1\text{ yr}$$
$$\text{RR} = 1.0$$
$$\text{ABS} = 1.0$$
$$\text{AT} = 1\text{ d}$$

For adults, the dose of chemical A is

$$\text{CR} = 2.0\text{ L/d}$$
$$\text{BW} = 70\text{ kg}$$
$$C_A = \left(10\,\frac{\mu g}{L}\right)\left(10^{-3}\,\frac{mg}{\mu g}\right) = 10 \times 10^{-3}\text{ mg/L}$$

$$I_N = \frac{C(\text{CR})(\text{EF})(\text{ED})(\text{RR})(\text{ABS})}{(\text{BW})(\text{AT})}$$

$$= \frac{\left(10 \times 10^{-3}\,\frac{mg}{L}\right)\left(2.0\,\frac{L}{d}\right)\left(1\,\frac{d}{yr}\right)}{\times (1\text{ yr})(1.0)(1.0)}$$
$$= 2.86 \times 10^{-4}\text{ mg/kg·d}$$

The hazard ratio is

$$\text{HR} = \frac{I_N}{\text{RfD}} = \frac{2.86 \times 10^{-4}\,\frac{mg}{kg\cdot d}}{5 \times 10^{-3}\,\frac{mg}{kg\cdot d}}$$
$$= 0.057$$

Similarly, the other pollutants contribute to the hazard ratio for adults.

pollutant	dose $\left(\frac{mg}{kg\cdot d}\right)$	RfD $\left(\frac{mg}{kg\cdot d}\right)$	hazard ratio
chemical A	2.86×10^{-4}	50×10^{-4}	0.057
metal B	7.14×10^{-4}	10×10^{-4}	0.714
chemical C	4.29×10^{-4}	360×10^{-4}	0.012
total			$\boxed{0.783}$

The answer is (B).

(b) For children, the dose of chemical A is

$$\text{CR} = 1.2\text{ L/d}$$
$$\text{BW} = 30\text{ kg}$$

$$I_N = \frac{C(\text{CR})(\text{EF})(\text{ED})(\text{RR})(\text{ABS})}{(\text{BW})(\text{AT})}$$

$$= \frac{\left(10 \times 10^{-3}\,\frac{mg}{L}\right)\left(1.2\,\frac{L}{d}\right)\left(1\,\frac{d}{yr}\right)}{\times (1\text{ yr})(1.0)(1.0)}$$
$$= 4 \times 10^{-4}\text{ mg/kg·d}$$

Find the hazard ratio.

$$\text{HR} = \frac{I_N}{\text{RfD}} = \frac{4 \times 10^{-4}\,\frac{mg}{kg\cdot d}}{5 \times 10^{-3}\,\frac{mg}{kg\cdot d}}$$
$$= 0.08$$

Similarly, the other pollutants contribute to the hazard ratio for children.

pollutant	dose $\left(\dfrac{mg}{kg \cdot d}\right)$	RfD $\left(\dfrac{mg}{kg \cdot d}\right)$	hazard ratio
chemical A	4×10^{-4}	50×10^{-4}	0.080
metal B	10×10^{-4}	10×10^{-4}	1.000
chemical C	6×10^{-4}	360×10^{-4}	0.017
total			$\boxed{1.097}$

The answer is (C).

(c) Because the hazard ratio is less than 1.0 for adults, there is no unacceptable risk for them. However, there is an unacceptable risk for children because the hazard ratio for them is greater than 1.0.

The answer is (C).

4. (a) Determine the mole fraction (in vapor phase) of each substance at NTP.

$$p = (1 \text{ atm})\left(\frac{101.325 \text{ kPa}}{1 \text{ atm}}\right) = 101.325 \text{ kPa}$$

$$x_v = \frac{p_{vp,\text{pure}} x_{li}}{p}$$

Because each flammable liquid is stored independently, the mole fraction (in liquid phase) of each $(x_{l,A})$ is 1.

For ethane,

$$x_{v,\text{ethane}} = \frac{(3850 \text{ kPa})(1)}{101.325 \text{ kPa}}$$
$$= 38.0 \quad (3800\%)$$

$x_{v,\text{ethane}}$ is greater than ethane's UFL of 12.5% and is too rich to form a flammable mixture in air.

For benzene,

$$x_{v,\text{benzene}} = \frac{(10 \text{ kPa})(1)}{101.325 \text{ kPa}}$$
$$= 0.0986 \quad (9.86\%)$$

$x_{v,\text{benzene}}$ is greater than benzene's UFL of 8.0% and is too rich to form a flammable mixture in air.

For n-heptane,

$$x_{v,n\text{-heptane}} = \frac{(4.6 \text{ kPa})(1)}{101.325 \text{ kPa}}$$
$$= 0.0454 \quad (4.54\%)$$

$x_{v,n\text{-heptane}}$ is between n-heptane's LFL and UFL (1.1% and 6.7%, respectively) and forms a flammable mixture in air.

The answer is (C).

(b) At NTP,

$$T_1 = 25°C + 273° = 298K$$

Determine the final temperature of each substance. Assume compression occurs adiabatically.

$$T_2 = T_1 \left(\frac{p_2}{p_1}\right)^{(k-1)/k}$$

For ethane,

$$T_{2,\text{ethane}} = (298K)\left(\frac{100 \text{ atm}}{1 \text{ atm}}\right)^{(1.206-1)/1.206}$$
$$= 654.4K - 273°$$
$$\approx 381°C$$

$T_{2,\text{ethane}}$ is less than 472°C, the autoignition temperature of ethane.

For benzene,

$$T_{2,\text{benzene}} = (298K)\left(\frac{100 \text{ atm}}{1 \text{ atm}}\right)^{(1.100-1)/1.100}$$
$$= 452.9K - 273°$$
$$\approx 180°C$$

$T_{2,\text{benzene}}$ is less than 500°C, the autoignition temperature of benzene.

For n-heptane,

$$T_{2,n\text{-heptane}} = (298K)\left(\frac{100 \text{ atm}}{1 \text{ atm}}\right)^{(1.4-1)/1.4}$$
$$= 1111K - 273°$$
$$\approx 838°C$$

$T_{2,n\text{-heptane}}$ is greater than 285°C, the autoignition temperature of n-heptane.

The answer is (C).

(c) For ethane,

$$x_{v,\text{ethane}} = \frac{n_{\text{ethane}}}{n_{\text{total}}} = \frac{2 \text{ mol}}{2 \text{ mol} + 4 \text{ mol} + 3 \text{ mol}}$$
$$= 0.22$$

For benzene,

$$x_{v,\text{benzene}} = \frac{n_{\text{benzene}}}{n_{\text{total}}} = \frac{4 \text{ mol}}{2 \text{ mol} + 4 \text{ mol} + 3 \text{ mol}}$$
$$= 0.44$$

For n-heptane,

$$x_{v,n\text{-heptane}} = \frac{n_{n\text{-heptane}}}{n_{\text{total}}} = \frac{3 \text{ mol}}{2 \text{ mol} + 4 \text{ mol} + 3 \text{ mol}}$$
$$= 0.33$$

$$\text{LFL}_{\text{mix}} = \frac{1}{\sum \left(\dfrac{x_{v,i}}{\text{LFL}_i} \right)} = \frac{1}{\dfrac{0.22}{3.0\%} + \dfrac{0.44}{1.2\%} + \dfrac{0.33}{1.1\%}}$$
$$= \boxed{1.34\%}$$

The answer is (B).

52 Electrical Systems and Equipment

PRACTICE PROBLEMS

Circuit Analysis

1. Most nearly, how much power is dissipated by the circuit shown?

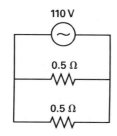

(A) 1.0 kW

(B) 3.0 kW

(C) 4.0 kW

(D) 12 kW

Transformers

2. An ideal step-down transformer has 200 primary coil turns and 50 secondary coil turns. 440 V are applied across the primary side. The resistance of an external secondary load is 5 Ω. The amount of heat dissipated in the resistance is most nearly

(A) 1200 W

(B) 2400 W

(C) 2600 W

(D) 3300 W

Induction Motors

3. The speed (in rpm) at which an induction motor rotates is

(A) $\dfrac{120(\text{frequency of the electric source, Hz})}{\text{number of poles}}$

(B) $\dfrac{120(\text{number of poles})}{\text{frequency of the electric source, Hz}}$

(C) $\dfrac{120(\text{voltage of the electric source, V})}{\text{frequency of the electric source, Hz}}$

(D) none of the above

4. The full-load phase current drawn by a 440 V (rms) 60 hp (total) three-phase induction motor having a full-load efficiency of 86% and a full-load power factor of 76% is most nearly

(A) 52 A

(B) 78 A

(C) 160 A

(D) 700 A

5. A 200 hp, three-phase, four-pole, 60 Hz, 440 V (rms) squirrel-cage induction motor operates at full load with an efficiency of 85%, power factor of 91%, and 3% slip.

(a) The speed is most nearly

(A) 1500 rpm

(B) 1700 rpm

(C) 2300 rpm

(D) 5800 rpm

(b) The torque developed is most nearly

(A) 180 ft-lbf

(B) 450 ft-lbf

(C) 600 ft-lbf

(D) 690 ft-lbf

(c) The line current is most nearly

(A) 34 A

(B) 150 A

(C) 250 A

(D) 280 A

6. A factory's induction motor load draws 550 kW at 82% power factor. What is most nearly the size of an additional synchronous motor required to produce 250 hp and raise the power factor to 95%? The line voltage is 220 V (rms).

(A) 140 kVA

(B) 230 kVA

(C) 240 kVA

(D) 330 kVA

7. The nameplate of an induction motor lists 960 rpm as the full-load speed. The frequency the motor was designed for is most nearly

(A) 24 Hz

(B) 34 Hz

(C) 48 Hz

(D) 50 Hz

Synchronous Motors

8. The speed (in rpm) at which a synchronous motor rotates is

(A) $\dfrac{120(\text{frequency of the electric source, Hz})}{\text{number of poles}}$

(B) $\dfrac{120(\text{number of poles})}{\text{frequency of the electric source, Hz}}$

(C) $\dfrac{120(\text{voltage of the electric source, V})}{\text{frequency of the electric source, Hz}}$

(D) none of the above

Three-Phase Power

9. The power triangle shown represents the total of all phases of a three-phase generator operating at 22 kV (rms). What is most nearly the rms line current for one phase?

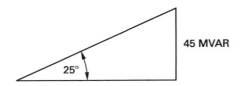

(A) 1.6 kA

(B) 2.2 kA

(C) 2.8 kA

(D) 4.8 kA

10. A 25 hp, three-phase motor draws 28 A (rms) at 480 V (rms). The motor efficiency is 92%. The total reactive load drawn by the motor is most nearly

(A) 8.6 kVAR

(B) 11 kVAR

(C) 16 kVAR

(D) 17 kVAR

Power Factor Correction

11. A machine shop that operates 14 hours per day, 349 days per year is considering some upgrades to decrease its energy usage. The shop currently uses electric motors generating a total of 600 hp. The upgrade would increase motor electrical-to-mechanical efficiencies from 86% to 95%. The shop is also considering replacing its 50 kW of incandescent lighting with 50 kW of fluorescent lighting. After the upgrade, the overall power factor of the shop will increase from 0.85 to 0.89. The cost for the upgrade is $160,000, and the electricity rate is $0.065/kVA-hr, based on apparent power. The simple payback period for this upgrade is most nearly

(A) 4.3 yr

(B) 5.9 yr

(C) 6.7 yr

(D) 8.1 yr

12. A manufacturing plant uses several large, 4160 V (rms), three-phase induction motors in its manufacturing process. The total motor power generated is 3960 hp, and the apparent drawn power is 3750 kVA. The motor electrical-to-mechanical efficiency is 90%, and the electricity cost is $0.045/kVA-hr, based on apparent power. The plant operates for 24 hours per day, seven days a week, but is shut down for maintenance two weeks per year. Several induction motors are replaced with synchronous motors to increase the plant power factor to 0.95. The synchronous motor efficiency is the same as the induction motor efficiency. Total plant motor power remains the same. The annual decrease in operating costs is most nearly

(A) $45,000

(B) $76,000

(C) $110,000

(D) $120,000

SOLUTIONS

1. The two resistors are in parallel. From Eq. 52.22, the equivalent resistance for the circuit is

$$\frac{1}{R_e} = \frac{1}{R_1} + \frac{1}{R_2} = \frac{1}{0.5\ \Omega} + \frac{1}{0.5\ \Omega}$$
$$R_e = 4\ \Omega$$

Although the voltage is AC, the circuit is purely resistive. The power dissipation is real power. The electrical power dissipated is

$$P = \frac{V^2}{R_e} = \frac{(110\ \text{V})^2}{(4\ \Omega)\left(1000\ \dfrac{\text{W}}{\text{kW}}\right)}$$
$$= \boxed{3.025\ \text{kW} \quad (3.0\ \text{kW})}$$

The answer is (B).

2. From Eq. 52.43, the transformer turns ratio is

$$a = \frac{N_p}{N_s} = \frac{200}{50} = 4$$

The secondary voltage is

$$V_s = \frac{V_p}{a} = \frac{440\ \text{V}}{4} = 110\ \text{V}$$

The secondary current is

$$I_s = \frac{V_s}{R_s} = \frac{110\ \text{V}}{5\ \Omega} = 22\ \text{A}$$

The heat dissipated in the resistance is

$$P_s = I_s^2 R_s = (22\ \text{A})^2 (5\ \Omega)$$
$$= \boxed{2420\ \text{W} \quad (2400\ \text{W})}$$

The answer is (B).

3. An induction motor is drawn forward (accelerated) by an induced current that decreases as the motor approaches the synchronous speed at which a synchronous motor would turn, option A. For this reason, an induction motor always turns a few percent slower than the synchronous speed. None of the options mention this "slip."

The answer is (D).

4. From Eq. 52.67, the full-load phase current is

$$I_p = \frac{P_p}{\eta\, V_p \cos\phi} = \frac{\left(\dfrac{60\ \text{hp}}{3}\right)\left(745.7\ \dfrac{\text{W}}{\text{hp}}\right)}{(0.86)(440\ \text{V})(0.76)}$$
$$= \boxed{51.86\ \text{A} \quad (52\ \text{A})}$$

The answer is (A).

5. (a) From Eq. 52.76 and Eq. 52.78,

$$n_r = n_{\text{synchronous}}(1 - s) = \left(\frac{120f}{p}\right)(1 - s)$$
$$= \left(\frac{(2)\left(60\ \dfrac{\text{sec}}{\text{min}}\right)(60\ \text{Hz})}{4}\right)(1 - 0.03)$$
$$= \boxed{1746\ \text{rpm} \quad (1700\ \text{rpm})}$$

The answer is (B).

(b) From Eq. 52.71(b), the torque developed is

$$T = \frac{5252P}{n_r} = \frac{\left(5252\ \dfrac{\text{ft-lbf}}{\text{hp-sec}}\right)(200\ \text{hp})}{1746\ \dfrac{\text{rev}}{\text{min}}}$$
$$= \boxed{602\ \text{ft-lbf} \quad (600\ \text{ft-lbf})}$$

The answer is (C).

(c) From Eq. 52.67, the line current is

$$I_l = \frac{P_t}{\sqrt{3}\,\eta\, V_l \cos\phi} = \frac{(200\ \text{hp})\left(745.7\ \dfrac{\text{W}}{\text{hp}}\right)}{(\sqrt{3})(0.85)(440\ \text{V})(0.91)}$$
$$= \boxed{253\ \text{A} \quad (250\ \text{A})}$$

The answer is (C).

6. Draw the power triangle.

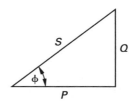

The original power angle is

$$\phi_i = \arccos 0.82 = 34.92°$$
$$P_1 = 550 \text{ kW}$$

From Eq. 52.56,

$$Q_i = P_1 \tan \phi_i = (550 \text{ kW})(\tan 34.92°) = 384.0 \text{ kVAR}$$

The new conditions are

$$P_2 = (250 \text{ hp})\left(0.7457 \frac{\text{kW}}{\text{hp}}\right) = 186.4 \text{ kW}$$
$$\phi_f = \arccos 0.95 = 18.19°$$

Since both motors perform real work,

$$P_f = P_1 + P_2 = 550 \text{ kW} + 186.4 \text{ kW} = 736.4 \text{ kW}$$

The new reactive power is

$$Q_f = P_f \tan \phi_f = (736.4 \text{ kW})(\tan 18.19°)$$
$$= 242.0 \text{ kVAR}$$

The change in reactive power is

$$\Delta Q = 384.0 \text{ kVAR} - 242.0 \text{ kVAR} = 142 \text{ kVAR}$$

Synchronous motors used for power factor correction are rated by apparent power.

$$S = \sqrt{(\Delta P)^2 + (\Delta Q)^2}$$
$$= \sqrt{(186.4 \text{ kW})^2 + (142 \text{ kVAR})^2}$$
$$= \boxed{234.3 \text{ kVA} \quad (230 \text{ kVA})}$$

The answer is (B).

7. From Eq. 52.76 and Eq. 52.78, the frequency is

$$f = \frac{pn_{\text{synchronous}}}{120} = \frac{pn_{\text{actual}}}{(2)\left(60 \frac{\text{sec}}{\text{min}}\right)(1-s)}$$

The slip and number of poles are unknown. Assume $s = 0$ and $p = 4$.

$$f = \frac{(4)\left(960 \frac{\text{rev}}{\text{min}}\right)}{(2)\left(60 \frac{\text{sec}}{\text{min}}\right)(1-0)} = 32 \text{ Hz}$$

32 Hz is not close to any frequency in commercial use. Try $p = 6$.

$$f = \frac{(6)\left(960 \frac{\text{rev}}{\text{min}}\right)}{(2)\left(60 \frac{\text{sec}}{\text{min}}\right)(1-0)} = 48 \text{ Hz}$$

With a 4% slip, $f = 50$ Hz.

$$\boxed{50 \text{ Hz (European)}}$$

The answer is (D).

8. A synchronous motor's speed is dependent on the speed of the stator rotating field. The motor can rotate at the speed of the applied alternating current or sub-multiples thereof. From Eq. 52.76,

$$n_{\text{synchronous}} = \frac{120f}{p}$$

The synchronous speed is 120 times the frequency of the electric source divided by the number of poles.

The answer is (A).

9. The total real power is

$$P_t = \frac{Q_t}{\tan \phi}$$
$$= \frac{(45 \text{ MVAR})\left(1000 \frac{\text{kVAR}}{\text{MVAR}}\right)}{\tan 25°}$$
$$= 96{,}502 \text{ kVAR}$$

From Eq. 52.67, the line current is

$$I_l = \frac{P_t}{\sqrt{3} V_l \cos \phi}$$
$$= \frac{96{,}502 \text{ kVAR}}{(\sqrt{3})(22 \text{ kV})\left(1000 \frac{\text{A}}{\text{kA}}\right)\cos 25°}$$
$$= \boxed{2.79 \text{ kA} \quad (2.8 \text{ kA})}$$

The answer is (C).

10. From Eq. 52.72, the total real electrical power drawn from the line by the motor is

$$P_{\text{electrical}} = \frac{P_{\text{rated}}}{\eta_m} = \frac{(25 \text{ hp})\left(745.7 \frac{\text{W}}{\text{hp}}\right)}{(0.92)\left(1000 \frac{\text{W}}{\text{kW}}\right)}$$
$$= 20.26 \text{ kW}$$

Rearrange Eq. 52.79 for the power factor.

$$\text{pf} = \frac{P_{\text{electrical}}}{\sqrt{3}\,V_l I_l} = \frac{(20.26 \text{ kW})\left(1000 \frac{\text{W}}{\text{kW}}\right)}{(\sqrt{3})(480 \text{ V})(28 \text{ A})}$$
$$= 0.870$$

The power factor angle is

$$\phi = \arccos \text{pf} = \arccos 0.870$$
$$= 29.5°$$

From Eq. 52.68, the total apparent power is

$$S_t = \sqrt{3}\,V_l I_l$$
$$= \frac{(\sqrt{3})(480 \text{ V})(28 \text{ A})}{1000 \frac{\text{VA}}{\text{kVA}}}$$
$$= 23.28 \text{ kVA}$$

Therefore, from Eq. 52.55, the total reactive load is

$$Q_t = S_t \sin \phi = (23.28 \text{ kVA})\sin 29.5°$$
$$= \boxed{11.46 \text{ kVAR} \quad (11 \text{ kVAR})}$$

The answer is (B).

11. The lighting changes do not decrease the lighting power draw, but they do affect the apparent power. The real and apparent powers before the motor upgrade are

$$P_{\text{before}} = \frac{P_m}{\eta_{\text{before}}} = \left(\frac{600 \text{ hp}}{0.86}\right)\left(0.7457 \frac{\text{kW}}{\text{hp}}\right) + 50 \text{ kW}$$
$$= 570.3 \text{ kW}$$

$$S_{\text{before}} = \frac{P_{\text{before}}}{\cos \phi} = \frac{P_{\text{before}}}{\text{pf}} = \frac{570.3 \text{ kW}}{0.85}$$
$$= 670.9 \text{ kVA}$$

The real and apparent powers after the motor upgrade are

$$P_{\text{after}} = \frac{P_m}{\eta_{\text{after}}} = \left(\frac{600 \text{ hp}}{0.95}\right)\left(0.7457 \frac{\text{kW}}{\text{hp}}\right) + 50 \text{ kW}$$
$$= 521.0 \text{ kW}$$
$$S_{\text{after}} = \frac{P_{\text{after}}}{\cos \phi} = \frac{P_{\text{after}}}{\text{pf}} = \frac{521.0 \text{ kW}}{0.89}$$
$$= 585.4 \text{ kVA}$$

The annual savings in cost from the upgrade, C_A, is

$$\Delta A = (S_{\text{before}} - S_{\text{after}})\left(\frac{\text{hr of operation}}{\text{yr}}\right) C_{\text{electricity}}$$
$$= (671.2 \text{ kVA} - 585.4 \text{ kVA})$$
$$\times \left(14 \frac{\text{hr}}{\text{day}}\right)\left(349 \frac{\text{day}}{\text{yr}}\right)\left(0.065 \frac{\$}{\text{kVA-hr}}\right)$$
$$= \$27,249/\text{yr}$$

The simple payback period is

$$\text{payback} = \frac{C_{\text{upgrade}}}{\Delta A} = \frac{\$160,000}{27,249 \frac{\$}{\text{yr}}}$$
$$= \boxed{5.87 \text{ yr} \quad (5.9 \text{ yr})}$$

The answer is (B).

12. From Eq. 52.72, the total real electrical power drawn by the motors from the line is

$$P_{\text{electrical}} = \frac{P_{\text{rated}}}{\eta_m} = \frac{3960 \text{ hp}}{0.90}$$
$$= 4400 \text{ hp}$$

Rearranging Eq. 52.57, the new total apparent power after the synchronous motors have been installed is

$$S_{\text{after}} = \frac{P_{\text{electrical}}}{\text{pf}} = \frac{(4400 \text{ hp})\left(745.7 \frac{\text{W}}{\text{hp}}\right)}{(0.95)\left(1000 \frac{\text{VA}}{\text{kVA}}\right)}$$
$$= 3454 \text{ kVA}$$

The annual decrease in operating cost is

$$\Delta A = (S_{\text{before}} - S_{\text{after}})\left(\frac{\text{hr of operation}}{\text{yr}}\right) C_{\text{electricity}}$$
$$= (3750 \text{ kVA} - 3454 \text{ kVA})$$
$$\times \left(24 \frac{\text{hr}}{\text{day}}\right)\left(365 \frac{\text{day}}{\text{yr}} - 14 \frac{\text{day}}{\text{yr}}\right)$$
$$\times \left(0.045 \frac{\$}{\text{kVA-hr}}\right)$$
$$= \boxed{\$112,207/\text{yr} \quad (\$110,000)}$$

The answer is (C).

53 Instrumentation and Digital Numbering Systems

PRACTICE PROBLEMS

Indicator Diagrams

1. Diesel engines in a waste-to-energy plant are powered by a gaseous mixture of methane and other digestion gases. The combustion conditions in each cylinder are continuously monitored by pressure and temperature transducers. A particular cylinder's indicator diagram is shown. The scale selected in the monitoring software is set to 111 kPa/mm. Most nearly, what is the mean effective pressure in the cylinder?

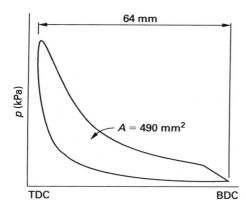

(A) 7.7 kPa

(B) 15 kPa

(C) 360 kPa

(D) 850 kPa

Variable Inductance Transformers

2. Which of the statements regarding the linear variable differential transformer (LVDT) shown is true?

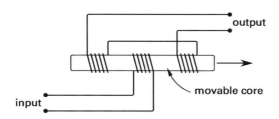

(A) The LVDT is a variable reluctance transducer, and its output voltage is directly proportional to its core movement.

(B) The LVDT is a variable reluctance transducer, and its output current is directly proportional to its core movement.

(C) The LVDT is a variable capacitance transducer, and its output current is directly proportional to core movement.

(D) The LVDT is a variable inductance transducer, and its output voltage is directly proportional to core movement.

Resistance Temperature Detectors

3. A resistance temperature detector (RTD) measures temperature in a bridge circuit. The resistances of resistors 1, 2, and 3 are 100 Ω, 200 Ω, and 50 Ω, respectively. The power supply voltage is 10 V. The reference resistance of the RTD is 100 Ω at 0°C. The alpha value for the RTD is 0.00392 1/°C. The variation of RTD resistance with temperature is linear. What is most nearly the temperature of the RTD when the voltmeter reads 4.167 V?

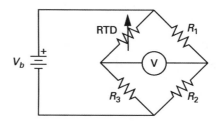

(A) 130°F

(B) 210°F

(C) 260°F

(D) 330°F

4. The temperature of a steel girder during a fire test is measured with a type 404 platinum RTD and a simple two-wire bridge, as shown. The characteristics of the RTD are: resistance, R, 100 Ω at 0°C; temperature coefficient, α, 0.00385 1/°C. The values of the bridge resistances are $R_1 = 1000\ \Omega$, and $R_3 = 1000\ \Omega$. When the meter is nulled out during a test, $R_2 = 376\ \Omega$. The lead resistances are each 100 Ω. Most nearly, what is the temperature of the beam during the test?

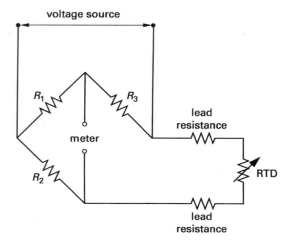

(A) 130°C

(B) 175°C

(C) 200°C

(D) 450°C

Piezoelectric Transducers

5. A compression-style crystal piezoelectric accelerometer has a sensitivity of 2.0 mV/g. It falls to the floor from a height of 3 ft and comes to rest 0.02 sec after contact with the floor. What is most nearly the average voltage output of the accelerometer during impact?

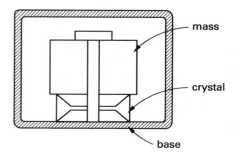

(A) 21 mV

(B) 32 mV

(C) 43 mV

(D) 57 mV

6. A piezoelectric transducer is connected to a charge amplifier as shown. What is the purpose of capacitor C_t?

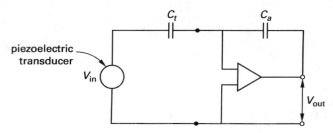

(A) C_t biases the transducer output.

(B) C_t isolates the transducer from inadvertent amplifier transients.

(C) C_t rectifies the transducer output.

(D) C_t prevents DC voltages from being passed to the amplifier.

Hydraulic Rams

7. A 2 ft diameter (outer dimension) oil-filled hydraulic scale has 2 in thick sidewalls. When the scale is unloaded, the pressure gauge reads zero. When the pressure gauge reads 50 psig, the weight of the object is most nearly

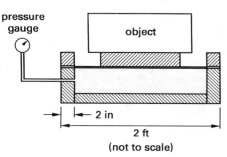

(A) 8000 lbf

(B) 16,000 lbf

(C) 23,000 lbf

(D) 63,000 lbf

Strain Gauges

8. A load cell measures tensile force with a steel bar and a bonded strain gauge. The modulus of elasticity of the steel is 30×10^6 lbf/in^2. The bar's tensile area is 1.0 in^2. The gage factor of the strain gauge is 6.0. The unloaded strain gauge resistance is 100 Ω, and the loaded resistance is 100.1 Ω. The applied force is most nearly

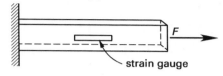

(A) 2500 lbf

(B) 5000 lbf

(C) 17,000 lbf

(D) 20,000 lbf

9. A strain gauge is bonded to each of the four vertical bars of a grain hopper scale. The hopper has a tare weight of 1500 lbf. The support bars have a modulus of elasticity of 20×10^6 lbf/in^2, and each bar has a cross-sectional area of 2 in^2. Each strain gauge has a gage factor of 3 and had an initial resistance of 300 Ω when originally applied to the unstressed bars. After the grain is loaded into the hopper, the strain gauge resistances are 300.05 Ω, 300.09 Ω, 300.11 Ω, and 300.03 Ω, respectively. What is most nearly the weight of the grain loaded into the hopper?

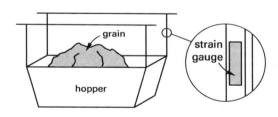

(A) 9500 lbf

(B) 11,000 lbf

(C) 12,000 lbf

(D) 14,000 lbf

10. A strain gauge with a gage factor of 2.0 and resistance of 120 Ω experiences a microstrain of 500. Most nearly, what is the percentage change in resistance?

(A) 0.00001%

(B) 0.001%

(C) 0.1%

(D) 1000%

11. A strain gauge has a gage factor of 3.0 and a nominal (strain-free) resistance of 350 Ω. The gauge is bonded to the top of a rectangular aluminum bar 12 cm long, 2 cm high, and 4 cm wide. The aluminum has a modulus of elasticity of 73 GPa. The bar is loaded as a cantilever beam. The distance from the applied load to the center of the strain gauge is 10 cm. The instrumentation consists of a simple Wheatstone bridge consisting of two 1000 Ω resistors, a variable resistor, and the strain gauge. Lead resistance is negligible. Most nearly, what is the change in resistance of the strain gauge when the free end of the aluminum bar is loaded vertically by a 1 kg mass?

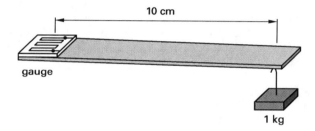

(A) 0.00054 Ω

(B) 0.0053 Ω

(C) 0.048 Ω

(D) 0.50 Ω

12. A cylindrical 2 in diameter solid shaft used to power an aerator blade is mounted as a stationary horizontal cantilever and instrumented with a strain gauge, R_1, bonded to its upper surface. The strain gauge is connected to an uncompensated bridge circuit where R_2 is 160 Ω and R_3 is 2500 Ω. R_4 is the adjustable resistance with a resistance of 2505 Ω when the shaft is unloaded, and 2511.7 Ω when the shaft tip 24 in from the midpoint of the strain gauge is loaded with vertical force of 1100 lbf. The gage factor of the strain gauge is 2.10. Disregard slip ring and lead resistances. Most nearly, what is the modulus of elasticity of the shaft?

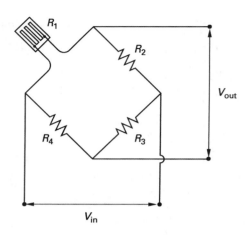

(A) 12×10^6 lbf/in^2

(B) 18×10^6 lbf/in^2

(C) 23×10^6 lbf/in^2

(D) 26×10^6 lbf/in^2

13. A 45° (rectangular) strain gauge rosette is bonded to the web of a large A36 steel built-up girder beam. The web's modulus of elasticity is 200 GPa, and Poisson's ratio is 0.3. The microstrains are $\epsilon_A = 650$, $\epsilon_B = -300$, and $\epsilon_C = 480$.

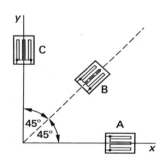

(a) Most nearly, what are the principal strains in the web?

(A) 840, 290

(B) 1400, −300

(C) 2300, 1200

(D) 2900, −610

(b) Most nearly, what are the principal stresses in the web?

(A) 250 MPa, 17 MPa

(B) 290 MPa, 57 MPa

(C) 300 MPa, 28 MPa

(D) 330 MPa, 57 MPa

Controllers and Control Logic

14. The schematic for a diesel generator's starter motor is shown. The table describes the conditions that affect the starting circuit's individual contacts.

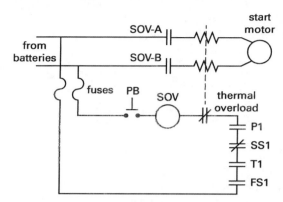

SOV = shutoff valve
PB = push button
P = pressure (switch)
SS = speed (switch)
T = temperature (switch)
FS = flow (switch)

contact	action
thermal overload	opens upon thermal overload
P1	closes when diesel oil pressure is greater than 3 psig
SS1	opens when diesel speed is greater than 100 rev/min
T1	closes when diesel crankcase temperature is greater than 45°F
FS1	closes when diesel cooling water flow is adequate

Which of the following conditions is NOT required for the diesel motor to run?

I.   fuses must not be blown

II.   push button must be pressed and held

III. no thermal overload on the starter motor

IV. diesel oil pressure must be greater than 3 psig

V.   diesel speed must be above 100 rev/min

VI. diesel crankcase temperature must be greater than 45°F

VII. diesel standby cooling water pump must be running

(A) V only

(B) VI only

(C) III and IV only

(D) IV and V only

15. An operator presses push button PB on the test circuit shown. Which statement best describes the condition of the circuit nine seconds after button PB is pressed?

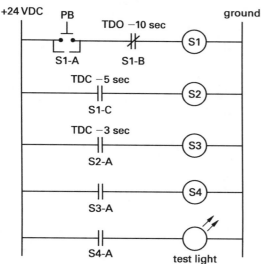

PB = push button
TDO = time-delayed opening
TDC = time-delayed closing
S = switch

(A) Nothing is energized, and the test light is not lit.

(B) S1, S2, and S3 are energized, and the test light is not lit.

(C) S1 and S2 are deenergized, S3 and S4 are energized, and the test light is lit.

(D) S1, S2, S3, and S4 are energized, and the test light is lit.

16. A room's temperature is held constant by using a proportional controller. The thermostat has a temperature range of 60–80°F, and the controller has a proportional range of 0–100%. When the thermostat and controller gain are set to 70°F and 50%, respectively, the temperature of the room is constant at 66°F. Which option will keep the temperature of the room at the setpoint?

(A) The gain on the proportional controller should be increased.

(B) The gain on the proportional controller should be decreased.

(C) The controller should be changed to a proportional-integral controller.

(D) The controller should be changed to a proportional-derivative controller.

17. Which statement about a proportional-integral-derivative (PID) controller is correct?

(A) The integral term ensures the controller controls at setpoint, and the derivative term increases controller damping and enhances controller stability.

(B) The derivative term ensures the PID controls at setpoint. The integral term increases controller damping and enhances controller stability.

(C) Adjusting the controller gain to the optimum setting will force the controller to control at setpoint.

(D) Controller stability is independent of the controller gain.

Thermocouples

18. The voltage across a copper-constantan thermocouple is 4.108 mV when it is placed in a 250°F oil bath. The temperature of the thermocouple reference junction is most nearly

(A) 32°F

(B) 65°F

(C) 80°F

(D) 85°F

19. Two iron-constantan thermocouples are wired in series to monitor the cylinder head temperature in an aircraft engine. The voltage across the thermocouples is read in the cockpit as 16.52 mV. The cockpit temperature is 70°F. What is most nearly the average temperature of the cylinder head?

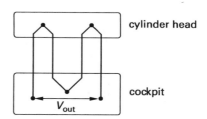

(A) 290°F

(B) 330°F

(C) 350°F

(D) 380°F

Systems, Mgmt., Professional

20. A copper-constantan thermocouple is used with an ice bath to measure the temperature of waste sludge during digestion. The voltage is proportional to the difference in temperature between the junctions. When the thermocouple is calibrated, it generates a voltage of 5.07×10^{-3} V when the temperature of the hot junction is 108.9°C and the reference temperature is at 0°C. Most nearly, what is the temperature of waste sludge when the voltage is 1.83×10^{-3} V?

(A) 40°C

(B) 45°C

(C) 50°C

(D) 55°C

Calculations in Other Numbering Systems

21. Perform the following binary (base-2) calculations.

(a) $101 + 011$

(A) 0110

(B) 1000

(C) 1001

(D) 1011

(b) $101 + 110$

(A) 0011

(B) 1010

(C) 1011

(D) 1100

(c) $101 + 100$

(A) 0100

(B) 0110

(C) 1000

(D) 1001

(d) $0100 - 1100$

(A) −1000

(B) −0100

(C) 0110

(D) 1000

(e) $1110 - 1000$

(A) 0010

(B) 0110

(C) 1100

(D) 1110

(f) $010 - 101$

(A) −100

(B) −011

(C) −010

(D) −001

(g) 111×11

(A) 10101

(B) 11011

(C) 11110

(D) 11111

(h) 100×11

(A) 0011

(B) 1000

(C) 1100

(D) 1110

(i) 1011×1101

(A) 10001111

(B) 11100111

(C) 11110000

(D) 11111111

22. Perform the following octal (base-8) calculations.

(a) $466 + 457$

(A) 923

(B) 1000

(C) 1105

(D) 1145

(b) $1007 + 6661$

(A) 7668

(B) 7670

(C) 7770

(D) 8670

(c) $321 + 465$

(A) 786

(B) 806

(C) 1006

(D) 1086

(d) $71 - 27$

   (A) 32

   (B) 42

   (C) 44

   (D) 52

(e) $1143 - 367$

   (A) 554

   (B) 754

   (C) 776

   (D) 876

(f) $646 - 677$

   (A) -131

   (B) -31

   (C) 31

   (D) 131

(g) 77×66

   (A) 5082

   (B) 6512

   (C) 6677

   (D) 7766

(h) 325×36

   (A) 11700

   (B) 12346

   (C) 14366

   (D) 14936

(i) 3251×161

   (A) 523411

   (B) 550111

   (C) 566011

   (D) 570231

23. Perform the following calculations involving hexadecimal (base-16) numbers.

(a) $BA + C$

   (A) B6

   (B) BA

   (C) C6

   (D) CA

(b) $BB + A$

   (A) B5

   (B) B6

   (C) C5

   (D) C6

(c) $BE + 10 + 1A$

   (A) B2

   (B) BF

   (C) E8

   (D) EC

(d) $FF - E$

   (A) F1

   (B) F7

   (C) FA

   (D) FB

(e) $74 - 4A$

   (A) 2A

   (B) 3A

   (C) 3B

   (D) 4B

(f) $FB - BF$

   (A) 3B

   (B) 3C

   (C) 4A

   (D) 4B

(g) $4A \times 3E$

   (A) 7F

   (B) 7AE

   (C) 11EC

   (D) 12EF

(h) $FE \times EF$

   (A) 22FE

   (B) ED22

   (C) EE11

   (D) EF22

(i) $17 \times 7A$

   (A) 8A

   (B) 6AE

   (C) AF6

   (D) FE8

Conversions Between Bases

24. Convert the following numbers to decimal (base-10) numbers.

(a) $(674)_8$

  (A) 136

  (B) 444

  (C) 1348

  (D) 3552

(b) $(101101)_2$

  (A) 45

  (B) 46

  (C) 48

  (D) 64

(c) $(734.262)_8$

  (A) 262.734

  (B) 349.674

  (C) 448.763

  (D) 476.348

(d) $(1011.11)_2$

  (A) 11.75

  (B) 14.75

  (C) 21.75

  (D) 38.75

25. Convert the following decimal (base-10) numbers to octal (base-8) numbers.

(a) $(75)_{10}$

  (A) $(113)_8$

  (B) $(131)_8$

  (C) $(311)_8$

  (D) $(331)_8$

(b) $(0.375)_{10}$

  (A) $(0.3)_8$

  (B) $(3)_8$

  (C) $(8)_8$

  (D) $(24)_8$

(c) $(121.875)_{10}$

  (A) $(151.7)_8$

  (B) $(171.7)_8$

  (C) $(177.1)_8$

  (D) $(182.6)_8$

(d) $(1011100.01110)_2$

  (A) $(134.34)_8$

  (B) $(143.43)_8$

  (C) $(341.41)_8$

  (D) $(413.14)_8$

26. Convert the following numbers to binary (base-2) numbers.

(a) $(83)_{10}$

  (A) $(1010001)_2$

  (B) $(1010011)_2$

  (C) $(1100100)_2$

  (D) $(1100101)_2$

(b) $(100.3)_{10}$

  (A) $(0001111.110111\cdots)_2$

  (B) $(0011011.100011\cdots)_2$

  (C) $(1000110.100010\cdots)_2$

  (D) $(1100100.010011\cdots)_2$

(c) $(0.97)_{10}$

  (A) $(0.101110\cdots)_2$

  (B) $(0.111110\cdots)_2$

  (C) $(110111.0)_2$

  (D) $(111011.1)_2$

(d) $(321.422)_8$

  (A) $(001100011.100001000)_2$

  (B) $(010010100.001001100)_2$

  (C) $(011000101.010010001)_2$

  (D) $(011010001.100010010)_2$

SOLUTIONS

1. The average height of the pressure plot is

$$\overline{h} = \frac{A}{w} = \frac{490 \text{ mm}^2}{64 \text{ mm}} = 7.656 \text{ mm}$$

The average pressure is

$$\overline{p} = k\overline{h} = \left(111 \ \frac{\text{kPa}}{\text{mm}}\right)(7.656 \text{ mm})$$

$$= \boxed{849.8 \text{ kPa} \quad (850 \text{ kPa})}$$

The answer is (D).

2. A coil of wire, particularly a coil with an iron core, behaves as an inductor in an AC circuit. In an LVDT, the movable transformer core changes the inductance between the primary and secondary windings. The transformer is constructed so that the output voltage of the LVDT is directly proportional to the movable core position. A linear variable differential transformer (LVDT) is a variable inductance transducer.

The answer is (D).

3. Since the voltage is not zero, the bridge is not balanced. Rearrange Eq. 53.16 to find the RTD resistance.

$$R_{\text{RTD}} = \frac{V_b R_1 R_3 + V(R_1 R_3 + R_2 R_3)}{V_b R_2 - V(R_1 + R_2)}$$

$$= \frac{\begin{array}{c} (10 \text{ V})(100 \ \Omega)(50 \ \Omega) \\ + (4.167 \text{ V})\left(\begin{array}{c}(100 \ \Omega)(50 \ \Omega) \\ + (200 \ \Omega)(50 \ \Omega)\end{array}\right)\end{array}}{\begin{array}{c}(10 \text{ V})(200 \ \Omega) - (4.167 \text{ V}) \\ \times (100 \ \Omega + 200 \ \Omega)\end{array}}$$

$$= 150.0 \ \Omega$$

Rearrange Eq. 53.3 to find the temperature. Since temperature varies linearly with resistance, only the alpha-value, α, is required.

$$T = \frac{R_{\text{ref}}(\alpha T_{\text{ref}} - 1) + R_{\text{RTD}}}{R_{\text{ref}} \alpha}$$

$$= \frac{(100 \ \Omega)\left(\left(0.00392 \ \frac{1}{^\circ\text{C}}\right)(0^\circ\text{C}) - 1\right) + 150.0 \ \Omega}{(100 \ \Omega)\left(0.00392 \ \frac{1}{^\circ\text{C}}\right)}$$

$$= 127.6^\circ\text{C}$$

Convert to the Fahrenheit scale.

$$T_{^\circ\text{F}} = 32^\circ + \tfrac{9}{5}T_{^\circ\text{C}} = 32^\circ\text{F} + \left(\tfrac{9}{5}\right)(127.6^\circ\text{C})$$

$$= \boxed{261.7^\circ\text{F} \quad (260^\circ\text{F})}$$

The answer is (C).

4. When the meter is nulled out, the resistance of the leg containing the RTD is

$$R_4 = R_{\text{RTD}} + 2R_{\text{lead}} = \frac{R_2 R_3}{R_1} = \frac{(376 \ \Omega)(1000 \ \Omega)}{1000 \ \Omega}$$

$$= 376 \ \Omega$$

The resistance of the RTD is

$$R_{\text{RTD}} = R_4 - 2R_{\text{lead}} = 376 \ \Omega - (2)(100 \ \Omega) = 176 \ \Omega$$

The relationship between temperature and resistance for an RTD is

$$R_T = R_0(1 + AT + BT^2)$$

For a platinum RTD with $\alpha = 0.00385 \ 1/^\circ\text{C}$, the resistance is

$$R_T = R_0 \left(\begin{array}{c} 1 + (3.9083 \times 10^{-3})T \\ + (-5.775 \times 10^{-7})T^2 \end{array}\right)$$

$$\approx R_0\left(1 + (3.9083 \times 10^{-3})T\right)$$

$$176 \ \Omega \approx (100 \ \Omega)\left(1 + (3.9083 \times 10^{-3})T\right)$$

$$T \approx \boxed{194.5^\circ\text{C} \quad (200^\circ\text{C})}$$

(If the squared term is kept, the temperature is 200.3°C.)

The answer is (C).

5. The displacement of the falling accelerometer is

$$x = x_0 + \text{v}_0 t - \tfrac{1}{2}gt^2$$

The initial displacement and the initial velocity are both zero. The time required to fall a distance of x is

$$t = \sqrt{\frac{2x}{g}}$$

The impact velocity is

$$\text{v} = gt = g\sqrt{\frac{2x}{g}} = \sqrt{2xg}$$

$$= \sqrt{(2)(3 \text{ ft})\left(32.2 \ \frac{\text{ft}}{\text{sec}^2}\right)}$$

$$= 13.9 \text{ ft/sec}$$

The average acceleration is

$$\bar{a} = \frac{v}{\Delta t} = \frac{13.9 \; \frac{ft}{sec}}{(0.02 \; sec)\left(32.2 \; \frac{ft}{sec^2 \text{-} g}\right)} = 21.58 \; g$$

The average voltage output is

$$\overline{V} = k\bar{a} = \left(2.0 \; \frac{mV}{g}\right)(21.58 \; g)$$

$$= \boxed{43.16 \; mV \quad (43 \; mV)}$$

The answer is (C).

6. The transducer capacitance, C_t, prevents DC voltages from being passed to the amplifier. The value of C_t does not influence the amplifier gain.

The answer is (D).

7. The weight of the object is equal to the force on the scale.

$$F = pA = p\left(\frac{\pi}{4}\right)(D - 2t)^2$$

$$= \left(50 \; \frac{lbf}{in^2}\right)\left(\frac{\pi}{4}\right)\left((2 \; ft)\left(12 \; \frac{in}{ft}\right) - (2)(2 \; in)\right)^2$$

$$= \boxed{15,708 \; lbf \quad (16,000 \; lbf)}$$

The answer is (B).

8. From Eq. 53.11, the strain is

$$\epsilon = \frac{\Delta R_g}{(GF)R_g} = \frac{100.1 \; \Omega - 100 \; \Omega}{(6.0)(100 \; \Omega)} = 1.67 \times 10^{-4}$$

From Eq. 53.24, the stress is

$$\sigma = E\epsilon = \left(30 \times 10^6 \; \frac{lbf}{in^2}\right)(1.67 \times 10^{-4})$$

$$= 5.0 \times 10^3 \; lbf/in^2$$

Rearranging Eq. 53.33, the applied force is

$$F = \sigma A = \left(5.0 \times 10^3 \; \frac{lbf}{in^2}\right)(1.0 \; in^2)$$

$$= \boxed{5.0 \times 10^3 \; lbf \quad (5000 \; lbf)}$$

The answer is (B).

9. Work with average values. The average strain gauge resistance is

$$R_{ave} = \frac{300.05 \; \Omega + 300.09 \; \Omega + 300.11 \; \Omega + 300.03 \; \Omega}{4}$$

$$= 300.07 \; \Omega$$

From Eq. 53.11, the average strain is

$$\epsilon_{ave} = \frac{\Delta R_g}{(GF)R_g} = \frac{300.07 \; \Omega - 300 \; \Omega}{(3)(300 \; \Omega)}$$

$$= 7.777 \times 10^{-5} \; in/in$$

From Eq. 53.24, the average stress is

$$\sigma_{ave} = E\epsilon_{ave} = \left(20 \times 10^6 \; \frac{lbf}{in^2}\right)\left(7.777 \times 10^{-5} \; \frac{in}{in}\right)$$

$$= 1555 \; lbf/in^2$$

Rearranging Eq. 53.33, the average force is

$$F_{ave} = \sigma_{ave} A = \left(1555 \; \frac{lbf}{in^2}\right)(2 \; in^2)$$

$$= 3110 \; lbf$$

The total weight of the hopper and grain is

$$W_t = 4F_{ave} = (4)(3110 \; lbf) = 12,440 \; lbf$$

Therefore, the weight of the grain is

$$W_g = W_t - W_h = 12,440 \; lbf - 1500 \; lbf$$

$$= \boxed{10,940 \; lbf \quad (11,000 \; lbf)}$$

The answer is (B).

10. Use Eq. 53.11.

$$\frac{\Delta R_g}{R_g} = (GF)\epsilon = (2.0)(500 \times 10^{-6})$$

$$= \boxed{0.001 \quad (0.1\%)}$$

The answer is (C).

11. The moment of inertia of the bar in bending is

$$I = \frac{bh^3}{12} = \frac{(4 \; cm)(2 \; cm)^3}{12} = 2.667 \; cm^4$$

The stress in the bar is

$$\sigma = \frac{Mc}{I} = \frac{mglc}{I}$$

$$= \frac{(1 \; kg)\left(9.81 \; \frac{m}{s^2}\right)(10 \; cm)\left(\frac{2 \; cm}{2}\right)\left(100 \; \frac{cm}{m}\right)^2}{2.667 \; cm^4}$$

$$= 367,829 \; Pa$$

The strain is

$$\epsilon = \frac{\sigma}{E} = \frac{367{,}829 \text{ Pa}}{(73 \text{ GPa})\left(10^9 \ \frac{\text{Pa}}{\text{GPa}}\right)} = 5.04 \times 10^{-6}$$

Use Eq. 53.11.

$$\Delta R_g = \epsilon(\text{GF})R_g = (5.04 \times 10^{-6})(3.0)(350 \ \Omega)$$
$$= \boxed{0.00529 \ \Omega \quad (0.0053 \ \Omega)}$$

The answer is (B).

12. First, use the information from the unloaded case to determine the unstressed resistance of the strain gauge. Since the bridge numbering does not correspond to Fig. 53.6, rewrite Eq. 53.16 using the ratio of the vertical legs, top to bottom.

$$\frac{R_1}{R_4} = \frac{R_2}{R_3}$$

The unstressed resistance of the strain gauge is

$$R_1 = \frac{R_4 R_2}{R_3} = \frac{(2505 \ \Omega)(160 \ \Omega)}{2500 \ \Omega} = 160.32 \ \Omega$$

The resistance of the strain gauge when the shaft is loaded is

$$R_1 = \frac{R_4 R_2}{R_3} = \frac{(2511.7 \ \Omega)(160 \ \Omega)}{2500 \ \Omega} = 160.7488 \ \Omega$$

Now, use Eq. 53.11 to determine the strain when the shaft is loaded.

$$\epsilon = \frac{\Delta R_g}{(\text{GF})R_g} = \frac{160.7488 \ \Omega - 160.32 \ \Omega}{(2.10)(160.32 \ \Omega)}$$
$$= 0.00127364$$

The shaft's moment of inertia as a beam is

$$I = \frac{\pi r^4}{4} = \frac{\pi \left(\frac{2 \text{ in}}{2}\right)^4}{4} = 0.7854 \text{ in}^4$$

The bending stress in the shaft is

$$\sigma = \frac{Mc}{I} = \frac{Flc}{I} = \frac{(1100 \text{ lbf})(24 \text{ in})\left(\frac{2 \text{ in}}{2}\right)}{0.7854 \text{ in}^4}$$
$$= 33{,}613 \text{ lbf/in}^2$$

The modulus of elasticity is

$$E = \frac{\sigma}{\epsilon} = \frac{33{,}613 \ \frac{\text{lbf}}{\text{in}^2}}{0.00127364}$$
$$= \boxed{26.39 \times 10^6 \text{ lbf/in}^2 \quad (26 \times 10^6 \text{ lbf/in}^2)}$$

The answer is (D).

13. (a) Use Table 53.6. (The strain gauge designations are changed.) Work in microstrains (μm/m). The principal strains are

$$\epsilon_p, \epsilon_q = \tfrac{1}{2}\left(\epsilon_A + \epsilon_C \pm \sqrt{2(\epsilon_A - \epsilon_B)^2 + 2(\epsilon_B - \epsilon_C)^2}\right)$$

$$= \left(\tfrac{1}{2}\right)\left(650 + 480 \pm \sqrt{\begin{array}{c}(2)\left(650 - (-300)\right)^2 \\ + (2)(-300 - 480)^2\end{array}}\right)$$

$$= \left(\tfrac{1}{2}\right)(1130 \pm 1738)$$

$$= \boxed{1434, -304 \quad (1400, -300)}$$

The answer is (B).

(b) Use Table 53.6. Work in microstrains (μm/m). The principal stresses are

$$\sigma_p, \sigma_q = \frac{E}{2}\left(\begin{array}{c}\dfrac{\epsilon_A + \epsilon_C}{1 - \nu} \pm \dfrac{1}{1 + \nu} \\ \times \sqrt{2(\epsilon_A - \epsilon_B)^2 + 2(\epsilon_B - \epsilon_C)^2}\end{array}\right)$$

$$= \left(\frac{200 \text{ GPa}}{2}\right)\left(10^9 \ \frac{\text{Pa}}{\text{GPa}}\right)$$

$$\times \left(\begin{array}{c}\dfrac{650 + 480}{1 - 0.3} \pm \left(\dfrac{1}{1 + 0.3}\right) \\ \times \sqrt{\begin{array}{c}(2)\left(650 - (-300)\right)^2 \\ + (2)(-300 - 480)^2\end{array}}\end{array}\right)$$

$$\times \left(10^{-6} \ \frac{\text{m}}{\mu\text{m}}\right)$$

$$= (1614 \pm 1337) \times 10^5 \text{ Pa}$$

$$= 295 \times 10^6 \text{ Pa}, 27.7 \times 10^6 \text{ Pa}$$

$$\boxed{\begin{array}{l}\text{The principal stresses are 295 MPa and} \\ \text{27.7 MPa (300 MPa and 28 MPa).}\end{array}}$$

The answer is (C).

14. In order for the starter motor to run, fuses must not be blown (condition I); the push button must be pressed and held (condition II); there can be no thermal overload on the starter motor (condition III); the diesel oil pressure must be greater than 3 psig (condition IV); the diesel speed must be less than 100 rev/min (condition V); and the diesel standby cooling water pump must be running (condition VII). The starter motor disengages as soon as the engine speed increases above 100 rev/min. Condition V is incorrect.

The answer is (A).

15. This is a ladder diagram, not a circuit diagram. All of the S1 elements are part of the same switch. Pressing the push button energizes S1 and causes S1-A to close to maintain power to S1. After a 5 sec delay, S1-C closes to energize S2. After an additional 3 sec delay, S2-A closes to energize S3. This causes S3 to close and immediately energize S4. S4-A then closes to illuminate the test light. The total time for this process is the sum of the delay times, which is 8 sec. At 10 sec, S1-B opens to de-energize the circuit.

Therefore, at 9 sec, S1, S2, S3, and S4 are energized, and the test light is lit.

The answer is (D).

16. The signal from a proportional-only controller is proportional to the difference between the actual temperature and the setpoint. The signal decreases as the temperature approaches the setpoint, so the temperature will approach, but never actually reach, the setpoint. Adjusting the gain on a proportional controller will not change this behavior and may make the controller unstable by causing the signal (and, hence, the temperature) to overshoot. A proportional-derivative (PD) controller responds to changing temperature (at any temperature), enhances stability, and increases damping of the controller. However, since the derivative of a constant is zero, a PD controller does not respond to a stable temperature and, therefore, cannot maintain the temperature at the setpoint. A proportional-integral (PI) controller eliminates offset and is able to control the temperature at the setpoint. Therefore, the controller should be changed to a PI controller.

The answer is (C).

17. Controller stability is affected by controller gain. If the gain is too high, the controller may become unstable; if the gain is too low, the controller response time may be too slow. The controller gain cannot be forced to operate at setpoint by adjusting the controller gain to an optimum setting. The integral term compensates for offset and ensures the controller controls at setpoint. The derivative term increases controller damping and enhances controller stability.

The answer is (A).

18. From App. 53.A, the voltage output from a copper-constantan thermocouple at 250°F is 5.280 mV. (Read horizontally on the 200°F line to the 50°F column.) The voltage corresponding to the reference junction temperature, V_{ref}, is

$$V_{ref} = V_{250°F} - V = 5.280 \text{ mV} - 4.108 \text{ mV}$$
$$= 1.172 \text{ mV}$$

By interpolation from App. 53.A, this corresponds to a reference junction temperature of $\boxed{85°F.}$

The answer is (D).

19. The average voltage across each thermocouple, V_{ave}, is

$$V_{ave} = \frac{V}{2} = \frac{16.52 \text{ mV}}{2}$$
$$= 8.26 \text{ mV}$$

The reference junction is at 70°F in the cockpit. From App. 53.A, the thermoelectric constant for an iron-constantan thermocouple at 70°F is 1.07 mV. This value, added to V_{ave}, represents the voltage that would be obtained if the cockpit were at the reference temperature of 32°F.

$$V = V_{ave} + 1.07 \text{ mV} = 8.26 \text{ mV} + 1.07 \text{ mV}$$
$$= 9.33 \text{ mV}$$

Interpolating from App. 53.A, the voltage yields an average cylinder head temperature of $\boxed{345°F \ (350°F).}$

The answer is (C).

20. During calibration, the temperature difference is $108.9°C - 0°C = 108.9°C$. The thermoelectric constant is

$$k_T = \frac{V}{\Delta T} = \frac{5.07 \times 10^{-3} \text{ V}}{108.9°C} = 4.656 \times 10^{-5} \text{ V/°C}$$

Use Eq. 53.9. The temperature of the waste sludge is

$$T = T_{ref} + \frac{V}{k_T}$$
$$= 0°C + \frac{1.83 \times 10^{-3} \text{ V}}{4.656 \times 10^{-5} \ \frac{\text{V}}{°C}}$$
$$= \boxed{39.3°C \quad (40°C)}$$

The answer is (A).

21. (a)

$$\begin{array}{r} 101 \\ + \ 011 \\ \hline \boxed{1000} \end{array}$$

The answer is (B).

(b)

$$\begin{array}{r} 101 \\ + \ 110 \\ \hline \boxed{1011} \end{array}$$

The answer is (C).

(c)

$$\begin{array}{r} 101 \\ + \ 100 \\ \hline \boxed{1001} \end{array}$$

The answer is (D).

(d)

$$-\left(\begin{array}{r} 1100 \\ - \ 0100 \\ \hline 1000 \end{array}\right) = \boxed{-1000}$$

The answer is (A).

(e)

$$\begin{array}{r} 1110 \\ - \ 1000 \\ \hline \boxed{0110} \end{array}$$

The answer is (B).

(f)

$$-\left(\begin{array}{r} 101 \\ - \ 010 \\ \hline 011 \end{array}\right) = \boxed{-011}$$

The answer is (B).

(g)

$$\begin{array}{r} 111 \\ \times \ 11 \\ \hline 111 \\ 111 \\ \hline \boxed{10101} \end{array}$$

The answer is (A).

(h)

$$\begin{array}{r} 100 \\ \times \ 11 \\ \hline 100 \\ 100 \\ \hline \boxed{1100} \end{array}$$

The answer is (C).

(i)

$$\begin{array}{r} 1011 \\ \times \ 1101 \\ \hline 1011 \\ 1011 \\ 1011 \\ \hline \boxed{10001111} \end{array}$$

The answer is (A).

22. (a)

$$\begin{array}{r} 466 \\ + \ 457 \\ \hline \boxed{1145} \end{array}$$

The answer is (D).

(b)

$$\begin{array}{r} 1007 \\ + \ 6661 \\ \hline \boxed{7670} \end{array}$$

The answer is (B).

(c)

$$\begin{array}{r} 321 \\ + \ 465 \\ \hline \boxed{1006} \end{array}$$

The answer is (C).

(d)

$$\begin{array}{r} 71 \\ - \ 27 \\ \hline \end{array} = \begin{array}{rr} 6 & 11 \\ - \ 2 & 7 \\ \hline \boxed{4} & \boxed{2} \end{array}$$

The answer is (B).

(e)

$$\begin{array}{r} 1143 \\ - \ 367 \\ \hline \end{array} = \begin{array}{rr} 113 & 13 \\ - \ 36 & 7 \\ \hline \end{array} = \begin{array}{rrr} 10 & 13 & 13 \\ - \ 3 & 6 & 7 \\ \hline \boxed{5} & \boxed{5} & \boxed{4} \end{array}$$

The answer is (A).

(f)

$$-\left(\begin{array}{r} 677 \\ -\ 646 \\ \hline 31 \end{array}\right) = \boxed{-31}$$

The answer is (B).

(g)

$$\begin{array}{r} 77 \\ \times\ 66 \\ \hline 572 \\ 572 \\ \hline \boxed{6512} \end{array}$$

The answer is (B).

(h)

$$\begin{array}{r} 325 \\ \times\ 36 \\ \hline 2376 \\ 1177 \\ \hline \boxed{14366} \end{array}$$

The answer is (C).

(i)

$$\begin{array}{r} 3251 \\ \times\ 161 \\ \hline 3251 \\ 23766 \\ 3251 \\ \hline \boxed{570231} \end{array}$$

The answer is (D).

23. (a)

$$\begin{array}{r} BA \\ +\ C \\ \hline \boxed{C6} \end{array}$$

The answer is (C).

(b)

$$\begin{array}{r} BB \\ +\ A \\ \hline \boxed{C5} \end{array}$$

The answer is (C).

(c)

$$\begin{array}{r} BE \\ 10 \\ +\ 1A \\ \hline \boxed{E8} \end{array}$$

The answer is (C).

(d)

$$\begin{array}{r} FF \\ -\ E \\ \hline \boxed{F1} \end{array}$$

The answer is (A).

(e)

$$\begin{array}{r} 74 \\ -\ 4A \\ \hline \end{array} = \begin{array}{rr} 6 & 14 \\ -\ 4 & A \\ \hline \boxed{2\ \ A} \end{array}$$

The answer is (A).

(f)

$$\begin{array}{r} FB \\ -\ BF \\ \hline \end{array} = \begin{array}{rr} E & 1B \\ -\ B & F \\ \hline \boxed{3\ \ C} \end{array}$$

The answer is (B).

(g)

$$\begin{array}{r} 4A \\ \times\ 3E \\ \hline 40C \\ DE \\ \hline \boxed{11EC} \end{array}$$

The answer is (C).

(h)

$$\begin{array}{r} FE \\ \times\ EF \\ \hline EE2 \\ DE4 \\ \hline \boxed{ED22} \end{array}$$

The answer is (B).

(i)

$$\begin{array}{r} 17 \\ \times\ 7A \\ \hline E6 \\ A1 \\ \hline \boxed{AF6} \end{array}$$

The answer is (C).

24. (a) $(674)_8 = (6)(8)^2 + (7)(8)^1 + (4)(8)^0$

$$= \boxed{444}$$

The answer is (B).

(b) $(101101)_2 = (1)(2)^5 + (0)(2)^4 + (1)(2)^3$
$$+ (1)(2)^2 + (0)(2)^1 + (1)(2)^0$$
$$= \boxed{45}$$

The answer is (A).

(c) $(734.262)_8 = (7)(8)^2 + (3)(8)^1 + (4)(8)^0$
$$+ (2)(8)^{-1} + (6)(8)^{-2} + (2)(8)^{-3}$$
$$= \boxed{476.348}$$

The answer is (D).

(d) $(1011.11)_2 = (1)(2)^3 + (0)(2)^2 + (1)(2)^1$
$$+ (1)(2)^0 + (1)(2)^{-1} + (1)(2)^{-2}$$
$$= \boxed{11.75}$$

The answer is (A).

25. (a)
$$75 \div 8 = 9 \quad \text{remainder } 3$$
$$9 \div 8 = 1 \quad \text{remainder } 1$$
$$1 \div 8 = 0 \quad \text{remainder } 1$$
$$(75)_{10} = \boxed{(113)_8}$$

The answer is (A).

(b)
$$0.375 \times 8 = 3 \quad \text{remainder } 0$$
$$(0.375)_{10} = \boxed{(0.3)_8}$$

The answer is (A).

(c)
$$121 \div 8 = 15 \quad \text{remainder } 1$$
$$15 \div 8 = 1 \quad \text{remainder } 7$$
$$1 \div 8 = 0 \quad \text{remainder } 1$$
$$0.875 \times 8 = 7 \quad \text{remainder } 0$$
$$(121.875)_{10} = \boxed{(171.7)_8}$$

The answer is (B).

(d) Since $(2)^3 = 8$, break the bits into groups of three starting at the decimal point and working outward in both directions.

$$001011100.011100 = 001\ 011\ 100.011\ 100$$

Convert each of the groups into its octal equivalent.

$$001\ 011\ 100.011\ 100 = 1\ 3\ 4.3\ 4$$
$$\boxed{(134.34)_8}$$

The answer is (A).

26. (a)
$$83 \div 2 = 41 \quad \text{remainder } 1$$
$$41 \div 2 = 20 \quad \text{remainder } 1$$
$$20 \div 2 = 10 \quad \text{remainder } 0$$
$$10 \div 2 = 5 \quad \text{remainder } 0$$
$$5 \div 2 = 2 \quad \text{remainder } 1$$
$$2 \div 2 = 1 \quad \text{remainder } 0$$
$$1 \div 2 = 0 \quad \text{remainder } 1$$
$$(83)_{10} = \boxed{(1010011)_2}$$

The answer is (B).

(b)
$$100 \div 2 = 50 \quad \text{remainder } 0$$
$$50 \div 2 = 25 \quad \text{remainder } 0$$
$$25 \div 2 = 12 \quad \text{remainder } 1$$
$$12 \div 2 = 6 \quad \text{remainder } 0$$
$$6 \div 2 = 3 \quad \text{remainder } 0$$
$$3 \div 2 = 1 \quad \text{remainder } 1$$
$$1 \div 2 = 0 \quad \text{remainder } 1$$
$$0.3 \times 2 = 0 \quad \text{remainder } 0.6$$
$$0.6 \times 2 = 1 \quad \text{remainder } 0.2$$
$$0.2 \times 2 = 0 \quad \text{remainder } 0.4$$
$$0.4 \times 2 = 0 \quad \text{remainder } 0.8$$
$$0.8 \times 2 = 1 \quad \text{remainder } 0.6$$
$$0.6 \times 2 = 1 \quad \text{remainder } 0.2$$
$$\vdots$$
$$(100.3)_{10} = \boxed{(1100100.010011\cdots)_2}$$

The answer is (D).

(c)

$$0.97 \times 2 = 1 \quad \text{remainder } 0.94$$
$$0.94 \times 2 = 1 \quad \text{remainder } 0.88$$
$$0.88 \times 2 = 1 \quad \text{remainder } 0.76$$
$$0.76 \times 2 = 1 \quad \text{remainder } 0.52$$
$$0.52 \times 2 = 1 \quad \text{remainder } 0.04$$
$$0.04 \times 2 = 0 \quad \text{remainder } 0.08$$
$$\vdots$$
$$(0.97)_{10} = \boxed{(0.111110\cdots)_2}$$

The answer is (B).

(d) Since $8 = (2)^3$, convert each octal digit into its binary equivalent.

$$321.422: \quad 3 = 011$$
$$2 = 010$$
$$1 = 001$$
$$4 = 100$$
$$2 = 010$$
$$2 = 010$$

$$\boxed{(321.422)_8 = (011010001.100010010)_2}$$

The answer is (D).

54 Engineering Economic Analysis

PRACTICE PROBLEMS

1. At 6% effective annual interest, approximately how much will be accumulated if $1000 is invested for ten years?

(A) $560

(B) $790

(C) $1600

(D) $1800

2. At 6% effective annual interest, the present worth of $2000 that becomes available in four years is most nearly

(A) $520

(B) $580

(C) $1600

(D) $2500

3. At 6% effective annual interest, approximately how much should be invested to accumulate $2000 in 20 years?

(A) $620

(B) $1400

(C) $4400

(D) $6400

4. At 6% effective annual interest, the year-end annual amount deposited over seven years that is equivalent to $500 invested now is most nearly

(A) $90

(B) $210

(C) $300

(D) $710

5. At 6% effective annual interest, the accumulated amount at the end of ten years if $50 is invested at the end of each year for ten years is most nearly

(A) $90

(B) $370

(C) $660

(D) $900

6. At 6% effective annual interest, approximately how much should be deposited at the start of each year for ten years (a total of 10 deposits) in order to empty the fund by drawing out $200 at the end of each year for ten years (a total of 10 withdrawals)?

(A) $190

(B) $210

(C) $220

(D) $250

7. At 6% effective annual interest, approximately how much should be deposited at the start of each year for five years to accumulate $2000 on the date of the last deposit?

(A) $350

(B) $470

(C) $510

(D) $680

8. At 6% effective annual interest, approximately how much will be accumulated in ten years if three payments of $100 are deposited every other year for four years, with the first payment occurring at $t = 0$?

(A) $180

(B) $480

(C) $510

(D) $540

9. $500 is compounded monthly at a 6% nominal annual interest rate. Approximately how much will have accumulated in five years?

(A) $515

(B) $530

(C) $675

(D) $690

10. The effective annual rate of return on an $80 investment that pays back $120 in seven years is most nearly

(A) 4.5%

(B) 5.0%

(C) 5.5%

(D) 6.0%

11. A new machine will cost $17,000 and will have a resale value of $14,000 after five years. Special tooling will cost $5000. The tooling will have a resale value of $2500 after five years. Maintenance will be $2000 per year. The effective annual interest rate is 6%. The average annual cost of ownership during the next five years will be most nearly

(A) $2000

(B) $2300

(C) $4300

(D) $5500

12. An old covered wooden bridge can be strengthened at a cost of $9000, or it can be replaced for $40,000. The present salvage value of the old bridge is $13,000. It is estimated that the reinforced bridge will last for 20 years, will have an annual cost of $500, and will have a salvage value of $10,000 at the end of 20 years. The estimated salvage value of the new bridge after 25 years is $15,000. Maintenance for the new bridge would cost $100 annually. The effective annual interest rate is 8%. Which is the best alternative?

(A) Strengthen the old bridge.

(B) Build the new bridge.

(C) Both options are economically identical.

(D) Not enough information is given.

13. A firm expects to receive $32,000 each year for 15 years from sales of a product. An initial investment of $150,000 will be required to manufacture the product. Expenses will run $7530 per year. Salvage value is zero, and straight-line depreciation is used. The income tax rate is 48%. The after-tax rate of return is most nearly

(A) 8.0%

(B) 9.0%

(C) 10%

(D) 11%

14. A public works project has initial costs of $1,000,000, benefits of $1,500,000, and disbenefits of $300,000.

(a) The benefit/cost ratio is most nearly

(A) 0.20

(B) 0.47

(C) 1.2

(D) 1.7

(b) The excess of benefits over costs is most nearly

(A) $200,000

(B) $500,000

(C) $700,000

(D) $800,000

15. A speculator in land pays $14,000 for property that he expects to hold for ten years. $1000 is spent in renovation, and a monthly rent of $75 is collected from the tenants. (Use the year-end convention.) Taxes are $150 per year, and maintenance costs are $250 per year. In ten years, the sale price needed to realize a 10% rate of return is most nearly

(A) $26,000

(B) $31,000

(C) $34,000

(D) $36,000

16. The effective annual interest rate for a payment plan of 30 equal payments of $89.30 per month when a lump sum payment of $2000 would have been an outright purchase is most nearly

(A) 27%

(B) 35%

(C) 43%

(D) 51%

17. A depreciable item is purchased for $500,000. The salvage value at the end of 25 years is estimated at $100,000.

(a) The depreciation in each of the first three years using the straight line method is most nearly

(A) $4000

(B) $16,000

(C) $20,000

(D) $24,000

(b) The depreciation in each of the first three years using the sum-of-the-years' digits method is most nearly

(A) $16,000; $16,000; $16,000

(B) $30,000; $28,000; $27,000

(C) $31,000; $30,000; $28,000

(D) $32,000; $31,000; $30,000

(c) The depreciation in each of the first three years using the double-declining balance method is most nearly

(A) $16,000; $16,000; $16,000

(B) $31,000; $30,000; $28,000

(C) $37,000; $34,000; $31,000

(D) $40,000; $37,000; $34,000

18. Equipment that is purchased for $12,000 now is expected to be sold after ten years for $2000. The estimated maintenance is $1000 for the first year, but it is expected to increase $200 each year thereafter. The effective annual interest rate is 10%.

(a) The present worth is most nearly

(A) $16,000

(B) $17,000

(C) $21,000

(D) $22,000

(b) The annual cost is most nearly

(A) $1100

(B) $2200

(C) $3600

(D) $3700

19. A new grain combine with a 20-year life can remove seven pounds of rocks from its harvest per hour. Any rocks left in its output hopper will cause $25,000 damage in subsequent processes. Several investments are available to increase the rock-removal capacity, as listed in the table. The effective annual interest rate is 10%. What should be done?

rock removal rate	probability of exceeding rock removal rate	required investment to achieve removal rate
7	0.15	0
8	0.10	$15,000
9	0.07	$20,000
10	0.03	$30,000

(A) Do nothing.

(B) Invest $15,000.

(C) Invest $20,000.

(D) Invest $30,000.

20. A mechanism that costs $10,000 has operating costs and salvage values as given. An effective annual interest rate of 20% is to be used.

year	operating cost	salvage value
1	$2000	$8000
2	$3000	$7000
3	$4000	$6000
4	$5000	$5000
5	$6000	$4000

(a) The cost of owning and operating the mechanism in year two is most nearly

(A) $3800

(B) $4700

(C) $5800

(D) $6000

(b) The cost of owning and operating the mechanism in year five is most nearly

(A) $5600

(B) $6500

(C) $7000

(D) $7700

(c) The economic life of the mechanism is most nearly

(A) one year

(B) two years

(C) three years

(D) five years

(d) Assuming that the mechanism has been owned and operated for four years already, the cost of owning and operating the mechanism for one more year is most nearly

(A) $6400

(B) $7200

(C) $8000

(D) $8200

21. (*Time limit: one hour*) A salesperson intends to purchase a car for $50,000 for personal use, driving 15,000 miles per year. Insurance for personal use costs $2000 per year, and maintenance costs $1500 per year. The car gets 15 miles per gallon, and gasoline costs $1.50 per gallon. The resale value after five years will be $10,000. The salesperson's employer has asked that the car be used for business driving of 50,000 miles per year and has offered a reimbursement of $0.30 per mile.

Systems, Mgmt., Professional

Using the car for business would increase the insurance cost to $3000 per year and maintenance to $2000 per year. The salvage value after five years would be reduced to $5000. If the employer purchased a car for the salesperson to use, the initial cost would be the same, but insurance, maintenance, and salvage would be $2500, $2000, and $8000, respectively. The salesperson's effective annual interest rate is 10%.

(a) Is the reimbursement offer adequate?

(b) With a reimbursement of $0.30 per mile, approximately how many miles must the car be driven per year to justify the employer buying the car for the salesperson to use?

(A) 20,000 mi

(B) 55,000 mi

(C) 82,000 mi

(D) 150,000 mi

22. Alternatives A and B are being evaluated. The effective annual interest rate is 10%.

	alternative A	alternative B
first cost	$80,000	$35,000
life	20 years	10 years
salvage value	$7000	0
annual costs		
years 1–5	$1000	$3000
years 6–10	$1500	$4000
years 11–20	$2000	0
additional cost		
year 10	$5000	0

(a) The present worth for alternative A is most nearly

(A) $91,000

(B) $93,000

(C) $100,000

(D) $120,000

(b) The equivalent uniform annual cost (EUAC) for alternative A is most nearly

(A) $10,000

(B) $11,000

(C) $12,000

(D) $14,000

(c) The present worth for alternative B is most nearly

(A) $56,000

(B) $62,000

(C) $70,000

(D) $78,000

(d) The EUAC for alternative B is most nearly

(A) $9100

(B) $10,000

(C) $11,000

(D) $13,000

(e) Which alternative is economically superior?

(A) Alternative A is economically superior.

(B) Alternative B is economically superior.

(C) Alternatives A and B are economically equivalent.

(D) Not enough information is provided.

23. A car is needed for three years. Plans A and B for acquiring the car are being evaluated. An effective annual interest rate of 10% is to be used.

Plan A: lease the car for $0.25/mile (all inclusive)

Plan B: purchase the car for $30,000; keep the car for three years; sell the car after three years for $7200; pay $0.14 per mile for oil and gas; pay other costs of $500 per year

(a) What is the annual mileage of plan A?

(b) What is the annual mileage of plan B?

(c) For an equal annual cost, what is the annual mileage?

(d) Which plan is economically superior?

(A) Plan A is economically superior.

(B) Plan B is economically superior.

(C) Plans A and B are economically equivalent.

(D) Not enough information is provided.

24. Two methods are being considered to meet strict air pollution control requirements over the next ten years. Method A uses equipment with a life of ten years. Method B uses equipment with a life of five years that will be replaced with new equipment with an additional life of five years. Capacities of the two methods are different, but operating costs do not depend on the throughput. Operation is 24 hours per day, 365 days per year. The effective annual interest rate for this evaluation is 7%.

	method A	method B	
	years 1–10	years 1–5	years 6–10
installation cost	$13,000	$6000	$7000
equipment cost	$10,000	$2000	$2200
operating cost per hour	$10.50	$8.00	$8.00
salvage value	$5000	$2000	$2000
capacity (tons/yr)	50	20	20
life	10 years	5 years	5 years

(a) The uniform annual cost per ton for method A is most nearly

(A) $1800

(B) $1900

(C) $2100

(D) $2200

(b) The uniform annual cost per ton for method B is most nearly

(A) $3500

(B) $3600

(C) $4200

(D) $4300

(c) Over what range of throughput (in units of tons/yr) does each method have the minimum cost?

25. A transit district has asked for assistance in determining the proper fare for its bus system. An effective annual interest rate of 7% is to be used. The following additional information was compiled.

cost per bus	$60,000
bus life	20 years
salvage value	$10,000
miles driven per year	37,440
number of passengers per year	80,000
operating cost	$1.00 per mile in the first year, increasing $0.10 per mile each year thereafter

(a) If the fare is to remain constant for the next 20 years, the break-even fare per passenger is most nearly

(A) $0.51/passenger

(B) $0.61/passenger

(C) $0.84/passenger

(D) $0.88/passenger

(b) If the transit district decides to set the per-passenger fare at $0.35 for the first year, approximately how much should the passenger fare go up each year thereafter such that the district can break even in 20 years?

(A) $0.022 increase per year

(B) $0.036 increase per year

(C) $0.067 increase per year

(D) $0.072 increase per year

(c) If the transit district decides to set the per-passenger fare at $0.35 for the first year and the per-passenger fare goes up $0.05 each year thereafter, the additional government subsidy (per passenger) needed for the district to break even in 20 years is most nearly

(A) $0.11

(B) $0.12

(C) $0.16

(D) $0.21

26. Make a recommendation to your client to accept one of the following alternatives. Use the present worth comparison method. (Initial costs are the same.)

Alternative A: a 25 year annuity paying $4800 at the end of each year, where the interest rate is a nominal 12% per annum

Alternative B: a 25 year annuity paying $1200 every quarter at 12% nominal annual interest

(A) Alternative A is economically superior.

(B) Alternative B is economically superior.

(C) Alternatives A and B are economically equivalent.

(D) Not enough information is provided.

27. A firm has two alternatives for improvement of its existing production line. The data are as follows.

	alternative A	alternative B
initial installment cost	$1500	$2500
annual operating cost	$800	$650
service life	5 years	8 years
salvage value	0	0

Determine the best alternative using an interest rate of 15%.

(A) Alternative A is economically superior.

(B) Alternative B is economically superior.

(C) Alternatives A and B are economically equivalent.

(D) Not enough information is provided.

28. Two mutually exclusive alternatives requiring different investments are being considered. The life of both alternatives is estimated at 20 years with no salvage values. The minimum rate of return that is considered acceptable is 4%. Which alternative is best?

	alternative A	alternative B
investment required	$70,000	$40,000
net income per year	$5620	$4075
rate of return on total investment	5%	8%

(A) Alternative A is economically superior.

(B) Alternative B is economically superior.

(C) Alternatives A and B are economically equivalent.

(D) Not enough information is provided.

29. Compare the costs of two plant renovation schemes, A and B. Assume equal lives of 25 years, no salvage values, and interest at 25%.

	alternative A	alternative B
first cost	$20,000	$25,000
annual expenditure	$3000	$2500

(a) Determine the best alternative using the present worth method.

   (A) Alternative A is economically superior.

   (B) Alternative B is economically superior.

   (C) Alternatives A and B are economically equivalent.

   (D) Not enough information is provided.

(b) Determine the best alternative using the capitalized cost comparison.

   (A) Alternative A is economically superior.

   (B) Alternative B is economically superior.

   (C) Alternatives A and B are economically equivalent.

   (D) Not enough information is provided.

(c) Determine the best alternative using the annual cost comparison.

   (A) Alternative A is economically superior.

   (B) Alternative B is economically superior.

   (C) Alternatives A and B are economically equivalent.

   (D) Not enough information is provided.

30. A machine costs $18,000 and has a salvage value of $2000. It has a useful life of 8 years. The interest rate is 8%.

(a) Using straight line depreciation, its book value at the end of 5 years is most nearly

   (A) $2000

   (B) $3000

   (C) $6000

   (D) $8000

(b) Using the sinking fund method, the depreciation in the third year is most nearly

   (A) $1500

   (B) $1600

   (C) $1800

   (D) $2000

(c) Repeat part (a) using double declining balance depreciation. The balance value at the fifth year is most nearly

   (A) $2000

   (B) $4000

   (C) $6000

   (D) $8000

31. A chemical pump motor unit is purchased for $14,000. The estimated life is 8 years, after which it will be sold for $1800. Find the depreciation in the first two years by the sum-of-the-years' digits method. The after-tax depreciation recovery using 15% interest with 52% income tax is most nearly

   (A) $3600

   (B) $3900

   (C) $4100

   (D) $6300

32. A soda ash plant has the water effluent from processing equipment treated in a large settling basin. The settling basin eventually discharges into a river that runs alongside the basin. Recently enacted environmental regulations require all rainfall on the plant to be diverted and treated in the settling basin. A heavy rainfall will cause the entire basin to overflow. An uncontrolled overflow will cause environmental damage and heavy fines. The construction of additional height on the existing basic walls is under consideration.

Data on the costs of construction and expected costs for environmental cleanup and fines are shown. Data on 50 typical winters have been collected. The soda ash plant management considers 12% to be their minimum rate of return, and it is felt that after 15 years the plant will be closed. The company wants to select the alternative that minimizes its total expected costs.

additional basin height (ft)	number of winters with basin overflow	expense for environmental clean up per year	construction cost
0	24	$550,000	0
5	14	$600,000	$600,000
10	8	$650,000	$710,000
15	3	$700,000	$900,000
20	1	$800,000	$1,000,000
	50		

The additional height the basin should be built to is most nearly

   (A) 5.0 ft

   (B) 10 ft

   (C) 15 ft

   (D) 20 ft

33. A wood processing plant installed a waste gas scrubber at a cost of $30,000 to remove pollutants from the exhaust discharged into the atmosphere. The scrubber has no salvage value and will cost $18,700 to operate next year, with operating costs expected to increase at the rate of $1200 per year thereafter. Money can be borrowed at 12%. Approximately when should the company consider replacing the scrubber?

(A) 3 yr

(B) 6 yr

(C) 8 yr

(D) 10 yr

34. Two alternative piping schemes are being considered by a water treatment facility. Head and horsepower are reflected in the hourly cost of operation. On the basis of a 10-year life and an interest rate of 12%, what is most nearly the number of hours of operation for which the two installations will be equivalent?

	alternative A	alternative B
pipe diameter	4 in	6 in
head loss for required flow	48 ft	26 ft
size motor required	20 hp	7 hp
energy cost per hour of operation	$0.30	$0.10
cost of motor installed	$3600	$2800
cost of pipes and fittings	$3050	$5010
salvage value at end of 10 years	$200	$280

(A) 1000 hr

(B) 3000 hr

(C) 5000 hr

(D) 6000 hr

35. An 88% learning curve is used with an item whose first production time was 6 weeks.

(a) Approximately how long will it take to produce the fourth item?

(A) 4.5 wk

(B) 5.0 wk

(C) 5.5 wk

(D) 6.0 wk

(b) Approximately how long will it take to produce the sixth through fourteenth items?

(A) 35 wk

(B) 40 wk

(C) 45 wk

(D) 50 wk

36. A company is considering two alternatives, only one of which can be selected.

alternative	initial investment	salvage value	annual net profit	life
A	$120,000	$15,000	$57,000	5 yr
B	$170,000	$20,000	$67,000	5 yr

The net profit is after operating and maintenance costs, but before taxes. The company pays 45% of its year-end profit as income taxes. Use straight line depreciation. Do not use investment tax credit.

(a) Determine whether each alternative has an ROR greater than the MARR.

(A) Alternative A has ROR > MARR.

(B) Alternative B has ROR > MARR.

(C) Both alternatives have ROR > MARR.

(D) Neither alternative has ROR > MARR.

(b) Find the best alternative if the company's minimum attractive rate of return is 15%.

(A) Alternative A is economically superior.

(B) Alternative B is economically superior.

(C) Alternatives A and B are economically equivalent.

(D) Not enough information is provided.

37. A company is considering the purchase of equipment to expand its capacity. The equipment cost is $300,000. The equipment is needed for 5 years, after which it will be sold for $50,000. The company's before-tax cash flow will be improved $90,000 annually by the purchase of the asset. The corporate tax rate is 48%, and straight line depreciation will be used. The company will take an investment tax credit of 6.67%. What is the after-tax rate of return associated with this equipment purchase?

(A) 10.9%

(B) 11.8%

(C) 12.2%

(D) 13.2%

38. A 120-room hotel is purchased for $2,500,000. A 25-year loan is available for 12%. The year-end convention applies to loan payments. A study was conducted to determine the various occupancy rates.

occupancy	probability
65% full	0.40
70%	0.30
75%	0.20
80%	0.10

The operating costs of the hotel are as follows.

taxes and insurance	$20,000 annually
maintenance	$50,000 annually
operating	$200,000 annually

The life of the hotel is figured to be 25 years when operating 365 days per year. The salvage value after 25 years is $500,000. Neglect tax credit and income taxes.

(a) The distributed profit is most nearly

(A) $300,000

(B) $320,000

(C) $340,000

(D) $380,000

(b) The annual daily receipts are most nearly

(A) $2300

(B) $2400

(C) $2500

(D) $2600

(c) The average occupancy is most nearly

(A) 0.65

(B) 0.70

(C) 0.75

(D) 0.80

(d) The average rate that should be charged per room per night to return 15% of the initial cost each year is most nearly

(A) $27

(B) $29

(C) $30

(D) $31

39. A company is insured for $3,500,000 against fire and the insurance rate is $0.69/$1000. The insurance company will decrease the rate to $0.47/$1000 if fire sprinklers are installed. The initial cost of the sprinklers is $7500. Annual costs are $200; additional taxes are $100 annually. The system life is 25 years.

(a) The annual savings is most nearly

(A) $300

(B) $470

(C) $570

(D) $770

(b) The rate of return is most nearly

(A) 3.8%

(B) 5.0%

(C) 13%

(D) 16%

40. Heat losses through the walls in an existing building cost a company $1,300,000 per year. This amount is considered excessive, and two alternatives are being evaluated. Neither of the alternatives will increase the life of the existing building beyond the current expected life of 6 years, and neither of the alternatives will produce a salvage value. Improvements can be depreciated.

Alternative A: Do nothing, and continue with current losses.

Alternative B: Spend $2,000,000 immediately to upgrade the building and reduce the loss by 80%. Annual maintenance will cost $150,000.

Alternative C: Spend $1,200,000 immediately. Then, repeat the $1,200,000 expenditure 3 years from now. Heat loss the first year will be reduced 80%. Due to deterioration, the reduction will be 55% and 20% in the second and third years. (The pattern is repeated starting after the second expenditure.) There are no maintenance costs.

All energy and maintenance costs are considered expenses for tax purposes. The company's tax rate is 48%, and straight line depreciation is used. 15% is regarded as the effective annual interest rate. Evaluate each alternative on an after-tax basis.

(a) The present worth of alternative A is most nearly

(A) −$5.9 million

(B) −$4.9 million

(C) −$2.6 million

(D) −$2.4 million

(b) The present worth of alternative B is most nearly

(A) −$3.4 million

(B) −$2.8 million

(C) −$2.6 million

(D) −$2.2 million

(c) The present worth of alternative C is most nearly

(A) −$3.2 million

(B) −$3.1 million

(C) −$2.4 million

(D) −$2.0 million

(d) Which alternative should be recommended?

(A) alternative A

(B) alternative B

(C) alternative C

(D) not enough information

41. You have been asked to determine if a 7-year-old machine should be replaced. Give a full explanation for your recommendation. Base your decision on a before-tax interest rate of 15%.

The existing machine is presumed to have a 10-year life. It has been depreciated on a straight line basis from its original value of $1,250,000 to a current book value of $620,000. Its ultimate salvage value was assumed to be $350,000 for purposes of depreciation. Its present salvage value is estimated at $400,000, and this is not expected to change over the next 3 years. The current operating costs are not expected to change from $200,000 per year.

A new machine costs $800,000, with operating costs of $40,000 the first year, and increasing by $30,000 each year thereafter. The new machine has an expected life of 10 years. The salvage value depends on the year the new machine is retired.

year retired	salvage
1	$600,000
2	$500,000
3	$450,000
4	$400,000
5	$350,000
6	$300,000
7	$250,000
8	$200,000
9	$150,000
10	$100,000

42. A company estimates that the demand for its product will be 500,000 units per year. The product incorporates three identical valves. The acquisition cost per valve is $6.50, and the cost of keeping a valve in storage per year is 40% of the acquisition cost per valve per year. It costs the company an average of $49.50 to process an order. The company receives no quantity discounts and does not use inventory safety stocks. The most economical quantity of valves to order is most nearly

(A) 2700 valves

(B) 4400 valves

(C) 4800 valves

(D) 7600 valves

43. As production facilities move toward just-in-time manufacturing, it is important to minimize

(A) demand rate

(B) production rate

(C) inventory carrying cost

(D) setup cost

SOLUTIONS

1.

$i = 6\%$ a year

Using the formula from Table 54.1,

$$F = P(1+i)^n = (\$1000)(1+0.06)^{10}$$
$$= \boxed{\$1790.85 \quad (\$1800)}$$

Using the factor from App. 54.B, $(F/P, i, n) = 1.7908$ for $i = 6\%$ a year and $n = 10$ years.

$$F = P(F/P, 6\%, 10) = (\$1000)(1.7908)$$
$$= \boxed{\$1790.80 \quad (\$1800)}$$

The answer is (D).

2.

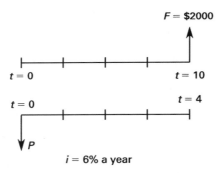

$i = 6\%$ a year

Using the formula from Table 54.1,

$$P = \frac{F}{(1+i)^n} = \frac{\$2000}{(1+0.06)^4}$$
$$= \boxed{\$1584.19 \quad (\$1600)}$$

Using the factor from App. 54.B, $(P/F, i, n) = 0.7921$ for $i = 6\%$ a year and $n = 4$ years.

$$P = F(P/F, 6\%, 4) = (\$2000)(0.7921)$$
$$= \boxed{\$1584.20 \quad (\$1600)}$$

The answer is (C).

3.

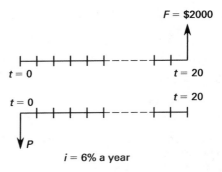

$i = 6\%$ a year

Using the formula from Table 54.1,

$$P = \frac{F}{(1+i)^n} = \frac{\$2000}{(1+0.06)^{20}}$$
$$= \boxed{\$623.61 \quad (\$620)}$$

Using the factor from App. 54.B, $(P/F, i, n) = 0.3118$ for $i = 6\%$ a year and $n = 20$ years.

$$P = F(P/F, 6\%, 20) = (\$2000)(0.3118)$$
$$= \boxed{\$623.60 \quad (\$620)}$$

The answer is (A).

4.

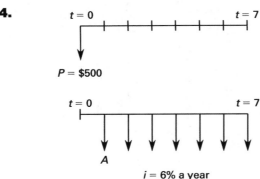

$i = 6\%$ a year

Using the formula from Table 54.1,

$$A = P\left(\frac{i(1+i)^n}{(1+i)^n - 1}\right) = (\$500)\left(\frac{(0.06)(1+0.06)^7}{(1+0.06)^7 - 1}\right)$$
$$= \boxed{\$89.57 \quad (\$90)}$$

Using the factor from App. 54.B, $(A/P, i, n) = 0.1791$ for $i = 6\%$ a year and $n = 7$ years.

$$A = P(A/P, 6\%, 7) = (\$500)(0.17914)$$
$$= \boxed{\$89.55 \quad (\$90)}$$

The answer is (A).

5.

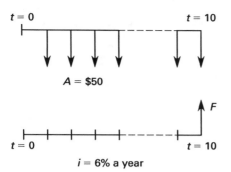

$$A = \$50$$

$$i = 6\% \text{ a year}$$

Using the formula from Table 54.1,

$$F = A\left(\frac{(1+i)^n - 1}{i}\right)$$

$$= (\$50)\left(\frac{(1+0.06)^{10} - 1}{0.06}\right)$$

$$= \boxed{\$659.04 \quad (\$660)}$$

Using the factor from App. 54.B, $(F/A, i, n) = 13.1808$ for $i = 6\%$ a year and $n = 10$ years.

$$F = A(F/A, 6\%, 10)$$

$$= (\$50)(13.1808)$$

$$= \boxed{\$659.04 \quad (\$660)}$$

The answer is (C).

6.

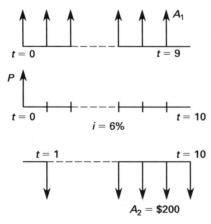

$$i = 6\%$$

$$A_2 = \$200$$

From Table 54.1, for each cash flow diagram,

$$P = A_1 + A_1\left(\frac{(1+0.06)^9 - 1}{(0.06)(1+0.06)^9}\right)$$

$$= A_2\left(\frac{(1+0.06)^{10} - 1}{(0.06)(1+0.06)^{10}}\right)$$

Therefore, for $A_2 = \$200$,

$$A_1 + A_1\left(\frac{(1+0.06)^9 - 1}{(0.06)(1+0.06)^9}\right)$$

$$= (\$200)\left(\frac{(1+0.06)^{10} - 1}{(0.06)(1+0.06)^{10}}\right)$$

$$7.80A_1 = \$1472.02$$

$$A_1 = \boxed{\$188.72 \quad (\$190)}$$

Using factors from App. 54.B,

$$(P/A, 6\%, 9) = 6.8017$$

$$(P/A, 6\%, 10) = 7.3601$$

$$A_1 + A_1(6.8017) = (\$200)(7.3601)$$

$$7.8017A_1 = \$1472.02$$

$$A_1 = \frac{\$1472.02}{7.8017}$$

$$= \boxed{\$188.68 \quad (\$190)}$$

The answer is (A).

7.

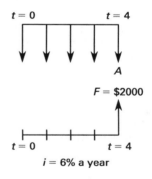

$$F = \$2000$$

$$i = 6\% \text{ a year}$$

From Table 54.1,

$$F = A\left(\frac{(1+i)^n - 1}{i}\right)$$

Since the deposits start at the beginning of each year, five deposits are made that contribute to the final amount. This is equivalent to a cash flow that starts at $t = -1$ without a deposit and has a duration (starting at $t = -1$) of 5 years.

$$F = A\left(\frac{(1+i)^n - 1}{i}\right) = \$2000$$

$$= A\left(\frac{(1+0.06)^5 - 1}{0.06}\right)$$

$$\$2000 = 5.6371A$$
$$A = \frac{\$2000}{5.6371}$$
$$= \boxed{\$354.79 \quad (\$350)}$$

Using factors from App. 54.B,

$$F = A\big((F/P, 6\%, 4) + (F/A, 6\%, 4)\big)$$
$$\$2000 = A(1.2625 + 4.3746)$$
$$A = \boxed{\$354.79 \quad (\$350)}$$

The answer is (A).

8.

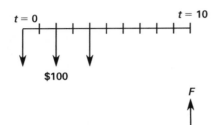

From Table 54.1, $F = P(1 + i)^n$. If each deposit is considered as P, each will accumulate interest for periods of 10, 8, and 6 years.

Therefore,

$$F = (\$100)(1 + 0.06)^{10} + (\$100)(1 + 0.06)^8$$
$$+ (\$100)(1 + 0.06)^6$$
$$= (\$100)(1.7908 + 1.5938 + 1.4185)$$
$$= \boxed{\$480.31 \quad (\$480)}$$

Using App. 54.B,

$$(F/P, i, n) = 1.7908 \text{ for } i = 6\% \text{ and } n = 10$$
$$= 1.5938 \text{ for } i = 6\% \text{ and } n = 8$$
$$= 1.4185 \text{ for } i = 6\% \text{ and } n = 6$$

By summation,

$$F = (\$100)(1.7908 + 1.5938 + 1.4185)$$
$$= \boxed{\$480.31 \quad (\$480)}$$

The answer is (B).

9.

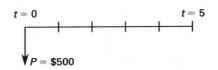

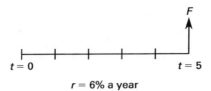

Since the deposit is compounded monthly, the effective interest rate should be calculated from Eq. 54.51.

$$i = \left(1 + \frac{r}{k}\right)^k - 1 = \left(1 + \frac{0.06}{12}\right)^{12} - 1$$
$$= 0.061678 \quad (6.1678\%)$$

From Table 54.1,

$$F = P(1 + i)^n = (\$500)(1 + 0.061678)^5$$
$$= \boxed{\$674.43 \quad (\$680)}$$

To use App. 54.B, interpolation is required.

$i\%$	factor F/P
6	1.3382
6.1678	desired
7	1.4026

$$\Delta(F/P) = \left(\frac{6.1678\% - 6\%}{7\% - 6\%}\right)(1.4026 - 1.3382)$$
$$= 0.0108$$

Therefore,

$$F/P = 1.3382 + 0.0108 = 1.3490$$
$$F = P(F/P, 6.1677\%, 5) = (\$500)(1.3490)$$
$$= \boxed{\$674.50 \quad (\$675)}$$

The answer is (C).

10.

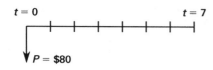

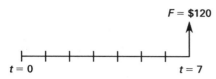

From Table 54.1,

$$F = P(1 + i)^n$$

Therefore,

$$(1 + i)^n = F/P$$

$$i = (F/P)^{1/n} - 1 = \left(\frac{\$120}{\$80}\right)^{1/7} - 1$$

$$= 0.0596 \approx \boxed{6.0\%}$$

From App. 54.B,

$$F = P(F/P, i\%, 7)$$

$$(F/P, i\%, 7) = F/P = \frac{\$120}{\$80} = 1.5$$

Searching App. 54.B,

$$(F/P, i\%, 7) = 1.4071 \text{ for } i = 5\%$$
$$= 1.5036 \text{ for } i = 6\%$$
$$= 1.6058 \text{ for } i = 7\%$$

Therefore, $i \approx \boxed{6.0\%.}$

The answer is (D).

11.

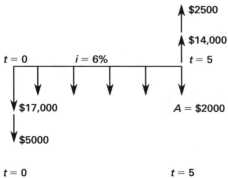

Annual cost of ownership, EUAC, can be obtained by the factors converting P to A and F to A.

$$P = \$17,000 + \$5000$$
$$= \$22,000$$
$$F = \$14,000 + \$2500$$
$$= \$16,500$$
$$\text{EUAC} = A + P(A/P, 6\%, 5) - F(A/F, 6\%, 5)$$
$$(A/P, 6\%, 5) = 0.2374$$
$$(A/F, 6\%, 5) = 0.1774$$
$$\text{EUAC} = \$2000 + (\$22,000)(0.2374)$$
$$\quad - (\$16,500)(0.1774)$$
$$= \boxed{\$4295.70 \quad (\$4300)}$$

The answer is (C).

12.

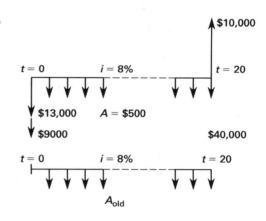

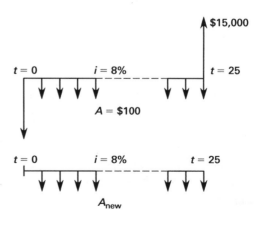

Consider the salvage value as a benefit lost (cost).

$$\text{EUAC}_{\text{old}} = \$500 + (\$9000 + \$13{,}000)(A/P, 8\%, 20)$$
$$- (\$10{,}000)(A/F, 8\%, 20)$$

$$(A/P, 8\%, 20) = 0.1019$$

$$(A/F, 8\%, 20) = 0.0219$$

$$\text{EUAC}_{\text{old}} = \$500 + (\$22{,}000)(0.1019)$$
$$- (\$10{,}000)(0.0219)$$
$$= \$2522.80$$

Similarly,

$$\text{EUAC}_{\text{new}} = \$100 + (\$40{,}000)(A/P, 8\%, 25)$$
$$- (\$15{,}000)(A/F, 8\%, 25)$$

$$(A/P, 8\%, 25) = 0.0937$$

$$(A/F, 8\%, 25) = 0.0137$$

$$\text{EUAC}_{\text{new}} = \$100 + (\$40{,}000)(0.0937)$$
$$- (\$15{,}000)(0.0137)$$
$$= \$3642.50$$

Therefore, the new bridge is more costly.

> The best alternative is to strengthen the old bridge.

The answer is (A).

13.

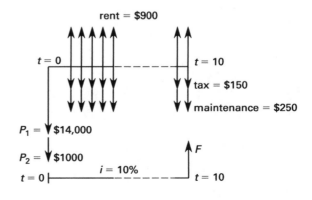

$32,000

t = 0   t = 15

$7530

$150,000

The annual depreciation is

$$D = \frac{C - S_n}{n} = \frac{\$150{,}000}{15} = \$10{,}000/\text{year}$$

The taxable income is

$$\$32{,}000 - \$7530 - \$10{,}000 = \$14{,}470/\text{year}$$

Taxes paid are

$$(\$14{,}470)(0.48) = \$6945.60/\text{year}$$

The after-tax cash flow is

$$\$32{,}000 - \$7530 - \$6945.60 = \$17{,}524.40$$

The present worth of the alternate is zero when evaluated at its ROR.

$$0 = -\$150{,}000 + (\$17{,}524.40)(P/A, i\%, 15)$$

Therefore,

$$(P/A, i\%, 15) = \frac{\$150{,}000}{\$17{,}524.40} = 8.55949$$

Searching App. 54.B, this factor matches $i = 8\%$.

> $\text{ROR} = 8.0\%$

The answer is (A).

14. (a) The conventional benefit/cost ratio is

$$B/C = \frac{B - D}{D}$$

The benefit/cost ratio will be

$$B/C = \frac{\$1{,}500{,}000 - \$300{,}000}{\$1{,}000{,}000} = \boxed{1.2}$$

The answer is (C).

(b) The excess of benefits over cost is $\boxed{\$200{,}000.}$

The answer is (A).

15. The annual rent is

$$(\$75)\left(12\ \frac{\text{months}}{\text{year}}\right) = \$900$$

$$P = P_1 + P_2 = \$15{,}000$$
$$A_1 = -\$900$$
$$A_2 = \$250 + \$150 = \$400$$

rent = $900

t = 0   t = 10

tax = $150

maintenance = $250

$P_1 = \$14{,}000$

$P_2 = \$1000$

t = 0   i = 10%   t = 10

F

Use App. 54.B.

$$F = (\$15,000)(F/P, 10\%, 10)$$
$$+ (\$400)(F/A, 10\%, 10)$$
$$- (\$900)(F/A, 10\%, 10)$$
$$(F/P, 10\%, 10) = 2.5937$$
$$(F/A, 10\%, 10) = 15.9374$$
$$F = (\$15,000)(2.5937) + (\$400)(15.9374)$$
$$- (\$900)(15.9374)$$
$$= \boxed{\$30,937 \quad (\$31,000)}$$

The answer is (B).

16.

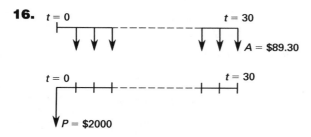

From Table 54.1,

$$P = A\left(\frac{(1+i)^n - 1}{i(1+i)^n}\right)$$

$$\frac{(1+i)^{30} - 1}{i(1+i)^{30}} = \frac{\$2000}{\$89.30} = 22.40$$

By trial and error,

$i\%$	$(1+i)^{30}$	$\dfrac{(1+i)^{30} - 1}{i(1+i)^{30}}$
10	17.45	9.43
6	5.74	13.76
4	3.24	17.29
2	1.81	22.40

2% per month is close.

$$i = (1 + 0.02)^{12} - 1 = \boxed{0.2682 \quad (27\%)}$$

The answer is (A).

17. (a) Use the straight line method, Eq. 54.25.

$$D = \frac{C - S_n}{n}$$

Each year depreciation will be the same.

$$D = \frac{\$500,000 - \$100,000}{25} = \boxed{\$16,000}$$

The answer is (B).

(b) Use Eq. 54.27.

$$T = \tfrac{1}{2}n(n + 1) = \left(\tfrac{1}{2}\right)(25)(25 + 1) = 325$$

Sum-of-the-years' digits (SOYD) depreciation can be calculated from Eq. 54.28.

$$D_j = \frac{(C - S_n)(n - j + 1)}{T}$$

$$D_1 = \frac{(\$500,000 - \$100,000)(25 - 1 + 1)}{325}$$
$$= \boxed{\$30,769 \quad (\$31,000)}$$

$$D_2 = \frac{(\$500,000 - \$100,000)(25 - 2 + 1)}{325}$$
$$= \boxed{\$29,538 \quad (\$30,000)}$$

$$D_3 = \frac{(\$500,000 - \$100,000)(25 - 3 + 1)}{325}$$
$$= \boxed{\$28,308 \quad (\$28,000)}$$

The answer is (C).

(c) The double declining balance (DDB) method can be used. By Eq. 54.32,

$$D_j = dC(1 - d)^{j-1}$$

Use Eq. 54.31.

$$d = \frac{2}{n} = \frac{2}{25}$$

$$D_1 = \left(\tfrac{2}{25}\right)(\$500,000)\left(1 - \tfrac{2}{25}\right)^0 = \boxed{\$40,000}$$

$$D_2 = \left(\tfrac{2}{25}\right)(\$500,000)\left(1 - \tfrac{2}{25}\right)^1 = \boxed{\$36,800 \quad (\$37,000)}$$

$$D_3 = \left(\tfrac{2}{25}\right)(\$500,000)\left(1 - \tfrac{2}{25}\right)^2 = \boxed{\$33,856 \quad (\$34,000)}$$

The answer is (D).

18.

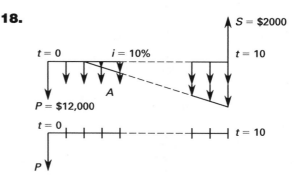

(a) A is \$1000, and G is \$200 for $t = n - 1 = 9$ years. With $F = S = \$2000$, the present worth is

$$P = \$12{,}000 + A(P/A, 10\%, 10) + G(P/G, 10\%, 10)$$
$$\quad - F(P/F, 10\%, 10)$$
$$= \$12{,}000 + (\$1000)(6.1446) + (\$200)(22.8913)$$
$$\quad - (\$2000)(0.3855)$$
$$= \boxed{\$21{,}952 \quad (\$22{,}000)}$$

The answer is (D).

(b) The annual cost is

$$A = (\$12{,}000)(A/P, 10\%, 10) + \$1000$$
$$\quad + (\$200)(A/G, 10\%, 10)$$
$$\quad - (\$2000)(A/F, 10\%, 10)$$
$$= (\$12{,}000)(0.1627) + \$1000 + (\$200)(3.7255)$$
$$\quad - (\$2000)(0.0627)$$
$$= \boxed{\$3572.10 \quad (\$3600)}$$

The answer is (C).

19. An increase in rock removal capacity can be achieved by a 20-year loan (investment). Different cases available can be compared by equivalent uniform annual cost (EUAC).

$$\text{EUAC} = \text{annual loan cost}$$
$$\quad + \text{expected annual damage}$$
$$= \text{cost}\,(A/P, 10\%, 20)$$
$$\quad + (\$25{,}000)(\text{probability})$$
$$(A/P, 10\%, 20) = 0.1175$$

A table can be prepared for different cases.

rock removal rate	cost (\$)	annual loan cost (\$)	expected annual damage (\$)	EUAC (\$)
7	0	0	3750	3750.00
8	15,000	1761.90	2500	4261.90
9	20,000	2349.20	1750	4099.20
10	30,000	3523.80	750	4273.80

$$\boxed{\text{It is cheapest to do nothing.}}$$

The answer is (A).

20. (a) Calculate the cost of owning and operating for years one and two.

$$A_1 = (\$10{,}000)(A/P, 20\%, 1) + \$2000$$
$$\quad - (\$8000)(A/F, 20\%, 1)$$
$$(A/P, 20\%, 1) = 1.2$$
$$(A/F, 20\%, 1) = 1.0$$
$$A_1 = (\$10{,}000)(1.2) + \$2000 - (\$8000)(1.0)$$
$$= \$6000$$
$$A_2 = (\$10{,}000)(A/P, 20\%, 2) + \$2000$$
$$\quad + (\$1000)(A/G, 20\%, 2)$$
$$\quad - (\$7000)(A/F, 20\%, 2)$$
$$(A/P, 20\%, 2) = 0.6545$$
$$(A/G, 20\%, 2) = 0.4545$$
$$(A/F, 20\%, 2) = 0.4545$$
$$A_2 = (\$10{,}000)(0.6545) + \$2000$$
$$\quad + (\$1000)(0.4545) - (\$7000)(0.4545)$$
$$= \boxed{\$5818 \quad (\$5800)}$$

The answer is (C).

(b) Calculate the cost of owning and operating for years three through five.

$$A_3 = (\$10{,}000)(A/P, 20\%, 3) + \$2000$$
$$\quad + (\$1000)(A/G, 20\%, 3)$$
$$\quad - (\$6000)(A/F, 20\%, 3)$$
$$(A/P, 20\%, 3) = 0.4747$$
$$(A/G, 20\%, 3) = 0.8791$$
$$(A/F, 20\%, 3) = 0.2747$$
$$A_3 = (\$10{,}000)(0.4747) + \$2000$$
$$\quad + (\$1000)(0.8791)$$
$$\quad - (\$6000)(0.2747)$$
$$= \$5977.90$$
$$A_4 = (\$10{,}000)(A/P, 20\%, 4)$$
$$\quad + \$2000 + (\$1000)(A/G, 20\%, 4)$$
$$\quad - (\$5000)(A/F, 20\%, 4)$$
$$(A/P, 20\%, 4) = 0.3863$$
$$(A/G, 20\%, 4) = 1.2742$$
$$(A/F, 20\%, 4) = 0.1863$$
$$A_4 = (\$10{,}000)(0.3863) + \$2000$$
$$\quad + (\$1000)(1.2742) - (\$5000)(0.1863)$$
$$= \$6205.70$$

$$A_5 = (\$10,000)(A/P, 20\%, 5) + \$2000$$
$$+ (\$1000)(A/G, 20\%, 5)$$
$$- (\$4000)(A/F, 20\%, 5)$$

$$(A/P, 20\%, 5) = 0.3344$$
$$(A/G, 20\%, 5) = 1.6405$$
$$(A/F, 20\%, 5) = 0.1344$$

$$A_5 = (\$10,000)(0.3344) + \$2000$$
$$+ (\$1000)(1.6405) - (\$4000)(0.1344)$$
$$= \boxed{\$6446.90 \quad (\$6500)}$$

The answer is (B).

(c) Since the annual owning and operating cost is smallest after two years of operation, it is advantageous to sell the mechanism after the second year.

$$\boxed{\text{The economic life is two years.}}$$

The answer is (B).

(d) After four years of operation, the owning and operating cost of the mechanism for one more year will be

$$A = \$6000 + (\$5000)(1 + i) - \$4000$$
$$i = 0.2 \quad (20\%)$$
$$A = \$6000 + (\$5000)(1.2) - \$4000$$
$$= \boxed{\$8000}$$

The answer is (C).

21. (a) To find out if the reimbursement is adequate, calculate the business-related expense.

Charge the company for business travel.

$$\begin{aligned}
\text{insurance:} &\quad \$3000 - \$2000 = \$1000 \\
\text{maintenance:} &\quad \$2000 - \$1500 = \$500 \\
\text{drop in salvage value:} &\quad \$10,000 - \$5000 = \$5000
\end{aligned}$$

The annual portion of the drop in salvage value is

$$A = (\$5000)(A/F, 10\%, 5)$$
$$(A/F, 10\%, 5) = 0.1638$$
$$A = (\$5000)(0.1638) = \$819/\text{yr}$$

The annual cost of gas is

$$\left(\frac{50,000 \text{ mi}}{15 \frac{\text{mi}}{\text{gal}}}\right)\left(\frac{\$1.50}{\text{gal}}\right) = \$5000$$

$$\text{EUAC per mile} = \frac{\$1000 + \$500 + \$819 + \$5000}{50,000 \text{ mi}}$$
$$= \$0.14638/\text{mi}$$

Since the reimbursement per mile was \$0.30 and since \$0.30 > \$0.14638, the reimbursement is $\boxed{\text{adequate.}}$

(b) Determine (with reimbursement) how many miles the car must be driven to break even.

If the car is driven M miles per year,

$$\left(\frac{\$0.30}{1 \text{ mi}}\right)M = (\$50,000)(A/P, 10\%, 5) + \$2500$$
$$+ \$2000 - (\$8000)(A/F, 10\%, 5)$$
$$+ \left(\frac{M}{15 \frac{\text{mi}}{\text{gal}}}\right)(\$1.50)$$

$$(A/P, 10\%, 5) = 0.2638$$
$$(A/F, 10\%, 5) = 0.1638$$
$$0.3M = (\$50,000)(0.2638) + \$2500 + \$2000$$
$$- (\$8000)(0.1638) + 0.1M$$
$$0.2M = \$16,379.60$$
$$M = \frac{\$16,379.60}{\frac{\$0.20}{1 \text{ mi}}}$$
$$= \boxed{81,898 \text{ mi} \quad (82,000 \text{ mi})}$$

The answer is (C).

22. The present worth of alternative A is

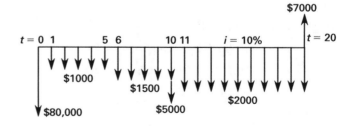

$$P_A = \$80,000 + (\$1000)(P/A, 10\%, 5)$$
$$+ (\$1500)(P/A, 10\%, 5)(P/F, 10\%, 5)$$
$$+ (\$2000)(P/A, 10\%, 10)(P/F, 10\%, 10)$$
$$+ (\$5000)(P/F, 10\%, 10)$$
$$- (\$7000)(P/F, 10\%, 20)$$

$(P/A, 10\%, 5) = 3.7908$

$(P/F, 10\%, 5) = 0.6209$

$(P/A, 10\%, 10) = 6.1446$

$(P/F, 10\%, 10) = 0.3855$

$(P/F, 10\%, 20) = 0.1486$

$$P_A = \$80,000 + (\$1000)(3.7908)$$
$$+ (\$1500)(3.7908)(0.6209)$$
$$+ (\$2000)(6.1446)(0.3855)$$
$$+ (\$5000)(0.3855)$$
$$- (\$7000)(0.1486)$$
$$= \boxed{\$92,946.15 \quad (\$93,000)}$$

The answer is (B).

(b) Since the lives are different, compare by EUAC.

$$\text{EUAC(A)} = (\$92,946.14)(A/P, 10\%, 20)$$
$$= (\$92,946.14)(0.1175)$$
$$= \boxed{\$10,921 \quad (\$11,000)}$$

The answer is (B).

(c) Evaluate alternative B.

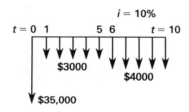

$$P_B = \$35,000 + (\$3000)(P/A, 10\%, 5)$$
$$+ (\$4000)(P/A, 10\%, 5)(P/F, 10\%, 5)$$

$(P/A, 10\%, 5) = 3.7908$

$(P/F, 10\%, 5) = 0.6209$

$$P_B = \$35,000 + (\$3000)(3.7908)$$
$$+ (\$4000)(3.7908)(0.6209)$$
$$= \boxed{\$55,787.23 \quad (\$56,000)}$$

The answer is (A).

(d) Since the lives are different, compare by EUAC.

$$\text{EUAC(B)} = (\$55,787.23)(A/P, 10\%, 10)$$
$$= (\$55,787.23)(0.1627)$$
$$= \boxed{\$9077 \quad (\$9100)}$$

The answer is (A).

(e) Since EUAC(B) < EUAC(A),

$$\boxed{\text{Alternative B is economically superior.}}$$

The answer is (B).

23. (a) If the annual cost is compared with a total annual mileage of M, for plan A,

$$A_A = \boxed{\$0.25M}$$

(b) For plan B,

$$A_B = (\$30,000)(A/P, 10\%, 3) + \$0.14M$$
$$+ \$500 - (\$7200)(A/F, 10\%, 3)$$

$(A/P, 10\%, 3) = 0.4021$

$(A/F, 10\%, 3) = 0.3021$

$$A_B = (\$30,000)(0.4021) + \$0.14M + \$500$$
$$- (\$7200)(0.3021)$$
$$= \boxed{\$10,387.88 + \$0.14M}$$

(c) For an equal annual cost $A_A = A_B$,

$$\$0.25M = \$10,387.88 + \$0.14M$$
$$\$0.11M = \$10,387.88$$
$$M = 94,435 \quad (94,000)$$

An annual mileage would be $\boxed{M = 94,000 \text{ mi.}}$

(d) For an annual mileage less than that, $A_A < A_B$.

$$\boxed{\begin{array}{l}\text{Plan A is economically superior until 94,000 mi is}\\\text{exceeded.}\end{array}}$$

The answer is (A).

24. (a) Method A:

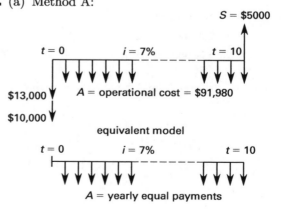

24 hr/day

365 days/yr

total of $(24)(365) = 8760$ hr/yr

$10.50 operational cost/hr

total of $(8760)(\$10.50) = \$91,980$ operational cost/yr

$$A = \$91,980 + (\$23,000)(A/P, 7\%, 10)$$
$$- (\$5000)(A/F, 7\%, 10)$$

$(A/P, 7\%, 10) = 0.1424$

$(A/F, 7\%, 10) = 0.0724$

$$A = \$91,980 + (\$23,000)(0.1424)$$
$$- (\$5000)(0.0724)$$
$$= \$94,893.20/\text{yr}$$

Therefore, the uniform annual cost per ton each year will be

$$\frac{\$94,893.20}{50 \text{ ton}} = \boxed{\$1897.86 \quad (\$1900)}$$

The answer is (B).

(b) Method B:

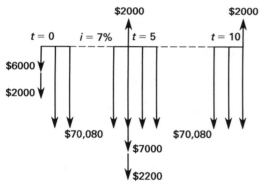

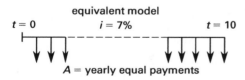

8760 hr/yr

$8 operational cost/hr

total of $70,080 operational cost/yr

$$A = \$70,080 + (\$6000 + \$2000)$$
$$\times (A/P, 7\%, 10)$$
$$+ (\$7000 + \$2200 - \$2000)$$
$$\times (P/F, 7\%, 5)(A/P, 7\%, 10)$$
$$- (\$2000)(A/F, 7\%, 10)$$

$(A/P, 7\%, 10) = 0.1424$

$(A/F, 7\%, 10) = 0.0724$

$(P/F, 7\%, 5) = 0.7130$

$$A = \$70,080 + (\$8000)(0.1424)$$
$$+ (\$7200)(0.7130)(0.1424)$$
$$- (\$2000)(0.0724)$$
$$= \$71,805.42/\text{yr}$$

Therefore, the uniform annual cost per ton each year will be

$$\frac{\$71,805.42}{20 \text{ ton}} = \boxed{\$3590.27 \quad (\$3600)}$$

The answer is (B).

(c) The following table can be used to determine which alternative is least expensive for each throughput range.

tons/yr	cost of using A		cost of using B		cheapest
0–20	$94,893	(1×)	$71,805	(1×)	B
20–40	$94,893	(1×)	$143,610	(2×)	A
40–50	$94,893	(1×)	$215,415	(3×)	A
50–60	$189,786	(2×)	$215,415	(3×)	A
60–80	$189,786	(2×)	$287,220	(4×)	A

25.
$$A_e = (\$60,000)(A/P, 7\%, 20) + A$$
$$+ G(P/G, 7\%, 20)(A/P, 7\%, 20)$$
$$- (\$10,000)(A/F, 7\%, 20)$$

$(A/P, 7\%, 20) = 0.0944$

$$A = (37{,}440 \text{ mi})\left(\frac{\$1.0}{1 \text{ mi}}\right) = \$37{,}440$$

$$G = 0.1A = (0.1)(\$37{,}440) = \$3744$$

$(P/G, 7\%, 20) = 77.5091$

$(A/F, 7\%, 20) = 0.0244$

$$A_e = (\$60{,}000)(0.0944) + \$37{,}440$$
$$+ (\$3744)(77.5091)(0.0944)$$
$$- (\$10{,}000)(0.0244)$$
$$= \$70{,}254.32$$

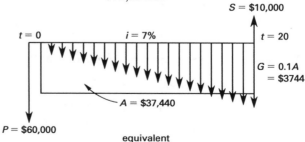

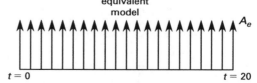

(a) With 80,000 passengers a year, the break-even fare per passenger would be

$$\text{fare} = \frac{A_e}{80,000} = \frac{\$70,254.32}{80,000}$$
$$= \boxed{\$0.878/\text{passenger} \quad (\$0.88/\text{passenger})}$$

The answer is (D).

(b) The passenger fare should go up each year by

$$\$0.878 = \$0.35 + G(A/G, 7\%, 20)$$
$$G = \frac{\$0.878 - \$0.35}{7.3163}$$
$$= \boxed{\$0.072 \text{ increase per year}}$$

The answer is (D).

(c) As in part (b), the subsidy should be

$$\text{subsidy} = \text{cost} - \text{revenue}$$
$$P = \$0.878 - \big(\$0.35 + (\$0.05)(A/G, 7\%, 20)\big)$$
$$= \$0.878 - \big(\$0.35 + (\$0.05)(7.3163)\big)$$
$$= \boxed{\$0.162 \quad (\$0.16)}$$

The answer is (C).

26. Use the present worth comparison method.

$$P(A) = (\$4800)(P/A, 12\%, 25)$$
$$= (\$4800)(7.8431)$$
$$= \$37,646.88$$

$$(4 \text{ quarters})(25 \text{ years}) = 100 \text{ compounding periods}$$
$$P(B) = (\$1200)(P/A, 3\%, 100)$$
$$= (\$1200)(31.5989)$$
$$= \$37,918.68$$
$$\boxed{\text{Alternative B is economically superior.}}$$

The answer is (B).

27. Use the equivalent uniform annual cost method.

$$EUAC(A) = (\$1500)(A/P, 15\%, 5) + \$800$$
$$= (\$1500)(0.2983) + \$800$$
$$= \$1247.45$$
$$EUAC(B) = (\$2500)(A/P, 15\%, 8) + \$650$$
$$= (\$2500)(0.2229) + \$650$$
$$= \$1207.25$$
$$\boxed{\text{Alternative B is economically superior.}}$$

The answer is (B).

28. The data given imply that both investments return 4% or more. However, the increased investment of $30,000 may not be cost effective. Do an incremental analysis.

$$\text{incremental cost} = \$70,000 - \$40,000 = \$30,000$$
$$\text{incremental income} = \$5620 - \$4075 = \$1545$$
$$0 = -\$30,000 + (\$1545)(P/A, i\%, 20)$$
$$(P/A, i\%, 20) = 19.417$$
$$i \approx 0.25\% < 4\%$$
$$\boxed{\text{Alternative B is economically superior.}}$$

(The same conclusion could be reached by taking the present worths of both alternatives at 4%.)

The answer is (B).

29. (a) The present worth comparison is

$$P(A) = (-\$3000)(P/A, 25\%, 25) - \$20,000$$
$$= (-\$3000)(3.9849) - \$20,000$$
$$= -\$31,954.70$$
$$P(B) = (-\$2500)(3.9849) - \$25,000$$
$$= -\$34,962.25$$
$$\boxed{\text{Alternative A is economically superior.}}$$

The answer is (A).

(b) The capitalized cost comparison is

$$CC(A) = \$20,000 + \frac{\$3000}{0.25} = \$32,000$$
$$CC(B) = \$25,000 + \frac{\$2500}{0.25} = \$35,000$$
$$\boxed{\text{Alternative A is economically superior.}}$$

The answer is (A).

(c) The annual cost comparison is

$$EUAC(A) = (\$20,000)(A/P, 25\%, 25) + \$3000$$
$$= (\$20,000)(0.2509) + \$3000$$
$$= \$8018.00$$
$$EUAC(B) = (\$25,000)(0.2509) + \$2500$$
$$= \$8772.50$$
$$\boxed{\text{Alternative A is economically superior.}}$$

The answer is (A).

30. (a) The depreciation in the first 5 years is

$$BV = \$18{,}000 - (5)\left(\frac{\$18{,}000 - \$2000}{8}\right)$$

$$= \boxed{\$8000}$$

The answer is (D).

(b) With the sinking fund method, the basis is

$$(\$18{,}000 - \$2000)(A/F, 8\%, 8)$$

$$= (\$18{,}000 - \$2000)(0.0940)$$

$$= \$1504$$

$$D_1 = (\$1504)(1.000) = \$1504$$

$$D_2 = (\$1504)(1.0800) = \$1624$$

$$D_3 = (\$1504)(1.0800)^2 = \boxed{\$1754 \quad (\$1800)}$$

The answer is (C).

(c) Using double declining balance depreciation, the first five years' depreciation is

$$D_1 = \left(\tfrac{2}{8}\right)(\$18{,}000) = \$4500$$

$$D_2 = \left(\tfrac{2}{8}\right)(\$18{,}000 - \$4500) = \$3375$$

$$D_3 = \left(\tfrac{2}{8}\right)(\$18{,}000 - \$4500 - \$3375) = \$2531$$

$$D_4 = \left(\tfrac{2}{8}\right)(\$18{,}000 - \$4500 - \$3375 - \$2531)$$

$$= \$1898$$

$$D_5 = \left(\tfrac{2}{8}\right)(\$18{,}000 - \$4500 - \$3375$$

$$- \$2531 - \$1898)$$

$$= \$1424$$

The balance value at the fifth year is

$$BV = \$18{,}000 - \$4500 - \$3375 - \$2531$$

$$- \$1898 - \$1424$$

$$= \boxed{\$4272 \quad (\$4000)}$$

The answer is (B).

31. The after-tax depreciation recovery is

$$T = \left(\tfrac{1}{2}\right)(8)(9) = 36$$

$$D_1 = \left(\tfrac{8}{36}\right)(\$14{,}000 - \$1800) = \$2711$$

$$\Delta D = \left(\tfrac{1}{36}\right)(\$14{,}000 - \$1800) = \$339$$

$$D_2 = \$2711 - \$339 = \$2372$$

$$DR = (0.52)(\$2711)(P/A, 15\%, 8)$$

$$- (0.52)(\$339)(P/G, 15\%, 8)$$

$$= (0.52)(\$2711)(4.4873)$$

$$- (0.52)(\$339)(12.4807)$$

$$= \boxed{\$4125.74 \quad (\$4100)}$$

The answer is (C).

32. Use the equivalent uniform annual cost method to find the best alternative.

$$(A/P, 12\%, 15) = 0.1468$$

$$\begin{aligned}
\text{EUAC}_{5\,\text{ft}} = {}& (\$600{,}000)(0.1468) \\
& + \left(\tfrac{14}{50}\right)(\$600{,}000) + \left(\tfrac{8}{50}\right)(\$650{,}000) \\
& + \left(\tfrac{3}{50}\right)(\$700{,}000) + \left(\tfrac{1}{50}\right)(\$800{,}000) \\
= {}& \$418{,}080
\end{aligned}$$

$$\begin{aligned}
\text{EUAC}_{10\,\text{ft}} = {}& (\$710{,}000)(0.1468) \\
& + \left(\tfrac{8}{50}\right)(\$650{,}000) + \left(\tfrac{3}{50}\right)(\$700{,}000) \\
& + \left(\tfrac{1}{50}\right)(\$800{,}000) \\
= {}& \$266{,}228
\end{aligned}$$

$$\begin{aligned}
\text{EUAC}_{15\,\text{ft}} = {}& (\$900{,}000)(0.1468) \\
& + \left(\tfrac{3}{50}\right)(\$700{,}000) + \left(\tfrac{1}{50}\right)(\$800{,}000) \\
= {}& \$190{,}120
\end{aligned}$$

$$\begin{aligned}
\text{EUAC}_{20\,\text{ft}} = {}& (\$1{,}000{,}000)(0.1468) \\
& + \left(\tfrac{1}{50}\right)(\$800{,}000) \\
= {}& \$162{,}800
\end{aligned}$$

$$\boxed{\text{Build to 20 ft.}}$$

The answer is (D).

33. Assume replacement after 1 year.

$$\begin{aligned}
\text{EUAC}(1) &= (\$30{,}000)(A/P, 12\%, 1) + \$18{,}700 \\
&= (\$30{,}000)(1.12) + \$18{,}700 \\
&= \$52{,}300
\end{aligned}$$

Assume replacement after 2 years.

$$\begin{aligned}
\text{EUAC}(2) = {}& (\$30{,}000)(A/P, 12\%, 2) \\
& + \$18{,}700 + (\$1200)(A/G, 12\%, 2) \\
= {}& (\$30{,}000)(0.5917) + \$18{,}700 \\
& + (\$1200)(0.4717) \\
= {}& \$37{,}017
\end{aligned}$$

Assume replacement after 3 years.

$$\text{EUAC}(3) = (\$30{,}000)(A/P, 12\%, 3)$$
$$+ \$18{,}700 + (\$1200)(A/G, 12\%, 3)$$
$$= (\$30{,}000)(0.4163) + \$18{,}700$$
$$+ (\$1200)(0.9246)$$
$$= \$32{,}299$$

Similarly, calculate to obtain the numbers in the following table.

years in service	EUAC
1	$52,300
2	$37,017
3	$32,299
4	$30,207
5	$29,152
6	$28,602
7	$28,335
8	$28,234
9	$28,240
10	$28,312

Replace after 8 yr.

The answer is (C).

34. Since the head and horsepower data are already reflected in the hourly operating costs, there is no need to work with head and horsepower.

Let N = no. of hours operated each year.

$$\text{EUAC}(A) = (\$3600 + \$3050)(A/P, 12\%, 10)$$
$$- (\$200)(A/F, 12\%, 10) + 0.30N$$
$$= (\$6650)(0.1770) - (\$200)(0.0570) + 0.30N$$
$$= 1165.65 + 0.30N$$
$$\text{EUAC}(B) = (\$2800 + \$5010)(A/P, 12\%, 10)$$
$$+ (\$280)(A/F, 12\%, 10) + 0.10N$$
$$= (\$7810)(0.1770) - (\$280)(0.0570) + 0.10N$$
$$= 1366.41 + 0.10N$$

$$\text{EUAC}(A) = \text{EUAC}(B)$$
$$1165.65 + 0.30N = 1366.41 + 0.10N$$
$$N = 1003.8 \text{ hr} \quad (1000 \text{ hr})$$

The answer is (A).

35. (a) From Eq. 54.88,

$$\frac{T_2}{T_1} = 0.88 = 2^{-b}$$
$$\log 0.88 = -b \log 2$$
$$-0.0555 = -(0.3010)b$$
$$b = 0.1843$$

$$T_4 = (6)(4)^{-0.1843} = \boxed{4.65 \text{ wk} \quad (4.5 \text{ wk})}$$

The answer is (A).

(b) From Eq. 54.89,

$$T_{6-14} = \left(\frac{T_1}{1-b}\right)\left(\left(n_2 + \tfrac{1}{2}\right)^{1-b} - \left(n_1 - \tfrac{1}{2}\right)^{1-b}\right)$$
$$= \left(\frac{6}{1 - 0.1843}\right)$$
$$\times \left(\left(14 + \tfrac{1}{2}\right)^{1-0.1843} - \left(6 - \tfrac{1}{2}\right)^{1-0.1843}\right)$$
$$= \left(\frac{6}{0.8157}\right)(8.857 - 4.017)$$
$$= \boxed{35.6 \text{ wk} \quad (35 \text{ wk})}$$

The answer is (A).

36. (a) First check that both alternatives have an ROR greater than the MARR. Work in thousands of dollars. Evaluate alternative A.

$$P(A) = -\$120 + (\$15)(P/F, i\%, 5)$$
$$+ (\$57)(P/A, i\%, 5)(1 - 0.45)$$
$$+ \left(\frac{\$120 - \$15}{5}\right)(P/A, i\%, 5)(0.45)$$
$$= -\$120 + (\$15)(P/F, i\%, 5)$$
$$+ (\$40.8)(P/A, i\%, 5)$$

Try 15%.

$$P(A) = -\$120 + (\$15)(0.4972) + (\$40.8)(3.3522)$$
$$= \$24.23$$

Try 25%.

$$P(A) = -\$120 + (\$15)(0.3277) + (\$40.8)(2.6893)$$
$$= -\$5.36$$

Since $P(A)$ goes through 0,

$$(\text{ROR})_A > \text{MARR} = 15\%$$

Next, evaluate alternative B.

$$P(B) = -\$170 + (\$20)(P/F, i\%, 5)$$
$$+ (\$67)(P/A, i\%, 5)(1 - 0.45)$$
$$+ \left(\frac{\$170 - \$20}{5}\right)(P/A, i\%, 5)(0.45)$$
$$= -\$170 + (\$20)(P/F, i\%, 5)$$
$$+ (\$50.35)(P/A, i\%, 5)$$

Try 15%.

$$P(B) = -\$170 + (\$20)(0.4972) + (\$50.35)(3.3522)$$
$$= \$8.73$$

Since $P(B) > 0$ and will decrease as i increases,

$$(ROR)_B > 15\%$$

ROR > MARR for both alternatives.

The answer is (C).

(b) Do an incremental analysis to see if it is worthwhile to invest the extra $\$170 - \$120 = \$50$.

$$P(B - A) = -\$50 + (\$20 - \$15)(P/F, i\%, 5)$$
$$+ (\$50.35 - \$40.8)(P/A, i\%, 5)$$

Try 15%.

$$P(B - A) = -\$50 + (\$5)(0.4972)$$
$$+ (\$9.55)(3.3522)$$
$$= -\$15.50$$

Since $P(B - A) < 0$ and would become more negative as i increases, the ROR of the added investment is greater than 15%.

Alternative A is superior.

The answer is (A).

37. Use the year-end convention with the tax credit. The purchase is made at $t = 0$. However, the credit is received at $t = 1$ and must be multiplied by $(P/F, i\%, 1)$.

$$P = -\$300,000 + (0.0667)(\$300,000)(P/F, i\%, 1)$$
$$+ (\$90,000)(P/A, i\%, 5)(1 - 0.48)$$
$$+ \left(\frac{\$300,000 - \$50,000}{5}\right)(P/A, i\%, 5)(0.48)$$
$$+ (\$50,000)(P/F, i\%, 5)$$
$$= -\$300,000 + (\$20,010)(P/F, i\%, 1)$$
$$+ (\$46,800)(P/A, i\%, 5)$$
$$+ (\$24,000)(P/A, i\%, 5)$$
$$+ (\$50,000)(P/F, i\%, 5)$$

By trial and error,

i	P
10%	$17,625
15%	-$20,409
12%	$1456
13%	-$6134
12$\frac{1}{4}$%	-$472

i is between 12% and 12$\frac{1}{4}$%.

The answer is (C).

38. (a) The distributed profit is

$$\text{distributed profit} = (0.15)(\$2,500,000)$$
$$= \boxed{\$375,000 \quad (\$380,000)}$$

The answer is (D).

(b) Find the annual loan payment.

$$\text{payment} = (\$2,500,000)(A/P, 12\%, 25)$$
$$= (\$2,500,000)(0.1275)$$
$$= \$318,750$$

After paying all expenses and distributing the 15% profit, the remainder should be 0.

$$0 = \text{EUAC}$$
$$= \$20,000 + \$50,000 + \$200,000$$
$$+ \$375,000 + \$318,750 - \text{annual receipts}$$
$$- (\$500,000)(A/F, 15\%, 25)$$
$$= \$963,750 - \text{annual receipts}$$
$$- (\$500,000)(0.0047)$$

This calculation assumes $i = 15\%$, which equals the desired return. However, this assumption only affects the salvage calculation, and since the number is so small, the analysis is not sensitive to the assumption.

$$\text{annual receipts} = \$961,400$$

The average daily receipts are

$$\frac{\$961,400}{365} = \boxed{\$2634 \quad (\$2600)}$$

The answer is (D).

(c) Use the expected value approach. The average occupancy is

$$(0.40)(0.65) + (0.30)(0.70) + (0.20)(0.75)$$
$$+ (0.10)(0.80) = \boxed{0.70}$$

The answer is (B).

(d) The average number of rooms occupied each night is

$$(0.70)(120 \text{ rooms}) = 84 \text{ rooms}$$

The minimum required average daily rate per room is

$$\frac{\$2634}{84} = \boxed{\$31.36 \quad (\$31)}$$

The answer is (D).

39. (a) The annual savings are

$$\frac{\text{annual}}{\text{savings}} = \left(\frac{0.69 - 0.47}{1000}\right)(\$3,500,000) = \boxed{\$770}$$

The answer is (D).

(b) $\qquad P = -\$7500 + (\$770 - \$200 - \$100)$
$$\times (P/A, i\%, 25) = 0$$
$$(P/A, i\%, 25) = 15.957$$

Searching the tables and interpolating, the rate of return is

$$i \approx \boxed{3.75\% \quad (3.8\%)}$$

The answer is (A).

40. (a) Evaluate alternative A, working in millions of dollars.

$$P(A) = -(\$1.3)(1 - 0.48)(P/A, 15\%, 6)$$
$$= -(\$1.3)(0.52)(3.7845)$$
$$= \boxed{-\$2.56 \quad (-\$2.6) \quad [\text{millions}]}$$

The answer is (C).

(b) Use straight line depreciation to evaluate alternative B.

$$D_j = \frac{\$2}{6} = \$0.333$$
$$P(B) = -\$2 - (0.20)(\$1.3)(1 - 0.48)(P/A, 15\%, 6)$$
$$\quad - (\$0.15)(1 - 0.48)(P/A, 15\%, 6)$$
$$\quad + (\$0.333)(0.48)(P/A, 15\%, 6)$$
$$= -\$2 - (0.20)(\$1.3)(0.52)(3.7845)$$
$$\quad - (\$0.15)(0.52)(3.7845)$$
$$\quad + (\$0.333)(0.48)(3.7845)$$
$$= \boxed{-\$2.202 \quad (-\$2.2) \quad [\text{millions}]}$$

The answer is (D).

(c) Evaluate alternative C.

$$D_j = \frac{1.2}{3} = 0.4$$
$$P(C) = -(\$1.2)\big(1 + (P/F, 15\%, 3)\big)$$
$$\quad - (\$0.20)(\$1.3)(1 - 0.48)$$
$$\quad \times \big((P/F, 15\%, 1) + (P/F, 15\%, 4)\big)$$
$$\quad - (\$0.45)(\$1.3)(1 - 0.48)$$
$$\quad \times \big((P/F, 15\%, 2) + (P/F, 15\%, 5)\big)$$
$$\quad - (\$0.80)(\$1.3)(1 - 0.48)$$
$$\quad \times \big((P/F, 15\%, 3) + (P/F, 15\%, 6)\big)$$
$$\quad + (\$0.4)(\$0.48)(P/A, 15\%, 6)$$
$$= -(\$1.2)(1.6575)$$
$$\quad - (\$0.20)(\$1.3)(0.52)(0.8696 + 0.5718)$$
$$\quad - (\$0.45)(\$1.3)(0.52)(0.7561 + 0.4972)$$
$$\quad - (\$0.80)(\$1.3)(0.52)(0.6575 + 0.4323)$$
$$\quad + (\$0.4)(0.48)(3.7845)$$
$$= \boxed{-\$2.428 \quad (-\$2.4) \quad [\text{millions}]}$$

The answer is (C).

(d) From parts (a) through (c), $\boxed{\text{alternative B is superior.}}$

The answer is (B).

41. This is a replacement study. Since production capacity and efficiency are not a problem with the defender, the only question is when to bring in the challenger.

Since this is a before-tax problem, depreciation is not a factor, nor is book value.

The cost of keeping the defender one more year is

$$\text{EUAC(defender)} = \$200,000 + (0.15)(\$400,000)$$
$$= \$260,000$$

For the challenger,

$$\text{EUAC(challenger)}$$
$$= (\$800,000)(A/P, 15\%, 10) + \$40,000$$
$$+ (\$30,000)(A/G, 15\%, 10)$$
$$- (\$100,000)(A/F, 15\%, 10)$$
$$= (\$800,000)(0.1993) + \$40,000$$
$$+ (\$30,000)(3.3832)$$
$$- (\$100,000)(0.0493)$$
$$= \$296,006$$

Since the defender is cheaper, keep it. The same analysis next year will give identical answers. Therefore, keep the defender for the next 3 years, at which time the decision to buy the challenger will be automatic.

Having determined that it is less expensive to keep the defender than to maintain the challenger for 10 years, determine whether the challenger is less expensive if retired before 10 years.

If retired in 9 years,

$$\text{EUAC(challenger)} = (\$800,000)(A/P, 15\%, 9) + \$40,000$$
$$+ (\$30,000)(A/G, 15\%, 9)$$
$$- (\$150,000)(A/F, 15\%, 9)$$
$$= (\$800,000)(0.2096)$$
$$+ \$40,000 + (\$30,000)(3.0922)$$
$$- (\$150,000)(0.0596)$$
$$= \$291,506$$

Similar calculations yield the following results for all the retirement dates.

n	EUAC
10	$296,000
9	$291,506
8	$287,179
7	$283,214
6	$280,016
5	$278,419
4	$279,909
3	$288,013
2	$313,483
1	$360,000

Since none of these equivalent uniform annual costs are less than that of the defender, it is not economical to buy and keep the challenger for any length of time.

Keep the defender.

42. The annual demand for valves is

$$D = (500,000 \text{ products})\left(3 \frac{\text{valves}}{\text{product}}\right)$$
$$= 1,500,000 \text{ valves}$$

The cost of holding (storing) a valve for one year is

$$h = \left(0.4 \frac{1}{\text{yr}}\right)(\$6.50) = \$2.60 \text{ 1/yr}$$

The most economical quantity of valves to order is

$$Q^* = \sqrt{\frac{2aK}{h}} = \sqrt{\frac{(2)\left(1,500,000 \frac{1}{\text{yr}}\right)(\$49.50)}{\$2.60 \frac{1}{\text{yr}}}}$$
$$= \boxed{7557 \text{ valves} \quad (7600 \text{ valves})}$$

The answer is (D).

43. With just-in-time manufacturing, production is one-at-a-time, according to demand. When one is needed, one is made. In order to make the EOQ approach zero, the setup cost must approach zero.

The answer is (D).

55 Project Management, Budgeting, and Scheduling

PRACTICE PROBLEMS

1. The activities that constitute a project are listed. The project starts at $t = 0$. (a) Draw an activity-on-node critical path network. (b) Indicate the critical path. (c) What is the earliest finish? (d) What is the latest finish? (e) What is the slack along the critical path? (f) What is the float along the critical path?

activity	predecessors	successors	duration
start	–	A	0
A	start	B, C, D	7
B	A	G	6
C	A	E, F	5
D	A	G	2
E	C	H	13
F	C	H, I	4
G	D, B	I	18
H	E, F	finish	7
I	F, G	finish	5
finish	H, I	–	0

2. PERT activities constituting a short project are listed with their characteristic completion times. If the project starts on day 15, what is the probability that the project will be completed on or before day 42?

activity	predecessors	successors	t_{min}	$t_{most\,likely}$	t_{max}
start	–	A	0	0	0
A	start	B, D	1	2	5
B	A	C	7	9	20
C	B	D	5	12	18
D	A, C	finish	2	4	7
finish	D	–	0	0	0

3. Listed is a set of activities and sequence requirements to start a warehouse construction project. Prepare an activity-on-arc project diagram.

activity	letter code	code of immediate predecessor
move-in	A	
job layout	B	A
excavations	C	B
make-up forms	D	A
shop drawing, order rebar	E	A
erect forms	F	C, D
rough in plumbing	G	F
install rebar	H	E, F
pour, finish concrete	I	G, H

4. Listed is a set of activities, sequence requirements, and estimated activity times required for the renewal of a pipeline.

(a) Prepare an activity-on-arc PERT project diagram.

activity	letter code	code of immediate predecessor	activity time requirement (days)
assemble crew for job	A		10
use old line to build inventory	B	D	28
measure and sketch old line	C	A	2
develop materials list	D	C	1
erect scaffold	E	D	2
procure pipe	F	D	30
procure valves	G	D	45
deactivate old line	H	B	1
remove old line	I	E	6
prefabricate new pipe	J	F	5
place valves	K	H, I	1
place new pipe	L	J	6
weld pipe	M	I, L	2
connect valves	N	G, K	1
insulate	O	M, N	4
pressure test	P	G, K	1
remove scaffold	Q	O	1
clean up and turn over to operating crew	R	P, Q	1

(b) There is additional information in the form of optimistic, most likely, and pessimistic time estimates for the project. Compute the expected mean time and the variance, σ^2, for the activities. Which activities have the greatest uncertainty in their completion schedules?

activity code	optimistic time, t_{min}	most likely time, $t_{most\ likely}$	pessimistic time, t_{max}
A	8	10	12
B	26	26.5	36
C	1	2	3
D	0.5	1	1.5
E	1.5	1.63	4
F	28	28	40
G	40	42.5	60
H	1	1	1
I	4	6	8
J	4	4.5	8
K	0.5	0.9	2
L	5	5.25	10
M	1	2	3
N	0.5	1	1.5
O	3	3.75	6
P	1	1	1
Q	1	1	1
R	1	1	1

(c) Suppose that, due to penalties in the contract, each day the pipeline renewal project can be shortened is worth $100. Which of the following possibilities would you follow and why?

- Shorten $t_{most\ likely}$ of activity B by 4 days at a cost of $100.

- Shorten $t_{most\ likely}$ of activity G by 5 days at a cost of $50.

- Shorten $t_{most\ likely}$ of activity O by 2 days at a cost of $150.

- Shorten $t_{most\ likely}$ of activity O by 2 days by drawing resources from activity N, thereby lengthening its $t_{most\ likely}$ by 2 days.

5. Activities constituting a bridge construction project are listed. (a) Draw an activity-on-node network showing the critical path. (b) Compute ES, EF, LS, and LF for each of the activities. Assume a target time for completing the project that, for the bridge, is 3 days after the EF time.

activity		immediate predecessors	duration (days)
A	start		0
B		A	4
C		B	2
D		C	4
E		D	6
F		C	1
G		F	2
H		F	3
I		D	2
J		D, G	4
K		I, J, H	10
L		K	3
M		L	1
N		L	2
O		L	3
P		E	2
Q		P	1
R		C	1
S		O, T	2
T		M, N	3
U		T	1
V		Q, R	2
W		V	5
X	finish	S, U, W	0

6. Prepare an activity-on-arc diagram for the activities and sequence requirements given in Prob. 5.

7. Which of the following describes a preparation difference between construction drawings (i.e., prepared by architects and used by bidders, contractors, and inspectors) and design drawings (prepared by engineers and designers)?

(A) Construction drawings are prepared on vellum, while design drawings are prepared on paper.

(B) Construction drawings are copies of originals, while design drawings are originals.

(C) Construction drawings are typically larger in size than design drawings.

(D) Construction drawings typically have blue lines (i.e., are prepared by the diazo blueprint process) while design drawings have black lines (i.e., are prepared by CAD and xerographic processes).

8. Record drawings (also known as "as-builts") of a buried sewer line installation that will be submitted to the client should be certified by the

(A) contractor

(B) building official

(C) architect

(D) engineer

9. Which of the following details would NOT normally be found in the specifications for a large roadway contract requiring subgrade preparation?

(A) proof testing

(B) limits and boundaries

(C) scarification

(D) moisture conditioning

10. In a bid document, the engineer allowed bidders to install either cast-in-place or precast concrete culverts and headers, with the requirement that submitted bids include a description and separate price (bid) for the method used. This is an example of bid

(A) alternatives

(B) enhancement

(C) diversification

(D) programming

11. Which of the following items identified in a bid take-off would normally NOT be bid on the basis of weight?

I. shotcrete or gunite

II. water pipe

III. riprap

IV. reinforcing steel

V. clearing and grubbing

VI. grout

(A) II, IV, and VI

(B) I, II, III, and VI

(C) I, II, V, and VI

(D) I, II, III, IV, and V

12. An inexperienced estimator wishes to determine the approximate cost of compacting soil by means of a sheepsfoot roller making 4 passes with 12 in lifts. How should the estimator proceed?

(A) contact the state contractor's license board

(B) survey equipment rental costs, labor rates, and cost-of-living indices

(C) use published resources and cost data

(D) contact the local chapter of the Associated General Contractors of America

13. The *Building Cost Index* (BCI) and *Construction Cost Index* (CCI) published by the Engineering News-Record are based on which cost elements?

I. carpentry labor

II. brick laying labor

III. structural steel shapes

IV. 2 × 4 lumber

V. portland cement

VI. heavy machinery

(A) II, V, and VI

(B) II, IV, V, and VI

(C) I, II, III, IV, and V

(D) I, II, III, IV, V, and VI

14. The table shown contains year-by-year cost indices. A job was bid in 1998. It was built over a period of three years, 2007, 2008, and 2009. What is the average escalation factor most nearly?

year	cost index
1997	1539
1998	1601
1999	1665
2000	1731
2001	1800
2002	1873
2003	1948
2004	2026
2005	2107
2006	2191
2007	2279
2008	2370
2009	2465

(A) 1.4

(B) 1.5

(C) 1.6

(D) 1.7

Systems, Mgmt., Professional

15. A contractor uses unbalanced bidding to bid a four-phase highway repaving project. The unbalanced bid results in phase billings of 38%, 12%, 42%, and 8%, respectively. The billings are roughly evenly spaced along the construction period. The increase in profit seen by the contractor from this practice is most nearly

(A) 0

(B) 0.7%

(C) 1.1%

(D) 1.6%

16. A construction firm has an annual fixed overhead (burden) cost of $500,000. During the year, the actual overhead was $550,000. The firm's gross billings during the year were $6,000,000. Each of the jobs billed had been estimated with a 10% allowance for overhead and profit. Assume the jobs were accurately bid and accurately managed. The firm's actual profit was most nearly

(A) $45,000

(B) $50,000

(C) $100,000

(D) $105,000

17. A project manager finds a loose sheet of binder paper at a construction job site. The paper contains various numbers, some with abbreviations of BF, CF, SF, WLD, and WHD. The manager should probably make an effort to give the sheet of paper to the

(A) engineer/architect

(B) contractor

(C) utility installer

(D) building inspector

18. A dump-hauler has a purchase price of $80,000. Freight for delivery is $1500. Tires are an additional $15,000. The hauler is expected to operate 1100 hours per year, and its useful life is estimated at 10 years. After 10 years, the salvage value is expected to be minimal. Interest, insurance, storage, fuel, oil, lubricants, filters, repairs, and sales taxes are estimated at $21,000 per year. Tires have an estimated life of 3500 hr. What is the before-tax estimated hourly cost of ownership and operation, excluding operator labor cost?

(A) $25/hr

(B) $31/hr

(C) $35/hr

(D) $40/hr

19. Which of the following financial ratios is stated INCORRECTLY?

(A) gross profit margin ratio = gross profit ÷ sales

(B) operating profit margin ratio = earnings after interest and taxes ÷ sales

(C) net profit margin ratio = net income ÷ sales

(D) return on assets = net income ÷ total assets

20. Given the following table of precedence relationships, how many nodes would be needed to form an activity-on-branch critical path network?

activity	description	predecessors
A	site clearing	–
B	removal of trees	–
C	general excavation	A
D	grading general area	A
E	excavation for utility trenches	B, C
F	placing formwork and reinforcement for concrete	B, C
G	installing sewer lines	D, E
H	installing other utilities	D, E
I	pouring concrete	F, G

(A) 6

(B) 7

(C) 9

(D) 11

21. Given the following table of precedence relationships, how many nodes would be needed to form an activity-on-node critical path network?

activity	description	predecessors
A	site clearing	–
B	removal of trees	–
C	general excavation	A
D	grading general area	A
E	excavation for utility trenches	B, C
F	placing formwork and reinforcement for concrete	B, C
G	installing sewer lines	D, E
H	installing other utilities	D, E
I	pouring concrete	F, G

(A) 6

(B) 7

(C) 9

(D) 11

22. A project is described by the following precedence table. The project manager wants to decrease the normal project time by 3 days. Most nearly, how much extra will it cost to reduce the project completion time by 3 days?

task	predecessors	normal time (days)	crash time	normal cost (per day)	crash cost (per day)
A	–	4	2	100	200
B	A	5	4	80	90
C	A	2	1	110	130
D	B	3	1	50	90
E	C	5	3	70	150
F	E	2	1	90	120
G	D, F	3	2	60	120

(A) 90

(B) 120

(C) 130

(D) 160

23. An engineering design firm has estimated the following billing scenarios for the following year.

billed amount	probability
$0–$500,000	0.60
$500,001–$1,000,000	0.25
$1,000,001–$1,500,000	0.10
$1,500,001–$2,000,000	0.05

What is the expected billing amount for the upcoming year?

(A) $550,000

(B) $625,000

(C) $690,000

(D) $740,000

24. An engineering design firm has a year-long lease on a bucket crane used to inspect construction details at second- and third-floor levels. Into which general ledger account should the cost of the lease be accumulated?

(A) rents

(B) amortization for leasehold

(C) depreciation

(D) overhead

25. A design firm contracts with a client using a standard 6% cost-plus arrangement to prepare a design for a footbridge and to manage the construction project. The owner has budgeted $750,000 for the design and construction. The design firm estimates that the bridge will cost the client $600,000. The bid-winning contractor estimates that the bridge materials and labor will cost $450,000 and bids the job at $550,000. The actual cost to the contractor to build the bridge as originally designed is $470,000. Along the way, the owner agrees to change orders, increasing the project billing by $40,000. Most nearly, what is the design firm's total billing to the owner?

(A) $570,000

(B) $610,000

(C) $630,000

(D) $650,000

26. What type of Gantt chart is illustrated?

	Jan	Feb	Mar	Apr	May	Jun	Jul	Aug	Sep
task 1	▬	▬							
task 2			▬	▬					
task 3			▬						
task 4					▬	▬	▬		
task 5				▬	▬				
task 6								▬	▬

(A) milestones Gantt

(B) Gantt with dependencies

(C) baseline Gantt

(D) timeline Gantt

27. A subcontractor's contract requires the general contractor to provide an area for lay-down. The lay-down area will be best used for

(A) parking heavy equipment and subcontractor employee vehicles

(B) receiving delivered materials to be used by the subcontractor

(C) space for subcontractor personnel during rest periods

(D) pre-installation assembly of overhead fire sprinkler pipes, electrical conduit, and standpipes

Systems, Mgmt., Professional

28. The main parties to a construction project contract include the

I.   client

II. engineer-architect

III. OSHA

IV. prime contractor

V.   subcontractors

VI. local building official

    (A) I, II, and IV

    (B) I, II, IV, and V

    (C) I, II, IV, V, and VI

    (D) I, II, III, IV, V, and VI

29. The term "design-build" means the

    (A) design firm designs the project and the client builds it

    (B) design firm both designs and builds the project

    (C) client designs the project and the contractor builds it

    (D) contractor designs the project and the subcontractors build it

30. The term "fast-track" refers to

    (A) constructing different areas of the project site in parallel

    (B) a project delivery method where the sequencing of construction activities allows some portions of the project to begin before the design is completed on other portions of the project

    (C) working multiple shifts, overtime, weekends, and holidays to decrease time to completion

    (D) providing financial incentives to the contractor and subcontractors to beat the published schedule

31. The term "turnkey" refers to

    (A) a contractor providing full service to the owner, including such things as obtaining financing, planning and costing, cost management, shell construction, and interior finishing

    (B) an OSHA inspection in which all areas of the jobsite are inspected

    (C) a responsibility-sharing partnership between an engineer, contractor, and architect

    (D) a financial lender's right to resell the construction loan to third party investors

32. Which of the following duties would normally NOT be a responsibility of the estimating department within a general contractor's organization?

    (A) obtaining bid documents

    (B) securing subcontractor/material quotations

    (C) project cost accounting

    (D) delivering competitive or negotiated proposals

33. Under normal circumstances, an architect-engineer will have all of the following authorities EXCEPT to

    (A) interpret contract questions from the prime contractor

    (B) judge a subcontractor's performance and condemn defective work

    (C) stop field operations

    (D) issue certificates of occupancy

34. What is the term used to describe a complete listing of all the materials and items of work that will be required for a project?

    (A) materials resource plan

    (B) specifications detail list

    (C) quantity survey

    (D) resource foundation accounting

35. Which two categories of projects have different bidding procedures, rules, and regulations?

    (A) commercial and residential

    (B) public and private

    (C) transportation (bridge and highway) and nontransportation

    (D) critical (school, hospital, emergency services, etc.) and noncritical

36. A contractor's decision to bid is affected by the bidding climate, a term that does NOT apply to

    (A) the contractor's bonding capacity

    (B) the owner's reputation and financial soundness

    (C) the completion date

    (D) construction loan interest rates

Systems, Mgmt., Professional

37. Which of the following is the most reliable source of productivity data for use in bidding by a contractor?

(A) U.S. Department of Labor

(B) professional construction/contractor associations

(C) compilations from RSMeans and Engineering News-Record

(D) internal historical data

38. Following the estimating process, a contractor normally adds a markup or margin. This is an allowance for

(A) profit

(B) general overhead

(C) contingency

(D) all of the above

39. All of the following are examples of indirect costs EXCEPT

(A) payroll taxes for construction labor

(B) equipment rentals for the project

(C) corporate income taxes

(D) worker's compensation insurance

40. Which statement about large construction equipment depreciation is true?

(A) Tires are normally depreciated separately from the chassis.

(B) The asset life selected must be 3, 5, or 15 years depending on the asset class.

(C) The salvage value must be included in the depreciation calculation.

(D) Repairs to the equipment increases the depreciation basis (i.e., is added to the depreciated cost).

41. What is the primary implication of the phrase, "50 minute hour"?

(A) Employees must be permitted to take their breaks and lunch.

(B) Heavy equipment is not 100% productive.

(C) Vacation and illness decrease employee productivity.

(D) Contractors must train their employees in safety procedures.

42. Contingency is something that should be added to a(n)

I. owner's construction loan amount

II. design engineering firm's fixed-fee design contract

III. prime's bid to an owner

IV. subcontractor's bid to the general contractor

V. inspector's schedule of inspection time

VI. lessee's move-in schedule

(A) I, III, and IV

(B) I, II, III, and IV

(C) II, III, IV, and V

(D) I, II, III, IV, V, and VI

43. When the bids are all opened, the owner will normally award the contract to the lowest

(A) available bidder

(B) qualified bidder

(C) responsible bidder

(D) bonded bidder

44. When a contractor is awarded a job on a cost-plus-fee basis, with the fee being 6% of the reimbursable expenses, what percentage of the contractor's total expenses will typically be reimbursed by the owner?

(A) 100%

(B) less than 100%

(C) more than 100%

(D) any of the above

45. The amount of a phase payment to the contractor withheld by the owner is known as

(A) retainage

(B) recovery

(C) incentive deposit

(D) overhead withholding

Systems, Mgmt.,
Professional

46. When a project is completed late, the amount of money specified by the contract to be paid to the owner by the contractor (or, withheld by the owner from the contractor's final payment) is known as

(A) responsibility penalties

(B) essence time payments

(C) completion recovery charges

(D) liquidated damages

47. A surety is

(A) an agreement that assigns liability for another party's debt, default, or failure in duty

(B) an agreement that outlines the conditions when liability for the debt, default, or failure in duty of another is assumed by the guarantor

(C) a party that assumes liability for the debt, default, or failure in duty of another party

(D) a deposit of money that is used to cover liability for the debt, default, or failure in duty of another party

48. Which of the following is NOT synonymous with surety bond?

(A) contract surety bond

(B) contract bond

(C) construction contract bond

(D) construction insurance

49. Once default has been established triggering a surety bond, the surety has an obligation to

(A) pay the face amount of the bond

(B) reimburse the owner's expenses in financing the remainder of the project

(C) pay the owner the balance of the construction contract amount not yet paid out

(D) do all steps necessary to complete the project

50. Which of the following is NOT a common type of bond on a construction project?

(A) bid bond

(B) performance bond

(C) payment bond

(D) schedule bond

51. A mechanic's lien is a

(A) claim on the recorded title of property

(B) public filing notice of a supplier's right to receive payment for material

(C) public filing that notifies the owner of the supplier's participation in the job

(D) public filing that notifies the owner of a default of the general contractor

52. Which of the following statements regarding the Miller Act is FALSE?

(A) The Miller Act was enacted in 1890.

(B) The Miller Act requires a performance and payment bond on all federal projects greater than $25,000.

(C) The Miller Act requires the bond to be 100% of the contract amount.

(D) The Miller Act protects first and second tier subcontractors only.

53. When the contract documents require a "Best Rating" of the surety, this means the

(A) contract is rated A-1

(B) surety must be rated A++

(C) surety must be rated by a company named "Best"

(D) surety must have the ability to bond 125% of the contract amount

54. Which of the following surety ratings means that the surety is in liquidation?

(A) C−

(B) E

(C) F

(D) S

55. Which of the following is NOT a characteristic of accrual accounting?

(A) income is counted when earned, not when received

(B) expenses are counted when paid, not when incurred

(C) income and expenses can be accumulated per project

(D) requires more detailed records and complex processes

56. Which one of the following financial equations is correct?

(A) assets = liabilities + net worth

(B) assets = cash + net worth

(C) net worth = net worth + liabilities

(D) assets = cash + accumulated depreciation

57. A contractor's liquidity ratio in three consecutive years is 1.01, 1.05, and 1.09, respectively. It is probably true that the contractor is

(A) becoming more profitable

(B) increasingly drawing on his/her line of credit

(C) accumulating cash

(D) receiving payments from clients increasingly earlier

58. Which of the following descriptions regarding materials purchasing and delivery to a job site is NOT correct?

(A) FOB (free on board) indicates that the seller puts the materials onboard the freight carrier free of expense to the buyer, with freight paid to the FOB point designated. Title to the material passes to the buyer when the seller consigns the material to the freight company.

(B) CIF (cost, insurance, freight) indicates that the purchase price includes the cost of goods, customary insurance, and freight to the buyer's destination. Title passes when the seller delivers the merchandise to the carrier and forwards to the buyer the bill of lading, insurance policy, and receipt showing payment of freight.

(C) COD (collect on delivery) indicates that the title passes to the buyer, if the buyer is to pay the transportation, at the time the goods are received by the carrier. However, the seller reserves the right to receive payment before surrender of possessions to the buyer.

(D) None of the above.

SOLUTIONS

1. (a) The activity-on-node critical path network is as follows.

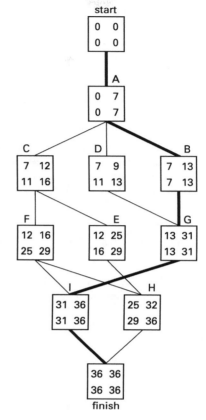

ES (earliest start) Rule: The earliest start time for an activity leaving a particular node is equal to the largest of the earliest finish times for all activities entering the node.

LF (latest finish) Rule: The latest finish time for an activity entering a particular node is equal to the smallest of the latest start times for all activities leaving the node.

The activity is critical if the earliest start equals the latest start.

(b) The critical path is $\boxed{\text{A-B-G-I.}}$

(c) The earliest finish is $\boxed{36.}$

(d) The latest finish is $\boxed{36.}$

(e) The slack along the critical path is $\boxed{0.}$

(f) The float along the critical path is $\boxed{0.}$

2. From Eq. 55.2,

$$\mu = \tfrac{1}{6}\left(t_{\min} + 4t_{\text{most likely}} + t_{\max}\right)$$

Use Eq. 55.3 to find the variance, which is the square of the standard deviation, σ.

$$\sigma^2 = \left(\tfrac{1}{6}(t_{\max} - t_{\min})\right)^2$$

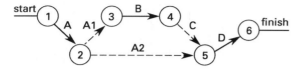

The critical path is A-A1-B-C-D.

The following probability calculations assume that all activities are independent. Use the following theorems for the sum of independent random variables and use the normal distribution for T (project time).

$$\mu_{\text{total}} = t_A + t_B + t_C + t_D$$
$$\sigma^2_{\text{total}} = \sigma^2_A + \sigma^2_B + \sigma^2_C + \sigma^2_D$$

The variance is 10.52778, and the standard deviation is 3.244654. (See the following table.)

$$\mu_{\text{total}} = 43.83333$$
$$\sigma^2_{\text{total}} = 10.52778$$
$$\sigma_{\text{total}} = 3.244654$$

$$z = \frac{t - \mu_{\text{total}}}{\sigma}$$
$$= \left|\frac{42 - 43.83333}{3.244654}\right|$$
$$= 0.565$$

From the normal table, the probability of finishing for $T \le 42$ is 0.286037 $\boxed{(28.6\%).}$

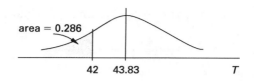

3. Draw an activity-on-arc diagram for the project.

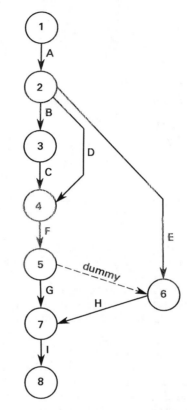

(A dummy activity cannot be replaced with any other activity letter without duplicating that activity.)

Table for Sol. 2

no.	name	activity exp. time	variance	earliest start	latest start	earliest finish	latest finish	slack LS − ES
1	A	2.33333	0.44444	15	15	17.3333	17.3333	0
2	A1	0	0	17.3333	17.3333	17.3333	17.3333	0
3	A2	0	0	17.3333	39.6667	17.3333	39.6667	22.3333
4	B	10.5000	4.69444	17.3333	17.3333	27.8333	27.8333	0
5	C	11.8333	4.69444	27.8333	27.8333	39.6667	39.6667	0
6	D	4.16667	0.69444	39.6667	39.6667	43.8333	43.8333	0

expected completion time = 43.83333 (43.83)

4. (a) Draw an activity-on-arc PERT diagram for the project. (See *Illustration for Sol. 4(a)*.)

(b) A table is the easiest way to determine which activities have the greatest uncertainty in their completion schedules.

activity	variance	mean time
A	0.44	10
B	2.77	28
C	0.11	2
D	0.027	1
E	0.17	2
F	4.0	30
G	11.11	45
H	0	1
I	0.44	6
J	0.44	5
K	0.0625	1
L	0.69	6
M	0.11	2
N	0.027	1
O	0.25	4
P	0	1
Q	0	1
R	0	1

Activities B, F, and G have the three largest variances, so these activities have the greatest uncertainties.

(c)

- Activity B is not on the critical path, so shortening it will not shorten the project.

- Activity G is on the critical path, so it should be shortened. Shortening it 5 days may change the critical path, however.

  path D-G-K: $1 + 45 = 6$ days

  path D-E-I-K: $1 + 2 + 6 = 9$ days

  path D-B-H-I-K: $1 + 28 + 1 + 6 = 36$ days

  Since the shortened path D-G-K has a length of 46 days − 5 days = 41 days and is still the longest path from D to K, it should be shortened.

- Activity O is on the critical path. The cost of the crash schedule is $150. Savings would be (2)($100) = $200. This activity should be shortened.

- The current path length is

  $$N\text{-}O\text{-}Q\text{-}R: 1 + 4 + 1 = 6 \text{ days}$$

  The new paths would be

  $$N\text{-}P\text{-}Q\text{-}R: 3 + 1 + 1 = 5 \text{ days}$$

  $$N\text{-}O\text{-}Q\text{-}R: 3 + 2 + 1 = 6 \text{ days}$$

  $$N\text{-}O\text{-}R: (3 + 2) + (4 - 2) = 7 \text{ days}$$

  N-O-R would become part of the critical path and would be longer than the original critical path. Therefore, it is not acceptable.

Illustration for Sol. 4(a)

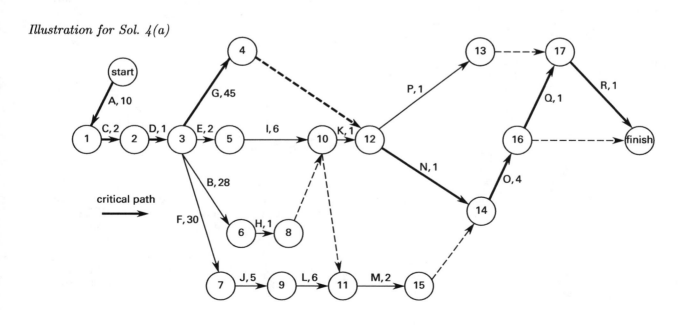

Systems, Mgmt., Professional

5. Draw the activity-on-node CPM network showing the critical path.

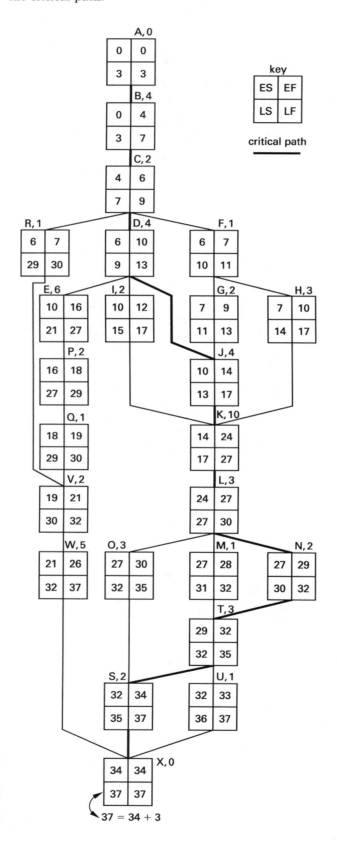

key

ES	EF
LS	LF

critical path

6. Draw a PERT diagram. (See *Illustration for Sol. 6.*)

7. The industry standard size for drawings prepared by architects is 36 in × 48 in (914 mm × 1219 mm), but this size sheet is rarely used in engineering design. The engineering design industry primarily uses 24 in × 36 in (610 mm × 914 mm), although 22 in × 34 in (559 × 864 mm) paper is also used. Half-size drawings are frequently used during reviews and submittals because they are less bulky (i.e., fit inside briefcases and file cabinets) and can be easily copied.

The answer is (C).

8. The engineer of record certifies engineering design. This is commonly referred to as "stamping" or "sealing" and includes dating and signing the design documents. Architects do not certify engineering work.

The answer is (D).

9. Construction documents, not the specifications, show the limits (in plan view) and boundaries of the subgrade preparation area. Other details of the preparation are found in the specifications.

The answer is (B).

10. *Bid alternatives* are used to give bidders the option of tailoring the methodology and materials to their expertise, experience, and capabilities.

The answer is (A).

11. Shotcrete (gunite) and clearing and grubbing are measured per square foot or square yard. Water pipe is measured by unit length or pieces. Grout is measured by volume. All of the other items are bid by weight.

The answer is (C).

12. Construction costs data are published regularly by several sources, including Engineering News-Record and RS Means. Other sources and some state transportation departments make similar data available.

The answer is (C).

13. The Engineering News-Record's (ENR's) *Building Cost Index* is developed from the costs of 68.38 hours of skilled labor at the 20-city average of bricklayers, carpenters, and structural ironworkers rates, plus 25 cwt of standard structural steel shapes at the mill price prior to 1996; as well as the fabricated 20-city price from 1996, plus 1.128 tons of portland cement at the 20-city price, plus 1088 board-ft of 2 × 4 lumber at the 20-city price.

Illustration for Sol. 6

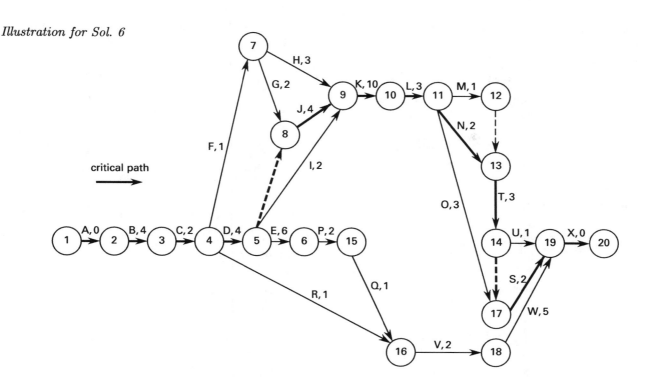

ENR's *Construction Cost Index* is developed from the costs of 200 hours of common labor at the 20-city average of common labor rates, plus 25 cwt of standard structural steel shapes at the mill price prior to 1996; as well as the fabricated 20-city price from 1996, plus 1.128 tons of portland cement at the 20-city price, plus 1088 board-ft of 2 × 4 lumber at the 20-city price. Neither index includes heavy machinery.

The answer is (C).

14. The average cost index over the construction period is

$$\frac{2279 + 2370 + 2465}{3} = 2371.3$$

Divide by the cost index from 1998 to determine the average escalation factor.

$$\frac{2371.3}{1601} = \boxed{1.48 \quad (1.5)}$$

The answer is (B).

15. Unbalanced bidding is used by contractors to improve cash flow by distorting cost in certain periods. While the overall bid remains competitive by virtue of the total cost, the payments are structured to suit the contractor's needs. Unbalanced bidding is not illegal, and it does not increase the contractor's gross profit, but it may be unethical as the contractor must make

untruthful cost statements in certain periods. The opportunity value of having the money earlier rather than later, as well as the time value of money, cannot be determined without knowing the contractor's marginal rate of return and the actual timing of the payments. In this problem, such information is clearly not available. Therefore, the question is answered from a general viewpoint—no increase in profit.

The answer is (A).

16. The jobs had been billed to create an allowance for profit and overhead of

$$(0.10)(\$6,000,000) = \$600,000$$

The actual profit is the allowance less the actual overhead cost.

$$\$600,000 - \$550,000 = \boxed{\$50,000}$$

The answer is (B).

17. The abbreviations most likely mean board feet (BF), cubic feet (CF), square feet (SF), width × length × depth (WLD), and width × height × depth (WHD). These would appear together on a materials or cost estimation sheet, probably for ordering concrete and formwork lumber. Such a sheet would have been prepared by the contractor.

The answer is (B).

Systems, Mgmt., Professional

18. The delivered price (less tires) is

$$C = \$80,000 + \$1500 = \$81,500$$

The hourly cost of ownership is

$$H_1 = \frac{\$81,500}{(10 \text{ yr})\left(1100 \dfrac{\text{hr}}{\text{yr}}\right)} = 7.41 \text{ \$/hr}$$

The hour cost of tire usage is

$$H_2 = \frac{\$15,000}{3500 \text{ hr}} = 4.29 \text{ \$/hr}$$

The hourly cost of operation is

$$H_3 = \frac{\$21,000}{1100 \text{ hr}} = 19.09 \text{ \$/hr}$$

The total hourly cost of ownership and operation is

$$
\begin{aligned}
H &= H_1 + H_2 + H_3 \\
&= 7.41 \frac{\$}{\text{hr}} + 4.29 \frac{\$}{\text{hr}} + 19.09 \frac{\$}{\text{hr}} \\
&= \boxed{30.79 \text{ \$/hr} \quad (\$31/\text{hr})}
\end{aligned}
$$

The answer is (B).

19. The operating profit margin is calculated before interest and taxes, not after.

The answer is (B).

20. The activity-on-branch representation of this project is

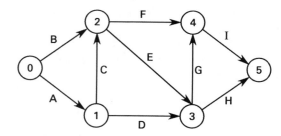

There are 6 nodes.

The answer is (A).

21. The activity-on-node representation of this project is

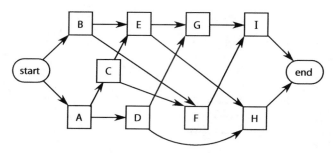

Counting the starting and ending dummy activities, there are 11 nodes.

The answer is (D).

22. Draw the activity-on-node network.

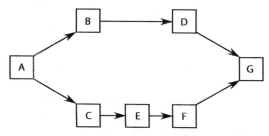

Path ABDG has a normal length of $4 + 5 + 3 + 3 = 15$.

Path ACEFG has a normal length of $4 + 2 + 5 + 2 + 3 = 16$.

Path ACEFG is the longest, and it is the critical path.

Determine the incremental costs of crashing.

activity	incremental cost (per day)
A	$200 - 100 = 100$
B	$90 - 80 = 10$
C	$130 - 110 = 20$
D	$90 - 50 = 40$
E	$150 - 70 = 80$
F	$120 - 90 = 30$
G	$120 - 60 = 60$

In order to shorten the project, an activity along the critical path must be crashed. Activity C has the lowest incremental cost of activities A, C, E, F, and G. For an incremental cost of 20, the duration can be reduced by 1.

At this point, both paths have the same length, and reducing one without reducing the other will not decrease the project duration. So, find the lowest incremental cost activity in each path. In path ABDG, activity B can reduce the path length by 1 at a cost of 10. In path ACEFG, with activity C already having been crashed, activity F can reduce the path length by 1 at a cost of 30.

Now, both paths have been reduced by 1 and have the same length. Each path must be reduced by one additional day. Crashing activity G at a cost of 60 will shorten both paths by 1.

The total cost of reducing the project duration by 3 days is

$$20 + 10 + 30 + 60 = \boxed{120}$$

The answer is (B).

23. Take the midpoint of each range.

billed amount	midpoint	probability
$0–$500,000	$250,000	0.60
$500,001–$1,000,000	$750,000	0.25
$1,000,001–$1,500,000	$1,250,000	0.10
$1,500,001–$2,000,000	$1,750,000	0.05

The expected billing amount is

$$
\begin{aligned}
E &= (\$250{,}000)(0.60) + (\$750{,}000)(0.25) \\
&\quad + (\$1{,}250{,}000)(0.10) + (\$1{,}750{,}000)(0.05) \\
&= \boxed{\$550{,}000}
\end{aligned}
$$

The answer is (A).

24. In a general ledger account, "rents" almost always refers to the expense of rental (leased) office space. Leasehold improvements are semi-permanent improvements made to rented (leased) office space, and their costs are amortized over time. Only an owner, not a lessee, can depreciate an asset. The crane is used for all projects, and like other generic costs, its expense should be accumulated into overhead. The design firm may also decide to allocate portions of the crane lease expense to the overhead accounts of each of the projects.

The answer is (D).

25. The total amount billed by the contractor through the design firm to the client is $550,000 + $40,000 = $590,000. This amount is the "cost" to the client, to which the design firm adds 6% for its efforts.

$$(1.06)(\$590{,}000) = \boxed{\$625{,}400 \quad (\$630{,}000)}$$

The answer is (C).

26. This is a basic timeline Gantt. No milestones, dependencies, or baselines are included.

The answer is (D).

27. Although a lay-down area could be used for almost anything, it is usually a storage area. Alternative B is the most general and, therefore, the best answer.

The answer is (B).

28. The client, engineer-architect, and the primary contractor are the main parties to a construction project contract. OSHA and the local building official have implicit authority over a project, but they are not parties to the contract. Subcontractors have contracts with the general contractor, but they are not usually parties to the project contract.

The answer is (A).

29. *Design-build* is a process where the client interacts only with a single entity, the "design-builder." While the design-builder is usually the general contractor, it can also be the design firm or a partnership consisting of the design firm and contractor.

The answer is (B).

30. *Fast-track* means that different phases of the design and construction overlap.

The answer is (B).

31. A *turnkey* is a contractor who provides full-service to the owner.

The answer is (A).

32. The estimation process does not include tracking costs or recording actual expenses.

The answer is (C).

33. The architect-engineer normally has the contractual authority to interpret the construction documents, judge quality, and stop work under unsafe and other conditions. Only the local building official can issue a certificate of occupancy.

The answer is (D).

34. The term *quantity survey* is synonymous with the terms *take-off survey* and *take-off report*.

The answer is (C).

Systems, Mgmt.,
Professional

35. Private projects have bidding procedures normally established by the owner and AE. Public projects follow various procurement statutes developed by federal, state, county, and municipal governments.

The answer is (B).

36. The *bidding climate* is affected by the following factors: bonding capacity considerations, location of project, severity of contractual terms (contractor responsibilities and liabilities), identity of the owner and owner's financial status, identity of the architect/ engineer, nature and size of project as it relates to company experience and equipment, labor conditions and availability, and completion date. Construction loan interest rates will affect an owner's decision to build the project, but the contractor does not generally care how the owner finances the project as long as it is not through nonpayment.

The answer is (D).

37. Productivity information is available from many sources; however, public and commercially available information is less reliable to bidding than internal historical data.

The answer is (D).

38. A contractor usually adds a markup or margin to the profit, general overhead, and contingency following the estimating process.

The answer is (D).

39. Indirect costs include such items as payroll taxes, insurance, employee fringe benefits (e.g., health, vacation, employee insurance, and 401K), employer contributions to social security, unemployment insurance, workers compensation insurance, and public liability and property damage insurance. Income taxes are incurred on profit, after all income and expenses (direct and indirect) have been accounted for.

The answer is (C).

40. Tires are a substantial part of the original cost of the equipment, but they do not last as long as the equipment. In fact, tires for some very large items of equipment (e.g., quarry trucks) have short life cycles. Therefore, tires are normally depreciated separately from the chassis.

The answer is (A).

41. On an average, heavy equipment is productive only approximately 50 minutes per hour. The remaining time is consumed by nonproductive activities such as greasing, fueling, minor repairs, maintenance, shifting position, operator delays, and waiting in line.

The answer is (B).

42. All of the items are subject to unforeseen delays; therefore, contingency should be applied to each.

The answer is (D).

43. In open competitive bidding by public and private owners, the bid will be awarded to the lowest responsible bidder.

The *lowest responsible bidder* is the lowest bidder whose offer best responds in quality, fitness, and capacity to fulfill the particular requirements of the proposed project, and who can fulfill these requirements with the qualifications needed to complete the job in accordance with the terms of the contract.

The answer is (C).

44. In a cost-plus-fee contract, 100% of a contractor's reimbursable expenses are paid to the contractor. However, it is very unlikely that all expenses incurred by the contractor will be reimbursable under the contract. For example, the contractor may incur an expense by flying first class to a distant meeting while the contract only provides for coach or economy seating. Therefore, only a portion of the expenses will be reimbursable.

The answer is (B).

45. *Retainage* is a percentage, usually 10%, of the payment to the contractor that is withheld from each payment. The owner may reduce retainage below 10% after substantial completion. The retainage is paid by owner to contractor after final completion. Retainage serves as an incentive for a contractor to complete all work per contract.

The answer is (A).

46. *Liquidated damages* are specified in the contract as a penalty for late completion. The amount is specified (e.g., per day of lateness) and is agreed to by the contractor at the time the agreement is entered into. The amount is established in advance in lieu of a determination of actual damages suffered by the owner. Liquidated damages must be a reasonable penalty based on a forecast of actual damages the owner would suffer if the contractor is in breach of contract by finishing late.

The answer is (D).

47. A *surety* is a party that assumes liability for the debt, default, or failure in duty of another.

The answer is (C).

48. A *surety bond* is not an insurance policy. A surety bond is a contract that outlines the conditions of the assumed liability by a surety.

The answer is (D).

49. If the contractor fails to fulfill his/her contractual obligations, the surety must assume the obligations of the contractor and see that the contract is completed, including paying all costs up to the face amount of the bond. This requires the surety to do more than merely provide money to get the project completed—the surety is responsible for finishing the contract. The contract can be completed in one of three ways: (1) Complete the contract itself through a completion contractor. (2) Select a new contractor to contract directly with the owner. (3) Allow the owner to complete the work with the surety paying the costs.

The answer is (D).

50. The contractor's obligation to maintain a reasonable schedule is contained in the contract agreement, and a separate bond is seldom, if ever, issued for that purpose. When the schedule is covered by a bond, it will be part of the performance bond. Other types of legitimate, but uncommon, bonds include bonds to release retainage, to discharge liens or claims (commonly referred to as bonds for "bonding over a lien"), and to indemnify owners against liens, license bonds, self-insurers' workers' compensation bonds, and union wage bonds.

The answer is (D).

51. A *mechanic's lien* is a right created by law to secure payment for work performed and material furnished in the improvement of real property. It is secured by the real property itself and will be recorded (e.g., with the county recorder) against the title or deed for a property.

The answer is (A).

52. The Miller Act was enacted in 1935. One of its provisions is that the subcontractor cannot claim on the payment bond until 90 days after the last day labor was performed on job.

The answer is (A).

53. When the contract documents require a "Best Rating" of the surety, the surety must be rated by the A. M. Best company. *A. M. Best Insurance Reports* contains financial ratings for insurance and surety companies.

The answer is (C).

54. The standard A. M. Best ratings for secure sureties are

- A++ and A+ (superior)
- A and A− (excellent)
- B++ and B+ (very good)

The standard A. M. Best ratings for vulnerable sureties are

- B and B− (fair)
- C++ and C+ (marginal)
- C and C− (weak)
- D (poor)
- E (under regulatory supervision)
- F (in liquidation)
- S (rating suspended)

The answer is (C).

55. Accrual accounting counts income and expenses when incurred, not when paid.

The answer is (B).

56. The net worth of a company is the difference between assets and liabilities.

$$\text{net worth} = \text{assets} - \text{liabilities}$$

This is algebraically the same as

$$\text{assets} = \text{liabilities} + \text{net worth}$$

The answer is (A).

57. Liquidity is a contractor's ability to meet short-term debt. One of the components of liquidity is cash, and another is assets that can quickly be converted into cash. Given the increasing liquidity ratio, it is probably true that the contractor is accumulating cash.

The answer is (C).

Systems, Mgmt., Professional

58. FOB indicates that the seller puts the goods on board the common carrier free of expense to the buyer, with freight paid to the FOB point designated (e.g., "FOB jobsite" or "FOB storage yard"). Title goes to the buyer when the carrier delivers the goods at the place indicated, not when the seller consigns the material to the freight company.

The answer is (A).

56 Professional Services, Contracts, and Engineering Law

PRACTICE PROBLEMS

1. List the different forms of company ownership. What are the advantages and disadvantages of each?

2. Define the requirements for a contract to be enforceable.

3. What standard features should a written contract include?

4. Describe the ways a consulting fee can be structured.

5. What is a retainer fee?

6. Which of the following organizations is NOT a contributor to the standard design and construction contract documents developed by the Engineers Joint Contract Documents Committee (EJCDC)?

(A) National Society of Professional Engineers

(B) Construction Specifications Institute

(C) Associated General Contractors of America

(D) American Institute of Architects

7. To be affected by the Fair Labor Standards Act (FLSA) and be required to pay minimum wage, construction firms working on bridges and highways must generally have

(A) 1 or more employees

(B) 2 or more employees and annual gross billings of $500,000

(C) 10 or more employees and be working on federally funded projects

(D) 50 or more employees and have been in business for longer than 6 months

8. A "double-breasted" design firm

(A) has errors and omissions as well as general liability insurance coverage

(B) is licensed to practice in both engineering and architecture

(C) serves both union and nonunion clients

(D) performs post-construction certification for projects it did not design

9. The phrase "without expressed authority" means which of the following when used in regards to partnerships of design professionals?

(A) Each full member of a partnership is a general agent of the partnership and has complete authority to make binding commitments, enter into contracts, and otherwise act for the partners within the scope of the business.

(B) The partnership may act in a manner that it considers best for the client, even though the client has not been consulted.

(C) Only plans, specifications, and documents that have been signed and stamped (sealed) by the authority of the licensed engineer may be relied upon.

(D) Only officers to the partnership may obligate the partnership.

10. A limited partnership has 1 managing general partner, 2 general partners, 1 silent partner, and 3 limited partners. If all partners cast a single vote when deciding on an issue, how many votes will be cast?

(A) 1

(B) 3

(C) 4

(D) 7

11. Which of the following are characteristics of a limited liability corporation (LLC)?

I.   limited liability for all members

II.   no taxation as an entity (no double taxation)

III. more than one class of stock

IV.   limited to fewer than 25 members

V.   fairly easy to establish

VI.   no "continuity of life" like regular corporation

(A) I, II, and IV

(B) I, II, III, and VI

(C) I, II, III, IV, and V

(D) I, III, IV, V, and VI

12. Which of the following statements is FALSE in regard to joint ventures?

(A) Members of a joint venture may be any combination of sole proprietorships, partnerships, and corporations.

(B) A joint venture is a business entity separate from its members.

(C) A joint venture spreads risk and rewards, and it pools expertise, experience, and resources. However, bonding capacity is not aggregated.

(D) A joint venture usually dissolves after the completion of a specific project.

13. Which of the following construction business types can have unlimited shareholders?

I.   S corporation

II.   LLC

III.   corporation

IV.   sole proprietorship

(A) II and III only

(B) I, II, and III

(C) I, III, and IV

(D) I, II, III, and IV

14. The phrase "or approved equal" allows a contractor to

(A) substitute one connection design for another

(B) substitute a more expensive feature for another

(C) replace an open-shop subcontractor with a union subcontractor

(D) install a product whose brand name and model number are not listed in the specifications

15. Cities, other municipalities, and departments of transportation often have standard specifications, in addition to the specifications issued as part of the construction document set, that cover such items as

(A) safety requirements

(B) environmental requirements

(C) concrete, fire hydrant, manhole structures, and curb requirements

(D) procurement and accounting requirements

16. What is intended to prevent a contractor from bidding on a project and subsequently backing out after being selected for the project?

(A) publically recorded bid

(B) property lien

(C) surety bond

(D) proposal bond

17. Which of the following is illegal, in addition to being unethical?

(A) bid shopping

(B) bid peddling

(C) bid rigging

(D) bid unbalancing

18. Which of the following is NOT normally part of a construction contract?

I.   invitation to bid

II.   instructions to bidders

III.   general conditions

IV.   supplementary conditions

V.   liability insurance policy

VI.   technical specifications

VII. drawings

VIII. addenda

IX.   proposals

X.   bid bond

XI.   agreement

XII. performance bond

XIII. labor and material payment bond

XIV. nondisclosure agreement

(A) I

(B) II

(C) V

(D) XIV

Systems, Mgmt., Professional

19. Once a contract has been signed by the owner and contractor, changes to the contract

(A) cannot be made

(B) can be made by the owner, but not by the contractor

(C) can be made by the contractor, but not by the owner

(D) can be made by both the owner and the contractor

20. A constructive change is a change to the contract that can legally be construed to have been made, even though the owner did not issue a specific, written change order. Which of the following situations is normally a constructive change?

(A) request by the engineer-architect to install OSHA-compliant safety features

(B) delay caused by the owner's failure to provide access

(C) rework mandated by the building official

(D) expense and delay due to adverse weather

21. A bid for foundation construction is based on owner supplied soil borings showing sandy clay to a depth of 12 ft. However, after the contract has been assigned and during construction, the backhoe encounters large pieces of concrete buried throughout the construction site. This situation would normally be referred to as

(A) concealed conditions

(B) unexplained features

(C) unexpected characteristics

(D) hidden detriment

22. If a contract has a value engineering clause and a contractor suggests to the owner that a feature or method be used to reduce the annual maintenance cost of the finished project, what will be the most likely outcome?

(A) The contractor will be able to share one time in the owner's expected cost savings.

(B) The contractor will be paid a fixed amount (specified by the contract) for making the suggestion, but only if the suggestion is accepted.

(C) The contract amount will be increased by some amount specified in the contract.

(D) The contractor will receive an annuity payment over some time period specified in the contract.

23. A contract has a value engineering clause that allows the parties to share in improvements that reduce cost. The contractor had originally planned to transport concrete on site for a small pour with motorized wheelbarrows. On the day of the pour, however, a concrete pump is available and is used, substantially reducing the contractor's labor cost for the day. This is an example of

(A) value engineering whose benefit will be shared by both contractor and owner

(B) efficient methodology whose benefit is to the contractor only

(C) value engineering whose benefit is to the owner only

(D) cost reduction whose benefit will be shared by both contractor and laborers

24. A material breach of contract occurs when the

(A) contractor uses material not approved by the contract to use

(B) contractor's material order arrives late

(C) owner becomes insolvent

(D) contractor installs a feature incorrectly

25. When an engineer stops work on a job site after noticing unsafe conditions, the engineer is acting as a(n)

I.   agent

II.   local official

III. OSHA safety inspector

IV.   competent person

(A) I only

(B) I and IV

(C) II and III

(D) III and IV

26. While performing duties pursuant to a contract for professional services with an owner/developer, a professional engineer gives erroneous instruction to a subcontractor hired by the prime contractor. To whom would the subcontractor look for financial relief?

I.   professional engineer

II.   owner/developer

III. subcontractor's bonding company

IV.   prime contractor

(A) III only

(B) IV only

(C) I and II only

(D) I, II, III, and IV

Systems, Mgmt., Professional

27. A professional engineer employed by a large, multinational corporation is the lead engineer in designing and producing a consumer product that proves injurious to some purchasers. What can the engineer expect in the future?

(A) legal action against him/her from consumers

(B) termination and legal action against him/her from his/her employer

(C) legal action against him/her from the U.S. Consumer Protection Agency

(D) thorough review of his/her work, legal support from the employer, and possible termination

28. A professional engineer in private consulting practice makes a calculation error that causes the collapse of a structure during the construction process. What can the engineer expect in the future?

(A) cancellation of his/her Errors & Omissions insurance

(B) criminal prosecution

(C) loss of his/her professional engineering license

(D) incarceration

29. A professional engineer is hired by a homeowner to design a septic tank and leach field. The septic tank fails after 18 years of operation. What sentence best describes what comes next?

(A) The homeowner could pursue a tort action claiming a septic tank should retain functionality longer than 18 years.

(B) The engineer will be protected by a statute of limitations law.

(C) The homeowner may file a claim with the engineer's original bonding company.

(D) The engineer is ethically bound to provide remediation services to the homeowner.

SOLUTIONS

1. The three different forms of company ownership are the (1) sole proprietorship, (2) partnership, and (3) corporation.

A *sole proprietor* is his or her own boss. This satisfies the proprietor's ego and facilitates quick decisions, but unless the proprietor is trained in business, the company will usually operate without the benefit of expert or mitigating advice. The sole proprietor also personally assumes all the debts and liabilities of the company. A sole proprietorship is terminated upon the death of the proprietor.

A *partnership* increases the capitalization and the knowledge base beyond that of a proprietorship, but offers little else in the way of improvement. In fact, the partnership creates an additional disadvantage of one partner's possible irresponsible actions creating debts and liabilities for the remaining partners.

A *corporation* has sizable capitalization (provided by the stockholders) and a vast knowledge base (provided by the board of directors). It keeps the company and owner liability separate. It also survives the death of any employee, officer, or director. Its major disadvantage is the administrative work required to establish and maintain the corporate structure.

2. To be legal, a contract must contain an *offer*, some form of *consideration* (which does not have to be equitable), and an *acceptance* by both parties. To be enforceable, the contract must be voluntarily entered into, both parties must be competent and of legal age, and the contract cannot be for illegal activities.

3. A written contract will identify both parties, state the purpose of the contract and the obligations of the parties, give specific details of the obligations (including relevant dates and deadlines), specify the consideration, state the boilerplate clauses to clarify the contract terms, and leave places for signatures.

4. A consultant will either charge a fixed fee, a variable fee, or some combination of the two. A one-time fixed fee is known as a *lump-sum fee*. In a *cost plus fixed fee* contract, the consultant will also pass on certain costs to the client. Some charges to the client may depend on other factors, such as the salary of the consultant's staff, the number of days the consultant works, or the eventual cost or value of an item being designed by the consultant.

5. A *retainer* is a (usually) nonreturnable advance paid by the client to the consultant. While the retainer may be intended to cover the consultant's initial expenses until the first big billing is sent out, there does not need to be any rational basis for the retainer. Often, a small retainer is used by the consultant to qualify the client (i.e., to make sure the client is not just shopping around and getting free initial consultations) and as a security deposit (to make sure the client does not change consultants after work begins).

6. The Engineers Joint Contract Documents Committee (EJCDC) consists of the National Society of Professional Engineers, the American Council of Engineering Companies (formerly the American Consulting Engineers Council), the American Society of Civil Engineers, Construction Specifications Institute, and the Associated General Contractors of America. The American Institute of Architects is not a member, and it has its own standardized contract documents.

The answer is (D).

7. A business in the construction industry must have two or more employees and a minimum annual gross sales volume of $500,000 to be subject to the Fair Labor Standards Act (FLSA). Individual coverage also applies to employees whose work regularly involves them in commerce between states (i.e., interstate commerce). Any person who works on, or otherwise handles, goods moving in interstate commerce, or who works on the expansion of existing facilities of commerce, is individually subject to the protection of the FLSA and the current minimum wage and overtime pay requirements, regardless of the sales volume of the employer.

The answer is (B).

8. A *double-breasted* design firm serves both union and nonunion clients. When union-affiliated companies find themselves uncompetitive in bidding on nonunion projects, the company owners may decide to form and operate a second company that is *open shop* (i.e., employees are not required to join a union as a condition of employment). Although there are some restrictions requiring independence of operation, the common ownership of two related firms is legal.

The answer is (C).

9. *Without expressed authority* means each member of a partnership has full authority to obligate the partnership (and the other partners).

The answer is (A).

10. Only general partners can vote in a partnership. Both silent and limited partners share in the profit and benefits of the partnership, but they only contribute financing and do not participate in the management. The identities of silent partners are often known only to a few, whereas limited partners are known to all.

The answer is (B).

11. Limited liability corporations have limited liability for all members, no double taxation, more than one type of stock, and no continuity of life. They are not limited in members, and they are comparatively fairly difficult to establish.

The answer is (B).

12. One of the reasons for forming joint ventures is that the bonding capacity is aggregated. Even if each contractor cannot individually meet the minimum bond requirements, the total of the bonding capacities may be sufficient.

The answer is (C).

13. Both normal corporations and limited liability corporations (LLCs) can have unlimited shareholders. S corporations and sole proprietorships are for individuals.

The answer is (A).

14. When the specifications include a nonstructural, brand-named article and the accompanying phrase "or approved equal," the contractor can substitute something with the same functionality, even though it is not the brand-named article. "Or approved equal" would not be used with a structural detail such as a connection design.

The answer is (D).

15. Municipalities that experience frequent construction projects within their boundaries have standard specifications that are included by reference in every project's construction document set. This document set would cover items such as concrete, fire hydrants, manhole structures, and curb requirements.

The answer is (C).

16. *Proposal bonds*, also known as *bid bonds*, are insurance policies payable to the owner in the event that the contractor backs out after submitting a qualified bid.

The answer is (D).

17. *Bid rigging*, also known as *price fixing*, is an illegal arrangement between contractors to control the bid prices of a construction project or to divide up customers or market areas. *Bid shopping* before or after the bid letting is where the general contractor tries to secure better subcontract proposals by negotiating with the subcontractors. *Bid peddling* is done by the subcontractor to try to lower its proposal below the lowest proposal. *Bid unbalancing* is where a contractor pushes the payment for some expense items to prior construction phases in order to improve cash flow.

The answer is (C).

18. The construction contract includes many items, some explicit only by reference. The contractor may be required to carry liability insurance, but the policy itself is between the contractor and its insurance company, and is not normally part of the construction contract.

The answer is (C).

19. Changes to the contract can be made by either the owner or the contractor. The method for making such changes is indicated in the contract. Almost always, the change must be agreed to by both parties in writing.

The answer is (D).

20. A *constructive change* to the contract is the result of an action or lack of action of the owner or its agent. If the project is delayed by the owner's failure to provide access, the owner has effectively changed the contract.

The answer is (B).

21. *Concealed conditions* are also known as *changed conditions* and *differing site conditions*. Most, but not all, contracts have provisions dealing with changed conditions. Some place the responsibility to confirm the site conditions before bidding on the contractor. Others detail the extent of changed conditions that will trigger a review of reimbursable expenses.

The answer is (A).

22. Changes to a structure's performance, safety, appearance, or maintenance that benefit the owner in the long run will be covered by the value engineering clause of a contract. Normally, the contractor is able to share in cost savings in some manner by receiving a payment or credit to the contract.

The answer is (A).

23. The problem gives an example of efficient methodology, where the benefit is to the contractor only. It is not an example of value engineering, as the change affects the contractor, not the owner. Performance, safety, appearance, and maintenance are unaffected.

The answer is (B).

24. A *material breach of the contract* is a significant event that is grounds for cancelling the contract entirely. Typical triggering events include failure of the owner to make payments, the owner causing delays, the owner declaring bankruptcy, the contractor abandoning the job, or the contractor getting substantially off schedule.

The answer is (C).

25. An engineer with the authority to stop work gets his/her authority from the agency clause of the contract for professional services with the owner/developer. It is unlikely that the local building department or OSHA would have authorized the engineer to act on their behalves. It is also possible that the engineer may have been designated as the jobsite's "competent person" (as required by OSHA) for one or more aspects of the job, with authority to implement corrective action.

The answer is (B).

26. In the absence of other information, the subcontractor only has a contractual relationship with the prime contractor, so the request for financial relief would initially go to the prime contractor. The prime contractor would ask for relief from the owner/developer, who in turn may want relief from the professional engineer. The subcontractor's bonding company's role is to guarantee the subcontractor's work product against subcontractor errors, not engineering errors.

The answer is (B).

27. Manufacturers of consumer products are generally held responsible for the safety of their products. Individual team members are seldom held personally accountable, and when they are, there is evidence of intentional wrongdoing and/or gross incompetence. The manufacturer will organize and absorb the costs of the legal defense. If the engineer is fired, it is likely this will occur after the case is closed.

The answer is (D).

28. A professional engineer is not held to the standard of perfection, so making a calculation error is neither a criminal act nor sufficient evidence of incompetence to result in loss of license. Without criminal prosecution for fraud or other wrongdoing, it is unlikely that the engineer would ever see the inside of a jail. However, it is likely that the engineer's bonding company will cancel his/her policy (after defending the case).

The answer is (A).

29. Satisfactory operation for 18 years is proof that the design was partially, if not completely, adequate. There is no ethical obligation to remediate normal wear-and-tear, particularly when the engineer was not involved in how the septic system was used or maintained. While the homeowner can threaten legal action and even file a complaint, in the absence of a warranty to the contrary, the engineer is probably well-protected by a statute of limitations.

The answer is (B).

57 Engineering Ethics

PRACTICE PROBLEMS

(Each problem has two parts. Determine whether the situation is (or can be) permitted legally. Then, determine whether the situation is permitted ethically.)

1. (a) Was it legal and/or ethical for an engineer to sign and seal plans that were not prepared by him or prepared under his responsible direction, supervision, or control?

(b) Was it legal and/or ethical for an engineer to sign and seal plans that were not prepared by him but were prepared under his responsible direction, supervision, and control?

2. Under what conditions would it be legal and/or ethical for an engineer to rely on the information (e.g., elevations and amounts of cuts and fills) furnished by a grading contractor?

3. Was it legal and/or ethical for an engineer to alter the soils report prepared by another engineer for his client?

4. Under what conditions would it be legal and/or ethical for an engineer to assign work called for in his contract to another engineer?

5. A licensed professional engineer was convicted of a felony totally unrelated to his consulting engineering practice.

(a) What actions would you recommend be taken by the state registration board?

(b) What actions would you recommend be taken by the professional or technical society (e.g., ASCE, ASME, IEEE, NSPE, and so on)?

6. An engineer came across some work of a predecessor. After verifying the validity and correctness of all assumptions and calculations, the engineer used the work. Under what conditions would such use be legal and/or ethical?

7. A building contractor made it a policy to provide cell phones to the engineers of projects he was working on. Under what conditions could the engineers accept the phones?

8. An engineer designed a tilt-up slab building for a client. The design engineer sent the design out to another engineer for checking. The checking engineer sent the plans to a concrete contractor for review. The concrete contractor made suggestions that were incorporated into the recommendations of the checking contractor. These recommendations were subsequently incorporated into the plans by the original design engineer. What steps must be taken to keep the design process legal and/or ethical?

9. A consulting engineer registered his corporation as "John Williams, P.E. and Associates, Inc." even though he had no associates. Under what conditions would this name be legal and/or ethical?

10. When it became known that a chemical plant was planning on producing a toxic product, an engineer employed by the plant wrote to the local newspaper condemning the chemical plant's action. Under what conditions would the engineer's action be legal and/or ethical?

11. An engineer signed a contract with a client. The fee the client agreed to pay was based on the engineer's estimate of time required. The engineer was able to complete the contract satisfactorily in half the time he expected. Under what conditions would it be legal and/or ethical for the engineer to keep the full fee?

12. After working on a project for a client, the engineer was asked by a competitor of the client to perform design services. Under what conditions would it be legal and/or ethical for the engineer to work for the competitor?

13. Two engineers submitted bids to a prospective client for a design project. The client told engineer A how much engineer B had bid and invited engineer A to beat the amount. Under what conditions could engineer A legally/ethically submit a lower bid?

14. A registered civil engineer specializing in well-drilling, irrigation pipelines, and farmhouse sanitary systems took a booth at a county fair located in a farming town. By a random drawing, the engineer's booth was located next to a hog-breeder's booth, complete with live (prize) hogs. The engineer gave away helium balloons with his name and phone number to all visitors to the booth. Did the engineer violate any laws/ethical guidelines?

15. While in a developing country supervising construction of a project an engineer designed, the engineer discovered the client's project manager was treating local workers in an unsafe and inhumane (but, for that country, legal) manner. When the engineer objected, the client told the engineer to mind his own business. Later, the local workers asked the engineer to participate in a walkout and strike with them.

(a) What legal/ethical positions should the engineer take?

(b) Should it have made any difference if the engineer had or had not yet accepted any money from the client?

16. While working for a client, an engineer learns confidential knowledge of a proprietary production process being used by the client's chemical plant. The process is clearly destructive to the environment, but the client will not listen to the objections of the engineer. To inform the proper authorities will require the engineer to release information that was gained in confidence. Is it legal and/or ethical for the engineer to expose the client?

17. While working for an engineering design firm, an engineer was moonlighting as a soils engineer. At night, the engineer used the employer's facilities to analyze and plot the results of soils tests. He then used his employer's computers to write his reports. The equipment and computers would otherwise be unused. Under what conditions could the engineer's actions be considered legal and/or ethical?

18. Ethical codes and state legislation forbidding competitive bidding by design engineers are

    (A) enforceable in some states

    (B) not enforceable on public (nonfederal) projects

    (C) enforceable for projects costing less than $5 million

    (D) not enforceable

SOLUTIONS

Introduction to the Solutions

Case studies in law and ethics can be interpreted in many ways. The problems presented are simple thumbnail outlines. In most real cases, there will be more facts to influence a determination than are presented in the case scenarios. In some cases, a state may have specific laws affecting the determination; in other cases, prior case law will have been established.

The determination of whether an action is legal can be made in two ways. The obvious interpretation of an illegal action is one that violates a specific law or statute. An action can also be *found to be illegal* if it is judged in court to be a breach of a written, verbal, or implied contract. Both of these approaches are used in the following solutions.

These answers have been developed to teach legal and ethical principles. While being realistic, they are not necessarily based on actual incidents or prior case law.

1. (a) Stamping plans for someone else is illegal. The registration laws of all states permit a registered engineer to stamp/sign/seal only plans that were prepared by him personally or were prepared under his direction, supervision, or control. This is sometimes called being in *responsible charge*. The stamping/signing/sealing, for a fee or gratis, of plans produced by another person, whether that person is registered or not and whether that person is an engineer or not, is illegal.

An illegal act, being a concealed act, is intrinsically unethical. In addition, stamping/signing/sealing plans that have not been checked violates the rule contained in all ethical codes that requires an engineer to protect the public.

(b) This is both ethical and legal. Consulting engineering firms typically operate in this manner, with a senior engineer being in responsible charge for the work of subordinate engineers.

2. Unless the engineer and contractor worked together such that the engineer had personal knowledge that the information was correct, accepting the contractor's information is illegal. Not only would using unverified data violate the state's registration law (for the same reason that stamping/signing/sealing unverified plans in Prob. 1 was illegal), but the engineer's contract clause dealing with assignment of work to others would probably be violated.

The act is unethical. An illegal act, being a concealed act, is intrinsically unethical. In addition, using unverified data violates the rule contained in all ethical codes that requires an engineer to protect the client.

3. It is illegal to alter a report to bring it "more into line" with what the client wants unless the alterations represent actual, verified changed conditions. Even when the alterations are warranted, however, use of the unverified remainder of the report is a violation of the state registration law requiring an engineer only to stamp/sign/seal plans developed by or under him. Furthermore, this would be a case of fraudulent misrepresentation unless the originating engineer's name was removed from the report.

Unless the engineer who wrote the original report has given permission for the modification, altering the report would be unethical.

4. Assignment of engineering work is legal (1) if the engineer's contract permitted assignment, (2) all prerequisites (e.g., notifying the client) were met, and (3) the work was performed under the direction of another licensed engineer.

Assignment of work is ethical (1) if it is not illegal, (2) if it is done with the awareness of the client, and (3) if the assignor has determined that the assignee is competent in the area of the assignment.

5. (a) The registration laws of many states require a hearing to be held when a licensee is found guilty of unrelated, but nevertheless unforgivable, felonies (e.g., moral turpitude). The specific action (e.g., suspension, revocation of license, public censure, and so on) taken depends on the customs of the state's registration board.

(b) By convention, it is not the responsibility of technical and professional organizations to monitor or judge the personal actions of their members. Such organizations do not have the authority to discipline members (other than to revoke membership), nor are they immune from possible retaliatory libel/slander lawsuits if they publicly censure a member.

6. The action is legal because, by verifying all the assumptions and checking all the calculations, the engineer effectively does the work. Very few engineering procedures are truly original; the fact that someone else's effort guided the analysis does not make the action illegal.

The action is probably ethical, particularly if the client and the predecessor are aware of what has happened (although it is not necessary for the predecessor to be told). It is unclear to what extent (if at all) the predecessor should be credited. There could be other extenuating circumstances that would make referring to the original work unethical.

7. Gifts, per se, are not illegal. Unless accepting the phones violates some public policy or other law, or is in some way an illegal bribe to induce the engineer to favor the contractor, it is probably legal to accept the phones.

Ethical acceptance of the phones requires (among other considerations) that (1) the phones be required for the job, (2) the phones be used for business only, (3) the phones are returned to the contractor at the end of the job, and (4) the contractor's and engineer's clients know and approve of the transaction.

8. There are two issues: (1) the assignment and (2) the incorporation of work done by another. To avoid a breach, the contracts of both the design and checking engineers must permit the assignments. To avoid a violation of the state registration law requiring engineers to be in responsible charge of the work they stamp/sign/seal, both the design and checking engineers must verify the validity of the changes.

To be ethical, the actions must be legal and all parties (including the design engineer's client) must be aware that the assignments have occurred and that the changes have been made.

9. The name is probably legal. If the name was accepted by the state's corporation registrar, it is a legally formatted name. However, some states have engineering registration laws that restrict what an engineering corporation may be named. For example, all individuals listed in the name (e.g., "Cooper, Williams, and Somerset—Consulting Engineers") may need to be registered. Whether having "Associates" in the name is legal depends on the state.

Using the name is unethical. It misleads the public and represents unfair competition with other engineers running one-person offices.

10. Unless the engineer's accusation is known to be false or exaggerated, or the engineer has signed an agreement (confidentiality, nondisclosure, and so on) with his employer forbidding the disclosure, the letter to the newspaper is probably not illegal.

The action is probably unethical. (If the letter to the newspaper is unsigned it is a concealed action and is definitely unethical.) While whistle-blowing to protect the public is implicitly an ethical procedure, unless the engineer is reasonably certain that manufacture of the toxic product represents a hazard to the public, he has a responsibility to the employer. Even then, the engineer should exhaust all possible remedies to render the manufacture nonhazardous before blowing the whistle. Of course, the engineer may quit working for the chemical plant and be as critical as the law allows without violating engineer-employer ethical considerations.

11. Unless the engineer's payment was explicitly linked in the contract to the amount of time spent on the job, taking the full fee would not be illegal or a breach of the contract.

An engineer has an obligation to be fair in estimates of cost, particularly when the engineer knows no one else is providing a competitive bid. Taking the full fee would be ethical if the original estimate was arrived at logically and was not meant to deceive or take advantage of the client. An engineer is permitted to take advantage of economies of scale, state-of-the-art techniques, and break-through methods. (Similarly, when a job costs more than the estimate, the engineer may be ethically bound to stick with the original estimate.)

12. In the absence of a nondisclosure or noncompetition agreement or similar contract clause, working for the competitor is probably legal.

Working for both clients is unethical. Even if both clients know and approve, it is difficult for the engineer not to "cross-pollinate" his work and improve one client's position with knowledge and insights gained at the expense of the other client. Furthermore, the mere appearance of a conflict of interest of this type is a violation of most ethical codes.

13. In the absence of a sealed-bid provision mandated by a public agency and requiring all bids to be opened simultaneously (and the award going to the lowest bidder), the action is probably legal.

It is unethical for an engineer to undercut the price of another engineer. Not only does this violate a standard of behavior expected of professionals, it unfairly benefits one engineer because a similar chance is not given to the other engineer. Even if both engineers are bidding openly against each other (in an auction format), the client must understand that a lower price means reduced service. Each reduction in price is an incentive to the engineer to reduce the quality or quantity of service.

14. It is generally legal for an engineer to advertise his services. Unless the state has relevant laws, the engineer probably did not engage in illegal actions.

Most ethical codes prohibit unprofessional advertising. The unfortunate location due to a random drawing might be excusable, but the engineer should probably refuse to participate. In any case, the balloons are a form of unprofessional advertising, and as such, are unethical.

15. (a) As stated in the scenario statement, the client's actions are legal for that country. The fact that the actions might be illegal in another country is irrelevant. Whether or not the strike is legal depends on the industry and the laws of the land. Some or all occupations (e.g., police and medical personnel) may be forbidden to strike. Assuming the engineer's contract does not prohibit participation, the engineer should determine the legality of the strike before making a decision to participate.

If the client's actions are inhumane, the engineer has an ethical obligation to withdraw from the project. Not doing so associates the profession of engineering with human misery.

(b) The engineer has a contract to complete the project for the client. (It is assumed that the contract between the engineer and client was negotiated in good faith, that the engineer had no knowledge of the work conditions prior to signing, and that the client did not falsely induce the engineer to sign.) Regardless of the reason for withdrawing, the engineer is breaching his contract. In the absence of proof of illegal actions by the client, withdrawal by the engineer requires a return of all fees received. Even if no fees have been received, withdrawal exposes the engineer to other delay-related claims by the client.

16. A contract for an illegal action cannot be enforced. Therefore, any confidentiality or nondisclosure agreement that the engineer has signed is unenforceable if the production process is illegal, uses illegal chemicals, or violates laws protecting the environment. If the production process is not illegal, it is not legal for the engineer to expose the client.

Society and the public are at the top of the hierarchy of an engineer's responsibilities. Obligations to the public take precedence over the client. If the production process is illegal, it would be ethical to expose the client.

17. It is probably legal for the engineer to use the facilities, particularly if the employer is aware of the use. (The question of whether the engineer is trespassing or violating a company policy cannot be answered without additional information.)

Moonlighting, in general, is not ethical. Most ethical codes prohibit running an engineering consulting business while receiving a salary from another employer. The rationale is that the moonlighting engineer is able to offer services at a much lower price, placing other consulting engineers at a competitive disadvantage. The use of someone else's equipment compounds the problem since the engineer does not have to pay for using the equipment, and so does not have to charge any clients for it. This places the engineer at an unfair competitive advantage compared to other consultants who have invested heavily in equipment.

18. Ethical bans on competitive bidding are not enforceable. The National Society of Professional Engineers' (NSPE) ethical ban on competitive bidding was struck down by the U.S. Supreme Court in 1978 as a violation of the Sherman Antitrust Act of 1890.

The answer is (D).

58 Engineering Licensing in the United States

PRACTICE PROBLEMS

There are no problems in this book corresponding to Chap. 58 of the *Environmental Engineering Reference Manual.*